STATISTICAL MECHANICS

STATISTICAL MECHANICS

Donald A. McQuarrie

University of California, Davis

University Science Books
Mill Valley, California

Mill Valley, California
www.uscibooks.com

Library of Congress Cataloging-in-Publication Data

McQuarrie, Donald A. (Donald Allan)
  Statistical mechanics / Donald A. McQuarrie
     p.   cm.
  Includes bibliographic references and index.
  ISBN 1-891389-15-7 (alk.paper)
  1. Statistical mechanics. I. Title

  QC174.8.M3 2000
  530.13–dc21   00–021962

Printed in the United States of America

10   9   8   7   6   5

CONTENTS

PREFACE

With this edition, University Science Books becomes the publisher of both my *Statistical Thermodynamics* and *Statistical Mechanics* books. I would like to thank the publisher, Bruce Armbruster, for his continual commitment to quality textbook publishing in the sciences.

Statistical Mechanics is the extended version of my earlier text, *Statistical Thermodynamics.* The present volume is intended primarily for a two-semester course or for a second one-semester course in statistical mechanics. Whereas *Statistical Thermodynamics* deals principally with equilibrium systems whose particles are either independent or effectively independent, *Statistical Mechanics* treats equilibrium systems whose particles are strongly interacting as well as nonequilibrium systems. The first twelve chapters of this book also form the first chapters in *Statistical Thermodynamics,* while the next ten chapters, 13-22, appear only in *Statistical Mechanics.* Chapter 13 deals with the radial distribution function approach to liquids, and Chapter 14 is a fairly detailed discussion of statistical mechanical perturbation theories of liquids. These theories were developed in the late 1960s and early 1970s and have brought the numerical calculation of the thermodynamic properties of simple dense fluids to a practical level. A number of the problems at the end of the Chapter 14 require the student to calculate such properties and compare the results to experimental data. Chapter 15, on ionic solutions, is the last chapter on equilibrium systems. Section 15-2 is an introduction to advances in ionic solution theory that were developed in the 1970s and that now allow one to calculate the thermodynamic properties of simple ionic solutions up to concentrations of 2 molar.

Chapters 16-22 treat systems that are not in equilibrium. Chapters 16 and 17 are meant to be somewhat of a review, although admittedly much of the material, particularly in Chapter 17, will be new. Nevertheless, these two chapters do serve as a background for the rest. Chapter 18 presents the rigorous kinetic theory of gases as formulated through the Boltzmann equation, the famous integro-differential equation whose solution gives the nonequilibrium distribution of a molecule in velocity space. The long-time or equilibrium solution of the Boltzmann equation is the well-known Maxwell-Boltzmann distribution (Chapter 7). Being an integro-differential equation, it is not surprising that its solution is fairly involved. We only outline the standard method of solution, called the Chapman-Enskog method, in Section 19-1, and the next two sections are a practical calculation of the transport properties of gases. In the last section of Chapter 19 we discuss Enskog's ad hoc extension of the

Boltzmann equation to dense hard-sphere fluids. Chapter 20, which presents the Langevin equation and the Fokker-Planck equation, again is somewhat of a digression but does serve as a background to Chapters 21 and 22.

The 1950s saw the beginning of the development of a new approach to transport processes that has grown into one of the most active and fruitful areas of nonequilibrium statistical mechanics. This work was initiated by Green and Kubo, who showed that the phenomenological coefficients describing many transport processes and time-dependent phenomena in general could be written as integrals over a certain type of function called a time-correlation function. The time-correlation function associated with some particular process is in a sense the analog of the partition function for equilibrium systems. Although both are difficult to evaluate exactly, the appropriate properties of the system of interest can be formally expressed in terms of these functions, and they serve as basic starting points for computationally convenient approximations. Before the development of the time-correlation function formalism, there was no single unifying approach to nonequilibrium statistical mechanics such as Gibbs had given to equilibrium statistical mechanics.

Chapters 21 and 22, two long chapters, introduce the time-correlation function approach. We have chosen to introduce the time-correlation function formalism through the absorption of electromagnetic radiation by a system of molecules since the application is of general interest and the derivation of the key formulas is quite pedagogical and requires no special techniques. After presenting a similar application to light scattering, we then develop the formalism in a more general way and apply the general formalism to dielectric relaxation, thermal transport, neutron scattering, light scattering, and several others.

Eleven appendixes are also included to supplement the textual material.

The intention here is to present a readable introduction to the topics covered rather than a rigorous, formal development. In addition, a great number of problems is included at the end of each chapter in order either to increase the student's understanding of the material or to introduce him or her to selected extensions.

As this printing goes to press, we are beginning to plan a new edition of *Statistical Mechanics*. We would be grateful for any and all feedback from loyal users intended to help us shape the next edition of this text.

I should like to acknowledge the people who have helped me in putting this book together: Donald DuPré and Stuart Rice for reviewing most of the manuscript; all my students, particularly Steve Brenner, Jim Ely, Dennis Isbister, Allen Nelson, and Bernard Pailthorpe, who had to suffer through several earlier versions of the final manuscript; Wilmer Olivares for reading all of the page proofs; and Lynn Johnson, who typed the final manuscript and is more than happy to be finished. After reading manuscript prefaces recently, it would seem appropriate at this point to thank my wife, Carole, for "encouragement and understanding during the writing of this book." It so happens, however, that she never did encourage me nor was she especially understanding, but she actually did something far better. As a biochemistry graduate student, she took both semesters of the course here (*not* from me) and found out for herself that statistical mechanics can be a rich and exciting field that has important applications not only in chemistry and physics but in biology as well. I hope that this book starts its readers along the same route.

Donald A. McQuarrie
March 2000
mcquarrie@mcn.org

STATISTICAL MECHANICS

CHAPTER 1

INTRODUCTION AND REVIEW

1–1 INTRODUCTION

Statistical mechanics is that branch of physics which studies macroscopic systems from a microscopic or molecular point of view. The goal of statistical mechanics is the understanding and prediction of macroscopic phenomena and the calculation of macroscopic properties from the properties of the individual molecules making up the system.

Present-day research in statistical mechanics varies from mathematically sophisticated discussions of general theorems to almost empirical calculations based upon simple, but nevertheless useful, molecular models. An example of the first type of research is the investigation of the question of whether statistical mechanics, as it is formulated today, is even capable of predicting the existence of a first-order phase transition. General questions of this kind are by their nature mathematically involved and are generally beyond the level of this book. We shall, however, discuss such questions to some extent later on. On the other hand, for many scientists statistical mechanics merely provides a recipe or prescription which allows them to calculate the properties of the physical systems which they are studying.

The techniques of statistical mechanics have been used in attacking a wide variety of physical problems. A quick glance through this text will show that statistical mechanics has been applied to gases, liquids, solutions, electrolytic solutions, polymers, adsorption, metals, spectroscopy, transport theory, the helix-coil transition of DNA, the electrical properties of matter, and cell membranes, among others.

Statistical mechanics may be broadly classified into two parts, one dealing with systems in equilibrium and the other with systems not in equilibrium. The treatment of systems in equilibrium is usually referred to as *statistical thermodynamics*, since it forms a bridge between thermodynamics (often called classical thermodynamics) and molecular physics.

Thermodynamics provides us with mathematical relations between the various experimental properties of macroscopic systems in equilibrium. An example of such a thermodynamic relation is that between the molar heat capacities at constant pressure and at constant volume,

$$C_p - C_v = \left[p + \left(\frac{\partial E}{\partial V} \right)_T \right] \left(\frac{\partial V}{\partial T} \right)_p \qquad (1\text{--}1)$$

Another, and one that we shall use in the next chapter, is

$$\left(\frac{\partial E}{\partial V} \right)_{N,T} - T \left(\frac{\partial p}{\partial T} \right)_{N,V} = -p \qquad (1\text{--}2)$$

Note that thermodynamics provides connections between many properties, but does not supply information concerning the magnitude of any one. Neither does it attempt to base any relation on molecular models or interpretations. This, in fact, is both the power and weakness of thermodynamics. It is a general discipline which does not need to recognize or rely upon the existence of atoms and molecules. Its many relations would remain valid even if matter were continuous. In addition, there are many systems (such as biological systems) which are too complicated to be described by an acceptable molecular theory, but here again the relations given by thermodynamics are exact. This great generality, however, is paid for by its inability to calculate physical properties separately or to supply physical interpretations of its equations. When one seeks a molecular theory which can do just this, one then enters the field of statistical thermodynamics. Thus thermodynamics and statistical thermodynamics treat the same systems. Thermodynamics provides general relations without the need of ever considering the ultimate constitution of matter, while statistical thermodynamics, on the other hand, assumes the existence of atoms and molecules to calculate and interpret thermodynamic quantities from a molecular point of view.

Statistical thermodynamics itself may be further divided into two areas: first, the study of systems of molecules in which molecular interactions may be neglected (such as dilute gases), and second, the study of systems in which the molecular interactions are of prime importance (such as liquids). We shall see that the neglect of intermolecular interactions enormously simplifies our problem. Chapters 4 through 11 of the book are devoted to the treatment of systems in which these interactions either may be ignored or highly simplified. This is the kind of statistical thermodynamics to which most undergraduates have been exposed and, to some extent, represents typical statistical thermodynamical research done in the 1930s. The more interesting and challenging problems, however, concern systems in which these molecular interactions cannot be neglected; Chapters 12 through 15 of the present volume are devoted to the study of such systems. It is in this latter area that a great deal of the research of the 1940s, 1950s, and 1960s was carried out. There are, of course, many important problems of this sort still awaiting attack. The theory of concentrated electrolyte solutions and the proof for the existence of first-order phase transitions are just two examples.

The most difficult branch of statistical mechanics, both mathematically and conceptually, is the study of systems not in equilibrium. This field is often referred to as *nonequilibrium statistical mechanics*. This is presently a very active area of research.

There are still some important unsolved conceptual problems in nonequilibrium statistical mechanics. Nevertheless, in the 1950s great strides were made toward the establishment of a firm basis for nonequilibrium statistical mechanics, commensurate with that of equilibrium statistical mechanics, or what we have called statistical thermodynamics. Chapters 16 through 22 of *Statistical Mechanics* present an introduction to some of the more elementary of these fairly new and useful concepts and techniques.

In Chapter 2 we shall introduce and discuss the basic concepts and assumptions of statistical thermodynamics. We shall present these ideas in terms of quantum mechanical properties such as energy states, wave functions, and degeneracy. Although it may appear at this point that quantum mechanics is a prerequisite for statistical thermodynamics, it will turn out that a satisfactory version of statistical thermodynamics can be presented by using only a few quantum mechanical ideas and results. We assume that the student is familiar with only the amount of quantum mechanics taught in most present-day physical chemistry courses. About the only requirement of the first few chapters is an understanding that the Schrödinger equation determines the possible energy values E_j available to the system and that these may have a degeneracy associated with them which we denote by $\Omega(E_j)$.

Before discussing the principles, however, we shall present in this chapter a discussion of some of the terms or concepts that are particularly useful in statistical thermodynamics. In Section 1–2 we shall treat classical mechanics, including an introduction to the Lagrangian and Hamiltonian formalisms. In Section 1–3 we shall briefly review the main features of quantum mechanics and give the solutions of the Schrödinger equation for some important systems. The only new material in this section to most students probably will be the discussion of the eigenvalues or energy levels of a many-body system. Then in Section 1–4 we shall review thermodynamics briefly, since it is assumed that the reader is familiar with the three laws of thermodynamics and the tedious manipulations of partial derivatives. Two important topics that are not usually discussed in elementary physical chemistry texts are introduced, however. These two topics are the Legendre transformation and Euler's theorem, both of which are useful in studying statistical thermodynamics. Finally, in Section 1–5 we shall discuss some mathematical techniques and results that are particularly useful in statistical thermodynamics. Much of this section may be new material to the reader.

1–2 CLASSICAL MECHANICS

NEWTONIAN APPROACH

Everyone knows the equation $F = ma$. What this equation really says is that the rate of change of momentum is equal to the applied force. If we denote the momentum by $\mathbf{p}$, we have then a more general version of Newton's second law, namely,

$$\frac{d\mathbf{p}}{dt} \equiv \dot{\mathbf{p}} = \mathbf{F} \tag{1-3}$$

If the mass is independent of time, then $d\mathbf{p}/dt = m\,d\dot{\mathbf{r}}/dt = m\ddot{\mathbf{r}} = m\mathbf{a}$. If $\mathbf{F}$ is given as a function of position $\mathbf{F}(x, y, z)$, then Eq. (1–3) represents a set of second-order differential equations in x, y, and z whose solutions give x, y, and z as a function of time if some initial conditions are known. Thus Eq. (1–3) is called an equation of motion. We shall consider three applications of this equation.

Example 1. Solve the equation of motion of a body of mass m shot vertically upward with an initial velocity v_0 in a gravitational field.

If we choose the x-axis (positive in the upward direction) to be the height of the body, then we have

$$m\ddot{x} = -mg$$

where mg is the magnitude of the force. The negative sign indicates that the force is acting in a downward direction. The solution to this differential equation for x is then

$$x(t) = -\tfrac{1}{2}gt^2 + v_0 t + x_0 \tag{1-4}$$

with x_0 equal to $x(0)$, which in our case is 0. This then gives the position of the body at any time after it was projected. The extension of this problem to two dimensions (i.e., a shell shot out of a cannon) and the inclusion of viscous drag on the body are discussed in Problems 1–1 and 1–2.

Example 2.   Set up and solve the equation of motion of a simple harmonic oscillator.

Let x_0 be the length of the unstrained spring. Hooke's law says that the force on the mass attached to the end of the spring is $F = -k(x - x_0)$. If we let $\xi = x - x_0$, we can write

$$\frac{d^2\xi}{dt^2} + \frac{k}{m}\xi = 0 \tag{1-5}$$

whose solution is

$$\xi(t) = A \sin \omega t + B \cos \omega t \tag{1-6}$$

The quantity

$$\omega = (k/m)^{1/2} \tag{1-7}$$

is the natural vibrational frequency of the system. Equation (1–6) can be written in an alternative form (see Problem 1–5)

$$\xi(t) = C \sin(\omega t + \phi) \tag{1-8}$$

This shows more clearly that the mass undergoes simple harmonic motion with frequency ω. Problems 1–3 through 1–5 illustrate some of the basic features of simple harmonic motion.

Example 3.   Two-dimensional motion of a body under coulombic attraction to a fixed center.

In this case the force is $\mathbf{F} = -K\mathbf{r}/r^3$, that is, it is of magnitude $-K/r^2$ and directed radially. Newton's equations become

$$m\ddot{x} = F_x = -\frac{Kx}{(x^2 + y^2)^{3/2}}$$

$$m\ddot{y} = F_y = -\frac{Ky}{(x^2 + y^2)^{3/2}} \tag{1-9}$$

Unlike our previous examples, these two equations are difficult to solve. Since the force depends, in a natural way, on the polar coordinates r and θ, it is more convenient for us to set the problem up in a polar coordinate system. Using then

$$x = r \cos \theta$$
$$y = r \sin \theta$$

and some straightforward differentiation, we get

$$\left\{ m(\ddot{r} - \dot{\theta}^2 r) + \frac{K}{r^2} \right\}\cos \theta - m(r\ddot{\theta} + 2\dot{\theta}\dot{r})\sin \theta = 0 \tag{1-10a}$$

$$\left\{ m(\ddot{r} - \dot{\theta}^2 r) + \frac{K}{r^2} \right\}\sin \theta + m(r\ddot{\theta} + 2\dot{\theta}\dot{r})\cos \theta = 0 \tag{1-10b}$$

By multiplying the first of these equations by $\cos\theta$ and the second by $\sin\theta$ and then adding the two, one gets

$$m(\ddot{r} - \dot{\theta}^2 r) + \frac{K}{r^2} = 0 \qquad (1\text{--}11)$$

But this is just the term in braces in Eq. (1–10), which leads us to the result that

$$m(r\ddot{\theta} + 2\dot{\theta}\dot{r}) = 0 \qquad (1\text{--}12)$$

as well. Equation (1–12) can be written in the form

$$\frac{1}{r}\frac{d}{dt}(mr^2\dot{\theta}) = 0 \qquad (1\text{--}13)$$

which implies that

$$mr^2\dot{\theta} = \text{constant} \qquad (1\text{--}14)$$

This quantity, $mr^2\dot{\theta}$, which maintains a fixed value during the motion of the particle, is called the angular momentum of the particle and is denoted by l. The angular momentum is always conserved if the force is central, that is, directed along $\mathbf{r}$ (see Problem 1–10).

Equation (1–14) can be used to eliminate $\dot{\theta}$ from Eq. (1–11) to give an equation in r alone, called the radial equation:

$$m\ddot{r} - \frac{l^2}{mr^3} + \frac{K}{r^2} = 0 \qquad (1\text{--}15)$$

This equation can be solved (at least numerically) to give $r(t)$, which together with Eq. (1–14) gives $\theta(t)$.

Even though the solution in this example is somewhat involved using polar coordinates, it is nevertheless much easier than if we had used Cartesian coordinates. This is just one example of many possibilities, which show that it is advantageous to recognize the symmetry of the problem by using the appropriate coordinate system.

This example was introduced, however, to illustrate another important point. Notice that Eq. (1–15) for r can be written as a Newtonian equation (i.e., in the form $F = ma$)

$$m\ddot{r} = -\frac{K}{r^2} + \frac{l^2}{mr^3}$$

if we interpret the term l^2/mr^3 as a force. This force is the well-known centrifugal force and must be introduced into the equation for $m\ddot{r}$.

This constitutes the main disadvantage of the Newtonian approach. The form of the equation $m\ddot{\eta} = F_\eta$ (where η is some general coordinate) is useful only in Cartesian systems, unless we are prepared to define additional forces, such as the centrifugal force in the above example. At times these necessary additional forces are fairly obscure.

There exist more convenient formulations of classical mechanics which are not tied to any one coordinate system. The two formulations that we are about to introduce are, in fact, independent of the coordinate system employed. These are the Lagrangian and the Hamiltonian formulations.

LAGRANGIAN APPROACH

Let K be the kinetic energy of a particle. In Cartesian coordinates

$$K(\dot{x}, \dot{y}, \dot{z}) = \frac{m}{2}(\dot{x}^2 + \dot{y}^2 + \dot{z}^2)$$

Let the potential energy be U. In many problems U is a function of position only, and so we write $U(x, y, z)$. Newton's equations are

$$m\ddot{x} = -\frac{\partial U}{\partial x}$$

with similar equations for y and z. Now introduce a new function

$$L(x, y, z, \dot{x}, \dot{y}, \dot{z}) \equiv K(\dot{x}, \dot{y}, \dot{z}) - U(x, y, z)$$

This function is called the Lagrangian of the system. In terms of L, we have

$$\frac{\partial L}{\partial \dot{x}} = \frac{\partial K}{\partial \dot{x}} = m\dot{x}$$

$$\frac{\partial L}{\partial x} = -\frac{\partial U}{\partial x}$$

$$\cdots$$

and we can write Newton's equations in the form

$$\frac{d}{dt}\left(\frac{\partial L}{\partial \dot{x}}\right) = \frac{\partial L}{\partial x} \tag{1-16}$$

with similar equations for y and z. These are Lagrange's equations of motion in Cartesian coordinates. The remarkable and useful property of Lagrange's equations is that they have the same form in any coordinate system. If the x, y, z are transformed into any other system, say q_1, q_2, q_3, Lagrange's equations take the form

$$\frac{d}{dt}\left(\frac{\partial L}{\partial \dot{q}_j}\right) = \frac{\partial L}{\partial q_j} \qquad j = 1, 2, 3 \tag{1-17}$$

This can be proved by writing $x = x(q_1, q_2, q_3)$, $y = y(q_1, q_2, q_3)$, and $z = z(q_1, q_2, q_3)$ and then transforming Eq. (1-16) into Eq. (1-17). (See Problem 1-13.)

Lagrange's equations are more useful than Newton's equations in many problems because it is usually much easier to write down an expression for the potential energy in some appropriate coordinate system than it is to recognize all the various forces. The Lagrangian formalism is based on the potential energy of the system, whereas the Newtonian approach is based on the forces acting on the system.

To illustrate the utility of the Lagrangian approach, we shall redo Example 3, the two-dimensional motion of a particle in a coulombic force field.

Example 3'.   The kinetic energy is

$$K = \frac{m}{2}(\dot{x}^2 + \dot{y}^2) = \frac{m}{2}(\dot{r}^2 + r^2\dot{\theta}^2)$$

and the potential energy is $U = -K/r$. The Lagrangian, then, is

$$L(r, \theta, \dot{r}, \dot{\theta}) = \frac{m}{2}(\dot{r}^2 + r^2\dot{\theta}^2) + \frac{K}{r} \tag{1-18}$$

The two Lagrangian equations of motion are

$$\frac{d}{dt}\left(\frac{\partial L}{\partial \dot{r}}\right) = \frac{\partial L}{\partial r}$$

$$\frac{d}{dt}\left(\frac{\partial L}{\partial \dot{\theta}}\right) = \frac{\partial L}{\partial \theta}$$

or using Eq. (1–18) for L,

$$\frac{d}{dt}(m\dot{r}) = mr\dot{\theta}^2 - \frac{K}{r^2}$$

$$\frac{d}{dt}(mr^2\dot{\theta}) = 0$$

These two equations are just Eqs. (1–11) and (1–13). Note, however, that they were obtained in a much more straightforward manner than were Eqs. (1–11) and (1–13). Problems 1–10 through 1–12 further illustrate the utility of the Lagrangian formulation. Other problems involve the motion of one and two particles in central force fields.

Equations (1–17) are three second-order ordinary differential equations. To completely specify the solutions, we need three initial velocities $\dot{q}_1(0), \dot{q}_2(0), \dot{q}_3(0)$, and three initial positions $q_1(0), q_2(0), q_3(0)$. These six initial conditions along with Lagrange's equations completely determine the future (and past) trajectory of the system. If there were N particles in the system, there would be $3N$ Lagrange equations and $6N$ initial conditions.

There is another formulation of classical mechanics that involves $6N$ first-order differential equations. Although this formulation is not as convenient as Lagrange's for solving problems, it is more convenient from a theoretical point of view, particularly in quantum mechanics and statistical mechanics. This is the Hamiltonian formulation.

HAMILTONIAN APPROACH

We define a generalized momentum by

$$p_j = \frac{\partial L}{\partial \dot{q}_j} \qquad j = 1, 2, \ldots, 3N \tag{1–19}$$

This generalized momentum is said to be *conjugate* to q_j. Note that Eq. (1–19) is simply $p_x = m\dot{x}$, and so on, in Cartesian coordinates.

We now define the Hamiltonian function for a system containing just one particle (for simplicity) by

$$H(p_1, p_2, p_3, q_1, q_2, q_3) = \sum_{j=1}^{3} p_j \dot{q}_j - L(\dot{q}_1, \dot{q}_2, \dot{q}_3, q_1, q_2, q_3) \tag{1–20}$$

It is understood here that the $\dot{q}_j$'s have been eliminated in favor of the p_j's by means of Eq. (1–19).

An important difference between the Lagrangian approach and the Hamiltonian approach is that the Lagrangian is considered to be a function of the generalized velocities $\dot{q}_j$ and the generalized coordinates q_j, whereas the Hamiltonian is considered to be a function of the generalized momenta p_j and the conjugate generalized coordinates q_j. This may appear to be a fine distinction at this point, but it will turn out to be important later on. It also may seem, at this time, that the definition Eq. (1–20) is rather obscure, but we shall give a motivation for its form in Section 1–4. (See Problem 1–38.)

For the kinds of systems that we shall treat in this book, the kinetic energy is of the form

$$K = \sum_{j=1}^{3N} a_j(q_1, q_2, \ldots, q_{3N}) \dot{q}_j^2 \tag{1–21}$$

that is, a quadratic function of the generalized velocities. The coefficients a_j are, in general, functions of generalized coordinates but not an explicit function of time. If, furthermore, the potential energy is a function only of the generalized coordinates, then the p_j occurring in Eq. (1–20) are given by

$$p_j = \frac{\partial L}{\partial \dot{q}_j} = \frac{\partial K}{\partial \dot{q}_j} = 2a_j \dot{q}_j$$

where the last equality comes from Eq. (1–21). Substituting this into Eq. (1–20) gives the important result

$$H = K + U = \text{total energy} \tag{1–22}$$

We shall now show that if the Lagrangian is not an explicit function of time, then $dH/dt = 0$. We begin with the definition of H, that is, Eq. (1–20).

$$dH = \sum_j \dot{q}_j \, dp_j + \sum_j p_j \, d\dot{q}_j - \sum \frac{\partial L}{\partial \dot{q}_j} d\dot{q}_j - \sum \frac{\partial L}{\partial q_j} dq_j$$

But if we use Eqs. (1–17) and (1–19), we see that

$$dH = \sum \dot{q}_j \, dp_j - \sum \dot{p}_j \, dq_j \tag{1–23}$$

The total derivative of H is (assuming no explicit dependence on time)

$$dH = \sum \left(\frac{\partial H}{\partial p_j} \right) dp_j + \sum \left(\frac{\partial H}{\partial q_j} \right) dq_j \tag{1–24}$$

Comparing Eqs. (1–23) and (1–24), we get Hamilton's equations of motion:

$$\frac{\partial H}{\partial p_j} = \dot{q}_j \qquad \frac{\partial H}{\partial q_j} = -\dot{p}_j \qquad j = 1, 2, \ldots, 3N \tag{1–25}$$

Hamilton's equations are $6N$ first-order differential equations. It is easy to show from Eqs. (1–24) and (1–25) that $dH/dt = 0$. (See Problem 1–14.) This along with Eq. (1–22) says that energy is conserved in such systems.

Since the Hamiltonian is so closely related to the energy, and it is the total energy which is usually the prime quantity in quantum and statistical mechanics, the Hamiltonian formalism will turn out to be the most useful from a conceptual point of view. Fortunately, however, we shall never have to solve the equations of motion for macroscopic systems. The role of statistical mechanics is to avoid doing just that.

1–3   QUANTUM MECHANICS

In the previous section we have seen that a knowledge of the initial velocities and coordinates of a particle or a system of particles was sufficient to determine the future course of the system if the equations of motion, essentially the potential field that the system experiences, are known. If the state (its velocities and coordinates) of the system is known at time t_0, then classical mechanics provides us with a method of calculating the state of the system at any other time t_1.

By the 1920s it was realized that such a calculation was too detailed in principle. The Heisenberg uncertainty principle states that it is impossible to precisely specify both the momentum and position of a particle simultaneously. Consequently, the prescription given by classical mechanics had to be modified to include the principle of uncertainty. This modification resulted in the development of quantum mechanics.

There are a number of levels of introducing the central ideas of quantum mechanics, but for most of the material in this text, we need consider only the most elementary. A fundamental concept of quantum mechanics is the so-called wave function $\Psi(\mathbf{q}, t)$, where $\mathbf{q}$ represents the set of coordinates necessary to describe the system. The wave function is given the physical interpretation that the probability that at time t the system is found between q_1 and $q_1 + dq_1$, q_2 and $q_2 + dq_2$, and so on, is

$$\Psi^*(\mathbf{q}, t)\Psi(\mathbf{q}, t)\, dq_1\, dq_2 \cdots dq_{3N}$$

We shall often write $dq_1 \cdots dq_{3N}$ as $d\mathbf{q}$. The uncertainty principle dictates that $\Psi(\mathbf{q}, t)$ is the most complete description of the system that can be obtained. Since the system is sure to be somewhere, we have

$$\int \Psi^*(\mathbf{q}, t)\Psi(\mathbf{q}, t)\, d\mathbf{q} = 1 \tag{1-26}$$

If Eq. (1–26) is satisfied, Ψ is said to be normalized.

A central problem of quantum mechanics is the calculation of $\Psi(\mathbf{q}, t)$ for any system of interest. We denote the time-independent part of $\Psi(\mathbf{q}, t)$ by $\psi(\mathbf{q})$. The state of the system described by a particular $\psi(\mathbf{q})$ is said to be a stationary state. Throughout this book we shall deal with stationary states only.

For our purpose, the wave function ψ is given as the solution of the Schrödinger equation

$$\mathscr{H}\psi = E\psi \tag{1-27}$$

where $\mathscr{H}$ is the Hamiltonian operator, and E is a scalar quantity corresponding to the energy of the system. The Hamiltonian operator is

$$\mathscr{H} = -\frac{\hbar^2}{2m}\left(\frac{\partial^2}{\partial x^2} + \frac{\partial^2}{\partial y^2} + \frac{\partial^2}{\partial z^2}\right) + U(x, y, z)$$

$$= -\frac{\hbar^2}{2m}\nabla^2 + U(x, y, z) \tag{1-28}$$

where $\hbar$ is $h/2\pi$, that is, Planck's constant divided by 2π. The first term here corresponds to the kinetic energy, and the second term is the potential energy. The Hamiltonian operator, then, corresponds to the total energy. There is a quantum mechanical operator and an equation similar to Eq. (1–27) corresponding to every quantity of classical mechanics, but we shall need only the one for the energy, namely, the Schrödinger equation.

Given certain physical boundary conditions of the system, a knowledge of $\mathscr{H}$ alone is sufficient to determine ψ and E. The wave function ψ is called an eigenfunction of the operator $\mathscr{H}$, and E is called an eigenvalue. There will usually be many ψ's and E's that satisfy Eq. (1–28), and this is indicated by labeling ψ and E with one or more subscripts. Generally, then, we have

$$\mathscr{H}\psi_j = E_j\psi_j \tag{1-29}$$

Equation (1–29) is a partial differential equation for ψ_j. The application of the boundary conditions often limits the values of E_j to only certain discrete values. Some simple examples are

1. a particle in a one-dimensional infinite well:

$$\mathcal{H} = -\frac{\hbar^2}{2m}\frac{\partial^2}{\partial x^2}$$

$$\varepsilon_n = \frac{h^2 n^2}{8ma^2} \qquad n = 1, 2, \ldots \tag{1–30}$$

2. a simple harmonic oscillator:

$$\mathcal{H} = -\frac{\hbar^2}{2m}\frac{\partial^2}{\partial x^2} + \frac{1}{2}kx^2$$

$$\varepsilon_n = (n + \tfrac{1}{2})\hbar\omega \qquad n = 0, 1, 2, \ldots \tag{1–31}$$

where $\omega = (k/m)^{1/2}$.

3. a rigid rotor (see Problem 1–21):

$$\mathcal{H} = -\frac{\hbar^2}{2I}\left\{\frac{1}{\sin\theta}\frac{\partial}{\partial\theta}\left(\sin\theta\,\frac{\partial}{\partial\theta}\right) + \frac{1}{\sin^2\theta}\frac{\partial^2}{\partial\phi^2}\right\}$$

$$\varepsilon_J = \frac{J(J+1)\hbar^2}{2I} \qquad J = 0, 1, 2, \ldots \tag{1–32}$$

Here I is the moment of inertia of the rotor (see Problem 1–15 for a treatment of the classical counterpart of this system).

The rigid rotor illustrates another important concept of quantum mechanics, namely, that of degeneracy. It happens that there may be a number of eigenfunctions or states of the system having the same eigenvalue or energy. The number of eigenfunctions having this energy is called the degeneracy of the system. For the rigid rotor, the degeneracy, ω_J, is $2J + 1$. The particle in a one-dimensional infinite well and the simple harmonic oscillator are nondegenerate, that is, the ω_n are unity. The concept of energy states and degeneracy plays an important role in statistical thermodynamics.

Consider the energy states of a particle in a three-dimensional infinite well. These are given by

$$\varepsilon_{n_x n_y n_z} = \frac{h^2}{8ma^2}(n_x^2 + n_y^2 + n_z^2) \qquad n_x, n_y, n_z = 1, 2, 3, \ldots \tag{1–33}$$

The degeneracy is given by the number of ways that the integer $M = 8ma^2\varepsilon/h^2$ can be written as the sum of the squares of three positive integers. In general, this is an erratic and discontinuous function of M (the number of ways will be zero for many values of M), but it becomes smooth for large M, and it is possible to derive a simple expression for it. Consider a three-dimensional space spanned by n_x, n_y, and n_z. There is a one-to-one correspondence between energy states given by Eq. (1–33) and the points in this n_x, n_y, n_z space with coordinates given by positive integers. Figure 1–1 shows a two-dimensional version of this space. Equation (1–33) is an equation for a sphere of radius $R = (8ma^2\varepsilon/h^2)^{1/2}$ in this space

$$n_x^2 + n_y^2 + n_z^2 = \frac{8ma^2\varepsilon}{h^2} = R^2$$

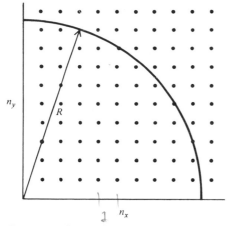

Figure 1–1. **A two-dimensional version of the** (n_x, n_y, n_z) **space, the space with the quantum numbers** $n_x, n_y,$ **and** n_z **as axes.**

We wish to calculate the number of lattice points that are at some fixed distance from the origin in this space. In general, this is very difficult, but for large R we can proceed as follows. We treat R or ε as a continuous variable and ask for the number of lattice points between ε and $\varepsilon + \Delta\varepsilon$. To calculate this quantity, it is convenient to first calculate the number of lattice points consistent with an energy $\leq \varepsilon$. For large ε, it is an excellent approximation to equate the number of lattice points consistent with an energy $\leq \varepsilon$ with the volume of one octant of a sphere of radius R. We take only one octant, because n_x, n_y, and n_z are restricted to be positive integers. If we denote the number of such states by $\Phi(\varepsilon)$, we can write

$$\Phi(\varepsilon) = \frac{1}{8}\left(\frac{4\pi R^3}{3}\right) = \frac{\pi}{6}\left(\frac{8ma^2\varepsilon}{h^2}\right)^{3/2} \tag{1–34}$$

The number of states between ε and $\varepsilon + \Delta\varepsilon$ $(\Delta\varepsilon/\varepsilon \ll 1)$ is

$$\omega(\varepsilon, \Delta\varepsilon) = \Phi(\varepsilon + \Delta\varepsilon) - \Phi(\varepsilon)$$

$$= \frac{\pi}{4}\left(\frac{8ma^2}{h^2}\right)^{3/2}\varepsilon^{1/2}\,\Delta\varepsilon + O((\Delta\varepsilon)^2) \tag{1–35}$$

If we take $\varepsilon = 3kT/2$, $T = 300°K$, $m = 10^{-22}$ g, $a = 10$ cm, and $\Delta\varepsilon$ to be 0.01ε (in other words a 1% band around ε), then $\omega(\varepsilon, \Delta\varepsilon)$ is $O(10^{28})$.* So even for a system as simple as a particle-in-a-box, the degeneracy can be very large at room temperature.

For an N-particle system, the degeneracy is tremendously greater than $O(10^{28})$. To see this, consider a system of N noninteracting particles in a cube. The energy of this system is

$$E = \frac{h^2}{8ma^2}\sum_{j=1}^{N}(n_{xj}^2 + n_{yj}^2 + n_{zj}^2) = \frac{h^2}{8ma^2}\sum_{j=1}^{3N}s_j^2$$

where n_{xj}, n_{yj}, n_{zj}, and s_j are positive integers. The degeneracy of this system can be calculated by generalizing the above derivation for one particle. Using the volume of

---

* We use the notation $O(10^{28})$ say, to mean of the order of magnitude 10^{28}. This differs from standard mathematical notation, but there should be no confusion.

an N-dimensional sphere from Problem 1–24, the number of states with energy $\leq E$ is*

$$\Phi(E) = \frac{1}{\Gamma(N+1)\Gamma[(3N/2)+1]}\left(\frac{2\pi ma^2 E}{h^2}\right)^{3N/2} \tag{1–36}$$

where $\Gamma(n)$ here is the gamma function. (See Problem 1–58.) The number of states between E and $E + \Delta E$ is

$$\Omega(E, \Delta E) = \frac{1}{\Gamma(N+1)\Gamma(3N/2)}\left(\frac{2\pi ma^2}{h^2}\right)^{3N/2} E^{(3N/2-1)}\,\Delta E \tag{1–37}$$

In this case, $E = 3NkT/2$. If we take $T = 300°K$, $m = 10^{-22}$ g, $a = 10$ cm, $N = 6.02 \times 10^{23}$, and ΔE equal to $0.01E$, we get $\Omega(E, \Delta E)$ to be $O(10^N)$ (see Problem 1–23), an extremely large number. This shows that as the number of particles in the system increases, the quantum mechanical degeneracy becomes enormous. Although we have shown this only for a system of noninteracting particles confined to a cubical box, that is, an ideal gas, the result is generally true. We shall see in the next chapter that the concept of the degeneracy of a macroscopic system is very important.

There is another quantum mechanical result that we shall use later on. It often happens that the Hamiltonian of a many-body system can be written either exactly or approximately as a summation of one-particle or few-particle Hamiltonians, that is,

$$\mathcal{H} = \mathcal{H}_\alpha + \mathcal{H}_\beta + \mathcal{H}_\gamma + \cdots \tag{1–38}$$

Let the eigenvalues of $\mathcal{H}_j$ be ε_j, and the eigenfunctions be ψ_j, where $j = \alpha, \beta, \gamma, \ldots$. To solve the many-body Schrödinger equation, we let $\psi = \psi_\alpha \psi_\beta \psi_\gamma \ldots$. Then

$$\begin{aligned}
\mathcal{H}\psi &= (\mathcal{H}_\alpha + \mathcal{H}_\beta + \mathcal{H}_\gamma + \cdots)\psi_\alpha \psi_\beta \psi_\gamma \cdots \\
&= \psi_\beta \psi_\gamma \cdots \mathcal{H}_\alpha \psi_\alpha + \psi_\alpha \psi_\gamma \cdots \mathcal{H}_\beta \psi_\beta + \cdots \\
&= \psi_\beta \psi_\gamma \varepsilon_\alpha \psi_\alpha + \psi_\alpha \psi_\gamma \varepsilon_\beta \psi_\beta + \cdots \\
&= (\varepsilon_\alpha + \varepsilon_\beta + \cdots)\psi_\alpha \psi_\beta \psi_\gamma \cdots = E\psi
\end{aligned} \tag{1–39}$$

In other words, the energy of the entire system is the sum of the energies of the individual particles if they do not interact. This is a very important result and will allow us to reduce a many-body problem to a one-body problem if the interactions are weak enough to ignore, such as in the case of a dilute gas. We shall see a number of cases where, even though the interactions are too strong to be ignored (such as in a solid), it is possible to formally or mathematically write the Hamiltonian in the form of Eq. (1–38). This will lead to defining quasi-particles like phonons and photons.

The last quantum mechanical topic we shall discuss here is that of the symmetry of wave functions with respect to the interchange of identical particles. Consider a system of N identical particles, described by a wave function $\psi(1, 2, 3, \ldots, N)$, where 1 denotes the coordinates of particle 1, and so on. If we interchange the position of any two of the particles, say particles 1 and 2, the wave function must either remain the same or change sign. (See Problem 1–26.) Thus if we let P_{12} be an operator that exchanges the two identical particles 1 and 2, then

$$\begin{aligned}
P_{12}\psi(1, 2, 3, \ldots, N) &= \psi(2, 1, 3, \ldots, N) \\
&= \pm\psi(1, 2, 3, \ldots, N)
\end{aligned} \tag{1–40}$$

---

* The extra factor of $\Gamma(N+1)$ occurs here because of the indistinguishability of the N particles. This will be discussed fully in Chapter 4.

It turns out that whether the wave function remains the same or changes sign is a function of the nature of the two identical particles that are exchanged. For particles with an integral spin (such as the He-4 nucleus, photons, ...), the wave function remains the same. In this case the wave function is called symmetric, and such particles are called bosons. For particles with half-integral spin (such as electrons, ...), the wave function is called antisymmetric, and the particles are called fermions. Chapter 4 considers the consequences of this symmetry requirement of wave functions.

1–4 THERMODYNAMICS

In this section we shall not attempt to review thermodynamics, but shall simply state the three laws and briefly discuss their consequences. Problems 1–27 through 1–36 review some of the equations and manipulations that arise in thermodynamics. Two topics that are not often treated in elementary physical chemistry are presented here, namely, Legendre transformations and Euler's theorem. Both of these topics will be used later on.

The pressure-volume work done by a system on its surroundings in going from state A to state B is

$$w = \int_A^B p\, dV$$

where p is the pressure exerted by the surroundings on the system. The differential quantity δw is positive if dV is positive.

The heat absorbed by the system from the surroundings during the change of the system from state A to state B is

$$q = \int_A^B \delta q$$

The first law of thermodynamics states that even though w and q depend upon the path taken from A to B, their difference does not. Their difference, then, is a function only of the two states A and B, or, namely, is a state function. This function is called the internal energy or thermodynamic energy and is denoted by E.

The first law of thermodynamics is

$$\Delta E = E_B - E_A = q - w$$

$$= \int_A^B \delta q - \int_A^B p\, dV \tag{1–41}$$

For simplicity, we consider only p–V work.

A reversible change is one in which the driving force (a difference in pressure, a difference in temperature, and so on) is infinitesimal. Any other change is called irreversible or spontaneous. Problem 1–27 asks the reader to show that for an isothermal process, $w_{rev} > w_{irrev}$ and $q_{rev} > q_{irrev}$.

The first law of thermodynamics is nothing but a statement of the law of conservation of energy. The second law is somewhat more abstract and can be stated in a number of equivalent ways. One of them is: There is a quantity S, called entropy, which is a state function. In an irreversible process, the entropy of the system and its surroundings increases. In a reversible process, the entropy of the system and its surroundings remains constant. The entropy of the system and its surroundings never decreases. The system and its surroundings are often referred to as the universe.

The mathematical expression for the difference in entropy between states A and B of a system is given by

$$\Delta S = \int_A^B \frac{dq_{rev}}{T} \tag{1-42}$$

Note that the heat appearing here is that associated with a reversible process. To compute ΔS between two states A and B, we must take the system from A to B in a reversible manner.

Another statement of the second law is: Along any reversible path, there exists an integrating factor T, common to all systems such that

$$dS = \frac{dq_{rev}}{T} \tag{1-43}$$

is an exact differential, that is, that S is a state function. Thus

$$\Delta S = \int_A^B \frac{dq_{rev}}{T}$$

For all other processes

$$\Delta S > \int_A^B \frac{dq}{T}$$

where T is the temperature of the surroundings.

The third law of thermodynamics states: If the entropy of each element in some crystalline state be taken as zero at the absolute zero of temperature, every substance has a finite positive entropy, but at the absolute zero of temperature, the entropy may become zero, and does become so in the case of perfect crystalline substances.

The second law is concerned with only the difference in the entropy between two states. The third law allows us to calculate the absolute entropy of a substance by means of the expressions

$$S - S_0 = \int_0^T \frac{dq_{rev}}{T} \quad \text{and} \quad S_0 = 0 \tag{1-44}$$

Problem 1–36 asks you to calculate the absolute entropy of gaseous nitromethane at its boiling point.

For simple one-component systems, the first law can be written in the form

$$dE = T\,dS - p\,dV \tag{1-45}$$

This implies that

$$\left(\frac{\partial E}{\partial S}\right)_V = T \quad \text{and} \quad \left(\frac{\partial E}{\partial V}\right)_S = -p \tag{1-46}$$

The simplicity of these partial derivatives implies that E is a "natural" function of S and V. For example, if we were to consider E to be a function of V and T (see Problem 1–30), we would get

$$dE = \left[T\left(\frac{\partial p}{\partial T}\right)_V - p\right]dV + C_V\,dT$$

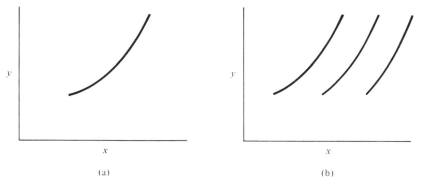

Figure 1–2. (a) **shows the function** $y(x)$ **and** (b) **shows a family of functions, all of which give the same value of** y **for any fixed value of** p.

Note that in this case the coefficients of dV and dT are not as simple as the coefficients of dV and dS obtained when E is expressed as a function of S and V. The " simplicity " of the expression $dE = T\,dS - p\,dV$ suggests that S and V are the " natural " variables for E. The quantities S and V (especially S) are difficult to control in the laboratory and consequently are not always the most desirable independent variables. A more useful pair might be (T, V) or (T, p). An important question that arises, then, is the existence of other thermodynamic state functions whose natural variables are (T, V) or (T, p), and so on. Furthermore, how would one find them if they do exist. This leads us to the topic of Legendre transformations.

We shall discuss a function of one variable in some detail and then simply present the generalization to a function of many variables. Consider a function $y = y(x)$, and let its slope be $p = p(x)$. We wish to describe the function $y(x)$ in terms of its slope. Figure 1–2, however, shows that the slope alone is not sufficient to completely specify $y(x)$. Figure 1–2(a) shows the curve $y(x)$, and Fig. 1–2(b) shows a family of curves, all of which give the same value of y for any one value of p. In order to uniquely describe the curve in Fig. 1–2(a), we must select one member of the family of curves in Fig. 1–2(b). We do this by specifying the intercepts of the tangent lines with the y-axis. Let the intercept be $\phi(p)$. Instead of describing the curve in Fig. 1–2(a) by y versus x, then, we can equally well represent it by specifying the slope at each point along with the intercept of the slope with the y-axis. Figure 1–3 shows these two representations. One sees that either representation can be used to describe the function. The relation between the two representations can be obtained by referring to Fig. 1–4. This figure shows that the slope p at any point is given by

$$p = \frac{y - \phi}{x - 0}$$

The result that we are after is

$$\phi(p) = y - px \tag{1–47}$$

The function $\phi(p)$ is the Legendre transformation of y. It is completely equivalent to $y(x)$, but considers p to be the independent variable instead of x. This may not be clear from the notation in Eq. (1–47), but it is understood there that y and x have been eliminated in favor of p by using the equations $y = y(x)$ and $x = x(p)$.

Let us apply this to the thermodynamic energy $E(S, V)$. We seek a function of T and V that is completely equivalent to E. Equation (1–46) shows that $T = (\partial E/\partial S)_V$,

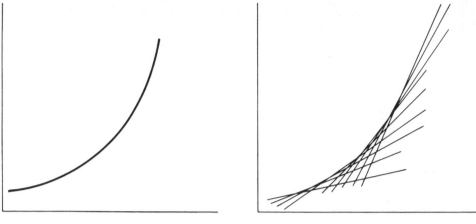

(a) y-x representation   (b) ϕ-p representation

Figure 1–3.   **In (a), the function is represented by the locus of points, y versus x. In (b), the same function is given by the envelope of its tangent curves.**

and so we are in a position to apply Eq. (1–47) directly. This can be treated as a one-variable problem, since V is held fixed throughout. Therefore the Legendre transformation of E that considers T and V to be the independent variables is $E - TS$. Of course, this is the Helmholtz free energy

$$A(T, V) = E - TS \tag{1–48}$$

whose differential form is

$$dA = -S\,dT - p\,dV \tag{1–49}$$

This shows that the natural variables of A are T and V. Another motivation for saying this is that the condition for equilibrium at constant T and V is that A assume its minimum value, or that $\Delta A \leq 0$ for a spontaneous process at constant T and V. To prove this, write

$$\begin{aligned} dA &= dE - T\,dS - S\,dT \\ &= \delta q - p\,dV - \delta q_{\text{rev}} - S\,dT \\ &= \delta q - \delta q_{\text{rev}} \end{aligned} \tag{1–50}$$

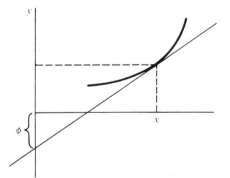

Figure 1–4.   **The diagram used to derive the connection between the y-x representation and the ϕ-p representation.**

at constant T and V. But $\delta q \leq \delta q_{\text{rev}}$ (see Problem 1–27), and so $\Delta A \leq 0$ at constant T and V.

In elementary physical chemistry, the function $A = E - TS$ is often presented as an a priori definition. But it should be apparent now that this form is dictated by the Legendre transformation if one specifies T and V to be the independent variables.

A function, whose natural variables are S and p, can be obtained in the same manner. Equation (1–46) shows that $p = -(\partial E/\partial V)_S$, so Eq. (1–47) gives that $E + pV$ is a thermodynamic state function, whose natural variables are S and p. This function is, of course, the enthalpy.

The generalization of Eq. (1–47) to more than one variable is simply

$$\phi(p) = y - \sum_j p_j x_j \tag{1–51}$$

where the x_j's are the independent variables of y, and $p_j = (\partial y/\partial x_j)$. We can use Eq. (1–51) to construct a thermodynamic state function, whose natural variables are T and p. Using Eqs. (1–46) and Eq. (1–51), we see that such a function is $E - TS + pV$, the Gibb's free energy. Its differential form $dG = -S\,dT + V\,dp$ and the fact that $\Delta G \leq 0$ for a spontaneous change at constant T and p suggest that T and p are the natural variables of G.

Up to this point we have considered only closed one-component systems. In general, E, H, A, and G depend upon the number of moles or molecules of each component. If we let N_j be the number of moles of component j, we have

$$dE = T\,dS - p\,dV + \sum_j \left(\frac{\partial E}{\partial N_j}\right)_{S,V,N_K,\,j \neq K} dN_j \tag{1–52}$$

$$= T\,dS - p\,dV + \sum_j \mu_j\,dN_j \tag{1–53}$$

where the second line defines μ_j. By adding $d(pV)$ to both sides of Eq. (1–53), we get

$$dH = T\,dS + V\,dp + \sum_j \mu_j\,dN_j \tag{1–54}$$

If we subtract $d(TS)$ from both sides of Eq. (1–53), we get

$$dA = -S\,dT - p\,dV + \sum_j \mu_j\,dN_j \tag{1–55}$$

Similar manipulations give

$$dG = -S\,dT + V\,dp + \sum_j \mu_j\,dN_j \tag{1–56}$$

Equations (1–52) through (1–56) show that

$$\mu_j = \left(\frac{\partial E}{\partial N_j}\right)_{S,V,\ldots} = \left(\frac{\partial H}{\partial N_j}\right)_{S,p,\ldots} = \left(\frac{\partial A}{\partial N_j}\right)_{V,T,\ldots} = \left(\frac{\partial G}{\partial N_j}\right)_{p,T,\ldots} \tag{1–57}$$

The quantity μ_j is called the chemical potential.

There is a mathematical theorem, called Euler's theorem, which is very useful in thermodynamics. Before discussing Euler's theorem, however, we must define extensive and intensive variables. Extensive properties are additive; their value for the whole system is equal to the sum of their values for the individual parts. Examples are the volume, mass, and entropy. Intensive properties are not additive. Examples are temperature and pressure. The temperature of any small part of a system in equilibrium is the same as the temperature of the whole system. Euler's theorem deals with extensive and intensive variables. If

$$f(\lambda x_1, \lambda x_2, \ldots, \lambda x_N) = \lambda^n f(x_1, x_2, \ldots, x_N) \tag{1-58}$$

f is said to be a homogeneous function of order n. The functions $f(x) = 3x^2$ and $f(x, y, z) = xy^2 + z^3 - 6x^4/y$ are homogeneous functions of degree 2 and 3, respectively, whereas $f(x) = x^2 + 2x - 3$ and $f(x, y) = xy - e^{xy}$ are not homogeneous. Euler's theorem states that if $f(x_1, \ldots, x_N)$ is a homogeneous function of order n, then

$$nf(x_1, \ldots, x_N) = x_1 \frac{\partial f}{\partial x_1} + x_2 \frac{\partial f}{\partial x_2} + \cdots + x_N \frac{\partial f}{\partial x_N} \tag{1-59}$$

The proof of Euler's theorem is simple. Differentiate Eq. (1–58) with respect to λ:

$$n\lambda^{n-1} f(x_1, x_2, \ldots, x_N) = \left(\frac{\partial f}{\partial \lambda x_1}\right)\left(\frac{\partial \lambda x_1}{\partial \lambda}\right) + \left(\frac{\partial f}{\partial \lambda x_2}\right)\left(\frac{\partial \lambda x_2}{\partial \lambda}\right) + \cdots + \left(\frac{\partial f}{\partial \lambda x_N}\right)\left(\frac{\partial \lambda x_N}{\partial \lambda}\right)$$

$$= x_1 \left(\frac{\partial f}{\partial \lambda x_1}\right) + x_2 \left(\frac{\partial f}{\partial \lambda x_2}\right) + \cdots + x_N \left(\frac{\partial f}{\partial \lambda x_N}\right)$$

Euler's theorem is proved by letting $\lambda = 1$.

Extensive thermodynamic variables are homogeneous of degree 1. Let us apply Euler's theorem to the Gibb's free energy.

$$G(T, p, \lambda N_1, \lambda N_2, \ldots) = \lambda G(T, p, N_1, N_2, \ldots)$$

The variables T and p here can be treated as constants. Equation (1–59) gives that

$$G = \sum_j N_j \left(\frac{\partial G}{\partial N_j}\right)_{T, p, \ldots} = \sum_j N_j \mu_j \tag{1-60}$$

Taking the derivative of this at constant T and p,

$$dG = \sum_j N_j \, d\mu_j + \sum_j \mu_j \, dN_j \qquad \text{(constant } T \text{ and } p)$$

But using Eq. (1–56) at constant T and p, we have

$$\sum_j N_j \, d\mu_j = 0 \qquad \text{(constant } T \text{ and } p) \tag{1-61}$$

This is called the Gibbs-Duhem equation and is very useful in the thermodynamic study of solutions. (See *Physical Chemistry*, 4th ed., by W. J. Moore, p. 235, under "Additional Reading," for a simple application of the Gibbs-Duhem equation.)

We shall conclude this section on thermodynamics with a brief discussion of the application of thermodynamics to chemical equilibria. Consider the general reaction

$$v_A A + v_B B + \cdots \rightleftharpoons v_D D + v_E E + \cdots \tag{1-62}$$

The capital letters represent the formulas of the compounds, and the v_j represent stoichiometric coefficients. It is more convenient to write Eq. (1–62) mathematically as

$$v_D D + v_E E + \cdots - v_A A - v_B B - \cdots = 0 \tag{1-63}$$

Define the extent of reaction λ, such that $dN_j = v_j \, d\lambda$ for all j, where the v's for products are positive, and those for reactants are negative.

At constant T and p, we have

$$dG = \sum_j \mu_j \, dN_j = \left(\sum_j \mu_j v_j\right) d\lambda \qquad \text{(constant } T \text{ and } p)$$

At equilibrium, G must be a minimum with respect to λ, so we write

$$dG = \sum_j \mu_j v_j = v_D \mu_D + v_E \mu_E + \cdots - v_A \mu_A - v_B \mu_B - \cdots = 0 \tag{1-64}$$

at equilibrium. The equilibrium between phases can be considered to be a chemical reaction of the form $A \rightleftharpoons B$, and so Eq. (1–64) gives that $\mu_A = \mu_B$ from the equilibrium condition between two pure phases.

Now consider the application of Eq. (1–64) to a chemical reaction between gases dilute enough to be considered ideal. Let the reaction be $v_A A + v_B B \rightleftharpoons v_C C + v_D D$. At constant temperature,

$$dG = V\,dp \qquad \text{(constant } T)$$

and so

$$G - G^0 = \int_{p_0}^{p} V\,dp = \int_{p_0}^{p} \frac{NkT}{p}\,dp = NkT \ln \frac{p}{p_0} \tag{1-65}$$

In this equation G^0 is the standard free energy of the gas, the standard state being the gas at a pressure p_0. Usually p_0 is taken to be 1 atmosphere. If we take N to be 1 mole, then G and G^0 become μ and μ^0. Each component in the reactive gas mixture will have an equation of the form of Eq. (1–65), and so we have

$$\mu_j(T, p) = \mu_j{}^0(T) + RT \ln \frac{p_j}{p_{0j}} \tag{1-66}$$

The total free energy change is

$$\Delta\mu = v_C \mu_C + v_D \mu_D - v_A \mu_A - v_B \mu_B$$

$$= \Delta\mu^0 + RT \ln \frac{(p_C')^{v_C}(p_D')^{v_D}}{(p_A')^{v_A}(p_B')^{v_B}} \tag{1-67}$$

In this equation the (p')'s are p/p_0, that is, they are the pressures relative to the standard states. These (p')'s are unitless. The argument of the logarithm here has the *form* of an equilibrium constant, but is not equal to the equilibrium constant unless the pressures are those which exist at chemical equilibrium. Equation (1–67) gives the change in free energy of the conversion of reactants at *arbitrary* pressures to products at *arbitrary* pressures.

At equilibrium, $\Delta\mu = 0$, and we have

$$\Delta\mu^0 = -RT \ln \left[\frac{(p_C')^{v_C}(p_D')^{v_D}}{(p_A')^{v_A}(p_B')^{v_B}} \right]_{\text{equilibrium}}$$

$$= -RT \ln K_p \tag{1-68}$$

There are extensive tabulations of μ^0's, and so $\Delta\mu^0$ is a simple matter to calculate. We see that if $\Delta\mu^0 < 0$, then $K_p > 1$, that is, the conversion of reactants *in their standard states* to products *in their standard states* proceeds spontaneously. On the other hand, if $\Delta\mu^0 > 0$, then $K_p < 1$, and we conclude that the reaction does not proceed spontaneously. It should be understood, however, that this applies only to reactants and products in their standard states. In general, it is $\Delta\mu$ along with Eq. (1–67) that determines the extent of a chemical reaction. (See Problem 1–34.)

1–5   MATHEMATICS

In this section we shall discuss several mathematical techniques or results that are repeatedly used in statistical thermodynamics. The topics we shall discuss here are random variables and distribution functions, Stirling's approximation, the binomial and multinomial coefficients, the Lagrange method of undetermined multipliers, and the behavior of binomial and multinomial coefficients for large numbers.

PROBABILITY DISTRIBUTIONS

Let u be a variable which can assume the M discrete values $u_1, u_2, \ldots, u_M$ with corresponding probabilities $p(u_1), p(u_2), \ldots, p(u_M)$. The variable u is said to be a discrete random variable, and $p(u)$ is said to be a discrete distribution. The mean, or average, value of u is

$$\bar{u} = \frac{\sum_{j=1}^{M} u_j\, p(u_j)}{\sum_{j=1}^{M} p(u_j)}$$

Since $p(u_j)$ is a probability, $p(u_j)$ must be normalized, that is, the summation in the denominator must equal unity. The mean of any function of u, $f(u)$, is given by

$$\overline{f(u)} = \sum_{j=1}^{M} f(u_j)p(u_j) \tag{1–69}$$

If $f(u) = u^m$, $\overline{f(u)}$ is called the mth moment of the distribution $p(u)$. If $f(u) = (u - \bar{u})^m$, $\overline{f(u)}$ is called the mth central moment, that is, the mth moment about the mean. In particular, the mean of $(u - \bar{u})^2$ is called the variance, and is a measure of the spread of the distribution. The square root of the variance is the standard deviation.

A very commonly occurring and useful discrete distribution is the Poisson distribution:

$$P(m) = \frac{a^m e^{-a}}{m!} \qquad m = 0, 1, 2, \ldots \tag{1–70}$$

This distribution has been applied to shot noise in electron tubes, the distribution of galaxies in space, aerial search, and many others.* (See Problem 1–42.)

If the random variable U is continuous rather than discrete, then we interpret $p(u)\,du$ as the probability that the random variable U lies between the values u and $u + du$. The mean of any function of U is

$$\overline{f(u)} = \int f(u)p(u)\, du \tag{1–71}$$

The limits of the integral are over the entire range of U.

The most important continuous probability distribution is the Gaussian distribution:

$$p(x) = \frac{1}{(2\pi\sigma^2)^{1/2}} \exp\left\{-\frac{(x - \bar{x})^2}{2\sigma^2}\right\} \qquad -\infty \le x \le \infty \tag{1–72}$$

The quantity σ^2, which is the variance, controls the width of the Gaussian distribution. The smaller the σ, the narrower the Gaussian distribution becomes. In the limit $\sigma \to 0$, Eq. (1–72) becomes a delta function (this is one representation of a delta function of Appendix B). Problems 1–43 through 1–45 involve some important results based on Eq. (1–72).

* See *Modern Probability Theory and Its Applications* by E. Parzen (New York: Wiley, 1960).

STIRLING'S APPROXIMATION

In statistical thermodynamics we often encounter factorials of very large numbers, such as Avogadro's number. The calculation and mathematical manipulation of factorials become awkward for large N. Therefore it is desirable to find an approximation for $N!$ for large N. Problems of this sort occur often in mathematics and are called asymptotic approximations, that is, an approximation to a function which improves as the argument of that function increases. Since $N!$ is actually a product, it is convenient to deal with $\ln N!$ because this is a sum. The asymptotic approximation to $\ln N!$ is called Stirling's approximation, which we now derive.

Since $N! = N(N - 1)(N - 2) \cdots (2)(1)$, $\ln N!$ is

$$\ln N! = \sum_{m=1}^{N} \ln m \tag{1-73}$$

Figure 1–5 shows $\ln x$ plotted versus x. The sum of the areas under these rectangles up to N is $\ln N!$. Figure 1–5 also shows the continuous curve $\ln x$ plotted on the same graph. Thus $\ln x$ is seen to form an envelope to the rectangles, and this envelope becomes a steadily smoother approximation to the rectangles as x increases. We can approximate the area under these rectangles by the integral of $\ln x$. The area under $\ln x$ will poorly approximate the rectangles only in the beginning. If N is large enough (we are deriving an asymptotic expansion), this area will make a negligible contribution to the total area. We may write, then,

$$\ln N! = \sum_{m=1}^{N} \ln m \approx \int_{1}^{N} \ln x \, dx = N \ln N - N \qquad (N \text{ large}) \tag{1-74}$$

which is Stirling's approximation to $\ln N!$. The lower limit could just as well have been taken as 0 in Eq. (1–74), since N is large. (Remember that $x \ln x \to 0$ as $x \to 0$.)

A more refined derivation of Stirling's approximation gives $\ln N! \approx N \ln N - N + \ln(2\pi N)^{1/2}$, but this additional term is seldom necessary. (See Problem 1–59.)

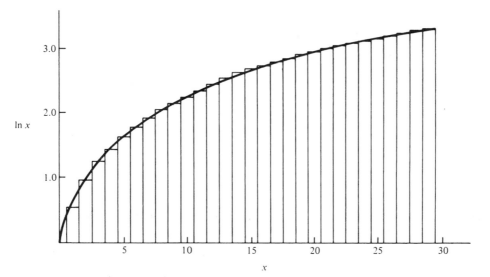

Figure 1–5. **A plot of $\ln x$ versus x, showing how the summation of $\ln m$ can be approximated by the integral of $\ln x$.**

BINOMIAL AND MULTINOMIAL DISTRIBUTION

During the course of our discussion of the canonical ensemble, we shall encounter the problem of determining how many ways it is possible to divide N distinguishable systems into groups such that there are n_1 systems in the first group, n_2 systems in the second group, and so on, and such that $n_1 + n_2 + \cdots = N$, that is, all the systems are accounted for. This is actually one of the easiest problems in combinatorial analysis. To solve this, we first calculate the number of permutations of N distinguishable objects, that is, the number of possible different arrangements or ways to order N distinguishable objects. Let us choose one of the N objects and place it in the first position, one of the $N - 1$ remaining objects and place it in the second position, and so on, until all N objects are ordered. Clearly there are N choices for the first position, $N - 1$ choices for the second position, and so on, until finally there is only one object left for the Nth position. The total number of ways of doing this is then the product of all the choices,

$$N(N - 1)(N - 2) \cdots (2)(1) \equiv N! \qquad \text{(distinguishable objects)}$$

Next we calculate the number of ways of dividing N distinguishable objects into two groups, one group containing N_1 objects, say, and the other containing the remaining $N - N_1$. There are $N(N - 1) \cdots (N - N_1 + 1)$ ways to form the first group, and $N_2! = (N - N_1)!$ ways to form the second group. The total number is, then, the product

$$N(N - 1) \cdots (N - N_1 + 1) \times (N - N_1)! = \frac{N!}{(N - N_1)!} \times (N - N_1)! = N!$$

But this has overcounted the situation drastically, since the order in which we place N_1 members in the first group and N_2 in the second group is immaterial to the problem as stated. All $N_1!$ orders of the first group and $N_2!$ orders of the second group correspond to just one division of N objects into N_1 objects and N_2 objects. Therefore the desired result is

$$\frac{N!}{N_1!(N - N_1)!} = \frac{N!}{N_1! N_2!} \tag{1-75}$$

Since the combination of factorials in Eq. (1-75) occurs in the binomial expansion,

$$(x + y)^N = \sum_{N_1 = 0}^{N} \frac{N! x^{N - N_1} y^{N_1}}{N_1!(N - N_1)!} = \sum_{N_1 N_2}^{*} \frac{N! x^{N_1} y^{N_2}}{N_1! N_2!} \tag{1-76}$$

$N!/N_1!(N - N_1)!$ is called a binomial coefficient. The asterisk on the second summation in Eq. (1-76) signifies the restriction $N_1 + N_2 = N$.

The generalization of Eq. (1-75) to the division of N into r groups, the first containing N_1, and so on, is easily seen to be

$$\frac{N!}{N_1! N_2! \cdots N_r!} = \frac{N!}{\prod_{j=1}^{r} N_j!} \tag{1-77}$$

where $N_1 + N_2 + \cdots + N_r = N$. This is known as a multinomial coefficient, since it occurs in the expansion

$$(x_1 + x_2 + \cdots + x_r)^N = \sum_{N_1 = 0}^{N} \sum_{N_2 = 0}^{N} \cdots \sum_{N_r = 0}^{N*} \frac{N! x_1^{N_1} \cdots x_r^{N_r}}{\prod_{j=1}^{r} N_j!} \tag{1-78}$$

where this time the asterisk signifies the restriction $N_1 + N_2 + \cdots + N_r = N$.

There are a number of other combinatorial formulas that are useful in statistical thermodynamics, but Eq. (1–77) is the most useful for our purposes. Combinatorial formulas can become rather demanding to derive. We refer to Appendix AVII of Mayer and Mayer* which contains a collection of formulas.

METHOD OF LAGRANGE MULTIPLIERS

It will be necessary, later, to maximize Eq. (1–77) with the constraint $N_1 + N_2 + \cdots + N_r = $ constant. This brings us to the mathematical problem of maximizing a function of several (or many) variables $f(x_1, x_2, \ldots, x_r)$ when the variables are connected by other equations, say $g_1(x_1, \ldots, x_r) = 0$, $g_2(x_1, \ldots, x_r) = 0$, and so on. This type of problem is readily handled by the method of Lagrange undetermined multipliers.

If it were not for the constraints, $g_j(x_1, x_2, \ldots, x_r) = 0$, the maximum of $f(x_1, \ldots, x_r)$ would be given by

$$\delta f = \sum_{j=1}^{r} \left(\frac{\partial f}{\partial x_j} \right)_0 \delta x_j = 0 \tag{1–79}$$

where the zero subscript indicates that this equation equals zero only when the r partial derivatives are evaluated at the maximum (or minimum) of f. Denote these values of x_j by $x_j{}^0$. If there were no constraints, each of the δx_j would be able to be varied independently and arbitrarily, and so we would conclude that $(\partial f/\partial x_j) = 0$ for every j, since δf must equal zero. This would give r equations from which the values of the $r x_j{}^0$ could be obtained.

On the other hand, if there is some other relation between the x's, such as $g(x_1, x_2, \ldots, x_r) = 0$, we have the additional equation

$$\delta g = \sum_{j=1}^{r} \left(\frac{\partial g}{\partial x_j} \right)_0 \delta x_j = 0 \tag{1–80}$$

This equation serves as a constraint that the δx_j must satisfy, thus making one of them depend upon the other $r - 1$. In the Lagrange method, one multiplies Eq. (1–80) by some parameter, say λ, and adds the result to Eq. (1–79) to get

$$\sum_{j=1}^{r} \left(\frac{\partial f}{\partial x_j} - \lambda \frac{\partial g}{\partial x_j} \right)_0 \delta x_j = 0 \tag{1–81}$$

The δx_j are still not independent, because of Eq. (1–80), and so they cannot be varied independently. Equation (1–80), however, can be treated as an equation giving one of the δx_j in terms of the other $r - 1$ independent ones. Pick any one of the $r \, \delta x_j$ as the dependent one. Let this be δx_μ.

The trick now is that we have not specified λ yet. We set it equal to $(\partial f/\partial x_\mu)_0/(\partial g/\partial x_\mu)_0$, making the coefficient of δx_μ in Eq. (1–81) vanish. The subscript zero here indicates that $(\partial f/\partial x_\mu)$ and $(\partial g/\partial x_\mu)$ are to be evaluated at values of the x_j such that f is at its maximum (or minimum) under the constraint of Eq. (1–80). Of course, we do not know these values of x_j yet, but we can nevertheless formally define λ in this manner. This leaves a sum of terms in Eq. (1–81) involving only the independent δx_j, which can be varied independently, yielding that

$$\left(\frac{\partial f}{\partial x_j} \right)_0 - \lambda \left(\frac{\partial g}{\partial x_j} \right)_0 = 0 \qquad j = 1, 2, \ldots, \mu - 1, \mu + 1, \ldots, r$$

---

* See Mayer and Mayer, *Statistical Mechanics* (New York: Wiley, 1940).

If we combine these $r - 1$ equations with our choice for λ, we have

$$\left(\frac{\partial f}{\partial x_j}\right)_0 - \lambda\left(\frac{\partial g}{\partial x_j}\right)_0 = 0 \tag{1-82}$$

for all j.

As we said above, the choice of λ here is certainly formal, since both $(\partial f/\partial x_\mu)_0$ and $(\partial g/\partial x_\mu)_0$ must be evaluated at these values of x_j which maximizes f, but these are known from Eq. (1–82) only in terms of λ. But this presents no difficulty, since in practice λ is determined by physical requirements. Examples of this will occur in the next two chapters.

Lagrange's method becomes no more difficult in the case in which there are several constraints. Let $g_1(x_1, \ldots, x_r), g_2(x_1, \ldots, x_r), \ldots$ be a set of constraints. We introduce a Lagrange multiplier for each $g_i(x_1, \ldots, x_r)$ and proceed as above to get

$$\frac{\partial f}{\partial x_j} - \lambda_1 \frac{\partial g_1}{\partial x_j} - \lambda_2 \frac{\partial g_2}{\partial x_j} - \cdots = 0 \tag{1-83}$$

BINOMIAL DISTRIBUTION FOR LARGE NUMBERS

Lastly, there is one other mathematical observation we need here in order to facilitate the discussion in the next chapter. This observation concerns the shape of the multinomial coefficient [Eq. (1–78)] as a function of the N_j's, as the N_j's become very large. To simplify notation, we shall consider only the binomial coefficient, but this will not affect our conclusions. Let us first find the value of N_1 for which $f(N_1) = N!/N_1!(N - N_1)!$ reaches its maximum value. Since N_1 and N are both very large, we treat them as continuous variables. Also since $\ln x$ is a monotonic function of x, we can maximize $f(N_1)$ by maximizing $\ln f(N_1)$. This allows us to use Stirling's approximation. The maximum of $f(N_1)$ is found, then, from

$$\frac{d \ln f(N_1)}{dN_1} = 0$$

to be located at $N_1{}^* = N/2$. Let us now expand $\ln f(N_1)$ about this point. The Taylor expansion is

$$\ln f(N_1) = \ln f(N_1{}^*) + \frac{1}{2}\left(\frac{d^2 \ln f(N_1)}{dN_1{}^2}\right)_{N_1 = N_1{}^*}(N_1 - N_1{}^*)^2 + \cdots \tag{1-84}$$

The linear term in $N_1 - N_1{}^*$ is missing, because the first derivative of $\ln f(N)$ is zero at $N_1 = N_1{}^*$. The second derivative appearing in Eq. (1–84) is equal to $-4/N$. Thus if we ignore higher-order terms (see Problem 1–53), Eq. (1–84) can be written in the form of a Gaussian curve

$$f(N_1) = f(N_1{}^*)\exp\left\{-\frac{2(N_1 - N_1{}^*)^2}{N}\right\} \tag{1-85}$$

Comparison of this with the standard form of the Gaussian function

$$f(x) = \frac{1}{(2\pi\sigma^2)^{1/2}}\exp\left\{-\frac{(x - x^*)^2}{2\sigma^2}\right\} \tag{1-86}$$

shows that the standard deviation is of the order of $N^{1/2}$. Equation (1–85) is, therefore, a bell-shaped function, centered at $N_1{}^* = N/2$ and having a width of a few multiples of $N^{1/2}$. Problem 1–43 establishes the well-known fact that a Gaussian function goes essentially to zero when x differs from x^* by a few σ's. Since we are interested only in large values of N_1 (or N), say numbers of the order of 10^{20}, we have a bell-shaped curve that is contained between $10^{20} \pm$ a few multiples of 10^{10}, which, if plotted, would for all practical purposes look like a delta function centered at $N_1^* = N/2$. Thus we have shown that the binomial coefficient peaks very strongly at the point $N_1 = N_2 = N/2$. This same behavior occurs for a multinomial coefficient as well. If there are s N_j's, the multinomial coefficient has a very sharp maximum at the point $N_1 = N_2 = \cdots = N_s = N/s$. (See Problem 1–50.) This peak becomes sharper as the N_j's become larger, and become a delta function in the limit $N_j \to \infty$ for all j.

MAXIMUM TERM METHOD

Another important result, which is a consequence of the large numbers encountered in statistical mechanics, is the *maximum-term method*. It says that under appropriate conditions the logarithm of a summation is essentially equal to the logarithm of the maximum term in the summation. To see how this goes, consider the sum

$$S = \sum_{N=1}^{M} T_N$$

where $T_N > 0$ for all N. Since all the terms are positive, the value of S must be greater than the value of the largest term, say T_{max}, and less than the product of the number of terms and the value of the largest term. Thus we can write

$$T_{max} \leq S \leq M T_{max}$$

Taking logarithms gives

$$\ln T_{max} \leq \ln S \leq \ln T_{max} + \ln M$$

We shall see that it is often the case in statistical mechanics that T_{max} will be $O(e^M)$. Thus we have

$$O(M) \leq \ln S \leq O(M) + \ln M$$

For large M, $\ln M$ is negligible with respect to M itself, and so we see that $\ln S$ is bounded from above and below by $\ln T_{max}$, and so

$$\ln S = \ln T_{max}$$

This is a rather remarkable theorem, and like a number of other theorems used in statistical mechanics, its validity results from the large numbers involved.

ADDITIONAL READING

1–2

DAVIDSON, N. 1962. *Statistical mechanics*. New York: McGraw-Hill. Chapter 2.
GOLDSTEIN, H. 1950. *Classical mechanics*. Reading, Mass.: Addison-Wesley.
SYMON, K. R. 1960. *Mechanics*. Reading, Mass.: Addison-Wesley.
TOLMAN, R. C. 1938. *Statistical mechanics*. London: Oxford University Press. Chapters 1 and 2.

1–3

HAMEKA, H. F. 1967. *Introduction to quantum theory*. New York: Harper & Row.
HANNA, M. W. 1969. *Quantum mechanics in chemistry*, 2nd ed. New York: Benjamin.
HILL, T. L. 1956. *Statistical mechanics*. New York: McGraw-Hill. Section 8.
KARPLUS, M., and PORTER, R. N. 1970. *Atoms and molecules*. New York: Benjamin.
KESTIN, J., and DORFMAN, J. R. 1971. *A course in statistical thermodynamics*. New York: Academic. Chapter 3.
PILAR, F. L. 1968. *Elementary quantum chemistry*. New York: McGraw-Hill.
TOLMAN, R. C. 1938. *Statistical mechanics*. London: Oxford University Press. Chapters 7 and 8.

1–4

ANDREWS, F. C. 1971. *Thermodynamics*. New York: Wiley.
CALLEN, H. B. 1960. *Thermodynamics*. New York: Wiley.
MAHAN, B. H. 1963. *Elementary chemical thermodynamics*. New York: Benjamin.
MOORE, W. J. 1972. *Physical chemistry*, 4th ed. Englewood Cliffs, N.J.: Prentice-Hall.
REISS, H. 1965. *Methods of thermodynamics*. Boston, Mass.: Blaisdell.
WASER, J. 1966. *Basic chemical thermodynamics*. New York: Benjamin.

1–5

ABRAMOWITZ, M., and STEGUN, I. A. "Handbook of Mathematical Functions," *Natl. Bur. Stan. Appl. Math. Series*, **55**, 1964.
ARFKEN, G., 1970. *Mathematical methods for physicists*, 2nd ed. New York: Academic.
KESTIN, J., and DORFMAN, J. R. 1971. *A course in statistical thermodynamics*. New York: Academic. Chapter 4.
KREYSZIG, E. 1962. *Advanced engineering mathematics*, 2nd ed. New York, Wiley.
REIF, F. 1965. *Statistical and thermal physics*. New York: McGraw-Hill, Chapter 1.

PROBLEMS

1–1. Solve the equation of motion of a body of mass m dropped from a height h. Assume that there exists a viscous drag on the body that is proportional to and in the opposite direction to the velocity of the body. (Let the proportionality constant be γ.) Solve for the so-called terminal velocity, that is, the limiting velocity as $t \to \infty$.

1–2. Calculate the trajectory of a shell shot out of a cannon with velocity v_0, assuming no aerodynamic resistance and that the cannon makes an angle θ with the horizontal axis.

1–3. Remembering that the potential energy is given by

$$V(x) = -\int_0^x F(\xi)\,d\xi = \tfrac{1}{2}kx^2$$

for a simple harmonic oscillator, derive an expression for the total energy as a function of time. Discuss how the kinetic and potential energy behave as a function of time.

1–4. Solve the equation for a harmonic oscillator of mass m and force constant k that is driven by an external force of the form $F(t) = F_0 \cos \omega_0 t$.

1–5. Show that

$$\xi(t) = A \sin \omega t + B \cos \omega t$$

can be written as

$$\xi(t) = C \sin(\omega t + \phi)$$

1–6. Show that the total linear momentum is conserved for a system of N particles with an interaction potential which depends only on the distance between particles.

1–7. When does $p = \partial L/\partial \dot{q}$ but $\neq \partial K/\partial \dot{q}$?

1–8. Consider a system of two-point particles with masses m_1 and m_2 moving in two dimensions. It is very common for their potential of interaction to depend upon their relative coordinates $(x_1 - x_2, y_1 - y_2)$ only. Thus the total energy is

$$E = \frac{m_1}{2}(\dot{x}_1{}^2 + \dot{y}_1{}^2) + \frac{m_2}{2}(\dot{x}_2{}^2 + \dot{y}_2{}^2) + U(x_1 - x_2, y_1 - y_2)$$

Now introduce four new variables

$$X = \frac{m_1 x_1 + m_2 x_2}{m_1 + m_2} \qquad Y = \frac{m_1 y_1 + m_2 y_2}{m_1 + m_2}$$

$$x_{12} = x_1 - x_2 \qquad y_{12} = y_1 - y_2$$

and show that this two-body problem can be reduced to two one-body problems, one involving the center of mass of the system and one involving the *relative* motion of the two particles.

Give a physical interpretation of the ratio $m_1m_2/(m_1 + m_2)$ that arises naturally in the relative motion. What is this quantity called? This result is easily extended to three dimensions.

1–9. Extend the development of Problem 1–8 to the case in which each particle also experiences an external potential energy, say $U(x_1, y_1, z_1)$ and $U(x_2, y_2, z_2)$. Interpret the resulting equations.

1–10. Derive Lagrange's equations for a particle moving in two dimensions under a central potential $u(r)$. Which of these equations illustrates the law of conservation of angular momentum? Is angular momentum conserved if the potential depends upon θ as well?

1–11. For a particle moving in three dimensions under the influence of a spherically symmetrical potential $U = U(r)$, write down the Lagrangian and the equations of motion in spherical coordinates (r, θ, ϕ). Show that $H = K + V$ from

$$H = \sum p_i \dot{q}_i - L$$

for this potential.

1–12. Solve the equation of motion of two masses m_1 and m_2 connected by a harmonic spring with force constant k.

1–13. Start with Lagrange's equations in Cartesian coordinates, that is,

$$\frac{d}{dt}\left(\frac{\partial L}{\partial \dot{x}}\right) = \frac{\partial L}{\partial x}$$

and so on. Now introduce three generalized coordinates q_1, q_2, and q_3 which are related to the Cartesian coordinates by $x = x(q_1, q_2, q_3)$, and so on. Show that by transforming Lagrange's equations from x, $\dot{x}$, y, $\dot{y}$, z, and $\dot{z}$ as independent variables to $q_1, \dot{q}_1, q_2, \dot{q}_2, q_3$, and $\dot{q}_3$ we get

$$\frac{d}{dt}\left(\frac{\partial L}{\partial \dot{q}_1}\right) = \frac{\partial L}{\partial q_1}$$

and so on.

1–14. If H, the classical Hamiltonian, does not depend explicitly on time, show that $dH/dt = 0$. What does this mean physically? Is this true if H does depend explicitly upon time?

1–15. Consider the rotation of a diatomic molecule with a fixed internuclear separation l and masses m_1 and m_2. By employing center of mass and relative coordinates, show that the rotational kinetic energy can be written in spherical coordinates as

$$\tfrac{1}{2}I(\dot{\theta}^2 + \dot{\phi}^2 \sin^2 \theta)$$

and from this derive the rotational Hamiltonian

$$H_{\text{rot}} = \frac{1}{2I}\left(p_\theta{}^2 + \frac{p_\phi{}^2}{\sin^2 \theta}\right)$$

In these equations, $I = \mu l^2$, where μ is the reduced mass. This Hamiltonian is useful for studying the rotation of diatomic molecules.

1–16. Show that the motion of a particle under a central force law takes place entirely in a single plane.

1–17. What is the expectation (average) value for the linear momentum p_x of a particle in a one-dimensional box $p_x{}^2$? Briefly discuss your results.

1–18. Show that the energy eigenvalues of a free particle confined to a cube of length a are given by

$$\varepsilon = \frac{h^2}{8ma^2}(n_x{}^2 + n_y{}^2 + n_z{}^2) \qquad n_x, n_y, n_z = 1, 2, \ldots$$

1–19. Show that the energy eigenvalues of a free particle confined to a rectangular parallelepiped of lengths a, b, and c are given by

$$\varepsilon = \frac{h^2}{8m}\left(\frac{n_x^2}{a^2} + \frac{n_y^2}{b^2} + \frac{n_z^2}{c^2}\right) \qquad n_x, n_y, n_z = 1, 2, \ldots$$

1–20. Calculate the energy eigenvalues of a particle confined to a ring of radius a.

1–21. Show that the Hamiltonian operator of a rigid rotor is given by Eq. (1–32).

• **1–22.** Calculate the degeneracy of the first few levels of a free particle confined to a cube of length a.

1–23. Verify the calculation that follows Eq. (1–37) which shows that the quantum mechanical degeneracy of a macroscopic system is $O(10^N)$.

1–24. We need to know the volume of an N-dimensional sphere in order to derive Eq. (1–36). This can be determined by the following device. Consider the integral

$$I = \int_{-\infty}^{\infty} \cdots \int e^{-(x_1^2 + x_2^2 + \cdots + x_N^2)}\, dx_1\, dx_2 \cdots dx_N$$

First show that $I = \pi^{N/2}$. Now one can formally transform the volume element $dx_1\, dx_2 \cdots dx_N$ to N-dimensional spherical (hyperspherical) coordinates to get

$$\int_{\text{angles}} dx_1\, dx_2 \cdots dx_N \to r^{N-1} S_N\, dr$$

where S_N is the factor that arises upon integration over the angles. Show that $S_2 = 2\pi$ and $S_3 = 4\pi$. S_N can be determined for any N by writing I in hyperspherical coordinates:

$$I = \int_0^\infty e^{-r^2} r^{N-1} S_N\, dr$$

Show that $I = S_N \Gamma(N/2)/2$, where $\Gamma(x)$ is the gamma function (see Problem 1–58). Equate these two values for I to get

$$S_N = \frac{2\pi^{N/2}}{\Gamma(N/2)}$$

Show that this reduces correctly for $N = 2$ and 3. Lastly now, convince yourself that the volume of an N-dimensional sphere of radius a is given by

$$V_N = \int_0^a S_N r^{N-1}\, dr = \frac{\pi^{N/2}}{\Gamma\left(\dfrac{N}{2} + 1\right)}\, a^N$$

and show that this reduces correctly for $N = 2$ and 3.

1–25. Derive an expression for the density of translational quantum states for a two-dimensional ideal gas.

1–26. Prove that a many-body wave function must be either symmetric or antisymmetric under the interchange of any two particles. Hint: Apply the exchange operation twice.

1–27. Show for an isothermal process that $w_{\text{rev}} > w_{\text{irrev}}$ and $q_{\text{rev}} > q_{\text{irrev}}$.

1–28. Derive the thermodynamic equation

$$C_p - C_V = \left[p + \left(\frac{\partial E}{\partial V}\right)_T\right]\left(\frac{\partial V}{\partial T}\right)_p$$

and evaluate this difference for an ideal gas and a gas that obeys the van der Waals equation.

• **1–29.** Derive the thermodynamic equation of state

$$\left(\frac{\partial E}{\partial V}\right)_T - T\left(\frac{\partial p}{\partial T}\right)_V = -p$$

1–30. Derive the equation

$$dE = \left[T\left(\frac{\partial p}{\partial T}\right)_V - p \right] dV + C_V \, dT$$

and from this show that $(\partial E/\partial V)_T = a/V^2$ for a van der Waals gas.

1–31. Show that

$$\left(\frac{\partial E}{\partial V}\right)_{\mu/T,\, 1/T} + \frac{1}{T}\left(\frac{\partial p}{\partial(1/T)}\right)_{\mu/T,\, V} = -p$$

1–32. Derive an expression for $\partial \ln K/\partial T$ in terms of ΔH, the heat of reaction, and in terms of C_p, the heat capacity at constant pressure.

1–33. Consider the "water-gas" reaction

$$CO + H_2O(g) \rightarrow H_2 + CO_2$$

where

$$K_p = \frac{P_{H_2} P_{CO_2}}{P_{CO} P_{H_2O}}$$

and given the following data:

Substance	(kcal/mole)	a	$b \times 10^3$	$c \times 10^7$	ΔH^0_{298}(kcal/m)
CO	−32.81	6.42	1.67	1.96	−26.4157
H$_2$O(g)	−54.64	7.26	2.30	2.83	−57.7979
CO$_2$	−94.26	6.21	10.40	−35.45	−94.0518
H$_2$	0.00	6.95	−0.20	4.81	0.00

where the heat capacity of the gases in cal deg^{-1} mole^{-1} is given by

$$C_p = a + bT + cT^2$$

Calculate K_p at 298°K and 800°K.

1–34. Calculate the free energy change at 700°C for the conversion of carbon monoxide at 10 atm and water vapor at 5 atm to carbon dioxide and hydrogen at partial pressures of 1.5 atm each. The equilibrium constant K_p for this reaction is 0.71. Is this process theoretically feasible?

1–35. It is illustrated in Chapter 17 that the speed of sound c_0 propagated through a gas is

$$c_0 = (m\rho\kappa_s)^{-1/2}$$

where κ_s is the adiabatic compressibility

$$\kappa_s = -\frac{1}{V}\left(\frac{\partial V}{\partial p}\right)_s$$

Show that this is equivalent to

$$c_0 = V\left\{ -\frac{\gamma}{M}\left(\frac{\partial p}{\partial V}\right)_T \right\}^{1/2}$$

where $\gamma = C_p/C_V$, and M is the molecular weight of the gas. Using the above result, show that

$$c_0 = \left(\gamma \frac{RT}{M}\right)^{1/2}$$

for an ideal gas.

1–36. Jones and Giauque obtained the following values for C_p of nitromethane.

°K	15	20	30	40	50	60	70	80	90	100
C_P	0.89	2.07	4.59	6.90	8.53	9.76	10.70	11.47	12.10	12.62

°K	120	140	160	180	200	220	240	260	280	300
C_P	13.56	14.45	15.31	16.19	17.08	17.98	18.88	25.01	25.17	25.35

The melting point is 244.7°K, heat of fusion 2319 cal/mole. The vapor pressure of the liquid at 298.1°K is 3.666 cm. The heat of vaporization at 298.0°K is 9147 cal/mole. Calculate the third-law entropy of CH_3NO_2 gas at 298.1°K and 1 atm pressure (assuming ideal gas behavior).

1–37. Derive the Legendre transformation of E in which $S \to T$ and $N \to \mu$.

1–38. Apply a Legendre transformation to the Lagrangian $L(q_J, \dot{q}_J)$ to eliminate the generalized velocities in favor of generalized momenta, defined by $p_J \equiv \partial L / \partial \dot{q}_J$. What function does this turn out to be?

1–39. Find the natural function of V, E, and μ. Hint: Start with the natural function of V, E, and N, namely, S, and transform $N \to \mu$.

1–40. Derive the Legendre transformation of E in which $S \to T$, $N \to \mu$, and $V \to p$. What peculiar thing happens when all the extensive variables are transformed out?

• **1–41.** Show that $\overline{(x - \bar{x})^2} = \overline{x^2} - \bar{x}^2$.

1–42. Show that the Poisson distribution $P(m) = a^m e^{-a}/m!$ is normalized. Calculate $\bar{m}$ and the variance. What is the significance of the parameter a?

• **1–43.** Sketch the Gaussian distribution as σ (or even $\sigma/\bar{x}$) becomes smaller and smaller. To what type of distribution does a Gaussian go in the limit $\sigma \to 0$. Discuss the meaning of this distribution.

• **1–44.** For the Gaussian distribution $p(x)$ show that

(a)

$$\int_{-\infty}^{\infty} p(x)\, dx = 1$$

(b) Calculate the nth central moment where $n = 0, 1, 2$, and 3.

(c) In the limit $\sigma \to 0$ what kind of distribution is approached where

$$p(x) = \frac{1}{\sigma\sqrt{2\pi}} \exp\left(-\frac{(x - \bar{x})^2}{2\sigma^2}\right)$$

1–45. The quantity $\overline{(x - \bar{x})^j}$ is called the jth central moment. Show that all odd central moments of a Gaussian vanish. What about the even ones? Relate the $j = 2$ central moment to the parameter σ.

1–46. Let $f(x, y)$ be a joint probability density, that is, $f(x, y)\, dx\, dy$ is the probability that X lies between x and $x + dx$ and Y lies between y and $y + dy$. If X and Y are *independent*, then

$$f(x, y)\, dx\, dy = f_1(x) f_2(y)\, dx\, dy$$

If X and Y are independent, show that the mean and variance of their sum is equal to the sum of the means and variances, respectively, of X and Y; that is, show that if $W = X + Y$, then

$$\overline{W} = \overline{X} + \overline{Y}$$

$$\overline{(W - \overline{W})^2} = \overline{(X - \overline{X})^2} + \overline{(Y - \overline{Y}^2)}$$

1–47. Let X be a random variable on the positive numbers, $0 \le x < \infty$, and let $p(x)$ be its probability density function. The function $\phi(s)$ defined by

$$\phi(s) = \int_0^\infty e^{-sx} p(x)\, dx$$

is called the characteristic function of $p(x)$. Find the relation between $\phi(s)$ and the moments of $p(x)$. Is knowledge of all the moments of $p(x)$ (assuming they exist) sufficient to specify $p(x)$ itself? Why or why not?

1–48. Show that the characteristic function of the density function of the sum of two independent random variables is the product of the characteristic functions of the densities of the two random variables themselves. What is the density function of $W = X + Y$?

1–49. Maximize

$$W(N_1, N_2, \ldots, N_M) = \frac{N!}{\prod_{j=1}^{M} N_j!}$$

with respect to each N_j under the constraints that

$$\sum N_j = N = \text{a fixed constant}$$
$$\sum E_j N_j = \mathscr{E} = \text{another fixed constant}$$

Hint: Consider the N_j's to be continuous, large enough to use Stirling's approximation of $N_j!$, and leave your answer in terms of the two undetermined multipliers.

1–50. Show that the maximum of a multinomial distribution is given when $N_1 = N_2 = \cdots = N_s = N/s$.

1–51. Use the method of undetermined multipliers to show that

$$-\sum_{j=1}^{N} P_j \ln P_j$$

subject to the condition

$$\sum_{j=1}^{N} P_j = 1$$

is a maximum when $P_j = \text{constant}$.

1–52. Consider the sum

$$\sum_{N=0}^{M} \frac{M! x^N}{N!(M-N)!}$$

where $x = O(1)$, and M and N are $O(10^{20})$. First show that $\ln \sum = M \ln(1 + x)$ *exactly*, and then calculate the logarithm of the maximum term. Hint: Remember the binomial expansion.

1–53. Show that the higher terms that were dropped in the expansion of $\ln f(N)$ in Eq. (1–84) are completely negligible for large values of N and M.

1–54. The Planck blackbody distribution law

$$\rho(\omega, T) d\omega = \frac{\hbar}{\pi^2 c^3} \frac{\omega^3 d\omega}{\exp(\beta\hbar\omega) - 1}$$

gives the blackbody radiation energy density between frequencies ω and $\omega + d\omega$. ($\hbar \equiv h/2\pi$, $\omega \equiv 2\pi\nu$, and $\varepsilon = h\nu = \hbar\omega$.) Substitute this into

$$\frac{E}{V} = \int_0^{\infty} \rho(\omega, T) d\omega$$

to derive the temperature dependence of E/V. Do this by expressing your result as a group of factors multiplying a dimensionless integral. You do not need to evaluate this integral.

1–55. Show that $e^x/(1 \pm e^x)^2$ is an even function of x.

1–56. The heat capacity of the Einstein model of a crystal is given by

$$C_V = 3Nk \left(\frac{\Theta_E}{T}\right)^2 \frac{e^{\Theta_E/T}}{(e^{\Theta_E/T} - 1)^2}$$

where Θ_E is the "characteristic temperature" of the crystal. Determine both the high- and low-temperature limiting expressions for the heat capacity. Do the same thing for the Debye model of crystals, in which

$$C_V = 9Nk \left(\frac{T}{\Theta_D}\right)^3 \int_0^{\Theta_D/T} \frac{x^4 e^x \, dx}{(e^x - 1)^2}$$

where Θ_D is the Debye temperature of the crystal.

1–57. Recognizing it as a geometric series, sum the following series in closed form:

$$S = \sum_{n=0}^{\infty} e^{-\alpha n}$$

Compare this result to

$$I = \int_0^{\infty} e^{-\alpha n} \, dn$$

Under what conditions are these two results the same?

1–58. One often encounters the gamma function in statistical thermodynamics. It was introduced by Euler as a function of x, which is continuous for positive values of x and which reduces to $n!$ when $x = n$, an integer. The gamma function $\Gamma(x)$ is defined by

$$\Gamma(x) = \int_0^{\infty} e^{-t} t^{x-1} \, dt$$

First show by integrating by parts that

$$\Gamma(x + 1) = x\Gamma(x)$$

Using this, show that $\Gamma(n + 1) = n!$ for n an integer. Show that

$$\Gamma(\tfrac{1}{2}) = \sqrt{\pi}$$

Evaluate $\Gamma(\tfrac{3}{2})$ using the recurrence formula $\Gamma(x + 1) = x\Gamma(x)$. Lastly show that

$$\Gamma\left(n + \frac{1}{2}\right) = \frac{1 \cdot 3 \cdots (2n - 1)}{2^n} \Gamma\left(\frac{1}{2}\right)$$

$$= \frac{(2n)!}{2^{2n}n!} \sqrt{\pi}$$

For a discussion of the gamma function, see G. Arfken, *Mathematical Methods for Physicists*, 2nd ed. (New York: Academic, 1970).

1–59. We can derive Stirling's approximation from an asymptotic approximation to the gamma function $\Gamma(x)$. From the previous problem

$$\Gamma(N + 1) = N! = \int_0^{\infty} e^{-x} x^N \, dx$$

$$= \int_0^{\infty} e^{Ng(x)} \, dx$$

where $g(x) = \ln x - x/N$. If $g(x)$ possesses a maximum at some point, say x_0, then for large N, $\exp(Ng(x))$ will be extremely sharply peaked at x_0. Under this condition, the integral for $N!$ will be dominated by the contribution of the integrand from the point x_0. First show that $g(x)$ does, in fact, possess at maximum at the point $x_0 = N$. Expand $g(x)$ about this point, keeping terms only up to and including $(x - N)^2$ to get

$$g(x) \approx g(N) - \frac{(x - N)^2}{2N^2} + \cdots$$

Why is there no linear term in $(x - N)$? Substitute this expression for $g(x)$ into the integral for $N!$ and derive the asymptotic formula

$$\ln N! \approx N \ln N - N + \ln(2\pi N)^{1/2}$$

1–60. Verify the energy conversion factors in Appendix A. (The one labeled "temperature" means that temperature required to give an energy equal to kT, where k is the Boltzmann constant.)

1–61. An integral that appears often in statistical mechanics and particularly in the kinetic theory of gases is

$$I_n = \int_0^\infty x^n e^{-ax^2}\, dx$$

This integral can be readily generated from two basic integrals. For even values of n, we first consider

$$I_0 = \int_0^\infty e^{-ax^2}\, dx$$

The standard trick to evaluate this integral is to square it, and then transform the variables into polar coordinates.

$$I_0{}^2 = \int_0^\infty \int_0^\infty e^{-ax^2} e^{-ay^2}\, dx\, dy$$

$$= \int_0^\infty \int_0^{\pi/2} e^{-ar^2} r\, dr\, d\theta$$

$$= \frac{\pi}{4a}$$

$$I_0 = \frac{1}{2}\left(\frac{\pi}{a}\right)^{1/2}$$

Using this result, show that for even n

$$I_n = \frac{1 \cdot 3 \cdot 5 \cdots (n-1)}{2(2a)^{n/2}}\left(\frac{\pi}{a}\right)^{1/2} \qquad n \text{ even}$$

For odd values of n, the basic integral I_1 is easy. Using I_1, show that

$$I_n = \frac{\Gamma\!\left(\dfrac{n+1}{2}\right)}{2a^{(n+1)/2}} \qquad n \text{ odd}$$

1–62. Show that a Gaussian distribution is extremely small beyond a few multiples of σ.

1–63. Another function that occurs frequently in statistical mechanics is the Riemann zeta function, defined by

$$\zeta(s) = \sum_{k=1}^\infty k^{-s}$$

First show that $\zeta(1) = \infty$, but that $\zeta(s)$ is finite for $s > 1$. Show that another definition of $\zeta(s)$ is

$$\zeta(s) = \frac{1}{\Gamma(s)} \int_0^\infty \frac{x^{s-1}\, dx}{(e^x - 1)}$$

that is, show that this is identical to the first definition. In addition, show that

$$\eta(s) = \sum_{k=1}^{\infty} (-1)^{k-1} k^{-s} = (1 - 2^{1-s})\zeta(s)$$

$$\lambda(s) = \sum_{k=0}^{\infty} (2k + 1)^{-s} = (1 - 2^{-s})\zeta(s)$$

The evaluation of $\zeta(s)$ for integral s can be done using Fourier series, and some results are $\zeta(2) = \pi^2/6$ and $\zeta(4) = \pi^4/90$.

For a discussion of the Riemann zeta function, see G. Arfken, *Mathematical Methods for Physicists*, 2nd ed. (New York: Academic, 1970).

CHAPTER 2

THE CANONICAL ENSEMBLE

In this chapter we shall introduce the basic concepts and assumptions of statistical thermodynamics, and then apply them to a system which has fixed values of V and N and is in thermal equilibrium with its environment. We shall derive the fundamental connection between the quantum mechanical energy levels available to an N-body system and its thermodynamic functions. This link is effected by a function, called the partition function, which is of central importance in statistical thermodynamics. In Section 2–4 we discuss the relevance of the statistical thermodynamic equations to the second and third laws of classical thermodynamics.

2–1 ENSEMBLE AVERAGES

Our goal is to calculate thermodynamic properties in terms of molecular properties. Given the structure of the individual molecules of our system and the form of the intermolecular potential, we wish to be able to calculate thermodynamic properties, such as entropy and free energy. We shall do this first with respect to mechanical properties (such as pressure, energy, volume), which are quantum mechanical or classical mechanical quantities, and then we shall bring nonmechanical thermodynamic variables (such as entropy, free energy) into our discussion by appealing to the equations of thermodynamics. One useful distinction between mechanical and nonmechanical properties is that mechanical properties are defined without appealing to the concept of temperature, whereas the definitions of nonmechanical properties involve the temperature.

Consider some macroscopic system of interest, such as a liter of water or a salt solution. From a macroscopic point of view, we can completely specify a system by a few parameters, say the volume, concentration or density, and temperature. Regardless of the complexity of the system, it requires only a small number of parameters to describe it. From a microscopic point of view, on the other hand, there will be an enormous number of quantum states consistent with the fixed macroscopic properties.

We saw in Chapter 1 that the degeneracy of an isolated N-body system is of the order of 10^N for all but the very lowest energies. This means that the liter of water or the salt solution could be in any one of the order of 10^N possible quantum states. It would be impossible for us to ever determine which of the order of 10^N possible states the system is in. The state of the system must be known, however, in order to calculate a mechanical thermodynamic property, such as the pressure, since the values of that property in each of the possible quantum states would, in general, be different. Thus we are faced with what appears to be an impossible task.

It is at this point that we appeal to the work of Maxwell, Boltzmann, and particularly Gibbs. The modern (postquantum) version of their approach is that in order to calculate the value of any mechanical thermodynamic property (say, the pressure), one calculates the value of that mechanical property in each and every one of the quantum states that is consistent with the few parameters necessary to specify the system in a macroscopic sense. The average of these mechanical properties is then taken, giving each possible quantum state the same weight. We then *postulate* that this average mechanical property corresponds to a parallel thermodynamic property. For example, we postulate that the average energy corresponds to the thermodynamic energy and that the average pressure corresponds to the thermodynamic pressure. It turns out that the calculation of a mechanical property averaged over all the consistent quantum states can be readily performed. Before doing this, however, we shall introduce some concepts that will make this procedure clearer.

We first discuss the concept of an ensemble of systems, first introduced by Gibbs. An ensemble is a (mental or virtual) collection of a very large number of systems, say $\mathscr{A}$, each constructed to be a replica on a thermodynamic (macroscopic) level of the particular thermodynamic system of interest. For example, suppose the system has a volume V, contains N molecules of a single component, and is known to have an energy E. That is, it is an isolated system with N, V, and E fixed. Then the ensemble would have a volume $\mathscr{A}V$, contain $\mathscr{A}N$ molecules, and have a total energy $\mathscr{E} = \mathscr{A}E$. Each of the systems in this ensemble is a quantum mechanical system of N interacting atoms or molecules in a container of volume V. The values of N and V, along with the force law between the molecules, are sufficient to determine the energy eigenvalues E_j of the Schrödinger equation along with their associated degeneracies $\Omega(E_j)$. These energies are the only energies available to the N-body system. Hence the fixed energy E must be one of these E_j's and, consequently, there is a degeneracy $\Omega(E)$. Note that there are $\Omega(E)$ different quantum states consistent with the only things we know about our macroscopic system of interest, namely, the values of N, V, and E. Although all the systems in the ensemble are identical from a thermodynamic point of view, they are not necessarily identical on a molecular level. So far we have said nothing about the distribution of the members of the ensemble with respect to the $\Omega(E)$ possible quantum states.

We shall further restrict our ensemble to obey the *principle of equal a priori probabilities*. That is to say, we require that each and every one of the $\Omega(E)$ quantum states is represented an equal number of times in the ensemble. Since we have no information to consider any one of the $\Omega(E)$ quantum states to be more important than any other, we must treat each of them equally, that is, we must utilize the principle of equal a priori probabilities. All of the $\Omega(E)$ quantum states are consistent with the given values of N, V, and E, the only information we have about the system. Clearly, the number of systems in the ensemble must be an integral multiple of $\Omega(E)$. The number of systems in an ensemble is a very large number and can be made arbitrarily large by simply doubling, tripling, and so on, the size of the ensemble. An alternative interpretation of the principle of equal a priori probabilities is that *an isolated system (N, V, and E fixed) is equally likely to be in any of its $\Omega(E)$ possible quantum states.*

We now define an ensemble average of a mechanical property as the average value of this property over all the members of the ensemble, utilizing the principle of equal a priori probabilities. We postulate that the ensemble average of a mechanical property can be equated to its corresponding thermodynamic property.

There are two complications in the above treatment that we should mention; neither of them, fortunately, is of any practical consequence. We have assumed that the isolated system that we have been using as an example has precisely the energy *E*. We know, however, from quantum mechanics that there always exists a small uncertainty ΔE in the value of *E*. For all thermodynamic purposes, this complication is completely inconsequential, and we shall therefore ignore it. The explanation of the other complication involves a greater knowledge of quantum mechanics than is generally required in this book. We have assumed that the systems of the ensemble are in one of the $\Omega(E)$ degenerate eigenstates having the eigenvalue *E*. The choice of these $\Omega(E)$ eigenfunctions, however, is somewhat arbitrary since any linear combination of these is also an eigenfunction with energy *E*. Moreover, a quantum mechanical system will, in general, not be in one of the $\Omega(E)$ selected states, but will be some linear combination of them. Thus we have tacitly assumed that any system with *N*, *V*, and *E* given will be a "pure state," whereas a system with *N*, *V*, and *E* given will most likely be in a "mixed state," that is, in a state described by a linear combination of the pure states we have chosen. In any event, this complication need not be considered, since the results do not differ appreciably from those obtained from the simpler and more naive point of view which we have presented above and now adopt.

Let us summarize this section by stating that we wish to calculate the ensemble average of some mechanical property, and then show that this can be set equal to the corresponding thermodynamic property. We have stated above that the calculation of the ensemble average is not difficult, and now we shall address ourselves to that problem. As Schrödinger says in his book:* "There is, essentially, only one problem in statistical thermodynamics, the distribution of a given amount of energy *E* over identical systems. Or perhaps better, to determine the distribution of an assembly of identical systems over the possible states in which the system can find itself, given that the energy of the assembly is a constant *E*."

So far, in this section, we have focused our attention on an ensemble whose members have *N*, *V*, and *E* fixed. This is called the *microcanonical ensemble* and is useful for theoretical discussions. For more practical applications, however, we consider not isolated systems, but those in which the temperature rather than the energy is fixed. The most commonly used ensemble in statistical thermodynamics is the *canonical ensemble*, in which the individual systems have *N*, *V*, and *T* fixed. The remainder of this chapter will deal with the canonical ensemble. There are many other types of ensembles, in fact, one for each set of thermodynamic variables that are used to specify an individual member of the ensemble. We shall discuss some of these other ensembles in the next chapter.

2–2 METHOD OF THE MOST PROBABLE DISTRIBUTION

Consider an experimental system with *N*, *V*, and *T* as its independent thermodynamic variables. We can mentally construct an ensemble of such systems in the following manner. We enclose each system in a container of volume *V* with walls that are heat

* E. Schrödinger, *Statistical Thermodynamics* (Cambridge: Cambridge University Press, 1952).

conducting but impermeable to the passage of molecules. The entire ensemble of systems is then placed in a very large heat bath at temperature T. When equilibrium is reached, the entire ensemble is at a uniform temperature T. Since the containing walls of each system are heat conducting, each and every system of the ensemble has the same fixed values of N, V, and T. Now, the entire ensemble is surrounded by thermal insulation, thus making the ensemble itself an isolated system with volume $\mathscr{A}V$, number of molecules $\mathscr{A}N$, and some total energy $\mathscr{E}$. (The actual value of $\mathscr{E}$ is not important.) Each of the $\mathscr{A}$ members of the canonical ensemble finds itself in a large heat bath at temperature T.

Because each of the systems of the canonical ensemble is not isolated but is at a fixed temperature, the energy of each system is not fixed at any set value. Thus we shall have to consider the entire spectrum of energy states for each member of the canonical ensemble. Let the energy eigenvalues of the quantum states of a system be $E_1(N, V)$, $E_2(N, V) \ldots$, ordered such that $E_{j+1} \geq E_j$. It is important to understand here that any particular energy, say E_i, is repeated according to its degeneracy, that is, occurs $\Omega(E_i)$ times. Any particular system might be found in any of these quantum states. We shall show later that the *average energy* or the *probability* that some system has a certain energy depends upon the temperature; however, any of the set of energies $\{E_j\}$ is possible, and so must be considered.

We can specify a state of the entire ensemble by saying that $a_1, a_2, a_3, \ldots$ of the systems are in states 1, 2, 3, $\ldots$, respectively, with energies $E_1, E_2, E_3, \ldots$. Thus we can describe any one state of the ensemble by writing

State No.	1,	2,	3,	$\ldots$,	$l \ldots$
Energy	E_1,	E_2,	E_3,	$\ldots$,	$E_l \ldots$
Occupation No.	a_1,	a_2,	a_3,	$\ldots$,	$a_l \ldots$

Occupation Number means the number of systems of the ensemble in that particular state. The set of occupation numbers is called a distribution. We shall often denote the set $\{a_j\}$ by **a**.

Of course, the occupation numbers satisfy the two conditions:

$$\sum_j a_j = \mathscr{A} \tag{2-1}$$

$$\sum_j a_j E_j = \mathscr{E} \tag{2-2}$$

The first condition simply accounts for all the members of the ensemble, and the second represents the fact that the entire canonical ensemble is an isolated system, and hence has some fixed energy $\mathscr{E}$.

Since the canonical ensemble has been isolated from its surroundings by thermal insulation, we can apply the principle of equal a priori probabilities to this isolated system. In the form that we wish to use here, the principle of equal a priori probabilities says that every possible state of the canonical ensemble, that is, every distribution of occupation numbers $a_1, a_2, \ldots$, consistent with Eqs. (2-1) and (2-2) is equally probable and must be given equal weight in performing ensemble averages.

The number of ways $W(\mathbf{a}) \equiv W(a_1, a_2, a_3, \ldots)$ that any particular distribution of the a_j's can be realized is the number of ways that $\mathscr{A}$ *distinguishable* objects can be arranged into groups, such that a_1 are in the first group, a_2 in the second, and so on [see Eq. (1-77)]:

$$W(\mathbf{a}) = \frac{\mathscr{A}!}{a_1! a_2! a_3! \cdots} = \frac{\mathscr{A}!}{\prod_k a_k!} \tag{2-3}$$

The systems are distinguishable since they are macroscopic systems, which we could, in principle, furnish with labels.

In general, there are very many distributions which are consistent with Eqs. (2–1) and (2–2). In any particular distribution, $a_j/\mathscr{A}$ is the fraction of systems or members of the canonical ensemble in the jth energy state (with energy E_j). The overall probability P_j that a system is in the jth quantum state is obtained by averaging $a_j/\mathscr{A}$ over all the allowed distributions, giving equal weight to each one according to the principle of equal a priori probabilities. Thus P_j is given by

$$P_j = \frac{\bar{a}_j}{\mathscr{A}} = \frac{1}{\mathscr{A}} \frac{\sum_{\mathbf{a}} W(\mathbf{a}) a_j(\mathbf{a})}{\sum_{\mathbf{a}} W(\mathbf{a})} \tag{2–4}$$

In Eq. (2–4), the notation $a_j(\mathbf{a})$ signifies that the value of a_j depends upon the distribution, and the summations are over all distributions that satisfy Eqs. (2–1) and (2–2). We shall later let $\mathscr{A} \to \infty$, but the ratio $\bar{a}_j/\mathscr{A}$ will remain finite since $\bar{a}_j \to \infty$ as well.

Given the probability that a system with fixed values of N, V, and T is in the jth quantum state, one can calculate the canonical ensemble average of any mechanical property from

$$\overline{M} = \sum_j M_j P_j \tag{2–5}$$

where M_j is the value of M in the jth quantum state. Thus the prescription for calculating the ensemble average of any mechanical property is given by Eqs. (2–4) and (2–5) and is, in principle, complete. The summations involved in Eq. (2–4), however, are very difficult to perform mathematically, and thus in practice Eqs. (2–4) and (2–5) are too complicated to use.

The fact that we can let $\mathscr{A} \to \infty$, however, allows us to appeal to the results of Section 1–5. We have seen there that multinomial coefficients, such as $W(\mathbf{a})$, are extremely peaked about their maximum value if all the variables a_j are large. In Eqs. (2–1) through (2–4), each of the a_j's can be made arbitrarily large since $\mathscr{A}$ can be made arbitrarily large. Thus we can use an argument here very similar to that used in Section 1–5. We need make only one modification or extension. We have shown that $W(\mathbf{a})$ is a maximum when all the a_j's are equal, under the one constraint Eq. (2–1). We have now an additional constraint Eq. (2–2) on the a_j's. So instead of peaking at the point at which all the a_j's are equal, it will peak at some other set of a_j's but the spread, nevertheless, will be arbitrarily small. We shall determine this set of a_j's shortly. Let us denote this distribution by $\mathbf{a}^* = \{a_j^*\}$.

The spread of $W(\mathbf{a})$ about its maximum value can be made arbitrarily narrow by taking the a_j, that is, $\mathscr{A}$, to be arbitrarily large. Thus the $W(\mathbf{a})$ in Eq. (2–4) at any set of a_j's other than the set $\mathbf{a}^*$, which maximizes $W(\mathbf{a})$, are completely negligible. We can replace the summations in Eq. (2–4) over all distributions by just one term, evaluated at $\mathbf{a}^*$. Thus we can write

$$P_j = \frac{1}{\mathscr{A}} \frac{\sum_{\mathbf{a}} W(\mathbf{a}) a_j(\mathbf{a})}{\sum_{\mathbf{a}} W(\mathbf{a})} = \frac{1}{\mathscr{A}} \frac{W(\mathbf{a}^*) a_j^*}{W(\mathbf{a}^*)} = \frac{a_j^*}{\mathscr{A}} \qquad (\lim a_j \to \infty) \tag{2–6}$$

where a_j^* is the value of a_j in that distribution that maximizes $W(\mathbf{a})$, that is, the most probable distribution. The name of this section, the method of the most probable distribution, is derived from Eq. (2–6). Comparing Eqs. (2–6) with (2–4), we have

$$P_j = \frac{\bar{a}_j}{\mathscr{A}} = \frac{a_j^*}{\mathscr{A}} \tag{2–7}$$

Thus, to calculate the probabilities to be used in ensemble averages, we need determine only that distribution **a*** that maximizes $W(\mathbf{a})$ under the two constraints Eqs. (2–1) and (2–2). This is the problem to which we now turn.

As this is a problem of maximizing a function of many variables with given constraints on the variables, we have a direct application of Lagrange's method of undetermined multipliers. Following Section 1–5, the set of a_j's that maximizes $W(\mathbf{a})$, subject to Eqs. (2–1) and (2–2), is found from

$$\frac{\partial}{\partial a_j}\left\{\ln W(\mathbf{a}) - \alpha \sum_k a_k - \beta \sum_k a_k E_k\right\} = 0, \qquad j = 1, 2, \dots \; . \tag{2–8}$$

where α and β are the undetermined multipliers. Using Eq. (2–3) for $W(\mathbf{a})$ along with Stirling's approximation (which is exact here since each of the a_j's can be made arbitrarily large), one gets

$$-\ln a_j^* - \alpha - 1 - \beta E_j = 0 \qquad\qquad j = 1, 2, \dots \tag{2–9}$$

or

$$a_j^* = e^{-\alpha'}e^{-\beta E_j} \qquad j = 1, 2, \dots \tag{2–10}$$

where $\alpha' = \alpha + 1$. (See Problem 2–3.) This gives us the most probable distribution in terms of α and β. We now evaluate α' and β by using Eqs. (2–1) and (2–2) along with physical arguments.

2–3   THE EVALUATION OF THE UNDETERMINED MULTIPLIERS, α AND β

We can obtain an expression for α (or α') in terms of β by summing both sides of Eq. (2–10) over j and using Eq. (2–1) to get

$$e^{\alpha'} = \frac{1}{\mathscr{A}}\sum_j e^{-\beta E_j} \tag{2–11}$$

Equation (2–7) thus becomes

$$P_j = \frac{a_j^*}{\mathscr{A}} = \frac{e^{-\beta E_j(N, V)}}{\sum_j e^{-\beta E_j(N, V)}} \tag{2–12}$$

Substituting this into Eq. (2–5), with E_j taken to be the mechanical property, gives

$$\bar{E} = \bar{E}(N, V, \beta) = \frac{\sum_j E_j(N, V)e^{-\beta E_j(N, V)}}{\sum_j e^{-\beta E_j(N, V)}} \tag{2–13}$$

According to the postulate of the ensemble method of Gibbs, this average energy $\bar{E}(N, V, \beta)$ corresponds to the thermodynamic energy E.

The pressure is another important mechanical variable. When a system is in the state j, $dE_j = -p_j\, dV$ is the work done on the system when its volume is increased by dV (keeping the number of particles in the system fixed). Thus the pressure in the state j is given by

$$p_j = -\left(\frac{\partial E_j}{\partial V}\right)_N \tag{2–14}$$

The canonical ensemble average of p_j is

$$\bar{p} = \sum_j p_j P_j = -\frac{\sum_j \left(\dfrac{\partial E_j}{\partial V}\right) e^{-\beta E_j}}{\sum_j e^{-\beta E_j}} \tag{2-15}$$

We postulate that $\bar{p}$ corresponds to the thermodynamic pressure.

The sum in the denominator of Eqs. (2–13) and (2–15) occurs throughout the equations of the canonical ensemble. Let this be denoted by $Q(N, V, \beta)$:

$$Q(N, V, \beta) = \sum_j e^{-\beta E_j(N, V)} \tag{2-16}$$

PARTITION FUNCTION

We shall see that this function $Q(N, V, \beta)$ is the central function of the canonical ensemble.

We have made two connections with thermodynamics:

$$\begin{array}{l} p \leftrightarrow \bar{p} \\ E \leftrightarrow \bar{E} \end{array} \quad \text{(ensemble postulate of Gibbs)}$$

Equation (2–13) gives E as a function of β. In principle, one could solve this equation for β as a function of E, but in practice this is not feasible. Fortunately β turns out to be a more convenient quantity than E, so much so that it is preferable to have E as a function of β rather than the inverse. We shall now evaluate β in two different ways.

We differentiate Eq. (2–13) with respect to V, keeping N and β fixed:

$$\left(\frac{\partial \bar{E}}{\partial V}\right)_{N,\beta} = -\bar{p} + \beta \overline{Ep} - \beta \bar{E}\bar{p} \tag{2-17}$$

In this equation,

$$\overline{Ep} = \frac{\sum_j p_j E_j e^{-\beta E_j}}{Q} = -\frac{\sum_j \left(\dfrac{\partial E_j}{\partial V}\right) E_j e^{-\beta E_j}}{Q}$$

and

$$\bar{E}\bar{p} = \frac{\sum_j E_j e^{-\beta E_j}}{Q} \cdot \frac{\sum_j p_j e^{-\beta E_j}}{Q}$$

Similarly, we can differentiate Eq. (2–15) to get

$$\left(\frac{\partial \bar{p}}{\partial \beta}\right)_{N,V} = \bar{E}\bar{p} - \overline{Ep} \tag{2-18}$$

From Eqs. (2–17) and (2–18) we get

$$\left(\frac{\partial \bar{E}}{\partial V}\right)_{N,\beta} + \beta \left(\frac{\partial \bar{p}}{\partial \beta}\right)_{N,V} = -\bar{p} \tag{2-19}$$

Note that $\bar{E}$ is a function of N, V, and β, whereas the E_j's are functions of N and V only. This is an important distinction, that should be clearly and completely understood.

Let us now compare Eq. (2–19) with the purely thermodynamic equation. (See Problem 1–29.)

$$\left(\frac{\partial E}{\partial V}\right)_{T,N} - T\left(\frac{\partial p}{\partial T}\right)_{N,V} = -p \tag{2-20}$$

which we rewrite in terms of $1/T$ instead of T:

$$\left(\frac{\partial E}{\partial V}\right)_{N,\, 1/T} + \frac{1}{T}\left(\frac{\partial p}{\partial 1/T}\right)_{N,\, V} = -p \tag{2-21}$$

A comparison of Eq. (2–19) with Eq. (2–21) allows us to deduce that $\beta = \text{const}/T$. It is customary to write $\beta = 1/kT$, where k is a constant, whose value could possibly vary from substance to substance. We shall show now, however, that k has the same value for all substances, that is, k is a universal constant.

Consider two closed systems A and B, each having its own kind of particles and energy states, but in thermal contact with each other and immersed in a heat bath of temperature T. We now construct a canonical ensemble of systems AB (as shown in Fig. 2–1) representative of a thermodynamic AB system at temperature T and apply the method of the most probable distribution to the AB system. Let the number of molecules and volume of the A and B systems be N_A, V_A, and N_B, V_B, respectively, and let their energy states be denoted by $\{E_{jA}\}$ and $\{E_{jB}\}$. If a_j denotes the number of A systems in state E_{jA}, and b_j denotes the number of B systems in state E_{jB}, then the number of states of the AB ensemble with compound distribution $\{a_j\}$ and $\{b_j\}$ is

$$W(\mathbf{a}, \mathbf{b}) = \frac{\mathscr{A}!}{\prod_j a_j!} \cdot \frac{\mathscr{B}!}{\prod_k b_k!} \tag{2-22}$$

where $\mathscr{A}$ and $\mathscr{B}$ ($\mathscr{A} = \mathscr{B}$) are the number of A and B system, respectively. Equation (2–22) turns out to be a product of the separate A and B factors, because we can arrange the A systems over their possible quantum states independently of the B systems, and vice versa. The a_j's and b_j's must satisfy the three relations:

$$\sum_j a_j = \mathscr{A}$$

$$\sum_j b_j = \mathscr{B} = \mathscr{A}$$

$$\sum_j (a_j E_{jA} + b_j E_{jB}) = \mathscr{E} \tag{2-23}$$

We now apply the method of the most probable distribution to Eqs. (2–22) and (2–23) to get Problem 2–9 for the simultaneous probability that the AB system has its A part in the ith quantum state and its B part in the jth quantum state:

$$P_{ij} = \frac{e^{-\beta E_{iA}}}{Q_A} \cdot \frac{e^{-\beta E_{jB}}}{Q_B} = P_{iA} P_{jB} \tag{2-24}$$

where

$$Q_A = \sum_k e^{-\beta E_{kA}} \quad \text{and} \quad Q_B = \sum_k e^{-\beta E_{kB}} \tag{2-25}$$

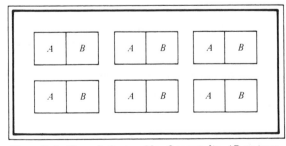

Figure 2–1. **Canonical ensemble of composite AB systems.**

Thus we have shown that two arbitrary systems in thermal contact have the same β. But we have seen from Eqs. (2–19) and (2–21) that $\beta = 1/kT$, and so the two systems must have the same value of k. Since the nature of the two systems is completely arbitrary, k must have the same value for all systems. Thus k is a universal constant and can therefore be evaluated using any convenient system. The most convenient is an ideal gas, and one can determine from the equation of state of an ideal gas [cf. Eq. (5–18)] that $k = 1.3806 \times 10^{-16}$ erg-deg^{-1}, where the temperature is in units of degrees Kelvin.

There is an alternative way to determine β which utilizes the fact that $1/T$ is an integrating factor of dq_{rev}. We shall present this argument here, since it will bring the nonmechanical property of entropy into our formalism.

The argument based around Fig. 2–1 shows that if two systems are in thermal contact at equilibrium, they have the same value of β. Since the two systems can be quite arbitrary, this implies that β must be some function of the temperature. We shall now show that $\beta \, dq_{rev}$ is an exact differential.

Consider the function $f = \ln Q$. We regard f as a function of β and all the E_j's:

$$f(\beta, E_1, E_2, \ldots) = \ln \left\{ \sum_j e^{-\beta E_j} \right\} \tag{2-26}$$

The total derivative f is

$$df = \left(\frac{\partial f}{\partial \beta} \right)_{E_j\text{'s}} d\beta + \sum_k \left(\frac{\partial f}{\partial E_k} \right)_{\beta, E_i\text{'s}} dE_k \tag{2-27}$$

The partial derivatives occurring here are determined from Eq. (2–26) to be

$$\left(\frac{\partial f}{\partial \beta} \right)_{E_j\text{'s}} = \frac{-\sum_j E_j e^{-\beta E_j}}{Q} = -\bar{E}$$

$$\left(\frac{\partial f}{\partial E_k} \right)_{\beta, E_j\text{'s}} = \frac{-\beta e^{-\beta E_k}}{Q} = -\beta P_k$$

Thus Eq. (2–27) becomes

$$df = -\bar{E} \, d\beta - \beta \sum_j P_j \, dE_j$$

which can be written as

$$d(f + \beta \bar{E}) = \beta \left(d\bar{E} - \sum_j P_j \, dE_j \right) \tag{2-28}$$

We now subject the ensemble of systems to the following physical process. We change the volume of all the systems by dV, changing, of course, the E_j's for all of them alike in order to still have an ensemble of macroscopically identical systems. We also change the temperature of the ensemble by dT by coupling it with a large heat bath (of the same temperature), changing the temperature slightly and then isolating the ensemble from the heat bath.

If initially there were a_j systems of the ensemble in the energy state j with energy E_j, then $a_j \, dE_j$ is the work done on all these systems in changing the energy from E_j to $E_j + dE_j$. The total work done on the ensemble is $\sum a_j \, dE_j$ and $\sum_j P_j \, dE_j$ is the ensemble average reversible work that we do on the systems. And since $d\bar{E}$ is the average energy increase, the term enclosed in parentheses on the right-hand side of Eq. (2–28) is the average reversible heat supplied to a system. Thus Eq. (2–28) is

$$d(f + \beta \bar{E}) = \beta \, \delta q_{rev} \tag{2-29}$$

which says that $\beta \, \delta q_{rev}$ is the derivative of a state function, that is, that β is an integrating factor of δq_{rev}. One statement of the second law of thermodynamics says that β must be equal to constant/T, or $1/kT$.

The left-hand side of Eq. (2–29), therefore, must be dS/k, and so we can write that

$$S = \frac{\bar{E}}{T} + k \ln Q + \text{constant} \qquad (2\text{–}30)$$

where the constant is independent of T and of the parameters (N, V, and so on) on which the E_j's depend. Since thermodynamics deals with ΔS only, the constant will always drop out of any calculations of entropy changes for chemical and/or physical changes. We shall, therefore, set this constant to zero and discuss the implications of this at the end of the chapter.

In the above argument that β was an integrating factor of δq_{rev}, we used the fact that the average work done on a system was $\sum_j P_j \, dE_j$. We do work, then, by changing the energies slightly, but keeping the population of these states fixed (the P_j's do not change). A molecular interpretation of thermodynamic work, then, is a change in the quantum mechanical energy states of the system, keeping the population over them fixed. That a molecular interpretation of the absorption of heat is the inverse of this can be seen from

$$d\bar{E} = \sum_j E_j \, dP_j + \sum_j P_j \, dE_j$$

$$= \delta q_{rev} - \delta \omega_{rev}$$

Thus when a small quantity of heat is absorbed from the surroundings, the energy states of the system do not change (N and V are fixed), but the population of these states does.

2–4   THERMODYNAMIC CONNECTION

We now complete the connection between thermodynamics and the canonical ensemble. Equation (2–13) for $\bar{E}$ can be written as (see Problem 2–10):

$$\bar{E} = kT^2 \left(\frac{\partial \ln Q}{\partial T} \right)_{N,V} \qquad (2\text{–}31)$$

and we can also easily derive (see Problem 2–10)

$$\bar{p} = kT \left(\frac{\partial \ln Q}{\partial V} \right)_{N,T} \qquad (2\text{–}32)$$

from Eq. (2–15). Equation (2–30) is an equation for the entropy S in terms of Q:

$$S = kT \left(\frac{\partial \ln Q}{\partial T} \right)_{N,V} + k \ln Q \qquad (2\text{–}33)$$

We have E, p, and S now as functions of Q, and so it is possible to derive expressions for all the thermodynamic functions in terms of Q. The function Q is the central statistical thermodynamic function of the canonical ensemble (N, V, and T fixed) and is called the *canonical (ensemble) partition function*:

$$Q(N, V, T) = \sum_j e^{-E_j(N,V)/kT} \qquad (2\text{–}34)$$

The partition function serves as a bridge between the quantum mechanical energy states of a macroscopic system and the thermodynamic properties of that system. If we can obtain Q as a function of N, V, and T, we can calculate thermodynamic properties in terms of quantum mechanical and molecular parameters. Although the E_j's are the energy states of an N-body system and consequently appear to be unobtainable in practice, we shall see that in a great many cases, we shall be able to reduce the N-body problem to a one-body, two-body, three-body problem, and so on, or approximate the system by classical mechanics. Both of these routes turn out to be very useful. For now, however, we need only assume that there is such a set of energies.

We can derive an equation for the Helmholtz free energy A in terms of Q by using Eqs. (2–31) and (2–33) along with the fact that $A = E - TS$. The result is

$$A(N, V, T) = -kT \ln Q(N, V, T) \tag{2–35}$$

Notice that of all the thermodynamic functions, it is A that is directly proportional to $\ln Q(N, V, T)$, and that A is the thermodynamic potential whose natural independent variables are those of the canonical ensemble. Equation (2–35) can be considered to be the most important connection between thermodynamics and the canonical partition function, since it is possible to derive many equations starting with its differential form (see Problem 2–11). Table 3–1 contains a summary of the formulas of the canonical ensemble.

In this chapter, we have developed the connection between thermodynamics and the quantum mechanical states available to a macroscopic system characterized by N, V, and T. This connection can be summarized by Eq. (2–35). Before concluding this chapter, we shall discuss the second and third laws of thermodynamics from a statistical thermodynamic point of view. A statement of the second law of thermodynamics for closed, isothermal systems is that $\Delta A < 0$ for a spontaneous process. We wish to derive this inequality starting with Eq. (2–35). To do this, we first write Eq. (2–34) in a slightly different form.

Consider Eq. (2–34) for $Q(N, V, T)$. The summation is over all the possible quantum states of the N-body system. In carrying out the summation, a particular value of $\exp(-E_j/kT)$ will occur $\Omega(E_j)$ times, where $\Omega(E_j)$ is the degeneracy. Instead of listing $\exp(-E_j/kT)\,\Omega(E_j)$ times, we could simply write $\Omega(E_j)\exp(-E_j/kT)$, and then sum over different values of E. If we do this, Eq. (2–34) is

$$Q(N, V, T) = \sum_E \Omega(N, V, E)e^{-E(N, V)/kT} \tag{2–36}$$

where we have dropped the no longer necessary j subscript of E_j. In Eq. (2–34), we sum over the *states* of the system. In Eq. (2–36) we sum over *levels*. Equation (2–36) is a more useful form for discussing the second law of thermodynamics.

Consider a typical spontaneous processes, such as the expansion of a gas into a vacuum. Figure 2–2 shows the initial and final states of such a process. For simplicity, we consider the entire system to be isolated. Initially the gas might be confined to one half of the container. After removing the barrier, the gas occupies the entire container.

Figure 2–2. **The initial and final states of the expansion of a gas into a vacuum.**

Equation (1–36) for $\Omega(N, V, E)$ of an ideal gas shows that the number of states is proportional to V^N. For the process illustrated in Fig. 2–2, the gas goes from a thermodynamic state of energy E, number of particles N, and volume $V/2$ to one with the same energy E (the system is isolated), the same number of particles N, but with volume V. Thus according to Eq. (1–36), the number of quantum states available or accessible to the system is increased.

Another example of a spontaneous process is the following. Initially we have an isolated system containing a mixture of hydrogen and oxygen gases. Although hydrogen and oxygen react to form water, in the absence of a catalyst the reaction is so slow that we can ignore it. Since the rate of this reaction (uncatalyzed) is very slow compared to any thermodynamic measurement, we can consider the mixture of hydrogen and oxygen to be simply a mixture of two gases in equilibrium. If we now add a small amount of catalyst to the system, the hydrogen and oxygen will readily form water, so that the system contains hydrogen, oxygen, and water. Thus the addition of a small amount of catalyst makes all the energy states associated with water molecules available or accessible to the system, and, hence, the system proceeds spontaneously to populate these states. Since the originally accessible states are still accessible (there is still some hydrogen and oxygen in the system), this spontaneous process is associated with an increase in the number of states accessible to the system by the removal of some constraint. In this case, the constraint was a high activation energy barrier, which was removed by the addition of the catalyst.

Both of the spontaneous processes that we have discussed occurred, because some restraint, inhibition, or barrier was removed which made additional quantum states accessible to the system. In general, any spontaneous process in an isolated system can be viewed in this manner. The removal of some constraint allows a greater number of quantum states to be accessible to the system, thus the "flow" of the system into these states is observed as a spontaneous process.

The above discussion is limited to isolated systems. In order to discuss the condition $\Delta A < 0$, we must now consider isothermal processes. When a system is in a heat bath rather than isolated, we must include all possible energy states or levels of the system. When a restraint is removed, the number of accessible quantum states of each and every energy E cannot decrease, and will usually increase, since the original states are still available. Thus we have that $\Omega_2(N, V, E) \geq \Omega_1(N, V, E)$ for all E, where the subscripts 1 and 2 denote the initial and final states, respectively. We now use this inequality along with Eq. (2–36) to show that $\Delta A < 0$. Since no term can be negative and many are positive, we have

$$Q_2 - Q_1 = \sum_E \{\Omega_2(N, V, E) - \Omega_1(N, V, E)\}e^{-E/kT} > 0 \qquad (2\text{–}37)$$

In Eq. (2–37) we sum over all the levels available to the final state. It follows immediately from the inequality in Eq. (2–37) that

$$\Delta A = A_2 - A_1 = -kT \ln \frac{Q_2}{Q_1} < 0 \qquad (2\text{–}38)$$

for a spontaneous isothermal process, and thus we have written the second law of thermodynamics in terms of Eq. (2–35).

Lastly, we consider the implications of putting the "constant" of Eq. (2–30) equal to zero. We shall see that this gives us a statistical thermodynamic version of the third law of thermodynamics. If we write Eq. (2–33) for S more explicitly, we get

$$S = k \ln \sum_j e^{-E_j/kT} + \frac{1}{T} \frac{\sum_j E_j e^{-E_j/kT}}{\sum_j e^{-E_j/kT}} \qquad (2\text{–}39)$$

We wish to study the behavior of this equation as $T \to 0$. Assume for generality that the first n states have the same energy ($E_1 = E_2 = \cdots = E_n$) and that the next m states have the same energy ($E_{n+1} = E_{n+2} = \cdots = E_{n+m}$), and so on. Then in the limit of small T, Eq. (2–39) becomes (see Problem 2–19):

$$S = k \ln n + \frac{km}{n} e^{-(E_{n+1}-E_1)/kT} + \frac{m}{nT} (E_{n+1} - E_1) e^{-(E_{n+1}-E_1)/kT}$$

and so

$$\lim_{T \to 0} S = k \ln n \qquad (2\text{–}40)$$

Thus as $T \to 0$, S is proportional to the logarithm of the degeneracy of the lowest level. Unless n is very large, Eq. (2–40) says that S is practically zero. For example, if the system were a gas of N-point particles, and the degeneracy of the lowest level were of the order of N, $k \ln N$ would be practically zero compared to a typical order of magnitude of the entropy, namely, Nk. Thus setting the "constant" of Eq. (2–30) equal to zero is equivalent to adopting the convention that the entropy of most systems is zero at the absolute zero of temperature [*cf.* Eq. (1–44)].

ADDITIONAL READING

General

ANDREWS, F. C. 1963. *Equilibrium statistical mechanics.* New York: Wiley. Chapters 9–11.

HILL, T. L. 1960. *Statistical thermodynamics.* Reading, Mass.: Addison-Wesley. Sections 1–1 through 1–5.

KESTIN, J., and DORFMAN, J. R. 1971. *A course in statistical thermodynamics.* New York: Academic. Sections 5–1 through 5–3 and 5–11 through 5–17.

KITTEL, C. 1969. *Thermal physics.* New York: Wiley. Chapters 1–4.

KNUTH, E. L. 1966. *Statistical thermodynamics.* New York: McGraw-Hill. Chapters 2, 4, and 5.

MAYER, J. E., and MAYER, M. G. 1940. *Statistical mechanics.* New York: Wiley. Chapters 3, 4, and 10.

SCHRÖDINGER, E. 1952. *Statistical thermodynamics.* Cambridge: Cambridge University Press. Chapters 1–3.

PROBLEMS

2–1. From statistical mechanics we have shown

$$\left(\frac{\partial \bar{E}}{\partial V}\right)_{N, \beta} + \beta \left(\frac{\partial \bar{p}}{\partial \beta}\right)_{N, V} = -\bar{p}$$

and from the thermodynamics we have

$$\left(\frac{\partial E}{\partial V}\right)_{N, T} - T \left(\frac{\partial p}{\partial T}\right)_{N, V} = -p$$

Why can't β be linearly proportional to the temperature? That is, $\beta = \text{constant} \times T$.

2–2. To investigate the replacement of $\bar{n}_j$ by $n_j{}^*$, that is, the replacement of the average number of systems in state j by the most probable number in state j, consider the simple example in which $\Omega(\mathbf{n})$ is just a binomial distribution

$$\Omega(\mathbf{n}) = \frac{n!}{n_1! (n - n_1)!}$$

and actually calculate $n_1{}^*$ and $\bar{n}_1$. Hint: Recall that

$$(1 + x)^n = \sum_{n_1=0}^{n} \frac{n! x^{n_1}}{n_1!(n - n_1)!}$$

2–3. Show that Eq. (2–9) follows from Eq. (2–8). Note that in deriving this result, we have written $\ln W(\mathbf{a})$ as $\mathscr{A} \ln \mathscr{A} - \mathscr{A} - \sum_j a_j \ln a_j + \mathscr{A}$ and have considered $\mathscr{A}$ to be a constant. Show that Eq. (2–10) is independent of this assumption, that is, derive Eq. (2–10) treating $\mathscr{A}$ as $\sum_j a_j$.

2–4. Starting with Eq. (2–31), prove that the Boltzmann constant k must be positive, using the fact that the heat capacity C_V is always positive.

2–5. Show that the entropy can be written as

$$S = -k \sum_j P_j \ln P_j$$

where P_j is given by Eq. (2–12).

2–6. Maximize the function defined as "information" in information theory.

$$I = \sum_j P_j \ln P_j$$

subject to the two constraints

$$\sum_j P_j = 1$$

and

$$\sum_j E_j P_j = E = \text{fixed}$$

Compare this result to that of Problem 1–51.

2–7. Obtain the most probable distribution of N molecules of an ideal gas contained in two equal and connected volumes at the same temperature by minimizing the Helmholtz free energy for the two systems.

2–8. Differentiate Eq. (2–16) with respect to β to derive Eq. (2–13).

2–9. Derive Eq. (2–24).

2–10. Derive Eqs. (2–31) and (2–32).

2–11. Derive Eqs. (2–31) through (2–33) by starting with $A = -kT \ln Q$.

2–12. We can derive Eq. (2–36) directly by the method of Lagrange multipliers. We label the *levels* rather than the states by a subscript l. The degeneracy of the lth level, whose energy is E_l, is Ω_l. The number of ways of distributing systems over levels, with degeneracy Ω_l, is

$$W(\mathbf{a}) = \frac{\mathscr{A}!}{\prod_l a_l!} \prod_l \Omega_l{}^{a_l}$$

where a_l is the number of systems in the lth level. Maximize this, subject to the constraints

$$\sum_l a_l = \mathscr{A}$$

$$\sum_l a_l E_l = \mathscr{E}$$

to get

$$a_l{}^* = \frac{\Omega_l e^{-E_l/kT}}{\sum_l \Omega_l e^{-E_l/kT}}$$

2–13. Show that for a particle confined to a cube of length a that

$$p_j = \frac{2}{3} \frac{E_j}{V}$$

By taking the ensemble average of both sides, we have

$$\bar{p} = \frac{2}{3} \frac{\bar{E}}{V}$$

If we use the fact that $\bar{E} = \frac{3}{2}NkT$ (to be proved in Chapter 5), we get the ideal gas equation of state.

⇒ **2-14.** We shall show in Chapter 5 that the partition function of a monatomic ideal gas is

$$Q(N, V, T) = \frac{1}{N!} \left(\frac{2\pi mkT}{h^2} \right)^{3N/2} V^N$$

Derive expressions for the pressure and the energy from this partition function. Also show that the ideal gas equation of state is obtained if Q is of the form $f(T)V^N$, where $f(T)$ is any function of temperature.

⇒ **2-15.** In Chapter 11 we shall approximate the partition function of a crystal by

$$Q = \left(\frac{e^{-h\nu/2kT}}{1 - e^{-h\nu/kT}} \right)^{3N} e^{U_0/kT}$$

where $h\nu/k \equiv \Theta_E$ is a constant characteristic of the crystal, and U_0 is the sublimation energy of the crystal. Calculate the heat capacity from this simple partition function and show that at high temperatures, one obtains the law of Dulong and Petit, namely, that $C_V \to 3Nk$ as $T \to \infty$.

⇒ **2-16.** In Chapter 13 of this author's textbook *Statistical Thermodynamics*, it is shown that the partition function of an ideal gas of diatomic molecules in an external electric field $\mathscr{E}$ is

$$Q(N, V, T, \mathscr{E}) = \frac{[q(V, T, \mathscr{E})]^N}{N!}$$

where

$$q(V, T, \mathscr{E}) = V \left(\frac{2\pi mkT}{h^2} \right)^{3/2} \left(\frac{8\pi^2 IkT}{h^2} \right) \frac{e^{-h\nu/2kT}}{(1 - e^{-h\nu/kT})} \left(\frac{kT}{\mu\mathscr{E}} \right) \sinh\left(\frac{\mu\mathscr{E}}{kT} \right)$$

Here I is the moment of inertia of the molecule; ν is its fundamental vibrational frequency; and μ is its dipole moment. Using this partition function along with the thermodynamic relation,

$$dA = -S\,dT - p\,dV - M\,d\mathscr{E}$$

where $M = N\bar{\mu}$, where $\bar{\mu}$ is the average dipole moment of a molecule in the direction of the external field $\mathscr{E}$, show that

$$\bar{\mu} = \mu \left[\coth\left(\frac{\mu\mathscr{E}}{kT} \right) - \frac{kT}{\mu\mathscr{E}} \right]$$

Sketch this result versus $\mathscr{E}$ from $\mathscr{E} = 0$ to $\mathscr{E} = \infty$ and interpret it.

2-17. In Chapter 14 we shall derive an *approximate* partition function for a dense gas, which is of the form

$$Q(N, V, T) = \frac{1}{N!} \left(\frac{2\pi mkT}{h^2} \right)^{3N/2} (V - Nb)^N e^{aN^2/VkT}$$

where a and b are constants that are given in terms of molecular parameters. Calculate the equation of state from this partition function. What equation of state is this? Calculate the thermodynamic energy and the heat capacity and compare it to Problem 1-30.

2-18. From electrostatics, the displacement vector D is given by $D = \mathscr{E} + 4\pi P$, where $\mathscr{E}$ is the electric field, and P is the polarization, i.e., the total dipole moment M per unit volume.

The dielectric constant ε is defined by $D = \varepsilon\mathscr{E}$. In the simple case of a parallel plate capacitor, D is the field produced by a set of charges on the plates, and so we can consider it to be external field; M is the total moment (both permanent and induced) of the substance between the plates; and $\mathscr{E}$ is the field between the plates, that is, the force that an infinitesimal change would feel. If only a vacuum existed between the plates, D and $\mathscr{E}$ would be the same. A real substance modifies D such that $D = \mathscr{E} + 4\pi(M/V)$, where V is the volume. Since D, $\mathscr{E}$, and M are all in the same direction (at least for simple fluids), D must be $\geq\mathscr{E}$, which says that $\varepsilon \geq 1$.

When an external electric field D is present, the first law of thermodynamics becomes

$$dE = T\,dS - p\,dV - M\,dD + \mu\,dN$$

Problem: Describe *concisely* how one would calculate ε (at least in principle) from statistical mechanics.

2–19. Derive equation 2–40.

CHAPTER 3

OTHER ENSEMBLES AND FLUCTUATIONS

In Chapter 2 we considered an ensemble in which N, V, and T are held fixed for each system. This ensemble is one of many possible ensembles that can be constructed. For example, if we allow the walls of the containers to be permeable to molecular transport, N is no longer fixed for each system, and we no longer have a canonical ensemble. The ensemble in this case is called a grand canonical ensemble and is discussed in Section 3-1. In Section 3-2 we discuss two other ensembles that are often used in statistical thermodynamics: the microcanonical ensemble, in which N, V, and E are fixed, and the isothermal-isobaric ensemble, in which N, T, and p are fixed. The last section, Section 3-3, is devoted to an investigation of fluctuations in statistical thermodynamics.

One of our basic assumptions is that the ensemble average of a mechanical property can be equated to the corresponding thermodynamic function; hence it is important that we investigate the expected spread about the mean value. We show in Section 3-3 that for macroscopic systems the probability distribution of observing some mechanical property is a very narrow Gaussian distribution whose mean is the ensemble average. One important deduction from this result is that the various ensembles are essentially equivalent and that one can choose to work with a partition function on the basis of mathematical convenience rather than on the basis of which thermodynamic variables are used to specify the system of interest.

3-1 GRAND CANONICAL ENSEMBLE

In the previous chapter we treated the canonical ensemble, in which each system is enclosed in a container whose walls are heat conducting, but impermeable to the passage of molecules. The entire ensemble is placed in a heat bath at temperature T until equilibrium is reached, and then is isolated from its surroundings. Each system of the ensemble is specified by N, V, and T. In this section we shall treat a grand canonical ensemble. In a grand canonical ensemble, each system is enclosed in a container whose walls are both heat conducting and permeable to the passage of molecules. The number of molecules in a system, therefore, can range over all possible values, that is, each system is open with respect to the transport of matter. We construct a grand canonical ensemble by placing a collection of such systems in a large heat bath at temperature T and a large reservoir of molecules. After equilibrium is reached, the entire ensemble is isolated from its surroundings. Since the entire ensemble is at equilibrium with respect to the transport of heat and matter, each system is specified by V, T, and μ, where μ is the chemical potential. (If there is more than one component, the chemical potential of each component is the same from system to system.) Figure 3–1 shows a schematic picture of a grand canonical ensemble.

We proceed now in the same manner as in the treatment of the canonical ensemble. In this case, however, we must specify a system not only by which quantum state it is in but also by the number of molecules in the system. For each value of N, there is a set of energy states $\{E_{Nj}(V)\}$. We let a_{Nj} be the number of systems in the ensemble that contain N molecules and are in the state j. Each value of N has a particular set of levels associated with it, so we first specify N and then j. The set of occupation numbers $\{a_{Nj}\}$ is a distribution. By the postulate of equal a priori probabilities, we assume that all states associated with all possible distributions are to be given equal weight or equal probability of occurrence in the ensemble. Each possible distribution must satisfy the following three conditions:

$$\sum_N \sum_j a_{Nj} = \mathscr{A} \tag{3-1}$$

$$\sum_N \sum_j a_{Nj} E_{Nj} = \mathscr{E} \tag{3-2}$$

$$\sum_N \sum_j a_{Nj} N = \mathscr{N} \tag{3-3}$$

The three symbols $\mathscr{A}$, $\mathscr{E}$, and $\mathscr{N}$ denote the number of systems in the ensemble, the total energy of the ensemble (the ensemble is isolated), and the total number of molecules in the ensemble.

For any possible distribution, the number of states is given by

$$W(\{a_{Nj}\}) = \frac{\mathscr{A}!}{\prod_N \prod_j a_{Nj}!} \tag{3-4}$$

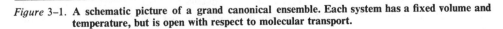

Figure 3–1. **A schematic picture of a grand canonical ensemble. Each system has a fixed volume and temperature, but is open with respect to molecular transport.**

As in the treatment of the canonical ensemble, the distribution that maximizes W subject to the appropriate constraints completely dominates all others. Thus we maximize Eq. (3–4) under the constraints of Eqs. (3–1) through (3–3), respectively, and we get (see Problem 3–1)

$$a_{Nj}^* = e^{-\alpha}e^{-\beta E_{Nj}(V)}e^{-\gamma N} \tag{3–5}$$

As before, the parameter α is easily determined in terms of the other parameter(s). We sum both sides of Eq. (3–5) over N and j and use Eq. (3–1) to get

$$P_{Nj}(V, \beta, \gamma) = \frac{a_{Nj}^*}{\mathscr{A}} = \frac{e^{-\beta E_{Nj}(V)}e^{-\gamma N}}{\sum_N \sum_j e^{-\beta E_{Nj}(V)}e^{-\gamma N}} \tag{3–6}$$

where $P_{Nj}(V, \beta, \gamma)$ is the probability that any randomly chosen system contains N molecules and be in the jth energy state, with energy $E_{Nj}(V)$.

The averages of the mechanical properties E, P, and N are

$$\bar{E}(V, \beta, \gamma) = \frac{1}{\Xi}\sum_N \sum_j E_{Nj}(V)e^{-\beta E_{Nj}(V)}e^{-\gamma N}$$

$$= -\left(\frac{\partial \ln \Xi}{\partial \beta}\right)_{V, \gamma} \tag{3–7}$$

$$p(V, \beta, \gamma) = \frac{1}{\Xi}\sum_N \sum_j \left(-\frac{\partial E_{Nj}}{\partial V}\right)e^{-\beta E_{Nj}(V)}e^{-\gamma N}$$

$$= \frac{1}{\beta}\left(\frac{\partial \ln \Xi}{\partial V}\right)_{\beta, \gamma} \tag{3–8}$$

$$\bar{N}(V, \beta, \gamma) = \frac{1}{\Xi}\sum_N \sum_j N e^{-\beta E_{Nj}(V)}e^{-\gamma N}$$

$$= -\left(\frac{\partial \ln \Xi}{\partial \gamma}\right)_{V, \beta} \tag{3–9}$$

where

$$\Xi(V, \beta, \gamma) = \sum_N \sum_j e^{-\beta E_{Nj}(V)}e^{-\gamma N} \tag{3–10}$$

We now determine β and γ. In our treatment of the canonical ensemble, one of the methods used to determine β was to derive an equation that related $(\partial \bar{E}/\partial V)_{N, \beta}$ to $(\partial \bar{p}/\partial \beta)_{N, V}$ and to compare this with a purely thermodynamic equation relating $(\partial E/\partial V)_{N, T}$ to $(\partial p/\partial T)_{N, V}$ [cf. Eqs. (2–17) to (2–21)]. This comparison suggested that β was proportional to $1/T$. We then showed that any two systems at the same temperature have the same value of β, thus proving that $\beta = 1/kT$, where k is a universal constant. We can do the same thing here (Problem 3–2), but it is not necessary.

A grand canonical ensemble can be considered to be a collection of canonical ensembles in thermal equilibrium with each other but with all possible values of N. Each of the systems has the same value of β, regardless of the number of molecules it contains. That β has the same value as in the canonical ensemble can be seen by imagining that we suddenly make the walls of the containers impermeable to the molecules but still heat conducting. This gives us a collection of canonical ensembles with V, N, and T fixed, and the arguments of Chapter 2 can be used to show that $\beta = 1/kT$.

The value of γ can be found by using the same method that we used in Chapter 2 to show that β was an integrating factor of δq_{rev}. Consider the function

$$f(\beta, \gamma, \{E_{Nj}(V)\}) = \ln \Xi = \ln \sum_N \sum_j e^{-\beta E_{Nj}(V)} e^{-\gamma N}$$

As the notation indicates, we regard f to be a function of β, γ, and the E_{Nj}'s. The total derivative of f is

$$df = \left(\frac{\partial f}{\partial \beta}\right)_{\gamma, \{E_{Nj}\}} d\beta + \left(\frac{\partial f}{\partial \gamma}\right)_{\beta, \{E_{Nj}\}} d\gamma + \sum_N \sum_j \left(\frac{\partial f}{\partial E_{Nj}}\right)_{\beta, \gamma, E_{Nj,s}} dE_{Nj}$$

Using Eqs. (3–6) through (3–10), we have

$$df = -\bar{E}\, d\beta - \bar{N}\, d\gamma - \beta \sum_N \sum_j P_{Nj}\, dE_{Nj}$$

The last term here is the ensemble average reversible work done by the systems. For simplicity, we assume only p–V work to get

$$df = -\bar{E}\, d\beta - \bar{N}\, d\gamma + \beta \bar{p}\, dV$$

Paralleling our development in Chapter 2, we add $d(\beta \bar{E}) + d(\gamma \bar{N})$ to both sides of this equation:

$$d(f + \beta \bar{E} + \gamma \bar{N}) = \beta\, d\bar{E} + \beta \bar{p}\, dV + \gamma\, d\bar{N}$$

If we compare this to the purely thermodynamic equation

$$T\, dS = dE + p\, dV - \mu\, dN$$

and use the fact that $\beta = 1/kT$, we can conclude that

$$\gamma = \frac{-\mu}{kT} \tag{3–11}$$

$$S = \frac{\bar{E}}{T} - \frac{\bar{N}\mu}{T} + k \ln \Xi \tag{3–12}$$

In Eq. (3–12), we have set the constant of integration equal to zero in accord with the third law of thermodynamics (see Problem 3–5).

We have now brought the entropy, a nonmechanical property, into our discussion. Equation (3–12), along with Eqs. (3–6) through (3–9), allows us to express any thermodynamic function of interest in a grand canonical ensemble in terms of $\Xi(V, T, \mu)$. This function is called the *grand (canonical ensemble) partition function*:

$$\Xi(V, T, \mu) = \sum_N \sum_j e^{-E_{Nj}(V)/kT} e^{\mu N/kT} \tag{3–13}$$

As the canonical partition function is the connection between thermodynamics and statistical thermodynamics for closed, isothermal systems (N, V, and T fixed), the grand partition function serves as the link for open, isothermal systems (V, T, and μ fixed). If we can determine Ξ for a system, we can calculate its thermodynamic properties.

By summing over j for fixed N in Eq. (3–13), we see that it is possible to write Ξ in the form

$$\Xi(V, T, \mu) = \sum_N Q(N, V, T) e^{\mu N/kT} \tag{3–14}$$

The term $e^{\mu/kT}$ is often denoted by λ. Since $\mu = kT \ln \lambda$, λ is an absolute activity, for the difference in chemical potentials between two states is given by $\Delta\mu = kT \ln(a_2/a_1)$, where a_1 and a_2 are activities.

Since we take the number of systems in an ensemble to be arbitrarily large, the number of particles in an ensemble becomes arbitrarily large, and hence the possible number of particles in any one system can approach infinity. Therefore the summation in Eq. (3–14) can be taken from 0 to ∞:

$$\Xi(V, T, \mu) = \sum_{N=0}^{\infty} Q(N, V, T)\lambda^N \tag{3-15}$$

Even though it may appear from Eq. (3–15) that Ξ would be more difficult to obtain than Q, it actually turns out in many problems that Ξ is easier to obtain, since the constraint of constant N is often mathematically awkward. This constraint can be avoided by using a grand partition function, that is, by summing over all values of N (see Section 4–2). Furthermore, there are many systems in which the many-body problem can be reduced to a one-body, two-body problem, and so on. In these cases, the grand partition function is particularly useful.

To complete our discussion of the grand canonical ensemble, we shall show that pV is the thermodynamic characteristic function of $\ln \Xi$. To see this, compare Eq. (3–12) with the thermodynamic equation

$$G = \mu N = E + pV - TS$$

Thus we have

$$pV = kT \ln \Xi(V, T, \mu) \tag{3-16}$$

Problem 1–37 shows that pV is the thermodynamic function whose natural variables are V, T, and μ. Equations (3–8), (3–9), and (3–12) can be derived from Eq. (3–16) and the thermodynamic equation $d(pV) = S\,dT + N\,d\mu + p\,dV$ (see Problem 3–6). Table 3–1 summarizes the formulas of the grand canonical ensemble.

3–2 OTHER ENSEMBLES

We could go on to consider other ensembles. For example, we could construct an ensemble of systems in which the containing walls of each system are heat conducting and flexible, so that each system of the ensemble is described by N, T, and p. The constraints would be on the total energy and total volume of the ensemble, and the partition function would turn out to be (see Problem 3–9):

$$\Delta(N, T, p) = \sum_{E} \sum_{V} \Omega(N, V, E)e^{-E/kT}e^{-pV/kT} \tag{3-17}$$

whose characteristic thermodynamic function is the Gibbs free energy, that is,

$$G = -kT \ln \Delta(N, T, p) \tag{3-18}$$

Equation (3–17) is called the isothermal-isobaric partition function. Notice that the natural variables of G are N, T, and p, the variables associated with this ensemble.

If we compare Eq. (3–17) with the two other partition functions that we have derived [Eqs. (2–34) and (3–13)], we see that all three can be obtained by starting with $\Omega(N, V, E)$, multiplying by some appropriate exponential, and summing over one or two of the variables N, V, and E. In a sense, $\Omega(N, V, E)$ is fundamental to all ensembles and, in fact, is itself the partition function for conceptually the most simple ensemble, the one representative of isolated systems. This is called the microcanonical ensemble.

We can apply the results of the previous section to a treatment of an isolated system. The grand canonical ensemble represents a collection of systems whose containing walls allow heat and molecules to pass freely from one system to another. From a physical point of view, the entire grand canonical ensemble is equivalent to one isolated system of volume $\mathscr{A}V$, containing $\mathscr{N}$ molecules and having energy $\mathscr{E}$. The partitions in Fig. 3–1 can be considered to be a conceptual division of one isolated system into $\mathscr{A}$ subsystems. The entropy of the entire ensemble S_e, considered as one isolated system, is $\mathscr{A}S$, where S is the entropy of each of the open, isothermal systems. This entropy is given by Eq. (3–12):

$$S = k(\beta \bar{E} + \gamma \bar{N} + \ln \Xi) \tag{3–19}$$

where we use the notation β and γ for convenience. We use Eqs. (3–7) and (3–9) for $\bar{E}$ and $\bar{N}$:

$$S = k \ln \Xi + k \left(\sum_{N,j} \beta E_{Nj} \frac{e^{-\beta E_{Nj}}e^{-\gamma N}}{\Xi} + \sum_{N,j} \gamma \frac{N e^{-\beta E_{Nj}}e^{-\gamma N}}{\Xi} \right)$$

$$= k \ln \Xi + k \sum_{N,j} (\beta E_{Nj} + \gamma N) \frac{e^{-\beta E_{Nj}}e^{-\gamma N}}{\Xi}$$

$$= k \ln \Xi - k \sum_{N,j} (\ln a_{Nj}{}^* + \ln \Xi - \ln \mathscr{A}) \frac{a_{Nj}{}^*}{\mathscr{A}} \tag{3–20}$$

where we have used Eq. (3–6) to write the last line. We can perform the summation over the second two terms in parentheses in Eq. (3–20):

$$S = k \ln \Xi - \frac{k}{\mathscr{A}} \sum_{N,j} a_{Nj}{}^* \ln a_{Nj}{}^* - k \ln \Xi + k \ln \mathscr{A}$$

or

$$S_e = \mathscr{A}S = k\mathscr{A} \ln \mathscr{A} - k \sum_{N,j} a_{Nj}{}^* \ln a_{Nj}{}^*$$

$$= k \ln W(\{a_{Nj}{}^*\}) \tag{3–21}$$

We see that for an isolated system, the entropy is proportional to the logarithm of the number of states available to the system. In another notation, we can write

$$S = k \ln \Omega(N, V, E) \tag{3–22}$$

Equation (3–22) shows that the more states there are available to an isolated system, the higher is its entropy. This equation serves as the basis for qualitative statements concerning entropy and disorder, randomness, and so on. In practice, Eq. (3–22) is not used for the calculation of thermodynamic functions since N, V, and E are all mechanical variables.

The argument leading to Eq. (2–38) incidentally can be immediately applied to Eq. (3–22). For any spontaneous process in an isolated system,

$$\Delta S = k \ln \frac{\Omega_2}{\Omega_1} > 0$$

where 1 and 2 represent the initial and final states, respectively.

Equation (3–22) is due to Boltzmann and is possibly the best-known equation in statistical thermodynamics, mainly for historical reasons. Of course, Boltzmann (1844–1906) did not express his famous equation in terms of quantum states, but rather in a classical mechanical framework. We shall take up classical statistical mechanics in Chapter 7. Boltzmann, in fact, was a great contributor to both equilibrium and nonequilibrium statistical mechanics. He was one of the first to see clearly how probability ideas could be combined with mechanics. Equation (3–22) is carved on his tombstone in the Zentralfriedhoff in Vienna, although the equation is not often used today. However, his contribution to nonequilibrium statistical mechanics is such that to this day the so-called Boltzmann equation (Chapters 18 and 19) still is the fundamental equation describing the transport of dilute gases. It is interesting to note that Boltzmann, who contributed so much to understanding macroscopic phenomena in terms of molecular mechanics, lived at a time when the atomic theory was not so generally accepted as it is today, and his work was severely criticized by some of the leading physicists of the day. He committed suicide in 1906 (for reasons not entirely clear) and never lived to see the full acceptance of his work in statistical mechanics.

Although $\Omega(N, V, E)$ is not generally available, we have determined it for an ideal gas in Section 1–3 [cf. Eq. (1–37)]. If we calculate $k \ln \Omega$, neglecting terms of order less than $O(N^{-1})$, we get (see Problem 3–11):

$$S = Nk \ln\left[\left(\frac{2\pi mkT}{h^2}\right)^{3/2} \frac{V e^{5/2}}{N}\right] \tag{3–23}$$

We shall see later that this equation gives excellent agreement with experiment, but now we simply show that if we use

$$dS = \frac{1}{T} dE + \frac{p}{T} dV - \frac{\mu}{T} dN \tag{3–24}$$

to get

$$\frac{p}{T} = \left(\frac{\partial S}{\partial V}\right)_{N,E}$$

and substitute Eq. (3–23) into this, we find

$$pV = NkT$$

which is the ideal gas equation of state. See Table 3–1 for a summary of the formulas related to the microcanonical ensemble.

It is possible to derive partition functions appropriate to other sets of independent variables, but the four that we have considered above are sufficient for most applications. We shall show that in the limit of large systems in equilibrium, one can choose an ensemble and its partition function on the basis of mathematical convenience rather than on the basis of which thermodynamic variables are used to describe the system. This result will come out of a study of fluctuations, which we turn to now.

3–3 FLUCTUATIONS

The methods that we have developed allow us to calculate ensemble averages of mechanical variables, which we then equate to thermodynamic functions. Equations such as Eq. (2–12) or (3–6) are the probability distributions over which these ensemble

Table 3–1. **A summary of formulas for several types of ensemble**

---

microcanonical ensemble, $\Omega(N, V, E)$

$$S = k \ln \Omega$$

$$dS = \frac{1}{T} dE + \frac{p}{T} dV - \frac{\mu}{T} dN$$

$$\frac{1}{kT} = \left(\frac{\partial \ln \Omega}{\partial E}\right)_{N, V} \tag{3–25}$$

$$\frac{p}{kT} = \left(\frac{\partial \ln \Omega}{\partial V}\right)_{N, E} \tag{3–26}$$

$$\frac{\mu}{kT} = -\left(\frac{\partial \ln \Omega}{\partial N}\right)_{V, E} \tag{3–27}$$

canonical ensemble, $Q(N, V, T)$

$$A = -kT \ln Q$$

$$dA = -S\, dT - p\, dV + \mu\, dN$$

$$S = k \ln Q + kT \left(\frac{\partial \ln Q}{\partial T}\right)_{N, V} \tag{3–28}$$

$$p = kT \left(\frac{\partial \ln Q}{\partial V}\right)_{N, T} \tag{3–29}$$

$$\mu = -kT \left(\frac{\partial \ln Q}{\partial N}\right)_{V, T} \tag{3–30}$$

$$E = kT^2 \left(\frac{\partial \ln Q}{\partial T}\right)_{N, V} \tag{3–31}$$

grand canonical ensemble, $\Xi(V, T, \mu)$

$$pV = kT \ln \Xi$$

$$d(pV) = S\, dT + N\, d\mu + p\, dV$$

$$S = k \ln \Xi + kT \left(\frac{\partial \ln \Xi}{\partial T}\right)_{V, \mu} \tag{3–32}$$

$$N = kT \left(\frac{\partial \ln \Xi}{\partial \mu}\right)_{V, T} \tag{3–33}$$

$$p = kT \left(\frac{\partial \ln \Xi}{\partial V}\right)_{\mu, T} = kT \frac{\ln \Xi}{V} \tag{3–34}$$

isothermal-isobaric ensemble, $\Delta(N, T, p)$

$$G = -kT \ln \Delta$$

$$dG = -S\, dT + V\, dp + \mu\, dN$$

$$S = k \ln \Delta + kT \left(\frac{\partial \ln \Delta}{\partial T}\right)_{N, p} \tag{3–35}$$

$$V = -kT \left(\frac{\partial \ln \Delta}{\partial p}\right)_{N, T} \tag{3–36}$$

$$\mu = -kT \left(\frac{\partial \ln \Delta}{\partial N}\right)_{T, p} \tag{3–37}$$

---

averages are taken. In Section 1–5 we saw that the average is the first of a family of moments. Another important moment is the second central moment or the variance, $\overline{(x - \bar{x})^2}$, which is a measure of the spread of a probability distribution about the mean value. Furthermore, we saw toward the end of Section 1–5 that the most meaningful measure of the spread of a distribution is the square root of the variance, that is, the standard deviation, relative to the mean value. A standard deviation of 10^{10} may be large as an absolute number, but it is extremely small if the mean of the probability distribution is 10^{20}. In this section we shall calculate the variances of several mechanical variables and compare these to the mean values.

Any deviation of a mechanical variable from its mean value is called a fluctuation, and the investigation of the probability of such deviations is called fluctuation theory. Fluctuation theory is important in statistical mechanics for a number of reasons. The most obvious reason is to determine to what extent we expect to observe deviations from the mean values that we calculate. If the spread about these is large, then experimentally we would observe a range of values, whose mean or average is given by statistical thermodynamics. We shall see, however, that the probability of observing any value other than the mean value is extremely remote. As a corollary to this important result, we shall see that all of the ensembles that we have considered earlier are equivalent for all practical purposes. In addition, there are several statistical thermodynamical theories of solutions and light scattering based on fluctuation theory, and one formulation of the statistical mechanical theory of transport focuses on the rate of decay of spontaneous fluctuations.

Let us consider first fluctuations in a canonical ensemble. In a canonical ensemble, N, V, and T are held fixed, and we can investigate fluctuations in the energy, pressure, and related properties since these are the ones that vary from system to system. It is important to be aware of the properties that can vary and those properties that are fixed in each ensemble. We shall consider fluctuations in the energy. Thus we use Eq. (2–12) for the probability distribution of the energy and write for the variance

$$\sigma_E{}^2 = \overline{(E - \bar{E})^2} = \overline{E^2} - \bar{E}^2$$
$$= \sum_j E_j{}^2 P_j - \bar{E}^2 \tag{3–38}$$

where

$$P_j = \frac{e^{-\beta E_j}}{Q(N, V, \beta)} \tag{3–39}$$

We can write Eq. (3–38) in a more convenient form by noting that

$$\sum_j E_j{}^2 P_j = \frac{1}{Q} \sum_j E_j{}^2 e^{-\beta E_j} = -\frac{1}{Q} \frac{\partial}{\partial \beta} \sum_j E_j e^{-\beta E_j}$$

$$= -\frac{1}{Q} \frac{\partial}{\partial \beta}(\bar{E}Q) = -\frac{\partial \bar{E}}{\partial \beta} - \bar{E}\frac{\partial \ln Q}{\partial \beta}$$

$$= kT^2 \frac{\partial \bar{E}}{\partial T} + \bar{E}^2 \tag{3–40}$$

Thus Eq. (3–38) becomes

$$\sigma_E{}^2 = kT^2 \left(\frac{\partial \bar{E}}{\partial T}\right)_{N,V} \tag{3–41}$$

and if we associate $\bar{E}$ with the thermodynamic energy, we have

$$\sigma_E{}^2 = kT^2 C_V \tag{3-42}$$

where C_V is the molar heat capacity.

To explore the *relative* magnitude of this spread, we look at

$$\frac{\sigma_E}{\bar{E}} = \frac{(kT^2 C_V)^{1/2}}{\bar{E}} \tag{3-43}$$

To get an order-of-magnitude estimate of this ratio, we use the values of $\bar{E}$ and C_V for an ideal gas, namely, $O(NkT)$ and $O(Nk)$, respectively. If we use these values in Eq. (3–43), we find that $\sigma_E/\bar{E}$ is $O(N^{-1/2})$, showing that in a typical macroscopic system, the relative deviations from the mean are extremely small. The probability distribution of the energy may, therefore, be regarded as a Gaussian distribution which is practically a delta function.

We can derive a Gaussian distribution approximation to $P(E)$, the probability of observing a particular value of E in a canonical ensemble. According to Eq. (2–36), $P(E)$ is given by $C\Omega(E)e^{-E/kT}$, where C is a normalization factor which is independent of E. Since $\Omega(E)$ is an increasing function of E, and $e^{-E/kT}$ is a decreasing function of E, their product $P(E)$ peaks at some value of E, say E^*. But we have just seen above that the spread about the maximum value is extremely small, and so E^* and $\bar{E}$ are essentially the same point. The width of $P(E)$ is $O(N^{-1/2})$, and so E^* and $\bar{E}$ differ by $O(N^{-1/2})$.

Let us now expand $P(E)$ in a Taylor series about E^*, or $\bar{E}$. As in Section 1–5, it is more convenient to work with $\ln P(E)$. From the definition of $E^*(\approx \bar{E})$ as the value of E at the maximum in $P(E)$,

$$\left(\frac{\partial \ln P}{\partial E}\right)_{E=E^*=\bar{E}} = \left(\frac{\partial \ln \Omega}{\partial E}\right)_{E=E^*=\bar{E}} - \beta = 0 \tag{3-44}$$

Equation (3–44) determines $\bar{E}$ as a function of β. The second derivative of $\ln P(E)$ is

$$\left(\frac{\partial^2 \ln P}{\partial E^2}\right) = \left(\frac{\partial^2 \ln \Omega}{\partial E^2}\right)$$

which is to be evaluated at $E = E^* = \bar{E}$. Since

$$\left(\frac{\partial^2 \ln \Omega}{\partial E^2}\right)_{E=\bar{E}} = \frac{\partial^2 \ln \Omega(\bar{E})}{\partial \bar{E}^2} = \frac{\partial}{\partial \bar{E}}\left(\frac{\partial \ln \Omega(\bar{E})}{\partial \bar{E}}\right)$$

$$= \frac{\partial}{\partial \bar{E}}\left(\frac{\partial \ln \Omega}{\partial E}\right)_{E=\bar{E}} = \frac{\partial \beta}{\partial \bar{E}}$$

where the last term follows from Eq. (3–44), we have

$$\left(\frac{\partial^2 \ln P}{\partial E^2}\right)_{E=E^*=\bar{E}} = \frac{\partial \beta}{\partial \bar{E}} = -\frac{1}{kT^2}\frac{\partial T}{\partial \bar{E}} = -\frac{1}{kT^2 C_V} \tag{3-45}$$

The Taylor expansion of $\ln P(E)$ through quadratic terms is

$$\ln P(E) = \ln P(\bar{E}) - \frac{(E-\bar{E})^2}{2kT^2 C_V} + \cdots \tag{3-46}$$

or

$$P(E) = P(\bar{E})\exp\left\{-\frac{(E-\bar{E})^2}{2kT^2 C_V}\right\} \tag{3-47}$$

Problem 3–16 involves showing that terms beyond the quadratic terms can be ignored in Eq. (3–46).

If we compare Eq. (3–47) to the standard form of a Gaussian distribution [Eq. (1–72)], we see that $\sigma_E^2 = kT^2C_V$ (in agreement with Eq. (3–42)] and that the normalization constant $P(\bar{E})$ is $(2\pi\sigma_E^2)^{-1/2}$. Equation (3–47) can be used to calculate the probability of observing a value of E that differs from $\bar{E}$. For example, the probability of observing an energy that differs by 0.1 percent from the average energy of 1 mole of an ideal gas $O(e^{-10^6})$, an extremely small number. (See Problem 3–12.)

Incidentally, the derivation of Eq. (3–47) is a case where we must be careful not to confuse the variable E in $P(E)$ with E^*, $\bar{E}$, or the thermodynamic quantity E, which unfortunately is also called " E." This is especially true of Eq. (3–44), where E is a variable, and $\bar{E}$ is that particular value of the variable for which the quantity $\partial \ln \Omega/\partial E$, a function of E, is equal to the preassigned value of β (N, V, T are given in a canonical ensemble).

We could also calculate the fluctuations in the pressure in a canonical ensemble, but this is left to Problem 3–18. Instead, we consider the fluctuations in a grand canonical ensemble. In a grand canonical ensemble, V, T, and μ are held fixed, while the energy and number of particles in each system are allowed to vary. We can calculate the fluctuation in the number of particles in the same manner as we treated the fluctuation in energy in a canonical ensemble. If σ_N^2 is the variance in the number of particles, then

$$\sigma_N^2 = \overline{N^2} - \overline{N}^2 = \sum_{N,j} N^2 P_{Nj} - \overline{N}^2 \qquad (3\text{–}48)$$

where

$$P_{Nj} = \frac{e^{-\beta E_{Nj}} e^{-\gamma N}}{\Xi(V, \beta, \gamma)}$$

We treat $\overline{N^2}$ in analogy to Eq. (3–40):

$$\sum_{N,j} N^2 P_{Nj} = \frac{1}{\Xi} \sum_{N,j} N^2 e^{-\beta E_{Nj}} e^{-\gamma N} = -\frac{1}{\Xi}\frac{\partial}{\partial \gamma} \sum_{N,j} N e^{-\beta E_{Nj}} e^{-\gamma N}$$

$$= -\frac{1}{\Xi}\frac{\partial}{\partial \gamma}(\overline{N}\Xi) = -\frac{\partial \overline{N}}{\partial \gamma} - \overline{N}\frac{\partial \ln \Xi}{\partial \gamma}$$

$$= kT\left(\frac{\partial \overline{N}}{\partial \mu}\right)_{V,T} + \overline{N}^2 \qquad (3\text{–}49)$$

Thus Eq. (3–48) becomes

$$\sigma_N^2 = kT\left(\frac{\partial \overline{N}}{\partial \mu}\right)_{V,T} \qquad (3\text{–}50)$$

The right-hand side of this equation can be written in a more familiar form by thermodynamic manipulations. Problem 3–26 proves that

$$\left(\frac{\partial \mu}{\partial \overline{N}}\right)_{V,T} = -\frac{V^2}{N^2}\left(\frac{\partial p}{\partial V}\right)_{N,T}$$

and so

$$\sigma_N^2 = \frac{\overline{N}^2 kT\kappa}{V} \qquad (3\text{–}51)$$

where κ is the isothermal compressibility

$$\kappa \equiv -\frac{1}{V}\left(\frac{\partial V}{\partial p}\right)_{N,T} \tag{3-52}$$

The value of σ_N relative to $\overline{N}$ is

$$\frac{\sigma_N}{\overline{N}} = \left(\frac{kT\kappa}{V}\right)^{1/2} \tag{3-53}$$

To get an order-of-magnitude estimate of this ratio, we use the fact that $\kappa = 1/p$ for an ideal gas to get $\sigma_N/\overline{N} = N^{-1/2}$. Again we find that relative deviations from the mean are very small. The result, $0(N^{-1/2})$, is typical of fluctuations in statistical thermodynamics.

Since V is fixed in the grand canonical ensemble, the fluctuation in the number of particles is proportional to the fluctuation in the density ρ, and so

$$\frac{\sigma_\rho}{\overline{\rho}} = \frac{\sigma_N}{\overline{N}} = \left(\frac{kT\kappa}{V}\right)^{1/2} \tag{3-54}$$

There is a condition under which the fluctuations in density are not negligible. At the critical point of a substance, $(\partial p/\partial V)_{N,T}$ is zero, and hence its isothermal compressibility is infinite. Thus there are large fluctuations in the density from point to point in a fluid at its critical point. This is observed macroscopically by the phenomenon of critical opalescence, in which a pure substance becomes turbid at its critical point.

We can also derive a Gaussian approximation to $P(N)$. Let $N^*(= \overline{N})$ be the value of N at the peak in $P(N)$. We have

$$P(N) = CQ(N, V, T)e^{\beta\mu N}$$

where C is a normalization constant. Then

$$\left(\frac{\partial \ln P}{\partial N}\right)_{N=N^*} = \left(\frac{\partial \ln Q}{\partial N}\right)_{N=\overline{N}} + \beta\mu = 0$$

This equation determines $\overline{N}$ as a function of $\beta\mu$. Also,

$$\left(\frac{\partial^2 \ln P}{\partial N^2}\right)_{N=\overline{N}} = \left(\frac{\partial^2 \ln Q}{\partial N^2}\right)_{N=\overline{N}} = \frac{\partial}{\partial \overline{N}}\frac{\partial \ln Q(\overline{N}, V, T)}{\partial \overline{N}} = -\frac{\partial \beta\mu}{\partial \overline{N}}$$

$$= -\frac{1}{kT(\partial \overline{N}/\partial \mu)_{V,T}}$$

Thus we find

$$P(N) = P(\overline{N})\exp\left[\frac{-(N - \overline{N})^2}{2kT(\partial \overline{N}/\partial \mu)_{V,T}}\right] \tag{3-55}$$

which gives the same expression for $\sigma_N{}^2$ as Eq. (3-50). Problems 3-19 through 3-20 involve the determination of fluctuations in the isothermal-isobaric ensemble.

An interesting application of the above fluctuation formulas is to the scattering of light by the atmosphere. It can be shown that if light of intensity I_0 is incident on a region of volume V with a dielectric constant ε, which differs from the average value of ε for the medium $\bar{\varepsilon}$, the intensity of light scattered at an angle θ at a distance R is

$$\frac{I(\theta)}{I_0} = \frac{\pi^2 V^2 \sigma_\varepsilon{}^2}{2\lambda^4}\frac{(1 + \cos^2 \theta)}{R^2} \tag{3-56}$$

where $\sigma_\varepsilon{}^2$ is the variance of ε, and λ is the wavelength of the incident light in vacuum.

This is called Rayleigh scattering. The dielectric constant ε is related to the density by the so-called Clausius-Mossotti equation

$$\frac{\varepsilon - 1}{\varepsilon + 2} = A\rho \tag{3-57}$$

which is derived and discussed in most physical chemistry texts. The quantity A is a constant, and ρ is the density. We can see from this equation that fluctuations in ρ lead to fluctuations in ε, and hence to Rayleigh scattering by Eq. (3–56). If we calculate σ_ε^2 in terms of σ_ρ^2 from Eq. (3–57), and use Eq. (3–54) for σ_ρ^2, we find (see Problem 3–21)

$$\frac{I(\theta)}{I_0} = \frac{\pi^2 kT}{18\lambda^4} \kappa(\varepsilon - 1)^2(\varepsilon + 2)^2 V \frac{(1 + \cos^2\theta)}{R^2}$$

where κ is the isothermal compressibility. By integrating this over the surface of a sphere of radius R, we obtain finally

$$\frac{I_{\text{scattered}}}{I_0} = \int \frac{I(\theta)}{I_0} R^2 \sin\theta \, d\theta \, d\phi$$

$$= \frac{8\pi^3}{27\lambda^4} kT\kappa(\varepsilon - 1)^2(\varepsilon + 2)^2 V$$

This equation shows that the blue color of the sky is due to fluctuations in the density of the atmosphere. The λ^4 in the denominator gives rise to a strong dependence on wavelength, so that the short wavelengths (blue) of the sun's light are scattered more than the red, and hence the sky appears blue. Similarly, red sunsets and sunrises are due to the fact that the long wavelengths (red) are not scattered as much as the blue.

There is one result of fluctuation theory which will be very useful to us. We have stated above that the various ensembles and their partition functions are essentially equivalent to each other, and that one can choose to work with a partition function on the basis of mathematical convenience. We now show why this is so.

Consider the canonical partition function:

$$Q(N, V, T) = \sum_E \Omega(N, V, E)e^{-E/kT} \tag{3-58}$$

We have seen in Eq. (3–47) that $P(E) = C\Omega(E)\exp(-E/kT)$ is an extremely narrow Gaussian function of E. In the limit of large N (and it is only in the limit of large N that classical thermodynamics is valid), only one value of E is important, namely, $E = E^* = \bar{E}$. Thus in the summation in Eq. (3–58), only the term with $E = \bar{E}$ contributes, and Eq. (3–58) becomes

$$Q(N, V, T) = \Omega(N, V, \bar{E})e^{-E/kT} \tag{3-59}$$

Although the systems of a canonical ensemble can, in principle, assume any value of E (as long as it is an eigenvalue of the N-particle Schrödinger equation), it happens that the energy of the entire ensemble is distributed uniformly throughout the ensemble, and each system is almost certain to be found with the average energy $\bar{E}$. A canonical ensemble degenerates, in a sense, to a microcanonical ensemble.

If we take the logarithm of Eq. (3–59) and use Eq. (2–35), we find that

$$A = \bar{E} - kT \ln \Omega(N, V, \bar{E})$$

or that

$$S = k \ln \Omega(N, V, \bar{E})$$

This is an alternative derivation of the fundamental relation between the entropy and the number of states accessible to the system.

The general results we have obtained here are also obtained for other ensembles. For example, although the systems of a grand canonical ensemble can assume any value of N and E, in practice it turns out that the total energy and the total number of molecules of the entire ensemble are distributed uniformly throughout the ensemble, and each system has the average energy and contains the average number of molecules. This, of course, is exactly what one expects intuitively, as long as the systems are of macroscopic size and the density is not *extremely* low.

These results can be used to write down, by inspection, the characteristic thermodynamic function of any partition function. Equations (3–58) and (3–59) are a good example. Suppose we did not know that $A = -kT \ln Q$. We do know that $S = k \ln \Omega$, however, and so if we take the logarithm of Eq. (3–59), we get that $\ln Q = S/k - \beta E$, which shows that $A = -kT \ln Q$. Since partition functions, in general, are a sum of $\Omega(N, V, E)$ multiplied by exponential factors, this method can always be used to determine the thermodynamic characteristic function. (See Problem 3–15.)

ADDITIONAL READING

General

HILL, T. L. 1960. *Statistical thermodynamics*. Reading, Mass.: Addison-Wesley. Sections 1–5 through 1–7 and Chapter 2.
———. 1956. *Statistical mechanics*. New York: McGraw-Hill. Sections 14, 15, and Chapter 4.
KESTIN, J., and DORFMAN, J. R. 1971. *A course in statistical thermodynamics*. New York: Academic. Section 5–18.
KNUTH, E. 1966. *Statistical thermodynamics*. New York: McGraw-Hill. Chapters 3, 4, and 5.
KUBO, R. 1965. *Statistical mechanics*. Amsterdam: North-Holland Publishing Co. Sections 1–12 to 1–14.
REIF, F. 1965. *Statistical and thermal physics*. New York: McGraw-Hill. Chapter 3.
RUSHBROOKE, G. S. 1949. *Statistical mechanics*. London: Oxford University Press. Chapters 15 and 17.

Fluctuations

ANDREWS, F. C. 1963. *Equilibrium statistical mechanics*. New York: Wiley. Chapter 33.
DAVIDSON, N. 1962. *Statistical mechanics*. New York: McGraw-Hill. Chapter 14.
KESTIN, J., and DORFMAN, J. R. 1971. *A course in statistical thermodynamics*. New York: Academic. Chapter 14.
LANDAU, L. D., and LIFSHITZ, E. M. 1958. *Statistical physics*. London: Pergamon Press. Chapter 12.
MÜNSTER, A. 1969. *Statistical thermodynamics*, Vol. I. Berlin: Springer-Verlag. Chapter 3.
SCHRÖDINGER, E. 1952. *Statistical thermodynamics*. Cambridge: Cambridge University Press. Chapter 5.
TOLMAN, R. C. 1938. *Statistical mechanics*. London: Oxford University Press. Section 141.

PROBLEMS

3–1. Derive Eq. (3–5).

3–2. Using a grand canonical formalism, show that any two systems at the same temperature have the same value of β.

3–3. For a grand canonical ensemble show that

$$\left(\frac{\partial \bar{E}}{\partial V}\right)_{\gamma, \beta} + \beta \left(\frac{\partial \bar{p}}{\partial \beta}\right)_{\gamma, V} = -\bar{p}$$

Compare this to the thermodynamic equation (see Problem 1–31)

$$\left(\frac{\partial E}{\partial V}\right)_{\mu/T, 1/T} + \frac{1}{T}\left(\frac{\partial p}{\partial(1/T)}\right)_{\mu/T, V} = -p$$

to suggest that $\beta = \text{const}/T$ for a grand canonical ensemble.

3–4. State and use Euler's theorem to show

$$p = kT \left(\frac{\partial \ln \Xi}{\partial V} \right)_{\mu, T} = kT \frac{\ln \Xi}{V}$$

3–5. Show that the entropy given by Eq. (3–12) goes to zero as T goes to zero.

3–6. Derive the principal thermodynamic connection formulas of the grand canonical ensemble starting from

$$pV = kT \ln \Xi$$

and

$$d(pV) = S \, dT + N \, d\mu + p \, dV$$

3–7. Show that for a two-component system

$$\Xi(\mu_1, \mu_2, T, V) = \sum_{N_1} \sum_{N_2} Q(N_1, N_2, V, T) \lambda_1{}^{N_1} \lambda_2{}^{N_2}$$

where $\lambda_i = e^{\mu_i/kT}$ ($i = 1, 2$). From this derive the corresponding thermodynamic connection formulas.

3–8. In the next chapter we shall see that the grand partition function of an ideal monatomic gas is

$$\Xi = e^{q\lambda}$$

where $q = (2\pi m k T/h^2)^{3/2} V$. Derive the thermodynamic properties of an ideal monatomic gas from Ξ.

3–9. Show that the partition function appropriate to an isothermal-isobaric ensemble is

$$\Delta(N, p, T) = \sum_E \sum_V \Omega(N, V, E) e^{-E/kT} e^{-pV/kT}$$

Derive the principal thermodynamic connection formulas for this ensemble.

3–10. In Problem 5–17 we shall show that the isothermal-isobaric partition function of an ideal monatomic gas is

$$\Delta = \left[\frac{(2\pi m)^{3/2}(kT)^{5/2}}{ph^3} \right]^N$$

Derive the thermodynamic properties of an ideal monatomic gas from Δ.

3–11. Derive Eq. (3–23) starting from Eq. (1–37).

3–12. Calculate the probability of observing an energy that differs by 10^{-4} percent from the average energy of 1 mole of an ideal gas.

3–13. Show that for macroscopic ideal systems, ones obtains the same result for the entropy whether one uses $S = k \ln \Phi(E)$, where Φ is the number of quantum states with energy $\leq E$ [Eq. (1–36)], or $S = k \ln \Omega(E, \Delta E)$, where $\Omega(E, \Delta E)$ is the number of quantum states within energy ΔE about E [Eq. (1–37)] as long as $\Delta E/E$ is small, but not zero. Show that S is insensitive to ΔE over a wide range of ΔE. The next problem discusses this remarkable result more generally.

3–14. Let $\Omega(E) \, dE$ be the number of quantum states between E and $E + dE$. In Chapter 1 we showed that $\Omega(E)$ is a monotonically increasing function of E (at least for an ideal gas). We can write two obvious inequalities for $\Omega(E)$:

$$\Phi(E) \equiv \int_0^E \Omega(E') \, dE' \geq \Omega(E) \, \Delta E$$

$$E\Omega(E) \geq \int_0^E \Omega(E') \, dE' = \Phi(E)$$

where ΔE is a small region surrounding E. By multiplying the second inequality by $\Delta E/E$, we get

$$\Omega(E)\,\Delta E \geq \frac{\Delta E}{E} \int_0^E \Omega(E')\,dE'$$

Combining this inequality with the first one above gives

$$\frac{\Delta E}{E} \int_0^E \Omega(E')\,dE' \leq \Omega(E)\,\Delta E \leq \int_0^E \Omega(E')\,dE'$$

Taking logarithms gives

$$\ln \Phi(E) - \ln \left(\frac{E}{\Delta E}\right) \leq \ln [\Omega(E)\,\Delta E] \leq \ln \Phi(E)$$

Now unless ΔE is extremely small, $\ln (E/\Delta E)$ is completely negligible compared to $\ln \Phi(E)$, since the total number of states with energies equal to or less than E is at least $0(e^N)$. Show that even if the energy could be measured to a millionth of a percent, $\ln (E/\Delta E) \approx 18$, which is completely negligible compared to N.

3–15. Fluctuation theory provides a simple method to determine the characteristic function associated with a particular partition function. Consider the canonical partition function

$$Q(N, V, T) = \sum_E \Omega(N, V, E)\, e^{-E/kT}$$

According to the theory of fluctuations, there is effectively only one term in this summation, and so we write

$$Q(N, V, T) = \Omega(N, V, \bar{E})\, e^{-E/kT}$$

Remembering that $S = k \ln \Omega$, we have, upon taking logarithms, that

$$\ln Q = \frac{S}{k} - \frac{E}{kT}$$

or that

$$\ln Q = \frac{-A}{kT}$$

Proceeding in a like manner, determine the characteristic thermodynamic function of the following partition functions:

$$\Xi(V, T, \mu) = \sum_N Q(N, V, T)\, e^{\beta \mu N}$$

$$\Delta(p, T, N) = \sum_V Q(N, V, T)\, e^{-\beta p V}$$

$$\phi(V, E, \beta\mu) = \sum_N \Omega(N, V, E)\, e^{\beta \mu N}$$

$$\Psi(V, T, \mu_1, N_2) = \sum_{N_1} Q(N_1, N_2, T, V)\, e^{\beta \mu N_1}$$

$$W(p, \gamma, T, N) = \sum_V \sum_{\mathscr{A}} Q(N, V, \mathscr{A}, T)\, e^{-\beta p V}\, e^{\beta \gamma \mathscr{A}}$$

where $\mathscr{A}$ is surface area, and γ is the surface tension.

3–16. When we derived the Gaussian expression for $P(E)$ in a canonical ensemble, we expanded $\ln P(E)$ in a Taylor expansion about $E = E^* \approx \bar{E}$, dropping terms after the quadratic term. Show that these terms are negligible.

3–17. Show that

$$\overline{(E - \bar{E})^3} = k^2 \left\{ T^4 \left(\frac{\partial C_V}{\partial T} \right) + 2T^3 C_V \right\}$$

and that

$$\frac{\overline{(E - \bar{E})^3}}{\bar{E}^3} = 0(N^{-2})$$

for a canonical ensemble.

3–18. Derive an expression for the fluctuation in the pressure in a canonical ensemble.

3–19. Show that for an isothermal-isobaric ensemble

$$P(V) = P(V^*) \exp \left\{ \frac{(V - \bar{V})^2}{2kT \left(\dfrac{\partial V}{\partial p} \right)_{N, T}} \right\}$$

3–20. Derive an equation for the fluctuation in the volume in an isothermal-isobaric ensemble. In other words, derive an equation for $\overline{V^2} - \bar{V}^2$. Express your answer in terms of the isothermal compressibility, defined by

$$\kappa = -\frac{1}{V} \left(\frac{\partial V}{\partial p} \right)_{N, T}$$

Show that $\sigma_v / \bar{V}$ is of the order of $N^{-1/2}$.

3–21. By calculating σ_ε^2 in terms of σ_ρ^2 from Eq. (3–57) and using Eq. (3–54) for σ_ρ^2, show that

$$\frac{I(\theta)}{I_0} = \frac{\pi^2 kT}{18\lambda^4} \kappa (\varepsilon - 1)^2 (\varepsilon + 2)^2 V \frac{(1 + \cos^2 \theta)}{R^2}$$

3–22. Show that the fluctuation in energy in a grand canonical ensemble is

$$\sigma_E^2 = (kT^2 C_V) + \left(\frac{\partial \bar{E}}{\partial \bar{N}} \right)_{T, V} \sigma_N^2$$

3–23. Show that in a two-component open, isothermal ensemble that

$$\overline{N_1 N_2} - \bar{N}_1 \bar{N}_2 = kT \left(\frac{\partial \bar{N}_1}{\partial \mu_2} \right)_{V, T, \mu_1}$$

$$= kT \left(\frac{\partial \bar{N}_2}{\partial \mu_1} \right)_{V, T, \mu_2}$$

3–24. Show that

$$\overline{H^2} - \bar{H}^2 = kT^2 C_p$$

in an N, p, T ensemble.

3–25. Use the formulas in Table 3–1 to derive expressions for any other thermodynamic functions for each of the four ensembles listed there.

3–26. Show that

$$\left(\frac{\partial \mu}{\partial N} \right)_{V, T} = -\frac{V^2}{N^2} \left(\frac{\partial p}{\partial V} \right)_{N, T}$$

CHAPTER 4

BOLTZMANN STATISTICS, FERMI-DIRAC STATISTICS, AND BOSE-EINSTEIN STATISTICS

The results that we have derived up to now are valid for macroscopic systems. In order to apply these equations, it is necessary to have the set of eigenvalues $\{E_j(N, V)\}$ of the N-body Schrödinger equation. In general, this is an impossible task. There are many important systems, however, in which the N-body Hamiltonian operator can be written as a sum of independent individual Hamiltonians. In such cases the total energy of the system can be written as a sum of individual energies. This leads to a great simplification of the partition function, and allows us to apply the results with relative ease.

We shall see that the final equations depend upon whether the individual particles of the system are fermions (that is, the N-body wave function is antisymmetric under the interchange of identical particles) or bosons (the N-body wave function is symmetric under the interchange of identical particles). These two types of particles obey different laws, called Fermi-Dirac or Bose-Einstein statistics. We shall show that under normal conditions (for example, sufficiently high temperatures), both of these distribution laws can be approximately reduced to an even simpler one, called Boltzmann statistics. The Boltzmann distribution law can also be derived from $Q(N, V, T)$ at high temperature without first deriving the Fermi-Dirac and Bose-Einstein distribution laws, and this is done in Section 4–1. We shall discuss in this section just what is meant by "normal" conditions or "sufficiently high" temperatures. Then in Section 4–2 we derive the two fundamental distribution laws, Fermi-Dirac and Bose-Einstein statistics, and show how both of them reduce to Boltzmann statistics in the appropriate limit.

4–1 THE SPECIAL CASE OF BOLTZMANN STATISTICS

In Section 1–3 it was shown that if the Hamiltonian of a many-body system can be written as a sum of one-body Hamiltonians, the energy of the system is the sum of individual energies, and the wave function is a product of the single-particle wave functions. In addition, the wave functions of a system of identical particles must satisfy certain symmetry requirements with respect to the interchange of the particles. All known particles fall into two classes: those whose wave function must be symmetric under the operation of the interchange of two identical particles, and those whose wave function must be antisymmetric under such an exchange. Particles belonging to the first class are called bosons, and the others are called fermions. There is no restriction of the distribution of bosons over their available energy states, but fermions have the very severe restriction that no two identical fermions can occupy the same single-particle energy state. This restriction follows immediately from the requirement that the wave function be antisymmetric (see Problem 1–26). These considerations become important in enumerating the many-body energy states available to the system.

There are many problems in which the Hamiltonian can be written as a sum of simpler Hamiltonians. The most obvious example perhaps is the case of a dilute gas, where the molecules are on the average far apart, and hence their intermolecular interactions can be neglected. Another example, which may be familiar from physical chemistry, is the decomposition of the Hamiltonian of a polyatomic molecule into its various degrees of freedom:

$$\mathscr{H} \approx \mathscr{H}_{\text{translational}} + \mathscr{H}_{\text{rotational}} + \mathscr{H}_{\text{vibrational}} + \mathscr{H}_{\text{electronic}} \tag{4–1}$$

Equation (4–1) is a good first approximation and can be systematically corrected by the introduction of small interaction terms.

There are many other problems in physics in which the Hamiltonian, by a proper and clever selection of variables, can be written as a sum of individual terms. Although these individual terms need not be Hamiltonians for actual individual molecules, they are nevertheless used to define the so-called quasi-particles, which mathematically behave like independent real particles. Some of the names of quasi-particles that are found in the literature are photons, phonons, plasmons, magnons, rotons, and other " ons." In spite of the apparent limitation of this requirement on the Hamiltonian, we can see that it is very useful and can be used to study solids (Chapter 11) and liquids (see Chapter 12 of *Statistical Thermodynamics*), systems in which the decomposition of a many-body Hamiltonian into a sum of independent terms would hardly appear to be justified. First let us consider the canonical partition function for a system of distinguishable particles, in which the Hamiltonian can be written as a sum of individual terms. Denote the individual energy states by $\{\varepsilon_j{}^a\}$, where the superscript denotes the particle (they are distinguishable), and the subscript denotes the state. In this case the canonical partition function becomes

$$
\begin{aligned}
Q(N, V, T) &= \sum_j e^{-E_j/kT} = \sum_{i,j,k,\ldots} e^{-(\varepsilon_i{}^a + \varepsilon_j{}^b + \varepsilon_k{}^c + \cdots)/kT} \\
&= \sum_i e^{-\varepsilon_i{}^a/kT} \sum_j e^{-\varepsilon_j{}^b/kT} \sum_k e^{-\varepsilon_k{}^c/kT} \cdots \\
&= q_a q_b q_c \cdots
\end{aligned}
\tag{4–2}
$$

where

$$q(V, T) = \sum_i e^{-\varepsilon_i/kT} \tag{4–3}$$

Equation (4–2) is a very important result. It shows that if we can write the N-particle Hamiltonian as a sum of independent terms, and if the particles are *distinguishable*, then the calculation of $Q(N, V, T)$ reduces to a calculation of $q(V, T)$. Since $q(V, T)$

requires a knowledge only of the energy values of an individual particle or quasi-particle, its evaluation is quite feasible. In most cases $\{\varepsilon_i\}$ is a set of molecular energy states; thus $q(V, T)$ is called a molecular partition function.

If the energy states of all the particles are the same, then Eq. (4–2) becomes

$$Q(N, V, T) = [q(V, T)]^N \qquad \text{(distinguishable particles)} \qquad (4\text{–}4)$$

Equation (4–4) shows that the original N-body problem (the evaluation of $Q(N, V, T)$) can be reduced to a one-body problem (the evaluation of $q(V, T)$) if the particles are independent and distinguishable. Although particles are certainly not distinguishable in general, there are many important cases where they can be treated as such. An excellent example of this is a perfect crystal. In a perfect crystal each atom is confined to one and only one lattice point, which we could, in principle, identify by a set of three numbers. Since each particle, then, is confined to a lattice point and the lattice points are distinguishable, the particles themselves are distinguishable. Furthermore, we shall see in Chapter 11 that although there are strong intermolecular interactions in crystals, we can treat the vibration of each particle about its lattice point as independent to a first approximation.

Another useful application of the separation indicated in Eq. (4–2) is to the molecular partition function itself. Equation (4–1) shows that the molecular Hamiltonian can be approximated by a sum of Hamiltonians for the various degrees of freedom of the molecule. Consequently we get the useful result that

$$q_{\text{molecule}} = q_{\text{translational}} q_{\text{rotational}} q_{\text{vibrational}} q_{\text{electronic}} \cdots \qquad (4\text{–}5)$$

where, for example,

$$q_{\text{translational}} = \sum_i e^{-\varepsilon_i^{\text{trans}}/kT} \qquad (4\text{–}6)$$

Thus not only can we reduce an N-body problem to a one-body problem, but it is possible to reduce it further into the individual degrees of freedom of the single particles.

Equation (4–4) is an attractive result, but atoms and molecules are, in general, not distinguishable; thus the utility of Eq. (4–4) is severely limited. The situation becomes more complicated when the inherent indistinguishability of atoms and molecules is considered. In this case, the N-body energy is

$$E_{ijkl\cdots} = \varepsilon_i + \varepsilon_j + \varepsilon_k + \varepsilon_l + \cdots \qquad (4\text{–}7)$$

and the partition function is

$$Q(N, V, T) = \sum_{i, j, k, l \cdots} e^{-(\varepsilon_i + \varepsilon_j + \varepsilon_k + \varepsilon_l + \cdots)/kT} \qquad (4\text{–}8)$$

Because the molecules are indistinguishable, one cannot sum over $i, j, k, l, \ldots$ separately as we did to get Eq. (4–2).

Consider, for example, the case of fermions. The antisymmetry of the wave function requires that no two identical fermions can occupy the same single-particle energy state. Thus in Eq. (4–8), terms in which two or more indices are the same cannot be included in the summation. The indices i, j, k, l, and so on, are not independent of one another, and a direct evaluation of $Q(N, V, T)$ for fermions by means of Eq. (4–8) is very difficult.

Bosons do not have the restriction that no two can occupy the same molecular state, but the summation in Eq. (4–8) is still complicated. Consider a term in Eq. (4–8) in which all of the indices are the same except one, that is, a term of the form $\varepsilon_i + \varepsilon_j + \varepsilon_j + \varepsilon_j + \cdots$ with $i \neq j$. Because the particles are indistinguishable, the position of ε_i is unimportant, and so this state is identical with $\varepsilon_j + \varepsilon_i + \varepsilon_j + \varepsilon_j + \cdots$ or $\varepsilon_j + \varepsilon_j + \varepsilon_i + \varepsilon_j + \varepsilon_j + \cdots$, and so on. Such a state should be included only once in Eq. (4–8), but an unrestricted summation over the indices in Eq. (4–8) would produce N terms of this type. Consider the other extreme in which all of the particles are in different molecular states, that is, the state with energy $\varepsilon_i + \varepsilon_j + \varepsilon_k + \cdots$ with $i \neq j \neq k \neq \cdots$. Because the particles are indistinguishable, the $N!$ states obtained by permuting the N different subscripts are identical and should occur only once in Eq. (4–8). Such terms will, of course, appear $N!$ times in an unrestricted summation. Consequently, a direct evaluation of Q for bosons by means of Eq. (4–8) also is difficult.

The terms that introduce complications are those in which two or more indices are the same. If it were not for this kind of term, one could carry out the summation in Eq. (4–8) in an unrestricted manner, and then correct the sum by dividing by $N!$ It turns out that this procedure yields an excellent approximation in many (most) cases for the following reason.

We showed in Section 1–3 that for a particle in a box, the number of molecular quantum states with energy $\leq \varepsilon$ is

$$\Phi(\varepsilon) = \frac{\pi}{6} \left(\frac{8ma^2\varepsilon}{h^2} \right)^{3/2} \tag{1 0}$$

For $m = 10^{-22}$ g, $a = 10$ cm, and $T = 300°$K, $\Phi(\varepsilon) = O(10^{30})$. Although this calculation is done for one particle in a cube (i.e., one molecule of an ideal gas), the order of magnitude of the result is general. Thus we see that the number of molecular quantum states available to a molecule at room temperature, say, is much greater than the number of molecules in the system for all but the most extreme densities. Since each particle has many individual states to choose from, it will be a rare event for two particles to be in the same molecular state. Therefore the vast majority of terms in Eq. (4–8) will have all different indices. This allows us to sum over all the indices unrestrictedly and divide by $N!$ to get

$$Q(N, V, T) = \frac{q^N}{N!} \qquad \text{(indistinguishable particles)} \tag{4–10}$$

with

$$q(V, T) = \sum_j e^{-\varepsilon_j/kT}$$

for a system of identical, indistinguishable particles satisfying the condition that the number of available molecular states is much greater than the number of particles.

Equation (4–10) is an extremely important result, since it reduces a many-body problem to a one-body problem. No longer is there a condition of distinguishability; the indistinguishability of the particles has been included by dividing by $N!$, a valid procedure for most systems under most conditions. We can investigate this condition in more detail using Eq. (4–9) for an ideal gas. Mathematically, we require that

$$\Phi(\varepsilon) \gg N$$

Using Eq. (4–9), we have the condition

$$\frac{\pi}{6} \left(\frac{12mkT}{h^2} \right)^{3/2} \gg \frac{N}{V} \tag{4–11}$$

where we have set $\varepsilon = 3kT/2$. Clearly this condition is favored by large mass, high temperature, and low density. Numerically it turns out that (4–11) is satisfied for all but the very lightest molecules at very low temperatures. Table 4–1 examines this condition for a number of systems. We see that the use of Eq. (4–10) is justified in most cases. We have examined (4–11) for only monatomic systems, but the results are valid for polyatomic molecules as well, since the translational energy states account for almost all of the energy states available to any molecule.

When Eq. (4–10) is valid, that is, when the number of available molecular states is much greater than the number of particles in the system, we say that the particles obey *Boltzmann statistics*. Boltzmann statistics is an approximation that becomes increasingly better at higher temperatures. We shall show in Chapter 7 that at high enough temperatures, one can describe the energy of a system by classical mechanics. Since the limiting case of Boltzmann statistics and the use of classical mechanics both require a high-temperature limit, Boltzmann statistics is also called the classical limit.

Let us examine Eq. (4–10). The total energy of the N-body system is

$$E = N\bar{\varepsilon} = kT^2 \left(\frac{\partial \ln Q}{\partial T}\right)_{N,V} = N \sum_j \varepsilon_j \frac{e^{-\varepsilon_j/kT}}{q} \tag{4–12}$$

The first equality is valid, because the molecules are assumed to be independent, and hence their energies are additive. We see from Eq. (4–12) that the average energy of a particle is

$$\bar{\varepsilon} = \sum_j \varepsilon_j \frac{e^{-\varepsilon_j/kT}}{q} \tag{4–13}$$

We can conclude from this equation that the probability that a molecule is in the jth energy state

$$\pi_j = \frac{e^{-\varepsilon_j kT}}{\sum_j e^{-\varepsilon_j kT}} = \frac{e^{-\varepsilon_j kT}}{q} \tag{4–14}$$

It is interesting to note that the fluctuations in ε are of the same order as ε itself (see Problems 4–18 and 4–19), that is, the probability distribution for single molecules is not sharp. A sharp probability distribution is a many-body effect.

Table 4–1. **The quantity** $(6N/\pi V)(h^2/12mkT)^{3/2}$ **for a number of simple systems***

	T (°K)	$\dfrac{6N}{\pi V}\left(\dfrac{h^2}{12mkT}\right)^{3/2}$
liquid helium	4	1.6
gaseous helium	4	0.11
gaseous helium	20	2.0×10^{-3}
gaseous helium	100	3.5×10^{-5}
liquid neon	27	1.1×10^{-2}
gaseous neon	27	8.2×10^{-5}
gaseous neon	100	3.1×10^{-6}
liquid argon	86	5.1×10^{-4}
gaseous argon	86	1.6×10^{-6}
liquid krypton	127	5.4×10^{-5}
gaseous krypton	127	2.0×10^{-7}
electrons in metals (sodium)	300	1465

* This quantity must be much less than unity for Eq. (4–10) to be valid. The temperatures associated with the liquid states are the normal boiling points [*cf*. Eq. (4–11)].

The similarity between Eq. (4–14) for molecular states and Eq. (2–12) for states of the entire N-body system is not fortuitous. Equation (4–14) can be derived by the same *mathematical* formalism of Chapter 2. The ensemble is considered to be the N actual molecules in thermal contact with each other. The number of molecules n_j in the state with energy ε_j is found by maximizing a combinatorial factor similar to Eq. (1–77). This point of view was the one originally proposed by Boltzmann. It is valid only for systems in which the total energy is a sum of individual molecular energies, that is, only for dilute gases. The conceptual generalization of these ideas by Gibbs was a magnificent achievement, which allowed statistical thermodynamics to be applicable to all physical systems. Furthermore, the derivation given in Chapter 2 is rigorous, since macroscopic systems can be labeled, and the size of the ensemble can be increased arbitrarily. This is not so for the Boltzmann approach, since the molecules cannot be labeled, and the system is finite.

Equation (4–14) can be reduced further if we assume that the energy of the molecule can be written in the form [*cf.* Eq. (4–1)]

$$\varepsilon = \varepsilon_i{}^{\text{trans}} + \varepsilon_j{}^{\text{rot}} + \varepsilon_k{}^{\text{vib}} + \varepsilon_l{}^{\text{elec}} + \cdots$$

Then Eq. (4–14) and Eq. (4–5) can be combined to give, for example,

$$\pi_j{}^{\text{vib}} = \frac{e^{-\varepsilon_j{}^{\text{vib}}}}{q_{\text{vib}}} \tag{4–15}$$

for the probability that a molecule is in the jth vibrational state irrespective of the other degrees of freedom.

Although Eq. (4–10) is applicable to most systems, it is important to complete the development of systems of independent, indistinguishable particles by evaluating Eq. (4–8) for the general case. The exact evaluation of Eq. (4–8) is necessary for several systems that we shall study. We must return, then, to a consideration of the effect of the symmetry requirements of N-body wave functions on the sum over states in Eq. (4–8).

4–2 FERMI-DIRAC AND BOSE-EINSTEIN STATISTICS

There are two cases to consider in the evaluation of Eq. (4–8). The resultant distribution function in the case of fermions is called Fermi-Dirac statistics, and that in the case of bosons is called Bose-Einstein statistics. Since all known particles are either fermions or bosons, these two "statistics" are the only exact distributions. We shall see, however, that in the case of high temperature and/or low density, both of these distributions go over into the Boltzmann or classical distribution.

It is most convenient to treat the general case by means of the grand canonical ensemble for reasons that we shall see shortly. Let $E_j(N, V)$ be the energy states available to a system containing N molecules. Let ε_k be the molecular quantum states. Finally, let $n_k = n_k(E_j)$ be the number of molecules in the kth molecular state when the system itself is in the quantum state with energy E_j. A quantum state of the entire system is specified by the set $\{n_k\}$. The energy of the system is

$$E_j = \sum_k \varepsilon_k n_k \tag{4–16}$$

and, of course,

$$N = \sum_k n_k \tag{4–17}$$

We can write $Q(N, V, T)$ as

$$Q(N, V, T) = \sum_j e^{-\beta E_j} = \sum_{\{n_k\}}{}^* e^{-\beta \sum_i \varepsilon_i n_i} \tag{4-18}$$

where the asterisk in the summation signifies the restriction that

$$\sum n_k = N$$

This restriction turns out to be mathematically awkward. We can avoid this restriction by using the grand canonical partition function instead. This will be an excellent example where one partition function is much easier to evaluate than another. Since we have demonstrated the equivalence of ensembles, we are free to make the choice strictly on mathematical convenience. We then use

$$\Xi(V, T, \mu) = \sum_{N=0}^{\infty} e^{\beta \mu N} Q(N, V, T)$$

We use Eq. (4–18) for $Q(N, V, T)$ and the absolute activity $\lambda = e^{\beta \mu}$ to get

$$\Xi(V, T, \mu) = \sum_{N=0}^{\infty} \lambda^N \sum_{\{n_k\}}{}^* e^{-\beta \sum_i \varepsilon_i n_i}$$

$$= \sum_{N=0}^{\infty} \sum_{\{n_k\}}{}^* \lambda^{\sum n_i} e^{-\beta \sum_j \varepsilon_j n_j}$$

$$= \sum_{N=0}^{\infty} \sum_{\{n_k\}}{}^* \prod_k (\lambda e^{-\beta \varepsilon_k})^{n_k} \tag{4-19}$$

Now comes the crucial step (which requires some thought). Since we are summing over all values of N, each n_k ranges over all possible values, and Eq. (4–19) can be written as (see Problem 4–6)

$$\Xi(V, T, \mu) = \sum_{n_1=0}^{n_1{}^{max}} \sum_{n_2=0}^{n_2{}^{max}} \cdots \prod_k (\lambda e^{-\beta \varepsilon_k})^{n_k} \tag{4-20}$$

Equations (4–19) and (4–20) are completely equivalent. Equation (4–20) can be written in a more lucid form:

$$\Xi(V, T, \mu) = \sum_{n_1=0}^{n_1{}^{max}} (\lambda e^{-\beta \varepsilon_1})^{n_1} \sum_{n_2=0}^{n_2{}^{max}} (\lambda e^{-\beta \varepsilon_2})^{n_2} \cdots$$

or

$$= \prod_k \sum_{n_k=0}^{n_k{}^{max}} (\lambda e^{-\beta \varepsilon_k})^{n_k} \tag{4-21}$$

Equation (4–21) is a simple product and is a general result. The crucial step in this series of equations is the step from Eq. (4–19) to Eq. (4–20), from which Eq. (4–21) follows immediately. The step from Eqs. (4–19) to (4–20) is possible only because we are summing over all values of N, or, in other words, since we are using the grand canonical partition function.

We now apply Eq. (4–21) to fermions and bosons. In *Fermi-Dirac* statistics, each of the n_k in Eq. (4–21) can be only either 0 or 1, since no two particles can be in the same quantum state. In this case $n_1{}^{max} = 1$, and Eq. (4–21) is simply

$$\Xi_{FD} = \prod_k (1 + \lambda e^{-\beta \varepsilon_k}) \tag{4-22}$$

where FD, of course, signifies Fermi-Dirac.

In *Bose-Einstein* statistics, on the other hand, the n_k can be 0, 1, 2, ..., since there is no restriction on the occupancy of each state. Therefore, $n_k{}^{max} = \infty$, and Eq. (4–21) becomes

$$\Xi_{BE} = \prod_k \sum_{n_k=0}^{\infty} (\lambda e^{-\beta\varepsilon_k})^{n_k} = \prod_k (1 - \lambda e^{-\beta\varepsilon_k})^{-1} \qquad \lambda e^{-\beta\varepsilon_k} < 1 \qquad (4\text{–}23)$$

To get Eq. (4–23), we have used the fact that

$$\sum_{j=0}^{\infty} x^j = (1 - x)^{-1}$$

for $x < 1$.

Equations (4–22) and (4–23) are the two fundamental distributions of the statistical thermodynamics of systems of independent particles. We can combine these two equations into

$$\Xi_{FD \atop BE} = \prod_k (1 \pm \lambda e^{-\beta\varepsilon_k})^{\pm 1} \qquad (4\text{–}24)$$

where as the notation indicates, the upper sign refers to Fermi-Dirac statistics, and the lower sign refers to Bose-Einstein statistics.

Using Eq. (3–33), we see that

$$\bar{N} = N = \sum_k \bar{n}_k = kT\left(\frac{\partial \ln \Xi}{\partial \mu}\right)_{V,T} = \lambda\left(\frac{\partial \ln \Xi}{\partial \lambda}\right)_{V,T} = \sum_k \frac{\lambda e^{-\beta\varepsilon_k}}{1 \pm \lambda e^{-\beta\varepsilon_k}} \qquad (4\text{–}25)$$

The average number of particles in the kth quantum state is

$$\bar{n}_k = \frac{\lambda e^{-\beta\varepsilon_k}}{1 \pm \lambda e^{-\beta\varepsilon_k}} \qquad (4\text{–}26)$$

Equation (4–26) is the quantum statistical counterpart of Eq. (4–14). We multiply Eq. (4–26) by ε_k and sum over k to get the quantum statistical version of Eq. (4–13).

$$\bar{E} = N\bar{\varepsilon} = \sum_k \bar{n}_k \varepsilon_k = \sum_k \frac{\lambda \varepsilon_k e^{-\beta\varepsilon_k}}{1 \pm \lambda e^{-\beta\varepsilon_k}} \qquad (4\text{–}27)$$

Lastly, Eq. (3–16) gives

$$pV = \pm kT \sum_k \ln[1 \pm \lambda e^{-\beta\varepsilon_k}] \qquad (4\text{–}28)$$

Equations (4–25) through (4–28) are the fundamental formulas of Fermi-Dirac ($+$) and Bose-Einstein ($-$) statistics. Note that the molecular partition function q is not a relevant quantity when we are dealing with quantum statistics, that is, Fermi-Dirac or Bose-Einstein statistics. In spite of the fact that we have neglected inter-molecular forces, the individual particles of the system are not independent because of the symmetry requirements of the wave functions.

We noted above that both kinds of statistics should go over into Boltzmann or classical statistics in the limit of high temperature or low density, where the number of available molecular quantum states is much greater than the number of particles. This condition implies that the average number of molecules in any state is very small, since most states will be unoccupied and those few states that are occupied will most likely contain only one molecule. This means that $\bar{n}_k \to 0$ in Eq. (4–26). This is achieved by letting $\lambda \to 0$. Thermodynamically, this means the limit of $N/V \to 0$ for fixed T, or $T \to \infty$ for fixed N/V. (See Problem 4–3.) For small λ, Eq. (4–26) becomes

$$\overline{n}_k = \lambda e^{-\beta\varepsilon_k} \qquad (\lambda \text{ small})$$

If we sum both sides of this equation over k to eliminate λ, we have

$$\frac{\bar{n}_k}{N} = \frac{e^{-\beta\varepsilon_k}}{q} \tag{4-29}$$

where

$$q = \sum_j e^{-\beta\varepsilon_j} \tag{4-30}$$

Equation (4–26) then goes over to the Boltzmann or classical limit for both Fermi-Dirac and Bose-Einstein statistics.

Equations (4–27) and (4–28) also reduce to the formulas of Section 4–1 as $\lambda \to 0$. Equation (4–27) becomes

$$\bar{E} \to \sum_j \lambda\varepsilon_j e^{-\beta\varepsilon_j}$$

and since $n_j \to \lambda e^{-\beta\varepsilon_j}$, we have

$$\bar{\varepsilon} = \frac{\bar{E}}{\bar{N}} \to \frac{\sum_j \varepsilon_j e^{-\beta\varepsilon_j}}{\sum_j e^{-\beta\varepsilon_j}} \tag{4-31}$$

This is the same as Eq. (4–13). Similarly, for small λ we can expand the logarithm in Eq. (4–28) to get

$$pV \to (\pm kT)\left(\pm\lambda \sum_j e^{-\beta\varepsilon_j}\right) \tag{4-32}$$

We have used the fact that $\ln(1 + x) \approx x$ for small x. Using Eq. (4–30), this becomes

$$pV = \lambda kT \sum_j e^{-\beta\varepsilon_j} = \lambda kT q \tag{4-33}$$

or

$$\beta pV = \ln \Xi = \lambda q \tag{4-34}$$

Equation (3–33) can be used to show that $\lambda q = N$, and so Eq. (4–34) is the perfect gas law as expected. Thus the formulas of Fermi-Dirac and Bose-Einstein statistics reduce to those of Boltzmann statistics in the classical limits.

We can also derive Eq. (4–10) directly from Eq. (4–34) for Ξ:

$$\Xi = e^{\lambda q} = \sum_{N=0}^{\infty} \frac{(\lambda q)^N}{N!}$$

If we compare this to Eq. (3–15), see that

$$Q(N, V, T) = \frac{q^N}{N!}$$

We shall defer a discussion of the equations of Fermi-Dirac and Bose-Einstein statistics to Chapter 10. There are a few systems such as electrons in metals, liquid helium, electromagnetic radiation, for which one must use quantum statistics. For most systems that we shall study in this book, however, we shall be able to use Boltzmann or classical statistics. In the next chapter we shall apply the limit of Boltzmann statistics to the simplest system, namely, a monatomic ideal gas.

ADDITIONAL READING

General

ANDREWS, F. C. 1963. *Equilibrium statistical mechanics*. New York: Wiley. Chapter 17.

HILL, T. L. 1960. *Statistical thermodynamics*. Reading, Mass.: Addison-Wesley. Chapter 3.

KITTEL, C. 1969. *Thermal physics*. New York: Wiley. Chapter 9.

KNUTH, E. 1966. *Statistical thermodynamics*. New York: McGraw-Hill. Chapter 6.

KUBO, R. 1965. *Statistical mechanics*. Amsterdam: North-Holland Publishing Co. Section 1–15 and Example 1–12.

MANDL, F. 1971. *Statistical physics*. New York: Wiley. Chapter 11.

REIF, F. 1965. *Statistical and thermal physics*. New York: McGraw-Hill. Sections 9–1 through 9–8.

RUSHBROOKE, G. S. 1949. *Statistical mechanics*. London: Oxford University Press. Chapters 2 and 3.

TOLMAN, R. C. 1938. *Statistical mechanics*. London: Oxford University Press. Chapters 10 and 14.

PROBLEMS

4–1. Calculate the temperature below which each of the substances listed below cannot be treated classically at 1 atmosphere. Compare this with the normal boiling temperature for each substance.

$$He, Ne, Ar, Kr, CO_2, N_2, H_2, Cl_2, H_2O$$

4–2. Show that the quantity

$$\frac{6N}{\pi V}\left(\frac{h^2}{12mkT}\right)^{3/2}$$

given in Table 4–1 is indeed very large for electrons in metals at room temperature.

4–3. Show that the condition that $\lambda \to 0$ corresponds thermodynamically to the limit $N/V \to 0$ for fixed T, or $T \to \infty$ for fixed density. Remember that $\lambda = e^{\beta\mu}$.

4–4. In deriving the limiting case of Boltzmann statistics, we claimed that if the number of quantum states M far exceeds the number of particles N, then the terms in the product of the molecular partition functions in which each particle is in a different quantum state constitute the overwhelming number of terms. Show, in fact, that the ratio of this type of term to the total number of terms approaches unity as $N/M \to 0$, N and M both large. Hint: remember that

$$\lim_{x \to \infty}\left(1 + \frac{a}{x}\right)^x = e^a$$

4–5. For an ideal gas, show that the relation

$$P = \frac{2}{3}\frac{E_{kin}}{V}$$

holds irrespective of its statistics, where E_{kin} is the total kinetic energy.

4–6. To convince yourself of the step leading from Eq. (4–19) to Eq. (4–20), consider the summation

$$S = \sum_{N=0}^{\infty} \sum_{(n_j)}^{*} x_1^{n_1} x_2^{n_2}$$

where n_1 and $n_2 = 0, 1,$ and 2. Show by directly expanding S for this simple case that this is equivalent to [Eq. (4–21)]

$$S = \prod_{K=1}^{2}(1 + x_K + x_K^2)$$

4–7. Recall that the equation of state for an ideal quantum gas is

$$pV = kT \ln \Xi = \pm kT \sum_{j} \ln [1 \pm \lambda e^{-\epsilon_j/kT}]$$

where $\lambda = e^{\mu/kT}$. Using the fact that the summation over states can be replaced by an integration over energy levels

$$\omega(\varepsilon) \, d\varepsilon = 2\pi \left(\frac{2m}{h^2}\right)^{3/2} V \varepsilon^{1/2} \, d\varepsilon$$

derive the quantum virial expansion

$$\frac{p}{kT} = \mp \frac{1}{\Lambda^3} \sum_{j=1}^{\infty} \frac{(\mp 1)^j \lambda^j}{j^{5/2}}$$

where $\Lambda = (h^2/2\pi m k T)^{1/2}$.

4–8. Show that the entropy of an ideal quantum gas can be written as

$$S = -k \sum_j [\bar{n}_j \ln \bar{n}_j \pm (1 \mp \bar{n}_j) \ln (1 \mp \bar{n}_j)]$$

where the upper (lower) sign denotes Fermi-Dirac (Bose-Einstein) statistics.

4–9. Show that $pV \geq \langle N \rangle kT$ for fermions, and $pV \leq \langle N \rangle kT$ for bosons.

4–10. Consider a system of independent, distinguishable particles, each of which has only two accessible states; a ground state of energy 0 and an excited state of energy ε. If the system is in equilibrium with a heat bath of temperature T, calculate A, E, S, and C_v. Sketch C_v versus T. Does the choice of the ground-state energy $= 0$ affect P, C_v, or S? How would your results change if ε_0 were added to both energy values?

4–11. Generalize Eq. (4–10) to the case of a mixture of several different species of noninteracting particles.

4–12. Consider a system of N distinguishable independent particles, each of which can be in the state $+\varepsilon_0$ or $-\varepsilon_0$. Let the number of particles with energy $\pm\varepsilon_0$ be $N_\pm$, so that the energy is

$$E = N_+ \varepsilon_0 - N_- \varepsilon_0 = 2N_+ \varepsilon_0 - N\varepsilon_0$$

Evaluate the partition function Q by summing $\exp(-E/kT)$ over levels and compare your result to $Q = q^N$. Do not forget the degeneracy of the levels, which in this case is the number of ways that N_+ particles out of N can be in the $+$ state. Calculate and plot the heat capacity C_V for this system.

4–13. The vibrational energy levels of a diatomic molecule can be approximated by a quantum mechanical harmonic oscillator. The fundamental vibrational frequency ν is $0(10^{13} \text{ sec}^{-1})$ for many diatomic molecules. Calculate the fraction of molecules in the first few vibrational levels in an ideal diatomic gas at 25°C. Derive a closed expression for the fraction of molecules in all excited states.

4–14. The rotational energy of diatomic molecules can be well approximated by a quantum mechanical rigid rotor. According to Eq. (1–32), the energy levels depend upon the moment of inertia, which for a diatomic molecule is $0(10^{-40} \text{ g-cm}^2)$. Calculate and plot the population of rotational levels of a diatomic ideal gas at 25°C. Do not forget to include the degeneracy $2J + 1$.

4–15. Show that $Q(N, V, T) = [q(V,T)]^N/N!$ implies that $q(V, T) = f(T)V$. Do this in both the canonical ensemble and grand canonical ensemble formalisms.

4–16. Show that the most probable distribution of $2N$ molecules of an ideal gas contained in two equal and connected volumes at the same temperature is N molecules in each volume.

4–17. In Fermi-Dirac statistics, the maximum occupancy of any state is 1, while in Bose-Einstein statistics, it is ∞. All particles appear to obey one of these two statistics. In 1940, however, Gentile* investigated the implications of an intermediate statistics, in which the maximum occupancy is m. Derive the distribution law for this case.

---

* G. Gentile, *Nuovo Cimento* **17**, p. 493, 1940.

4-18. Derive the equation

$$C_V = \frac{N}{kT^2} \left[\overline{\varepsilon^2} - \overline{\varepsilon}^2 \right]$$

for independent particles and show that fluctuations of molecular energies are not at all negligible.

4-19. Starting from Eqs. (4-13) and (4-14), show that the fluctuations in ε, the energy of a single particle, are not small, and in fact, are given by

$$\frac{\sigma_\varepsilon}{\overline{\varepsilon}} = N^{1/2} \frac{\sigma_E}{\overline{E}}$$

4-20. Consider a gas in equilibrium with the surface of a solid. Some of the molecules of the gas will be adsorbed onto the surface, and the number adsorbed will be a function of the pressure of the gas. A simple statistical mechanical model for this system is to picture the solid surface to be a two-dimensional lattice of M sites. Each of these sites can be either unoccupied, or occupied by at most one of the molecules of the gas. Let the partition function of an unoccupied site be 1 and that of an occupied site be $q(T)$. (We do not need to know $q(T)$ here.) Assuming that molecules adsorbed onto the lattice sites do not interact with each other, the partition function of N molecules adsorbed onto M sites is then

$$Q(N, M, T) = \frac{M!}{N!(M-N)!} [q(T)]^N$$

The binomial coefficient accounts for the number of ways of distributing the N molecules over the M sites. By using the fact the adsorbed molecules are in equilibrium with the gas phase molecules (considered to be an ideal gas), derive an expression for the fractional coverage, $\theta \equiv N/M$, as a function of the pressure of the gas. Such an expression, that is, $\theta(p)$, is called an adsorption isotherm, and this model gives the so-called Langmuir adsorption isotherm.

4-21. Consider a lattice of M equivalent noninteracting magnetic dipoles, μ (associated, say, with electron or nuclear spins). When placed in a magnetic field H, each dipole can orient itself either in the same direction, $\uparrow$, or opposed to, $\downarrow$, the field. The energy of a dipole is $-\mu H$ if oriented with the field, and $+\mu H$ if oriented against the field. Let N be the number of $\downarrow$ states and $M - N$ the number of $\uparrow$ states. For a given value of N, the total energy is

$$\mu H N - \mu H (M - N) = (2N - M)\mu H$$

The total magnetic moment I is

$$I = (M - 2\overline{N})\mu$$

where $\overline{N}$ is the average value of N for a given $M, H,$ and T. The work necessary to increase H by dH is $-IdH$. Find the specific heat C and the total magnetic moment for this system, and sketch both I versus $\mu H/kT$, that is, the total magnetization versus the applied field, and C/Nk versus $kT/\mu H$.

4-22. (a) Consider a system of M independent and distinguishable macromolecules on which any number from 0 to m small molecules may bind. Let $q(j)$ be the macromolecular partition function when j molecules are bound. If there are N small molecules (or ions) and M macromolecules (say proteins), then

$$Q(N, M, T) = \sum_{a}^{*} \frac{M! \, q(0)^{a_0} q(1)^{a_1} \cdots q(m)^{a_m}}{a_0! a_1! \cdots a_m!}$$

where the number of macromolecules having j bound molecules is a_j, and where the asterisk indicates the restrictions

$$\sum_{j=0}^{m} a_j = M \qquad \sum_{j=0}^{m} ja_j = N$$

Show that the grand partition function for this system can be written in the form

$$\Xi(M, T, \mu) = \xi(\mu, T)^M$$

where

$$\xi(\mu, T) = q(0) + q(1)\lambda + \cdots + q(m)\lambda^m$$

Interpret this result.

(b) Extend this result to the case in which the macromolecules are not distinguishable.

CHAPTER 5

IDEAL MONATOMIC GAS

In this chapter, we shall apply the general results of the preceding chapters to an ideal monatomic gas. By ideal, we mean a gas dilute enough that intermolecular interactions can be neglected. The results that we derive here will be applicable to real monatomic gases at pressures and temperatures for which the equation of state is well represented by $pV = NkT$, that is, pressures below 1 atmosphere and temperatures greater than room temperature.

low P, high T

We have shown in Section 4–1 that the number of available quantum states far exceeds the number of particles for an ideal gas. Thus we can write the partition function of the entire system in terms of the individual atomic partition functions:

$$Q(N, V, T) = \frac{[q(V, T)]^N}{N!} \tag{5-1}$$

A monatomic gas has translational, electronic, and nuclear degrees of freedom. The translational Hamiltonian is separable from the electronic and nuclear degrees of freedom, and the electronic and nuclear Hamiltonians are separable to a very good approximation. Thus we have

$$q(V, T) = q_{\text{trans}} q_{\text{elect}} q_{\text{nucl}} \tag{5-2}$$

We shall study each of these factors separately in the following sections of this chapter.

5–1 THE TRANSLATIONAL PARTITION FUNCTION

In this section we shall evaluate the translational partition function. The energy states are given by

$$\varepsilon_{n_x n_y n_z} = \frac{h^2}{8ma^2} (n_x{}^2 + n_y{}^2 + n_z{}^2) \qquad n_x, n_y, n_z = 1, 2, \ldots \tag{5-3}$$

We substitute this into q_{trans} to get

$$q_{trans} = \sum_{n_x, n_y, n_z = 1}^{\infty} e^{-\beta \varepsilon_{n_x n_y n_z}}$$

$$= \sum_{n_x = 1}^{\infty} \exp\left(-\frac{\beta h^2 n_x^2}{8ma^2}\right) \sum_{n_y = 1}^{\infty} \exp\left(-\frac{\beta h^2 n_y^2}{8ma^2}\right) \sum_{n_z = 1}^{\infty} \exp\left(-\frac{\beta h^2 n_z^2}{8ma^2}\right)$$

$$= \left(\sum_{n=1}^{\infty} \exp\left(-\frac{\beta h^2 n^2}{8ma^2}\right)\right)^3 \tag{5-4}$$

This summation cannot be evaluated in closed form, that is, it cannot be expressed in terms of any simple analytic function. This does not present any difficulty, however, for the following reason. The successive terms in these summations differ so little from each other that the terms vary essentially continuously, and so the summation can, for all practical purposes, be replaced by an integral. To prove this, we show that the argument of the exponential changes little in going from n_x to $n_x + 1$. This difference, Δ, is given by

$$\Delta = \frac{\beta h^2 (n_x + 1)^2}{8ma^2} - \frac{\beta h^2 n_x^2}{8ma^2} = \frac{\beta h^2 (2n_x + 1)}{8ma^2}$$

At room temperature, for $m = 10^{-22}$ g and $a = 10$ cm, this difference is

$$\Delta \approx (2n_x + 1) \times 10^{-20}$$

A typical value of n_x at room temperature is $O(10^{10})$ (see Problem 5–3), so Δ is indeed very small for all but very large values of n_x. A value of n_x for which Δ is as large as 10^{-5} would correspond to an energy of $(10^{10}kT)$, an extremely improbable energy. Thus we can replace the summation in Eq. (5–4) by an integration:

$$q_{trans}(V, T) = \left(\int_0^{\infty} e^{-\beta h^2 n^2 / 8ma^2} \, dn\right)^3 = \left(\frac{2\pi mkT}{h^2}\right)^{3/2} V \tag{5-5}$$

where we have written V for a^3.

It is instructive to evaluate q_{trans} in another way. Equation (5–4) is a sum over the states of the system. We could also write q_{trans} as a sum over levels. Recognizing that the levels are very densely distributed, we can write q_{trans} as an integral:

$$q_{trans} = \int_0^{\infty} \omega(\varepsilon) e^{-\beta \varepsilon} \, d\varepsilon \tag{5-6}$$

The function $\omega(\varepsilon)$ is the number of energy states between ε and $\varepsilon + d\varepsilon$, or, in other words, the effective degeneracy. Equation (5–6) is simply a continuous form of a sum over levels rather than a sum over states. We have already evaluated $\omega(\varepsilon)$ in Section 1–3. It is given by Eq. (1–35)

$$\omega(\varepsilon) \, d\varepsilon = \frac{\pi}{4} \left(\frac{8ma^2}{h^2}\right)^{3/2} \varepsilon^{1/2} \, d\varepsilon \tag{5-7}$$

If we substitute this into Eq. (5–6), we get

$$q_{trans} = \frac{\pi}{4} \left(\frac{8ma^2}{h^2}\right)^{3/2} \int_0^{\infty} \varepsilon^{1/2} e^{-\beta \varepsilon} \, d\varepsilon$$

$$= \left(\frac{2\pi mkT}{h^2}\right)^{3/2} V \tag{5-8}$$

Of course, we obtain the same result as Eq. (5–5).

The factor $(h^2/2\pi mkT)^{1/2}$ that occurs in the translational partition function has units of length and is usually denoted by Λ. In this notation, Eq. (5–5) or (5–8) read

$$q_{trans} = \frac{V}{\Lambda^3} \tag{5-9}$$

The quantity Λ can be given the following interpretation. The average translational or kinetic energy of an ideal gas molecule can be calculated immediately from Eq. (5–8) and Eq. (4–13), which in terms of q_{trans} is

$$\bar{\varepsilon}_{trans} = kT^2 \left(\frac{\partial \ln q_{trans}}{\partial T} \right)$$

We find that $\bar{\varepsilon}_{trans} = \frac{3}{2}kT$, and since $\varepsilon_{trans} = p^2/2m$, where p^2 is the momentum of the particle, we can say that the average momentum is essentially $(mkT)^{1/2}$. Thus Λ is essentially h/p, which is equal to the De Broglie wavelength of the particle. Consequently, Λ is called the thermal De Broglie wavelength. The condition for the applicability of classical or Boltzmann statistics is equivalent to the condition that $\Lambda^3/V \ll 1$, which physically says that the thermal De Broglie wavelength must be small compared to the dimensions of the container. Such a condition is similar to the condition that quantum effects decrease as the De Broglie wavelength becomes small (*cf.* Table 4–1).

5–2 THE ELECTRONIC AND NUCLEAR PARTITION FUNCTIONS

In this section we shall investigate the electronic and nuclear contributions to q.

It is more convenient to write the electronic partition function as a sum of levels rather than a sum over states. We have, then,

$$q_{elect} = \sum \omega_{ei} e^{-\beta \varepsilon_i} \tag{5-10}$$

where ω_{ei} is the degeneracy, and ε_i the energy of the ith electronic level. We first fix the arbitrary zero of energy such that $\varepsilon_1 = 0$, that is, we shall measure all of our electronic energies relative to the ground state. The electronic contribution to q can then be written as

$$q_{elect} = \omega_{e1} + \omega_{e2} e^{-\beta \Delta \varepsilon_{12}} + \cdots \tag{5-11}$$

where $\Delta \varepsilon_{1j}$ is the energy of the jth electronic level relative to the ground state. These $\Delta\varepsilon$'s are typically of the order of electron volts, and so $\beta\Delta\varepsilon$ is typically quite large at ordinary temperatures (see Problem 5–10). Therefore at ordinary temperatures, only the first term in the summation for q_{elect} is significantly different from zero. However, there are some cases, such as the halogen atoms, where the first excited state lies only a fraction of an electron volt above the ground state, so that several terms in q_{elect} are necessary. Even in these cases the sum converges extremely rapidly.

The electronic energies of atoms and ions are determined by atomic spectroscopy and are well tabulated. The standard reference is the tables of Moore* which list the energy levels and energies of many atoms and ions. Table 5–1 lists the first few levels for H, He, Li, O, and F. A look at this table will indicate that electronic states are labeled or characterized by a so-called term symbol, which is briefly explained in Section 5–4. (A knowledge of the meaning of atomic term symbols is not necessary for the calculation of q_{elect}, but they are explained in Section 5–4 for completeness.)

* See Table 5–1.

Table 5–1. **Atomic energy states**

atom	electron configuration	term symbol	degeneracy $g = 2J + 1$	energy (cm^{-1})	energy (eV)
H	$1s$	$^2S_{1/2}$	2	0	0
	$2p$	$^2P_{1/2}$	2	82258.907	10.20
	$2s$	$^2S_{1/2}$	2	82258.942	
	$2p$	$^2P_{3/2}$	4	82259.272	
He	$1s^2$	$^1S_0$	1	0	
	$1s2s$	$^3S_1$	3	159850.318	19.82
		$^1S_0$	1	166271.70	
Li	$1s^22s$	$^2S_{1/2}$	2	0	
	$1s^22p$	$^2P_{1/2}$	2	14903.66	1.85
		$^2P_{3/2}$	4	14904.00	
	$1s^23s$	$^2S_{1/2}$	2	27206.12	
O	$1s^22s^22p^4$	$^3P_2$	5	0	
		$^3P_1$	3	158.5	0.02
		$^3P_0$	1	226.5	0.03
		$^1D_2$	5	15867.7	1.97
		$^1S_0$	1	33792.4	4.19
F	$1s^22s^22p^5$	$^2P_{3/2}$	4	0	
		$^2P_{1/2}$	2	404.0	0.05
	$1s^22s^22p^43s$	$^4P_{5/2}$	6	102406.50	12.70
		$^4P_{3/2}$	4	102681.24	
		$^4P_{1/2}$	2	102841.20	
		$^2P_{3/2}$	4	104731.86	
		$^2P_{1/2}$	2	105057.10	

Source: C. E. Moore, "Atomic Energy States," *Natl. Bur. Standards, Circ.*, **1**, p. 467, 1949.

Some general observations about Table 5–1 are: All the rare gases have a ground state $^1S_0$ (called a singlet S) with the first excited state O(10 eV) higher; the alkali metals have a $^2S_{1/2}$ (called a doublet S) ground state with the next state O(1 eV) higher; the halogen atoms have a $^2P_{3/2}$ (called a doublet P) ground state with the next one, a $^2P_{1/2}$ (also a doublet P) only O(0.1 eV) higher. Thus at ordinary temperatures the electronic partition function of the rare gases is essentially unity and that of the alkali metals is 2, while those for halogen atoms consist of two terms.

Using the data in Table 5–1, we can now calculate the fraction of He atoms in the lowest triplet state, $^3S_1$. This fraction is given by

$$f_2 = \frac{\omega_{e2}\, e^{-\beta\Delta\varepsilon_{12}}}{\omega_{e1} + \omega_{e2}\, e^{-\beta\Delta\varepsilon_{12}} + \omega_{e3}\, e^{-\beta\Delta\varepsilon_{13}} + \cdots}$$

$$= \frac{3e^{-\beta\Delta\varepsilon_{12}}}{1 + 3e^{-\beta\Delta\varepsilon_{12}} + \omega_{e3}\, e^{-\beta\Delta\varepsilon_{13}} + \cdots} \tag{5–12}$$

At 300°K, $\beta\,\Delta\varepsilon_{12} = 770$, and so $f_2 \approx 10^{-334}$. Even at 3000°K, $f_2 \approx 10^{-33}$. This is typical of the rare gases. The energy separation must be less than a few hundred cm^{-1} or so before any population of that level is significant. Incidentally, it is useful to know that Boltzmann's constant in units of cm^{-1}/deg-molecule is 0.695 (almost ln 2), and 1 eV = 8065.73 cm^{-1}.

Table 5–2 gives the fraction of fluorine atoms in the first excited electronic state as a function of temperature. It can be seen that fluorine is a case where it is necessary to use two terms in q_{elec}.

We shall write the electronic partition function as

$$q_{elec}(T) \approx \omega_{e1} + \omega_{e2}\, e^{-\beta\varepsilon_{12}} \tag{5–13}$$

Table 5–2. **The fraction of fluorine atoms in the first excited electronic state as a function of temperature**

$T(°K)$	f_2
200	0.027
400	0.105
600	0.160
800	0.195
1000	0.219
1200	0.236
2000	0.272

but at temperatures at which the second term is not negligible with respect to the first term, we must check the possible contribution of higher terms as well. This will rarely be necessary, however.

We now consider the nuclear partition function. The nuclear partition function has a form similar to that of the electronic partition function. Nuclear energy levels are separated by millions of electron volts, however, which means that it requires temperatures of the order of $10^{10}°K$ to produce excited nuclei. At terrestrial temperatures then, we need consider only the first term, that is, the degeneracy of the ground nuclear state ω_{n1}. We take our zero of nuclear energy states to be the ground state. Note that we have taken the overall atomic ground state to be the atom in its ground translational, electronic, and nuclear states. The nuclear partition function, $q_n = \omega_{n1}$, then contributes only a multiplicative constant to Q, and hence affects only the entropy and free energies by a constant additive factor. Since the nuclear state is rarely altered in any chemical process, it does not contribute to thermodynamic changes, and so we shall usually not include it in q. We cannot do this in the case of the electronic contribution since there are many chemical processes in which the electronic states change.

This completes the partition function of monatomic ideal gases. In summary then, we have

$$Q = \frac{(q_{\text{trans}} q_{\text{elec}} q_{\text{nucl}})^N}{N!} \tag{5–14}$$

where

$$q_{\text{trans}} = \left(\frac{2\pi mkT}{h^2}\right)^{3/2} V = \frac{V}{\Lambda^3}$$

$$q_{\text{elect}} = \omega_{e1} + \omega_{e2} e^{-\beta \Delta\varepsilon_{12}} + \cdots$$

$$q_{\text{nucl}} = \omega_{n1} + \cdots \tag{5–15}$$

The nuclear partition function, although not always equal to unity, is usually omitted. We can now calculate thermodynamic properties of a monatomic ideal gas.

5–3 THERMODYNAMIC FUNCTIONS

The Helmholtz free energy is given by

$$A(N, V, T) = -kT \ln Q = -NkT \ln\left[\left(\frac{2\pi mkT}{h^2}\right)^{3/2} \frac{Ve}{N}\right]$$

$$- NkT \ln(\omega_{e1} + \omega_{e2} e^{-\beta\Delta\varepsilon_{12}}) \tag{5–16}$$

The argument of the first logarithm here is much larger than the argument of the second logarithm, and so the electronic contribution to A is quite small.

The thermodynamic energy is

$$E = kT^2 \left(\frac{\partial \ln Q}{\partial T}\right)_{N,V} = \tfrac{3}{2} NkT + \frac{N\omega_{e2}\,\Delta\varepsilon_{12}\,e^{-\beta\Delta\varepsilon_{12}}}{q_{elec}} + \cdots \qquad (5\text{--}17)$$

The contribution of the electronic degrees of freedom to the energy is small at ordinary temperatures (see Problem 5–12). Since we have neglected the contribution of the intermolecular potential to the total energy of the gas, the first term of Eq. (5–17) represents only kinetic energy. Furthermore, each atom has an average kinetic energy $3kT/2$, or $kT/2$ for each degree of translational freedom. We shall give an interesting interpretation of this result when we study classical statistical mechanics in Chapter 7. If we ignore the very small contribution from the electronic degrees of freedom, the molar heat capacity at constant volume is $3Nk/2$, a well-known experimental result for dilute gases.

The pressure is

$$p = kT \left(\frac{\partial \ln Q}{\partial V}\right)_{N,T} = \frac{NkT}{V} \qquad (5\text{--}18)$$

Note that Eq. (5–18) results because $q(V, T)$ is of the form $f(T)V$, and the only contribution to the pressure is from the translational energy of the atoms. This is what one expects intuitively, since the pressure is due to bombardment of the walls of the container by the atoms and molecules of the gas.

The entropy is given by

$$S = \tfrac{3}{2} Nk + Nk \ln\left[\left(\frac{2\pi m kT}{h^2}\right)^{3/2}\frac{Ve}{N}\right] + Nk \ln(\omega_{e1} + \omega_{e2} e^{-\beta\Delta\varepsilon_{12}})$$

$$+ \frac{Nk\omega_{e2}\,\beta\Delta\varepsilon_{12}\,e^{-\beta\Delta\varepsilon_{12}}}{q_{elec}} \qquad (5\text{--}19)$$

$$= Nk \ln\left[\left(\frac{2\pi m kT}{h^2}\right)^{3/2}\frac{Ve^{5/2}}{N}\right] + S_{elec} \qquad (5\text{--}20)$$

In Eq. (5–20), S_{elec} denotes the last two terms of Eq. (5–19). Equation (5–20) is called the Sackur-Tetrode equation. Table 5–3 compares the results of this equation with experimental values for several monatomic gases.

Table 5–3. **Comparison of experimental entropies at 1 atm and $T = 298°K$ to those calculated from the statistical thermodynamical equation for the entropy of an ideal monatomic gas***

	exp. (e.u.)	calc. (e.u.)
He	30.13	30.11
Ne	34.95	34.94
Ar	36.98	36.97
Kr	39.19	39.18
Xe	40.53	40.52
C	37.76	37.76
Na	36.72	36.70
Al	39.30	39.36
Ag	41.32	41.31
Hg	41.8	41.78

* The experimental values have been corrected for any nonideal gas behavior.

The chemical potential is

$$\mu(T, p) = -kT\left(\frac{\partial \ln Q}{\partial N}\right)_{V,T} = -kT \ln \frac{q}{N}$$

$$\boxed{Q = \frac{q^N}{N!}}$$ · approx. for many indistinguishable parts.

$$= -kT \ln\left[\left(\frac{2\pi mkT}{h^2}\right)^{3/2}\frac{V}{N}\right] - kT \ln q_e q_n$$

$$= -kT \ln\left[\left(\frac{2\pi mkT}{h^2}\right)^{3/2}\frac{kT}{p}\right] - kT \ln q_e q_n$$

$$= -kT \ln\left[\left(\frac{2\pi mkT}{h^2}\right)^{3/2}kT\right] - kT \ln q_e q_n + kT \ln p$$

$$= \mu_0(T) + kT \ln p \tag{5-21}$$

where the last line is the thermodynamic equation for $\mu(T, p)$ for an ideal gas [cf. Eq. (1–66)]. Thus statistical thermodynamics yields an expression for $\mu_0(T)$:

$$\mu_0(T) = -kT \ln\left[\left(\frac{2\pi mkT}{h^2}\right)^{3/2}kT\right] - kT \ln q_e q_n \tag{5-22}$$

The argument of the first logarithm here has units of pressure, but remember that there is a $kT \ln p$ term in Eq. (5–21) so that $\mu(T, p)$ itself has units of energy and is $O(kT)$.

5–4 A DIGRESSION ON ATOMIC TERM SYMBOLS

The electronic state of an atom is designated by a so-called atomic term symbol. Since one encounters atomic term symbols in the calculation of electronic partition functions, we discuss them in this section. (The quantum mechanical level of this section is above that of most of the book, and it is not necessary to read this section on first reading.)

In addition to the usual kinetic energy and electrostatic terms in the Hamiltonian of a many-electron atom, there are a number of magnetic and spin terms. The most important of these is the spin-orbit interaction term, which represents the interaction of the magnetic moment associated with the spin of the electron with the magnetic field generated by the electric current produced by its own orbital motion. There are other terms such as spin–spin and orbit–orbit interaction terms, but these are numerically much less important. The Hamiltonian can then be written as

$$H = -\frac{\hbar^2}{2m}\sum_j \nabla_j^2 - \sum_j \frac{Ze^2}{r_j} + \sum_{i<j}\frac{e^2}{r_{ij}} + \sum_j \xi(r_j)\mathbf{1}_j \cdot \mathbf{s}_j \tag{5-23}$$

where $\mathbf{1}_j$ and $\mathbf{s}_j$ are the individual electronic orbital and spin angular momenta, respectively, and $\xi(r_j)$ is a scalar function of r_j, whose form is not necessary here. We can abbreviate this equation by writing

$$H = H_0 + H_{ee} + H_{so} \tag{5-24}$$

where H_0 represents the first two terms (no interelectronic interactions), H_{ee} the third, and H_{so} the fourth.

For light atoms ($Z < 40$), H_{so} is small enough to be considered a small perturbation. If H_{so} is neglected altogether, it can be shown that the total orbital angular momentum **L** and the total spin angular momentum **S** are conserved (i.e., yield "good" quantum numbers, or commute with $H_0 + H_{ee}$). However, in this case, the *individual* orbital and spin angular momenta are not conserved; hence they are not useful concepts. The eigenvalues of the square of the total orbital angular momentum operator $\hat{L}^2$ and the square of the total spin angular momentum operator $\hat{S}^2$ are $L(L+1)\hbar^2$ and $S(S+1)\hbar^2$, respectively. One often interprets these eigenvalues by saying that the orbital angular momentum has the value L or that the total spin is S, but it should always be borne in mind that the orbital or spin angular momentum itself is not an eigenoperator in quantum mechanics.

The quantities $\mathbf{L}$ and $\mathbf{S}$ are the vector sums of the individual orbital $\mathbf{1}_j$ and spin $\mathbf{s}_j$, angular momenta. The possible ways of adding these $\mathbf{1}_j$ or $\mathbf{s}_j$ are governed by quantum mechanics with the result being that only certain values of the quantum numbers L and S are allowed. In the case of two electrons, L can take on only the values $1_1 + 1_2, 1_1 + 1_2 - 1, \dots |1_1 - 1_2|$, with a similar result for S. What we really mean by this, of course, is that the only allowed values of the eigenvalues of L^2 are $(1_1 + 1_2)(1_1 + 1_2 + 1)\hbar^2$, $(1_1 + 1_2 - 1)(1_1 + 1_2)\hbar^2$, $\dots (|1_1 - 1_2|)(|1_1 - 1_2| + 1)\hbar^2$. The addition of electronic angular momenta to obtain $\mathbf{L}$ for cases involving more than two electrons can be accomplished using a scheme which is a straightforward but rather tedious electron-by-electron extension of the above two electron systems. Actually, rather specialized and advanced techniques have been developed to handle this problem, but we need not be concerned with them here.

The electronic energy states are designated by a term symbol, part of which is given by ^{2S+1}L. Terms with $L = 0, 1, 2, \dots$ are denoted by $S, P, D, \dots$.

When the spin–orbit term is taken into account, $\mathbf{L}$ and $\mathbf{S}$ are no longer conserved (that is, do not commute with the total H), and only the total angular momentum, $\mathbf{J} = \mathbf{L} + \mathbf{S}$, is conserved. The eigenvalues of $\hat{J}^2 = (\hat{L} + \hat{S})^2$ are $J(J+1)\hbar^2$, with a degeneracy $2J + 1$, corresponding to the $2J + 1$ eigenvalues of $\hat{J}_z$, namely, $J\hbar, (J-1)\hbar, \dots, -J\hbar$. Just as in the addition of $\mathbf{1}_1$ and $\mathbf{1}_2$ above, the allowed values of J are $L + S, L + S - 1, \dots, |L - S|$. The spin–orbit term causes each of these values of J to have a slightly different energy, and so the value of J is included in the term symbol as a subscript to give $^{2S+1}L_J$ as a characterization of the electronic state of an atom.

Table 5–1 lists the first few electronic states for some of the first row atoms.

This light atom approximate coupling scheme, in which L and S are almost good quantum numbers (that is, not good quantum numbers because of the small spin–orbit perturbation term) and in which the total angular momentum $\mathbf{J}$ is found by adding $\mathbf{L}$ and $\mathbf{S}$, is called Russell-Saunders or L–S coupling. As the atomic number of the atom becomes larger, the spin–orbit term becomes larger than the interelectronic repulsion term, and H_{ee} can be considered to be a small perturbation on the others. In this case L and S are no longer useful, and the individual total angular momenta, $\mathbf{j}_i = \mathbf{1}_i + \mathbf{s}_i$, become the approximately conserved quantities. One then couples the j's to get the total angular momentum. This scheme is called j–j coupling and is applicable to heavier atoms. In spite of the deterioration of L–S coupling as Z increases, it is still approximately useful, and so the electronic states of even heavy atoms are designated by term symbols of the form $^{2S+1}L_J$.

We are ready to discuss an ideal gas of diatomic molecules. In addition to having translational and electronic degrees of freedom, diatomic molecules have rotational and vibrational degrees of freedom as well. It should be apparent at this point that the additional input into our statistical thermodynamical equations will be the rotation-vibration energy levels.

We could also leave the study of gases for a while and apply our general results to other systems such as solids and liquids. This would involve no more effort than continuing with ideal gases and perhaps would be a change of pace. We could go directly to Chapter 11, for example, but we shall finish gases before treating other systems.

ADDITIONAL READING

General

ANDREWS, F. C. 1963. *Equilibrium statistical mechanics.* New York: Wiley. Chapter 18.
HILL, T. L. 1960. *Statistical thermodynamics.* Reading, Mass.: Addison-Wesley. Chapter 4.
KESTIN, J., and DORFMAN, J. R. 1971. *A course in statistical thermodynamics.* New York: Academic. Sections 6–7 through 6–8.
KITTEL, C. 1969. *Thermal physics.* New York: Wiley. Chapters 11 and 12.
KNUTH, E. 1966. *Statistical thermodynamics.* New York: McGraw-Hill. Chapter 8.
MÜNSTER, A. 1969. *Statistical thermodynamics*, Vol. I. Berlin: Springer-Verlag. Sections 6–1 and 6–2.

Term symbols

STRAUSS, H. L. 1968. *Quantum mechanics.* Englewood Cliffs, N.J.: Prentice-Hall. Chapter 8.

PROBLEMS

5–1. Convert Boltzmann's constant $k = 1.3806 \times 10^{-16}$ ergs-molecule^{-1} deg^{-1} to cm^{-1}-molecule^{-1}-deg^{-1} and to eV-molecule^{-1}-deg^{-1}.

5–2. By considering the special case of an ideal gas, determine the order of magnitude of E, A, G, S, C_V, and μ. Express your answers in terms of N, kT, or Nk, whichever is appropriate.

5–3. Calculate the value of n_x, n_y, and n_z for the case $n_x = n_y = n_z$ for a hydrogen atom (atomic weight 1.00) in a box of dimensions 1 cc if the particle has a kinetic energy $3kT/2$, for $T = 27°C$. What significant fact does this calculation illustrate?

5–4. Calculate the entropy of Ne at 300°K and 1 atmosphere. Use the entropy, in turn, to estimate the "translational" degeneracy of the gas.

5–5. Calculate the entropy of 1 mole of argon at 298°K and 1 atm and compare this to Table 5–3.

5–6. The quantum mechanical energy of a particle confined to a rectangular parallelepiped of lengths a, b, and c is

$$\varepsilon_{n_x n_y n_z} = \frac{h^2}{8m} \left(\frac{n_x^2}{a^2} + \frac{n_y^2}{b^2} + \frac{n_z^2}{c^2} \right)$$

Show that the translational partition function for this geometry is the same as that of a cube of the same volume.

5–7. Given that the quantum mechanical energy levels of a particle in a two-dimensional box are

$$\varepsilon = \frac{h^2}{8ma^2} (s_x^2 + s_y^2) \qquad s_x, s_y = 1, 2, \ldots$$

First calculate the density of states $\omega(\varepsilon)\, d\varepsilon$, that is, the number of states between ε and $\varepsilon + d\varepsilon$, and use this to find the translational partition function of a two-dimensional ideal gas. Then find the partition function by another method. And finally find the equation of state, the thermodynamic energy E, the heat capacity C_A, and the entropy. This is a model for a gas adsorbed onto a surface or for long-chain fatty acids on the surface of water, say, as long as the number of molecules per unit area is small enough.

5–8. Calculate the entropy of a mixture of 50 percent neon and 50 percent argon at 500°K and 10 atm, assuming ideal behavior.

5–9. Calculate the De Broglie wavelength of an argon atom at 25°C and compare this with the average interatomic spacing at 1 atm.

5–10. Evaluate $\beta\Delta\varepsilon_{elec}$ at room temperature, given that electronic energy levels are usually separated by energies of the order of electron volts.

5–11. Using the data in Table 5–1, calculate the population of the first few electronic energy levels of an oxygen atom at room temperature.

5–12. Show that the contribution of the electronic degrees of freedom to the total energy is small at ordinary temperatures [*cf.* Eq. (5–17)].

5–13. Generalize the results of this chapter to an ideal binary mixture. In particular, show that

$$Q = \frac{q_1^{N_1} q_2^{N_2}}{N_1! N_2!}$$

$$E = \tfrac{3}{2}(N_1 + N_2)kT$$

and

$$S = N_1 k \ln\left(\frac{Ve^{5/2}}{\Lambda_1^3 N_1}\right) + N_2 k \ln\left(\frac{Ve^{5/2}}{\Lambda_2^3 N_2}\right)$$

if we ignore q_{elec} and q_{nucl}.

5–14. Derive the standard thermodynamic formula for the entropy of mixing by starting with Eq. (5–20).

5–15. Calculate A, E, μ, C_V, and S for 1 mole of Kr at 25°C and 1 atm (assuming ideal behavior).

5–16. Show that the most probable distribution of $2N$ molecules of an ideal gas contained in two equal and connected volumes at the same temperature is N molecules in each volume.

5–17. Evaluate the isothermal-isobaric partition function of a monatomic ideal gas by converting the summation over V in Eq. (3–17) to an integral. The result is

$$\Delta(N, p, T) = \left(\frac{kT}{p\Lambda^3}\right)^N$$

Using the fact that $G = -kT \ln \Delta$, derive expressions for S and V.

5–18. Consider a monatomic ideal gas of N particles in a volume V. Show that the number n of particles in some small subvolume v is given by the Poisson distribution

$$P_n = (\lambda q)^n \frac{e^{-\lambda q}}{n!}$$

$$= (\bar{n})^n \frac{e^{-\bar{n}}}{n!}$$

Hint: Use the grand canonical ensemble and particularly the result that $\Xi = \exp(\lambda q)$.

5–19. Calculate q_{elec} for a hydrogen atom. The energy levels are given by

$$E_n = -\frac{2\pi^2 me^4}{n^2 h^2} \qquad n = 1, 2, \ldots$$

and the degeneracy is $2n^2$. How is this seemingly paradoxical result explained? (See S. J. Strickler, *J. Chem. Educ.*, **43**, p. 364, 1966.)

CHAPTER 6

IDEAL DIATOMIC GAS

In this chapter we shall treat an ideal gas composed of diatomic molecules. In addition to translational and electronic degrees of freedom, diatomic molecules possess vibrational and rotational degrees of freedom as well. The general procedure would be to set up the Schrödinger equation for two nuclei and n electrons and to solve this equation for the set of eigenvalues of the diatomic molecule. Such a general exact approach is very difficult and has been done only for H_2. Fortunately, a series of very good approximations can be used to reduce this complicated two-nuclei, n-electron problem to a set of simpler problems. The simplest of these approximations is the rigid rotor-harmonic oscillator approximation. In Section 6–1 we shall discuss this approximation, and then in Sections 6–2 and 6–3 we discuss the vibrational and rotational partition functions within this approximation. Section 6–4 is a discussion of the symmetry of the wave functions of homonuclear diatomic molecules under the interchange of the two nuclei, and Section 6–5 is an application of these results to the rotational partition function of homonuclear diatomic molecules. This section contains a detailed discussion of ortho- and para-hydrogen. Section 6–6 summarizes the thermodynamic functions under the rigid rotor–harmonic oscillator approximation.

6–1 THE RIGID ROTOR–HARMONIC OSCILLATOR APPROXIMATION

We first make the Born-Oppenheimer approximation. The physical basis of the Born-Oppenheimer approximation is that the nuclei are much more massive than the electrons, and thus move slowly relative to the electrons. Therefore the electrons can be considered to move in a field produced by the nuclei fixed at some internuclear separation. Mathematically, the Schrödinger equation approximately separates into two simpler equations. One equation describes the motion of the electrons in the field of the fixed nuclei. We denote the eigenvalues of this equation by $u_j(r)$, where r is the internuclear separation. The other equation describes the motion of the nuclei in the electronic potential $u_j(r)$, that is, the potential set up by the electrons in the electronic state j. Each electronic state of the molecule creates its own characteristic internuclear potential. As in the atomic case, the first excited electronic state usually lies several electron volts above the ground state, and so only the ground electronic state potential is necessary. The calculation of $u_j(r)$ for even the ground state is a difficult n-electron calculation, and so semiempirical approximations such as the Morse potential are often used. Figure 6–1 illustrates a typical internuclear potential. Given $u_0(r)$, we treat the motion of the two nuclei in this potential.

Problem 1–8 shows that the motion of two masses in a spherically symmetric potential can be rigorously separated into two separate problems by the introduction of center of mass and relative coordinates. The center-of-mass motion is that of a freely

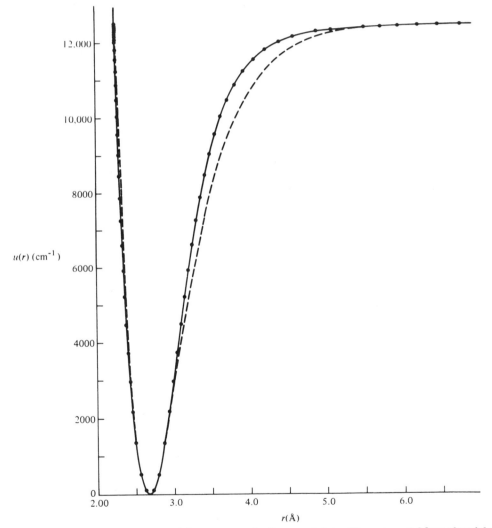

Figure 6–1. The internuclear potential energy curve for the ground state of I_2 as computed from ultraviolet spectroscopy. The dashed curve is the Morse curve. (From R. D. Verma, *J. Chem. Phys.*, **32**, 738, 1960.)

translating point of mass $m_1 + m_2$ situated at the center of mass. The other problem is that of the *relative* motion of the two bodies, which can be interpreted as one body of reduced mass $\mu = m_1 m_2/(m_1 + m_2)$ moving about the other one fixed at the origin.

The Hamiltonian can be written as

$$\mathcal{H} = \mathcal{H}_{\text{trans}} + \mathcal{H}_{\text{int}}$$

with eigenvalues

$$\varepsilon = \varepsilon_{\text{trans}} + \varepsilon_{\text{int}}$$

The partition function of the diatomic molecule, therefore, is

$$q = q_{\text{trans}} q_{\text{int}}$$

where

$$q_{\text{trans}} = \left[\frac{2\pi(m_1 + m_2)kT}{h^2} \right]^{3/2} V \tag{6–1}$$

The density of translational states alone is so great that we can write

$$Q(N, V, T) = \frac{q_{\text{trans}}^N q_{\text{int}}^N}{N!} \tag{6–2}$$

Thus we must investigate q_{int} to complete our treatment of diatomic molecules.

The relative motion of the two nuclei in the potential $u(r)$ consists of rotary motion about the center of mass and relative vibratory motion of the two nuclei. It turns out that the amplitude of the vibratory motion is very small, and so it is a good approximation to consider the angular motion to be that of a rigid dumbbell of fixed internuclear distance r_e. In addition, the internuclear potential $u(r)$ can be expanded about r_e:

$$u(r) = u(r_e) + (r - r_e)\left(\frac{du}{dr}\right)_{r=r_e} + \tfrac{1}{2}(r - r_e)^2\left(\frac{d^2 u}{dr^2}\right)_{r=r_e} + \cdots$$
$$= u(r_e) + \tfrac{1}{2}k(r - r_e)^2 + \cdots \tag{6–3}$$

The linear term vanishes because (du/dr) is zero at the minimum of $u(r)$. The parameter k is a measure of the curvature of the potential at the minimum and is called the force constant. A large value of k implies a stiff bond; a small value implies a loose bond.

The approximation introduced in the previous paragraph is called the rigid rotor–harmonic oscillator approximation. It allows the Hamiltonian of the relative motion of the nuclei to be written as

$$\mathcal{H}_{\text{rot, vib}} = \mathcal{H}_{\text{rot}} + \mathcal{H}_{\text{vib}} \qquad \text{(rigid rotor–harmonic oscillator} \tag{6–4}$$
$$\text{approximation)}$$

and

$$\varepsilon_{\text{rot, vib}} = \varepsilon_{\text{rot}} + \varepsilon_{\text{vib}} \tag{6–5}$$

The partition function $q_{\text{rot, vib}}$, then, is

$$q_{\text{rot, vib}} = q_{\text{rot}} q_{\text{vib}} \qquad \text{(rigid rotor-harmonic oscillator approximation)} \tag{6–6}$$

The energy eigenvalues and the degeneracy of a rigid rotor are given in Eq. (1–32)

$$\varepsilon_J = \frac{\hbar^2 J(J+1)}{2I} \qquad J = 0, 1, 2, \ldots$$

$$\omega_J = 2J + 1 \tag{6–7}$$

where I is the moment of inertia, μr_e^2, of the molecule. The energy and degeneracy of an harmonic oscillator are [cf. Eq. (1–31)]

$$\varepsilon_{\text{vib}} = h\nu(n + \tfrac{1}{2}) \qquad n = 0, 1, 2, \ldots$$

$$\omega_n = 1 \quad \text{for all } n \tag{6–8}$$

where

$$\nu = \frac{1}{2\pi} \left(\frac{k}{\mu}\right)^{1/2} \tag{6–9}$$

Transitions from one rotational level to another can be induced by electromagnetic radiation. The selection rules for this are: (1) The molecule must have a permanent dipole moment, and (2) $\Delta J = \pm 1$. The frequency of radiation absorbed in the process of going from a level J to $J + 1$ is given by

$$\nu = \frac{\varepsilon_{J+1} - \varepsilon_J}{h} = \frac{h}{4\pi^2 I}(J + 1) \qquad J = 0, 1, 2, \ldots \tag{6–10}$$

We thus expect absorption of radiation at frequencies given by multiples of $h/4\pi^2 I$ and should observe a set of equally spaced spectral lines, which for typical molecular values of μ and r_e^2 will be found in the microwave region. Experimentally one does see a series of almost equally spaced lines in the microwave spectra of linear molecules. The usual units of frequency in this region are wave numbers, or reciprocal wavelengths.

$$\bar{\omega}(\text{cm}^{-1}) = \frac{1}{\lambda} = \frac{\nu}{c} \tag{6–11}$$

Microwave spectroscopists define the rotational constant $\bar{B}$ by $h/8\pi^2 Ic$ (units of cm^{-1}), so that the energy of rigid rotor (in cm^{-1}) becomes

$$\bar{\varepsilon}_J = \bar{B}J(J + 1) \tag{6–12}$$

Table 6–1 lists the values of $\bar{B}$ for several diatomic molecules.

For a molecule to change its vibrational state by absorbing radiation it must (1) change its dipole moment when vibrating and (2) obey the selection rule $\Delta n = \pm 1$. The frequency of absorption is, then, seen to be

$$\nu = \frac{\varepsilon_{n+1} - \varepsilon_n}{h} = \frac{1}{2\pi}\left(\frac{k}{\mu}\right)^{1/2} \tag{6–13}$$

Equation (6–13) predicts that the vibrational spectrum of a diatomic molecule will consist of just one line. This line occurs in the infrared, typically around 1000 cm^{-1}, giving force constants k of the order of 10^5 or 10^6 dynes/cm. (See Problem 6–5.) Table 6–1 gives the force constants of a number of diatomic molecules.

Table 6-1. **Molecular constants for several diatomic molecules***

molecule	electronic state	$\bar{\omega}$ (cm^{-1})	Θ_v (°K)	$\bar{B}$ (cm^{-1})	Θ_r (°K)	$k \times 10^{-5}$ (dynes/cm)	D_0 (kcal/mole)
H$_2$	$^1\Sigma_g^+$	4320	6215	59.3	85.3	5.5	103.2
D$_2$	$^1\Sigma_g^+$	3054	4394	29.9	42.7	5.5	104.6
Cl$_2$	$^1\Sigma_g^+$	561	808	0.244	0.351	3.2	57.1
Br$_2$	$^1\Sigma_g^+$	322	463	0.0809	0.116	2.4	45.4
I$_2$	$^1\Sigma_g^+$	214	308	0.0373	0.0537	1.7	35.6
O$_2$	$^3\Sigma_g^-$	1568	2256	1.437	2.07	11.6	118.0
N$_2$	$^1\Sigma_g^+$	2345	3374	2.001	2.88	22.6	225.1
CO	$^1\Sigma^+$	2157	3103	1.925	2.77	18.7	255.8
NO	$^2\Pi_{1/2}$	1890	2719	1.695	2.45	15.7	150.0
HCl	$^1\Sigma^+$	2938	4227	10.44	15.02	4.9	102.2
HBr	$^1\Sigma^+$	2640	3787	8.36	12.02	3.9	82.4
HI	$^1\Sigma^+$	2270	3266	6.46	9.06	3.0	70.5
Na$_2$	$^1\Sigma_g^+$	159	229	0.154	0.221	0.17	17.3
K$_2$	$^1\Sigma_g^+$	92.3	133	0.0561	0.081	0.10	11.8

* These parameters were obtained from a variety of sources and do not necessarily represent the most accurate values since they are obtained under the rigid rotor–harmonic oscillator approximation.

We furthermore assume that the electronic and nuclear degrees of freedom can be written separately, and thus we have

$$\mathcal{H} = \mathcal{H}_{trans} + \mathcal{H}_{rot} + \mathcal{H}_{vib} + \mathcal{H}_{elec} + \mathcal{H}_{nucl} \tag{6-14}$$

which implies that

$$\varepsilon = \varepsilon_{trans} + \varepsilon_{rot} + \varepsilon_{vib} + \varepsilon_{elec} + \varepsilon_{nucl} \tag{6-15}$$

and

$$q = q_{trans} q_{rot} q_{vib} q_{elec} q_{nucl} \tag{6-16}$$

The translational partition function is given by Eq. (6–1); the electronic partition function will be similar to Eq. (5–11) for a monatomic gas; and we shall usually adopt the convention that $q_{nucl} = 1$. Although Eq. (6–14) is not exact, it serves as a useful approximation. Within this approximation, the partition function of the gas itself is given by

$$Q(N, V, T) = \frac{(q_{trans} q_{rot} q_{vib} q_{elec} q_{nucl})^N}{N!} \tag{6-17}$$

We shall introduce several corrections to Eqs. (6–14) through (6–17) in Problems 6–23 and 6–24.

Only the vibrational and rotational contributions to the partition function are not known in Eq. (6–17), and we shall discuss these contributions in the next few sections. Before discussing these, however, we must choose a zero of energy for the rotational and vibrational states. The zero of rotational energy will usually be taken to be the $J = 0$ state. In the vibrational case we have two choices. One is to take the zero of vibrational energy to be that of the ground state, and the other is to take the zero to be the bottom of the internuclear potential well. In the first case, the energy of the ground vibrational state is zero, and in the second case it is $h\nu/2$. We shall choose the zero of

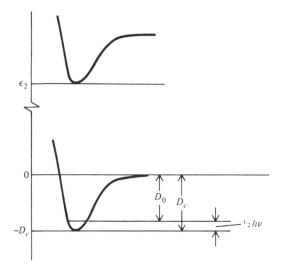

Figure 6–2. **The ground and first excited electronic states as a function of the internuclear separation** *r*, **illustrating the quantities** D_0, D_e, **and** ε_2.

vibrational energy to be the bottom of the internuclear potential well of the lowest electronic state. Lastly, we take the zero of the electronic energy to be the separated, electronically unexcited atoms at rest. If we denote the depth of the ground electronic state potential well by D_e, the energy of the ground electronic state is $-D_e$, and the electronic partition function is

$$q_{elec} = \omega_{e1} e^{D_e/kT} + \omega_{e2} e^{-\varepsilon_2/kT} + \cdots \qquad (6\text{–}18)$$

where D_e and ε_2 are shown in Fig. 6–2. We also define a quantity D_0 by $D_e - \frac{1}{2}h\nu$. As Fig. 6–2 shows, D_0 is the energy difference between the lowest vibrational state and the dissociated molecule. The quantity D_0 can be measured spectroscopically (by predissociation spectra, for example) or calorimetrically from the heat of reaction at any one temperature and the heat capacities from $0°K$ to that temperature. The values of D_0 for a number of diatomic molecules are given in Table 6–1.

6–2 THE VIBRATIONAL PARTITION FUNCTION

Since we are measuring the vibrational energy levels relative to the bottom of the internuclear potential well, we have

$$\varepsilon_n = (n + \tfrac{1}{2})h\nu \qquad n = 0, 1, 2, \ldots \qquad (6\text{–}19)$$

with $\nu = (k/\mu)^{1/2}/2\pi$, where k is the force constant of the molecule, and μ is its reduced mass [*cf*. Eq. (6–9)]. The vibrational partition function q_{vib}, then, becomes

$$\begin{aligned}
q_{vib}(T) &= \sum_n e^{-\beta\varepsilon_n} \\
&= e^{-\beta h\nu/2} \sum_{n=0}^{\infty} e^{-\beta h\nu n} \\
&= \frac{e^{-\beta h\nu/2}}{1 - e^{-\beta h\nu}}
\end{aligned} \qquad (6\text{–}20)$$

where we have recognized the summation above as a geometric series. This is one of the rare cases in which q can be summed directly without having to approximate it by an integral, as we did in the translational case in Chapter 5 and shall do shortly in the rotational case. The quantity $\beta h\nu$ is ordinarily larger than 1, but if the temperature is high enough, $\beta h\nu \ll 1$, and we can replace the sum in Eq. (6–20) by an integral to get

$$q_{\text{vib}}(T) = e^{-\beta h\nu/2} \int_0^\infty e^{-\beta h\nu n}\, dn = \frac{kT}{h\nu} \qquad (kT \gg h\nu) \tag{6–21}$$

which we see is what results from Eq. (6–20) if $\beta h\nu \ll 1$. Although we shall rarely use this approximation since we have $q_{\text{vib}}(T)$ exactly, it will be interesting to compare this limit to some others which we shall derive later on. From $q_{\text{vib}}(T)$ we can calculate the vibrational contribution to the thermodynamic energy

$$E_v = NkT^2 \frac{d\ln q_v}{dT} = Nk\left(\frac{\Theta_v}{2} + \frac{\Theta_v}{e^{\Theta_v/T} - 1}\right) \tag{6–22}$$

where $\Theta_v \equiv h\nu/k$ and is called the vibrational temperature. Table 6–1 gives Θ_v for a number of diatomic molecules. The vibrational contribution to the heat capacity is

$$\left(\frac{\partial E_v}{\partial T}\right)_N = Nk\left(\frac{\Theta_v}{T}\right)^2 \frac{e^{\Theta_v/T}}{(e^{\Theta_v/T} - 1)^2} \tag{6–23}$$

Notice that as $T \to \infty$, $E_v \to NkT$ and $C_v \to Nk$, a result given in many physical chemistry courses and one whose significance we shall understand more fully when we discuss equipartition of energy.

Figure 6–3 shows the vibrational contribution of an ideal diatomic gas to the molar heat capacity as a function of temperature.

An interesting quantity to calculate is the fraction of molecules in excited vibrational states. The fraction of molecules in the vibrational state designated by n is

$$f_n = \frac{e^{-\beta h\nu(n + 1/2)}}{q_{\text{vib}}} \tag{6–24}$$

This equation is shown in Fig. 6–4 for Br_2 at $300°K$. Notice that most molecules are in the ground vibrational state and that the population of the higher vibrational states

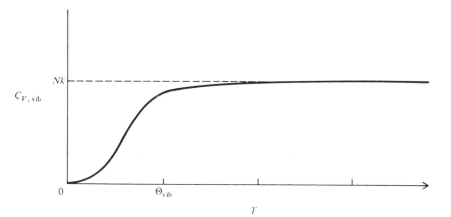

Figure 6–3. **The vibrational contribution of an ideal diatomic gas to the molar heat capacity as a function of temperature. Room temperature is typically $O(0.1\Theta_v)(Cf.$ Table 6–1).**

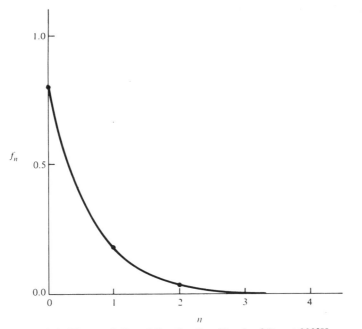

Figure 6–4. **The population of the vibrational levels of Br_2 at 300°K.**

decreases exponentially. Bromine has a force constant smaller than most molecules, however (*cf.* Table 6–1), and so the population of excited vibrational levels of Br_2 is greater than most other molecules. Table 6–2 gives the fraction of molecules in all excited states for a number of molecules. This fraction is given by

$$f_{n>0} = \sum_{n=1}^{\infty} \frac{e^{-\beta h\nu(n+1/2)}}{q_{\text{vib}}} = 1 - f_0 = e^{-\beta h\nu} = e^{-\Theta_v/T} \tag{6-25}$$

Table 6–2. **The fraction of molecules in excited vibrational states at 300°K and 1000°K**

gas	Θ_v, °K	$e^{-\Theta_v/T}$ 300°K	1000°K
H_2	6215	1.04×10^{-9}	2.03×10^{-3}
HCl	4227	1.02×10^{-6}	1.59×10^{-2}
N_2	3374	1.51×10^{-5}	3.55×10^{-2}
CO	3100	3.71×10^{-5}	4.65×10^{-2}
Cl_2	810	6.72×10^{-2}	4.45×10^{-1}
I_2	310	3.56×10^{-1}	7.33×10^{-1}

6–3 THE ROTATIONAL PARTITION FUNCTION OF A HETERONUCLEAR DIATOMIC MOLECULE

For heteronuclear diatomic molecules, the calculation of the rotational partition function is straightforward. The rotational partition function is given by

$$q_{\text{rot}}(T) = \sum_{J=0}^{\infty} (2J + 1)e^{-\beta BJ(J+1)} \tag{6-26}$$

In the nomenclature of Chapter 2, Eq. (6–26) is a summation over levels rather than over states.

The ratio $\bar{B}/k$ is denoted by Θ_r and is called the characteristic temperature of rotation. This is given in Table 6–1 for a number of molecules. Unlike the vibrational case, this sum cannot be written in closed form. However, because Θ_r/T is quite small at ordinary temperatures for most molecules, we can approximate this sum by an integral. (It is really $\Delta\varepsilon/kT = 2\Theta_r(J + 1)/T$ that must be small compared to one, and this of course cannot be true as J increases. However, by the time J is large enough to contradict this, the terms are so small that it makes no difference.)

At high enough temperatures, then,

$$q_{\text{rot}}(T) = \int_0^\infty (2J + 1)e^{-\Theta_r J(J+1)/T}\,dJ \tag{6–27}$$

$$= \int_0^\infty e^{-\Theta_r J(J+1)/T}\,d\{J(J + 1)\} = \frac{T}{\Theta_r} \tag{6–28}$$

$$= \frac{8\pi^2 I k T}{h^2} \qquad \Theta_r \ll T \tag{6–29}$$

This result improves as the temperature increases and is called the high-temperature limit. For low temperatures or for molecules with large values of Θ_r, say HD with $\Theta_r = 42.7°\text{K}$, one can use the sum directly. For example,

$$q_{\text{rot}}(T) = 1 + 3e^{-2\Theta_r/T} + 5e^{-6\Theta_r/T} + 7e^{-12\Theta_r/T} \tag{6–30}$$

is sufficient to give the sum to within 0.1 percent for $\Theta_r > 0.7T$. For Θ_r less than $0.7T$ but not small enough for the integral to give a good approximation, we need some intermediate approximation.

The replacement of a sum by an integral can be viewed as the first of a sequence of approximations. The full scheme is a standard result of the field of the calculus of finite differences and is called the Euler-MacLaurin summation formula. It states that if $f(n)$ is a function defined on the integers and continuous in between, then

$$\sum_{n=a}^b f(n) = \int_a^b f(n)\,dn + \tfrac{1}{2}\{f(b) + f(a)\}$$

$$+ \sum_{j=1}^\infty (-)^j \frac{B_j}{(2j)!}\{f^{(2j-1)}(a) - f^{(2j-1)}(b)\} \tag{6–31}$$

where $f^{(k)}(a)$ is the kth derivative of f evaluated at a. The B_j's are the Bernoulli numbers, $B_1 = \tfrac{1}{6}$, $B_2 = \tfrac{1}{30}$, $B_3 = \tfrac{1}{42}$, …. Before applying this to $q_{\text{rot}}(T)$, let us apply it first to a case we can do exactly. Consider the sum [cf. Eq. (6–20)]

$$\sum_{j=0}^\infty e^{-\alpha j} = \frac{1}{1 - e^{-\alpha}} \tag{6–32}$$

Applying the Euler-MacLaurin summation formula, we get

$$\sum_{j=0}^\infty e^{-\alpha j} = \frac{1}{\alpha} + \frac{1}{2} + \frac{\alpha}{12} - \frac{\alpha^3}{720} + \cdots \tag{6–33}$$

The expansion of $(1 - e^{-\alpha})^{-1}$ is

$$\frac{1}{1 - e^{-\alpha}} = \frac{1}{\alpha - \dfrac{\alpha^2}{2} + \dfrac{\alpha^3}{6} - \cdots} = \frac{1}{\alpha} + \frac{1}{2} + \frac{\alpha}{12} - \frac{\alpha^3}{720} + \cdots \tag{6–34}$$

We see that these two expansions are the same. If α is large, we can use the first few terms of Eq. (6–33); otherwise, we use the Euler-MacLaurin expansion in α.

Applying this formula to $q_{rot}(T)$ gives (see Problem 6–9):

$$q_{rot}(T) = \frac{T}{\Theta_r}\left\{1 + \frac{1}{3}\left(\frac{\Theta_r}{T}\right) + \frac{1}{15}\left(\frac{\Theta_r}{T}\right)^2 + \frac{4}{315}\left(\frac{\Theta_r}{T}\right)^3 + \cdots\right\} \qquad (6\text{–}35)$$

which is good to within one percent for $\Theta_r < T$. For simplicity we shall use only the high-temperature limit in what we do here since Θ_r is $\ll T$ for most molecules at room temperature (*cf.* Table 6–1).

The rotational contribution to the thermodynamic energy is

$$E_{rot} = NkT^2\left(\frac{\partial \ln q_{rot}}{\partial T}\right) = NkT + \cdots \qquad (6\text{–}36)$$

and the contribution to the heat capacity is

$$C_{V,rot} = Nk + \cdots \qquad (6\text{–}37)$$

The fraction of molecules in the Jth rotational state is

$$\frac{N_J}{N} = \frac{(2J + 1)e^{-\Theta_r J(J+1)/T}}{q_{rot}(T)} \qquad (6\text{–}38)$$

Figure 6–5 shows this fraction for HCl at 300°K. Contrary to the vibrational case, most molecules are in excited rotational levels at ordinary temperatures. We can find the maximum of this curve by differentiating Eq. (6–38) with respect to J to get

$$J_{max} = \left(\frac{kT}{2\bar{B}}\right)^{1/2} - \frac{1}{2} \approx \left(\frac{T}{2\Theta_r}\right)^{1/2} = \left(\frac{kT}{2\bar{B}}\right)^{1/2}$$

We see then that J_{max} increases with T and is inversely related to $\bar{B}$, and so increases with the moment of inertia of the molecule since $\bar{B} \propto 1/I$.

The next two sections deal with the rotational partition function of homonuclear diatomic molecules. The wave function of a homonuclear diatomic molecule must possess a certain symmetry with respect to the interchange of the two identical nuclei in the molecule. In particular, if the two nuclei have integral spins, the wave function must be symmetric with respect to an interchange; if the nuclei have half odd integer

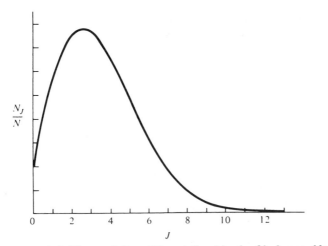

Figure 6–5. **The population of the rotational levels of hydrogen chloride at 300°K.**

spin, the wave function must be antisymmetric. This symmetry requirement has a profound effect on the rotational energy levels of a homonuclear diatomic molecule, which can be understood only by understanding the general symmetry properties of a homonuclear diatomic molecule. This is discussed in the next section. Then in Section 6–5 we apply these results to the rotational partition function. We shall see there that at low temperatures, the symmetry properties have an important effect on the thermodynamic properties of certain molecules, and in particular we shall discuss ortho- and para-hydrogen.

The discussion of the symmetry requirement is somewhat involved, however, and so we present here a summary of the high-temperature limit, which for most systems is completely adequate. At temperatures such that the summation in Eq. (6–26) can be approximated by an integral or even the Euler-MacLaurin expansion, the result for a homonuclear diatomic molecule is

$$q_{rot} = \frac{T}{2\Theta_r}\left\{1 + \frac{\Theta_r}{3T} + \frac{1}{15}\left(\frac{\Theta_r}{T}\right)^2 + \cdots\right\} \tag{6–39}$$

Note that this equation is the same as Eq. (6–35) except for the factor of 2 in the denominator. This factor is due to the symmetry of the homonuclear diatomic molecule and, in particular, is due to the fact that there are two indistinguishable orientations of a homonuclear diatomic molecule. There is a two-fold axis of symmetry perpendicular to the internuclear axis.

Equations (6–35) and (6–39) can be written as one equation by introducing a factor σ into the denominator of Eq. (6–35). If $\sigma = 1$, we have Eq. (6–35), and if $\sigma = 2$, we have Eq. (6–39). The factor σ is called the *symmetry number* of the molecule and represents the number of indistinguishable orientations that the molecule can have. The exact origin of σ can only be completely understood from the arguments presented in Sections 6–4 and 6–5, but on first reading it is possible to accept the factor of σ and proceed directly to Eq. (6–47) from here.

6–4 THE SYMMETRY REQUIREMENT OF THE TOTAL WAVE FUNCTION OF A HOMONUCLEAR DIATOMIC MOLECULE

The calculation of the rotational partition function is not quite so straightforward for homonuclear diatomic molecules. The total wave function of the molecule, that is, the electronic, vibrational, rotational, translational, and nuclear wave function, must be either symmetric or antisymmetric under the interchange of the two identical nuclei. It must be symmetric if the nuclei have integral spins (bosons), or antisymmetric if they have half-integral spins (fermions). This symmetry requirement has profound consequences on the thermodynamic properties of homonuclear diatomic molecules at low temperatures. We shall discuss the interchange of the two identical nuclei of a homonuclear diatomic molecule in this section, and then apply the results to the calculation of q_{rot} in the next section.

It is convenient to imagine this interchange as a result of (1) an inversion of all the particles, electrons and nuclei, through the origin, and then (2) an inversion of just the electrons back through the origin. This two-step process is equivalent to an exchange of the nuclei. Let us write ψ_{total} *exclusive* of the nuclear part as

$$\psi'_{total} = \psi_{trans}\psi_{rot}\psi_{vib}\psi_{elec}$$

where the prime on ψ_{total} indicates that we are ignoring the nuclear contribution for now. The translational wave function depends only upon the coordinates of the center of mass of the molecule, and so this factor is not affected by inversion. Furthermore, ψ_{vib} depends only upon the magnitude of $(r - r_e)$, and so this part of the total wave function is unaffected by any inversion operation. Therefore, we concentrate on ψ_{elec} and ψ_{rot}.

The property of ψ_{elec} under the inversions in both Steps (1) and (2) above depends upon the symmetry of the ground electronic state of the molecule. The ground electronic state of most molecules is symmetric under both of these operations. Such a state is designated by the term symbol $\sum_g^+$. Thus it is ψ_{rot} that controls the symmetry of ψ_{total}.

Only Step (1) above, the inversion of both electrons and nuclei through the origin, affects ψ_{rot}. The effect of this inversion is to change the coordinates (r, θ, ϕ) that describe the orientation of the diatomic molecule into $(r, \pi - \theta, \phi + \pi)$. One can see this either analytically from the eigenfunctions themselves or pictorially from the rotational wave functions shown in Fig. 6–6. Notice that the rigid rotor wave functions are the same functions as the angular functions of the hydrogen atom.

The net result then, when the ground electronic state is symmetric, that is, $\sum_g^+$ is that ψ_{total} remains unchanged for even J and changes sign for odd J. This result applies to the total wave function, exclusive of nuclear spin.

Now consider a molecule such as H_2, whose nuclei have a spin of $\frac{1}{2}$. Just as in the case of the two electrons in the helium atom, the two nuclei of spin $\frac{1}{2}$ have three symmetric spin functions $\alpha\alpha$, $\beta\beta$, and $2^{-1/2}(\alpha\beta + \beta\alpha)$, and one antisymmetric spin function $2^{-1/2}(\alpha\beta - \beta\alpha)$. Since nuclei with spin $\frac{1}{2}$ act as fermions, the total wave function must be antisymmetric in the exchange of these two nuclei. Now states with both even and odd values of J can be brought to the required antisymmetry by coupling them with the right spin functions. Since three symmetric nuclear spin functions can be combined with the odd J levels to achieve the correct overall antisymmetry for $^1\sum_g^+$ electronic states, we see that the odd J levels have a statistical weight of 3, compared to a weight of 1 for even J levels. This leads to the existence of ortho- (parallel nuclear spins) states and para- (opposed nuclear spins) states in H_2. This weighting of the rotational states will be seen shortly to have a profound effect on the low-temperature thermodynamics of H_2.

More generally, for nuclei of spin I, there are $2I + 1$ spin states for each nucleus. Let the eigenfunctions of these spin states be denoted by $\alpha_1, \alpha_2, \ldots, \alpha_{2I+1}$. There are $(2I + 1)^2$ nuclear wave functions to include in ψ_{total}. (In the case of H_2, $I = \frac{1}{2}$, there are two spin states α and β, and there are four nuclear spin functions, three of which are symmetric and one of which is antisymmetric.) The antisymmetric nuclear spin functions are of the form $\alpha_i(1)\alpha_j(2) - \alpha_i(2)\alpha_j(1)$, $1 \leq i, j \leq 2I + 1$. There are $(2I + 1)(2I)/2$ such combinations, and so this is the number of antisymmetric nuclear spin functions. (For H_2, we find that there is only one antisymmetric choice in agreement with the above paragraph.) All the remaining $(2I + 1)^2$ total nuclear spin functions are symmetric, and so their number is $(2I + 1)^2 - I(2I + 1) = (I + 1)(2I + 1)$. Thus we can write the following summary for $\sum_g^+$ states;

integral spin
$I(2I + 1)$ antisymmetric nuclear spin functions couple with odd J
$(I + 1)(2I + 1)$ symmetric nuclear spin functions couple with even J
half-integral spin
$I(2I + 1)$ antisymmetric nuclear spin functions couple with even J
$(I + 1)(2I + 1)$ symmetric nuclear spin functions couple with odd J

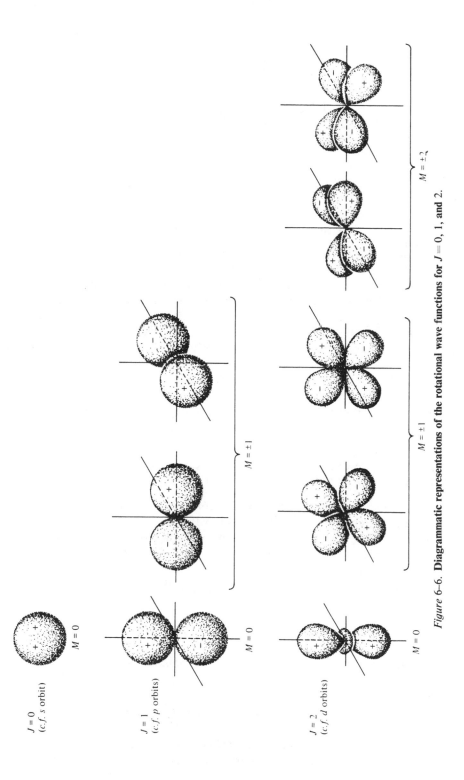

Figure 6-6. **Diagrammatic representations of the rotational wave functions for** $J = 0$, 1, and 2.

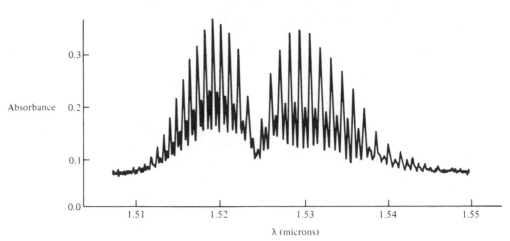

Figure 6–7. **The vibration–rotation spectrum of acetylene. This represents one vibrational line. The alternation in the intensity of the lines is due to the statistical weights of the rotational levels.** (From L. W. Richards, *J. Chem. Ed.*, **43**, p. 645, 1966.)

These combinations of nuclear and rotational wave functions produce the correct symmetry required of the total wave function under interchange of identical nuclei. Remember that all of these conclusions are for $\sum_g^+$ electronic states, the most commonly occurring ground state. (See Problem 6–26 for a discussion of O_2.)

Even though we have considered only diatomic molecules here, the results of this section apply also to linear polyatomic molecules such as CO_2, H_2C_2. For example, the molecules $HC^{12}C^{12}H$ and $DC^{12}C^{12}D$ have their rotational states weighted in a similar way as H_2 and D_2. Figure 6–7 shows the vibration-rotation spectrum of H_2C_2. The alternation in the intensity of these rotational lines due to the statistical weights is very apparent.

6–5 THE ROTATIONAL PARTITION FUNCTION OF A HOMONUCLEAR DIATOMIC MOLECULE

The results of the previous section show that for homonuclear diatomic molecules with nuclei having integral spin, rotational levels with odd values of J must be coupled with the $I(2I + 1)$ antisymmetric nuclear spin functions, and that rotational levels with even values of J must be coupled with the $(I + 1)(2I + 1)$ symmetric nuclear spin functions. Thus we write

$$q_{rot, nucl}(T) = (I + 1)(2I + 1) \sum_{J \text{ even}} (2J + 1)e^{-\Theta_r J(J + 1)/T}$$
$$+ I(2I + 1) \sum_{J \text{ odd}} (2J + 1)e^{-\Theta_r J(J + 1)/T} \qquad (6–40)$$

Likewise, for molecules with nuclei with half-integer spins, we have

$$q_{rot, nucl}(T) = I(2I + 1) \sum_{J \text{ even}} (2J + 1)e^{-\Theta_r J(J + 1)/T}$$
$$+ (I + 1)(2I + 1) \sum_{J \text{ odd}} (2J + 1)e^{-\Theta_r J(J + 1)/T} \qquad (6–41)$$

Notice that in this case the combined rotational and nuclear partition function *does not factor* into $q_{rot}q_{nucl}$. This is a situation in which we cannot ignore q_{nucl}. For most molecules at ordinary temperatures, $\Theta_r \ll T$, and we can replace the sum by an integral. We see then that

$$\sum_{J \text{ even}} \approx \sum_{J \text{ odd}} \approx \frac{1}{2} \sum_{\text{all } J} \approx \frac{1}{2} \int_0^\infty (2J + 1)e^{-\Theta_r J(J+1)/T} \, dJ = \frac{T}{2\Theta_r} \tag{6-42}$$

and so both Eqs. (6–40) and (6–41) become

$$q_{\text{rot, nucl}}(T) = \frac{(2I + 1)^2 T}{2\Theta_r} \tag{6-43}$$

which can be written as $q_{\text{rot}}(T)q_{\text{nucl}}$ where

$$q_{\text{rot}}(T) = \frac{T}{2\Theta_r} \quad \text{and} \quad q_{\text{nucl}} = (2I + 1)^2 \tag{6-44}$$

For Eq. (6–42) to be valid, Θ_r/T must be less than about 0.20.

This result is to be compared to the result for a heteronuclear diatomic molecule, namely, $q_{\text{rot}}(T) = T/\Theta_r$. The factor of 2 that appears above in the high-temperature limit takes into account that the molecule is homonuclear, and so its rotational partition function is given by Eq. (6–40) or (6–41) instead of (6–26). This factor of 2 is called the symmetry number and is denoted by σ. It legitimately appears only when Θ_r is less than approximately $0.2T$, since only then can we use Eq. (6–42). Understanding the origin of this fact then, we can write

$$q_{\text{rot}}(T) \approx \frac{8\pi^2 I k T}{\sigma h^2} \approx \frac{1}{\sigma} \sum_{J=0}^\infty (2J + 1)e^{-\Theta_r J(J+1)/T} \qquad \Theta_r \ll T \tag{6-45}$$

where $\sigma = 1$ for heteronuclear molecules, and $\sigma = 2$ for homonuclear diatomic molecules. Remember that this is applicable only to the high-temperature limit or its Euler-MacLaurin correction. A similar factor will appear for polyatomic molecules also.

There are some interesting systems in which Θ_r/T is not small. Hydrogen is one of the most important such cases. Each nucleus in H_2 has nuclear spin $\frac{1}{2}$, and so

$$q_{\text{rot, nucl}} = \sum_{J \text{ even}} (2J + 1)e^{-\Theta_r J(J+1)/T} + 3 \sum_{J \text{ odd}} (2J + 1)e^{-\Theta_r J(J+1)/T} \tag{6-46}$$

The hydrogen with only even rotational levels allowed (antisymmetric nuclear spin function or "opposite" nuclear spins) is called para-hydrogen; that with only odd rotational levels allowed (symmetric nuclear spin function or "parallel" nuclear spins) is called ortho-hydrogen. The ratio of the number of ortho-H_2 molecules to the number of para-H_2 molecules is

$$\frac{N_{\text{ortho}}}{N_{\text{para}}} = \frac{3 \sum\limits_{J \text{ odd}} (2J + 1)e^{-\Theta_r J(J+1)/T}}{\sum\limits_{J \text{ even}} (2J + 1)e^{-\Theta_r J(J+1)/T}}$$

Figure 6–8 shows the percentage of p–H_2 versus temperature in an equilibrium mixture of ortho- and para-hydrogen. Note that the system is all para- at $0°K$ and 25 percent para- at high temperatures.

Figure 6–9 illustrates an interesting situation that occurs with low-temperature heat capacity measurements on H_2. Equation (6–46) can be used to calculate the heat capacity of H_2, and this is plotted in Fig. 6–9, along with the experimental results. It can be seen that the two curves are in great disagreement. These calculations and

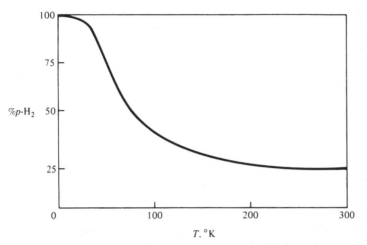

Figure 6–8. **The percentage of para-hydrogen in an equilibrium mixture as a function of temperature.**

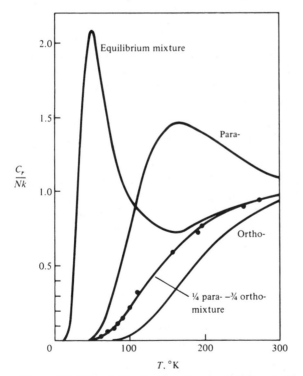

Figure 6–9. **The rotational and nuclear contribution to the molar heat capacity for ortho-hydrogen, para-hydrogen, an equilibrium mixture of ortho- and para-hydrogen, a metastable 75 percent ortho- and 25 percent para- mixture, and the experimental data.** (From K. F. Bonhoeffer and P. Harteck, *Z. Physikal. Chem.*, **4B**, p. 113, 1929.)

measurements were made at a time when quantum mechanics was being developed, and was not accepted by all scientists. For a while, the disagreement illustrated in Fig. 6–9 was a blow to the proponents of the new quantum mechanics. It was Dennison[*] who finally realized that the conversion between ortho- and para-hydrogen is extremely slow in the absence of a catalyst, and so when hydrogen is prepared in the laboratory at room temperature and then cooled down for the low-temperature heat capacity measurements, the room-temperature composition persists instead of the equilibrium composition. Thus the experimental data illustrated in Fig. 6–9 are not for an equilibrium system of ortho- and para-hydrogen, but for a metastable system whose ortho-para composition is that of equilibrium room-temperature hydrogen, namely, 75 percent ortho- and 25 percent para-. If one calculates the heat capacity of such a system, according to

$$C_V = \tfrac{3}{4}C_V(\text{ortho-}) + \tfrac{1}{4}C_V(\text{para-})$$

where $C_V(\text{ortho-})$ is obtained from just the second term of Eq. (6–46), and $C_V(\text{para-})$ is obtained from the first term of Eq. (6–46), one obtains excellent agreement with the experimental curve. A clever confirmation of this explanation was shortly after obtained by Bonhoeffer and Harteck,[†] who performed heat capacity measurements on hydrogen in the presence of activated charcoal, a catalyst for the ortho–para conversion. This produces an equilibrium system at each temperature. The experimental data are in excellent agreement with the equilibrium calculation in Fig. 6–9.

The explanation of the heat capacity of H_2 was one of the great triumphs of post-quantum mechanical statistical mechanics. You should be able to go through a similar argument for D_2, sketching the equilibrium heat capacity, the pure ortho- and para-heat capacity, and finally what you should expect the experimental curve to be for D_2 prepared at room temperature and at some other temperature, say 20°K. (See Problem 6–17.)

In principle, such nuclear spin effects should be observable in other homonuclear molecules, but a glance at Table 6–1 shows that the characteristic rotational temperatures for all the other molecules are so small that these molecules reach the "high-temperature limit" while still in the solid state. Hydrogen is somewhat unusual in that its rotational constant is so much greater than its boiling point.

For most cases then, we can use Eq. (6–45) which, when we use the Euler-MacLaurin expansion, becomes

$$q_{\text{rot}}(T) = \frac{T}{\sigma\Theta_r}\left\{1 + \frac{\Theta_r}{3T} + \frac{1}{15}\left(\frac{\Theta_r}{T}\right)^2 + \frac{4}{315}\left(\frac{\Theta_r}{T}\right)^3 + \cdots\right\} \tag{6–47}$$

Usually only the first term of this is necessary. Some of the thermodynamic functions are

$$E_{\text{rot}} = NkT\left\{1 - \frac{\Theta_r}{3T} - \frac{1}{45}\left(\frac{\Theta_r}{T}\right)^2 + \cdots\right\} \tag{6–48}$$

$$C_{\text{rot}} = Nk\left\{1 + \frac{1}{45}\left(\frac{\Theta_r}{T}\right)^2 + \cdots\right\} \tag{6–49}$$

[*] D. M. Dennison, *Proc. Roy. Soc.* **A115**, 483, 1927.
[†] K. F. Bonhoeffer and P. Harteck, *Z. Phys. Chem.*, **4B**, 113, 1926.

$$S_{\text{rot}} = Nk\left\{1 - \ln\left(\frac{\sigma\Theta_r}{T}\right) - \frac{1}{90}\left(\frac{\Theta_r}{T}\right)^2 + \cdots\right\} \tag{6-50}$$

where all of these formulas are valid in the same region, in which σ itself is a meaningful concept, that is, $\Theta_r < 0.2T$. The terms in Θ_r/T and its higher powers are usually not necessary. Note that Eq. (6–47) is identical to Eq. (6–35) except for the occurrence of the symmetry number in Eq. (6–47).

6-6 THERMODYNAMIC FUNCTIONS

Having studied each contribution to the total partition function q in Eq. (6–17), we can write in the harmonic oscillator–rigid rotor approximation

$$q(V, T) = \left(\frac{2\pi mkT}{h^2}\right)^{3/2} V \frac{8\pi^2 IkT}{\sigma h^2} e^{-\beta h\nu/2}(1 - e^{-\beta h\nu})^{-1}\omega_{e_1}e^{D_e/kT} \tag{6-51}$$

Remember that this requires that $\Theta_r \ll T$, that only the ground electronic state is important, and that the zero of energy is taken to be the separated states at rest in their ground electronic states. Note that only q_{trans} is a function of V, and this is of the form $f(T)V$ which, we have seen before, is responsible for the ideal gas equation of state. The thermodynamic functions associated with Eq. (6–51) are

$$\frac{E}{NkT} = \frac{5}{2} + \frac{h\nu}{2kT} + \frac{h\nu/kT}{e^{h\nu/kT} - 1} - \frac{D_e}{kT} \tag{6-52}$$

$$\frac{C_V}{Nk} = \frac{5}{2} + \left(\frac{h\nu}{kT}\right)^2 \frac{e^{h\nu/kT}}{(e^{h\nu/kT} - 1)^2} \tag{6-53}$$

$$\frac{S}{Nk} = \ln\left[\frac{2\pi(m_1 + m_2)kT}{h^2}\right]^{3/2} \frac{Ve^{5/2}}{N} + \ln\frac{8\pi^2 IkTe}{\sigma h^2}$$
$$+ \frac{h\nu/kT}{e^{h\nu/kT} - 1} - \ln(1 - e^{-h\nu/kT}) + \ln\omega_{e1} \tag{6-54}$$

$$pV = NkT \tag{6-55}$$

$$\frac{\mu^0(T)}{kT} = -\ln\left[\frac{2\pi(m_1 + m_2)kT}{h^2}\right]^{3/2} kT - \ln\frac{8\pi^2 Ik\,T}{\sigma h^2} + \frac{h\nu}{2kT}$$
$$+ \ln(1 - e^{-h\nu/kT}) - \frac{D_e}{kT} - \ln\omega_{e1} \tag{6-56}$$

Table 6–1 contains the characteristic rotational temperatures, the characteristic vibrational temperatures, and $D_0 = D_e - \frac{1}{2}h\nu$ for a number of diatomic molecules.

Table 6–3 presents a comparison of Eq. (6–54) with experimental data. It can be seen that the agreement is quite good and is typical of that found for the other thermodynamic functions. It is possible to improve the agreement considerably by including the first corrections to the rigid rotor–harmonic oscillator model. These include centrifugal distortion effects, anharmonic effects, and other extensions. The consideration of these effects introduces a new set of molecular constants, all of which are determined spectroscopically and are well tabulated. (See Problem 6–24.) The use of such

Table 6–3. **The entropies of some diatomic molecules calculated according to Eq. (6–54) compared to the experimental values at 1-atm pressure and 25°C***

	S(calc.) (e.u.)	S(exp.) (e.u.)
H_2	31.1	31.2
O_2	49.0	49.0
N_2	45.7	45.7
Cl_2	53.2	53.3
HCl	44.6	44.6
HBr	47.4	47.4
HI	49.4	49.3
CO	47.2	46.2

* The experimental values have been corrected for any nonideal gas behavior.

additional parameters from spectroscopic data can give calculated values of the entropy and heat capacity that are actually more accurate than experimental ones.

It should be pointed out, however, that extremely accurate calculations can require a sophisticated knowledge of molecular spectroscopy. For example, we said above that the electronic partition function was similar to that in the atomic case. This, however, is not entirely true. For molecules in states other than a $\sum$ state (which has zero total angular momentum), the total electronic angular momentum must be coupled with the overall rotational angular momentum, and this coupling must be treated in a detailed quantum mechanical way. This is too specialized to discuss here, but the result of this coupling is that the electronic and rotational partition functions do not separate. When $T \gg \Theta_r$, however, the molecules are in states with large enough rotational quantum numbers [*cf.* Eq. (6–39)] that the angular momentum coupling is no longer important, and the rotational-electronic partition function separates into a rotational part and an electronic part. Since we have chosen the zero of energy to be the separated electronically unexcited atoms at rest, the electronic partition function is

$$q_e(T) = \omega_{e1} e^{D_e/kT} + \omega_{e2}\, e^{-\varepsilon_2/kT} + \cdots$$

where the ω_{ej} are the degeneracies, and the ε_j's are measured relative to the ground electronic state of the molecule. Keep in mind, however, that for some molecules, such as NO, this equation is valid only at high temperatures, and that the low-temperature partition function requires a fairly specialized knowledge of the coupling of electronic and rotational angular momenta. See Herzberg under "Additional Reading" for a thorough discussion of this complication.

It seems logical at this time to go on to a discussion of polyatomic molecules in much the same manner as we have for diatomics. We would see very quickly, however, that unless the molecule possesses a certain degree of symmetry, it is impossible to write down any closed-form expression for its rotational energy levels. This means that a calculation of $q_{rot}(T)$ is at best a complicated numerical problem. This would appear to imply that we have come to the end of the line for statistical thermodynamical applications, and we have not even begun to consider *interactions* between molecules! Even just *two* simple molecules, to say nothing of N particles, interacting through any kind of realistic interatomic potential becomes an extremely complicated quantum mechanical problem. At this point we must go back and reconsider some of the results we have derived up to now.

ADDITIONAL READING

General

DAVIDSON, N. 1962. *Statistical mechanics*. New York: McGraw-Hill. Chapters 8 and 9.

FOWLER, R. H., and GUGGENHEIM, E. A. 1956. *Statistical thermodynamics*. Cambridge: Cambridge University Press. Sections 312 through 325.

HILL, T. L. 1960. *Statistical thermodynamics*. Reading, Mass.: Addison-Wesley. Chapter 8.

KESTIN, J., and DORFMAN, J. R. 1971. *A course in statistical thermodynamics*. New York: Academic. Sections 6–9, 6–10, and 6–13.

KNUTH, E. 1966. *Statistical and thermodynamics*. New York: McGraw-Hill. Chapter 13.

KUBO, R. 1965. *Statistical mechanics*. Amsterdam: North-Holland Publishing Co. Chapter 3 and Example 3–1.

MAYER, J. E., and MAYER, M. G. 1940. *Statistical mechanics*. New York: Wiley. Chapter 7.

MÜNSTER, A. 1969. *Statistical thermodynamics*. Vol. I. Berlin: Springer-Verlag. Sections 6–3 through 6–7.

RUSHBROOKE, G. S. 1949. *Statistical mechanics*. London: Oxford University Press. Chapters 6, 7, and 8.

Spectroscopy

BARROW, G. 1962. *Introduction to molecular spectroscopy*. New York: McGraw-Hill.

DUNFORD, H. B. 1968. *Elements of diatomic molecular spectra*. Reading, Mass.: Addison-Wesley.

HERZBERG, G. 1950. *Spectra of diatomic molecules*, 2nd ed. New York: Van Nostrand.

KING, G. W. 1964. *Spectroscopy and molecular structure*. New York: Holt, Rinehart and Winston.

PROBLEMS

6–1. The Morse potential is

$$U(r) = D_e(1 - e^{-\beta(r-r_e)})^2$$

Show that $\beta = \nu(2\pi^2\mu/D_e)^{1/2}$.

6–2. The dissociation energy D_0 of H_2 is 103.2 kcal/mole, and its fundamental vibrational frequency $\bar{\omega}$ is 4320 cm^{-1}. From this information, calculate D_0 and $\bar{\omega}$ for D_2, T_2, and HD, assuming the Born-Oppenheimer approximation.

6–3. Given that D_0 for H_2 is 103.2 kcal/mole and that Θ_v is 6215°K, calculate D_0 for both D_2 and T_2.

6–4. Show that the moment of inertia of a diatomic molecule is μr_e^2, where μ is the reduced mass, and r_e is the equilibrium separation.

6–5. Show that the force constants in Table 6–1 are consistent with the frequencies given there.

6–6. Using the data in Table 6–1, calculate the frequencies that are expected to be found in the rotational spectrum of HCl.

6–7. In the far infrared spectrum of HBr, there is a series of lines separated by 16.72 cm^{-1}. Calculate the moment of inertia and internuclear separation in HBr.

6–8. Show that the vibrational contribution to the heat capacity C_V of a diatomic molecule is Nk as $T \to \infty$.

6–9. Derive Eq. (6–35) from the Euler–MacLaurin summation formula.

6–10. Show that the rotational level that is most populated is given by $J_{max} = (kT/2\bar{B})^{1/2}$. Calculate J_{max} for CO_2 and H_2 at room temperature.

6–11. The rotational constant $\bar{B}$ for $HC^{12}N^{14}$ is 44,315.97 MHz (megahertz) and $DC^{12}N^{14}$ for 36,207.40 MHz. Deduce the moments of inertia for these molecules. Assuming that the bond lengths are independent of isotopic substitution, calculate the H–C and C–N bond length.

6–12. Given that the values of Θ_r and Θ_v for H_2 are 85.3°K and 6215°K, respectively, calculate these quantities for HD and D_2.

6–13. What is the most probable value of the rotational quantum number J of a gas phase N_2 molecule at 300°K? What is the most probable vibrational quantum number n for this same situation?

6–14. Using the Euler-MacLaurin expansion, derive the second- and third-order corrections to the (first-order) high-temperature limit of E_r and C_{V_r}. Express your result in terms of a power series of Θ_r/T.

6–15. Calculate the rotational contribution to the entropy of HD at 20°K, 100°K, and 300°K, using the formulas appropriate for each particular temperature, and estimate the error involved in each.

6–16. Discuss the statistical weights of a hypothetical diatomic molecule X_2 with a ground electronic state $\sum_u^-$, supposing the X nuclei have integral spin (bosons) and half-integral spin (fermions). Derive the rotational partition function and the rotational contribution to the heat capacity C_V for each case.

6–17. Calculate the percent of para-D_2 as a function of temperature (assuming equilibrium) and also calculate the heat capacity of the equilibrium mixture, para-D_2, ortho-D_2, and finally what you expect would be the experimental heat capacity.

6–18. Why does one not see discussions in the literature concerning the ortho-para forms of fluorine?

6–19. Show that the thermodynamic quantities p and C_V are independent of the choice of a zero of energy.

6–20. In the far infrared spectrum of HCl, there is a series of lines with an almost constant spacing of 20.7 cm^{-1}. In the near infrared spectrum, there is one intense band at 3.46 microns. Use these data to calculate the entropy of HCl at 300°K and 1 atm (assuming ideal behavior).

6–21. Molecular nitrogen is heated in an electric arc, and it is found spectroscopically that the relative populations of excited vibrational levels is

n	0	1	2	3	4	$\cdots$
$\dfrac{f_n}{f_0}$	1.000	0.200	0.040	0.008	0.002	$\cdots$

Is the nitrogen in thermodynamic equilibrium with respect to vibrational energy? What is the vibrational temperature of the gas? Is this necessarily the same as the translational temperature?

6–22. Without looking anything up, put in order of decreasing magnitudes the following "temperatures":

$$\Theta_v^{H2}, \ \Theta_r^{H2}, \ \Theta_v^{Cl2}, \ \Theta_r^{Cl2}, \ \Theta_v^{HCl}, \ \Theta_r^{HCl}$$

6–23. A more accurate expression for the vibrational energy of a diatomic molecule is

$$\varepsilon_n = (n + \tfrac{1}{2})h\nu - x_e(n + \tfrac{1}{2})^2 h\nu$$

where x_e is called the anharmonicity constant. The additional term here represents the first deviations from strictly harmonic behavior. Treating x_e as a small parameter, calculate the anharmonic effect on the various thermodynamic functions at least to first order in x_e.

6–24. The model of a diatomic molecule presented in this chapter is called the rigid rotor–harmonic oscillator model. The rotational-vibrational energy in this approximation is

$$\varepsilon_{vr} = (n + \tfrac{1}{2})h\nu + \bar{B}J(J + 1)$$

This expression can be improved in a number of ways. The harmonic oscillator approximation can be modified to include terms that reflect the deviations from harmonic behavior (anharmonicity) as the vibrational energy of the molecule increases. This is done by quantum mechanical perturbation theory, which gives

$$\varepsilon_v = (n + \tfrac{1}{2})h\nu - x_e(n + \tfrac{1}{2})^2 h\nu + \cdots$$

where x_e is a small constant called the anharmonicity constant. In addition to this, there is a correction due to the fact that the molecule is not a rigid rotor and, in fact, stretches some as the molecule rotates with greater energy. This is also handled by perturbation theory and gives

$$\varepsilon_r = \bar{B}J(J+1) - \bar{D}J^2(J+1)^2 + \cdots$$

where $\bar{D}$ is a small constant called the centrifugal distortion constant. Lastly, there exists a coupling between the rotational and vibrational modes of the molecule, since its moment of inertia changes as the molecule vibrates. Putting all this together gives

$$\varepsilon_{vr} = (n + \tfrac{1}{2})h\nu + \bar{B}J(J+1) - x_e(n + \tfrac{1}{2})^2 h\nu - \bar{D}J^2(J+1)^2 - \alpha(n + \tfrac{1}{2})J(J+1)$$

where α is the rotation-vibration coupling constant. These terms, which correct the rigid rotor–harmonic oscillator approximation, are usually quite small. Using this more rigorous expression for ε_{vr}, show that the molecular partition function can be written in the form

$$q(V, T) = q_{rr-ho}q_{corr}$$

where

$$q_{corr} = 1 + \frac{2kT}{\bar{B}}\left(\frac{\bar{D}}{\bar{B}}\right) + \frac{1}{e^{\beta h\nu} - 1}\left(\frac{\alpha}{\bar{B}}\right)$$

$$+ \frac{2\beta h\nu}{(e^{\beta h\nu} - 1)^2}x_e + \text{higher-order terms in } \bar{D}, \alpha, \text{ and } x_e$$

Calculate the effect of q_{corr} on E and C_V for O_2 at 300°K, given the following values of the spectroscopic parameters: $x_e = 0.0076$, $\bar{D} = 4.8 \times 10^{-6}$ cm^{-1}, and $\alpha = 0.016$ cm^{-1}.

6–25. Consider a system of independent diatomic molecules constrained to move in a plane, that is, a two-dimensional ideal diatomic gas. How many degrees of freedom does a two-dimensional diatomic molecule have? Given that the energy eigenvalues of a two-dimensional rigid rotor are

$$\varepsilon_J = \frac{\hbar^2 J^2}{2I} \qquad J = 0, 1, 2, \ldots$$

with a degeneracy $\omega_J = 2$ for all J except $J = 0$, for which $\omega_J = 1$, calculate the rotational partition function. I is the moment of inertia of the molecule. The vibrational partition function is the same as for a three-dimensional diatomic gas. Write out

$$q(T) = q_{trans}(T)q_{rot}(T)q_{vib}(T)$$

and derive an expression for the average energy of this two-dimensional ideal diatomic gas.

6–26. Show that the molecule $O^{16}O^{16}$ has only odd rotational levels in its ground electronic state.

CHAPTER 7

CLASSICAL STATISTICAL MECHANICS

So far we have been able to derive translational, rotational, and vibrational partition functions for linear molecules. In each case we saw that if the temperature were high enough we could replace sums by integrals and obtain high-temperature limits. These prove to be numerically satisfactory for most gases at ordinary temperatures (with the exception perhaps of the vibrational case).

As the temperature increases, the average energy per molecule increases, and so in a quantum mechanical sense, the quantum numbers describing this motion (n_x for translational, J for rotational, etc.) also increase, meaning that the molecules are in the high quantum number limit. For example at room temperature translational quantum numbers are typically 10^8. (See Problem 5–3.) It is the recognition of this fact that will point the way to a solution to the problem discussed in the last paragraph of Chapter 6.

It is one of the fundamental principles of quantum mechanics that classical behavior is obtained in the limit of large quantum numbers. So we see that up to now our procedure has been to solve a particular quantum mechanical problem, use this result in the molecular partition function, use a high-temperature approximation, and then find that this high-temperature limit is satisfactory. In other words, we were starting with a quantum mechanical solution and then taking the classical limit at a later stage. It is natural to seek a procedure in which we can use classical mechanics throughout, and such an approach is developed in this chapter. For simplicity we shall first consider only molecular partition functions, although we shall generalize our results afterward.

7–1 THE CLASSICAL PARTITION FUNCTION

Consider the molecular partition function

$$q = \sum_j e^{-\beta \varepsilon_j} \tag{7–1}$$

This is of the form of a sum of $e^{-\beta(\text{energy})}$ over all possible quantum states. It is natural to assume that the corresponding classical expression is a similar sum, or since the energy in the classical sense is a continuous function of the momenta p_j and coordinates q_j, this sum would become an *integral* over all the possible classical "states" of the system. Since the classical energy is the Hamiltonian function $H(p, q)$, the molecular partition function $q(V, T)$ becomes

$$q_{\text{class}} \sim \int \cdots \int e^{-\beta H(p, q)} \, dp \, dq \qquad (7\text{–}2)$$

In Eq. (7–2) the notation (p, q) denotes all the momenta and coordinates on which H depends; dp stands for $dp_1 \, dp_2 \cdots dp_s$ and dq for $dq_1 \cdots dq_s$, where s is the number of momenta or coordinates necessary to completely specify the motion or position of the molecule. The quantity s represents the number of degrees of freedom of the molecule. The set of coordinates $\{q_j\}$ does not necessarily have to be a set of Cartesian coordinates, and more usually represents a set of generalized coordinates, that is, any set of coordinates that conveniently specifies the position of the molecule. For a mass point, for example, the generalized coordinates might be simply x, y, and z; for a rigid rotor, we might choose the two angles θ and ϕ needed to specify the orientation of the molecule. Usually the choice of generalized coordinates is obvious. The momenta $\{p_j\}$ in Eq. (7–2) are the generalized momenta conjugate to the $\{q_j\}$ [cf. Eq. (1–19)].

At this stage Eq. (7–2) is just a plausible conjecture. Let us now pursue this idea by considering a monatomic ideal gas once again. From Eq. (5–8), we have

$$q_{\text{trans}}(V, T) = \left(\frac{2\pi m k T}{h^2}\right)^{3/2} V$$

The classical Hamiltonian of one atom of a monatomic ideal gas is simply the kinetic energy:

$$H = \frac{1}{2m}(p_x{}^2 + p_y{}^2 + p_z{}^2)$$

According to Eq. (7–2), then,

$$q_{\text{class}} \sim \int \cdots \int \exp\left\{-\frac{\beta(p_x{}^2 + p_y{}^2 + p_z{}^2)}{2m}\right\} \, dp_x \, dp_y \, dp_z \, dx \, dy \, dz \qquad (7\text{–}3)$$

Notice here that since it takes three coordinates to specify the position of a point particle, q_{class} is a six-fold integral. The integral over $dx \, dy \, dz$ simply yields the volume of the container V, and so we have

$$q_{\text{class}} \sim V \left\{\int_{-\infty}^{\infty} e^{-\beta p^2/2m} \, dp\right\}^3 = (2\pi m k T)^{3/2} V \qquad (7\text{–}4)$$

We see that except for a factor of Planck's constant cubed, this is just the translational partition function that we obtained before. Of course, we cannot expect to derive a purely classical expression that contains h, and so although our conjecture may be incomplete, there seems to be some element of truth to it.

Let us see how this procedure works for the other partition functions that we have evaluated. For the rigid rotor, the Hamiltonian is

$$H = \frac{1}{2I}\left(p_\theta{}^2 + \frac{p_\phi{}^2}{\sin^2 \theta}\right)$$

where I is the moment of inertia of the molecule. The generalized coordinates and momenta in this case are θ, ϕ, p_θ, and p_ϕ, and so Eq. (7–2) is

$$q_{rot} \sim \int_{-\infty}^{\infty} \int dp_\theta\, dp_\phi \int_0^{2\pi} d\phi \int_0^\pi d\theta\, e^{-\beta H} = (8\pi^2 IkT) \tag{7–5}$$

For the classical harmonic oscillator,

$$H = \frac{p^2}{2\mu} + \frac{k}{2} x^2 \tag{7–6}$$

and

$$q_{vib} \sim \int_{-\infty}^{\infty} dp \int_{-\infty}^{\infty} dx\, e^{-\beta H} = \frac{kT}{\nu} \tag{7–7}$$

where

$$\nu = \frac{1}{2\pi} \left(\frac{k}{\mu} \right)^{1/2}$$

We can see from these three examples that the translational partition function is incorrect by a factor of h^3; the rotational partition function is incorrect by a factor of h^2; and the vibrational partition function is incorrect by a factor of h. It appears that a factor of h results for each product $dp_j\, dq_j$ occurring in q_{class}. Since partition functions are dimensionless, and h has units of momentum times length, we see that this at least automatically satisfies a dimensional requirement. We shall therefore *assume* that

$$q = \sum_j e^{-\beta \varepsilon_j} \rightarrow \frac{1}{h^s} \int \cdots \int e^{-\beta H} \prod_{j=1}^{s} dp_j\, dq_j \tag{7–8}$$

We now shall extend this assumption to systems of molecules. Equation (4–10) says that at high enough temperatures, we can write for a system of N independent indistinguishable particles

$$Q = \frac{q^N}{N!}$$

$$= \frac{1}{N!} \prod_{j=1}^{N} \left\{ \frac{1}{h^s} \int \cdots \int e^{-\beta H_j} \prod_{i=1}^{s} dp_{ji}\, dq_{ji} \right\}$$

where H_j is the Hamiltonian of the jth molecule and is a function of $p_{j1}, \ldots, p_{js}$, $q_{j1}, \ldots, q_{js}$. We now simply relabel the momenta and coordinates such that p_1 through p_s represent p_{11} through p_{1s}; p_{s+1} through p_{s+s} represent p_{21} through p_{2s}, and so on, and write

$$Q = \frac{1}{N!\, h^{sN}} \int \cdots \int e^{-\beta \Sigma_j H_j} \prod_{i=1}^{sN} dp_i\, dq_i$$

$$= \frac{1}{N!\, h^{sN}} \int \cdots \int e^{-\beta H} \prod_{i=1}^{sN} dp_i\, dq_i$$

where H is the Hamiltonian of the N-body system. This form suggests the classical limit of Q for systems of interacting particles. We *conjecture* that

$$Q = \frac{1}{N!\, h^{sN}} \int \cdots \int e^{-\beta H(p,q)}\, dp\, dq \tag{7–9}$$

where $H(p, q)$ is the classical N-body Hamiltonian for *interacting* particles. The notation (p, q) represents the set of p_j's and q_j's that describes the entire system, and $dp\,dq$ represents

$$\prod_{j=1}^{sN} dp_j \, dq_j$$

We have assumed then that the classical limit of $Q(N, V, T)$ is given by

$$Q = \sum_j e^{-\beta E_j} \to \frac{1}{N! \, h^{sN}} \int \cdots \int e^{-\beta H(p, q)} \, dp \, dq \tag{7-10}$$

For a monatomic gas, for example,

$$H(p, q) = \frac{1}{2m} \sum_{j=1}^{N} (p_{xj}{}^2 + p_{yj}{}^2 + p_{zj}{}^2) + U(x_1, y_1, z_1, \ldots, x_N, y_N, z_N) \tag{7-11}$$

Equation (7–10) is, in fact, the correct classical limit of Q, although we have not proved it here. It is actually possible to start with the quantum mechanical sum in Eq. (7–10) and to derive the integral as the classical limit, that is, the limiting result as $h \to 0$ (*cf.* Section 10–7).

If we substitute Eq. (7–11) into Eq. (7–10), the momentum integrations can be done easily, and we get

$$Q_{\text{class}} = \frac{1}{N!} \left(\frac{2\pi mkT}{h^2} \right)^{3N/2} Z_N \tag{7-12}$$

where

$$Z_N = \int_V e^{-U(x_1, \ldots, z_N)/kT} \, dx_1 \cdots dz_N \tag{7-13}$$

In Eq. (7–13), Z_N is called the *classical configuration integral*. Since the intermolecular forces depend upon the relative distances between molecules, this integral is, in general, extremely difficult and is essentially responsible for the research in equilibrium statistical mechanics. In the absence of intermolecular forces, $U = 0$ and $Z_N = V^N$. Equations (7–12) and (7–13) are fundamental equations in the study of monatomic, classical, imperfect gases and liquids.

It often happens that not all of the degrees of freedom of a molecule can be treated classically. For example, we have seen that the spacing between translational and rotational levels is small enough that the sum over states or levels can be replaced by an integral, that is, these degrees of freedom can be treated classically. This is not the case, however, with the vibrational degrees of freedom, and these degrees of freedom must be treated quantum mechanically.

Suppose, then, that the Hamiltonian of a molecule can be written as

$$H = H_{\text{class}} + H_{\text{quant}} \tag{7-14}$$

where H_{class} refers to the s degrees of freedom that can be treated classically, and H_{quant} refers to the degrees of freedom that cannot be treated classically. Then

$$q = q_{\text{class}} q_{\text{quant}} \tag{7-15}$$

where

$$q_{\text{class}} = \frac{1}{h^s} \int e^{-H_{\text{class}}(p, q)/kT} \, dp_1 \, dq_1 \cdots dp_s \, dq_s \tag{7-16}$$

Note that Eq. (6–51) is of the form of Eq. (7–15), where the translational and rotational degrees of freedom are treated classically, and the vibrational and electronic degrees of freedom are treated quantum mechanically.

Equations (7–14) to (7–16) are immediately generalizable to a system of interacting molecules. If the Hamiltonian of the entire system is separable into a classical part and a quantum part, then

$$H = H_{\text{class}} + H_{\text{quant}} \tag{7–17}$$

$$Q = Q_{\text{class}} Q_{\text{quant}} \tag{7–18}$$

$$= \frac{Q_{\text{quant}}}{N! \, h^{sN}} \int e^{-H_{\text{class}}/kT} \, dp_{\text{class}} \, dq_{\text{class}} \tag{7–19}$$

7–2   PHASE SPACE AND THE LIOUVILLE EQUATION

Until now our approach has been to go to the classical limit only when it was necessary. Historically, however, statistical mechanics was originally formulated by Boltzmann, Maxwell, and Gibbs in the nineteenth century before the evolution of quantum mechanics. Their formulation, therefore, was based on classical mechanics, and since this still is a most useful limit, we shall now discuss the classical mechanical formulation of statistical mechanics. This formalism forms the basis of most of the work involving interacting systems in equilibrium and nonequilibrium statistical mechanics that is done today.

Consider any classical system containing N (interacting) molecules. Let each molecule have s degrees of freedom, that is, each molecule requires s coordinates to completely describe its position. Let the number of coordinates necessary to describe the positions of all N molecules be $l = sN$. The l coordinates, $q_1, q_2, \ldots, q_l$, then completely describe the spatial orientation of the entire N-body system. To each of these l coordinates, there corresponds a conjugate momentum p_j, say, defined by Eq. (1–19). The l spatial coordinates $\{q_j\}$ and the l momenta $\{p_j\}$ completely specify the classical mechanical state of the N-body system. These $2l$ coordinates, along with the equations of motion of the system, completely determine the future and past course of the system.

We now construct a conceptual Euclidean space of $2l$ dimensions, with $2l$-rectangular axes, one for each of the spatial coordinates $q_1, \ldots, q_l$ and one for each of the momenta $p_1, \ldots, p_l$. Following Gibbs, we speak of such a conceptual space as a *phase space* for the system under consideration. The state of the classical N-body system at any time t is completely specified by the location of *one* point in phase space. Such a point is called a *phase point*. As the system evolves in time, its dynamics is completely described by the motion or trajectory of the phase point through phase space. The trajectory of the phase point is given by Hamilton's equations of motion:

$$\dot{q}_j = \frac{\partial H}{\partial p_j} \quad \text{and} \quad \dot{p}_j = -\frac{\partial H}{\partial q_j} \qquad j = 1, 2, \ldots, l = sN \tag{7–20}$$

In principle, these $2l$ equations can be integrated to give $\{q_j(t)\}$ and $\{p_j(t)\}$. For notational simplicity, we shall denote the set of l q's by $q(t)$, and the set of l p's by $p(t)$. The $2l$ constants of integration can be fixed by the location of the phase point at some initial time, say t_0. Of course, in practice, such an integration is not feasible.

We now introduce the concept of an ensemble of systems in phase space. For simplicity we shall consider a microcanonical ensemble, that is, an ensemble representative of an isolated system. Consider a large number $\mathscr{A}$ of isolated systems, each of which having the same values of macroscopic variables N, V, and E.

The detailed classical state of each system in the ensemble has a representative phase point in the *same* phase space. The entire ensemble then appears as a cloud of points in phase space. As time evolves, each point will trace out its *independent* trajectory. The trajectories are independent, since each one represents an isolated system and is, therefore, independent of all the others. The postulate of equal a priori probabilities requires that there is a representative phase point in phase space for each and every set of coordinates and momenta consistent with the few fixed macroscopic variables. In particular, the postulate of equal a priori probabilities states that for a microcanonical ensemble, the density is *uniform* over the constant energy "surface" in phase space, where the value of the energy on the surface is that of the isolated system. We consider all parts of phase space equally important, as long as the (p, q)'s are consistent with all that we know macroscopically about the system, that is, consistent with the values of N, V, and E for the system that the ensemble represents. Just as every quantum state was equally likely before, now we consider every classical state to be equally probable.

This cloud of points is very dense then, and we can define a number density $f(p, q, t)$, such that the number of systems in the ensemble that have phase points in $dp\, dq$ about the point p, q at time t is $f(p, q, t)\, dp\, dq$. Clearly we must have

$$\int \cdots \int f(p, q)\, dp\, dq = \mathscr{A} \tag{7-21}$$

The ensemble average of any function, say $\phi(p, q)$, of the momenta and coordinates of the system is defined as

$$\bar{\phi} = \frac{1}{\mathscr{A}} \int \cdots \int \phi(p, q) f(p, q, t)\, dp\, dq \tag{7-22}$$

It is Gibbs' postulate to equate this ensemble average to the corresponding thermodynamic function. Note the similarity between this equation and Eq. (2–5), its quantum mechanical analog.

Since the equations of motion determine the trajectory of each phase point, they must also determine the density $f(p, q, t)$ at any time if the dependence of f on p and q is known at some initial time t_0. The time dependence of f is thus controlled by the laws of mechanics and is not arbitrary. The time dependence of f is given by the Liouville equation, which we now derive.

Consider the small volume element $\delta p_1 \cdots \delta p_l\, \delta q_1 \cdots \delta q_l$ about the point $p_1, \cdots p_l, q_1, \cdots q_l$. The number of phase points inside this volume at any instant is

$$\delta N = f(p_1, \ldots, p_l, q_1, \ldots, q_l, t)\, \delta p_1 \cdots \delta p_l\, \delta q_1 \cdots \delta q_l$$

This number will, in general, change with time since the natural trajectories of phase points will take them into and out of this volume element, and the number passing through any one "face" will, in general, be different from the number passing through the opposite "face." Let us calculate the number entering one face and leaving through the opposite. Consider two faces perpendicular to the q_1-axis and located at q_1 and $q_1 + \delta q_1$. The number of phase points entering the first of these faces per unit time is

$$f\dot{q}_1\, \delta q_2 \cdots \delta q_l\, \delta p_1 \cdots \delta p_l \tag{7-23}$$

(See Problem 7–33.) The number passing through the other face per unit time is

$$f(q_1 + \delta q_1, q_2, \ldots, q_l, p_1, \ldots, p_l)$$
$$\times \dot{q}_1(q_1 + \delta q_1, q_2, \ldots, q_l, p_1, \ldots, p_l)\,\delta q_2 \cdots \delta q_l\,\delta p_1 \cdots \delta p_l$$

which, if we expand f and $\dot{q}_1$ to linear terms in δq_1, gives

$$\left(f + \frac{\partial f}{\partial q_1}\delta q_1\right)\left(\dot{q}_1 + \frac{\partial \dot{q}_1}{\partial q_1}\delta q_1\right)\delta q_2 \cdots \delta q_l\,\delta p_1 \cdots \delta p_l + \cdots \tag{7-24}$$

Subtracting (7–24) from (7–23), we get the *net* flow of phase points in the q_1-direction into the volume element $\delta q_1 \cdots \delta q_l\,\delta p_1 \cdots \delta p_l$:

$$\text{net flow} = -\left(\frac{\partial f}{\partial q_1}\dot{q}_1 + f\frac{\partial \dot{q}_1}{\partial q_1}\right)\delta p_1 \cdots \delta p_l\,\delta q_1 \cdots \delta q_l$$

in the q_1-direction. In a similar manner, the net flow in the p_1-direction is (remember that momenta and spatial coordinates have equal status in phase space):

$$-\left(\frac{\partial f}{\partial p_1}\dot{p}_1 + f\frac{\partial \dot{p}_1}{\partial p_1}\right)\delta p_1 \cdots \delta p_l\,\delta q_1 \cdots \delta q_l$$

Thus the change in the number of phase points through all the faces is

$$-\sum_{j=1}^{l}\left(\frac{\partial f}{\partial q_j}\dot{q}_j + f\frac{\partial \dot{q}_j}{\partial q_j} + \frac{\partial f}{\partial p_j}\dot{p}_j + f\frac{\partial \dot{p}_j}{\partial p_j}\right)\delta p_1 \cdots \delta p_l\,\delta q_1 \cdots \delta q_l.$$

This must be equal to the change of δN with time, and so we have

$$\frac{d(\delta N)}{dt} = -\sum_{j=1}^{l}\left[f\left(\frac{\partial \dot{q}_j}{\partial q_j} + \frac{\partial \dot{p}_j}{\partial p_j}\right) + \left(\frac{\partial f}{\partial q_j}\dot{q}_j + \frac{\partial f}{\partial p_j}\dot{p}_j\right)\right]\delta p_1 \cdots \delta p_l\,\delta q_1 \cdots \delta q_l \tag{7-25}$$

This result can be immediately simplified. Since

$$\dot{q}_j = \frac{\partial H}{\partial p_j} \qquad \dot{p}_j = -\frac{\partial H}{\partial q_j} \tag{7-26}$$

the first term in parentheses in Eq. (7–25) is

$$\frac{\partial \dot{q}_j}{\partial q_j} + \frac{\partial \dot{p}_j}{\partial p_j} = 0 \tag{7-27}$$

Furthermore, we divide Eq. (7–25) by the volume element $\delta p_1 \cdots \delta p_l\,\delta q_1 \cdots \delta q_l$. This gives the rate of change in the density itself around the point $p_1, \ldots p_l, q_1, \ldots q_l$, so that we can write

$$\frac{\partial f}{\partial t} = -\sum_{j=1}^{l}\left(\frac{\partial f}{\partial q_j}\dot{q}_j + \frac{\partial f}{\partial p_j}\dot{p}_j\right) \tag{7-28}$$

where we have written $\partial f/\partial t$ to indicate that we have fixed our attention on a given stationary point in the phase space.

Equation (7–28) can be written in a more conventional form by using Eqs. (7–26) for $\dot{q}_j$ and $\dot{p}_j$. The result is

$$\frac{\partial f}{\partial t} = -\sum_{j=1}^{l}\left(\frac{\partial H}{\partial p_j}\frac{\partial f}{\partial q_j} - \frac{\partial H}{\partial q_j}\frac{\partial f}{\partial p_j}\right) \tag{7-29}$$

This is the Liouville equation, the most fundamental equation of classical statistical mechanics. In fact, it can be shown that the Liouville equation is equivalent to the $6N$ Hamiltonian equations of motion of the N-body system. See Mazo under "Additional Reading" for a proof of this. In Cartesian coordinates, the Liouville equation for N point masses is (see Problem 7–11):

$$\frac{\partial f}{\partial t} + \sum_{j=1}^{N} \frac{\mathbf{p}_j}{m_j} \cdot \nabla_{\mathbf{r}_j} f + \sum_{j=1}^{N} \mathbf{F}_j \cdot \nabla_{\mathbf{p}_j} f = 0 \tag{7-30}$$

In this equation $\nabla_{\mathbf{r}_j}$ denotes the gradient with respect to the spatial variables in f; $\nabla_{\mathbf{p}_j}$ denotes the gradient with respect to the momentum variables in f; and $\mathbf{F}_j$ is the total force on the jth particle. The Liouville equation forms the starting point of most theories of nonequilibrium statistical mechanics.

There are several interesting deductions from the Liouville equation which we now discuss. Consider Eq. (7–28)

$$\frac{\partial f}{\partial t} + \sum_{j=1}^{l} \left(\frac{\partial f}{\partial p_j}\right) \dot{p}_j + \sum_{j=1}^{l} \left(\frac{\partial f}{\partial q_j}\right) \dot{q}_j = 0 \tag{7-31}$$

Since $f = f(p, q, t)$, this equation is equivalent to

$$\frac{df}{dt} = 0 \tag{7-32}$$

Physically, this equation says that the density in the neighborhood of any selected moving phase point is a constant along the trajectory of that phase point. Thus the cloud of phase points behaves as an incompressable fluid. Gibbs called this the principle of the conservation of density in phase. An equivalent statement of this is that if p, q are the coordinates of a phase point at time t, which at time t_0 were (p_0, q_0), then Liouville's equation implies that (see Problem 7–12)

$$f(p, q; t) = f(p_0, q_0; t_0) \tag{7-33}$$

Because of the equations of motions, the point (p, q) should be considered a function of the initial point (p_0, q_0) and the elapsed time t. That is

$$p = p(p_0, q_0; t)$$
$$q = q(p_0, q_0; t)$$

Now let us select a small element of volume at (p_0, q_0) at time t_0. At a later time, $t_0 + t$, the phase points originally on the surface of this volume element will have formed a new surface enclosing a volume element of different shape at the phase point (p, q). The volume element at (p, q) must contain the same number of phase points as the original volume element at (p_0, q_0). This follows because a phase point outside or inside the volume element can never cross the surface as the element moves through phase space, for otherwise there would be two different trajectories through the same point in phase space. This is impossible, however, because of the uniqueness of the equations of motion of a phase point. Trajectories of phase points can never cross. Now since the density and number of phase points in the volume element are the same at p_0, q_0 and p, q, it follows that although the shape of this volume may change and contort itself as it moves through phase space, its volume remains constant. Gibbs called this result *conservation of extension in phase space*. This fact is expressed mathematically by writing

$$\delta p \, \delta q = \delta p_0 \, \delta q_0 \qquad \text{for all } t \tag{7-34}$$

Another way of expressing this is to say that the Jacobian of the set (p, q) to (p_0, q_0) is unity. This can be proved directly from the equations of motion of the system. See Mazo under "Additional Reading."

A corollary of this theorem, whose proof demands a more extensive knowledge of classical mechanics, is that if we are given two sets of coordinates and their conjugate momenta, say,

$$q_1, q_2, \ldots, q_{3n}, p_1, p_2, \ldots, p_{3n}$$
$$Q_1, Q_2, \ldots, Q_{3n}, P_1, P_2, \ldots, P_{3n}$$

which can describe a system in phase space equally well, then

$$dq_1 \, dq_2 \cdots dq_{3n} \, dp_1 \cdots dp_{3n} = dQ_1 \cdots dQ_{3n} \, dP_1 \cdots dP_{3n}$$

For example, a single particle in three dimensions may be described by the coordinates (x, y, z) or the spherical coordinates (r, θ, ϕ). It is straightforward, albeit lengthy to show that

$$dp_x \, dp_y \, dp_z \, dx \, dy \, dz = dp_r \, dp_\theta \, dp_\phi \, dr \, d\theta \, d\phi \tag{7-35}$$

Notice that although the volume elements in ordinary coordinate space are $dx \, dy \, dz$ and $r^2 \sin \theta \, dr \, d\theta \, d\phi$, the $r^2 \sin \theta$ factor does not occur in the phase space transformation. These simple volume element transformations would not generally be true if we have chosen the generalized coordinates and velocities instead of momenta This is one reason why momenta and not velocities are used to describe classical systems.

7–3 EQUIPARTITION OF ENERGY

We have seen that classical statistical mechanics is applicable when the temperature is high enough to replace the quantum statistical summation by an integral. Under these conditions, it is not necessary to know the eigenvalues of the quantum mechanical problem, only the classical Hamiltonian is required. There is an interesting theorem of classical statistical mechanics which can be used to understand more fully some of the results of the last two chapters.

Consider the expression for the average energy of a molecule in a system of independent molecules,

$$\bar{\varepsilon} = \frac{\iint H e^{-\beta H} \, dp_1 \cdots dq_s}{\iint e^{-\beta H} \, dp_1 \cdots dq_s} \tag{7-36}$$

which can be evaluated in principle for any known dependence of H on the p's and the q's. Multiplying by the total number of molecules gives an expression for the total energy of the system, and by differentiating with respect to T, we obtain an expression for its heat capacity at constant volume.

If it so happens that the Hamiltonian is of the form

$$H(p_1, p_2, \ldots, q_s) = \sum_{j=1}^{m} a_j p_j^2 + \sum_{j=1}^{n} b_j q_j^2 + H(p_{m+1}, \ldots, p_s, q_{n+1}, \ldots, q_s) \tag{7-37}$$

where the a_j and b_j are constants, then it is easy to show that each of these quadratic terms will contribute $kT/2$ to the energy and $k/2$ to the heat capacity. (See Problem 7–29.) This result is called the principle of equipartition of energy. It should be emphasized that the principle is a consequence of the quadratic form of terms in the Hamiltonian, rather than a general consequence of classical statistical mechanics.

Let us apply this general theorem to some of the cases we have treated in Chapters 5 and 6. For instance, for a monatomic ideal gas, the Hamiltonian is

$$H = \frac{p_x^2 + p_y^2 + p_z^2}{2m} \tag{7-38}$$

Since there are three quadratic terms, each atom contributes $3kT/2$ to the total energy and so $3k/2$ to the constant volume heat capacity. This is exactly our result in Chapter 5. For the case of a rigid rotor, the Hamiltonian is

$$H = \frac{1}{2I}\left(p_\theta^2 + \frac{p_\phi^2}{\sin^2\theta}\right)$$

The $\sin^2\theta$ in the p_ϕ^2 term would seem to exclude the p_ϕ^2 term from the principle of equipartition, since Eq. (7-37) requires that the coefficients a_j be constants. There is a more general version, however, that allows the a_j and b_j to be functions of the momenta and coordinates not involved in the quadratic terms, that is, to be functions of $p_{m+1}, \ldots, p_s$ and $q_{n+1}, \ldots, q_s$ in Eq. (7-37). The proof of this is more difficult than the proof of the simpler version. (See either Problem 7-30 or Tolman under "Additional Reading.") Because of this, each quadratic term above still contributes its equipartition value, and so the rotational contribution of a rigid rotor to the energy is kT per molecule, just as we obtained in Chapter 6 [cf. Eq. (6-36)].

Note that equipartition is a classical concept, that is, the degree of freedom contributing must be such that $\Delta\varepsilon/kT$ is small in passing from one level to another. We have seen that this is true for translational and rotational degrees of freedom at ordinary temperatures, but not vibrational degrees of freedom. The heat capacity for an ideal diatomic gas in the rigid rotor–harmonic oscillator approximation is [cf. Eq. (6-53)]

$$C_V = \tfrac{5}{2}Nk + \frac{Nk(\Theta_v/T)^2 e^{\Theta v/T}}{(e^{\Theta v/T} - 1)^2} \tag{7-39}$$

where the $\tfrac{5}{2}Nk$ comes from the translational plus rotational degrees of freedom which, we have seen, are excited enough to be treated classically. The second term is the vibrational contribution, which reaches its expected classical limit of Nk, since the classical Hamiltonian for a harmonic oscillator is $(p^2/2m) + (k/2)x^2$, when Θ_v/T becomes small, which is far above room temperature for most molecules. A value of the vibrational contribution to C_V differing from Nk is thus a quantum mechanical result.

There are more general formulations of the principle of equipartition of energy than we have given here, but they are not necessary for most purposes. In fact, the principle itself is perhaps more of historical interest today than actual practical interest. It is interesting to note in this regard that when the electronic structure of atoms and metals evolved toward the end of the nineteenth century, it was of great concern to Gibbs that the electrons contributed only a very small fraction of their equipartition value to the heat capacities of metals. He did not live to see this anomalous result completely explained by quantum statistics. Since electrons have such a small mass, they behave not at all classically and should, therefore, not be governed by the equipartition of energy (cf. Section 10-2).

We have made this long detour through phase space for more than just historical reasons. As we said earlier, most of the systems of interest to chemists can be treated very satisfactorily by classical methods. In fact, the quantum statistical theories of systems of interacting particles are quite a demanding and specialized subject whose techniques are still being developed. Fortunately, being chemists, we are spared from having to master these techniques. Even today the classical Liouville equation forms the starting point for most of the rigorous approaches to nonequilibrium statistical mechanics. We shall now discuss the problem that sent us here in the first place, namely, the study of ideal polyatomic gases.

ADDITIONAL READING

General

EYRING, H., HENDERSON, D., STOVER, B. J., and EYRING, E. M. 1964. *Statistical mechanics and dynamics.* New York: Wiley. Chapter 7.

GIBBS, J. W. 1960. *Elementary principles in statistical mechanics.* New York: Dover.

HILL, T. L. 1956. *Statistical mechanics.* New York: McGraw-Hill. Chapter 1.

HUANG, K. 1963. *Statistical mechanics.* New York: Wiley. Chapter 7.

KHINCHIN, A. I. 1949. *Mathematical foundations of statistical mechanics.* New York: Dover.

KILPATRICK, J. E. 1967. In *Physical chemistry, an advanced treatise,* Vol. II, ed. by H. Eyring, D. Henderson, and W. Jost. New York: Academic.

KUBO, R. 1965. *Statistical mechanics.* Amsterdam: North-Holland Publishing Co. Sections 1–1 to 1–14.

MAZO, R. M. 1967. *Statistical mechanical theories of transport processes.* New York: Pergamon. Chapter 7.

MÜNSTER, A. 1969. *Statistical thermodynamics,* Vol. I. Berlin: Springer-Verlag. Chapter 1.

RUSHBROOKE, G. S. 1949. *Statistical mechanics.* London: Oxford University Press. Chapter 4.

TER HAAR, D. 1966. *Elements of thermostatistics.* London: Oxford University Press. Chapter 5.

———. 1955. *Rev. Mod. Phys.* **27**, p. 289, 1955.

TOLMAN, R. C. 1938. *Statistical mechanics.* London: Oxford University Press. Chapter 3.

PROBLEMS

7–1. Show that at room temperature the translational quantum numbers are typically around 10^8 or so.

7–2. What is the constant energy surface in phase space for a simple linear harmonic oscillator? What is it for a single-point mass? What is it for an ideal gas of N-point masses?

7–3. Convince yourself that trajectories in phase space can never cross, also that surfaces (really hypersurfaces) of constant energy can never intersect if the energies are different.

7–4. Consider a classical ideal gas enclosed in an infinitely tall cylinder in a gravitational field. Assuming that the temperature is uniform up the cylinder, derive the barometric formula

$$p(z) = p(0)\exp\left(\frac{-mgz}{kT}\right)$$

From this calculate the atmospheric pressure at the top of Mt. Everest.

7–5. An ideal gas consisting of N particles of mass m is enclosed in an infinitely tall cylindrical container placed in a uniform gravitational field, and is in thermal equilibrium. Calculate the classical partition function, Helmholtz free energy, mean energy, and heat capacity of this system.

7–6. Consider a perfect gas of molecules with permanent electric dipole moments μ in an electric field $\mathscr{E}$. Neglecting the polarizability of the molecules, the potential energy is

$$U = -\mu\mathscr{E}\cos\theta$$

where θ is the angle between μ and $\mathscr{E}$. Using classical mechanics, derive an expression for the additional effect of $\mathscr{E}$ on the energy E and heat capacity of the gas.

7–7. The potential energy of N molecules in a container V can often be fairly well approximated by a sum of pair-wise potentials:

$$U(\mathbf{r}_1, \ldots, \mathbf{r}_N) = \sum_{i<j} u(\mathbf{r}_i, \mathbf{r}_j)$$

In addition, the pair-wise potentials $u(\mathbf{r}_i, \mathbf{r}_j)$ are often assumed to depend only upon the distance $r_{ij} = |\mathbf{r}_i - \mathbf{r}_j|$ between the two molecules. Thus one often writes

$$U(\mathbf{r}_1, \ldots, \mathbf{r}_N) = \sum_{i<j} u(r_{ij})$$

Convince yourself that even these two simplifications of U do not help in trying to evaluate the configuration integral Z.

7–8. It is possible to determine the value of Boltzmann's constant by observing the distribution of suspended Brownian particles in a gravitational field as a function of their height z. Given that the particles have a mass of 1.0×10^{-14} g, that the temperature is 300°K, and the following data:

z(cm)	Number of particles
0.0000	100
0.0025	55
0.0050	31
0.0075	17
0.0100	9

calculate the value of the Boltzmann constant.

7–9. We can calculate the microcanonical ensemble partition function for a classical monatomic ideal gas in the following way. This partition function is given by

$$\Omega(E, \Delta E) = \frac{1}{N!h^{3N}} \int \cdots \int^* dp_1 \, dp_2 \cdots dq_{3N}$$

where the asterisk indicates that one integrates over the region of phase space such that

$$E - \Delta E \leq \frac{1}{2m} \sum_{j=1}^{3N} p_j^2 \leq E$$

We have seen in the quantum mechanical case that the thermodynamic consequences of this equation are remarkably insensitive to the value of ΔE. (See Problem 3–14.) We can find Ω most readily by first evaluating

$$I(E) = \frac{1}{N!h^{3N}} \int \cdots \int^* dp_1 \, dp_2 \cdots dq_{3N}$$

where now the asterisk signifies the constraint

$$0 \leq \frac{1}{2m} \sum_{j=1}^{3N} p_j^2 \leq E$$

Note that $\Omega(E, \Delta E)$ is given by $I(E) - I(E - \Delta E)$. The integration of $dq_1 \cdots dq_{3N}$ in $I(E)$ immediately gives V^N, and the remaining integration over the momenta is just the volume of a $3N$-dimensional sphere of radius $(2mE)^{1/2}$. The volume of a $3N$-dimensional sphere of radius R is (see Problem 1–24)

$$\frac{\pi^{3N/2}}{(3N/2)!} R^{3N}$$

(Note that this reduces correctly when $3N = 2$ and 3.) Using this formula then, show that

$$I(E) = \frac{\pi^{3N/2} V^N (2mE)^{3N/2}}{N! \, h^{3N} (3N/2)!}$$

is in agreement with Eq. (1–36).

7–10. In Problem 3–14 we showed that the entropy could be calculated from $k \ln \Omega(E)\Delta E$ or $k \ln \Phi(E)$, where $\Omega(E)\Delta E$ is the number of states with energies between E and $E + \Delta E$, and $\Phi(E)$ is the total number of states with energies less than E. In addition to this, we showed that the result is remarkably insensitive to the choice of ΔE. We shall now discuss the classical analog of this. In particular, this problem involves showing that the volume of an N-dimensional sphere is essentially the same as the volume of the hypershell of thickness s. First write the volume of the hypersphere as

$$V_{\text{sphere}}(R) = \text{const} \times R^N$$

Now show that if N is large enough such that $sN \gg R$, then

$$\begin{aligned} V_{\text{shell}} &= V(R) - V(R - s) \\ &= \text{const} \times R^N (1 - e^{-sN/R}) \\ &\approx V_{\text{sphere}} \end{aligned}$$

7–11. Show that in Cartesian coordinates, the Liouville equation takes the form of Eq. (7–30).

7–12. Convince yourself that a corollary of Liouville's equation is

$$f(p, q; t) = f(p_0, q_0, t_0)$$

Although we did not discuss it explicitly, much of the kinetic theory of gases is contained in this chapter. Problems 7–13 through 7–25 develop some of the kinetic theory of gases.

7–13. Consider a system of N interacting molecules, whose vibrational degrees of freedom are treated quantum mechanically and whose translational and rotational degrees of freedom are treated classically with Hamiltonian

$$H_{\text{class}} = K_{\text{trans}} + K_{\text{rot}} + U$$

where K represents kinetic energy, and U represents potential energy. Substitute this into Eq. (7–19); integrate over all the coordinates except the $3N$ translational momentum coordinates; and derive

$$\text{prob}\{K_{\text{trans}}\} = \frac{e^{-K_{\text{trans}}/kT} \, dp_{\text{trans}}}{\int e^{-K_{\text{trans}}/kT} \, dp_{\text{trans}}}$$

Now realize that

$$K_{\text{trans}} = \sum_{j=1}^{N} \frac{1}{2m} (p_{xj}^2 + p_{yj}^2 + p_{zj}^2)$$

and derive the normalized Maxwell-Boltzmann distribution, namely,

$$f(p_x, p_y, p_z) \, dp_x \, dp_y \, dp_z = (2\pi mkT)^{-3/2} \, e^{-(p_x^2 + p_y^2 + p_z^2)/2mkT} \, dp_x \, dp_y \, dp_z \tag{7–40}$$

One can derive all of the usual expressions of the kinetic theory of gases from this.

7–14. An integral that appears often in statistical mechanics and particularly in the kinetic theory of gases is

$$I_n = \int_0^\infty x^n e^{-ax^2} \, dx$$

This integral can be readily generated from two basic integrals. For even values of n, we first consider

$$I_0 = \int_0^\infty e^{-ax^2} \, dx$$

The standard trick to evaluate this integral is to square it, and then transform the variables into polar coordinates:

$$I_0^2 = \int_0^\infty \int_0^\infty e^{-ax^2} e^{-ay^2} \, dx \, dy$$

$$= \int_0^\infty \int_0^{\pi/2} e^{-ar^2} r \, dr \, d\theta$$

$$= \frac{\pi}{4a}$$

$$I_0 = \frac{1}{2} \left(\frac{\pi}{a}\right)^{1/2}$$

Using this result, show that for even n

$$I_n = \frac{1 \cdot 3 \cdot 5 \cdots (n-1)}{2(2a)^{n/2}} \left(\frac{\pi}{a}\right)^{1/2} \qquad n \text{ even}$$

For odd values of n, the basic integral I_1 is easy. Using I_1, show that

$$I_n = \frac{\Gamma\left(\dfrac{n+1}{2}\right)}{2a^{(n+1)/2}} \qquad n \text{ odd}$$

7–15. Convert Eq. (7–40) (see Problem 7–13) from a Cartesian coordinate to a spherical coordinate representation by writing

$$p^2 = p_x^2 + p_y^2 + p_z^2$$
$$p_z = p \cos \theta$$
$$p_x = p \sin \theta \cos \phi$$
$$p_y = p \sin \theta \sin \phi$$
$$dp_x \, dp_y \, dp_z \to p^2 \sin \theta \, dp \, d\theta \, d\phi$$

and integrating over θ and ϕ to get

$$f(p) \, dp = 4\pi (2\pi m k T)^{-3/2} p^2 \, e^{-p^2/2mkT} \, dp$$

for the fraction of molecules with momentum between p and $p + dp$. By substituting $p = mv$, we get the fraction of molecules with *speeds* between v and $v + dv$:

$$f(v) \, dv = 4\pi \left(\frac{m}{2\pi k T}\right)^{3/2} v^2 e^{-mv^2/2kT} \, dv$$

7–16. Prove that the most probable molecular speed is $v_{mp} = (2kT/m)^{1/2}$, that the mean speed is $\langle v \rangle = (8kT/\pi m)^{1/2}$, and that the root-mean-square speed is $\langle v^2 \rangle^{1/2} = (3kT/m)^{1/2}$. Evaluate these for H_2 and N_2 at 25°C.

7–17. Show that the mean-square fluctuation of the velocity of the Maxwell-Boltzmann distribution is

$$\overline{v^2} - \bar{v}^2 = \frac{kT}{m} \left(3 - \frac{8}{\pi}\right)$$

7–18. Show that the average velocity in any direction (say x, y, or z) vanishes. What does this mean?

7–19. Derive an expression for the fraction of molecules with translational energy between ε and $\varepsilon + d\varepsilon$ from both Eqs. (7–40) and (5–7).

7–20. According to Problem 1–35, the speed of sound in an ideal gas is given by

$$c_0 = \left(\gamma \frac{RT}{M}\right)^{1/2}$$

where M is the molecular weight of the gas, and $\gamma = C_p/C_V$. Show that $c_0 = 0.81\bar{v}$ for an ideal monatomic gas.

7–21. Calculate the probability that two molecules will have a total kinetic energy between ε and $\varepsilon + d\varepsilon$.

7–22. Calculate the fraction of molecules with x-component of velocity between $\pm n(2kT/m)^{1/2}$, where $n = 1$, 2, and 3. Remember that the integral of $\exp(-x^2)$ with finite limits cannot be evaluated in closed form and is expressed in terms of the error function $\text{erf}(x)$ by

$$\text{erf}(x) = \frac{2}{\pi^{1/2}} \int_0^x e^{-t^2}\, dt$$

7–23. What is the average kinetic energy $\bar{\varepsilon}$ and the most probable kinetic energy ε_{mp} of a gas molecule?

7–24. Show that the number of molecules striking a unit area per unit time is $\rho\bar{v}/4$, where $\rho = N/V$.

7–25. How would you interpret the velocity distribution

$$\phi(\mathbf{v}) = \left(\frac{m}{2\pi kT}\right)^{3/2} \exp\left\{-\frac{m}{2kT}\left[(v_x - a)^2 + (v_y - b)^2 + (v_z - c)^2\right]\right\}$$

in which a, b, and c are constants?

7–26. The relativistic dependence of the kinetic energy on momentum is

$$\varepsilon = c(p_x^2 + p_y^2 + p_z^2 + m_0^2 c^2)^{1/2}$$

where m_0 is the rest mass of the particle, and c is the speed of light. Determine the thermodynamic properties of an ideal gas in the extreme relativistic limit, where $p \gg m_0 c$.

7–27. If an atom is radiating light of wavelength λ_0, the wavelength measured by an observer will be

$$\lambda = \lambda_0\left(1 + \frac{v_z}{c}\right)$$

if moving away from or toward the observer with velocity v_z. In this equation c is the speed of light. This is known as the Doppler effect. If one observes the radiation emitted from a gas at temperature T, it is found that the line at λ_0 will be spread out by the Maxwellian distribution of velocities v_z of the molecules emitting the radiation. Show that $I(\lambda)\, d\lambda$, the intensity of radiation observed between wavelengths λ and $\lambda + d\lambda$, is

$$I(\lambda) \propto \exp\left\{-\frac{mc^2(\lambda - \lambda_0)^2}{2\lambda_0^2 kT}\right\}$$

This spreading about the line at λ_0 is known as Doppler broadening. Estimate the Doppler line width for HCl radiating microwave radiation at room temperature.

7–28. Plot C_V in Eq. (7–39) versus temperature and see that the vibrational contribution does not contribute until the temperature approaches Θ_V.

7–29. Prove that if the Hamiltonian is given by Eq. (7–37), then each of the quadratic terms will contribute $kT/2$ to the average molecular energy and $k/2$ to the molecular heat capacity.

7–30. Prove that even if the a_j and b_j in the Hamiltonian of Eq. (7–37) are functions of the momenta and coordinates not involved in the quadratic terms, the law of equipartition still applies. In particular, show how this more general version of the law of equipartition applies to the rigid rotor Hamiltonian.

7–31. Let $H(p, q)$ be the classical Hamiltonian for a classical system of N interacting particles. Let x_j be one of the $3N$ momentum components or one of the $3N$ spatial coordinates. Prove the generalized equipartition theorem, namely, that

$$\left\langle x_i \frac{\partial H}{\partial x_j} \right\rangle = kT\delta_{ij}$$

and from this derive the principle of equipartition of energy that we discussed earlier. Hint: Realize that the potential $U \to \infty$ at the walls of the container.

7–32. Consider a two-dimensional harmonic oscillator with Hamiltonian

$$H = \frac{1}{2m}(p_x^2 + p_y^2) + \frac{k}{2}(x^2 + y^2)$$

According to the principle of equipartition of energy, the average energy will be $2kT$. Now transform this Hamiltonian to plane polar coordinates to get

$$H = \frac{1}{2m}\left(p_r^2 + \frac{p_\theta^2}{r^2}\right) + \frac{k}{2}r^2$$

What would you predict for the average energy now? Show by direct integration in plane polar coordinates that $\bar\varepsilon = 2kT$. Is anything wrong here? Why not?

7–33. Convince yourself that the number of phase points passing through a face perpendicular to q_1 per unit time is

$$f\dot{q}_1\, \delta q_2\, \delta q_3 \cdots \delta p_l$$

CHAPTER 8

IDEAL POLYATOMIC GAS

In this chapter we make a direct extension to polyatomic molecules of the methods and approximations used in the previous chapters for diatomic molecules and classical statistical mechanics. The introduction of the concept of normal coordinates will allow us to treat the vibrational problem of polyatomic molecules as a simple extension of the vibration of diatomic molecules (Section 8–1). We shall see, however, that the rotational problem must be treated by the method of classical statistical mechanics since, for most polyatomic molecules, the quantum mechanical rotational energy levels cannot be written in any convenient closed form, and most temperatures are such that classical statistical mechanics is applicable anyway (Section 8–2). One new feature that arises with polyatomic molecules is hindered internal rotation (in molecules such as ethane). This type of consideration is important in organic chemistry, where the results can be directly applied to a determination of the various conformers that exist in certain molecules. Hindered rotation is treated in Section 8–4.

The discussion in Section 6–1 for diatomic molecules applies equally well to polyatomic molecules. After making the Born-Oppenheimer approximation, we transform the Schrödinger equation that describes the motion of the nuclei in the Born-Oppenheimer potential into center of mass and relative coordinates. This allows us to write

$$\mathcal{H} = \mathcal{H}_{trans} + \mathcal{H}_{int}$$

$$\varepsilon = \varepsilon_{trans} + \varepsilon_{int}$$

$$q = q_{trans} q_{int}$$

where

$$q_{trans} = \left[\frac{2\pi M k T}{h^2}\right]^{3/2} V \tag{8-1}$$

and M is the total mass of the molecule. As before, the density of translational energy states alone is sufficient to guarantee that the number of energy states available to any molecule is much greater than the number of molecules in the system, and so

$$Q(N, V, T) = \frac{q_{\text{trans}}^N q_{\text{int}}^N}{N!}$$

The Born-Oppenheimer potential, that is, the potential set up by the electrons moving in the field of the fixed nuclei, depends upon a number of internuclear distances. Each of the n atoms in a polyatomic molecule requires three coordinates to locate it. Thus it takes $3n$ coordinates to specify the polyatomic molecule itself. Of these $3n$ coordinates, 3 are needed to specify the center of mass of the molecule. In addition, it requires two coordinates (θ and ϕ) to specify its orientation if it is linear, and three coordinates if it is nonlinear. The remaining $3n - 5$ or $3n - 6$ coordinates are internal coordinates that are required to specify the *relative* location of the n nuclei. Thus the Born-Oppenheimer potential of a polyatomic molecule depends upon $3n - 5$ (linear) or $3n - 6$ (nonlinear) coordinates. Since these coordinates specify the relative locations of the n nuclei, they are referred to as vibrational degrees of freedom. Similarly, since the two (linear) or three (nonlinear) orientation angles are used to specify the orientation of the molecule about its center of mass, they represent rotational degrees of freedom. The three coordinates used to locate the center of mass of the molecule are translational degrees of freedom.

As in the case of diatomic molecules, we use a rigid rotor-harmonic oscillator approximation. This allows us to separate the rotational motion from the vibrational motion of the molecule, and we can treat each one separately. Both problems are somewhat more complicated for polyatomic molecules than for diatomic molecules. Nevertheless, we can write the polyatomic analog of Eqs. (6–14) to (6–17):

$$Q(N, V, T) = \frac{(q_{\text{trans}} q_{\text{rot}} q_{\text{vib}} q_{\text{elec}} q_{\text{nucl}})^N}{N!} \tag{8–2}$$

We choose as the zero of energy all n atoms completely separated in their ground electronic states. Thus the energy of the ground electronic state is $-D_e$. As before then, the electronic partition function is

$$q_{\text{elect}} = \omega_{e1} e^{D_e/kT} + \cdots \tag{8–3}$$

and we set $q_{\text{nucl}} = 1$.

To calculate $Q(N, V, T)$ then, we must investigate q_{vib} and q_{rot}, and this is done in the next two sections.

8–1 THE VIBRATIONAL PARTITION FUNCTION

For a polyatomic molecule, the potential in which the nuclei vibrate is a function of $3n - 5$ or $3n - 6$ relative coordinates. This potential is a generalization of Fig. 6–1 to $3n - 5$ or $3n - 6$ dimensions, and thus is a complicated energy surface. As in the case of diatomic molecules, however, the amplitude of the nuclear vibrations is very small, and we can expand the potential function about its stable or equilibrium configuration. Let the Cartesian coordinates of each nucleus in the molecule be $x_1, y_1, z_1, x_2, \ldots,$ x_n, y_n, z_n. For small vibrations about the equilibrium configuration, we have a generalization of Eq. (6–3) to $3n - 5$ or $3n - 6$ relative coordinates, and the potential energy will be a quadratic function of $3n - 5$ or $3n - 6$ relative coordinates such as $x_2 - x_1, x_3 - x_1, y_2 - y_1,$ and so on. When terms such as $(x_2 - x_1)^2$ are multiplied out, there result cross terms of the form $x_1 x_2$, $x_1 x_3$, $y_1 y_2$, and so on. The presence of such cross terms makes the potential energy and hence the Hamiltonian a complicated mixture of x_1, x_2, ... By the introduction of certain linear combinations of the Cartesian coordinates x_1, x_2, ..., it is possible to eliminate the cross terms in the potential and to write it as a sum of squares of these new coordinates, say Q_1, Q_2, ... In other words, it is possible to transform a Hamiltonian which contains cross terms in terms of x_1, x_2, ..., y_1, y_2, ..., to a Hamiltonian, which when written in terms of Q_1, Q_2, ..., becomes

$$\mathscr{H} = -\sum_{j=1}^{\alpha} \frac{\hbar^2}{2\mu_j} \frac{\partial^2}{\partial Q_j{}^2} + \sum_{j=1}^{\alpha} \frac{k_j}{2} Q_j{}^2 \tag{8–4}$$

In this equation α is the number of vibrational degrees of freedom, that is, $3n - 5$ for a linear molecule and $3n - 6$ for a nonlinear molecule, and μ_j and k_j are effective reduced masses and force constants.

Equation (8–4) is the Hamiltonian of a sum of *independent* harmonic oscillators, and so the total energy is of the form

$$\varepsilon = \sum_{j=1}^{\alpha} (n_j + \tfrac{1}{2}) h v_j \qquad n_j = 0, 1, 2, \ldots \tag{8–5}$$

where

$$v_j = \frac{1}{2\pi} \left(\frac{k_j}{\mu_j} \right)^{1/2}$$

There are straightforward methods to determine the Q_j's that allow the Hamiltonian to be written as a sum of independent terms as in Eq. (8–4). The Q_j's are called normal coordinates, and their determination for any particular molecule is called a normal coordinate analysis. The α fundamental frequencies v_j are obtained automatically in a normal coordinate analysis, but in practice they are usually determined spectroscopically. The normal coordinates for a linear triatomic molecule (CO_2) are as shown in the following figure.

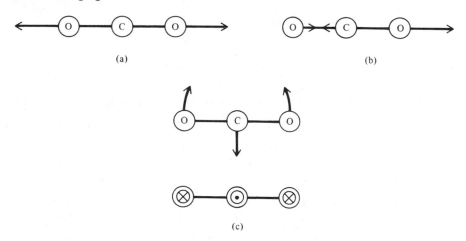

(a)   (b)

(c)

The mode labeled (a) is a symmetric stretch; the one labeled (b) is an asymmetric stretch; and the one labeled (c) is a bending mode. This mode is doubly degenerate, with one of the modes being in the plane of the page and the other being perpendicular to the page.

For a nonlinear triatomic molecule such as water, we have the following three modes.

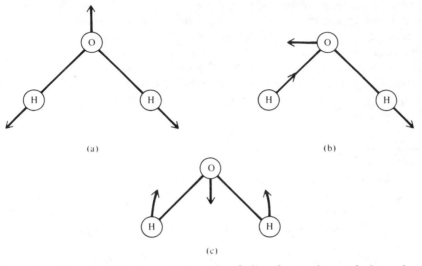

(a)   (b)

(c)

Because of Eq. (8–4), each normal mode of vibration makes an independent contribution to thermodynamic functions such as E, C_V, S, and so on. If the normal frequencies ν_1, ν_2, ..., ν_α, where $\alpha = 3n - 5$ or $3n - 6$, are known, we have immediately [see Eqs. (6–22) and (6–23)] equations such as

$$q_{\text{vib}} = \prod_{j=1}^{\alpha} \frac{e^{-\Theta_{vj}/2T}}{(1 - e^{-\Theta_{vj}/T})} \tag{8–6}$$

$$E_{\text{vib}} = Nk \sum_{j=1}^{\alpha} \left(\frac{\Theta_{vj}}{2} + \frac{\Theta_{vj} e^{-\Theta_{vj}/T}}{1 - e^{-\Theta_{vj}/T}} \right) \tag{8–7}$$

$$C_{V,\text{vib}} = Nk \sum_{j=1}^{\alpha} \left[\left(\frac{\Theta_{vj}}{T} \right)^2 \frac{e^{-\Theta_{vj}/T}}{(1 - e^{-\Theta_{vj}/T})^2} \right] \tag{8–8}$$

where

$$\Theta_{vj} = \frac{h\nu_j}{k}$$

Table 8–1 contains values of Θ_{vj} for a number of molecules.

Table 8–1. **Values of the characteristic rotational temperatures, the characteristic vibrational temperatures, and D_0 for polyatomic molecules***

molecule	$\Theta_{\text{rot}}(°K)$			$\Theta_{\text{vib}}(°K)$	D_0(kcal/mole)
CO_2		0.561		3360, 954(2), 1890	381.5
H_2O	40.1	20.9	13.4	5360, 5160, 2290	219.3
NH_3	13.6	13.6	8.92	4800, 1360, 4880(2), 2330(2)	276.8
ClO_2	2.50	0.478	0.400	1360, 640, 1600	90.4
SO_2	2.92	0.495	0.422	1660, 750, 1960	254.0
N_2O		0.603		3200, 850(2), 1840	263.8
NO_2	11.5	0.624	0.590	1900, 1980, 2330	221.8
CH_4	7.54	7.54	7.54	4170, 2180(2), 4320(3), 1870(3)	392.1
CH_3Cl	7.32	0.637	0.637	4270, 1950, 1050, 4380(2), 2140(2), 1460(2)	370.7
CCl_4	0.0823	0.0823	0.0823	660, 310(2), 1120(3), 450(3)	308.8

* These parameters were obtained from a variety of sources and do not necessarily represent the most accurate values since they are obtained under the rigid rotor–harmonic oscillator approximation.

8-2 THE ROTATIONAL PARTITION FUNCTION

The rotational properties of a polyatomic molecule depend upon the general shape of the molecule. If the molecule is linear, such as CO_2 and C_2H_2, the problem is exactly the same as for a diatomic molecule. The energy levels are given by

$$\varepsilon_J = \frac{J(J+1)h^2}{8\pi^2 I} \qquad J = 0, 1, 2, \ldots$$

$$\omega_J = 2J + 1 \tag{8-9}$$

where, in this case, the moment of inertia I is

$$I = \sum_{j=1}^{n} m_j d_j^2 \tag{8-10}$$

where d_j is the distance of the jth nucleus from the center of mass of the molecule. Recall that the coordinates of the center of mass of a molecule are given by

$$x_{cm} = \frac{1}{M} \sum_{j=1}^{n} m_j x_j$$

$$y_{cm} = \frac{1}{M} \sum_{j=1}^{n} m_j y_j$$

$$z_{cm} = \frac{1}{M} \sum_{j=1}^{n} m_j z_j \tag{8-11}$$

where x_j, y_j, and z_j are the Cartesian coordinates of the jth nucleus in an arbitrary coordinate system, and $M = m_1 + m_2 + \cdots + m_n$. (See Problem 8-1.)

The rotational partition function of a linear polyatomic molecule is

$$q_{\text{rot}} = \frac{8\pi^2 IkT}{\sigma h^2} = \frac{T}{\sigma \Theta_r} \tag{8-12}$$

As before, we have introduced a symmetry number, which is unity for unsymmetrical molecules such as N_2O and COS and equal to two for symmetrical molecules such as CO_2 and C_2H_2. The symmetry number is the number of different ways the molecule can be rotated into a configuration indistinguishable from the original. Classically, it is a factor introduced to avoid over counting indistinguishable configurations in phase space. Table 8-1 gives Θ_r for several linear polyatomic molecules.

The energy calculated from Eq. (8-12) is kT, in accord with equipartition of energy, since there are two degrees of rotational freedom of a linear molecule.

The moment of inertia is a fundamental property of rigid bodies. The rotational properties of a rigid body are characterized by the *principal moments* of the body, which are defined in the following way. Choose any set of Cartesian axes with origin at the center of mass of the body. The moments of inertia about these three axes are

$$I_{xx} = \sum_{j=1}^{n} m_j[(y_j - y_{cm})^2 + (z_j - z_{cm})^2]$$

$$I_{yy} = \sum_{j=1}^{n} m_j[(x_j - x_{cm})^2 + (z_y - z_{cm})^2]$$

$$I_{zz} = \sum_{j=1}^{n} m_j[(x_j - x_{cm})^2 + (y_j - y_{cm})^2]$$

In addition to these, there are also products of inertia, such as

$$I_{xy} = \sum_{j=1}^{n} m_j(x_j - x_{cm})(y_j - y_{cm}) \cdots$$

Now there is a theorem of rigid body motion that says that there always exists a particular set of Cartesian coordinates X, Y, Z, called the principal axes, passing through the center of mass of the body such that all the products of inertia vanish. The moments of inertia about these axes I_{XX}, I_{YY}, and I_{ZZ} are called the principal moments of inertia. The principal moments of inertia are customarily denoted by I_A, I_B, and I_C.

If the molecule possesses any degree of symmetry, the principal axes are simple to find. For example, if the molecule is planar, one of the principal axes will be perpendicular to the plane. Usually an axis of symmetry of the molecule will be a principal axis. The C–H bond of $CHCl_3$ is a three-fold axis of symmetry and also a principal axis. In general, however, it is not often necessary to calculate the principal moments of inertia, since there are extensive tables in the literature. They are usually given in terms of rotational constants in units of cm^{-1}, defined by

$$\bar{A} = \frac{h}{8\pi I_A c}$$

$$\bar{B} = \frac{h}{8\pi I_B c}$$

$$\bar{C} = \frac{h}{8\pi I_C c} \tag{8–13}$$

from which it is an easy matter to calculate the corresponding rotational temperatures Θ_A, Θ_B, and Θ_C. In these quantities, c is the speed of light. Table 8–1 contains values of Θ_A, Θ_B, and Θ_C for a number of polyatomic molecules.

The relative magnitudes of the three principal moments of inertia are used to characterize the rigid body. If all three are equal, the body is called a spherical top; if only two are equal, the body is called a symmetric top; and if all three are different, the body is called an asymmetric top. Table 8–1 shows that CH_4 and CCl_4 are spherical tops; CH_3Cl and NH_3 are symmetric tops; and H_2O and NO_2 are asymmetric tops. The quantum mechanical problem of the rotation of spherical tops and symmetric tops can be readily solved, but the rotation of an asymmetric top is fairly involved.

We shall start with the easiest example, namely, that of spherical top $I_A = I_B = I_C$. The quantum mechanical problem of a spherical top is readily solvable, having energy levels ε_J and degeneracy ω_J given by

$$\varepsilon_J = \frac{J(J + 1)\hbar^2}{2I} \qquad J = 0, 1, 2, \ldots \tag{8–14}$$

$$\omega_J = (2J + 1)^2$$

The high-temperature limit of the partition function is

$$q_{rot} = \frac{1}{\sigma} \int_0^{\infty} (2J + 1)^2 e^{-J(J+1)\hbar^2/2IkT} \, dJ \tag{8–15}$$

Here again we have introduced a symmetry number σ. The symmetry number for a polyatomic molecule is simply the number of ways that the molecule can be rotated "into itself." For example, $\sigma = 2$ for H_2O and $\sigma = 3$ for NH_3. For methane, $\sigma = 12$ since there is three-fold symmetry about each of the four carbon–hydrogen bonds. Similarly, $\sigma = 4$ for ethylene and 12 for benzene. Classically, the symmetry number avoids overcounting indistinguishable configurations in phase space. For those readers who know some group theory, σ is the number of pure rotational elements (including the identity) in the point group of a nonlinear molecule. Since high temperature means that high values of J are important, we may neglect 1 compared to J in Eq. (8–15) and write

$$q_{rot} = \frac{1}{\sigma} \int_0^{\infty} 4J^2 e^{-J^2 \hbar^2 / 2IkT} \, dJ = \frac{\pi^{1/2}}{\sigma} \left(\frac{8\pi^2 IkT}{h^2} \right)^{3/2} \tag{8–16}$$

Problem 8–2 involves the derivation of Eq. (8–16) by evaluating the integral in Eq. (8–15) without neglecting 1 compared to J.

The quantum mechanical problem of a symmetric top ($I_A = I_B \neq I_C$) is also solvable in closed form. In this case the energy levels depend upon two quantum numbers, one of which is a measure of the total rotational angular momentum of the molecule J, and the other a measure of the component of the rotational angular momentum along the unique axis of the symmetric top K, that is, the axis having the unique moment of inertia (customarily denoted by I_C). It might be pointed out here that any molecule with an n-fold axis of symmetry, with $n \geq 3$, is at least a symmetric top. The expression for the energy levels is

$$\varepsilon_{JK} = \frac{\hbar^2}{2} \left\{ \frac{J(J+1)}{I_A} + K^2 \left(\frac{1}{I_C} - \frac{1}{I_A} \right) \right\}$$

where $J = 0, 1, 2, \ldots$; $K = J, J - 1, \ldots, -J$; and the degeneracy is

$$\omega_{JK} = (2J + 1)$$

The partition function is, then,

$$q_{rot} = \frac{1}{\sigma} \sum_{J=0}^{\infty} (2J + 1) e^{-\alpha_A J(J+1)} \sum_{K=-J}^{+J} e^{-(\alpha_C - \alpha_A)K^2}$$

where

$$\alpha_j = \frac{\hbar^2}{2I_j kT} \qquad j = A \text{ or } C$$

Problem 8–3 converts this to a double integral over J and K, which results in

$$q_{rot} = \frac{\pi^{1/2}}{\sigma} \left(\frac{8\pi^2 I_A kT}{h^2} \right) \left(\frac{8\pi^2 I_C kT}{h^2} \right)^{1/2} \tag{8–17}$$

Notice that this reduces to the rotational partition of a spherical top [Eq. (8–16)] when $I_A = I_C$.

The next case is that of an asymmetric top $I_A \neq I_B \neq I_C$. This is the most commonly occurring type of molecule. The quantum mechanical problem of the rotational levels of an asymmetric top is a fairly involved problem and must be solved numerically. Consequently a quantum-statistical treatment is awkward, and it is desirable to use classical mechanics. Even the classical Hamiltonian of a rigid asymmetrical rotor is quite complicated. We confine ourselves, therefore, to the statement that insertion of the classical Hamiltonian into the classical phase integral leads, after a rather long but straightforward integration, to (see Problem 8–16)

$$q_{rot} = \frac{\pi^{1/2}}{\sigma} \left(\frac{8\pi^2 I_A kT}{h^2}\right)^{1/2} \left(\frac{8\pi^2 I_B kT}{h^2}\right)^{1/2} \left(\frac{8\pi^2 I_C kT}{h^2}\right)^{1/2} \tag{8–18}$$

Notice that this is a generalization of Eqs. (8–16) and (8–17).

If we introduce the characteristic rotational temperatures into Eq. (8–18), we have

$$q_{rot} = \frac{\pi^{1/2}}{\sigma} \left(\frac{T^3}{\Theta_A \Theta_B \Theta_C}\right)^{1/2} \tag{8–19}$$

Table 8–1 contains Θ_A, Θ_B, and Θ_C for several molecules. The rotational contributions to some thermodynamic functions are

$$E_{rot} = \tfrac{3}{2} NkT \tag{8–20}$$

$$C_{V,\,rot} = \tfrac{3}{2} Nk \tag{8–21}$$

$$S_{rot} = Nk \, \ln\left[\frac{\pi^{1/2}}{\sigma} \left(\frac{T^3 e^3}{\Theta_A \Theta_B \Theta_C}\right)^{1/2}\right] \tag{8–22}$$

Note that since there are three degrees of rotational freedom, the rotational kinetic energy here is $3NkT/2$, in accord with equipartition of energy.

8–3   THERMODYNAMIC FUNCTIONS

We can now use the results of Sections 8–1 and 8–2 to construct $q(V, T)$. We get for linear polyatomic molecules

$$q = \left(\frac{2\pi MkT}{h^2}\right)^{3/2} V \cdot \frac{T}{\sigma \Theta_r} \cdot \left\{\prod_{j=1}^{3n-5} \frac{e^{-\Theta_{vj}/2T}}{(1 - e^{-\Theta_{vj}/T})}\right\} \omega_{e1} e^{D_e/kT} \tag{8–23}$$

$$-\frac{A}{NkT} = \ln\left[\left(\frac{2\pi MkT}{h^2}\right)^{3/2} \frac{Ve}{N}\right] + \ln\left(\frac{T}{\sigma \Theta_r}\right)$$
$$- \sum_{j=1}^{3n-5} \left[\frac{\Theta_{vj}}{2T} + \ln(1 - e^{-\Theta_{vj}/T})\right] + \frac{D_e}{kT} + \ln \omega_{e1} \tag{8–24}$$

$$\frac{E}{NkT} = \tfrac{3}{2} + \tfrac{2}{2} + \sum_{j=1}^{3n-5} \left[\left(\frac{\Theta_{vj}}{2T}\right) + \frac{\Theta_{vj}/T}{(e^{\Theta_{vj}/T} - 1)}\right] - \frac{D_e}{kT} \tag{8–25}$$

$$\frac{C_V}{Nk} = \tfrac{3}{2} + \tfrac{2}{2} + \sum_{j=1}^{3n-5} \left(\frac{\Theta_{vj}}{T}\right)^2 \frac{e^{\Theta_{vj}/T}}{(e^{\Theta_{vj}/T} - 1)^2} \tag{8–26}$$

$$\frac{S}{Nk} = \ln\left[\left(\frac{2\pi MkT}{h^2}\right)^{3/2} \frac{Ve^{5/2}}{N}\right] + \ln\left(\frac{Te}{\sigma \Theta_r}\right)$$
$$+ \sum_{j=1}^{3n-5} \left[\frac{\Theta_{vj}/T}{e^{\Theta_{vj}/T} - 1} - \ln(1 - e^{-\Theta_{vj}/T})\right] + \ln \omega_{e1} \tag{8–27}$$

$$pV = NkT \tag{8–28}$$

and for nonlinear polyatomic molecules:

$$q = \left(\frac{2\pi MkT}{h^2}\right)^{3/2} V \cdot \frac{\pi^{1/2}}{\sigma} \left(\frac{T^3}{\Theta_A \Theta_B \Theta_C}\right)^{1/2} \cdot \left\{\prod_{j=1}^{3n-6} \frac{e^{-\Theta_{vj}/2T}}{(1 - e^{-\Theta_{vj}/T})}\right\} \omega_{e1} e^{D_e/kT}$$

(8–29)

$$-\frac{A}{NkT} = \ln\left[\frac{2\pi MkT}{h^2}\right]^{3/2} \frac{Ve}{N} + \ln \frac{\pi^{1/2}}{\sigma} \left(\frac{T^3}{\Theta_A \Theta_B \Theta_C}\right)^{1/2}$$

$$- \sum_{j=1}^{3n-6} \left[\frac{\Theta_{vj}}{2T} + \ln(1 - e^{-\Theta_{vj}/T})\right] + \frac{D_e}{kT} + \ln \omega_{e1}$$

(8–30)

$$\frac{E}{NkT} = \tfrac{3}{2} + \tfrac{3}{2} + \sum_{j=1}^{3n-6} \left(\frac{\Theta_{vj}}{2T} + \frac{\Theta_{vj}/T}{e^{\Theta_{vj}/T} - 1}\right) - \frac{D_e}{kT}$$

(8–31)

$$\frac{C_V}{Nk} = \tfrac{3}{2} + \tfrac{3}{2} + \sum_{j=1}^{3n-6} \left(\frac{\Theta_{vj}}{T}\right)^2 \frac{e^{\Theta_{vj}/T}}{(e^{\Theta_{vj}/T} - 1)^2}$$

(8–32)

$$\frac{S}{Nk} = \ln\left[\frac{2\pi MkT}{h^2}\right]^{3/2} \frac{Ve^{5/2}}{N} + \ln \frac{\pi^{1/2}e^{3/2}}{\sigma} \left(\frac{T^3}{\Theta_A \Theta_B \Theta_C}\right)^{1/2}$$

$$+ \sum_{j=1}^{3n-6} \left[\frac{\Theta_{vj}/T}{e^{\Theta_{vj}/T} - 1} - \ln(1 - e^{-\Theta_{vj}/T})\right] + \ln \omega_{e1}$$

(8–33)

$$pV = NkT$$

(8–34)

Table 8–1 contains the characteristic rotational temperatures, the characteristic vibrational temperatures, and

$$D_0 = D_e - \sum_j \tfrac{1}{2}h\nu_j$$

for a number of polyatomic molecules. See Herzberg under "Additional Reading" for an excellent chapter dealing with practical statistical thermodynamical calculations for polyatomic molecules. He includes a number of corrections to the above formulas and discusses them numerically.

Table 8–2 gives the vibrational contribution to the heat capacity for a variety of molecules of different shapes. It can be seen that the vibrational contributions are far

Table 8–2. **Vibrational contribution to heat capacities of some polyatomic molecules at 300°K**

molecule						C^{vib}/Nk	total C_V/Nk (calc.)
CO_2, linear	$\Theta_v(°K)$	1890	3360	954			
	degeneracy	1	1	2			
	contribution to C_V/Nk	0.073	0.000	0.458		0.99	3.49
N_2O, linear	$\Theta_v(°K)$	1840	3200	850			
	degeneracy	1	1	2			
	contribution to C_V/Nk	0.082	0.003	0.533		1.15	2.65
NH_3, pyramidal	$\Theta_v(°K)$	4800	1360	4880	2330		
	degeneracy	1	1	2	2		
	contribution to C_V/Nk	0.000	0.226	0.000	0.026	0.28	3.28
CH_4, tetrahedron	$\Theta_v(°K)$	4170	2180	4320	1870		
	degeneracy	1	2	3	3		
	contribution to C_V/Nk	0.000	0.037	0.000	0.077	0.30	3.30
H_2O, isosceles triangle	$\Theta_v(°K)$	2290	5160	5360			
	degeneracy	1	1	1			
	contribution to C_V/Nk	0.028	0.000	0.000		0.03	3.03

from their equipartition values at the temperatures listed and that the agreement between the calculated and experimental values of C_V/Nk is excellent. A calculation for more complicated molecules would show similar agreement between the calculated values and the experimental data.

Table 8–3 compares calculated values of the entropy to those measured calorimetrically. It can be seen again that the agreement with experiment is quite good. In fact, calculated values of the entropy are often more accurate than measured values, provided sophisticated enough spectroscopic models are used.

Table 8–3. **The entropy of several polyatomic gases at 25°C and 1-atm pressure***

	S(calc.) (e.u.)	S(exp.) (e.u.)
CO_2	51.1	51.0
NH_3	46.1	46.0
NO_2	57.5	57.5
ClO_2	59.4	59.6
CH_4	44.5	44.5
CH_3Cl	55.8	56.0
CCl_4	74.0	73.9
C_6H_6	64.5	64.4

* The experimental values have been corrected for nonideal gas behavior.

There is, however, a class of molecules for which the type of agreement in Table 8–3 is not found. For example, it is found that for carbon monoxide, $S_{calc} = 47.3$ e.u. and $S_{exp} = 46.2$, for a discrepancy of 1.1 e.u. Other such discrepancies are found, and in all cases $S_{calc} > S_{exp}$. This difference is often referred to as *residual* entropy. The explanation of this is the following. Carbon monoxide has a very small dipole moment, and so when carbon monoxide is crystallized, the molecules do not have a strong tendency to line up in an energetically favorable way. The resultant crystal, then, is a random mixture of the two possible orientations CO and OC. As the crystal is cooled down toward 0°K, each molecule gets locked into its orientation and cannot realize the state of lowest energy with $\Omega = 1$, that is, all the molecules oriented in the same direction. Instead, the number of configurations Ω of the crystal is 2^N, since each of the N molecules exists equally likely (almost equally likely since the dipole moment is so small) in two states. Thus the entropy of the crystal at 0°K is $S = k \ln \Omega = Nk \ln 2$ instead of zero. If $Nk \ln 2 = 1.4$, entropy units are added to the experimental entropy, the agreement in the case of carbon monoxide becomes satisfactory. If it were possible to obtain carbon monoxide in its true equilibrium state at $T = 0$, this discrepancy would not occur. A similar situation occurs with nitrous oxide. For H_3CD, the residual entropy is 2.8 esu, and this is explained by realizing that each molecule of monodeuterated methane can assume four different orientations in the low-temperature crystal, and so $S_{residual} = Nk \ln 4 = 2.7$ esu, in very close agreement with the experimental value.

8–4 HINDERED ROTATION

There is one extension or modification of the partition function of polyatomic molecules that we shall discuss in this section. In molecules such as ethane, one of the most important internal degrees of freedom is a rotation about the single carbon–carbon bond. Because of the interactions between the hydrogen atoms on each carbon, this rotation is not free, but is said to be restricted or hindered. As the two methyl

Figure 8–1. **Potential energy of internal rotation in ethane. This curve can be represented approximately by** $\frac{1}{2}V_0(1 - \cos 3\phi)$.

groups rotate about the carbon–carbon bond, the hydrogen atoms become alternately eclipsed (directly opposite each other) and staggered. The potential energy associated with this rotation is shown in Fig. 8–1. The maxima correspond to the configuration where the hydrogen atoms are eclipsed, and the minima correspond to the configuration in which they are staggered. At temperatures such that $kT \gg V_0$, the internal rotation is essentially free and can be treated by methods similar to the rigid rotor. At temperatures such that $kT \ll V_0$, the molecule is trapped at the bottom of the wells in Fig. 8–1, and the motion is that of a simple torsional vibration, which can be treated by a method similar to that used for the simple harmonic oscillator. Typical values of V_0 are such, however, that at ordinary temperatures, the motion is intermediate between that of free rotation and torsional vibration.

It is necessary to solve the Schrödinger equation for the potential shown in Fig. 8–1. This potential can be approximately represented by $\frac{1}{2}V_0(1 - \cos 3\phi)$, for which the Schrödinger equation is

$$-\frac{h^2}{8\pi^2 I_r}\frac{\partial^2 \psi}{\partial \phi^2} + \frac{1}{2}V_0(1 - \cos 3\phi)\psi = \varepsilon\psi$$

where I_r is an effective moment of inertia whose precise form we shall not need. This differential equation is difficult to solve analytically, but the eigenvalues have been tabulated numerically as a function of V_0. These can be used to compute a partition function for the restricted rotation, which can then be used to compute thermodynamic properties. There are extensive tables of the various thermodynamic functions as a function of V_0/kT.

Figure 8–2 shows a sketch of the contribution of internal rotation to the heat capacity in an ethanelike molecule. One can use curves such as these to fit heat capacity

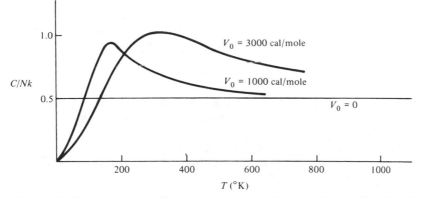

Figure 8–2. **The contribution of internal rotation to the heat capacity as a function of barrier height V_0.**

Table 8–4. **Potential barriers of some molecules**

molecule	V_0(kcal/mole)
CH_3CH_3	2.7–3.0
CH_3CCl_3	2.7
CH_3CH_2Cl	2.7–4.7
CH_3OH	1.1–1.6
CH_3SH	1.3–1.5
CH_3NH_2	1.9
$CH_3CH_2CH_3$	3.3
CH_3CHCCH_2	1.59–1.65

data and thus obtain a thermodynamic estimate of V_0. One can also generate curves for the entropy as a function of V_0 and hence determine V_0 by comparing the results to experimental values of the entropy. The values of V_0 determined from heat capacity data and entropy data are in fair agreement. There are a number of other ways of determining V_0 (see e.g. Section 9–2(F)) and the reader is referred to a review article by Wilson* for a general discussion of the problem of barriers to internal rotation in molecules and a comparison of the various methods for determining V_0 experimentally. Table 8–4 lists potential barriers for several molecules. The range of values given for each molecule is an indication of the agreement between different experimental methods of determining V_0.

ADDITIONAL READING

General

DAVIDSON, N. 1962. *Statistical mechanics*. New York: McGraw-Hill. Chapter 11.
FOWLER, R. H., and GUGGENHEIM, E. A. 1956. *Statistical thermodynamics*. Cambridge: Cambridge University Press. Sections 326 through 331.
HILL, T. L. 1952. *Statistical thermodynamics*. Cambridge: Cambridge University Press. Chapter 9.
KUBO, R. 1965. *Statistical mechanics*. Amsterdam: North-Holland Publishing Co. Sections 3–1 to 3–3. Example 3–2.
MAYER, J. E., and MAYER, M. G. 1940. *Statistical mechanics*. New York: Wiley. Chapter 8.
MÜNSTER, A. 1969. *Statistical thermodynamics*, Vol. I. Berlin: Springer-Verlag. Sections 6–8 through 6–11.
RUSHBROOKE, G. S. 1949. *Statistical mechanics*. London: Oxford University Press. Chapter 9.

Spectroscopy

BARROW, G. M. 1962. *Introduction to molecular spectroscopy*. New York: McGraw-Hill.
COSTAIN, C. C. 1970. In *Physical chemistry: an advanced treatise*, Vol. IV, ed. by H. Eyring, D. Henderson, and W. Jost. New York: Academic.
HALL, J. R. 1970. In *Physical chemistry, an advanced treatise*, Vol. IV, ed. by H. Eyring, D. Henderson, and W. Jost. New York: Academic.
HERZBERG, G. 1945. *Infrared and Raman spectra of polyatomic molecules*. New York: Van Nostrand.
KING, G. W. 1970. In *Physical chemistry, an advanced treatise*, Vol. IV, ed. by H. Eyring, D. Henderson, and W. Jost. New York: Academic.
———. 1964. *Spectroscopy and molecular structure*. New York: Holt, Rinehart & Winston.

PROBLEMS

8–1. The HOH bond angle in water is 104°, and the OH bond length is 0.96Å. Calculate the center of mass and the three moments of inertia of water. From this verify the results for Θ_{rot} for H_2O given in Table 8–1.

8–2. Evaluate Eq. (8–16) without neglecting 1 compared to J in Eq. (8–15).

8–3. Derive Eq. (8–17) from its corresponding summation by converting the sum to an integral.

---

* E. B. Wilson, Jr., *Adv. Chem. Phys.*, **2**, p. 367, 1959.

8–4. Verify the calculated entries in Table 8–3 for CH_4, using the data in Table 8–1.

8–5. Use the data in Table 8–1 to calculate the entropy of CO_2 at 25°C and 1 atm. Compare your result to that in Table 8–3.

8–6. The same as Problem 8–5, but for H_2O.

8–7. The same as Problem 8–5, but for CH_4.

8–8. What molar heat capacities would you expect under classical conditions for the following gases: (a) Ne, (b) O_2, (c) H_2O, (d) CO_2, and (e) $CHCl_3$?

8–9. Verify the results for methane in Table 8–3.

8–10. Verify the results for ammonia and water in Table 8–2.

8–11. Calculate the entropy of ClO_2 at 298°K and compare this to the experimental value of 61 e.u.

8–12. Verify from group theoretic character tables that the symmetry number is equal to the number of pure rotational elements (including the identity) in the point group of the molecule.

8–13. Show that C_V for NH_3 at 300°K is $3.3Nk$.

8–14. In Problem 1–36, heat-capacity data were listed for a calculation of the third-law entropy of nitromethane. From the following molecular data, calculate the statistical entropy S^0_{298}. Bond distances (Å): N–O 1.21; C–N 1.46; C–H 1.09. Bond angles: O–N–O 127°; H–C–N 109½°. From these distances, calculate the principal moments of inertia, $I = 67.2$, 76.0, 137.9 × 10^{-40} g-cm². The fundamental vibration frequencies in cm^{-1} are 476, 599, 647, 921, 1097, 1153, 1384, 1413, 1449, 1488, 1582, 2905, 3048(2). One of the torsional vibrations has become a free rotation around the C–N bond with $I = 4.86 \times 10^{-40}$.

8–15. The classical rotational kinetic energy of a symmetric top molecule is

$$K = \frac{p_\theta^2}{2I_A} + \frac{(p_\phi - p_\psi \cos\theta)^2}{2I_A \sin^2\theta} + \frac{p_\psi^2}{2I_C}$$

where I_A, I_A, and I_C are the principal moments of inertia, and θ, ϕ, and ψ are the three Euler angles. Derive the classical limit of the rotational partition function for a symmetric top molecule. Hint: Recall that the Euler angles have the ranges:

$$0 \le \theta \le \pi$$
$$0 \le \phi \le 2\pi$$
$$0 \le \psi \le 2\pi$$

8–16. The classical Hamiltonian for an asymmetric top molecule with principal moments of inertia I_A, I_B, and I_C is given by

$$H = \frac{1}{2I_A \sin^2\theta} \{(p_\phi - p_\psi \cos\theta)\cos\psi - p_\theta \sin\theta \sin\psi\}^2$$

$$+ \frac{1}{2I_B \sin^2\theta} \{(p_\phi - p_\psi \cos\theta)\sin\psi + p_\theta \sin\theta \cos\psi\}^2 + \frac{1}{2I_C} p_\psi^2$$

Derive the classical limit of the rotational partition function for an asymmetric top molecule. Hint: It may help to rearrange the Hamiltonian and integrate over p_θ, p_ψ, p_ψ in that order.

CHAPTER 9

CHEMICAL EQUILIBRIUM

One of the most important chemical applications of statistical thermodynamics is the calculation of equilibrium constants in terms of molecular parameters. Often the results of such calculations are more accurate than the experimental values. In Section 9–1 we derive the basic equations, giving the equilibrium constant in terms of partition functions. This is an easy derivation, since Eq. (1–68) gives the thermodynamic equilibrium condition in terms of chemical potentials, and in Chapters 5, 6, and 8 we derived thermodynamic functions such as the chemical potential in terms of partition functions. Thus there are no new principles introduced in this chapter. In a sense, it is simply a numerical application of the results of the previous chapters. The bulk of this chapter is a numerical discussion of the few equations of Section 9–1. In Section 9–2 we discuss six types of reactions: (1) the association of atoms or molecules in a vapor, (2) a simple isotopic exchange reaction, (3) a more complicated isotopic exchange reaction, (4) a chemical reaction involving only diatomic molecules, (5) a chemical reaction involving polyatomic molecules, and (6) a chemical reaction involving a molecule with restricted internal rotation. Then in Section 9–3 we discuss the use of thermodynamic tables to calculate equilibrium constants. We shall see that thermodynamic tables are equivalent to a tabulation of the partition function at various temperatures.

9–1 THE EQUILIBRIUM CONSTANT IN TERMS OF PARTITION FUNCTIONS

We shall consider the general homogeneous gas phase chemical reaction

$$v_A A + v_B B \rightleftharpoons v_C C + v_D D \tag{9–1}$$

at equilibrium in a closed thermostated vessel. The v's are stoichiometric coefficients and A, B, and so on, represent the reactants and products. The thermodynamic condition for chemical equilibrium is derived as follows.

We first write Eq. (9–1) algebraically as

$$v_C C + v_D D - v_A A - v_B B = 0 \tag{9–2}$$

We then define a variable λ such that $dN_j = v_j \, d\lambda$, where $j = A, B, C,$ or D and where v_j is taken to be positive for products and negative for reactants. The Helmholtz free energy of the system is

$$dA = -S \, dT - p \, dV + \sum_j \mu_j \, dN_j$$

For a reaction vessel at fixed volume and temperature,

$$dA = \sum_j \mu_j \, dN_j = \left(\sum_j v_j \mu_j \right) d\lambda \qquad \text{(constant } T \text{ and } V\text{)} \tag{9–3}$$

For a system at equilibrium, the free energy must be a minimum with respect to all possible changes $d\lambda$, and so $(\partial A / \partial \lambda)_{T, V} = 0$. From Eq. (9–3) then, we have the condition for chemical equilibrium:

$$\sum_j v_j \mu_j = v_C \mu_C + v_D \mu_D - v_A \mu_A - v_B \mu_B = 0 \tag{9–4}$$

This is the general thermodynamic equation of chemical equilibrium. We shall now introduce statistical thermodynamics through the relation between chemical potential and partition functions. In a mixture of ideal gases, the species are independent and distinguishable, and so the partition function of the mixture is a product of the partition functions of the individual components. Thus

$$Q(N_A, N_B, N_C, N_D, V, T) = Q(N_A, V, T)Q(N_B, V, T)Q(N_C, V, T)Q(N_D, V, T)$$

$$= \frac{q_A(V, T)^{N_A}}{N_A!} \frac{q_B(V, T)^{N_B}}{N_B!} \frac{q_C(V, T)^{N_C}}{N_C!} \frac{q_D(V, T)^{N_D}}{N_D!} \tag{9–5}$$

The chemical potential of each species is given by an equation such as

$$\mu_A = -kT \left(\frac{\partial \ln Q}{\partial N_A} \right)_{N_j, V, T} = -kT \ln \frac{q_A(V, T)}{N_A} \tag{9–6}$$

where Stirling's approximation has been used for $N_A!$. The N_j subscript on the partial derivative indicates that the numbers of particles of the other species are held fixed. Equation (9–6) simply says that the chemical potential of one species of an ideal gas mixture is calculated as if the other species were not present. This, of course, is obvious for an ideal gas mixture.

If we substitute Eq. (9–6) into Eq. (9–4), we get

$$\frac{N_C{}^{v_C} N_D{}^{v_D}}{N_A{}^{v_A} N_B{}^{v_B}} = \frac{q_C{}^{v_C} q_D{}^{v_D}}{q_A{}^{v_A} q_B{}^{v_B}} \tag{9–7}$$

For an ideal gas, the molecular partition function is of the form $f(T)V$ (see Problem 4–15), so that q/V is a function of temperature only. This allows us to write

$$K_c(T) = \frac{\rho_C{}^{v_C} \rho_D{}^{v_D}}{\rho_A{}^{v_A} \rho_B{}^{v_B}} = \frac{(q_C/V)^{v_C}(q_D/V)^{v_D}}{(q_A/V)^{v_A}(q_B/V)^{v_B}} \tag{9–8}$$

where $K_c(T)$ is the equilibrium constant of the reaction. For an ideal system, K_c is a function of temperature only.

Another commonly used equilibrium constant is $K_p(T)$, which is expressed in terms of partial pressures rather than concentrations. We can derive the equation for $K_p(T)$ by substituting $p_j = \rho_j kT$ into Eq. (9–8):

$$K_p(T) = \frac{p_C{}^{\nu_C} p_D{}^{\nu_D}}{p_A{}^{\nu_A} p_B{}^{\nu_B}} = (kT)^{\nu_C + \nu_D - \nu_A - \nu_B} K_c(T) \tag{9–9}$$

By means of Eq. (9–8) or (9–9), along with the results of Chapters 5 through 8, it is a simple matter to calculate equilibrium constants in terms of molecular parameters. This is best illustrated by means of examples.

9–2 EXAMPLES OF THE CALCULATION OF EQUILIBRIUM CONSTANTS

A. THE ASSOCIATION OF ALKALI METAL VAPORS

We use, as an example, the reaction

$$2\text{Na} \rightleftharpoons \text{Na}_2$$

The equilibrium constant for this reaction can be written as

$$K_p(T) = \frac{p_{\text{dimer}}}{p^2{}_{\text{monomer}}} \tag{9–10}$$

$$= (kT)^{-1} \frac{(q_{\text{Na}_2}/V)}{(q_{\text{Na}}/V)^2} \tag{9–11}$$

The equation for the partition function of a monatomic ideal gas [Eq. (5–15)] and a diatomic molecule [Eq. (6–51)] are

$$q_{\text{Na}}(T, V) = \left(\frac{2\pi m_{\text{Na}} kT}{h^2}\right)^{3/2} V q_{\text{elec}}(T) \tag{9–12}$$

$$q_{\text{Na}_2}(T, V) = \left(\frac{2\pi m_{\text{Na}_2} kT}{h^2}\right)^{3/2} V \frac{8\pi^2 IkT}{2h^2} \frac{e^{-\beta h\nu/2}}{(1 - e^{-\beta h\nu})} \omega_{1e} e^{D_e/kT}$$

$$= \left(\frac{2\pi m_{\text{Na}_2} kT}{h^2}\right)^{3/2} V \left(\frac{T}{2\Theta_r}\right)(1 - e^{-\Theta_v/T})^{-1} e^{D_0/kT} \tag{9–13}$$

Note that we have introduced $D_0 = D_e - \frac{1}{2}h\nu$ into Eq. (9–13). We shall usually do this since there are extensive tables of D_0. Tables 6–1 and 8–1 contain values of D_0 for diatomic and polyatomic molecules. In addition to values of D_0, Table 6–1 contains the other parameters needed in Eq. (9–13). From Table 6–1, $\Theta_v = 229°\text{K}$, $\Theta_r = 0.221°\text{K}$, and $D_0 = 17.3$ Kcal/mole. In addition, we need to know that the ground electronic state of a sodium atom is $^2S_{1/2}$, and that the next electronic state lies approximately 16,000 cm^{-1} above the ground state. At 1000°K, then

$$\frac{q_{\text{Na}}}{V} = \left(\frac{2\pi \times 23 \times 1.66 \times 10^{-24} \times 1.38 \times 10^{-16} \times 10^3}{6.626 \times 6.626 \times 10^{-54}}\right)^{3/2} q_{\text{elec}}$$

$$= (6.54 \times 10^{26}) \times 2 = 1.31 \times 10^{27} \tag{9–14}$$

$$\frac{q_{\text{Na}_2}}{V} = (1.85 \times 10^{27}) \times (2.26 \times 10^3)(4.88)(5.96 \times 10^3)$$

$$= (1.22 \times 10^{35}) \tag{9–15}$$

and the equilibrium constant is

$$K_p(T) = \frac{1.22 \times 10^{35}}{(1.38 \times 10^{-16})(1000)(1.72 \times 10^{54})}$$

$$= 0.50 \times 10^{-6}(\text{dyne/cm}^2)^{-1} = 0.50 \text{ atm}^{-1} \tag{9-16}$$

The experimental value is 0.475 atm^{-1}. We have used the fact that $1 \text{ atm} = 1.01 \times 10^6$ dynes/cm^2. This can easily be calculated from the gas constant or the Boltzmann constant in ergs/deg-mole and liter-atm/deg-mole. Note that K_p has units of 1/(pressure), even though it is a function only of the temperature for an ideal system. Table 9-1 gives $K_p(T)$ at a number of other temperatures. The agreement is seen to be good.

Table 9-1. **A comparison of the experimental values of the equilibrium constant for the reaction** $2\text{Na} \rightleftharpoons \text{Na}_2$ **with the values calculated from Eq. (9-11)**

$T(^\circ\text{K})$	$K_p(\text{calc.})(\text{atm})^{-1}$	$K_p(\text{exp.})^*(\text{atm})^{-1}$
900	1.44	1.32
1000	0.50	0.47
1100	0.22	0.21
1200	0.11	0.10

* See C. T. Ewing, *et al.*, *J. Chem. Phys.*, **71**, 473, 1967.

A general principle of dissociation or association reactions of this type is that the dimer is favored by the energetics of the reaction but the monomers are favored by the entropy. Problems 9-2 and 9-3 involve similar calculations for the association of potassium vapors and the dissociation of I_2.

B. AN ISOTOPIC EXCHANGE REACTION

We consider the isotopic exchange reaction

$$\text{H}_2 + \text{D}_2 \rightleftharpoons 2\text{HD}$$

We can use this simple reaction to illustrate the consequences of the Born-Oppenheimer approximation in isotopic exchange reactions. The Born-Oppenheimer approximation is based upon the approximation that the nuclei are so much more massive than the electrons that it is legitimate to calculate the electronic state of a molecule in a field of fixed nuclei. Thus H_2, D_2, and HD all have the same internuclear potential function, and therefore have the same force constant k, the same depth of the potential D_e, and the same internuclear separation. This leads to a great deal of canceling between the numerator and denominator in the ratio of partition functions in the equilibrium constant.

The equilibrium constant is

$$K(T) = K_p(T) = K_c(T) = \frac{\rho_{\text{HD}}^2}{\rho_{\text{H}_2}\rho_{\text{D}_2}} = \frac{p_{\text{HD}}^2}{p_{\text{H}_2}p_{\text{D}_2}} = \frac{q_{\text{HD}}^2}{q_{\text{H}_2}q_{\text{D}_2}}$$

$$= \frac{\left(\dfrac{2\pi m_{\text{HD}}kT}{h^2}\right)^3 \left(\dfrac{T}{\Theta_{r,\text{HD}}}\right)^2 \left(\dfrac{e^{-\Theta_{v,\text{HD}}/2T}}{1-e^{-\Theta_{v,\text{HD}}/T}}\right)^2 e^{2D_e/kT}}{\left(\dfrac{2\pi m_{\text{H}_2}kT}{h^2}\right)^{3/2}\left(\dfrac{2\pi m_{\text{D}_2}kT}{h^2}\right)^{3/2}\left(\dfrac{T^2}{4\Theta_{r,\text{H}_2}\Theta_{r,\text{D}_2}}\right)\left(\dfrac{e^{-\Theta_{v,\text{H}_2}/2T}}{1-e^{-\Theta_{v,\text{H}_2}/T}}\right)\left(\dfrac{e^{-\Theta_{v,\text{D}_2}/2T}}{1-e^{-\Theta_{v,\text{D}_2}/T}}\right)e^{2D_e/kT}}$$

$$= \frac{m_{\text{HD}}^3}{(m_{\text{H}_2}m_{\text{D}_2})^{3/2}} \frac{4\Theta_{r,\text{H}_2}\Theta_{r,\text{D}_2}}{\Theta_{r,\text{HD}}^2} \frac{(1-e^{-\Theta_{v,\text{H}_2}/T})(1-e^{-\Theta_{v,\text{D}_2}/T})}{(1-e^{-\Theta_{v,\text{HD}}/T})^2} e^{-(2\Theta_{v,\text{HD}}-\Theta_{v,\text{H}_2}-\Theta_{v,\text{D}_2})/2T}$$

$$\tag{9-17}$$

From Table 6–1, $\Theta_{r,H_2} = 85.3°K$, $\Theta_{r,D_2} = 42.7°K$, $\Theta_{v,H_2} = 6215°K$, and $\Theta_{v,D_2} = 4394°K$. The values for HD are easy to calculate from these, since $\Theta_v = hv/k$, and $v = (k/\mu)^{1/2}/2\pi$, where μ is the reduced mass of the molecule. Since the force constants of all three molecules are equal under the Born-Oppenheimer approximation, we have

$$\frac{v_{HD}}{v_{H_2}} = \left(\frac{\mu_{H_2}}{\mu_{HD}}\right)^{1/2} \quad \text{or} \quad \frac{\Theta_{v,HD}}{\Theta_{v,H_2}} = \left(\frac{\mu_{H_2}}{\mu_{HD}}\right)^{1/2} \tag{9-18}$$

which gives

$$\Theta_{v,HD} = \left(\tfrac{3}{4}\right)^{1/2}\Theta_{v,H_2} \tag{9-19}$$

Of course, we also have the relations

$$\Theta_{v,HD} = \left(\tfrac{3}{2}\right)^{1/2}\Theta_{v,D_2} \quad \text{and} \quad \Theta_{v,D_2} = \left(\tfrac{1}{2}\right)^{1/2}\Theta_{v,H_2} \tag{9-20}$$

Similarly, since $\Theta_r = h^2/8\pi^2 I k$ and $I = \mu r_e^2$, where r_e is the internuclear separation, we can write

$$\frac{\Theta_{r,HD}}{\Theta_{r,H_2}} = \frac{\mu_{H_2}}{\mu_{HD}} = \frac{3}{4} \tag{9-21}$$

or

$$\frac{\Theta_{r,HD}}{\Theta_{r,D_2}} = \frac{3}{2} \tag{9-22}$$

Note that $\Theta_{r,H_2} = 2\Theta_{r,D_2}$. We shall calculate $K(T)$ between 195°K and 741°K. At these temperatures, the $(1 - e^{-\Theta_v/T})$ factors in Eq. (9–17) are approximately unity.

Substituting all of these reduction formulas into Eq. (9–17), we get

$$K(T) = 4\left(\tfrac{9}{8}\right)^{3/2}\left(\tfrac{8}{9}\right)e^{-[3^{1/2}-1-(1/2)^{1/2}]\Theta_{v,H_2}/2T}$$

$$= 4(1.06)\exp\left\{\frac{-77.7}{T}\right\} \tag{9-23}$$

Table 9–2 compares Eq. (9–23) with the experimental data of Urey and Rittenberg, obtained during Urey's early investigations on heavy hydrogen. Equation (9–23) shows that the product or the symmetry numbers of the H_2 and D_2 predominates in the calculation of the equilibrium constant. All of the other factors are close to unity.

The hydrogen–deuterium exchange reaction, in fact, shows an unusually large deviation from $K = 4$, since the percentage mass difference between hydrogen and

Table 9–2. **A comparison of the experimental values of the equilibrium constant of the reaction** $H_2 + D_2 \rightleftharpoons 2HD$ **with the values calculated from Eq. (9–23)**

$T(°K)$	K(calc.)	K(exp.)
195	2.84*	2.92
273	3.18	3.24
298	3.26	3.28
383	3.46	3.50
543	3.67	3.85
670	3.77	3.8
741	3.81	3.75

* Quantum effects begin to be important at these low temperatures. If these effects are included, a value of 2.87 is obtained.

Source: D. Rittenberg, W. Bleakney, and H. C. Urey, *J. Chem. Phys.*, **2**, p. 362, 1934, and A. J. Gould, W. Bleakney, and H. S. Taylor, *J. Chem. Phys.*, **2**, p. 362, 1934.

deuterium is large. For other diatomic isotopic exchange reactions such as $N_2^{14} + N_2^{15} \rightleftharpoons 2N^{14}N^{15}$, the equilibrium constant differs very little from the value 4. It is easy to derive a general expression for the equilibrium constant of such reactions by expanding the complete expression for $K(T)$ in a power series in Δ/M, where Δ is the difference in atomic mass between the two isotopes, and M is the atomic mass of the heavier isotope. To order $(\Delta/M)^2$, the result is (see Problem 9–6)

$$K(T) = 4\left(1 + \frac{\Delta^2}{8M^2}\right)e^{-\Delta^2\Theta_{M,\text{vib}}/32M^2T} \tag{9-24}$$

where $\Theta_{M,\text{vib}}$ is the characteristic vibrational temperature of the diatomic molecule containing two heavier nuclei. For the $N_2^{14} + N_2^{15} \rightleftharpoons 2N^{14}N^{15}$ reaction, $\Delta/M = 1/15$ and $\Theta_{M,\text{vib}} = 0.97$ times the value in Table 6–1, or $\Theta_{M,\text{vib}} = 3260°K$. Equation (9–24) becomes

$$K(T) = 4(1.0005)e^{-0.44/T} \tag{9-25}$$

which is essentially equal to 4 for all temperatures greater than 100°K, say. Thus the equilibrium constant is completely determined by the symmetry of the molecules involved in the reaction. This is not necessarily so for more complicated molecules, however, as the next section shows.

C. A MORE COMPLICATED ISOTOPIC EXCHANGE REACTION: AN ILLUSTRATION OF THE TELLER-REDLICH PRODUCT RULE

Consider the exchange reaction

$$CH_4 + DBr \rightleftharpoons CH_3D + HBr$$

Because of the Born-Oppenheimer approximation, the internuclear potential surfaces of CH_4 and CH_3D are the same, and the internuclear potential energy curves of HBr and DBr are the same. As before, this leads to a great deal of simplification in the ratio of partition functions. The equilibrium constant for this reaction is

$$K(T) = \frac{\rho(CH_3D)\rho(HBr)}{\rho(CH_4)\rho(Br)} = \frac{q(CH_3D)q(HBr)}{q(CH_4)q(DBr)}$$

$$= \frac{\sigma_{CH_4}\sigma_{DBr}}{\sigma_{CH_3D}\sigma_{HBr}}\left(\frac{M_{CH_3D}M_{HBr}}{M_{CH_4}M_{DBr}}\right)^{3/2}\frac{I_{HBr}}{I_{DBr}}\frac{(I_AI_BI_C)_{CH_3D}^{1/2}}{(I_AI_BI_C)_{CH_4}^{1/2}}\frac{q_{\text{vib},CH_3D}\,q_{\text{vib},HBr}}{q_{\text{vib},CH_4}\,q_{\text{vib},DBr}} \tag{9-26}$$

The fundamental vibrational frequencies of CH_4, CH_3D, HBr, and DBr are given in Table 9–3. All of these frequencies are high enough that the factors $(1 - e^{-\Theta_v/T})$ in

Table 9–3. **Fundamental vibrational frequencies of CH_4, CH_3D, HBr, and DBr**

CH₄		CH₃D	
frequency	degeneracy	frequency	degeneracy
2917 cm⁻¹	1	2200 cm⁻¹	1
1534	2	2945	1
3019	3	1310	1
1306	3	1471	2
		3021	2
		1155	2

HBr		DBr	
frequency	degeneracy	frequency	degeneracy
2650 cm⁻¹	1	1880 cm⁻¹	1

the vibrational partition functions are essentially unity at room temperature. The ratio of the vibrational partition functions in Eq. (9–26) includes then only the zero-point vibrational energy terms and is

$$\frac{q_{\text{vib,CH}_3\text{D}}\, q_{\text{vib,HBr}}}{q_{\text{vib,CH}_4}\, q_{\text{vib,DBr}}} \approx \exp\left\{ -\sum_j \frac{(\Theta_{vj}^{\text{CH}_3\text{D}} + \Theta_{vj}^{\text{HBr}} - \Theta_{vj}^{\text{CH}_4} - \Theta_{vj}^{\text{DBr}})}{2T} \right\}$$

$$= \exp\left\{ \frac{220\ \text{cm}^{-1}}{kT} \right\} \tag{9–27}$$

using the frequencies given in Table 9–3.

The ratio of the symmetry numbers in Eq. (9–26) is $12 \times 1/(3 \times 1) = 4$. To calculate the remainder of Eq. (9–26), we could look up all the moments of inertia we need in the literature, but this is not necessary. We can avoid it by using the *Teller-Redlich product rule* for isotopically substituted compounds, which says that if molecules A and A' differ only by isotopic substitution, then

$$\left(\frac{M'}{M} \right)^{3/2} \frac{I'}{I} = \prod_{i=1}^{n} \left(\frac{m_i'}{m_i} \right)^{3/2} \prod_{j=1}^{3n-5} \frac{v_j'}{v_j} \qquad \text{(linear)} \tag{9–28}$$

for linear molecules and

$$\left(\frac{M'}{M} \right)^{3/2} \frac{(I_A' I_B' I_C')^{1/2}}{(I_A I_B I_C)^{1/2}} = \prod_{i=1}^{n} \left(\frac{m_i'}{m_i} \right)^{3/2} \prod_{j=1}^{3n-6} \frac{v_j'}{v_j} \qquad \text{(nonlinear)} \tag{9–29}$$

for nonlinear molecules. In these two equations, M denotes the total mass of molecule A; m_j is the mass of the jth atom; and v_j is the j fundamental vibrational frequency. The proof of the Teller-Redlich product involves a detailed consideration of the normal coordinates of molecules, and so we shall not prove it here. It is easy to show, however, that it is valid for a diatomic molecule under the Born-Oppenheimer approximation (see Problem 9–7).

The Teller-Redlich product rule is useful to us, since we already have the v_j in Table 9–3. Furthermore, when this is used in Eq. (9–26), the products of the atomic masses cancel, and we are left with the simple result:

$$K(T) = 4e^{317/T}\, \frac{v_{\text{HBr}}}{v_{\text{DBr}}} \prod_{j=1}^{9} \left(\frac{v_{j\,\text{CH}_3\text{D}}}{v_{j\,\text{CH}_4}} \right) \tag{9–30}$$

$$= 2.99 e^{317/T}$$

$$= 8.65 \qquad \text{at } 298°\text{K} \tag{9–31}$$

D. A CHEMICAL REACTION INVOLVING DIATOMIC MOLECULES

We shall calculate the equilibrium constant for the reaction

$$\text{H}_2 + \text{I}_2 \rightleftharpoons 2\text{HI}$$

from 500°K to 1000°K. The equilibrium constant is given by

$$K(T) = \frac{(q_{\text{HI}}/V)^2}{(q_{\text{H}_2}/V)(q_{\text{I}_2}/V)} = \frac{q_{\text{HI}}^2}{q_{\text{H}_2}\, q_{\text{I}_2}}$$

$$= \left(\frac{m_{\text{HI}}^2}{m_{\text{H}_2}\, m_{\text{I}_2}} \right)^{3/2} \left(\frac{4\Theta_r^{\text{H}_2}\Theta_r^{\text{I}_2}}{(\Theta_r^{\text{HI}})^2} \right) \frac{(1 - e^{-\Theta_v^{\text{H}_2}/T})(1 - e^{-\Theta_v^{\text{I}_2}/T})}{(1 - e^{-\Theta_v^{\text{HI}}/T})^2}$$

$$\times \exp \frac{(2\text{D}_0^{\text{HI}} - \text{D}_0^{\text{H}_2} - \text{D}_0^{\text{I}_2})}{RT} \tag{9–32}$$

All of the necessary parameters are given in Table 6–1. Figure 9–1 shows ln K plotted

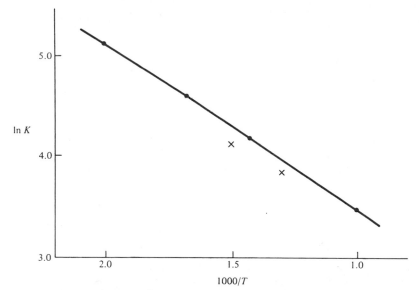

Figure 9–1. **The logarithm of the equilibrium constant versus** $1/T$ **for the reaction** $H_2 + I_2 \rightleftharpoons 2HI$. **The points are calculated from Eq. (9–32) and the crosses are the experimental values.** (From A. H. Taylor and R. H. Crist, *J. Am. Chem, Soc.*, **63**, 1377, 1941.)

versus $1/T$. The thermodynamic equation for the variation of the equilibrium constant with temperature is

$$d(\ln K) = -\frac{\Delta H}{R} d\left(\frac{1}{T}\right) \tag{9–33}$$

where ΔH is the heat of reaction. From Fig. 9–1 we get $\Delta H = -3100$ cal/mole, compared to the experimental value of -2950 cal/mole.

Problems 9–8 and 9–9 involve the calculation of the equilibrium constant for some other reactions involving only diatomic molecules.

E. A REACTION INVOLVING POLYATOMIC MOLECULES

As an example of a reaction involving a polyatomic molecule, consider the reaction

$$H_2 + \tfrac{1}{2}O_2 \rightleftharpoons H_2O$$

whose equilibrium constant is

$$K_p(T) = \frac{(q_{H_2O}/V)}{(kT)^{1/2}(q_{H_2}/V)(q_{O_2}/V)^{1/2}} \tag{9–34}$$

It is almost as convenient to calculate each partition function separately as to substitute them into K_p first. Furthermore, we shall use the values in the next section.

The necessary parameters are given in Tables 6–1 and 8–1. At 1500°K, the three partition functions are

$$\frac{q_{H_2}}{V} = \left(\frac{2\pi m_{H_2} kT}{h^2}\right)^{3/2} \left(\frac{T}{2\Theta_r^{H_2}}\right)(1 - e^{-\Theta_v^{H_2}/T})^{-1} e^{D_0^{H_2}/RT}$$

$$= 2.80 \times 10^{26} e^{D_0^{H_2}/RT} \tag{9–35}$$

$$\frac{q_{O_2}}{V} = \left(\frac{2\pi m_{O_2} kT}{h^2}\right)^{3/2} \left(\frac{T}{2\Theta_r^{O_2}}\right)(1 - e^{-\Theta_v^{O_2}/T})^{-1} 3 e^{D_0^{O_2}/RT}$$

$$= 2.79 \times 10^{30} e^{D_0^{O_2}/RT} \tag{9–36}$$

and

$$\frac{q_{H_2O}}{V} = \left(\frac{2\pi m_{H_2O} kT}{h^2}\right)^{3/2} \frac{\pi^{1/2}}{\sigma} \left(\frac{T^3}{\Theta_A{}^{H_2O}\Theta_B{}^{H_2O}\Theta_C{}^{H_2O}}\right)^{1/2} \prod_{j=1}^{3}(1 - e^{-\Theta_{vj}{}^{H_2O}/T})^{-1} e^{D_0{}^{H_2O}/RT}$$

$$= 5.33 \times 10^{29} e^{D_0{}^{H_2O}/RT} \tag{9-37}$$

The factor of 3 occurs in q_{O_2}/V because the ground state of O_2 is $^3\sum_g^-$. (See Table 6–1.) At 1500°K, the equilibrium constant is

$$K_p = 4.77 \times 10^5 (\text{atm})^{-1/2} \tag{9-38}$$

Table 9–4 compares the calculated values of $\log K_p$ with the experimental data of

Table 9–4. **The logarithm of the equilibrium constant for the reaction $H_2 + \frac{1}{2}O_2 \rightleftharpoons H_2O$**

$T(°K)$	$\log K$(calc.)	$\log K$(exp.)*
1000	10.2	10.1
1500	5.7	5.7
2000	3.7	3.5

* See H. Zeiss, *Z. Electrochem.*, **43**, p. 706, 1937.

Zeiss. The agreement can be considerably improved by using more sophisticated spectroscopic models. At high temperatures, the rotational energies of the molecules are high enough to warrant centrifugal distortion effects and other extensions of the simple rigid rotor–harmonic oscillator approximation. Problems 9–10 through 9–12 involve similar calculations.

F. A CHEMICAL REACTION INVOLVING A MOLECULE WITH RESTRICTED INTERNAL ROTATION, ETHYLENE–ETHANE EQUILIBRIUM

As a final example, we discuss the equilibrium constant for the reaction

$$C_2H_4 + H_2 \rightleftharpoons C_2H_6$$

This calculation is more complicated than the previous ones because of the hindered rotation in ethane. The equilibrium constant is calculated for various values of the barrier height V_0, and V_0 is chosen to give the best agreement with experiment. Table 9–5 gives some values for the equilibrium constant K_p for this reaction at various

Table 9–5. **The equilibrium constant for the reaction $C_2H_4 + H_2 \rightleftharpoons C_2H_6$ calculated with $V_0 = 3100$ kcal/mole**

$T(°K)$	$K_p(\text{atm}^{-1})$ calc.	obs.
673	1.1×10^4	1.2×10^4
773	4.3×10^2	4.2×10^2
873	33.5	32.0
973	4.5	5.0

Source: E. A. Guggenheim, *Trans. Faraday Soc.*, **37**, p. 97, 1941.

temperatures as calculated by Guggenheim.* The value of V_0 that seemed to give the best agreement was 3.1 kcal/mole. This value agrees fairly well with other determinations of V_0 for ethane. This is another method for determining barrier heights and some values determined by this method are included in Table 8–4.

* E. A. Guggenheim, *Trans. Faraday Soc.*, **37**, 97, 1941.

9–3 THERMODYNAMIC TABLES

In the previous section we have seen that the rigid rotor-harmonic oscillator approximation can be used to calculate equilibrium constants in reasonably good agreement with experiment, and because of the simplicity of the model, the calculations involved are not extensive. If greater accuracy is desired, however, one must include corrections to the rigid rotor–harmonic oscillator model, and the calculations become increasingly more laborious. It is natural, then, that a number of numerical tables of partition functions has evolved, and in this section we shall discuss the use of these tables. These tables are actually much more extensive than a compilation of partition functions. They include many experimentally determined values of thermodynamic properties, often complemented by theoretical calculations. The thermodynamic tables we are about to discuss in this section, then, represent a collection of the thermodynamic and/or statistical thermodynamic properties of many substances. In order to fully appreciate the use of these tables, it is necessary to express the equilibrium constant in terms of the standard chemical potential $\mu_0(T)$ and then to discuss at length the problem, or rather the convention, of the choice of the zero of energy.

First we shall derive the relation between the equilibrium constant and the standard chemical potential [cf. Eq. (1–68)]. Equation (9–4) gives the thermodynamic condition for chemical equilibrium:

$$v_C \mu_C + v_D \mu_D - v_A \mu_A - v_B \mu_B = 0$$

For an ideal gas, the chemical potential is of the form [cf. Eq. (1–66) or Eq. (5–21)]

$$\mu(T, p) = \mu_0(T) + kT \ln p \tag{9–39}$$

If we substitute this into Eq. (9–4), we get

$$\ln K_p = -\frac{\Delta \mu_0}{kT} \tag{9–40}$$

where

$$\Delta \mu_0 = v_C \mu_{0C} + v_D \mu_{0D} - v_A \mu_{0A} - v_B \mu_{0B} \tag{9–41}$$

The μ_0 are related to the molecular partition function in the following way. For an ideal gas, $Q = q^N/N!$. The chemical potential is given by $-kT(\partial \ln Q/\partial N)_{V,T}$, so

$$\mu = -kT \ln\left(\frac{q}{N}\right) \tag{9–42}$$

For an ideal gas, the molecular partition function is a function of temperature times the volume, and so q/V is a function of temperature only. We can rewrite Eq. (9–42) as

$$\mu = -kT \ln\left\{\left(\frac{q}{V}\right)\frac{V}{N}\right\}$$

and use the ideal gas equation of state for V/N to get

$$\mu = -kT \ln\left\{\left(\frac{q}{V}\right)kT\right\} + kT \ln p \tag{9–43}$$

This equation is of the form of Eq. (9–39) with $\mu_0(T)$ given by

$$\mu_0(T) = -kT \ln\left\{\left(\frac{q}{V}\right)kT\right\} \tag{9–44}$$

The argument of the logarithm here has units of pressure, but of course the units cancel when this term is combined with $kT \ln p$. Nevertheless, the numerical value of $\mu_0(T)$ does depend upon the units of pressure. In anticipation of the convention used in the thermodynamic tables, we shall express the pressure in atmospheres. If we use the fact that 1 atm $= 1.01 \times 10^6$ dynes/cm^2, we can write Eq. (9–43) in the form

$$\mu = -kT \ln\left\{\left(\frac{q}{V}\right)\frac{kT}{1.01 \times 10^6}\right\} + kT \ln p(\text{atm}) \tag{9–45}$$

where (q/V) and kT in the $\mu_0(T)$ term are expressed in cgs units. The equilibrium constant calculated with these values of $\mu_0(T)$, however, will be in atmospheres, since we have explicitly included the conversion factor of 1.01×10^6. We shall see below that the thermodynamic tables are given in pressure units of atmospheres.

In principle, then, we could calculate $\mu_0(T)$ over a range of temperature and tabulate the results for future use in Eq. (9–40). This is almost the content of thermodynamic tables, but there is one last convention we must discuss, namely, the convention of the zero of energy. Throughout this book we have calculated partition functions in terms of energies relative to a zero of energy being taken as the infinitely separated atoms in their lowest energy states. From a molecular point of view, this is a satisfactory choice and is conceptually the most convenient choice. It is experimentally difficult to determine the energy required to separate a molecule into its constituent atoms, and so the thermodynamic tables utilize another convention which we describe below. Of course, it makes no difference what we choose as the zero of energy as long as we are consistent and use the same convention on both sides of a chemical equation. Let us go back now and reexamine at what point this arbitrariness in the choice of zero energy enters the calculation of partition functions.

We have seen that to a good approximation, the molecular partition function can be written as

$$q(V, T) = q_{\text{trans}}(V, T)q_{\text{rot}}(T)q_{\text{vib}}(T)q_{\text{elec}}(T) \tag{9–46}$$

where, of course, q_{rot} and q_{vib} do not appear in the case of an atom. If we recall the discussion centering around Fig. 6–2, it is in the calculation of $q_{\text{elec}}(T)$ that the zero-of-energy convention was used. The electronic partition function is

$$q_{\text{elec}}(T) = \omega_{e1}e^{-\varepsilon_{e1}/kT} + \omega_{e2}e^{-\varepsilon_{e2}/kT} + \cdots \tag{9–47}$$

If one chooses the separated atoms in their ground states as the zero of energy, then (cf. Fig. 6–2 or Fig. 9–2)

$$\varepsilon_{e1} = -D_e$$
$$\varepsilon_{e2} = \varepsilon_{e1} + (\varepsilon_{e2} - \varepsilon_{e1}) = \varepsilon_{e1} + \Delta\varepsilon_{12}$$
$$= -D_e + \Delta\varepsilon_{12}$$
$$\cdots \tag{9–48}$$

and

$$q_{\text{elec}}(T) = \omega_{e1}e^{D_e/kT} + \omega_{e2}e^{D_e/kT}e^{-\Delta\varepsilon_{12}/kT} + \cdots$$
$$= e^{D_e/kT}(\omega_{e1} + \omega_{e2}e^{-\Delta\varepsilon_{12}/kT} + \cdots)$$
$$= e^{D_e/kT}q^0_{\text{elec}}(T) \tag{9–49}$$

The superscript on the $q^0_{\text{elec}}(T)$ in the last line indicates that this is the electronic partition function of a molecule with its ground electronic state taken to be zero.

If we substitute Eq. (9–49) into Eq. (9–46), we get

$$q(V, T) = q_{\text{trans}}(V, T)q_{\text{rot}}(T)q_{\text{vib}}(T)q^0_{\text{elec}}(T)e^{D_e/kT} \tag{9–50}$$

The vibrational partition function is calculated using energies whose zero is taken as the bottom of the potential well, so that there is an energy of $\frac{1}{2}h\nu(\sum_j \frac{1}{2}h\nu_j$ in a polyatomic molecule) in the ground vibrational state. It is this zero-point vibrational energy that leads to the numerator in the vibrational partition functions of Chapters 6 and 8, that is, to the $e^{-\Theta_{vj}/2T}$ in

$$q_{\text{vib}}(T) = \prod_j \frac{e^{-\Theta_{vj}/2T}}{(1 - e^{-\Theta_{vj}/T})} \tag{9–51}$$

If we substitute this into Eq. (9–50), we have

$$q(V, T) = q_{\text{trans}}(V, T)q_{\text{rot}}(T)\left\{\prod_j (1 - e^{-\Theta_{vj}/T})^{-1}\right\}q^0_{\text{elec}}(T)e^{(D_e - (1/2)\sum_j h\nu_j/kT)} \tag{9–52}$$

$$= q_{\text{trans}}(V, T)q_{\text{rot}}(T)q^0_{\text{vib}}(T)q^0_{\text{elec}}(T)e^{D_0/kT} \tag{9–53}$$

where we have written D_0 for $D_e - \frac{1}{2}\sum_j h\nu_j$ (cf. Fig. 9–2) and have superscripted q_{vib} with a ^{0} since the term in brackets in Eq. (9 52) is a vibrational partition function whose lowest vibrational state is taken to be zero. In addition, we could have superscripted q_{trans} and q_{rot} since these partition functions are calculated on the basis of the lowest translational and rotational states being zero. Therefore we can write the molecular partition function as

$$q(V, T) = q^0(V, T)e^{D_0/kT} \tag{9–54}$$

where $q^0(V, T)$ indicates that the partition function is calculated on the basis that the ground state of the entire molecule is taken to be zero. The factor of $\exp(D_0/kT)$ "scales" the partition function to the zero of energy being the separated atoms in their ground states.

Equation (9–54) can be interpreted as saying that the partition function can be written as the product of an internal part $q^0(V, T)$, and a scaling factor that accounts for the arbitrary zero of energy. Figure 9–2(a) shows the energy D_0, and Fig. 9–2(b)

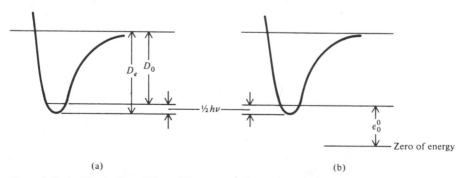

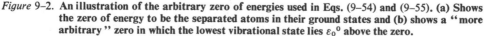

(a)   (b)

Figure 9–2. **An illustration of the arbitrary zero of energies used in Eqs. (9–54) and (9–55). (a) Shows the zero of energy to be the separated atoms in their ground states and (b) shows a "more arbitrary" zero in which the lowest vibrational state lies ε_0^0 above the zero.**

shows a completely arbitrary zero of energy. In terms of this other zero of energy, the partition function is

$$q(V, T) = q^0(V, T)e^{-\varepsilon_0{}^0/kT} \tag{9-55}$$

where $q^0(V, T)$ is the same partition function that occurs in Eq. (9–54). Note that if $\varepsilon_0{}^0 = -D_0$, Eq. (9–55) reduces to Eq. (9–54). In a sense, Eq. (9–55) is more arbitrary than Eq. (9–54) since the zero of energy has no physical basis and is completely arbitrary. If we substitute this into Eq. (9–45), we get

$$\mu - \varepsilon_0{}^0 = -kT \ln\left\{\left(\frac{q^0}{V}\right)\frac{kT}{1.01 \times 10^6}\right\} + kT \ln p(\text{atm}) \tag{9-56}$$

Equation (9–56) clearly displays the fact that the chemical potential is calculated relative to some zero of energy. As $T \to 0$, the molecule is found in its lowest energy state, which according to Fig. 9–2(b) is $\varepsilon_0{}^0$. This is the significance of the zero subscript on $\varepsilon_0{}^0$. It is easy to show mathematically that at 1 atm, $\mu \to \varepsilon_0{}^0$ as $T \to 0$ since $T \ln T \to 0$ as $T \to 0$.

We are now ready to introduce the notation and the conventions used in most thermodynamic tables. We shall adopt the convention that the energy of an element is zero at 0°K if the element is in the physical state (i.e., gas, liquid, or solid) characteristic of 25°C and 1 atm pressure. All energies, therefore, are calculated relative to this convention. We shall, in addition, multiply Eq. (9–56) by Avogadro's number and recognize that $N\mu$ is the Gibbs free energy per mole. We shall write Eq. (9–56) as

$$G^0 - E_0{}^0 = -RT \ln\left\{\left(\frac{q^0}{V}\right)\frac{kT}{1.01 \times 10^6}\right\} \tag{9-57}$$

The significance of the ^{0} superscript on G^0 indicates that this is the Gibbs free energy per mole at 1 atm pressure relative to the zero-of-energy convention adopted above. It is clear from Eq. (9–57) that $G^0 \to E_0{}^0$ as $T \to 0$, and so $E_0{}^0$ is also the standard free energy at 0°K. Thus we could write $G^0 - G_0{}^0$ instead of $G^0 - E_0{}^0$. This notation is often found in thermodynamic tables. Furthermore, since the enthalpy $H = E + pV$, and $p \to 0$ as $T \to 0$, we also have $H_0{}^0 = E_0{}^0$, and one also finds the notation $G^0 - H_0{}^0$. Since $G_0{}^0 = H_0{}^0 = E_0{}^0$, the quantities $G^0 - G_0{}^0$, $G^0 - E_0{}^0$, and $G^0 - H_0{}^0$ are equivalent. The important point is that $G^0 - E_0{}^0$ can be calculated without assuming any convention concerning the zero of energy, since this represents the standard Gibbs free energy relative to arbitrary zero, $E_0{}^0$. The right-hand side of Eq. (9–57) is based upon the energy of a molecule in its lowest state, that is, at $T = 0$°K, being zero, and hence can be calculated for any molecule independent of the choice of $E_0{}^0$.

Table 9–6 gives $-(G^0 - E_0{}^0)/T$ for a number of substances. The ratio $(G^0 - E_0{}^0)/T$ rather than $(G^0 - E_0{}^0)$ is given since $(G^0 - E_0{}^0)/T$ varies more slowly with temperature, and hence the tables are easier to interpolate. In addition to $-(G^0 - E_0{}^0)/T$, it is necessary to tabulate $E_0{}^0$ according to our chosen convention that the energy of an element is zero at 0°K if the element is in the physical state characteristic of 25°C and 1 atm. For a molecule, $E_0{}^0$ represents the energy of a molecule at 0°K relative to the elements, and therefore it also represents the heat of formation at 0°K. Consequently,

Table 9–6. **Thermodynamic functions of some selected substances**

| | \multicolumn{5}{c}{$-(G^0 - E_0^{\,0})/T$ cal/deg-mole} | | |
	298.15°K	500°K	1000°K	1500°K	2000°K	$H_{298}^0 - E_0^{\,0}$	$E_0^{\,0}$(kcal/mole)
Ne	29.98	32.56	36.00	38.00	39.44	1.481	0
Ar	32.01	34.59	38.04	40.04	41.48	1.481	0
Kr	34.22	36.80	40.24	42.25	43.69	1.481	0
H(g)	22.42	24.99	28.44	30.45	31.88	1.481	51.62
Cl(g)	34.43	37.06	40.69	42.83	44.34	1.499	28.54
Br(g)	36.84	39.41	42.85	44.89	46.36	1.481	26.90
I(g)	38.22	40.78	44.23	46.24	47.68	1.481	25.61
O(g)	33.08	35.84	39.46	41.54	43.00	1.607	58.98
N(g)	31.65	34.22	37.66	39.67	41.10	1.481	112.54
H_2(g)	24.42	27.95	32.74	35.59	37.67	2.024	0
Cl_2(g)	45.93	49.85	55.43	58.85	61.34	2.194	0
Br_2(g)	50.85	54.99	60.80	64.31	66.83	2.325	8.37
I_2(g)	54.18	58.46	64.40	67.96	70.52	2.148	15.66
O_2(g)	42.06	45.68	50.70	53.81	56.10	2.07	0
N_2(g)	38.82	42.42	47.31	50.28	52.48	2.072	0
CO(g)	40.25	43.86	48.77	51.78	53.99	2.073	−27.202
NO(g)	42.98	46.76	51.86	54.96	57.24	2.194	21.48
HCl(g)	37.72	41.31	46.16	49.08	51.23	2.065	−22.019
HBr(g)	40.53	44.12	48.99	51.95	54.13	2.067	−8.1
HI(g)	42.40	45.99	50.90	53.90	56.11	2.069	6.7
CO_2(g)	43.56	47.67	54.11	58.48	61.85	2.238	−93.969
H_2O(g)	37.17	41.29	47.01	50.60	53.32	2.368	−57.107
NH_3(g)	37.99	42.28	48.63	53.03	56.56	2.37	−9.37
Cl_2O(g)	54.52	59.49	67.04	71.91	—	2.719	18.61
ClO_2(g)	52.79	57.48	64.65	69.33	—	2.577	25.59
N_2O(g)	44.89	49.11	55.76	60.27	—	2.291	20.31
NO_2(g)	49.19	53.60	60.23	64.58	67.88	2.465	8.68
CH_4(g)	36.46	40.75	47.65	52.84	57.1	2.397	−15.99
CH_3Cl(g)	47.45	52.06	59.78	65.54	—	2.489	−17.7
CH_2Cl_2(g)	55.08	60.36	69.58	76.17	—	2.834	−19
$CHCl_3$(g)	59.29	65.81	76.78	84.63	—	3.390	−23
CCl_4(g)	60.15	68.12	81.41	89.96	—	4.111	−25
C_2H_6(g)	45.27	50.77	61.11	69.46	—	2.856	−16.52
C_2H_4(g)	43.98	48.74	57.29	63.94	69.46	2.525	14.52
CH_2O(g)	44.25	48.54	55.11	59.81	63.58	2.393	−26.8
Na_2(g)	46.65	51.04	57.14	60.77	—	2.484	—
K_2(g)	51.06	55.57	61.76	65.47	—	2.566	—

Source: G. N. Lewis and M. Randall, *Thermodynamics*, revised by K. S. Pitzer and L. Brewer (New York: McGraw-Hill, 1961).

one often finds the notation $\Delta H_0^{\,0}$ instead of $E_0^{\,0}$. Thus we can calculate equilibrium constants from the tables since Eq. (9–40), is, in the notation introduced above,

$$- R \ln K_p = \frac{\Delta E_0^{\,0}}{T} + \Delta \left(\frac{G^0 - E_0^{\,0}}{T} \right) \tag{9–58}$$

where Δ has the same significance here as in Eq. (9–41).

We shall illustrate the use of these tables by means of examples. First let us show how the tables are consistent with our previous calculations. Consider the reaction $H_2 + I_2 \rightleftharpoons 2HI$. For hydrogen,

$$q^0(V, T) = \left(\frac{2\pi m k T}{h^2} \right)^{3/2} V \left(\frac{T}{2\Theta_r} \right) (1 - e^{-\Theta_v/T})^{-1}$$

$$= 9.95 \times 10^{25} \, V \quad \text{at } 1000°K$$

According to Eq. (9–57),

$$-\left(\frac{G^0 - E_0{}^0}{T}\right) = R \ln\left\{\frac{9.95 \times 10^{25} \times 1.38 \times 10^{-13}}{1.01 \times 10^6}\right\}$$

$$= 32.87 \qquad \text{at } 1000°K$$

in good agreement with Table 9–6. Of course, the value given in Table 9–6 must be considered to be more accurate than the value we have calculated here. Since hydrogen is a gas at 25°C and 1 atm, we find from Table 9–6 that $E_0{}^0 = 0$. For I_2, the parameters in Table 6–1 give $-(G^0 - E_0{}^0)/T = 64.40$ given in Table 9–6. Since I_2 is a solid at 25°C and 1 atm, $E_0{}^0$ is not zero, but is 15.66 kcal/mole, the heat of sublimation of I_2 at 0°K. If we substitute the values from Table 9–6 into Eq. (9–58), we get $K_p = 32.5$, compared to 30.9 from the simpler equations of Chapter 6.

For the reaction $H_2 + \frac{1}{2}O_2 \rightleftharpoons H_2O$ at 1000°K, the tables give $\log K_p = 10.00$ compared to the experimental value of 10.06 and the less accurate value 10.2 calculated in the previous section on the basis of the rigid rotor–harmonic oscillator model. Note that K_p is expressed in atmospheres, since the tables are based on the standard pressure being 1 atm.

Table 9–6 also gives $H_{298}^0 - E_0{}^0$, which is the enthalpy at 298°K and 1 atm pressure relative to the zero of energy already explained. One often finds $H_{298}^0 - H_0{}^0$, which, of course, is equivalent to $H_{298}^0 - E_0{}^0$. For 1 mole of an ideal gas,

$$H = E + pV = E + RT$$

$$= RT^2\left(\frac{\partial \ln Q}{\partial T}\right)_{N,V} + RT \tag{9–59}$$

and so at temperatures high enough that the rotation is classical, we have

$$H_T{}^0 - E_0{}^0 = \tfrac{3}{2}RT + \tfrac{2}{2}RT + \frac{R\Theta_v}{e^{\Theta_v/T} - 1} + RT \qquad \text{(diatomic molecule)} \tag{9–60}$$

$$= \tfrac{3}{2}RT + \tfrac{2}{2}RT + \sum_j \frac{R\Theta_{vj}}{e^{\Theta_{vj}/T} - 1} + RT \qquad \text{(linear polyatomic molecule)}$$
$$\tag{9–61}$$

$$= \tfrac{3}{2}RT + \tfrac{3}{2}RT + \sum_j \frac{R\Theta_{vj}}{e^{\Theta_{vj}/T} - 1} + RT \qquad \text{(nonlinear polyatomic molecule)}$$
$$\tag{9–62}$$

Note that there is no contribution from the zero-point vibrational energy here as there is in Eq. (6–22), since we have taken the energy of the ground vibrational state to be zero. A value of 2.07 kcal/mole for a diatomic molecule comes from the $7RT/2$. Values higher than this are due to the vibrational contributions; a value lower than this (such as H_2) indicates that the rotation is not completely classical, and Eq. (6–35) should be used.

Although entropies are not given in Table 9–6, they can be easily obtained from the data given there since

$$S_{298}^0 = \left(\frac{H_{298} - E_0{}^0}{T}\right) - \left(\frac{G_{298}^0 - E_0{}^0}{T}\right) \tag{9–63}$$

The values of the entropy obtained in this way are in excellent agreement with those given in Tables 6–3 and 8–3.

Table 9–6 can also be used to calculate values of D_0 for various molecules. For example, consider D_0 for H_2. Table 9–6 shows that the energy of a hydrogen atom relative to a hydrogen molecule (both at $0°K$) is 51.62 kcal/mole. The value given in Table 6–1 is twice this.

For H_2O, we have

$$H_2 + \tfrac{1}{2}O_2 \longrightarrow H_2O + 57.107 \text{ kcal} \qquad \text{at } 0°K$$

We add to this the following two reactions

$$\left.\begin{array}{rcl} 2H & \longrightarrow & H_2 + 103.24 \text{ kcal} \\ O & \longrightarrow & \tfrac{1}{2}O_2 + 58.98 \text{ kcal} \end{array}\right\} \quad \text{at } 0°K$$

to get

$$2H + O \longrightarrow H_2O + 219.3 \text{ kcal} \qquad \text{at } 0°K$$

At $0°K$, both reactants and products are in their ground states, and so the 219.3 kcal represents D_0 for H_2O, in agreement with Table 8–1.

As a last example, we calculate D_0 for HI. This case is slightly different since iodine is a solid at $25°C$ and 1 atm. Table 9–6 gives that

$$\tfrac{1}{2}H_2(g) + \tfrac{1}{2}I_2(s) \longrightarrow HI - 6.7 \text{ kcal} \qquad \text{at } 0°K$$

We add the following equations to this:

$$H \longrightarrow \tfrac{1}{2}H_2(g) + 51.62 \text{ kcal} \qquad \text{at } 0°K$$

$$I(g) \longrightarrow \tfrac{1}{2}I_2(s) + 25.61 \text{ kcal} \qquad \text{at } 0°K$$

The result is

$$H + I(g) \longrightarrow HI + 70.53 \text{ kcal} \qquad \text{at } 0°K$$

and hence D_0 for HI is 70.53 kcal/mole.

Thermodynamic tables contain a great deal of thermodynamic and/or statistical thermodynamic data. Their use requires some amount of practice, but it is well worth the effort. Problems 9–17 through 9–26 are meant to supply this practice.

ADDITIONAL READING

General

FOWLER, R. H., and GUGGENHEIM, E. A. 1956. *Statistical thermodynamics*. Cambridge: Cambridge University Press. Chapter 5.

KITTEL, C. 1969. *Thermal physics*. New York: Wiley. Chapter 21.

MAYER, J. E., and MAYER, M. G. 1940. *Statistical mechanics*. New York: Wiley. Chapter 9.

MÜNSTER, A. 1969. *Statistical thermodynamics*, Vol. 1. Berlin: Springer-Verlag. Chapter 7.

RUSHBROOKE, G. S. 1949. *Statistical mechanics*. London: Oxford University Press. Chapters 11 and 12.

WIBERG, K. B. 1964. *Physical organic chemistry*. New York: Wiley. Part 2.

WILSON, Jr., E. B. 1940. *Chem. Rev.*, **27**, 17.

Thermodynamic Tables

LEWIS, G. N., and RANDALL, M. 1961. *Thermodynamics*, rev. ed. by K. S. Pitzer and L. Brewer. New York: McGraw-Hill.

PROBLEMS

9–1. Consider the reaction $A \rightleftharpoons 2B$. The canonical ensemble partition function for an ideal binary mixture is

$$Q(N_A, N_B, V, T) = \frac{q_A{}^{N_A} q_B{}^{N_B}}{N_A! N_B!}$$

Minimize the Helmholtz free energy with the stoichiometric constraint $2N_A + N_B = $ constant to show that

$$\frac{N_B^{*2}}{N_A^*} = \frac{q_B^2}{q_A}$$

where N_A^* and N_B^* are the equilibrium numbers of A and B. Can you generalize this approach to the reaction $\nu_A A + \nu_B B \rightleftharpoons \nu_C C + \nu_D D$ and derive Eq. (9–7)?

9–2. Using the data in Table 6–1 (plus atomic spectroscopic tables), calculate the equilibrium constant for the reaction $2K \rightleftharpoons K_2$ at 800°K and 1000°K. The experimental values* at these temperatures are 0.673 and 0.123, respectively.

9–3. Calculate the equilibrium constant K_p for the reaction $I_2 \rightleftharpoons 2I$, using the data in Table 6–1 and the fact that the ground electronic state of the iodine atom is $^2P_{3/2}$ and that the first excited electronic state lies 0.94 eV higher. The experimental values† are

$T(°K)$	800	900	1000	1100	1200
K_p	1.14×10^{-2}	4.74×10^{-2}	0.165	0.492	1.23

9–4. Show that the equilibrium constant for a reaction such as

$$HCl + DBr \rightleftharpoons DCl + HBr$$

approaches unity at sufficiently high temperatures.

9–5. Calculate the equilibrium constant for the reaction $N_2^{14} + N_2^{15} \rightleftharpoons 2N^{14}N^{15}$ and compare your results to the experimental values of Joris and Taylor.‡

9–6. Derive Eq. (9–24).

9–7. Show that the Teller-Redlich product rule is valid for a diatomic molecule under the Born-Oppenheimer approximation.

9–8. Calculate the enthalpy of reaction for $H_2 + I_2 \rightleftharpoons 2HI$ around 300°K.

9–9. Calculate the equilibrium constant for the reaction $\frac{1}{2}N_2 + \frac{1}{2}O_2 \rightleftharpoons NO$, using the data in Table 6–1. The observed values are

$T(°K)$	1500	2000	2500
K_p	2.4×10^{-3}	1.5×10^{-2}	4.5×10^{-2}

Why is the agreement not as good as you obtained in the other problems in this chapter?

9–10. Calculate the equilibrium constant for the reaction $CO_2 \rightleftharpoons CO + \frac{1}{2}O_2$ at 3000°K. The experimental value is 0.378 (atm)$^{1/2}$.

9–11. Using the data in Tables 6–1 and 8–1, calculate the equilibrium constant for the water gas reaction $CO_2 + H_2 \rightleftharpoons CO + H_2O$ at 900°K and 1200°K. The experimental values at these two temperatures are 0.46 and 1.37, respectively.

9–12. Using the data in Tables 6–1 and 8–1, calculate the equilibrium constant of the reaction $3H_2 + N_2 \rightleftharpoons 2NH_3$ at 400°C. The accepted value is 3.3×10^{-4} (atm)$^{-2}$.

9–13. Calculate the temperature at which molecular nitrogen is 99 percent dissociated at 100 atm, 1 atm, and 0.1 atm.

9–14. For the two ionization processes

$$H \rightleftharpoons H^+ + e^-$$
$$Cs \rightleftharpoons Cs^+ + e^-$$

derive an expression for the ratio of the equilibrium constants K_H/K_{Cs} where

$$K_H = \frac{\rho_{H^+}\rho_{e^-}}{\rho_H} \quad \text{and} \quad K_{Cs} = \frac{\rho_{Cs^+}\rho_{e^-}}{\rho_{Cs}}$$

* See C. T. Ewing, *et al., J. Chem. Phys.*, **71**, p. 473, 1967.
† See Perlman and Rollefson, *J. Chem. Phys.*, **9**, p. 362, 1941.
‡ See G. C. Joris and H. S. Taylor, *J. Chem. Phys.*, **7**, p. 893, 1939.

in terms of the respective ionization potentials I_H and I_{Cs}. What is the high-temperature limit of K_H/K_{Cs}? At room temperature do you expect K_H/K_{Cs} to be greater than, less than, or approximately equal to one?

9–15. Calculate the equilibrium constant for the first ionization of argon at 10,000°K. Use the data in A. B. Cambel, *Plasma Physics and Magnetofluid Dynamics* (New York: McGraw-Hill, 1963, p. 119).

9–16. Estimate the temperature at which gaseous atomic hydrogen would be at least 99 percent dissociated at 10^{-3} torr pressure into electrons and protons. Assume classical statistics, but remember that the electron has a spin degeneracy of 2.

9–17. Calculate the equilibrium constants of Problem 9–2, using the data in Table 9–6 and the *Handbook of Chemistry and Physics*.

9–18. Calculate the equilibrium constants of Problem 9–8, using the data in Table 9–6.

9–19. Calculate the equilibrium constants of Problem 9–9, using the data in Table 9–6.

9–20. Calculate the equilibrium constants of Problem 9–10, using the data in Table 9–6.

9–21. Calculate the equilibrium constants of Problem 9–11, using the data in Table 9–6.

9–22. Calculate the equilibrium constants of Problem 9–12, using the data in Table 9–6.

9–23. Verify the results in Table 9–6 for Ne, using the results of Chapter 5.

9–24. Verify the results in Table 9–6 for HBr, using the results of Chapter 6.

9–25. Verify the results of Table 9–6 for CH_3Cl, using the results of Chapter 8.

9–26. Verify the results of Table 9–6 for NO_2, using the results of Chapter 8.

CHAPTER 10

QUANTUM STATISTICS

In Chapter 4 we derived the two fundamental distribution laws of statistical mechanics. One, the Fermi-Dirac distribution, applies to systems whose N-body wave function is antisymmetric with respect to an interchange of any two identical particles; the other, Bose-Einstein statistics, applies to systems whose N-body wave function is symmetric under such an interchange. All elementary particles that have a half-odd-integral spin, such as the electron and the proton, obey Fermi-Dirac statistics and are called fermions; elementary particles that have an integral spin, such as the deuteron and the photon, obey Bose-Einstein statistics and are called bosons. The classification of compound particles can become a delicate problem at times, but as long as the binding energy of the compound particle is large compared to all other energies in the problem (which will be so for all cases that we shall discuss), one can say that a compound particle containing an odd number of fermions, such as He-3, will obey Fermi-Dirac statistics, and one with an even number, such as He-4, will obey Bose-Einstein statistics. Incidentally, it should be pointed out that it is not obvious that all particles occurring in nature are necessarily fermions or bosons, but there appear to be no known exceptions.

The basic equations associated with the two fundamental distribution laws are [Eqs. (4–24) through (4–28)]

$$\Xi(V, T, \lambda) = \prod_k (1 \pm \lambda e^{-\beta \varepsilon_k})^{\pm 1} \tag{10–1}$$

$$N = \sum_k \frac{\lambda e^{-\beta \varepsilon_k}}{1 \pm \lambda e^{-\beta \varepsilon_k}} \tag{10–2}$$

$$\bar{n}_k = \frac{\lambda e^{-\beta \varepsilon_k}}{1 \pm \lambda e^{-\beta \varepsilon_k}} \tag{10–3}$$

$$E = \sum_k \frac{\lambda \varepsilon_k e^{-\beta \varepsilon_k}}{1 \pm \lambda e^{-\beta \varepsilon_k}} \tag{10–4}$$

and

$$pV = \pm kT \sum_k \ln(1 \pm \lambda e^{-\beta \varepsilon_k}) \qquad (10\text{--}5)$$

where $\lambda = \exp(\mu/kT)$. In these equations the upper sign $(+)$ corresponds to Fermi-Dirac statistics, and the lower sign $(-)$ corresponds to Bose-Einstein statistics. In order to discuss the thermodynamic properties given by these equations, it is necessary to solve Eq. (10–2) for λ in terms of N and the $\beta\varepsilon$'s, and since the ε_k's are functions of V, this procedure gives λ as a function of N, V, and T. This solution for λ is then substituted into Eqs. (10–4) and (10–5) to give E and p, and hence other thermodynamic functions, in terms of N, V, and T. (See Problem 10–1.)

The difficulty is that it is not possible to solve Eq. (10–2) analytically for all values of λ. In Chapter 4 we showed that if λ is small, both Fermi-Dirac and Bose-Einstein reduce to Boltzmann, or classical, statistics. In this case $\lambda = N/q$ [*cf.* Eq. (10–2)], and all the mathematical manipulations are easy to perform. Thus the magnitude of λ can be regarded as a measure of the degree of quantum behavior of the system. Small values of λ correspond to classical or near-classical behavior, and large values of λ apparently correspond to quantum statistical behavior. Table 4–1 and the discussion surrounding it show that quantum effects become important for low temperatures and high densities. Thus we may expect λ to be small for high temperatures and low densities and large for low temperatures and high densities. We shall verify this later in the chapter.

Fortunately most systems of interest can be described by Boltzmann statistics and hence are characterized by small values of λ. This led us to the detailed discussion of classical ideal gases in Chapters 5 to 9.

There are several important and interesting systems, however (see Table 4–1), which cannot be described by classical statistics, and hence for which λ is not small. In this chapter we shall discuss these applications.

In Section 10–1 we shall study an ideal Fermi-Dirac gas for values of λ such that a series expansion of Eqs. (10–2) through (10–5) is useful. The first terms of these expansions represent the Boltzmann, or classical, limit, and so these expansions are useful in a temperature and density region where there are only small deviations from classical behavior, that is, high temperature and low density. We shall find the interesting result that even though we neglect the intermolecular forces in an ideal gas model, the equation of state will no longer be $p = \rho kT$, but will be a virial expansion, that is, a power series in the density. In Section 10–2 we shall examine the case for large values of λ (low values of the temperature or high values of the density). In this region we expect large deviations from classical behavior. Such systems are said to have large quantum effects or to be strongly *degenerate*. We shall see that although the results of these two sections are derived for an ideal gas, they serve as an interesting model for the electrons in a metal, and, in fact, Section 10–2 considers this model specifically.

Then in Section 10–3 we shall treat Bose-Einstein statistics in the range of λ, where a series expansion of Eqs. (10–2) through (10–5) is useful. The results are very similar to those of Section 10–1. In Section 10–4 we examine the low-temperature or high-density limit of Eqs. (10–2) through (10–5). This section, although more mathematically involved than the others, is very interesting, since we find that a system of ideal bosons undergoes a kind of phase transition as the temperature is lowered. We shall discuss the implications of this result with respect to the well-known λ-transition that occurs in He-4 at 2.18°K. In Section 10–5 we shall treat an ideal gas of photons or, in reality, electromagnetic radiation in thermal equilibrium. In this section we shall derive the fundamental blackbody radiation distribution law first derived by Planck and which led to the development of quantum mechanics.

Section 10–6 is devoted to an alternative formulation of quantum statistics, namely, through the density matrix, which is the quantum statistical analog of the density of phase points in classical phase space. Lastly, Section 10–7 is a rigorous discussion of the classical limit of Q.

10–1 A WEAKLY DEGENERATE IDEAL FERMI-DIRAC GAS

As the title of this section indicates, we shall consider an ideal gas of fermions in a region where λ is small enough that we may represent the deviations from classical behavior by a series expansion in λ. Consider Eqs. (10–2) and (10–4) with the upper signs:

$$N = \sum_k \frac{\lambda e^{-\beta \varepsilon_k}}{1 + \lambda e^{-\beta \varepsilon_k}} \tag{10–6}$$

$$pV = kT \sum_k \ln(1 + \lambda e^{-\beta \varepsilon_k}) \tag{10–7}$$

The ε_k's in these equations are the eigenvalues of a particle in a cube [cf. Eq. (5–3)], and the one index k really stands for n_x, n_y, and n_z:

$$\varepsilon_{n_x n_y n_z} = \frac{h^2}{8mV^{2/3}} (n_x{}^2 + n_y{}^2 + n_z{}^2) \qquad n_x, n_y, n_z = 1, 2, \ldots \tag{10–8}$$

where we have replaced a^2 by $V^{2/3}$. In Chapter 5 we converted sums over the energy states to integrals over energy levels. We used the argument that at room temperature, successive values of ε/kT differed so little from each other that ε_k ($\varepsilon_{n_x n_y n_z}$) was essentially a continuous function. We derived an expression for the density of states $\omega(\varepsilon)$ between ε and $\varepsilon + d\varepsilon$ on the same basis. In this chapter we do not wish to restrict our discussion to high temperatures, but the very same type of argument applies if we consider the thermodynamic limit of a large volume [cf. Eq. (10–8)]. Thus we may consider the energy states in Eq. (10–8) to be continuous and write

$$N = 2\pi \left(\frac{2m}{h^2}\right)^{3/2} V \int_0^\infty \frac{\lambda \varepsilon^{1/2} e^{-\beta \varepsilon} \, d\varepsilon}{1 + \lambda e^{-\beta \varepsilon}} \tag{10–9}$$

$$pV = 2\pi kT \left(\frac{2m}{h^2}\right)^{3/2} V \int_0^\infty \varepsilon^{1/2} \ln(1 + \lambda e^{-\beta \varepsilon}) \, d\varepsilon \tag{10–10}$$

This procedure is valid only because the summands in Eqs. (10–6) and (10–7) are finite continuous functions of λ for all the ε_k. Equations (10–6) and (10–7) indicate that the range of λ is $0 \leq \lambda < \infty$. Because of the minus sign in the denominator of the expressions for Bose-Einstein statistics, such a conversion of a summation into an integration must be done with care.

The integrals occurring in Eqs. (10–9) and (10–10) cannot be evaluated in closed form, but they can be written as power series in λ by expanding the denominator of Eq. (10–9) and the logarithm in Eq. (10–10) in a power series in λ and integrating term by term (see Problem 10–4):

$$\rho = \frac{1}{\Lambda^3} \sum_{l=1}^\infty \frac{(-1)^{l+1} \lambda^l}{l^{3/2}} \tag{10–11}$$

and

$$\frac{p}{kT} = \frac{1}{\Lambda^3} \sum_{l=1}^{\infty} \frac{(-1)^{l+1} \lambda^l}{l^{5/2}} \qquad (10\text{--}12)$$

These series are valid only for small values of λ. Equation (10–11) gives ρ as a power series in λ, and Eq. (10–12) gives p/kT as a power series in λ. We wish to solve Eq. (10–11) for λ as a function of ρ, and then substitute this into Eq. (10–12) to give p/kT as a function of ρ.

The problem of solving an equation like Eq. (10–11) is called *reversion* of a series and can be done in general. The general solution is involved, however, and since we are interested only in small deviations from classical behavior in this section, we can use a straightforward algebraic method that readily gives the first few terms. We assume that λ is a power series in ρ and write

$$\lambda = a_0 + a_1\rho + a_2\rho^2 + \cdots$$

We then substitute this into Eq. (10–11) and equate coefficients of like powers of ρ on both sides of the equation to get

$$a_0 = 0$$

$$a_1 = \Lambda^3$$

$$a_2 - \frac{a_1{}^2}{2^{3/2}} = 0$$

$$a_3 - \frac{a_1 a_2}{2^{1/2}} + \frac{a_1{}^3}{3^{3/2}} = 0$$

$$\cdots \qquad (10\text{--}13)$$

Thus λ is

$$\lambda = \rho\Lambda^3 + \frac{1}{2^{3/2}}(\rho\Lambda^3)^2 + \left(\frac{1}{4} - \frac{1}{3^{3/2}}\right)(\rho\Lambda^3)^3 + \cdots \qquad (10\text{--}14)$$

We now substitute this into Eq. (10–12) and get

$$\frac{p}{kT} = \rho + \frac{\Lambda^3}{2^{5/2}}\rho^2 + \left(\frac{1}{8} - \frac{2}{3^{5/2}}\right)\Lambda^6\rho^3 + \cdots \qquad (10\text{--}15)$$

This equation is in the form of a virial expansion for the pressure, which is customarily written in the form

$$\frac{p}{kT} = \rho + B_2(T)\rho^2 + B_3(T)\rho^3 + \cdots \qquad (10\text{--}16)$$

where $B_j(T)$ is a function of only the temperature and is called the *j*th virial coefficient. Virial coefficients reflect deviations from ideality, or intermolecular interactions. Although there are no intermolecular forces in this case, the particles, nevertheless, experience an effective interaction through the symmetry requirement of the N-body wave function. In the case of fermions, this interaction may be said to be repulsive since the first correction to ideal behavior, $B_2(T)$, is positive and hence increases the pressure above that of a classical ideal gas under the same conditions. Note also that $B_2(T)$ is $0(\Lambda^3)$. But Λ is just the thermal De Broglie wavelength, and so we see that quantum statistical effects decrease as the thermal De Broglie wavelength decreases In fact, if Eq. (10–15) is written in a dimensionless form, we see once again that it is the dimensionless ratio Λ^3/V that is a measure of the quantum effects.

The Fermi-Dirac equation for $\bar{E}$, Eq. (10–4) with the positive sign, also can be converted to an integral and then written as a series expansion in λ:

$$E = \tfrac{3}{2}VkT\frac{1}{\Lambda^3}\sum_{l=1}^{\infty}\frac{(-1)^{l+1}\lambda^l}{l^{5/2}} \tag{10–17}$$

If Eq. (10–14) for λ is substituted into this, we get

$$E = \tfrac{3}{2}NkT\left(1 + \frac{\Lambda^3}{2^{5/2}}\rho + \cdots\right) \tag{10–18}$$

All other thermodynamic functions can be obtained from Eqs. (10–14), (10–15), and (10–16). For example, the chemical potential μ follows immediately from Eq. (10–14) since $\lambda = \exp(\mu/kT)$. The entropy can be obtained from $G = \mu N = E - TS + pV$.

These series expansions for the thermodynamic functions are valid only for small values of λ, or for temperatures and densities such that quantum effects are a small correction to the classical limit. The leading terms in all these expansions are the classical limits obtained in Chapter 5. Now we shall discuss the case for large values of λ.

10–2   A STRONGLY DEGENERATE IDEAL FERMI-DIRAC GAS

In this section we shall treat the case of an ideal Fermi-Dirac gas at low temperature and/or high density. For concreteness we shall develop the results in terms of the free electron model of metals, where the valence electrons of the atoms of the metal are represented by an ideal gas of electrons. It is possible to understand a number of important physical properties of some metals, in particular the simple monovalent metals, in terms of the free electron model. There are several reasons why such a simple model can be used. One is that although the electrons do indeed interact with each other and with the atomic cores through a Coulombic potential, this potential is so long range ($1/r$) that the total electronic potential that any one electron "sees" is almost constant from point to point in the metallic crystal. Furthermore, in general many of the physically observable properties of a metallic crystal are due more to quantum statistical effects than to the details of the electron–electron and electron–ionic core interactions.

Consider Eq. (10–3) (with the positive sign in the denominator), which gives the number of particles in the molecular energy state k:

$$\bar{n}_k = \frac{\lambda e^{-\beta\varepsilon_k}}{1 + \lambda e^{-\beta\varepsilon_k}} = \frac{1}{1 + e^{\beta(\varepsilon_k - \mu)}} \tag{10–19}$$

As in the previous section, ε_k is essentially a continuous parameter, and we can write

$$f(\varepsilon) = \frac{1}{1 + e^{\beta(\varepsilon - \mu)}} \tag{10–20}$$

where $f(\varepsilon)$ is the probability that a given state is occupied. This equation is plotted in Fig. 10–1, where, in particular, $f(\varepsilon)$ is plotted versus ε/μ for fixed values of $\beta\mu$. In the extreme limit of low temperatures, that is, $T = 0$, the distribution is unity for energies

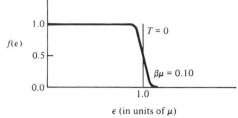

Figure 10-1. The Fermi-Dirac distribution as a function of ε (in units of μ) for $T = 0$ and $\beta\mu = 0.10$.

less than μ, and zero for energies greater than μ. The quantity μ is, in general, a function of temperature, and so we indicate the value of μ at $T = 0$ by μ_0.

The zero-temperature limit of $f(\varepsilon)$ is very simple, and it is instructive to understand its behavior. At the absolute zero of temperature, all the states with energy less than μ_0 are occupied and those with energy greater than μ_0 unoccupied. Thus μ_0 has the property of being a cutoff energy. According to Eq. (1–35), the number of states with energy between ε and $\varepsilon + d\varepsilon$ is

$$\omega(\varepsilon)\,d\varepsilon = 4\pi\left(\frac{2m}{h^2}\right)^{3/2} V\varepsilon^{1/2}\,d\varepsilon \tag{10-21}$$

where we have introduced a factor of 2, since an electron has two spin states ($\pm \frac{1}{2}$) associated with each translational state. We can find μ_0 immediately from the fact that all the states below $\varepsilon = \mu_0$ are occupied and all these above are unoccupied. Thus if N is the number of valence electrons,

$$N = 4\pi\left(\frac{2m}{h^2}\right)^{3/2} V \int_0^{\mu_0} \varepsilon^{1/2}\,d\varepsilon$$

$$= \frac{8\pi}{3}\left(\frac{2m}{h^2}\right)^{3/2} V(\mu_0)^{3/2} \tag{10-22}$$

from which we write

$$\mu_0 = \frac{h^2}{2m}\left(\frac{3}{8\pi}\right)^{2/3}\left(\frac{N}{V}\right)^{2/3} \tag{10-23}$$

Using the fact that the molar volume of Na, say, is 23.7 cm³/mole, and that a sodium atom has one valence electron, the quantity μ_0 is 3.1 eV. We see, then, that at $T = 0°K$, the conduction electrons in sodium metal fill all the energy states up to 3.1 eV according to the Pauli exclusion principle of allowing only two electrons to occupy each state. The quantity μ_0 is called the Fermi energy of a metal and is typically of the order of 1–5 eV (see Problem 10–6).

This is a very significant result since a plot as in Fig. 10–1 shows that at room temperature, where $\beta\mu_0$ is of order of 10^2, the distribution $f(\varepsilon)$ is still essentially a step function like at $T = 0°K$. Compared to a characteristic temperature μ_0/k, room temperature may be considered to be zero, and it is an excellent first approximation to use the distribution

$$\begin{aligned} f(\varepsilon) &= 1 \qquad \varepsilon < \mu_0 \\ &= 0 \qquad \varepsilon > \mu_0 \end{aligned} \tag{10-24}$$

at room temperature. The quantity μ_0/k is called the Fermi temperature and is denoted by T_F. Fermi temperatures are typically of the order of thousands of degrees Kelvin. In this approximation, the thermodynamic energy is [cf. Eq. (10–4)]

$$E_0 = 4\pi \left(\frac{2m}{h^2}\right)^{3/2} V \int_0^{\mu_0} \varepsilon^{3/2} \, d\varepsilon$$
$$= \tfrac{3}{5} N \mu_0 \qquad\qquad (10\text{–}25)$$

where we have written E_0 to emphasize that this is a $T = 0°\text{K}$ result. The energy E_0 is the zero-point energy of a Fermi-Dirac gas. Equation (10–25) implies that the contribution of the conduction electrons to the heat capacity is zero, in sharp contrast to the equipartition value of $3k/2$ for each electron. The physical explanation of this is that in order to contribute to the heat capacity, the electrons must be excited to higher quantum states, but since μ_0 is so large compared to kT, only a small fraction of all the particles will be within kT from the top of the distribution (near μ_0), where there are vacant states lying above. Thus a very small fraction of all the electrons can contribute to the heat capacity, and so the experimental heat capacity is almost zero.

The pressure is given by [cf. Eq. (10–10) with the additional factor of 2 due to the two allowed spin orientations of an electron]:

$$p = 4\pi kT \left(\frac{2m}{h^2}\right)^{3/2} \int_0^{\mu_0} \varepsilon^{1/2} \ln(1 + e^{\beta(\mu_0 - \varepsilon)}) \, d\varepsilon$$

In order to evaluate this integral, one can neglect unity compared to $e^{\beta(\mu_0 - \varepsilon)}$ since $\beta(\mu_0 - \varepsilon)$ is much larger than one over most of the range of integration. Therefore

$$p_0 = 4\pi \left(\frac{2m}{h^2}\right)^{3/2} \int_0^{\mu_0} \varepsilon^{1/2}(\mu_0 - \varepsilon) \, d\varepsilon$$
$$= \tfrac{2}{5} N \mu_0 / V \qquad\qquad (10\text{–}26)$$

This " zero-point pressure " is $0(10^6)$ atm. The occurrence of h in this expression shows that the zero-point pressure is a quantum effect. It follows from $G = N\mu = E - TS + pV$ and Eqs. (10–23), (10–25), and (10–26) that $S_0 = 0$. This is to be expected, since there is only one way to put the N indistinguishable particles into the lowest possible quantum states.

Equations (10–22) through (10–26) are based on the zero-temperature distribution function, Eq. (10–24). It is not difficult to calculate corrections to these zero-temperature results as an expansion in powers of a parameter $\eta = kT/\mu_0$. At room temperature, $\eta = 0(10^{-2})$, so that such an expansion converges quickly. To determine this expansion, first notice that all the thermodynamic quantities $N, E, p, \ldots$ can be written as

$$I = \int_0^{\infty} f(\varepsilon)h(\varepsilon) \, d\varepsilon \qquad\qquad (10\text{–}27)$$

where the following table gives examples of I and $h(\varepsilon)$:

I	$h(\varepsilon)$
N	$4\pi \left(\frac{2m}{h^2}\right)^{3/2} V \varepsilon^{1/2}$
E	$4\pi \left(\frac{2m}{h^2}\right)^{3/2} V \varepsilon^{3/2}$

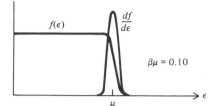

Figure 10–2. **The Fermi-Dirac distribution function and its derivative for a value of** $\beta\mu = 0.10$.

Figure 10–2 shows $f(\varepsilon)$ and the derivative of $f(\varepsilon)$, $f'(\varepsilon)$, versus ε for $\beta\mu = 0.10$. This shows that even at $\beta\mu = 0.10$ (which, we shall show below, corresponds to a temperature at which most metals are liquid), $f(\varepsilon)$ is a step function with rounded edges, and its derivative therefore is a function that is zero everywhere except around the region in which $\varepsilon = \mu$. Thus it is convenient to express the integral I in Eq. (10–27) in terms of $f'(\varepsilon)$. This is done by an integration by parts:

$$I = - \int_0^\infty f'(\varepsilon)H(\varepsilon)\, d\varepsilon \tag{10–28}$$

where

$$H(\varepsilon) = \int_0^\varepsilon h(x)\, dx \tag{10–29}$$

and we have used the fact that $h(\varepsilon) = 0$ at $\varepsilon = 0$. Now since $f'(\varepsilon)$ is nonzero only for some small region around $\varepsilon = \mu$, the only important values of $H(\varepsilon)$ will be for values of ε around $\varepsilon = \mu$. Thus we expand $H(\varepsilon)$ in a Taylor series about $\varepsilon = \mu$:

$$H(\varepsilon) = H(\mu) + (\varepsilon - \mu)\left(\frac{dH}{d\varepsilon}\right)_{\varepsilon = \mu} + \tfrac{1}{2}(\varepsilon - \mu)^2 \left(\frac{d^2 H}{d\varepsilon^2}\right)_{\varepsilon = \mu} + \cdots \tag{10–30}$$

If we substitute this into Eq. (10–28), we get

$$I = H(\mu)L_0 + \left(\frac{dH}{d\varepsilon}\right)_{\varepsilon = \mu} L_1 + \frac{1}{2}\left(\frac{d^2 H}{d\varepsilon^2}\right)_{\varepsilon = \mu} L_2 + \cdots \tag{10–31}$$

where

$$L_j \equiv - \int_0^\infty (\varepsilon - \mu)^j f'(\varepsilon)\, d\varepsilon \tag{10–32}$$

The first of these integrals, L_0, is simply unity since it is equal to $f(0) - f(\infty)$. In the others, such as L_1 and L_2, we may replace the lower limit by $-\infty$ since $f'(\varepsilon)$ is so small from $-\infty$ to 0 that there is no contribution to the integral. If we let $x = \beta(\varepsilon - \mu)$, we have

$$L_j = \frac{1}{\beta^j} \int_{-\infty}^\infty \frac{x^j e^x}{(1 + e^x)^2}\, dx \qquad j = 0, 1, 2, \ldots . \tag{10–33}$$

All of these integrals with odd values of j vanish, since except for the x^j, the integrands are even functions of x. (See Problem 10–10.) The integral for L_2 is standard and may be found in tables:

$$\int_{-\infty}^\infty \frac{x^2 e^x\, dx}{(1 + e^x)^2} = \frac{\pi^2}{3}$$

This gives $L_2 = \pi^2(kT)^2/3$ and, by Eq. (10–31),

$$I = H(\mu) + \frac{\pi^2}{6}(kT)^2 H''(\mu) + \cdots \tag{10–34}$$

As an example of the use of this equation, we calculate N. In this case $h(\varepsilon) = 4\pi(2m/h^2)^{3/2}V\varepsilon^{1/2}$, and so

$$N = \frac{8\pi}{3}\left(\frac{2m}{h^2}\right)^{3/2}V\mu^{3/2}\left[1 + \frac{\pi^2}{8}(\beta\mu)^{-2} + \cdots\right] \tag{10–35}$$

Using Eq. (10–23) for μ_0, this becomes

$$\mu_0 = \mu\left[1 + \frac{\pi^2}{8}(\beta\mu)^{-2} + \cdots\right]^{2/3}$$

$$= \mu\left[1 + \frac{\pi^2}{12}(\beta\mu)^{-2} + \cdots\right]$$

This gives μ_0/μ as a power series in $(\beta\mu)^{-2}$. We can get μ/μ_0 as a power series in $\eta = (\beta\mu_0)^{-1}$ by taking the reciprocal of this equation

$$\frac{\mu}{\mu_0} = 1 - \frac{\pi^2}{12}(\beta\mu)^{-2} + \cdots$$

and then substituting $\mu = \mu_0(1 - \pi^2/12(\beta\mu)^2 + \cdots)$ into the right-hand side to get

$$\mu = \mu_0\left[1 - \frac{\pi^2}{12}\eta^2 + \cdots\right] \tag{10–36}$$

This equation shows that μ changes slowly with temperature and is approximately μ_0 throughout the entire solid-state range of a metal.

We can use Eq. (10–34) to calculate other thermodynamic properties such as E. This gives

$$E = \frac{8\pi}{5}\left(\frac{2m}{h^2}\right)^{3/2}V\mu^{5/2}[1 + \tfrac{5}{8}\pi^2(\beta\mu)^{-2} + \cdots]$$

$$= E_0\left(\frac{\mu}{\mu_0}\right)^{5/2}[1 + \tfrac{5}{8}\pi^2(\beta\mu)^{-2} + \cdots]$$

We can use Eq. (10–36) now to write E as a power series in η:

$$E = E_0\left[1 + \frac{5\pi^2}{12}\eta^2 + \cdots\right] \tag{10–37}$$

The contribution of the conductance electrons to the heat capacity is

$$C_V = \frac{\pi^2 NkT}{2(\mu_0/k)} = \frac{\pi^2}{2}Nk\left(\frac{T}{T_F}\right) \tag{10–38}$$

where we have introduced the Fermi temperature T_F. This equation predicts the molar electronic heat capacity of metals to be of the order of $10^{-4}T$ cal/deg-mole, and this is observed for many, but not all, metals to which the free electron model is expected to be applicable.

Fermi-Dirac statistics have been applied to other physical systems besides the electrons in metals. Two such applications are the theory of white dwarf stars and "nuclear" gases.

10–3   A WEAKLY DEGENERATE IDEAL BOSE-EINSTEIN GAS

In this section we shall treat an ideal Bose-Einstein gas in a region where λ is small enough that we may use series expansions in λ as we did in Section 10–1 for an ideal gas of fermions. We shall see that the results of this section are very similar to those of Section 10–1. In this case we use Eqs. (10–2) and (10–4) with the lower signs:

$$N = \sum_k \frac{\lambda e^{-\beta \varepsilon_k}}{(1 - \lambda e^{-\beta \varepsilon_k})} \tag{10-39}$$

$$pV = -kT \sum_k \ln(1 - \lambda e^{-\beta \varepsilon_k}) \tag{10-40}$$

The energy values here are also given by Eq. (10–8).

In Section 10–1 we converted these summations to integrals, since for large V the ε_k's vary essentially continuously. The discussion following Eq. (10–10) states that such a procedure is valid only because the summands are finite continuous functions of λ for all ε_k. If we let ε_0 be the ground translational energy state, we see from Eq. (10–39) that λ is restricted to the values $0 \leq \lambda < e^{\beta \varepsilon_0}$. Thus if we wish our results to be valid for all values of λ, we must take care in converting the summations in Eqs. (10–39) and (10–40) to integrals. In particular, we must single out the ground state and write

$$N = \frac{\lambda e^{-\beta \varepsilon_0}}{1 - \lambda e^{-\beta \varepsilon_0}} + \sum_{k \neq 0} \frac{\lambda e^{-\beta \varepsilon_k}}{1 - \lambda e^{-\beta \varepsilon_k}} \tag{10-41}$$

for Eq. (10–39), where the $k \neq 0$ means a summation over all states other than the ground state. Now since $0 \leq \lambda < e^{\beta \varepsilon_0}$, and $\varepsilon_k > \varepsilon_0$ for all $k \neq 0$, the summation here can be converted to an integral by the same method that we used in Section 10–1. Equation (10–41) becomes

$$N = \frac{\lambda e^{-\beta \varepsilon_0}}{1 - \lambda e^{-\beta \varepsilon_0}} + 2\pi \left(\frac{2m}{h^2}\right)^{3/2} V \int_{\varepsilon > \varepsilon_0}^{\infty} \frac{\lambda \varepsilon^{1/2} e^{-\beta \varepsilon} \, d\varepsilon}{1 - \lambda e^{-\beta \varepsilon}} \tag{10-42}$$

According to Eq. (10–8), the ground state $\varepsilon_0 = 3h^2/8mV^{2/3}$, but we can always redefine the energy such that $\varepsilon_0 = 0$. Physical results must be independent of where one sets the zero of energy of a system, and the following equations are simpler if we choose the energy such that $\varepsilon_0 = 0$. Equation (10–42) becomes

$$\rho = \frac{N}{V} = 2\pi \left(\frac{2m}{h^2}\right)^{3/2} \int_{\varepsilon > 0}^{\infty} \frac{\lambda \varepsilon^{1/2} e^{-\beta \varepsilon} \, d\varepsilon}{1 - \lambda e^{-\beta \varepsilon}} + \frac{\lambda}{V(1 - \lambda)} \tag{10-43}$$

Similarly, Eq. (10–40) is

$$\frac{p}{kT} = -2\pi \left(\frac{2m}{h^2}\right)^{3/2} \int_{\varepsilon > 0}^{\infty} \varepsilon^{1/2} \ln(1 - \lambda e^{-\beta \varepsilon}) \, d\varepsilon - \frac{1}{V} \ln(1 - \lambda) \tag{10-44}$$

where in both Eqs. (10–43) and (10–44), $0 \leq \lambda < 1$.

The second terms of both equations contain a factor of $1/V$. Ordinarily it is legitimate to ignore such terms, since we are always interested only in the thermodynamic limit, where $V \to \infty$. We must be careful here, however, since if $\lambda \to 1$, the term $\lambda/(1 - \lambda)$ becomes very large, and $\lambda/(1 - \lambda)$ divided by V is not necessarily negligible compared to the first term in Eq. (10–43). In this section we shall consider only the case where λ is close to zero, and so the $1/V$ terms in Eqs. (10–43) and (10–44) are not important. But in the next section we shall consider the strongly degenerate case in which $\lambda \to 1$, and we shall see that in this case the $1/V$ terms contribute in an important way to the thermodynamic functions.

The integrals occurring in Eqs. (10–43) and (10–44) can be evaluated as power series in λ in much the same way as Eqs. (10–9) and (10–10). The result is

$$\rho = \frac{1}{\Lambda^3} g_{3/2}(\lambda) \qquad (10\text{–}45)$$

and

$$\frac{p}{kT} = \frac{1}{\Lambda^3} g_{5/2}(\lambda) \qquad (10\text{–}46)$$

where

$$g_n(\lambda) = \sum_{l=1}^{\infty} \frac{\lambda^l}{l^n} \qquad (10\text{–}47)$$

Equation (10–45) gives ρ as a power series in λ, and Eq. (10–46) gives p/kT as a power series in λ. We can derive an expression for p/kT as an expansion in the density by *reverting* Eq. (10–45) and substituting this result into Eq. (10–46):

$$\frac{p}{\rho kT} = 1 - \frac{\Lambda^3}{2^{5/2}} \rho + \cdots \qquad (10\text{–}48)$$

The second virial coefficient in this case is negative, implying that the effective interaction between ideal bosons is attractive, in contrast to the case for fermions.

The Bose-Einstein equation for E, Eq. (10–4) with the negative sign, can be converted to an integral and then written as a series expansion in λ:

$$E = \tfrac{3}{2} V k T \frac{1}{\Lambda^3} g_{5/2}(\lambda)$$

In terms of ρ, this becomes

$$E = \tfrac{3}{2} N k T \left(1 - \frac{\Lambda^3}{2^{5/2}} \rho + \cdots \right) \qquad (10\text{–}49)$$

All other thermodynamic functions for a weakly degenerate ideal gas of bosons follow in a similar way. Just as in Section 10–1, these virial expansion expressions for the thermodynamic functions are useful only for small values of λ or ρ and represent small quantum corrections to the limiting classical results. In the next section we shall discuss the region of strong degeneracy, where λ approaches its largest allowed value ($\lambda \to 1$), and the $1/V$ terms in Eqs. (10–43) and (10–44) cannot be ignored.

10–4   A STRONGLY DEGENERATE IDEAL BOSE-EINSTEIN GAS

We now consider the situation when λ is not necessarily small. Let us return to Eqs. (10–43) and (10–44) for the density and pressure of an ideal Bose-Einstein gas

$$\rho = 2\pi \left(\frac{2m}{h^2}\right)^{3/2} \int_0^\infty \frac{\lambda \varepsilon^{1/2} e^{-\varepsilon/kT}}{1 - \lambda e^{-\varepsilon/kT}} \, d\varepsilon + \frac{\lambda}{V(1 - \lambda)}$$

$$\frac{p}{kT} = -2\pi \left(\frac{2m}{h^2}\right)^{3/2} \int_0^\infty \varepsilon^{1/2} \ln(1 - \lambda e^{-\varepsilon/kT}) \, d\varepsilon - \frac{1}{V} \ln(1 - \lambda)$$

or

$$\rho = \frac{1}{\Lambda^3} g_{3/2}(\lambda) + \frac{\lambda}{V(1 - \lambda)} \tag{10–50}$$

$$\frac{p}{kT} = \frac{1}{\Lambda^3} g_{5/2}(\lambda) - \frac{1}{V} \ln(1 - \lambda) \tag{10–51}$$

where $g_{3/2}(\lambda)$ and $g_{5/2}(\lambda)$ are defined by Eq. (10–47). Since λ can approach unity, we cannot ignore the $1/V$ terms in these equations.

Equation (10–3) shows that the average number of particles in their ground state is

$$\bar{n}_0 = \frac{\lambda}{1 - \lambda}$$

and so it is clear that $0 \leq \lambda < 1$. Note that there is no such restriction on the range of λ in the Fermi-Dirac case. There $0 \leq \lambda < \infty$. The function $g_{3/2}(\lambda)$ in Eq. (10–50) is a bounded, positive, and monotonically increasing function of λ in the range $0 \leq \lambda < 1$. At $\lambda = 1$ the first derivative diverges, but $g_{3/2}$ itself is finite since

$$g_{3/2}(1) = \sum_{l=1}^\infty \frac{1}{l^{3/2}} = \zeta(\tfrac{3}{2}) = 2.612 \ldots \tag{10–52}$$

where $\zeta(n)$ is the Riemann zeta function, defined by (see Problem 1–63)

$$\zeta(n) = \sum_{l=1}^\infty \frac{1}{l^n} \tag{10–53}$$

The function $g_{3/2}(\lambda)$ is plotted in Fig. 10–3.

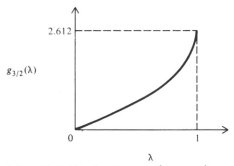

Figure 10–3. **The function** $g_{3/2}(\lambda)$ **versus** λ.

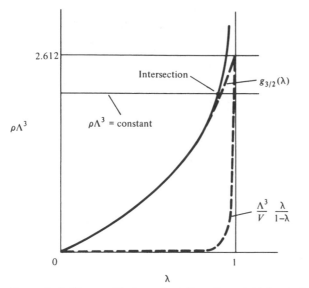

Figure 10–4. **The graphical solution of Eq. (10–50) for λ as a function of ρ and T. The dotted lines are $g_{3/2}(\lambda)$ and $\Lambda^3\lambda/V(1-\lambda)$. The solid lines are $\rho\Lambda^3$ = constant and $g_{3/2}(\lambda)+\Lambda^3\lambda/V(1-\lambda)$.**

In order to determine the equation of state, we must determine λ as a function of ρ and T by solving Eq. (10–50) for λ, and then substituting this into Eq. (10–51) for p/kT. Equation (10–50) cannot be solved analytically, but can be solved graphically by plotting both sides of the equation on the same graph and picking off the intersection. In particular, we shall plot $\rho\Lambda^3$ and $g_{3/2}(\lambda) + \Lambda^3\lambda/V(1 - \lambda)$ versus λ. (See Fig. 10–4.) For fixed values of ρ and T, the function $\rho\Lambda^3$ is just a constant and so appears as a horizontal line. The function $g_{3/2}(\lambda) + \Lambda^3\lambda/V(1 - \lambda)$, on the other hand, deserves some thought. Since V is large, the term $\Lambda^3\lambda/V(1 - \lambda)$ is small for all values of λ except those where λ is very close to unity. In fact, this term is negligible compared to $g_{3/2}(\lambda)$ *except* when $\lambda = 1 - 0(1/V)$, say $1 - a/V$, where a is some positive number (remember that λ must be less than 1). Thus a graph of $g_{3/2}(\lambda) + \Lambda^3\lambda/V(1 - \lambda)$ is indistinguishable from a graph of $g_{3/2}(\lambda)$ everywhere except where $\lambda \approx 1$. This sum $g_{3/2}(\lambda) + \Lambda^3\lambda/V(1 - \lambda)$, along with its two contributions, is plotted in Fig. 10–4. The dashes are the two separate terms, and the solid line is their sum. Note that the value of λ given by the intersection of the curves $\rho\Lambda^3$ = constant and $g_{3/2}(\lambda) + \Lambda^3\lambda/V(1 - \lambda)$, and the value of λ given by the intersection of $\rho\Lambda^3$ = constant and $g_{3/2}(\lambda)$ differ by only $0(1/V)$. This is so since $g_{3/2}(\lambda)$ and $g_{3/2}(\lambda) + \Lambda^3\lambda/V(1 - \lambda)$ differ only when $\lambda = 1 - 0(1/V)$ in the first place.

Figure 10–5 shows the set of intersections obtained by varying $\rho\Lambda^3$, which may be done by varying the density at constant temperature or by varying the temperature at constant density. Note that above $\rho\Lambda^3 = 2.612$, λ is essentially equal to unity, being different from unity by $0(1/V)$. This may be immediately seen by noting that the solid curve in Fig. 10–4 lies very close to the line $\lambda = 1$ for values of $\rho\Lambda^3$ greater than 2.612. The point $\rho\Lambda^3 = 2.612$ represents some sort of a critical point as we shall see below.

We can summarize the results of Figs. 10–4 and 10–5 mathematically in the following way. We wish to solve Eq. (10–50) for λ as a function of ρ and T, that is, we wish to solve

$$\rho\Lambda^3 = g_{3/2}(\lambda) + \frac{\Lambda^3}{V}\frac{\lambda}{1 - \lambda} \qquad (10\text{–}54)$$

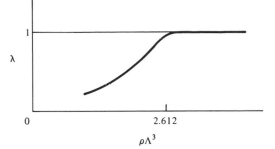

Figure 10-5. **A plot of λ versus $\rho\Lambda^3$, obtained from the locus of intercepts of plots such as the one in Figure 10-4.**

for λ. Figures 10-4 and 10-5 show that for values of $\rho\Lambda^3$ less than 2.612, λ is given to within $0(1/V)$ by the root of $g_{3/2}(\lambda) = \rho\Lambda^3 =$ constant. For values of $\rho\Lambda^3$ greater than 2.612, on the other hand, $\lambda = 1 - 0(1/V) = 1 - a/V$, where a is some positive number. We can determine a by substituting $\lambda = 1 - a/V$ into Eq. (10-54) and solving for a:

$$a = \frac{\Lambda^3}{\rho\Lambda^3 - g_{3/2}(1)} \qquad (\rho\Lambda^3 > 2.612) \tag{10-55}$$

where we have written $g_{3/2}(1)$ for $g_{3/2}(1 - 0(1/V))$ since $g_{3/2}(\lambda)$ is a continuous function of λ. Thus we have

$$\lambda = 1 - \frac{a}{V} \qquad \text{if } \rho\Lambda^3 > g_{3/2}(1)$$

$$= \text{the root of } g_{3/2}(\lambda) = \rho\Lambda^3 \qquad \text{if } \rho\Lambda^3 < g_{3/2}(1) \tag{10-56}$$

where a is given by Eq. (10-55). In the thermodynamic limit $V \to \infty$,

$$\lambda = 1 \qquad \text{if } \rho\Lambda^3 > g_{3/2}(1)$$
$$= \text{the root of } g_{3/2}(\lambda) = \rho\Lambda^3 \qquad \text{if } \rho\Lambda^3 < g_{3/2}(1) \tag{10-57}$$

Clearly the point $\rho\Lambda^3 = g_{3/2}(1) = 2.612$ is a special point. To explore its physical significance, consider $\rho\Lambda^3 = \rho(h^2/2\pi mkT)^{3/2}$ to be a function of temperature for a fixed density.

At high temperatures such that $\rho\Lambda^3 < 2.612$, λ must be determined numerically from the equation $g_{3/2}(\lambda) = \rho\Lambda^3$. But at low temperatures such that $\rho\Lambda^3 > 2.612$, $\lambda = 1 - a/V$ where a is given by Eq. (10-55). The quantity $\lambda/(1 - \lambda)$ is the average number of particles in their ground state, and for temperatures such that $\rho\Lambda^3 > 2.612$, we have

$$\bar{n}_0 = \frac{\lambda}{1 - \lambda} = \frac{V}{a} = \frac{V}{\Lambda^3}(\rho\Lambda^3 - g_{3/2}(1)) \qquad (\rho\Lambda^3 > 2.612) \tag{10-58}$$

We can write this in a more instructive form by defining a temperature T_0 by

$$\rho\Lambda_0^3 = \rho\left(\frac{h^2}{2\pi mkT_0}\right)^{3/2} = g_{3/2}(1) \tag{10-59}$$

In terms of this T_0 then, Eq. (10-58) becomes

$$\frac{\bar{n}_0}{N} = 1 - \left(\frac{T}{T_0}\right)^{3/2} \qquad T < T_0 \tag{10-60}$$

For temperatures greater than T_0, the value of λ determined from $g_{3/2}(\lambda) = \rho\Lambda^3$ will not be in the neighborhood of $\lambda = 1$ (unless $T = T_0 + 0(1/V)$, which is, of course, of no practical interest). Let this value of λ be denoted by λ_0. Then $\bar{n}_0 = \lambda_0/(1 - \lambda_0)$ and so $\bar{n}_0/N$ would be vanishingly small. Thus, at temperatures above T_0,

$$\frac{\bar{n}_0}{N} = 0 \qquad T > T_0 \tag{10-61}$$

Equations (10–60) and (10–61) for $\bar{n}_0/N$ are plotted in Fig. 10–6.

It can be seen that when $T > T_0$, the fraction of molecules in their ground state is essentially zero. This is the normal situation, where the molecules are distributed smoothly over the many molecular quantum states available to each one. However, as the temperature is lowered past T_0, suddenly the ground state begins to be appreciably populated, and the population increases until at $T = 0$ all the molecules are in their ground state. The fact that one state (the ground state) out of the many available to each molecule starts to become greatly preferred abruptly at $T = T_0$ is analogous to an ordinary phase transition. This "condensation" of the molecules into their ground states is called *Bose-Einstein condensation*.

A similar result is found if the temperature is held fixed and the density is allowed to vary. There is a critical density, above which the ground states of the molecules are not preferred, and below which the molecules tend to populate their ground state. The analog of Eqs. (10–60) and (10–61) is (see Problem 10–36)

$$\frac{\bar{n}_0}{N} = 1 - \frac{\rho_0}{\rho} \qquad \rho > \rho_0$$
$$= 0 \qquad \rho < \rho_0 \tag{10-62}$$

To determine the properties of a Bose-Einstein condensation, consider the equation of state. From Eq. (10–51), the pressure is given by

$$\frac{p}{kT} = \frac{1}{\Lambda^3} g_{5/2}(\lambda) - \frac{1}{V}\ln(1 - \lambda) \tag{10-63}$$

We may neglect the logarithm term here, since its greatest value is achieved when $1 - \lambda = a/V$. In this case the second term is $(1/V)\ln(1/V)$, and this vanishes as $V \to \infty$ since $x \ln x \to 0$ as $x \to 0$. Thus Eq. (10–63) is simply

$$\frac{p}{kT} = \frac{1}{\Lambda^3} g_{5/2}(\lambda) \tag{10-64}$$

where λ is determined in terms of ρ and T by the arguments surrounding Figs. 10–4 and 10–5, which are summarized in Eq. (10–56).

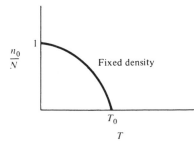

Figure 10–6. **The fraction of particles in their ground state as a function of temperature.**

If we consider p/kT to be a function of the density for fixed temperature, we can write

$$\frac{p}{kT} = \frac{1}{\Lambda^3} g_{5/2}(\lambda) \qquad \rho < \rho_0$$

$$= \frac{1}{\Lambda^3} g_{5/2}(1) \qquad \rho > \rho_0$$

(10–65)

where $g_{5/2}(1) = \zeta(\tfrac{5}{2}) = 1.342 \ldots$. The crucial point is that for $\rho > \rho_0$, p/kT is independent of the density and hence appears as a horizontal line when p/kT is plotted versus ρ. For $\rho < \rho_0$, on the other hand, p/kT is a function of ρ that can be determined numerically.

It is customary to plot isotherms of p versus v rather than p versus ρ, and these are shown in Fig. 10–7. Note that these isotherms are very similar to the isotherms observed for real gases.

The horizontal lines represent that region in which the system is a mixture of two phases. The points A and B correspond to the two phases in equilibrium: the condensed phase (A) and the dilute phase (B). The dilute phase has a specific volume v_0, and the condensed phase has a specific volume 0. The two phases have the same pressure, namely, the vapor pressure, which is given by

$$p_0(T) = \frac{kT}{\Lambda^3} g_{5/2}(1)$$

(10–66)

If we differentiate this and compare it to the Clapeyron equation,

$$\frac{dp}{dT} = \frac{\Delta H_{\text{cond}}}{T \, \Delta V}$$

(10–67)

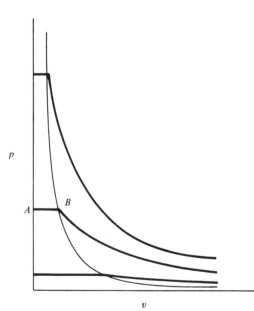

Figure 10–7. **The pressure-volume isotherms for an ideal Bose-Einstein gas. The points A and B correspond to the two phases in equilibrium.**

we see that there is a heat of transition associated with this process that is given by (see Problem 10–20)

$$\Delta H_{\text{cond}} = \tfrac{5}{2} kT \, \frac{g_{5/2}(1)}{g_{3/2}(1)} \tag{10-68}$$

Therefore the Bose-Einstein condensation is a first-order process. This is a very unusual first-order transition, however, since the condensed phase has no volume, and the system therefore has a uniform macroscopic density rather than the two different densities that are usually associated with first-order phase transitions. This is often interpreted by saying that the condensation occurs in momentum space rather than coordinate space, particularly since from a classical point of view, the particles in the condensed phase are found in the same region of momentum space, namely, zero momentum.

One observes similar behavior in the other thermodynamic functions. The heat capacity has a particularly interesting property. The thermodynamic energy is given by

$$E = \frac{3}{2} \frac{kTV}{\Lambda^3} g_{5/2}(\lambda) \tag{10-69}$$

which becomes

$$\frac{E}{N} = \frac{3}{2} \frac{kTv}{\Lambda^3} g_{5/2}(\lambda) \qquad T > T_0$$
$$= \frac{3}{2} \frac{kTv}{\Lambda^3} g_{5/2}(1) \qquad T < T_0 \tag{10-70}$$

We differentiate these with respect to T at constant N and V, we get (see Problem 10–19)

$$\frac{C_V}{Nk} = \frac{15}{4} \frac{v}{\Lambda^3} g_{5/2}(\lambda) - \frac{9}{4} \frac{g_{3/2}(\lambda)}{g_{1/2}(\lambda)} \qquad T > T_0$$
$$= \frac{15}{4} \frac{v}{\Lambda^3} g_{5/2}(1) \qquad\qquad T < T_0 \tag{10-71}$$

Again, λ must be determined numerically from Eq. (10–50), and the result of this is shown in Fig. 10–8(a), where C_V is plotted against T. There is no discontinuity in C_V at T_0, but there is a discontinuity in the slope there.

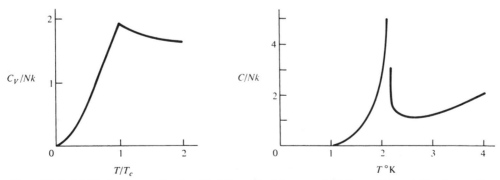

Figure 10–8. **(a) The heat capacity of an ideal Bose-Einstein gas and (b) the experimental heat capacity of liquid helium under its saturated vapor.**

Bose-Einstein condensation is an interesting phenomenon since it occurs even though the particles do not interact with each other through an intermolecular potential. Indeed, however, there is an effective interaction through the symmetry requirement of the N-body wave function of the system, and it is this effective interaction that leads to the condensation. Even though the results derived here are valid only for an ideal gas, there is a real system to which they are approximately applicable. Helium exists in the form of two isotopes: He-3 and He-4. The thermodynamic properties of the pure isotopes have been extensively studied, and it turns out that He-4, which has a spin of zero and therefore must obey Bose-Einstein statistics, exhibits many remarkable properties, one of which being the heat capacity curve shown in Fig. 10–8(b). The similarity between this experimental curve and the Greek letter λ has led the transition to be referred to as a "lambda transition." Although the heat capacity appears to diverge logarithmically at $T = 2.18°$K, the similarity between it and the heat capacity curve of an ideal Bose-Einstein gas shown in Fig. 10–8(a) is striking. One cannot expect complete agreement since liquid He-4 is an inextricable combination of quantum statistics *and* intermolecular interactions, but it appears that the experimental heat capacity is due in part to the quantum statistics of the Bose-Einstein He-4 system. Furthermore, liquid He-3, which obeys Fermi-Dirac statistics, does not have any unusual behavior in its heat capacity curve, just as the heat capacity curve of an ideal Fermi-Dirac gas is "normal." Most intriguing, however, is that if we calculate the value of T_0 from Eq. (10–59) (using the density of liquid helium = 0.145 g/cm^3), we find that $T_0 = 3.14°$K, which is the right order of magnitude. One important difference between the λ-transition in He-4 and the Bose-Einstein condensation is that the λ-transition is not a first-order transition. Nevertheless, Bose-Einstein statistics seem to play an important role in the λ-transition in liquid He-4.

10–5 AN IDEAL GAS OF PHOTONS (BLACKBODY RADIATION)

In this section we shall apply statistical thermodynamics to electromagnetic radiation enclosed in a fixed volume V and at a fixed temperature T. The experimental system is obtained by making a cavity in any material, evacuating the cavity, and then heating the material to the temperature T. The atoms of the walls of the cavity constantly emit and absorb radiation, and so at equilibrium we will have a cavity filled with electromagnetic radiation. Such a cavity is called a blackbody cavity, and the radiation within the cavity is called blackbody radiation.

The quantum mechanical theory of electromagnetic radiation tells us that an electromagnetic wave may be regarded as a massless particle of spin angular momentum $\hbar = h/2\pi$ and with a momentum and energy that are functions of the wavelength. These massless particles are called photons. Since photons have a spin of 1 (in units of $\hbar$), they form an ideal Bose-Einstein gas. A new feature here is that since the walls of the blackbody cavity are constantly emitting and absorbing photons, the number of photons is not fixed at any instant, and thus N is not an independent thermodynamic variable. Thermodynamically, the blackbody cavity is described by V and T.

Let us briefly review some of the results of vibratory motion. In what follows, we shall consider only the "electro" part of an electromagnetic wave, since this is several orders of magnitude more important for our purposes. Consider a harmonic electromagnetic wave of unit amplitude traveling with velocity c in the positive x-direction. Mathematically, this is described by

$$E(x, t) = \sin\left[\frac{2\pi}{\lambda}(x - ct)\right] \tag{10–72}$$

and is called a traveling wave. The symbol λ is the wavelength of the wave. Physicists customarily write this equation in the form

$$E(x, t) = \sin(kx - \omega t) \tag{10-73}$$

where $k = 2\pi/\lambda$ and is called the wave vector, and $\omega = 2\pi\nu$. In some sense, this wave also describes a photon with energy and momentum given by

$$\varepsilon = h\nu = \hbar\omega = \hbar c k$$

$$\text{momentum} = \frac{h}{\lambda} = \hbar k \tag{10-74}$$

We shall consider blackbody radiation at equilibrium to be a system of standing waves set up in the cavity. It is not necessary to do this, since as the volume becomes large, the boundary conditions chosen have no effect on the thermodynamic properties, but imposing a boundary condition of standing waves is as convenient as any. Consider the superposition of two traveling harmonic waves of unit amplitude and traveling in opposite directions. This is given by

$$\phi(x, t) = \sin(kx - \omega t) + \sin(kx + \omega t)$$
$$= 2 \sin kx \cos \omega t \tag{10-75}$$

This new wave does not move either backward or forward. It vanishes at the values of x for which $\sin kx = 0$ for all values of t. These points are called nodes. They occur at the values of x given by $kx = n\pi$, where $n = 1, 2, \ldots$. In between the nodes, the disturbance vibrates harmonically with time. Figure 10–9 shows the first few standing waves that can be set up between 0 and L. The appropriate boundary condition is that the standing wave be fixed at the end points 0 and L, which implies that the wave vector k be given by

$$k = \frac{n\pi}{L} \qquad n = 1, 2, \ldots \tag{10-76}$$

Note that this is equivalent to saying that there must be an integral number of half wavelengths between 0 and L, that is, $n(\lambda/2) = L$.

In order to discuss three-dimensional waves, it is convenient to use an exponential representation rather than a sine or a cosine. In one dimension, a wave of unit amplitude can be described by

$$E(x, t) = e^{i(kx - \omega t)} \tag{10-77}$$

Equation (10–73) is obtained by taking the imaginary part of this. This is also a harmonic wave traveling in the positive x-direction. A standing wave in this representation is given by the imaginary part of

$$E(x, t) = e^{i(kx - \omega t)} + e^{i(kx + \omega t)}$$
$$= 2e^{ikx} \cos \omega t \tag{10-78}$$

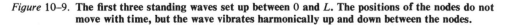

Figure 10–9. **The first three standing waves set up between 0 and L. The positions of the nodes do not move with time, but the wave vibrates harmonically up and down between the nodes.**

The boundary condition given in Eq. (10–76) is obtained from this by requiring that the imaginary part of Eq. (10–78) vanish at the endpoints 0 and L. Remember that the imaginary part of $e^{i\theta}$ vanishes whenever θ is an integral multiple of π.

In three dimensions, an electromagnetic wave is described by

$$\mathbf{E}(\mathbf{r}, t) = \boldsymbol{\sigma}e^{i(\mathbf{k} \cdot \mathbf{r} - \omega t)} \tag{10–79}$$

This represents a traveling wave propagated in the direction of the wave vector $\mathbf{k}$. The wavelength λ is given by $|\mathbf{k}| = 2\pi/\lambda$. The wave propagates in the direction of $\mathbf{k}$, but the direction of $\mathbf{E}$ itself is in the direction of $\boldsymbol{\sigma}$. The vector $\boldsymbol{\sigma}$ is called the polarization vector and is perpendicular to $\mathbf{k}$. Thus $\mathbf{E}$ represents a *transverse* wave, that is, one in which the disturbance vibrates in a direction perpendicular to the direction of propagation. In fact, $\mathbf{E}$ vibrates in the plane perpendicular to $\mathbf{k}$, and hence there are two and only two independent polarization vectors.

A standing wave can be formed from Eq. (10–79) (where it is understood that we take the imaginary part):

$$\mathbf{E}(\mathbf{r}, t) = 2\boldsymbol{\sigma}e^{i\mathbf{k} \cdot \mathbf{r}} \cos \omega t \tag{10–80}$$

If we let the components of the wave vector $\mathbf{k}$ be k_x, k_y, and k_z, the imaginary part of this vanishes when $k_x L = n_x \pi$, $k_y L = n_y \pi$, and $k_z L = n_z \pi$, or in vector notation, when

$$\mathbf{k} = \frac{\pi}{L}\mathbf{n} \qquad n_x, n_y, n_z = 1, 2, \dots \tag{10–81}$$

We have assumed for simplicity here that the volume is a cube of length L. The energy and momentum are given by

$$\begin{aligned} \varepsilon &= \hbar c |\mathbf{k}| = \hbar c k \\ \text{momentum} &= \hbar |\mathbf{k}| = \hbar k \end{aligned} \tag{10–82}$$

Note that the energy depends only upon $|\mathbf{k}|$. We are going to need an expression for the number of standing waves with energy between ε and $\varepsilon + d\varepsilon$. This is found by the same method that we used to find the number of translational energy states between ε and $\varepsilon + d\varepsilon$ in Section 1–3. The square of $\mathbf{k}$ is

$$k^2 = \frac{\pi^2}{L^2}(n_x^2 + n_y^2 + n_z^2) \qquad n_x, n_y, n_z = 1, 2, \dots \tag{10–83}$$

Thus the number of standing waves with the magnitude of the vector less than k is

$$\Phi(k) = \frac{\pi}{6}\left(\frac{Lk}{\pi}\right)^3 = \frac{L^3 k^3}{6\pi^2} = \frac{V k^3}{6\pi^2} \tag{10–84}$$

and the number between k and k and $k + dk$ is

$$\omega(k)\,dk = \frac{d\Phi}{dk}\,dk = \frac{V k^2\,dk}{2\pi^2} \tag{10–85}$$

According to Eq. (10–82), the energy $\varepsilon = \hbar c k$. Furthermore, there are two polarizations with the energy ε, and so we can write

$$\omega(\varepsilon)\,d\varepsilon = \frac{V\varepsilon^2\,d\varepsilon}{\pi^2 c^3 \hbar^3} \tag{10–86}$$

for the number of standing waves with energy between ε and $\varepsilon + d\varepsilon$.

The total energy of the system is given by

$$E(\{n_k\}) = \sum_k \varepsilon_k n_k \tag{10-87}$$

where the number of "particles" with energy ε_k (actually $\hbar ck$) is $n_k = 0, 1, 2, \ldots$ since photons are bosons. The partition function is

$$Q(V, T) = \sum_{\{n_k\}} e^{-\beta E(\{n_k\})} = \sum_{\{n_k\}} e^{-\beta \sum_k \varepsilon_k n_k} \tag{10-88}$$

This summation is similar to the one in Eq. (4–18), but in this case there is no restriction on the set $\{n_k\}$ since the number of photons is not conserved. This is why Q in Eq. (10–88) is a function of only V and T rather than the usual N, V, and T. The quantity N is not a thermodynamic variable in this case, since it is not fixed.

Without the restriction, a summation like the one in Eq. (10–88) is easy since

$$Q(V, T) = \prod_k \left(\sum_{n=0}^{\infty} e^{-\beta \varepsilon_k n} \right)$$

$$= \prod_k \frac{1}{1 - e^{-\beta \varepsilon_k}} \tag{10-89}$$

All thermodynamic functions are expressed in terms of $\ln Q$, which is

$$\ln Q = -\sum_k \ln(1 - e^{-\beta \varepsilon_k}) = -\sum_\varepsilon \ln(1 - e^{-\beta \varepsilon}) \tag{10-90}$$

As usual, this summation can be converted to an integral by introducing the density of states and treating ε_k to be a continuous variable. Thus

$$\ln Q = -\frac{V}{\pi^2 c^3 \hbar^3} \int_0^\infty \varepsilon^2 \ln(1 - e^{-\beta \varepsilon}) \, d\varepsilon \tag{10-91}$$

This integral may be readily evaluated by expanding the logarithm and integrating term by term:

$$\ln Q = \frac{V}{\pi^2 c^3 \hbar^3} \sum_{n=1}^{\infty} \frac{1}{n} \int_0^\infty \varepsilon^2 e^{-n\beta \varepsilon} \, d\varepsilon$$

$$= \frac{V}{\pi^2 c^3 \hbar^3} \frac{2}{\beta^3} \sum_{n=1}^{\infty} \frac{1}{n^4} = \frac{2V}{\pi^2 (c\hbar\beta)^3} \zeta(4) \tag{10-92}$$

where $\zeta(4)$ is the summation, which is a Riemann zeta function [cf. Eq. (10–52)] and is equal to $\pi^4/90$. (See Problem 1–63.)

The thermodynamic energy is given by

$$E = kT^2 \left(\frac{\partial \ln Q}{\partial T} \right)_V = \frac{\pi^2 V (kT)^4}{15(\hbar c)^3} \tag{10-93}$$

This result can be used to derive the Stefan-Boltzmann law. In Problem 7–24 it is shown that the number of gas molecules striking a surface per unit area per unit time is $\rho \bar{v}/4$, where ρ is the number density, and $\bar{v}$ is the average velocity. By analogy, then, $cE(T)/4V$ is the energy incident per unit area per unit time on the wall of the enclosure containing the radiation. Thus if one cuts a small hole of unit area in the wall, the energy radiated per unit time

$$R = \frac{cE}{4V} = \frac{\pi^2 (kT)^4}{60\hbar^3 c^2} \equiv \sigma T^4 \tag{10-94}$$

This is known as the Stefan-Boltzmann law, and σ is known as the Stefan-Boltzmann constant.

The pressure of the blackbody radiation is

$$p = kT\left(\frac{\partial \ln Q}{\partial V}\right)_T = \frac{2(kT)^4}{\pi^2(\hbar c)^3}\,\zeta(4) = \frac{\pi^2(kT)^4}{45(\hbar c)^3} \qquad (10\text{--}95)$$

The pressure due to the radiation is negligible except for the highest temperatures.

It is instructive to go on and calculate the entropy and the chemical potential. The entropy is given by [*cf*. Eq. (2–33)]

$$\begin{aligned}
S &= k \ln Q + kT\left(\frac{\partial \ln Q}{\partial T}\right)_V \\
&= \frac{4\pi^2 Vk(kT)^3}{45(\hbar c)^3}
\end{aligned} \qquad (10\text{--}96)$$

If we calculate $G = \bar{N}\mu$ from the equation $\bar{N}\mu = E - TS + pV$, we find that

$$\bar{N}\mu = 0$$

and since $\bar{N}$ is not zero, this implies that the chemical potential equals zero for an ideal gas of photons. We can prove on thermodynamic grounds that this is true for any system in which the number of particles is not conserved. In such a system we can write the chemical reaction $mA \rightleftharpoons nA$, where m and n are arbitrary integers. If we apply the criterion that the change in chemical potential $\Delta\mu = 0$ for a system in equilibrium, we find that $\Delta\mu = m\mu - n\mu = (m - n)\mu = 0$. But since $m - n$ does not equal zero, this implies that $\mu = 0$. Since $\lambda = \exp(\beta\mu)$, this also says that we could have derived all of the above results by setting $\lambda = 1$ in the Bose-Einstein formulas (see Problem 10–33).

Before leaving this section, we wish to calculate the energy density at each frequency. We can do this by noting that Eq. (10–90) is equivalent to

$$\ln Q = -\sum_\omega \ln(1 - e^{-\beta\hbar\omega})$$

where $\omega = \varepsilon/\hbar = ck$. If we use this directly in Eq. (10–93), we find that

$$E = \sum_\omega \frac{\hbar\omega e^{-\beta\hbar\omega}}{1 - e^{-\beta\hbar\omega}} \qquad (10\text{--}97)$$

This can be converted to an integral over ω by introducing the number of states between ω and $\omega + d\omega$ from Eq. (10–86) and $\omega = \varepsilon/\hbar$. (It is unfortunate that the same notation is utilized for the density of states and the frequency, but since the density of states always occurs with an argument, such as $\omega(\varepsilon)$ or $\omega(k)$, there should be no confusion. We hesitate writing $\omega(\omega)$, however.) If we introduce the density of frequencies, Eq. (10–97) becomes

$$E = \frac{V\hbar}{\pi^2 c^3} \int_0^\infty \frac{\omega^3}{e^{\beta\hbar\omega} - 1}\, d\omega \qquad (10\text{--}98)$$

The thermodynamic energy per unit volume can be written in terms of the energy density of each frequency by

$$\frac{E}{V} = \int_0^\infty \rho(\omega, T)\, d\omega$$

where

$$\rho(\omega, T)\, d\omega = \frac{\hbar}{\pi^2 c^3} \frac{\omega^3}{e^{\beta\hbar\omega} - 1}\, d\omega \tag{10-99}$$

This is the famous blackbody distribution law first derived by Planck in 1901. Clearly he did not derive this formula in the way that we have since his work preceded, and, in fact, led to the development of quantum mechanics. Problems 10–25 through 10–27 are involved with some well-known results of blackbody radiation.

10–6 THE DENSITY MATRIX

In the previous sections of this chapter, we have treated ideal Fermi-Dirac and Bose-Einstein gases by applying the results of Chapter 4: The basic equations there are derived on the basis of independent particles and so are not applicable when inter-molecular forces are present. In this section we introduce a formalism of quantum statistical mechanics, which can be generalized to study nonideal quantum systems.

Let us return now to the completely general equations

$$Q = \sum_j e^{-\beta E_j} \tag{10-100}$$

$$\overline{M} = \frac{1}{Q} \sum_j M_j e^{-\beta E_j} \tag{10-101}$$

where $\overline{M}$ is the ensemble average of the mechanical property, and M_j is the quantum mechanical expectation value of the operator $\hat{M}$ in the jth quantum state.

We wish to express Q and $\overline{M}$ in terms of the quantum mechanical operators $\mathscr{H}$ and $\hat{M}$. Let $\{\psi_j\}$ be the set of normalized eigenfunctions of $\mathscr{H}$ and $\{E_j\}$ be the corresponding eigenvalues, that is,

$$\mathscr{H}\psi_j = E_j\psi_j \tag{10-102}$$

Since ψ_j is an eigenfunction of $\mathscr{H}$, we can also write

$$\mathscr{H}^n\psi_j = E_j^n\psi_j \tag{10-103}$$

for integral n. This allows us to define the result of an analytic function of $\mathscr{H}$ acting on ψ_j. In particular, we have

$$e^{-\beta\mathscr{H}}\psi_j = \left(\sum_{n=0}^{\infty} \frac{(-\beta)^n}{n!} \mathscr{H}^n\right)\psi_j = \sum_{n=0}^{\infty} \frac{(-\beta)^n}{n!} \mathscr{H}^n\psi_j$$

$$= \sum_{n=0}^{\infty} \frac{(-\beta)^n}{n!} E_j^n\psi_j = e^{-\beta E_j}\psi_j \tag{10-104}$$

(In general, functions of operators are defined through their MacLaurin expansions.) Thus we have

$$e^{-\beta\mathscr{H}}\psi_j = e^{-\beta E_j}\psi_j \tag{10-105}$$

We multiply both sides of this equation by $\psi_j{}^*$ and integrate over all the coordinates involved to get

$$e^{-\beta E_j} = \int \psi_j{}^* e^{-\beta\mathscr{H}}\psi_j\, d\tau \tag{10-106}$$

where we have used the fact that $\exp(-\beta E_j)$ is just a number and that ψ_j is normalized. The symbol $d\tau$ represents an integration over all the coordinates on which ψ_j depends. Using Eq. (10–106) then, Q becomes

$$Q = \sum_j e^{-\beta E_j} = \sum_j \int \psi_j^* e^{-\beta \mathcal{H}} \psi_j \, d\tau \tag{10–107}$$

Integrals of quantum mechanical operators, such as $\exp(-\beta \mathcal{H})$ above, between wave functions occur naturally in the matrix formulation of quantum mechanics and are called matrix elements. They can be represented by the notation

$$\int \psi_i^* e^{-\beta \mathcal{H}} \psi_j \, d\tau = (e^{-\beta \mathcal{H}})_{ij} \tag{10–108}$$

In this notation, the canonical partition function is

$$Q = \sum_j (e^{-\beta \mathcal{H}})_{jj} = \text{Tr}(e^{-\beta \mathcal{H}}) \tag{10–109}$$

where we have employed the notation of representing a summation over the diagonal elements of a matrix by Tr, which stands for "trace."

Now there is a standard theorem of matrix algebra which says that the trace of a matrix is independent of the particular function ψ_i in Eq. (10–108) used to calculate the matrix elements. This theorem is easy to prove. Let $\{\psi_j\}$ be the orthonormal set of eigenfunctions of $\mathcal{H}$ and let $\{\phi_n\}$ be any orthonormal set of functions that can be expanded in terms of the ψ_j as

$$\phi_j = \sum_n a_{jn} \psi_n \tag{10–110}$$

where the a_{jn}'s are constant. We can calculate the a_{jn}'s by multiplying both sides of eq. (10–110) by ψ_k^*, integrating over all values of the coordinates, and using the fact that the ψ_j's are orthonormal:

$$a_{jn} = \int \psi_n^* \phi_j \, d\tau \tag{10–111}$$

Furthermore, since the ϕ_j's are normalized, we have that the a_{jn} must satisfy the conditions (see Problem 10–34)

$$\sum_n a_{jn}^* a_{jn} = 1 \tag{10–112}$$

We shall also need the coefficients for the expansion of ψ_j in terms of the ϕ's. Write

$$\psi_s = \sum_t b_{st} \phi_t \tag{10–113}$$

The b_{st} can be related to the a_{jn} by multiplying both sides of this equation by ϕ_t^* and using the fact that the ϕ_t's are orthonormal:

$$b_{st} = \int \phi_t^* \psi_s \, d\tau = a_{ts}^* \tag{10–114}$$

where we have used Eq. (10–111) to write the last equality. Thus b_{ij}'s are obtained from the a_{ij}'s by reversing the subscripts and taking the complex conjugate. Since the ψ's are normalized, we have

$$\sum_n b_{jn}^* b_{jn} = \sum_n a_{nj}^* a_{nj} = 1 \tag{10–115}$$

Note that this equation is similar to Eq. (10–112). Here we sum over the first subscript, and there we sum over the second. We shall need this relation to prove the theorem referred to above, namely, that

$$Q = \sum_j \int \psi_j{}^* e^{-\beta\mathscr{H}} \psi_j \, d\tau = \sum_j \int \phi_j{}^* e^{-\beta\mathscr{H}} \phi_j \, d\tau \tag{10–116}$$

To prove this, we substitute Eq. (10–110) into Eq. (10–116) to get

$$\int \phi_j{}^* e^{-\beta\mathscr{H}} \phi_j \, d\tau = \sum_{m,n} a_{jm}{}^* a_{jn} e^{-\beta E_n} \int \psi_m{}^* \psi_n \, d\tau \tag{10–117}$$

But the integral is just a Kroenecker delta δ_{mn} since the set $\{\psi_j\}$ is orthonormal. Thus the double summation over m and n becomes a single summation, and we have

$$\int \phi_j{}^* e^{-\beta\mathscr{H}} \phi_j \, d\tau = \sum_n a_{jn}{}^* a_{jn} e^{-\beta E_n} \tag{10–118}$$

We now sum both sides over j and use Eq. (10–115)

$$\sum_j \int \phi_j{}^* e^{-\beta\mathscr{H}} \phi_j \, d\tau = \sum_n e^{-\beta E_n} = \sum_j e^{-\beta E_j} = Q \tag{10–119}$$

which shows that the trace of $\exp(-\beta\mathscr{H})$ is independent of the particular orthonormal set of functions used to compute the matrix elements.

Equation (10–109) gives Q in terms of the quantum mechanical operator $\mathscr{H}$. We now wish to express Eq. (10–101) for $\bar{M}$ in terms of quantum mechanical operators. The ensemble average of M is

$$\bar{M} = \frac{\sum_j M_j e^{-\beta E_j}}{\sum_j e^{-\beta E_j}} \tag{10–120}$$

We first use the fact that M_j is given by

$$M_j = \int \psi_j{}^* \hat{M} \psi_j \, d\tau \tag{10–121}$$

where $\hat{M}$ denotes the quantum mechanical operator corresponding to M, and ψ_j is the eigenfunction of $\mathscr{H}$ in the jth state. We substitute this into the numerator of Eq. (10–120) and perform a series of elementary manipulations:

$$\sum_j M_j e^{-\beta E_j} = \sum_j e^{-\beta E_j} \int \psi_j{}^* \hat{M} \psi_j \, d\tau$$

$$= \sum_j \int \psi_j{}^* \hat{M} e^{-\beta E_j} \psi_j \, d\tau = \sum_j \int \psi_j{}^* \hat{M} e^{-\beta\mathscr{H}} \psi_j \, d\tau$$

$$= \sum_j (\hat{M} e^{-\beta\mathscr{H}})_{jj} = \mathrm{Tr}(\hat{M} e^{-\beta\mathscr{H}}) \tag{10–122}$$

The invariance of the trace with respect to the functions used to calculate the matrix elements says that any convenient orthonormal set of functions could have been used to derive this result.

Using this result for the numerator in Eq. (10–120) gives

$$\bar{M} = \frac{\mathrm{Tr}(\hat{M} e^{-\beta\mathscr{H}})}{\mathrm{Tr}(e^{-\beta\mathscr{H}})} \tag{10–123}$$

The denominator of this expression is a scalar, and hence we can define a new operator $\hat{\rho}$ by

$$\hat{\rho} = \frac{e^{-\beta\mathscr{H}}}{\mathrm{Tr}(e^{-\beta\mathscr{H}})} \tag{10–124}$$

and write for Eq. (10–123)

$$\overline{M} = \mathrm{Tr}(\hat{M}\hat{\rho}) \tag{10–125}$$

The operator $\hat{\rho}$ is the quantum mechanical analog of the equilibrium density of points in the phase space for an canonical ensemble. The matrix corresponding to $\hat{\rho}$ is called the *density matrix*. Equation (10–123) corresponds to the classical expression

$$\overline{M} = \frac{\int \cdots \int dp\, dq\, M(p,q) e^{-\beta\mathscr{H}(p,q)}}{\int \cdots \int dp\, dq\, e^{-\beta\mathscr{H}(p,q)}} \tag{10–126}$$

One can prove that the trace operation in Eq. (10–123) goes into the phase space integration in Eq. (10–126) as $h \to 0$.

We have derived the above equations and defined the density matrix ρ only for a canonical ensemble. The density matrix can be defined in a much more general manner, but since we shall need only the results for a canonical ensemble in this book we simply refer to Tolman (see "Additional Reading"), who gives a complete discussion of the density matrix. Tolman shows that ρ can be defined for nonequilibrium systems, giving the quantum mechanical correspondence to the density of phase points in phase space. There is, for example, a quantum mechanical analog of the Liouville equation.

10–7 THE CLASSICAL LIMIT FROM THE QUANTUM MECHANICAL EXPRESSION FOR Q

In this section we shall derive the classical mechanical form of the canonical partition function (an integration over phase space) directly from its quantum mechanical form (a summation over energy states). In other words, we shall show that

$$Q = \sum_j e^{-\beta E_j} \xrightarrow{\hbar \to 0} \frac{1}{N!h^{3N}} \int \cdots \int e^{-\beta H}\, d\mathbf{p}_1 \cdots d\mathbf{p}_N\, d\mathbf{q}_1 \cdots d\mathbf{q}_N$$

In Chapter 7 we presented an argument that was meant to make this correspondence plausible, but here we shall present a rigorous treatment due originally to Kirkwood.* In particular, we shall show how to derive an expansion for Q in powers of $\hbar$, with the leading term being the classical limit. Such expansions were derived in Sections 10–1 and 10–3, but there the results were valid only for ideal systems. Here we shall relax this restriction. The quantum mechanical level of this section is somewhat higher than most of the others in this book, and this section can be omitted on first reading, since the material presented here is not necessary in the development of later sections. For those who do choose to read this section, however, Appendix B presents a brief discussion of Fourier transforms, delta functions, and so on.

We start with Eq. (10–116) for a system of N monatomic particles, namely,

$$Q = \sum_j \int \phi_j{}^* e^{-\beta\mathscr{H}} \phi_j\, d\mathbf{r} \tag{10–127}$$

* J. G. Kirkwood, *Phys. Rev.*, **44** p. 31, 1933; **45**, p. 116, 1934. See also Hill in "Additional Reading."

where we have written $dr = dx_1 \, dy_1 \cdots dz_N$ for $d\tau$. The functions $\phi_j(\mathbf{r}_1, \ldots, \mathbf{r}_N)$ in this case are taken to be the eigenfunctions of the Hamiltonian operator and form a complete orthonormal set. The Hamiltonian operator $\mathscr{H}$ is

$$\mathscr{H} = \mathscr{K} + U = -\frac{h^2}{2m} \sum_{l=1}^{N} \nabla_l^2 + U(\mathbf{r}_1, \ldots, \mathbf{r}_N)$$

We wish to transform Eq. (10–127) into an integration over phase space. To accomplish this, we introduce the eigenfunctions $u(\mathbf{p}_1, \ldots, \mathbf{r}_N)$ of the momentum operator $-i\hbar\nabla$:

$$u(\mathbf{p}_1, \ldots, \mathbf{r}_N) = \exp\left[\frac{i}{\hbar} \sum_{k=1}^{N} \mathbf{p}_k \cdot \mathbf{r}_k\right] \tag{10–128}$$

Note that $-i\hbar\nabla_i u = \mathbf{p}_i u$. We now expand the ϕ_j in Eq. (10–127) in terms of the momentum eigenfunctions

$$\phi_j(\mathbf{r}_1, \ldots, \mathbf{r}_N) = \int \cdots \int A_j(\mathbf{p}_1, \ldots, \mathbf{p}_N) \exp\left[\frac{i}{\hbar} \sum_{k=1}^{N} \mathbf{p}_k \cdot \mathbf{r}_k\right] d\mathbf{p}_1 \cdots d\mathbf{p}_N \tag{10–129}$$

The ϕ_j in Eq. (10–127), being eigenfunctions of $\mathscr{H}$, are either symmetric or antisymmetric in the coordinates of the N particles. Thus they should be expanded in terms of linear combinations of the $u(\mathbf{p}_1, \ldots, \mathbf{r}_N)$ in Eq. (10–128) that themselves are symmetric or antisymmetric in the coordinates of the N particles. We could do this by introducing a permutation operator $\mathscr{P}$, but the resulting equations become fairly involved. For simplicity, we shall ignore this complication and simply use Eq. (10–129). The result of this is that we shall not derive the factor of $N!$ that occurs in the classical partition function. Thus the central result of this section, Eq. (10–133) with Eqs. (10–135), (10–138), through (10–140), is lacking an $N!$ in its denominator. This factor is included, however, in Kirkwood's original paper and in the more pedagogical discussion of Hill.

Equation (10–129) shows $\phi_j(\mathbf{r}_1, \ldots, \mathbf{r}_N)$ as a Fourier transform of A_j, and so we can use the inversion theorem of Fourier transforms to write

$$A_j(\mathbf{p}_1, \ldots, \mathbf{p}_N) = \frac{1}{(2\pi\hbar)^{3N}} \int \cdots \int \phi_j(\mathbf{r}_1, \ldots, \mathbf{r}_N) \exp\left[-\frac{i}{\hbar} \sum_{k=1}^{N} \mathbf{p}_k \cdot \mathbf{r}_k\right] d\mathbf{r}_1 \cdots d\mathbf{r}_N \tag{10–130}$$

We now substitute Eq. (10–129) into Eq. (10–127) to get

$$Q = \sum_j \int \cdots \int \phi_j^*(\mathbf{r}_1, \ldots, \mathbf{r}_N) A_j(\mathbf{p}_1, \ldots, \mathbf{p}_N) e^{-\beta\mathscr{H}} \exp\left[\frac{i}{\hbar} \sum_{k=1}^{N} \mathbf{p}_k \cdot \mathbf{r}_k\right] d\mathbf{p}_1 \cdots d\mathbf{r}_N$$

and then use Eq. (10–130) for A_j (with $\mathbf{r}_1', \ldots, \mathbf{r}_N'$ replacing $\mathbf{r}_1, \ldots, \mathbf{r}_N$ as the variables of integration) to give

$$Q = \frac{1}{h^{3N}} \int \cdots \int \left\{\sum_j \phi_j^*(\mathbf{r}_1, \ldots, \mathbf{r}_N)\phi_j(\mathbf{r}_1', \ldots, \mathbf{r}_N')\right\} \exp\left[-\frac{i}{\hbar} \sum_k \mathbf{p}_k \cdot \mathbf{r}_k'\right]$$

$$\times \exp(-\beta\mathscr{H}) \exp\left[\frac{i}{\hbar} \sum_k \mathbf{p}_k \cdot \mathbf{r}_k\right] d\mathbf{p}_1 \cdots d\mathbf{r}_1 \cdots d\mathbf{r}_1' \cdots d\mathbf{r}_N'$$

The summation over j in the braces is $\delta(\mathbf{r}_1 - \mathbf{r}_1', \ldots, \mathbf{r}_N - \mathbf{r}_N')$ (cf. Problem B–7), and so the integration over the primed variables simply gives

$$Q = \frac{1}{h^{3N}} \int \cdots \int \exp\left[-\frac{i}{\hbar}\sum_k \mathbf{p}_k \cdot \mathbf{r}_k\right] \exp(-\beta\mathcal{H}) \exp\left[\frac{i}{\hbar}\sum_k \mathbf{p}_k \cdot \mathbf{r}_k\right] d\mathbf{p}_1 \cdots d\mathbf{r}_N$$

$$(10\text{–}131)$$

Note that $\mathcal{H}$ contains the operators ∇_l^2 and so does not commute with

$$\exp\left[\pm\frac{i}{\hbar}\sum_k \mathbf{p}_k \cdot \mathbf{r}_k\right]$$

Thus the order of the terms in the integrand of Eq. (10–131) is important.

Equation (10–131) is in the desired form of an integral over phase space, but the integrand contains the quantum mechanical operator $\exp(-\beta\mathcal{H})$ instead of the classical function $\exp(-\beta H)$. Kirkwood defines a function $w(\mathbf{p}_1, \ldots, \mathbf{r}_N, \beta)$ by the relation

$$\exp(-\beta\mathcal{H})\exp\left[\frac{i}{\hbar}\sum_k \mathbf{p}_k \cdot \mathbf{r}_k\right] = \exp(-\beta H)\exp\left[\frac{i}{\hbar}\sum_k \mathbf{p}_k \cdot \mathbf{r}_k\right] w(\mathbf{p}_1, \ldots, \mathbf{r}_N, \beta)$$

$$= F(\mathbf{p}_1, \ldots, \mathbf{r}_N, \beta) \qquad (10\text{–}132)$$

Note that this function $w(\mathbf{p}_1, \ldots, \mathbf{r}_N, \beta)$ has been defined such that Eq. (10–131) becomes

$$Q = \frac{1}{h^{3N}} \int \cdots \int \exp(-\beta H) w(\mathbf{p}_1, \ldots, \mathbf{r}_N, \beta)\, d\mathbf{p}_1 \cdots d\mathbf{r}_N \qquad (10\text{–}133)$$

Remember that this expression for Q is missing a factor of $N!$ in the denominator since we have chosen to ignore the symmetry properties of the ϕ_j in Eq. (10–127). This equation for Q shows that the quantum corrections to the classical partition function lie in the function $w(\mathbf{p}_1, \ldots, \mathbf{r}_N, \beta)$, and so we must now investigate this function. In particular, we wish to show that $w(\mathbf{p}_1, \ldots, \mathbf{r}_N, \beta) \to 1$ as $\hbar \to 0$.

It is possible to evaluate w by carrying out the operation on the left-hand side of Eq. (10–132) and comparing the result to the right-hand side, but this turns out to be an extremely tedious route. A more convenient way is to differentiate Eq. (10–132) with respect to β to get

$$\frac{\partial F}{\partial \beta} = \frac{\partial}{\partial \beta}\exp(-\beta\mathcal{H})\exp\left[\frac{i}{\hbar}\sum_k \mathbf{p}_k \cdot \mathbf{r}_k\right]$$

$$= \frac{\partial}{\partial \beta}\left(1 - \beta\mathcal{H} + \frac{\beta^2}{2}\mathcal{H}^2 + \cdots\right)\exp\left[\frac{i}{\hbar}\sum_k \mathbf{p}_k \cdot \mathbf{r}_k\right]$$

$$= -\mathcal{H}F \qquad (10\text{–}134)$$

This differential equation is called a Bloch differential equation and in this case has the boundary condition

$$F(\beta = 0) = \exp\left[\frac{i}{\hbar}\sum_k \mathbf{p}_k \cdot \mathbf{r}_k\right]$$

It is not possible to solve this equation in general (note its similarity with the Schrödinger equation), but it is fairly straightforward to determine the first few coefficients in an expansion of F in powers of $\hbar$. We do this by actually expanding not F but w according to

$$w(\mathbf{p}_1, \ldots, \mathbf{r}_N, \beta) = \sum_{l=0}^{\infty} \hbar^l w_l(\mathbf{p}_1, \ldots, \mathbf{r}_N, \beta) \tag{10-135}$$

This equation defines the functions w_l. We now substitute this expansion into the right-hand side of Eq. (10–132) and that result into Eq. (10–134), and after some amount of work and cancellation of $\exp(-\beta H)$ and

$$\exp\left(\frac{i}{\hbar} \sum_k \mathbf{p}_k \cdot \mathbf{r}_k\right)$$

we get

$$\left\{\frac{\partial w_0}{\partial \beta} + \frac{\hbar}{\partial \beta}\frac{\partial w_1}{\partial \beta} + O(\hbar^2)\right\} - H\{w_0 + \hbar w_1 + O(\hbar^2)\} = -U\{w_0 + \hbar w_1 + O(\hbar^2)\}$$

$$- K\{w_0 + \hbar w_1 + O(\hbar^2)\} + \frac{i\hbar}{m}\left\{\sum_{i=1}^{N} \mathbf{p}_i \cdot \nabla_i w_0 - \beta w_0 \sum_{i=1}^{N} \mathbf{p}_i \cdot \nabla_i U\right\} + O(\hbar^2) \tag{10-136}$$

where K is the sum of the kinetic energies of all the particles. To arrive at this result, we have used the following relations:

$$\nabla e^{-\beta H} = e^{-\beta K} \nabla e^{-\beta U} = -\beta e^{-\beta H} \nabla U$$

$$\nabla^2 e^{-\beta H} = -\beta e^{-\beta H} \nabla^2 U + \beta^2 e^{-\beta H}(\nabla U)^2$$

$$\nabla_j \exp\left[\frac{i}{\hbar} \sum_k \mathbf{p}_k \cdot \mathbf{r}_k\right] = \frac{i}{\hbar} \mathbf{p}_j \exp\left[\frac{i}{\hbar} \sum_k \mathbf{p}_k \cdot \mathbf{r}_k\right]$$

$$\nabla_j^2 \exp\left[\frac{i}{\hbar} \sum_k \mathbf{p}_k \cdot \mathbf{r}_k\right] = -\frac{2m}{\hbar^2} K_j \exp\left[\frac{i}{\hbar} \sum_k \mathbf{p}_k \cdot \mathbf{r}_k\right]$$

where K_j is the kinetic energy of the jth particle.

The coefficient of $\hbar$ to the zero power in Eq. (10–136) gives $\partial w_0/\partial \beta = 0$ or $w_0 = $ constant. The value of the "constant" can be found from the boundary condition

$$F(\beta = 0) = \exp\left[\frac{i}{\hbar} \sum_k \mathbf{p}_k \cdot \mathbf{r}_k\right]$$

From the defining equation for $F(\beta)$, that is, Eq. (10–132), we see that this is equivalent to the condition

$$w(\mathbf{p}_1, \ldots, \mathbf{r}_N, \beta = 0) = 1$$

and so Eq. (10–135) reads

$$1 = w_0(\mathbf{p}_1, \ldots, \mathbf{r}_N, \beta = 0) + \hbar w_1(\mathbf{p}_1, \ldots, \mathbf{r}_N, \beta = 0) + \hbar^2 w_2(\mathbf{p}_1, \ldots, \mathbf{r}_N, \beta = 0) + \cdots$$

Since the w_l are independent of $\hbar$, this implies that

$$w_0(\mathbf{p}_1, \ldots, \mathbf{r}_N, \beta = 0) = 1$$
$$w_l(\mathbf{p}_1, \ldots, \mathbf{r}_N, \beta = 0) = 0 \qquad l \geq 2 \tag{10-137}$$

But we have seen that w_0 is a constant independent of β, and so we have generally that

$$w_0 = 1 \tag{10-138}$$

and a formal proof that the quantum mechanical partition function goes over into the classical limit according to Eq. (7–9). Note that the classical limit is obtained in two ways. The more obvious one is to let $\hbar \rightarrow 0$, but it is also obtained from the limit $\beta = 0$ [*cf*. Eqs. (10–137)].

We can calculate the first quantum correction to $w(\mathbf{p}_1, \ldots, \mathbf{r}_N, \beta)$ by comparing the coefficients of $\hbar$ to the first power in Eq. (10–136), namely,

$$\frac{\partial w_1}{\partial \beta} = -\frac{i\beta}{m} \sum_{j=1}^{N} \mathbf{p}_j \cdot \nabla_j U$$

which upon integration gives

$$w_1 = -\frac{i\beta^2}{2m} \sum_{j=1}^{N} \mathbf{p}_j \cdot \nabla_j U \tag{10–139}$$

Since this term is odd in the momenta, its contribution to Q according to Eq. (10–133) will vanish. (See Problem 10–35.) The contribution from w_2 does not vanish, however, and represents the first correction to Q. The evaluation of w_2 follows along the same lines as the evaluation for w_0 and w_1, but is quite a bit more lengthy. The result is

$$w_2 = -\frac{1}{2m} \left\{ \frac{\beta^2}{2} \nabla^2 U - \frac{\beta^3}{3} \left[(\nabla U)^2 + \frac{1}{m} (\mathbf{p} \cdot \nabla)^2 U \right] + \frac{\beta^4}{4m} (\mathbf{p} \cdot \nabla U)^2 \right\} \tag{10–140}$$

where we have used the abbreviated notation

$$\mathbf{p} \cdot \mathbf{a} \equiv \sum_{j=1}^{N} \mathbf{p}_j \cdot \mathbf{a}_j$$

In Chapter 15, Eq. (10–140) is used to calculate the first quantum correction to the thermodynamic properties of imperfect gases.

ADDITIONAL READING

General

HUANG, K. 1963. *Statistical mechanics*. New York: Wiley. Chapters 9, 11, and 12.

ISIHARA, A. 1971. *Statistical physics*. New York: Academic. Chapter 4.

KESTIN, J., and DORFMAN, J. R. 1971. *A course in statistical thermodynamics*. New York: Academic. Chapter 8.

KITTEL, C. 1969. *Thermal physics*. New York: Wiley. Chapters 14, 15, and 17.

KUBO, R. 1965. *Statistical mechanics*. Amsterdam: North-Holland Publishing Co. Chapter 4.

LANDAU, L. D., and LIFSHITZ, E. M. 1958. *Statistical physics*. Oxford: Pergamon. Chapter 5.

MAYER, J. E., and MAYER, M. G. 1940. *Statistical mechanics*. New York: Wiley. Chapter 16.

MÜNSTER, A. 1969. *Statistical thermodynamics*, Vol. I. Berlin: Springer-Verlag. Chapter 2.

SCHRÖDINGER, E. 1952. *Statistical thermodynamics*. Cambridge: Cambridge University Press. Chapter 8.

TER HAAR, D. 1966. *Elements of thermostatics*. London: Oxford University Press. Chapter 4.

TOLMAN, R. C. 1938. *Statistical mechanics*. London: Oxford University Press. Chapter 10.

Fermi-Dirac statistics (electrons in metals)

FOWLER, R. H., and GUGGENHEIM, E. A. 1956. *Statistical thermodynamics*. (Cambridge: Cambridge University Press. Chapter 11.

KITTEL, C. 1967. *Solid state physics*, 3rd ed. New York: Wiley. Chapter 7.

REIF, F. 1965. *Statistical and thermal physics*. New York: McGraw-Hill Book Co. Sections 9–16 and 9–17.

Bose-Einstein statistics

LONDON, F. 1954. *Superfluids* Vol. II. New York: Wiley.

Blackbody radiation

DAVIDSON, N. 1962. *Statistical mechanics*. New York: McGraw-Hill. Chapter 12.
EYRING, H., HENDERSON, D., STOVER, B. J., and EYRING, E. M. 1964. *Statistical mechanics and dynamics*. New York: Wiley. Chapters 5 and 7.
KESTIN, J., and DORFMAN, J. R. 1971. *A course in statistical thermodynamics*. New York: Academic. Chapter 10.
KNUTH, E. 1966. *Statistical thermodynamics*. New York: McGraw-Hill. Chapter 10.
MANDL, F. 1971. *Statistical physics*. New York: Wiley. Chapter 10.
REIF, F. 1965. *Statistical and thermal physics*. New York: McGraw-Hill. Sections 9–13 through 9–15.

Density matrix

EYRING, H., HENDERSON, D., STOVER, B. J., and EYRING, E. M. 1964. *Statistical mechanics and dynamics*. New York: Wiley. Chapter 5.
HILL, T. L. 1956. *Statistical mechanics*. New York: McGraw-Hill. Sections 11 and 12.
ISIHARA, A. 1971. *Statistical physics*. New York: Academic. Chapter 10.
KUBO, R. 1965. *Statistical mechanics*. Amsterdam: North-Holland Publishing Co. Section 2–7 and Example 2–7.
TER HAAR, D. 1966. *Elements of thermostatics*. London: Oxford University Press. Chapter 6.
TOLMAN, R. C. 1938. *Statistical mechanics*. London: Oxford University Press. Chapter 9.

Classical limit of Q

HILL, T. L. 1956. *Statistical mechanics*. New York: McGraw-Hill. Section 16.
MÜNSTER, A. 1969. *Statistical thermodynamics*, Vol. I. Berlin: Springer-Verlag. Section 216.

PROBLEMS

10–1. Referring to the discussion following Eq. (10–5), derive expressions for the thermodynamic functions A, G, μ, and S from E and p as functions of N, V, and T.

10–2. Derive a virial expansion for E, A, and S for a Fermi-Dirac ideal gas.

10–3. Derive a virial expansion for E, A, and S for a Bose-Einstein ideal gas.

10–4. Derive Eqs. (10–11) and (10–12) from Eqs. (10–9) and (10–10).

10–5. Consider the power series

$$z = x + \frac{x^2}{2!} + \frac{x^3}{3!} + \cdots$$

Invert this to find x as a power series in z. Compare your result to the expansion of $\ln(1 + z)$, since z above is actually $e^x - 1$.

10–6. Give numerical estimates of the Fermi energy, μ_0, and of $T_F = \mu_0/k$ for (a) electrons in a typical metal such as Ag or Cu, (b) nucleons, for example, neutrons and protons, in a heavy nucleus, (c) He^3 atoms in He^3 gas, in which the volume available to each atom is 50Å^3. Treat the particles as free fermions.

10–7. The density of sodium metal at room temperature is 0.95 g/cm³. Assuming that there is one conduction electron per sodium atom, calculate the Fermi energy and Fermi temperature of sodium.

10–8. Estimate the pressure of an ideal Fermi-Dirac gas of electrons at 0°K.

10–9. Show that the derivative of $f(\varepsilon)$ in Eq. (10–20) is symmetric about μ and that

$$\int_{-\infty}^{\infty} f'(\varepsilon)\, d\varepsilon = -1$$

where $f'(\varepsilon) = df/d\varepsilon$.

10–10. Prove that the expression $e^x/(1 + e^x)^2$ is an even function of x.

10–11. Prove that

$$\int_{-\infty}^{\infty} \frac{x^j e^x\, dx}{(1 + e^x)^2} = 0 \qquad j \text{ odd}$$

$$= -2(j!) \sum_{n=1}^{\infty} \frac{(-1)^n}{n^j} \qquad j \text{ even}$$

$$\equiv 2(j!)\eta(j)$$

The summation $\eta(j)$ is closely related to the Riemann zeta function (*cf.* Abromowitz and Stegun) and has the values $\eta(2) = \pi^2/12$, $\eta(4) = 7\pi^4/720$, ...

10–12. Carry Eq. (10–36) one term further and show that

$$\mu = \mu_0 \left\{ 1 - \frac{\pi^2}{12} \left(\frac{kT}{\mu_0} \right)^2 - \frac{\pi^4}{80} \left(\frac{kT}{\mu_0} \right)^4 + \cdots \right\}$$

10–13. Show that for an ideal Fermi-Dirac gas that

$$p = \frac{2}{5} \frac{N\mu_0}{V} \left\{ 1 + \frac{5\pi^2}{12} \left(\frac{kT}{\mu_0} \right)^2 - \frac{\pi^4}{16} \left(\frac{kT}{\mu_0} \right)^4 + \cdots \right\}$$

10–14. Show that at $3000°$K, μ for aluminum differs from μ_0 by less than 0.1 percent ($\mu_0 = 11.7$ eV).

10–15. Show that for an ideal Fermi-Dirac gas, the Helmholtz free energy is

$$A = \tfrac{3}{5} N\mu_0 \left\{ 1 - \frac{5\pi^2}{12} \left(\frac{kT}{\mu_0} \right)^2 + \frac{\pi^4}{48} \left(\frac{kT}{\mu_0} \right)^4 + \cdots \right\}$$

and that the entropy is

$$S = \frac{N\mu_0}{T} \left[\frac{\pi^2}{2} \left(\frac{kT}{\mu_0} \right)^2 - \frac{\pi^2}{20} \left(\frac{kT}{\mu_0} \right)^4 + \cdots \right]$$

10–16. Take Eq. (10–37) one term further and show that

$$E = \tfrac{3}{5} N\mu_0 \left\{ 1 + \frac{5\pi^2}{12} \left(\frac{kT}{\mu_0} \right)^2 - \frac{\pi^4}{16} \left(\frac{kT}{\mu_0} \right)^4 + \cdots \right\}$$

10–17. Show that

$$C_V = \frac{\pi^2}{3} k^2 T f(\mu_0)$$

is the constant volume heat capacity of an ideal Fermi-Dirac gas if $\mu_0 \gg kT$, where $f(\varepsilon)$ is the density of states.

10–18. Consider a system in which the density of states of the electrons $f(\varepsilon)$ is

$$f(\varepsilon) = \text{constant} = D \qquad \varepsilon > 0$$
$$= 0 \qquad \varepsilon < 0$$

Calculate the Fermi energy for this system; determine the condition for the system being highly degenerate; and then show that the heat capacity is proportional to T for the highly degenerate case.

10–19. Derive Eq. (10–71) for the constant volume heat capacity of an ideal Bose-Einstein gas.

10–20. Prove that the heat of transition associated with Bose-Einstein condensation is given by Eq. (10–68).

10–21. Show that $g_n(\lambda)$ defined by Eq. (10–47) obeys the following recursion formula:

$$g_{n-1} = \frac{\partial g_n}{\partial(\ln \lambda)}$$

Also show that for λ close to unity, that

$$g_{5/2}(\lambda) = 2.363(-\ln \lambda)^{3/2} + 1.342 + 2.612 \ln \lambda - 0.730 (\ln \lambda)^2 + \cdots$$

From these two results, show that the discontinuity of $(\partial C_V/\partial T)$ at $T = T_c$ for an ideal Bose-Einstein gas is

$$\left(\frac{\partial C_V}{\partial T} \right)_{T \to T_c + \varepsilon} - \left(\frac{\partial C_V}{\partial T} \right)_{T \to T_c - \varepsilon} = \frac{3.66 Nk}{T_c}$$

10–22. Consider an ideal Bose-Einstein gas in which the particles have internal degrees of freedom. Assume for simplicity that only the first excited state need be considered and that this has an energy ε relative to the ground state, taken to be zero. Show that the Bose-Einstein condensation temperature of this system is given by

$$T_c = T_c^0 \{1 - 0.255 e^{-\varepsilon/kT_c^0} + \cdots\}$$

assuming that $e^{-\varepsilon/kT_c^0} \ll 1$. How are the thermodynamic functions affected by ε.

10–23. Does a two-dimensional Bose-Einstein ideal gas display a condensation as it does in three dimensions?

10–24. Show that in two dimensions the heat capacity C_V of an ideal Fermi-Dirac and Bose-Einstein is the same.

10–25. Derive the Rayleigh-Jeans law

$$\rho(\nu, T)\, d\nu = \frac{8\pi\nu^2}{c^3}\, kT\, d\nu$$

from the Planck radiation law by considering the limit $h\nu \ll kT$. Derive the Wien empirical distribution law

$$\rho(\nu, T)\, d\nu = \frac{8\pi h\nu^3}{c^3}\, e^{-h\nu/kT}\, d\nu$$

by considering the high-frequency limit.

10–26. Show that the Planck blackbody distribution can be written in terms of wavelengths λ rather than frequency:

$$\rho(\lambda, T)\, d\lambda = \frac{8\pi hc}{\lambda^5}\, \frac{d\lambda}{e^{hc/\lambda kT} - 1}$$

where $\rho(\lambda, T)\, d\lambda$ is the amount of energy between wavelength λ and $\lambda + d\lambda$.

10–27. If ω_{max} is the frequency at which $\rho(\omega, T)$ is a maximum, illustrate by maximizing $\ln \rho(\omega, T)$ that ω_{max} is given by

$$\frac{\hbar\omega_{max}}{kT} = 3(1 - e^{-\hbar\omega_{max}/kT})$$

and so

$$\frac{\hbar\omega_{max}}{kT} = 2.82$$

Similarly show that

$$\lambda_{max} T = 0.290 \text{ cm-deg}$$

Calculate the temperature for which λ_{max} is in the red region of the spectrum.

10–28. Derive Eq. (10–93) by evaluating the integral in Eq. (10–98).

10–29. In Problem 7–24, it was shown that the number of molecules striking a surface per unit area per unit time is $\rho\bar{v}/4$. By a similar approach, show that the total energy flux radiated by a blackbody is

$$e(T) = \frac{c}{4}\frac{E}{V} = \sigma T^4$$

where $\sigma = 2\pi^5 k^4/15h^3 c^3$. This result is known as the Stefan-Boltzmann law, and σ is the Stefan-Boltzmann constant. Verify that σ, a universal constant, equals 5.669×10^{-5} erg/cm²-deg⁴-sec.

10–30. Show that

$$p = \frac{1}{3}\frac{E}{V}$$

for a photon gas and compare this to the analogous result for bosons and fermions with nonzero rest mass.

10–31. It has been stated that in the early stages, a nuclear fission explosion generates a temperature of the order of a million degrees Kelvin over a sphere 10 cm in diameter. Assuming this to be true, estimate the total rate of radiation emitted from the surface of this sphere, the radiation flux a few miles away, and the wavelength corresponding to the maximum in the radiated power spectrum.

10–32. Derive an expression for Planck's blackbody distribution law and Stefan's radiation law for a two-dimensional world.

10–33. Derive all of the principal results for a photon gas by setting $\lambda = 1$ in the Bose-Einstein formulas.

10–34. Prove Eq. (10–112).

10–35. Show that the contribution of the term w_1 [Eq. (10–139)] to the partition function Q vanishes, since it is odd in the momenta.

10–36. Derive Eqs. (10–62).

CHAPTER 11

CRYSTALS

In this chapter we shall discuss the application of statistical thermodynamics to the calculation of the thermodynamic properties of crystals. Unlike dilute gases, the interatomic interactions in a crystal are not negligible, but we shall see that the concept of normal coordinates allows us to treat a crystal as a system of independent "particles." In Section 11–1 we show that all of the thermodynamic properties of a crystal can be expressed in terms of the distribution of its normal vibrational frequencies. This distribution function is difficult to calculate exactly, but in Sections 11–2 and 11–3 we discuss two well-known simple approximations, the first due to Einstein (Section 11–2) and the other due to Debye (Section 11–3). In Section 11–4 we turn to the problem of an exact vibrational analysis of a crystalline solid. We shall determine the exact vibrational spectrum of two types of one-dimensional lattices. Although the results for one-dimensional lattices are not directly applicable to real crystals, the basic ideas and techniques associated with such a calculation serve as an introduction to the field of lattice dynamics, that is, the calculation of the vibrational spectrum of more realistic lattices. The final two sections of the chapter contain a discussion of two important topics in the statistical thermodynamics of crystals. In Section 11–5 we introduce the concept of a phonon, and in Section 11–6 we discuss several of the most important types of defects or imperfections that occur in real crystals.

11–1 THE VIBRATIONAL SPECTRUM OF A MONATOMIC CRYSTAL

In this section we shall derive the partition function for a monatomic crystal. Although we can hardly ignore the interatomic (or intermolecular) interactions in a solid, we shall see that it is, nevertheless, possible to treat a crystalline solid as a system of independent "particles." The crucial point here is the existence of normal coordinates. For many purposes, a crystal may be represented by a system of regularly spaced masses and springs, as illustrated two dimensionally in Fig. 11–1. The springs represent the resultant interatomic force that each atom "sees" about its lattice point. Each

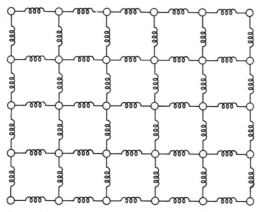

Figure 11–1. **A two-dimensional version of a mass and spring model of a crystalline lattice.**

atom sits in a potential well whose minimum is at a lattice point, and as the atom moves from its equilibrium position, the net force restores it to its equilibrium position. Thus the effect of the intermolecular forces between all the atoms in a crystalline solid may be represented by a system of springs as illustrated in Fig. 11–1. Clearly the force constants that we assign to these springs reflect the interatomic forces involved.

In a typical crystal, the potential well that each atom " sees " is very steep, and so each atom vibrates about its equilibrium lattice point with a small amplitude. This allows us to expand the interatomic potential of the entire crystal in a Taylor series. Consider the one-dimensional example shown in Fig. 11–2. The total potential energy is a function of the displacements of the N atoms from their equilibrium positions, that is, $U = U(\xi_1, \xi_2, \ldots, \xi_N)$. Since the atoms vibrate with small amplitude about their equilibrium positions, we write

$$U(\xi_1, \xi_2, \ldots, \xi_N) = U(0, 0, \ldots, 0)$$

$$+ \sum_{j=1}^{N} \left(\frac{\partial U}{\partial \xi_j}\right)_0 \xi_j + \frac{1}{2} \sum_{i=1}^{N} \sum_{j=1}^{N} \left(\frac{\partial^2 U}{\partial \xi_i \, \partial \xi_j}\right)_0 \xi_i \xi_j + \cdots \quad (11\text{–}1)$$

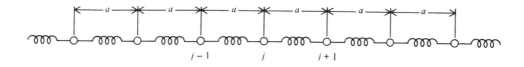

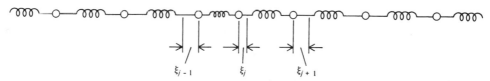

Figure 11–2. **A one-dimensional system of masses and springs. The upper system is in its equilibrium configuration, where all the masses are separated by a distance a, the unstrained length of the spring. The lower curve shows an arbitrary configuration which is described by the set of displacements $\{\xi_j\}$ of the atoms from their equilibrium positions.**

where the zero subscript on the derivatives indicated that they are to be evaluated at the point at which all the ξ_j equal zero. Since U is a minimum when the $\xi_j = 0$, the first derivatives in Eq. (11–1) are zero, and so we have

$$U(\xi_1, \xi_2, \ldots, \xi_N) = U(0, 0, \ldots, 0) + \frac{1}{2} \sum_{i,j} \left(\frac{\partial^2 U}{\partial \xi_i \xi_j} \right)_0 \xi_i \xi_j + \cdots$$

$$= U(0, 0, \ldots, 0) + \frac{1}{2} \sum_{i,j} k_{ij} \xi_i \xi_j + \cdots \qquad (11\text{–}2)$$

where we have introduced the set of force constants k_{ij}. The important result here is that $U(\xi_1, \xi_2, \ldots, \xi_N)$ is a *quadratic* function of the displacements. Note that $U(0, 0, \ldots, 0)$ is a function of the lattice spacing, which in turn is a function of the ratio V/N, or the density. To emphasize this, we shall write $U(0, 0, \ldots, 0)$ as $U(\mathbf{0}; \rho)$. We are taking the zero of energy to be the separated atoms at rest. In addition, the curvature of $U(\xi_1, \xi_2, \ldots, \xi_N)$ at the minimum is a function only of V/N, and so the force constants are functions of V/N only.

Equation (11–2) represents a system of *coupled* harmonic oscillators. We say that they are coupled because of the cross terms in Eq. (11–2). If it were not for these cross terms, Eq. (11–2) would be a sum of independent squared terms, and the Lagrange equations of motion (classically) or the Schrödinger equation (quantum mechanically) would yield N separate or uncoupled harmonic oscillators. This is similar to the situation that occurred when we treated the vibration of a polyatomic molecule in Section 8–1, and, in fact, a crystal containing N atoms can be considered to be just a large polyatomic molecule. We saw in Chapter 8 that the vibrational motion of a polyatomic molecule can be rigorously decomposed into a set of *independent* harmonic oscillators by introducing normal coordinates. In principle, we can apply such a normal coordinate analysis to an entire crystal. If there are N atoms in a monatomic crystal, there are $3N$ degrees of freedom, of which three are associated with the translational motion of the whole crystal, and three more are concerned with the rotation of the crystal. There are then $3N - 6$ vibrational degrees of freedom. But with $N = O(10^{20})$, we can take this number of vibrational degrees to be $3N$, without noticeable error. The result of a normal coordinate analysis would yield $3N - 6 \approx 3N$ vibrational frequencies

$$\nu_j = \frac{1}{2\pi} \left(\frac{k_j}{\mu_j} \right)^{1/2} \qquad j = 1, 2, \ldots, 3N - 6 \approx 3N \qquad (11\text{–}3)$$

where k_j and μ_j are an effective force constant and an effective reduced mass, respectively. The precise form of k_j and μ_j are not important for our purposes. The important point is that the complicated general vibrational problem can be mathematically reduced to $3N$ (really $3N - 6$) independent harmonic oscillators, each with its own frequency, which is a complicated function of the masses, force constants, and geometry of the lattice. Since the k_j in Eq. (11–3) depends upon the k_{ij} in Eq. (11–2), the k_j and the frequencies ν_j in Eq. (11–3) depend upon V/N rather than V or N separately.

Since the crystal does not translate or rotate,* the complete partition function is given by

$$Q\left(\frac{V}{N}, T \right) = e^{-U(\mathbf{0}; \rho)/kT} \prod_{j=1}^{3N-6} q_{\text{vib}, j} \qquad (11\text{–}4)$$

---

* More precisely, the translational and rotational degrees of freedom of the crystal contribute negligibly to the partition function on a per molecule basis.

where $q_{\text{vib},j}$ is the vibrational partition function associated with the jth vibrational frequency. Note that there is no factor of $N!$ in the denominator of Q. Since each molecule is restricted to the neighborhood of its lattice point, and the lattice points could, in principle, be labeled, the atoms themselves must be considered to be distinguishable. We have already evaluated q_{vib} in our treatment of diatomic gases and found that

$$q_{\text{vib}} = \frac{e^{-hv/2kT}}{1 - e^{-hv/kT}} \tag{11-5}$$

Therefore the total partition function is given by

$$Q = \prod_{j=1}^{3N} \left(\frac{e^{-hv_j/2kT}}{1 - e^{-hv_j/kT}} \right) e^{-U(0;\rho)/kT} \tag{11-6}$$

Since there are $3N$ normal frequencies, they are essentially continuously distributed, and we can introduce a function $g(v)\, dv$, which gives the number of normal frequencies between v and $v + dv$. If we introduce this into the logarithm of Eq. (11-6), we have

$$-\ln Q = \frac{U(0;\rho)}{kT} + \int_0^\infty \left[\ln(1 - e^{-hv/kT}) + \frac{hv}{2kT} \right] g(v)\, dv \tag{11-7}$$

where

$$\int_0^\infty g(v)\, dv = 3N \tag{11-8}$$

since there are $3N$ normal frequencies in all. If we can determine the function $g(v)$, we can then calculate the thermodynamic properties of the crystal. For example, we have

$$E = U(0;\rho) + \int_0^\infty \left[\frac{hv e^{-hv/kT}}{(1 - e^{-hv/kT})} + \frac{hv}{2} \right] g(v)\, dv \tag{11-9}$$

and

$$C_V = k \int_0^\infty \frac{(hv/kT)^2 e^{-hv/kT} g(v)\, dv}{(1 - e^{-hv/kT})^2} \tag{11-10}$$

Equations (11-7) through (11-10) are essentially exact. In order to use them, we must know the function $g(v)$, and this is, of course, where the difficulty lies. The function $g(v)$ is easier to determine than the entire set of individual frequencies, but it is, nevertheless, a very difficult problem. We shall discuss some exact calculations of $g(v)$ in Section 11-4, but before that, we shall introduce two useful and well-known approximations to $g(v)$. One of these is due to Einstein and says that all the normal frequencies are the same; the other is due to Debye, who treated a crystal as a continuous elastic medium and calculated $g(v)$ by studying the elastic waves that can be set up in such a body.

11-2   THE EINSTEIN THEORY OF THE SPECIFIC HEAT OF CRYSTALS

In this section we shall discuss an extremely simple model for the vibrational character of a crystal. It is so simple, in fact, that we should go back in time to the beginning of this century in order to appreciate its great insight and impact. Classical statistical thermodynamics was fairly well developed by the end of the nineteenth century. If it is considered that the N atoms of a crystalline solid behave as harmonic oscillators about their equilibrium positions, classical theory (equipartition) predicts that each atom would contribute R cal/deg-mole for each of its three vibrational degrees of freedom, or that the molar heat capacity at constant volume would be $3Nk = 3R = 6$ cal/deg-mole. This prediction, which is known as the law of Dulong and Petit, is in good agreement with the observed heat capacity of many crystals at high enough temperatures, and often down to room temperature, but the agreement fails completely at low temperatures. For example, the heat capacity C_V of silver is shown in Fig. 11–3. It can be seen that the Dulong and Petit value is approached asymptotically, but that the curve falls rapidly to zero as $T \to 0$. This behavior is observed quite generally, and experimentally it is found that the heat capacity goes to zero as T^3 as $T \to 0$. This is known as the T^3-law and is an experimental observation which any successful theory must reproduce.

At the beginning of this century, deviations from predicted classical behavior were being discovered regularly, and each one was a severe challenge to the physical theories of the time. Einstein was the first to present a theoretical explanation of the low-temperature heat capacity of solids by applying the revolutionary blackbody radiation theory ideas of Planck to the vibrations of atoms in crystals. Einstein assumed that each atom in the crystal vibrates about its equilibrium configuration as a simple harmonic oscillator, so that the entire crystal could be considered to be a set of $3N$ independent harmonic oscillators, each oscillator having the same frequency v. Physically then, he assumed that each atom of the crystal sees the same environment as any other, and so all N atoms could be treated as independent oscillators in the x-, y-, and z-directions. Classically such an assumption leads to the Dulong and Petit value of $3R$ cal/deg-mole, but Einstein's great contribution (in 1907) was to say that the energy of each of these $3N$ independent oscillators had to be quantized according to the procedure developed by Planck. Thus, with our advantage of using a formalism and notation developed long after the turbulent years of the beginning of the century, we can say that Einstein assumed that the frequency spectrum $g(v)$ was a delta function at one frequency

$$g(v) = 3N\delta(v - v_E) \tag{11–11}$$

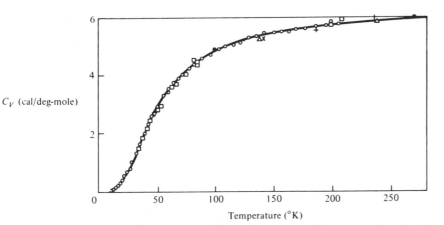

Figure 11–3. **The molar heat capacity at constant volume of metallic silver as a function of temperature.** (From C. Kittel, *Solid State Physics, 2nd ed.* New York: Wiley, 1956.)

where the factor $3N$ is included, so that Eq. (11–8) is satisfied, and v_E is the single frequency assigned to all $3N$ independent oscillators of the crystal. The value of the Einstein frequency v_E varies from substance to substance and, in some way, reflects the nature of the interatomic interactions for the particular crystal. In the light of the organized development presented in Section 11–1, this may appear to be a terribly gross assumption, but it was a major step forward, and "in a sense the final step,"* in the understanding of the heat capacity of solids.

If we substitute Eq. (11–11) into Eq. (11–10), we find

$$C_V = 3Nk\left(\frac{hv_E}{kT}\right)^2 \frac{e^{-hv_E/kT}}{(1 - e^{-hv_E/kT})^2} \tag{11–12}$$

for the heat capacity. It is customary to define a quantity Θ_E by hv_E/k, which has units of temperature and is called the Einstein temperature of the crystal. In terms of the Einstein temperature, Eq. (11–12) is

$$C_V = 3Nk\left(\frac{\Theta_E}{T}\right)^2 \frac{e^{-\Theta_E/T}}{(1 - e^{-\Theta_E/T})^2} \tag{11–13}$$

Equation (11–13) contains one adjustable parameter to fit the entire heat capacity curve shown in Fig. 11–3. Figure 11–4 shows a comparison of Eq. (11–13) versus the experimental heat capacity of diamond. This figure is taken from Einstein's original paper and shows the success of such a simple theory. It is easy to show from Eq. (11–13) that C_V approaches the Dulong and Petit value of $3Nk = 3R$ as $T \to \infty$. (See Problem 11–2.)

Although Fig. 11–4 shows that the Einstein model of a crystal is capable of giving an impressive qualitative agreement with experiment, it is not in quantitative agreement. In particular, Eq. (11–13) predicts that the low-temperature heat capacity goes as

$$C_V \to 3Nk\left(\frac{\Theta_E}{T}\right)^2 e^{-\Theta_E/T} \tag{11–14}$$

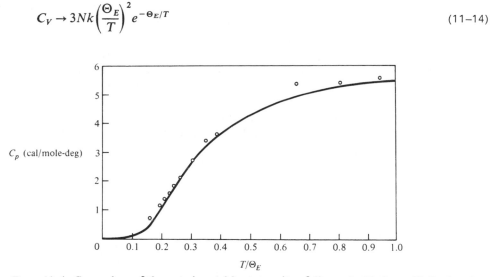

Figure 11–4. **Comparison of the experimental heat capacity of diamond with the prediction based on the Einstein theory with** $\Theta_E = 1320°K$. (From C. Kittel, *Solid State Physics*, 3rd ed. New York: Wiley, 1967., after A. Einstein, *Ann. Physik.*, **22**, 180, 1907.)

* From Blackman in "Additional Reading."

instead of T^3 as the experimental data. The low-temperature heat capacity predicted by Eq. (11–14) falls to its zero value more rapidly than T^3-law. We shall see in the next section that the Debye theory, which came a few years after Einstein's, gives a T^3-law as $T \to 0$.

Before going on to the Debye theory, we point out an important feature of Eq. (11–13), which is also exhibited in more rigorous theories. Equation (11–13) predicts that C_V is the same function for all substances if it is plotted versus T/Θ_E. When this happens, we say that C_V is a universal function of T/Θ_E, and that the various crystals obey a *law of corresponding states*. Once the temperature is scaled or "reduced" by a quantity that depends upon the particular substance, the heat capacity versus the reduced temperature will superimpose for all crystals. Although the Einstein model does not quantitatively reproduce experimental data, its prediction of a law of corresponding states is, in fact, correct.

11–3 THE DEBYE THEORY OF THE HEAT CAPACITY OF CRYSTALS

According to the ideas of Planck, the energy of an oscillator is proportional to the frequency, and since it is the lower energies that are populated at low temperatures, we can reason that it is the low-frequency or long-wavelength modes that are most important at low temperatures. The success of the Debye theory is that it treats the long-wavelength frequencies of a crystal in an exact manner, and hence is able to predict the low-temperature heat capacity.

The normal frequencies of a crystal vary from essentially zero to some value of the order of 10^{13} cycles/sec (Hz) or so. Normal frequencies are not due to the vibrations of single atoms, but are a concerted harmonic motion of all the atoms. This concerted motion is called a normal coordinate or a normal mode. Note, for example, that the normal coordinates of CO_2 and H_2O involve the synchronous motion of all the atoms in each molecule. Two extremes of normal modes of a one-dimensional crystal are shown in Fig. 11–5. The upper mode is one in which the atoms vibrate against each other and has a wavelength of $2a$. The lower one is one in which a long row of atoms moves smoothly together to produce a long-wavelength mode of low frequency. It is the long-wavelength modes that Debye was able to treat in a clever manner.

Debye reasoned that those normal modes whose wavelengths are long compared to the atomic spacing do not depend upon the detailed atomic character of the solid and could be calculated by assuming that the crystal is a continuous elastic body. The approximation of the Debye theory is that it treats all the normal frequencies from this point of view.

The distribution of frequencies that can be set up in a solid body is calculated in almost the same way that we calculated the set of standing waves in a blackbody cavity.

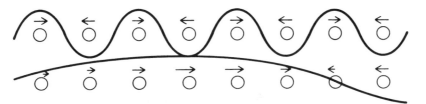

Figure 11–5. **Two types of normal modes in a one-dimensional crystal. The upper one is a high-frequency mode and the bottom is a low-frequency mode.**

The first part of Section 10–5 can be read independently of the rest of the chapter. We state there that the imaginary part of

$$u(\mathbf{r}, t) = A e^{i(\mathbf{k}\cdot\mathbf{r} - \omega t)} \tag{11-15}$$

represents a wave of amplitude A traveling through a medium in the direction of $\mathbf{k}$ and with frequency $\omega = 2\pi v$. The quantity $\mathbf{k}$ is called the wave vector, and its magnitude is $2\pi/\lambda$. The velocity of this wave is given by $v = \omega/k = v\lambda$. A standing wave can be obtained by superimposing two waves traveling in opposite directions. This gives

$$u = 2A e^{i\mathbf{k}\cdot\mathbf{r}} \cos \omega t$$

In order that this be a standing wave, we require that the imaginary part of this vanish at the edges of the crystal. If we assume for simplicity that the crystal is a cube of length L, this boundary condition gives that $k_x L = n_x \pi$, $k_y L = n_y \pi$, and $k_z L = n_z \pi$, where k_x, and so on, are the components of $\mathbf{k}$, and n_x, n_y, and n_z are positive integers. In vector notation, we have

$$\mathbf{k} = \frac{\pi}{L}\,\mathbf{n} \tag{11-16}$$

The frequency v depends upon only the magnitude of $\mathbf{k}$ (through the relation $\omega = vk$), which is given by

$$k^2 = \left(\frac{\pi}{L}\right)^2 (n_x{}^2 + n_y{}^2 + n_z{}^2) \tag{11-17}$$

The number of standing waves with wave number between k and $k + dk$ is found by the same method that we used to find the number of translational energy states between ε and $\varepsilon + d\varepsilon$ in Section 1–3. Using Eq. (11–17), the number of standing waves with the wave vector of magnitude less than k is [cf. Eq. (10–84)]

$$\Phi(k) = \frac{\pi}{6}\left(\frac{Lk}{\pi}\right)^3 = \frac{L^3 k^3}{6\pi^2} = \frac{V k^3}{6\pi^2}$$

and the number between k and $k + dk$ is [cf. Eq. (10–85)]

$$\omega(k)\,dk = \frac{d\Phi}{dk}\,dk = \frac{V k^2\,dk}{2\pi^2} \tag{11-18}$$

We can convert this into $g(v)\,dv$ by using the relation $v = v/\lambda = vk/2\pi$:

$$g(v)\,dv = \frac{4\pi V v^2}{v^3}\,dv \tag{11-19}$$

This is almost the desired result. We must recognize that there are two kinds of waves that can propagate through a continuous medium. These are transverse waves, in which the medium vibrates perpendicular to the direction of propagation (the direction of $\mathbf{k}$), and longitudinal waves, in which the medium vibrates in the same direction as the wave is propagated. Since it is possible to draw two independent vectors perpendicular to $\mathbf{k}$ and only one parallel to $\mathbf{k}$, there are two transverse waves and one longitudinal wave. The three of these contribute to $g(v)$, and we finally have the complete expression for the Debye approximation to $g(v)$, namely,

$$g(v)\,dv = \left(\frac{2}{v_t{}^3} + \frac{1}{v_l{}^3}\right) 4\pi V v^2\,dv \tag{11-20}$$

In this expression, v_t and v_l are the transverse and longitudinal velocities, respectively. It is conventional to introduce a kind of average velocity by means of

$$\frac{3}{v_0{}^3} \equiv \frac{2}{v_t{}^3} + \frac{1}{v_l{}^3} \tag{11-21}$$

so that Eq. (11–20) can be written in the form

$$g(v)\,dv = \frac{12\pi V}{v_0{}^3}\,v^2\,dv \tag{11-22}$$

This expression is exact in the limit of low frequencies or long wavelengths, where the atomic nature of the solid is not important, and the crystal can, in fact, be treated as a continuous elastic body. The Debye theory uses Eq. (11–22) for all the normal frequencies, however. The total number of normal frequencies is $3N$, and so Debye defined a maximum frequency v_D such that the integral of $g(v)\,dv$ from 0 to v_D equals $3N$. Thus

$$\int_0^{v_D} g(v)\,dv = 3N \tag{11-23}$$

which, when Eq. (11–22) is used for $g(v)$, gives

$$v_D = \left(\frac{3N}{4\pi V}\right)^{1/3} v_0 \tag{11-24}$$

The frequency v_D is called the Debye frequency. In terms of v_D, the distribution function $g(v)\,dv$ is

$$g(v)\,dv = \frac{9N}{v_D{}^3}\,v^2\,dv \qquad 0 \le v \le v_D$$

$$= 0 \qquad\qquad v > v_D \tag{11-25}$$

This summarizes the Debye theory of crystals.

We can now substitute Eq. (11–25) for $g(v)\,dv$ into Eqs. (11–7) through (11–10) to calculate the thermodynamic properties of a crystal according to the Debye theory. The most interesting thermodynamic function is the heat capacity C_V, given by Eq. (11–10) with Eq. (11–25) for $g(v)\,dv$:

$$C_V = 9Nk\left(\frac{T}{\Theta_D}\right)^3 \int_0^{\Theta_D/T} \frac{x^4 e^x}{(e^x - 1)^2}\,dx \tag{11-26}$$

where we have let $x = hv/kT$ and have defined the Debye temperature by

$$\Theta_D = \frac{hv_D}{k} \tag{11-27}$$

The integral in Eq. (11–26) cannot be evaluated in terms of simple functions and must be evaluated numerically. Note that the integral is a function of only the upper limit of the integral, that is, a function of Θ_D/T. It is customary to define a function $D(T/\Theta_D)$ by

$$D\left(\frac{T}{\Theta_D}\right) = 3\left(\frac{T}{\Theta_D}\right)^3 \int_0^{\Theta_D/T} \frac{x^4 e^x}{(e^x - 1)^2}\,dx \tag{11-28}$$

so that the heat capacity is

$$C_V = 3Nk D\left(\frac{T}{\Theta_D}\right) \tag{11-29}$$

The function $D(T/\Theta_D)$ is called the Debye function. It is a well-tabulated function of T/Θ_D. (See Appendix C.) Figure 11–6 is from Debye's original paper and shows two comparisons of Eq. (11–29) with experimental data. It can be seen that the agreement is very good. The values of Θ_D are those which give the best overall fit to the data. Table 11–1 gives the Debye temperatures for many monatomic solids. Note that most of these values are of the order of a few hundred degrees Kelvin.

Table 11–1. **The Debye temperature of various monatomic solids**

solid	$\Theta_D(°K)$	solid	$\Theta_D(°K)$
Na	150	Fe	420
K	100	Co	385
Cu	315	Ni	375
Ag	215	Al	390
Au	170	Ge	290
Be	1000	Sn	260
Mg	290	Pb	88
Zn	250	Pt	225
Cd	172	C (diam)	1860

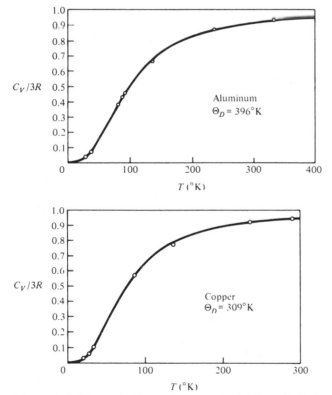

Figure 11–6. **Observed values and curves calculated on the Debye model for the heat capacity of aluminum and copper, taking $\Theta_D = 396°K$ and $309°K$, respectively.** (After P. Debye, *Ann. Physik*, **39**, 789, 1912. From C. Kittel, *Solid State Physics*, 2nd ed., New York: Wiley, 1956.)

Although Eq. (11–29) must be evaluated numerically for arbitrary values of T/Θ_D, it is easy to investigate its high- and low-temperature behavior. At high temperatures, the upper limit in the integral in Eq. (11–28) becomes very small. Hence the range of x is very small and it is legitimate to expand the integrand to get

$$\int_0^{\Theta_D/T} \frac{x^4 e^x}{(e^x - 1)^2}\, dx \to \int_0^{\Theta_D/T} \frac{x^4 (1 + x + \cdots)}{(1 + x + \cdots - 1)^2}\, dx = \int_0^{\Theta_D/T} x^2\, dx = \frac{1}{3}\left(\frac{\Theta_D}{T}\right)^3$$

Therefore $D(T/\Theta_D) \to 1$, and Eq. (11–29) for C_V becomes

$$C_V \to 3Nk = 3R = 6 \text{ cal/deg-mole} \tag{11–30}$$

which is the classical limiting law of Dulong and Petit.

The low-temperature limit of C_V is more interesting. This can be obtained by letting the upper limit of the integral in $D(T/\Theta_D)$ go to infinity. Then

$$D\left(\frac{T}{\Theta_D}\right) \to 3\left(\frac{T}{\Theta_D}\right)^3 \int_0^\infty \frac{x^4 e^x}{(e^x - 1)^2}\, dx$$

The integral here is standard and equals $4\pi^4/15$. (See Problem 11–6.) The low-temperature limit of C_V then is

$$C_V \to \frac{12\pi^4}{5} Nk \left(\frac{T}{\Theta_D}\right)^3 \tag{11–31}$$

which is the famous T^3-law. This was the great triumph of the Debye theory. Although it is not obvious from Fig. 11–4, the Einstein heat capacity curve falls much too rapidly as $T \to 0$, and the agreement at low temperatures is very poor. This is more readily seen in Table 11–2, where the Einstein and Debye theories are compared to experimental data for silver. It can be seen that although both theories agree for temperatures greater than approximately 100°K, only the Debye theory is able to be used at lower temperature.

There are several important features of the Debye theory. The one that we have just discussed in some detail is that it predicts a T^3-law for the low-temperature heat capacity. Another is that it predicts a law of corresponding states for the heat capacity. Equation (11–29) clearly shows that if C_V is plotted versus T/Θ_D, all substances will lie on the one curve. Another way of saying this is that C_V is a universal function for all substances, determined by one parameter Θ_D in the form of T/Θ_D. Figure 11–7 shows the heat capacity data for a number of substances plotted on the same graph of C_V versus T/Θ_D. Note that the Debye curve fits all the points over the entire temperature range.

An interesting consequence of Debye's approach is that it is possible to calculate Θ_D in terms of the elastic constants of the solid. We shall not prove it here, but it should be clear that such a thing is possible since the Debye theory is based upon treating a crystal as a continuous elastic body. The elastic constants of a body are quantities such as the compressibility and Young's modulus. Table 11–3 compares the Debye temperatures determined by fitting heat capacity data with those calculated from the elastic properties of the solid. The agreement, although not perfect, is quite good.

Table 11–2. **Heat capacity of silver at different temperatures**

temperatures (°K)	C_V(obs.) (cal/mole-deg)	C_V calculated	
		Einstein	Debye
1.35	0.000254	8.76×10^{-49}	—
2	0.000626	1.39×10^{-32}	—
3	0.00157	6.16×10^{-20}	—
4	0.00303	5.92×10^{-15}	—
5	0.00509	1.62×10^{-11}	—
6	0.00891	3.24×10^{-9}	—
7	0.0151	1.30×10^{-7}	0.0172
8	0.0236	2.00×10^{-6}	0.0257
10	0.0475	1.27×10^{-4}	0.0502
12	0.0830	0.0010	0.0870
14	0.1336	0.0052	0.137
16	0.2020	0.0180	0.207
20	0.3995	0.0945	0.394
28.56	1.027	0.579	1.014
36.16	1.694	1.252	1.69
47.09	2.582	2.272	2.60
55.88	3.186	2.946	3.22
65.19	3.673	3.521	3.73
74.56	4.039	3.976	4.13
83.91	4.326	4.309	4.45
103.14	4.797	4.795	4.86
124.20	5.084	5.124	5.17
144.38	5.373	5.323	5.37
166.78	5.463	5.476	5.51
190.17	5.578	5.581	5.61
205.30	5.605	5.633	5.66

Source: C. Kittel, *Solid State Physics*, 2nd ed. New York: Wiley, 1956.

One cannot expect perfect agreement in Table 11–3 since the Debye theory is, of course, an approximate theory. In fact, as more experimental data became available in the 1920s, the Debye theory was subjected to more and more severe tests, and discrepancies began to appear. For example, one can calculate Θ_D at any particular temperature from Eq. (11–29) if the experimental heat capacity of that temperature is known. If the Debye theory were exact, the value of Θ_D obtained would be independent of the temperature used to calculate it. However, it turns out that the value

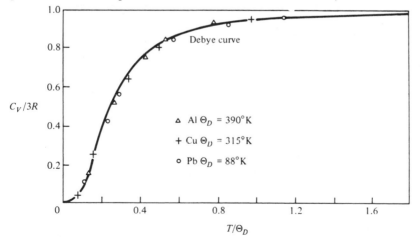

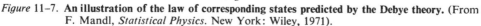

Figure 11–7. **An illustration of the law of corresponding states predicted by the Debye theory.** (From F. Mandl, *Statistical Physics*. New York: Wiley, 1971).

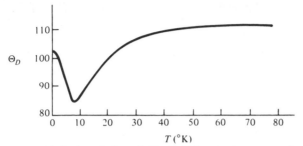

Figure 11–8. **A typical result for the Debye temperature as function of *T* for a monatomic solid.**

Table 11–3. **Comparison of Debye temperatures determined from elastic constants or heat capacity measurements**

substance	Θ (elastic) °K	Θ (heat capacity) °K
Al	399	396
Cu	329	313
Ag	212	220
Au	166	186
Cd	168	164
Sn	185	165
Pb	72	86
Bi	111	111
Pt	226	220

of Θ_D does depend upon the temperature at which it is evaluated. This can be shown most clearly on a Θ_D versus T plot, where the value of Θ_D calculated at some temperature is plotted against the temperature. Such a Θ_D–T plot is shown in Fig. 11–8. If the Debye theory were exact, Θ_D versus T would be a horizontal straight line, and so a deviation from such a straight line indicates a failing of the Debye theory. In the main, however, these deviations do not exceed 10 percent in most cases, and so we can say that the Debye theory is a successful theory of the thermodynamic properties of crystal lattices. In the next section we shall present the basic ideas of a more rigorous approach to the dynamical properties of lattices, which has been highly developed since the 1930s and constitutes the presently active field of lattice dynamics.

11–4 INTRODUCTION TO LATTICE DYNAMICS

In this section we shall introduce the basic ideas of lattice dynamics by calculating the frequency distribution of two types of one-dimensional lattices: one where all the masses are the same (a one-dimensional model for an elemental crystal such as Zn) and one with two alternating different masses (a one-dimensional model of NaCl, say). These two models have quite different vibrational spectra, both of which are qualitatively observed in real crystals. At the end of the section, we shall simply state the types of results found in two and three dimensions.

The Hamiltonian for the first case is (see Fig. 11–2)

$$H = \sum_{j=1}^{N} \frac{m}{2} \dot{\xi}_j{}^2 + \sum_{j=2}^{N} \frac{f}{2} (\xi_j - \xi_{j-1})^2 \qquad (11\text{–}32)$$

where we are using f here for the force constants. The equations of motion corresponding to this are (see Problem 11–14)

$$m\ddot{\xi}_j = f(\xi_{j+1} + \xi_{j-1} - 2\xi_j) \tag{11-33}$$

This set of equations represents a set of *coupled* harmonic oscillators. We assume that the time dependence is harmonic and let

$$\xi_j(t) = e^{i\omega t}y_j \tag{11-34}$$

where $\omega = 2\pi v$, and y_j is independent of time. Substituting this into Eq. (11–33) gives

$$-m\omega^2 y_j = f(y_{j+1} + y_{j-1} - 2y_j) \tag{11-35}$$

This type of equation is called a difference equation (as opposed to a differential equation); it is a linear difference equation with constant coefficients. Linear difference equations with constant coefficients are solved by $y_j = A^j$ (compare to $y(x) = e^{nx}$ as solutions to linear differential equations with constant coefficients). A little experience with equations of this type would suggest letting A be $e^{i\phi}$. Substituting this into Eq. (11–35) gives

$$-m\omega^2 = f(e^{i\phi} + e^{-i\phi} - 2)$$
$$= f(2\cos\phi - 2)$$

or

$$\omega^2 = \frac{4f}{m}\sin^2\left(\frac{\phi}{2}\right) \tag{11-36}$$

If we note that the maximum value that ω^2 can have is $4f/m$ (since the maximum value that $\sin^2(\phi/2)$ can have is 1), then we can write

$$\omega = \omega_{max}\left|\sin\left(\frac{\phi}{2}\right)\right| \tag{11-37}$$

The solution to Eq. (11–33) is then

$$\xi_j(t) = e^{i(\omega t + j\phi)}$$

where ϕ is as yet undetermined.

Since this functional form for $\xi_j(t)$ repeats for every $\Delta j = 2\pi/\phi$, there is a wavelength λ equal to $a\,\Delta j = 2\pi a/\phi$, where a is the lattice spacing of the one-dimensional chain. From this, ϕ is given by

$$\phi = \frac{2\pi a}{\lambda} \equiv ka \tag{11-38}$$

where k, which equals $2\pi/\lambda$, is the wave vector of the motion of the chain. We see then that

$$\xi_j(t) = e^{i(jka + \omega t)} \tag{11-39}$$

represents a wave of wavelength $2\pi/k$ and frequency ω traveling along the chain. From the De Broglie relation, $\hbar k$ is the momentum of this wave, or in the language of the next section, $\hbar k$ is the momentum of a phonon associated with this frequency.

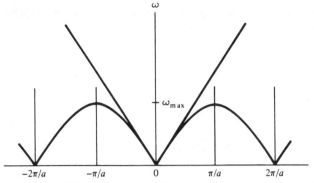

Figure. 11–9 **The dispersion curve for a one-dimensional monatomic lattice. [Eq. (11–40)]. The straight lines correspond to a continuous string.**

The relation between the frequency and the wave vector is called a dispersion curve. The dispersion curve for our simple one-dimensional lattice is [*cf.* Eqs. (11–37) and (11–38)]

$$\omega = \omega_{max} \left| \sin\left(\frac{ka}{2}\right) \right| \tag{11–40}$$

This is plotted in Fig. 11–9. In the limit of small values of ka (wavelength long compared to the lattice spacing), we have essentially a continuous chain as in the Debye theory, and the dispersion curve is

$$\omega = \omega_{max} \frac{ka}{2} \tag{11–41}$$

which we write as

$$\frac{\omega}{k} = \lambda v = \frac{a\omega_{max}}{2} = \text{constant velocity} \tag{11–42}$$

Typical lattice spacings are $O(10^{-8}\text{cm})$, and typical maximum frequencies are $O(10^{13} \text{ sec}^{-1})$, giving $O(10^5 \text{ cm/sec})$ for the velocity of the wave. This is the order of magnitude for the velocity of sound in solids.

Note that, in general, however, λv (or ω/k) is *not* a constant. From Eq. (11–40),

$$\lambda v = \frac{\omega}{k} = \frac{\omega_{max}}{k} \left| \sin\left(\frac{ka}{2}\right) \right| = c(k) \tag{11–43}$$

which shows that the velocity is, in fact, a function of k or λ. Waves actually have two types of velocities associated with them, and the velocity $c(k)$ defined in Eq. (11–43) is called the phase velocity. It is the fact that different wavelengths travel with different velocities, which leads to the dispersion of waves as they pass through a prism, and so $\omega(k)$ is called a dispersion curve. Dispersion curves can be determined experimentally by inelastic neutron scattering measurements.

An important property of Eq. (11–39) for $\xi_j(t)$ is that the substitution $k \rightarrow k_n = k + 2\pi n/a$ with $n = \pm 1, \pm 2, \ldots$ leaves $\xi_j(t)$ unchanged. Furthermore, this substitution leaves Eq. (11–40) for ω unchanged. In other words, there is no physical difference between states corresponding to wave vectors k or $k \mp 2\pi n/a$. In order to obtain a unique relationship between the state of vibration of the lattice and the wave vector k, k must be restricted to a range of values $2\pi/a$. Usually one chooses

$$-\frac{\pi}{a} \le k \le \frac{\pi}{a}$$

We now show how to express the thermodynamic properties of the lattice in terms of the dispersion curve. Before doing this, however, we must consider the boundary conditions in our lattice. We shall use *periodic boundary conditions*, namely, that $\xi_j(t) = \xi_{j+N}(t)$. One way of thinking about these boundary conditions is to picture the linear chain of N atoms to be bent around into a circle. Clearly, if the chain is long enough, joining its two ends has a negligible effect on the thermodynamic properties of the chain. Applying these boundary conditions to the $\xi_j(t)$ [Eq. (11–39)] gives $\exp(iNka) = 1$, which implies that $k = 2\pi j/Na$, where j is an integer. Now because k is restricted to lie between $\pm\pi/a$, the possible values for j are ± 1, ± 2, ..., $\pm N/2$. The thermodynamic energy of the crystal (without the zero-point energy) is then

$$E = \sum_j \frac{\hbar\omega_j}{\exp(\beta\hbar\omega_j) - 1}$$

$$= \frac{Na}{\pi} \int_0^{\pi/a} \frac{\hbar\omega(k)\, dk}{\exp(\beta\hbar\omega(k)) - 1} \tag{11–44}$$

where the summation over the possible wave vectors defined above has been approximated by an integral and the fact that $\omega(k) = \omega(|k|)$ has been used. We see, then, that if $\omega(k)$, that is, the dispersion curve of the lattice, is known, we can calculate E and other thermodynamic properties of the lattice.

This integral over k can be converted to an integral over ω itself by means of the dispersion curve.

$$dk = \frac{dk}{d\omega}\, d\omega = \frac{d}{d\omega}\left\{\frac{2}{a} \sin^{-1}\left(\frac{\omega}{\omega_{max}}\right)\right\} d\omega$$

$$= \frac{2\, d\omega}{a(\omega_{max}^2 - \omega^2)^{1/2}} \tag{11–45}$$

Substituting this into Eq. (11–44) gives

$$E = \frac{2N}{\pi} \int_0^{\omega_{max}} \frac{\hbar\omega\, d\omega}{[\exp(\beta\hbar\omega) - 1][\omega_{max}^2 - \omega^2]^{1/2}} \tag{11–46}$$

If we compare this to Eq. (11–9) (without the zero-point energy $h\nu/2$), we see that

$$g(\nu) = \frac{2N}{\pi} \frac{1}{(\nu_{max}^2 - \nu^2)^{1/2}} \tag{11–47}$$

Problem 11–33 is involved with showing that a one-dimensional Debye approximation agrees with Eq. (11–47) as $\nu \to 0$.

In one dimension, the distribution of frequencies is related to the dispersion curve [*cf.* Eq. (11–47) and Eq. (11–45)] by

$$g(\nu) = \frac{Na}{\pi} \frac{1}{d\nu/dk}$$

Notice that $g(\nu)$ has singularities at the points where $d\nu/dk$ equals zero. For a continuum, $d\nu/dk = $ velocity = constant. In general, however, $d\nu/dk$ is not constant. The derivative dv/dk is called the group velocity of the wave and physically represents the rate-of-energy transmission of the wave. For a continuum, the group velocity and phase velocity [Eq. (11–43)] are both equal to the same constant value.

Now let us consider a one-dimensional lattice with two alternating kinds of atoms with masses m_1 and m_2. This represents a one-dimensional analog of a crystal like sodium chloride. The Hamiltonian for this lattice is

$$H = \sum_{j=1}^{N} \left\{ \frac{m_1}{2} \dot{\xi}_{2j}^2 + \frac{m_2}{2} \dot{\xi}_{2j-1}^2 \right\} + \frac{f}{2} \sum_{j=1}^{N} \{ (\xi_{2j} - \xi_{2j-1})^2 + (\xi_{2j+1} - \xi_{2j})^2 \}$$

In this case one gets two sets of equations of motion, one for the masses m_1 and one for the masses m_2, which eventually yield (see Problem 11–34)

$$\omega^2 = \omega_0^2 \left\{ 1 \pm \left(1 - \frac{4m_1 m_2 \sin^2 \phi}{(m_1 + m_2)^2} \right)^{1/2} \right\} \tag{11–48}$$

where

$$\omega_0^2 = \frac{f}{\mu} \tag{11–49}$$

In these equations, ϕ = multiple of π/N, and μ is the reduced mass of m_1 and m_2.

According to which sign is chosen, the dispersion curve of Eq. (11–48) yields two branches: a high-frequency branch called the optical branch and a low-frequency branch called the acoustical branch. Figure 11–10(a) shows these two dispersion curves, and Fig. 11–10(b) shows the corresponding frequency distribution $g(v)$. In the normal modes belonging to the acoustical branch, neighboring atoms are displaced in the same direction (as in the lower curve in Fig. 11–5) to produce a long wavelength mode. In the optical branch, neighboring atoms are displaced in opposite directions (as in the upper curve in Fig. 11–5).

If the two masses of our lattice are ions of opposite sign (such as NaCl), the vibrational motion in the optical modes produces oscillating dipole moments. An oscillating dipole moment leads to an absorption of infrared radiation. Figure 11–11 shows the infrared spectrum of NaCl. Furthermore, the larger the reduced mass, the lower

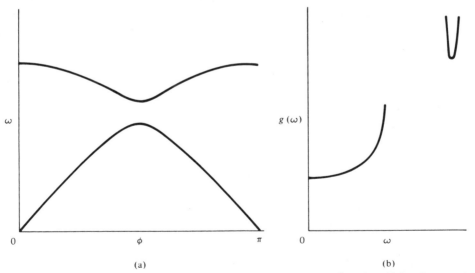

(a)   (b)

Figure 11–10. **(a) The angular frequency for a diatomic chain as a function of the phase angle ϕ. (b) Density of normal vibrations of a linear chain with $m_1/m_2 = 3$.**

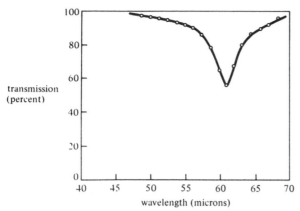

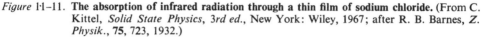

Figure 1·1–11. **The absorption of infrared radiation through a thin film of sodium chloride.** (From C. Kittel, *Solid State Physics*, 3rd ed., New York: Wiley, 1967; after R. B. Barnes, *Z. Physik.*, **75**, 723, 1932.)

the frequency of the optical modes, and hence the longer the wavelengths at which the infrared absorption occurs. Figure 11–12 shows the onset of absorption in various alkali halides. Evidently, if we wish to get good transmission in the infrared, we must use a crystal in which the ions are as heavy as possible. Such considerations are important when designing prisms for infrared spectrometers.

The exact lattice dynamic calculations of two- and three-dimensional lattices proceeds in much the same way that we did in the previous section, but the actual calculations are much more difficult. Figure 11–13 shows two experimentally determined vibrational spectra to give an idea of the complexity of the spectra that are obtained for three-dimensional lattices. There have been several exact or near-exact calculations of such spectra, and there is now an extensive literature on lattice dynamics calculations (see "Additional Reading").

The last two sections of this chapter deal with two special topics, namely, the concept of phonons (Section 11–5) and the concept of defects in crystal lattices (Section 11–6).

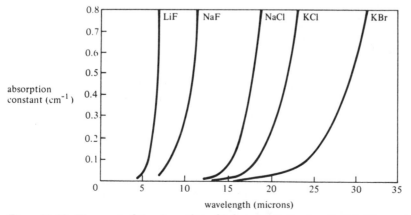

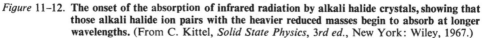

Figure 11–12. **The onset of the absorption of infrared radiation by alkali halide crystals, showing that those alkali halide ion pairs with the heavier reduced masses begin to absorb at longer wavelengths.** (From C. Kittel, *Solid State Physics*, 3rd ed., New York: Wiley, 1967.)

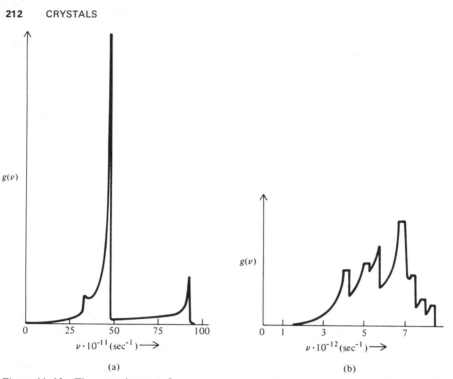

Figure 11–13. **The experimental frequency spectrum of (a) aluminum and (b) iron.** (From M. Blackman in *Encyclopedia of Physics*, vol. VII, pt. 1, ed. by S. Flügge. Berlin: Springer-Verlag, 1955.)

11–5 PHONONS

We have seen in Section 11–1 that if a crystal has N atoms, it has $3N$ normal coordinates, each with its own characteristic frequency $v_1, v_2, \ldots, v_{3N}$. The total energy of the crystal is

$$E(\{n_j\}) = \sum_{j=1}^{3N} h v_j (n_j + \tfrac{1}{2}) \tag{11–50}$$

$$= \sum_{j=1}^{3N} h v_j n_j + \sum_{j=1}^{3N} \frac{h v_j}{2}$$

$$= \sum_{j=1}^{3N} h v_j n_j + E_0 \tag{11–51}$$

where we have set E_0 equal to the total zero-point vibrational energy. Since E is a sum of terms, this expression for E can be *interpreted* as the energy of a system of *independent* particles, which occupy the states 1, 2, ..., $3N$ with corresponding energies $h v_1, h v_2, \ldots, h v_{3N}$ with n_1 particles in the first state, n_2 in the second, and so on. Note that the set of numbers $\{n_j\}$, the occupation numbers of the $3N$ states, completely specifies the state of the system. Since it is only the number of "particles" in each state that specifies the system, we can consider these "particles" to be indistinguishable, and furthermore, since there is no restriction on the numbers n_j, that is, $n_j = 0, 1, \ldots, 3N$, these "particles" are bosons.

This is actually a very useful interpretation, since it gives us a system of *noninteracting* bosons, and we can use the Bose-Einstein formulas of Chapter 4 or

Chapter 10. As far as the vibrations of the lattice are concerned then, we shall treat it as if it were an *ideal* Bose-Einstein gas. The "particles" we have invented here are examples of the quasi-particles we mentioned at the beginning of Chapter 4 and are called *phonons*. Phonons are essentially quanta of lattice vibrations, just as photons are quanta of electromagnetic vibrations. Since lattice vibrations are closely related to sound waves passing through the crystals, phonons can be thought of as quanta of sound waves.

We can directly apply the basic Bose-Einstein equations of Chapter 4 or even the formulas developed in Section 10–5 for a photon gas. For instance, the equation for the average occupation of the jth state is [Eq. (4–26)]

$$\bar{n}_j = \frac{\lambda e^{-\beta \varepsilon_j}}{1 - \lambda e^{-\beta \varepsilon_j}} = \frac{1}{\lambda^{-1} e^{\beta \varepsilon_j} - 1} \tag{11–52}$$

As we saw in Chapter 10, these quantum statistical equations are awkward to use because of the presence of λ, but in the case of a phonon gas, we can evaluate λ easily. The reason for this is that the number of phonons is, in fact, not fixed, since clearly it is possible to have a number of sets $\{n_j\}$ in which $E(\{n_j\})$ is fixed but $n = \sum_j n_j$ is different. Thermodynamically, the system is characterized by E and V only. A system of phonons, then, is mathematically identical to a photon gas (photons obey Bose-Einstein statistics and their number is not conserved), and in Chapter 10 we showed that $\mu = 0$ or $\lambda = 1$ for this case. In Section 10–5 we gave a simple thermodynamic proof that $\mu = 0$. In brief, the phonons are likened to a chemical equilibrium of the type $nA \rightleftharpoons mA$, where n and m are integers. Since the number of phonons is not conserved, $n \neq m$. The condition for equilibrium (Chapter 9) is that $(m - n)\mu = 0$, and thus we have $\mu = 0$ and $\lambda = 1$.

With λ set equal to 1, then we have

$$\bar{n}_j = \frac{1}{e^{\beta \varepsilon_j} - 1} \tag{11–53}$$

and

$$\bar{E} = \sum_{j=1}^{3N} \bar{n}_j h v_j + E_0 = \sum_{j=1}^{3N} \frac{h v_j}{e^{\beta v_j} - 1} + E_0 \tag{11–54}$$

If we introduce $g(v)$, we can write

$$\bar{E} = E_0 + \int_0^\infty \frac{g(v) h v \, dv}{e^{\beta h v} - 1} \tag{11–55}$$

which is the same as Eq. (11–9). Equation (11–54) is the same as Eq. (11–44) without the zero-point energy. The derivative of this gives Eq. (11–10) for the heat capacity. We can thus derive all of the results of Sections 11–2 and 11–3 by treating the lattice vibrations of a crystal as a gas of noninteracting phonons.

Actually, the concept of phonons is much more useful than this. For instance, phonons can be assigned a momentum (just as a photon can), which must be conserved in collisions. The collision of the phonons and the electrons in a metal lead to the electrical resistance of a metal. The inelastic scattering of phonons with photons is known as Brillouin scattering. The vibrational frequency spectrum of a crystal can be determined experimentally by inelastic neutron scattering, which is treated theoretically by phonon–neutron scattering.

11–6 POINT DEFECTS IN SOLIDS

Up to now we have assumed that every atom or ion of a crystal is situated at a lattice site and that every lattice site is occupied by one and only one particle. Such a perfect periodic arrangement is called a perfect crystal. We shall show below, however, that such a perfect arrangement is thermodynamically unattainable. Any deviation from such perfect behavior is called an imperfection or a defect. There are quite a variety of defects that exist in crystals, but in this section we shall study only the most common *point* defects, namely, vacant lattice sites and interstitial atoms (extra atoms not at lattice sites). Although it will turn out that the concentration of such defects will be fairly small, they nevertheless have a profound effect on the properties of crystals. The conductivity of some semiconductors is due entirely to trace amounts of chemical impurities. The color of many crystals is due to imperfections, and the mechanical and elastic properties of crystals depend strongly on the number and type of defects present. The diffusion of atoms through solids is another property that is dependent upon defects.

The simplest type of defect is a lattice vacancy. Such a missing atom or ion is known as a *Schottky defect*. A Schottky defect can be created by transferring an atom or ion from the body of the crystal to its surface. Although it requires energy to do this, there is an increase of entropy. The final equilibrium concentration is found by minimizing the free energy. If we assume that it takes an energy ε_v to bring an atom from an interior lattice site to a surface lattice site and also assume that the concentration of defects is small enough to consider them to be independent, then we can write

$$A(n) = E - TS$$
$$= n\varepsilon_v - kT \ln \frac{N!}{n!(N-n)!} \tag{11–56}$$

The quantity n is the number of vacancies, and the combinatorial factor is just the number of ways of distributing n vacancies over N sites. We now minimize $A(n)$ with respect to n to find the number of vacancies expected. Setting $(\partial A/\partial n)_T$ equal to zero and solving for n give

$$n \simeq Ne^{-\varepsilon_v/kT} \tag{11–57}$$

where we have neglected n compared to N. If ε_v is 1 eV and $T = 300°$K, then $n/N \approx 10^{-17}$. At 1000°K, $n/N \approx 10^{-5}$ (see Problem 11–19). (In ionic crystals such as NaCl, it is usually favorable to form roughly equal numbers of positive and negative defects. This keeps the crystal electrostatically neutral on a local scale.)

The other common type of defect that we shall discuss here is a *Frenkel defect*, in which an atom is displaced from a lattice position to an interstitial position. If we let ε_I be the energy it takes to do this, let N be the number of lattice sites, and let N' be the number of possible interstitial sites, then

$$A(n) = n\varepsilon_I - kT \ln\left\{\frac{N!}{n!(N-n)!} \cdot \frac{N'!}{n!(N'-n)!}\right\} \tag{11–58}$$

In this case the combinatorial is a product of the number of ways of choosing n out of N lattice sites and the number of ways of distributing the n chosen atoms over the N' available interstitial sites. Minimizing $A(n)$ gives (see Problem 11–20)

$$n \approx (NN')^{1/2}e^{-\varepsilon_I/2kT} \tag{11–59}$$

It turns out that the most common type of point defect in alkali halides are Schottky defects, and the most common defect in silver halides are Frenkel defects. This is probably due to the fact that silver ions are smaller than alkali ions and so can fit into interstitial positions more easily. It is possible to determine if one type of defect is predominant by careful measurements of the density. The formation of Schottky defects lowers the density of the crystal since the volume is increased with no change in mass. On the other hand, the formation of Frenkel defects does not change the volume, and so there is no change in density. It is also possible to use ionic conductivity measurements to differentiate between Schottky and Frenkel defects.

Lattice vacancies in controlled concentrations can be produced by the addition of divalent ions. For instance, if a crystal of KCl is grown with controlled amounts of $CaCl_2$, the Ca^{2+} enters the lattice at a normal K^+ site, and the two Cl^- enter at two normal Cl^- sites. The net result of this is the production of vacant positive ion sites. This is shown in Fig. 11-14. The formation of these lattice vacancies can be observed by measuring the density of $CaCl_2$-KCl mixtures. The volume of such a mixture is actually larger than the volume of the separate components.

We said earlier in this section that the rate of diffusion in solids is greatly affected by the presence of defects. As an atom diffuses through a crystal, it must surmount a series of energy barriers presented by its neighbors as it moves from lattice site to lattice site or from interstitial position to interstitial position. Let us consider interstitial diffusion of impurities. If this barrier height is ε, then $\exp(-\varepsilon/kT)$ can be thought of as the fraction of time that the atom will have an energy exceeding ε. If v is the frequency with which the diffusing atom vibrates around its interstitial position, then the probability per unit time that the atom will be able to pass over the barrier is

$$p \approx v e^{-\varepsilon/kT} \qquad \qquad \text{(11–60)}$$

attempt frequency   successful jumps/second

Now consider two parallel planes of impurity atoms in interstitial sites. The planes are separated by the lattice constant a. There will be c impurity atoms on one plane and $c + a(dc/dx)$ on the other. The net number of atoms crossing between these planes per unit time is $pa(dc/dx)$. If n is the concentration of impurity atoms, then $c = an$. The diffusion flux is then

$$j = -pa^2 \frac{\partial n}{\partial x} \qquad \qquad \text{(11–61)}$$

Remember that a flux of a quantity Ψ is the rate of flow of Ψ through a unit area of surface per second. The minus sign occurs because the direction of the diffusion is

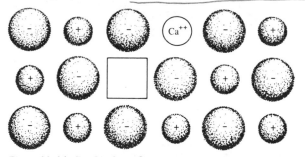

Figure 11–14. **Production of a lattice vacancy by the solution of $CaCl_2$ in KCl: to ensure electrical neutrality a positive ion vacancy is introduced into the lattice with each divalent cation Ca^{2+}. The two Cl^- ions of $CaCl_2$ enter normal negative ion sites.** (From C. Kittel, *Solid State Physics*, 3rd ed. New York: Wiley, 1967.)

opposed to the concentration gradient. We compare this to Fick's law, which simply states that a diffusion flux is proportional to the concentration gradient of the diffusing substance, that is,

$$j = -D\frac{\partial n}{\partial x} \qquad (11\text{--}62)$$

where D is the diffusion constant. We see then that

$$D = pa^2 \approx va^2 e^{-\varepsilon/kT} \qquad (11\text{--}63)$$

Figure 11–15 shows the experimentally determined temperature dependence of the diffusion coefficient of carbon in iron. If we let v be 10^{14} sec^{-1}, a be 3×10^{-8} cm, and ε be 1 eV, then $D \approx 10^{-18}$ cm^2/sec at 300°K and 10^{-6} cm^2/sec at 1000°K. (See Problem 11–23.)

Another property that is much affected by defects is the color of slightly impure (doped) alkali halide crystals. Pure alkali halides are transparent in the visible region, but if a sodium chloride crystal is heated in sodium vapor, it becomes yellow. Similarly, if KCl is heated in potassium vapor, it takes on a magenta color. These same effects can be produced by X-ray, neutron, or electron bombardment or by electrolysis. The color of the doped crystals is due to defects which absorb light and are called *color centers.* A number of different kinds of color centers have been discovered, but here we shall mention only the first discovered and perhaps the simplest, namely, an *F center.* The name comes from the German word for color, *farbe. F* centers have been identified by electron spin resonance to be electrons bound to a negative ion vacancy. This is consistent with the fact that they can be produced by heating the pure crystal in an excess of the metal vapor. When this is done, the metal atoms of the vapor are incorporated into the crystal lattice. The metal atom loses its electron, and the positive ion takes up a lattice site while the electron associates itself with a vacant negative

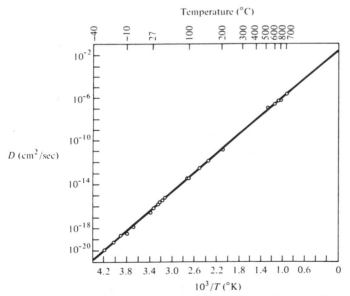

Figure 11–15. **Diffusion coefficient of carbon in iron.** (From C. Kittel, *Solid State Physics,* 3rd ed. New York: Wiley, 1967.)

ion lattice site which is lacking an anion. A vacant negative ion lattice site has an effective positive charge, and so the electron behaves somewhat like the electron bound to a nucleus and can absorb light. The model is consistent with a number of experimental facts. For instance, F band absorption is characteristic of the crystal and not of the alkali metal used in the vapor; that is, NaCl will become yellow whether heated in sodium vapor or any other alkali metal vapor. Furthermore, the absorption corresponds quantitatively to the amount of excess metal, and it has been determined that colored alkali halide crystals are less dense than pure alkali halide crystals.

ADDITIONAL READING

General

BLACKMAN, M. 1955. In *Encyclopaedia of physics*, Vol. VII, pt. 1, ed. by S. Flügge. Berlin: Springer-Verlag.

DEKKER, A. J. 1957. *Solid state physics*. Englewood Cliffs, N.J.: Prentice-Hall.

FOWLER, R. H., and GUGGENHEIM, E. A. 1956. *Statistical thermodynamics*. Cambridge: Cambridge University Press. Chapter 4.

KESTIN, J., and DORFMAN, J. R. 1971. *A course in statistical thermodynamics*. New York: Academic. Chapter 9.

KITTEL, C. 1967. *Solid state physics*, 3rd ed. New York: Wiley.

MAYER, J. E., and MAYER, M. G. 1940. *Statistical Mechanics*. New York: Wiley. Chapter 11.

MOORE, W. J. 1967. *Seven solid states*. New York: Benjamin.

WANNIER, G. H. 1966. *Statistical physics*. New York: Wiley. Chapter 13.

Einstein and Debye theories

DAVIDSON, N. 1962. *Statistical mechanics*. New York: McGraw-Hill. Chapter 16.

EYRING, H., HENDERSON, D., STOVER, B. J., and EYRING, E. M. 1964. *Statistical mechanics and dynamics*. New York: Wiley. Chapter 7.

HILL, T. L. 1960. *Statistical thermodynamics*. Reading, Mass.: Addison-Wesley. Chapter 5.

KNUTH, E. 1966. *Statistical thermodynamics*. New York: McGraw-Hill. Chapters 9 and 14.

MANDL, F. 1971. *Statistical physics*. New York: Wiley. Chapter 6.

REIF, F. 1965. *Statistical and thermal physics*. New York: McGraw-Hill. Sections 10–1 and 10–2.

Lattice dynamics

DE LAUNAY, J. 1956. *Solid state physics*, Vol. 2, ed. by F. Seitz and D. Turnbull. New York: Academic.

MARADUDIN, A., *Ann. Rev. Phys. Chem.*, **14**, p. 89, 1963.

MARADUDIN, A. A., MONTROLL, E. W., and WEISS, G. H. 1963. *Theory of lattice dynamics in the harmonic oscillator approximation*. New York: Academic.

MITRA, S. S. 1962. *Solid state physics*, Vol. 13, ed. by F. Seitz and D. Turnbull. New York: Academic.

Defects in crystals

BARR, L. W., and LIDIARD, A. B. 1970. In *Physical chemistry, an advanced treatise*, Vol. X, ed. by H. Eyring, D. Henderson, and W. Jost. New York: Academic.

PROBLEMS

11–1. The difference between the constant pressure and constant volume heat capacities is

$$C_p - C_V = -T\left(\frac{\partial V}{\partial T}\right)_p^2 \left(\frac{\partial p}{\partial V}\right)_T$$

In terms of the volume expansion coefficient $\alpha_V = (1/V)(\partial V/\partial T)_p$ and the isothermal compressibility $\kappa = -(1/V)(\partial V/\partial p)_T$, this difference is

$$C_p - C_V = \alpha_V^2 \frac{TV}{\kappa}$$

which is often rewritten as

$$C_p = C_V(1 + \gamma \alpha_V T)$$

where $\gamma = \alpha_V V / \kappa C_V$ is practically independent of temperature. This is called the Grüneisen constant. Calculate the Grüneisen constant given that $\alpha_V = 6.22 \times 10^{-5}$ deg^{-1} and $\kappa = 12.3 \times 10^{-12}$ cm²-dyne^{-1} at room temperature.

11–2. Find both the high- and low-temperature limiting forms of the heat capacity according to the Einstein model.

11–3. Derive an expression for the Einstein specific heat for a two-dimensional crystal.

11–4. Determine the various thermodynamic properties of an Einstein crystal.

11–5. Derive Eq. (11–24).

11–6. Prove that

$$\int_0^\infty \frac{x^4 e^x \, dx}{(e^x - 1)^2} = \frac{4\pi^4}{15}$$

Hint: See Problem 1–63.

11–7. Derive an expression for the heat capacity of a Debye crystal as a power series expansion in Θ_D / T.

11–8. Why is the heat capacity of diamond at room temperature far below its Dulong-Petit value?

11–9. Prove that for a monatomic crystal that

$$\int_0^\infty [3Nk - C_V(T)] \, dt = E(0)$$

where $E(0)$ is the energy of the crystal at 0°K. Give a graphical interpretation of this equation.

11–10. Starting with a two-dimensional wave equation, derive a Debye type equation for the heat capacity C_V for a two-dimensional crystal, assuming that the transverse and longitudinal velocities are the same.

11–11. Show that in a monatomic crystal, the high-temperature limiting form of the heat capacity depends only on the existence of a cutoff frequency and is given by the law of Dulong and Petit. That is, assume only that the distribution of frequencies

$$g(v) = 0 \qquad \text{for } v > v_{\max}$$

11–12. The potential energy of the atoms of a solid of density N/V in their equilibrium positions is denoted by $U(0; N/V)$. The normal frequencies of vibration of the atoms near their equilibrium positions are functions of the density $v_j(N/V)$ where $j = 1, 2, \ldots, 3N - 6$. It is a good approximation that

$$\frac{\partial \log v_j}{\partial \log V} = -\gamma \qquad (j = 1, 2, \ldots, 3N - 6)$$

for all frequencies. The constant γ is called the Grüneisen constant. Show that under this approximation, the pressure of the solid is given by

$$p = -\frac{\partial U}{\partial V} + \gamma \frac{E}{V}$$

This equation of state is known as the Mie-Grüneisen equation.

11–13. A modification of the Debye theory was introduced by Born, who proposed a different cutoff for the spectrum of vibrational modes. He proposed that the cutoff be made such that both the longitudinal and transverse modes have a common minimum wavelength. If we denote this common minimum wavelength by λ_m, then $\lambda_m v_{\text{long}} = c_{\text{long}}$ and $\lambda_m v_{\text{trans}} = c_{\text{trans}}$, Equation (11–23) now becomes

$$4\pi V \left\{ \int_0^{v_t} \frac{2}{c_t{}^3} v^2 \, dv + \int_0^{v_l} \frac{v^2 \, dv}{c_l{}^3} \right\} = 3N$$

Show that this leads to the following expression for the specific heat:

$$C_V = R\left[D\left(\frac{\Theta_l}{T}\right) + 2D\left(\frac{\Theta_t}{T}\right)\right]$$

where $D(x)$ is the Debye function

$$D(x) = \frac{3}{x^3}\int_0^x \frac{e^z z^4\,dz}{(e^z - 1)^2}$$

11–14. Derive Eq. (11–33) from (11–32).

11–15. The Debye theory treats the crystal as a continuum body, and hence the dispersion relation for the phonons is $\omega = ck$, where c is the speed of sound in the body. In a ferromagnetic solid at low temperatures, there exist quantized waves of magnetization called spin waves, and the dispersion relation for these type of waves is $\omega \propto k^2$. Find the low-temperature heat capacity due to spin waves.

11–16. Suppose that wavelike quasi-particles, having the dispersion relation $\omega = Ak^n$, exist in a solid and yield a specific heat when they are excited as thermal motion. Using the relation

$$E = \sum_k \frac{\hbar\omega_k}{\exp(\beta\hbar\omega_k) - 1}$$

$$= \frac{V}{(2\pi)^3}\int \frac{\hbar\omega(\mathbf{k})\,d\mathbf{k}}{\exp(\beta\hbar\omega(\mathbf{k})) - 1} \qquad \text{(three dimensions)}$$

show that the heat capacity is proportional to $T^{3/n}$ at low temperatures. Note that for $n = 1$, that is, $\omega/k = \lambda\nu = A = $ constant, one has the Debye theory.

11–17. Consider a planar square lattice of identical atoms which vibrate perpendicular to the plane of the lattice. If we let u_{lm} be the displacement of the atom in the lth column and mth row, show that the equation of motion is

$$m\left(\frac{d^2 u_{lm}}{dt^2}\right) = f[(u_{l+1,m} + u_{l-1,m} - 2u_{lm}) + (u_{l,m+1} + u_{l,m-1} - 2u_{lm})]$$

where m is the mass of an atom, and f is the force constant. By analogy with Eq. (11–39), assume a solution of the form

$$u_{lm} = \exp[i(lk_y a + mk_x a + \omega t)]$$

where a is the nearest-neighbor lattice spacing. Show that the dispersion relation for this system is

$$\omega^2 = \frac{2f}{m}(2 - \cos k_x a - \cos k_y a)$$

For $ka = (k_x^2 + k_y^2)^{1/2}a$ where $a \ll 1$, show that

$$\omega = \left(\frac{fa^2}{m}\right)^{1/2} k$$

11–18. Use the equations $E = h\nu$ and $p = h/\lambda$ to show that the group velocity for a free particle of mass m is p/m.

11–19. Calculate the number of Schottky defects per mole of crystal at $300°K$ and $1000°K$ given that it takes 1.0 eV to bring an atom or ion from an interior lattice site to a surface lattice site.

11–20. Show that the number n of Frenkel defects in equilibrium in a crystal having N lattice points and N' possible interstitial positions is given by the equation

$$\varepsilon_1 = k_B T \log \left[\frac{(N-n)(N'-n)}{n^2} \right]$$

which, for $n \ll N, N'$, gives

$$n \simeq (NN')^{1/2} \exp\left(\frac{-\varepsilon_1}{2k_B T}\right)$$

Here ε_1 is the energy necessary to remove an atom from a lattice site to an interstitial position.

11–21. (a) Assuming a simple coulombic interaction between positive and negative ion vacancies, calculate the binding energy between a pair of oppositely charged vacancies in NaCl.

(b) If n_1 represents the number of single positive or negative ion vacancies and n_2 the number of pairs of vacancies, derive an expression for the Helmholtz free energy and show by minimizing this with respect to n_1 and n_2 that

$$\frac{n_2}{n_1} = 6 \exp \left[\frac{(\varepsilon - \tfrac{1}{2}\phi)}{kT} \right]$$

where ε is the binding energy of a pair of oppositely charged vacancies, and ϕ is the energy required to produce a single positive and negative ion vacancy.

11–22. Treat an F center as a free electron moving in the field of a point charge e in a medium of dielectric constant $\varepsilon = n^2$ where n is the index of refraction. What is the $1s - 2p$ energy difference of F centers in NaCl?

11–23. If a sodium atom next to a vacancy has to move over a potential hill of 0.5 eV, and the atomic vibration frequency is 10^{12} Hz, estimate the diffusion coefficient at room temperature for radioactive sodium in normal sodium. Assume a lattice spacing of 4 Å.

11–24. Consider a closed container containing a small solid and its vapor in equilibrium at temperature T. Assume that the volume of the solid, v, is much less than the volume of the container, V. Let the partition function of the solid be of the form $Q = q_s(T)^{N_s}$, and let there be N_g molecules in the vapor phase. Show that the equilibrium condition is given by

$$N_g = \frac{q_g}{q_s}$$

where q_g is the partition function of a vapor phase molecule. Hint: Minimize the total free energy of the system, that is, $A_g(T, V, N_g) + A_s(T, N_s)$, with respect to N_g keeping $N_g + N_s = N = $ constant.

11–25. Using the result of Problem 11–24, show that the vapor pressure of an Einstein crystal is

$$p = kT \left(\frac{2\pi mkT}{h^2} \right)^{3/2} \left(2 \sinh \frac{\Theta_E}{2T} \right)^3 e^{-\phi/kT}$$

where ϕ is given by $U(0; \rho) = -N_s \phi$.

11–26. The pV term in $G = A + pV$ can be neglected for a condensed phase. Using this fact, the chemical potential of a Debye crystal can be well approximated by $\mu = A/N$ rather than the correct G/N. Show that the vapor pressure of a Debye crystal is given by

$$\ln p = \ln \left[\left(\frac{2\pi mkT}{h^2} \right)^{3/2} kT \right] + \left(\frac{\phi(0)/2 + 9k\Theta_D/8}{kT} \right) - \frac{\pi^4}{5} \left(\frac{T}{\Theta_D} \right)^3$$

Identify the quantity $\phi(0)/2 + 9K\Theta_D/8$.

11–27. Consider a gas in equilibrium with the surface of a solid. Some of the molecules of the gas will be adsorbed onto the surface, and the number adsorbed will be a function of the pressure of the gas. A simple statistical mechanical model for this system is to picture the solid surface to be a two-dimensional lattice of M sites. Each of these sites can be either unoccupied or occupied by at most one of the molecules of the gas. Let the partition function of an unoccupied site be 1 and that of an occupied site be $q(T)$. (We do not need to know $q(T)$ here.) Assuming that molecules adsorbed onto the lattice sites do not interact with each other, the partition function of N molecules adsorbed onto M sites is then

$$Q(N, M, T) = \frac{M!}{N!(M-N)!} [q(T)]^N$$

The binomial coefficient accounts for the number of ways of distributing the N molecules over the M sites. By using the fact the adsorbed molecules are in equilibrium with the gas phase molecules (considered to be an ideal gas), derive an expression for the fractional coverage, $\theta \equiv N/M$, as a function of the pressure of the gas. Such an expression, that is, $\theta(p)$, is called an adsorption isotherm, and this model gives the so-called Langmuir adsorption isotherm.

11–28. The low-temperature constant volume heat capacity of many metals can be written in the form

$$C_V = \gamma T + AT^3$$

where the T^3 term is from the lattice vibrations and the linear term is due to the electrons. In the previous chapter it was shown that C_V for a free electron gas is given by $C_V = \pi^2 NkT/2T_F$, where T_F is the Fermi temperature. Given that the Fermi temperatures of Na and Cu are $3.7 \times 10^4 °K$ and $8.2 \times 10^4 °K$, respectively, calculate γ and compare it to the experimental values.

11–29. Compare the contributions of the electrons and the lattice vibrations to the heat capacity of sodium at low temperatures.

11–30. Show that the bulk modulus $B = V(\partial P/\partial V)_T$ of an electron gas at $0°K$ is $B = \frac{5}{3}P = 10E_0/9V$. The valence electron density of potassium is about 1.40×10^{22} cm^{-3}. Calculate B for potassium and compare with the experimental value of 3.66×10^{10} dyne/cm^2 at $4°K$.

11–31. Show that the entropy of a Debye crystal at low temperature is given by

$$S = \frac{4\pi^4 Nk}{5} \left(\frac{T}{\Theta_D}\right)^3$$

11–32. The heat capacity of copper at $100°K$ is 3.85 cal/mole-deg. Using this information, calculate the value of Θ_E and Θ_D for copper. Now calculate the heat capacity at $25°K$ and compare it to the experimental value of 0.23 cal/mole-deg. Which model gives better results at low temperatures?

11–33. Show that $g(\nu)$ of a one-dimensional Debye crystal agrees with Eq. (11–47) as $\nu \to 0$.

11–34. Derive Eq. (11–48).

CHAPTER 12

IMPERFECT GASES

In the limit of low densities, all gases approach perfect-gas behavior, or in other words, they obey the well-known equation of state

$$p = \rho k T \tag{12-1}$$

This equation was derived in Chapter 5 for a monatomic gas in which the intermolecular potential could be ignored. Physically, this means that the particles spend most of their time far away from each other and so do not "feel" any intermolecular potential. To see again how Eq. (12–1) arises from statistical thermodynamics, consider the classical canonical partition function of N monatomic particles contained in a volume V at temperature T:

$$Q = \frac{1}{N! \, h^{3N}} \int \cdots \int e^{-\beta H} \, d\mathbf{p}_1 \cdots d\mathbf{p}_N \, d\mathbf{r}_1 \cdots d\mathbf{r}_N \tag{12-2}$$

Since H is of the form

$$H = \frac{1}{2m} \sum_{n=1}^{N} (p_{xn}^2 + p_{yn}^2 + p_{zn}^2) + U(x_1, y_1, \ldots, z_N)$$

we can immediately integrate over the momenta to get

$$Q = \frac{1}{N!} \left(\frac{2\pi m k T}{h^2} \right)^{3N/2} Z_N \tag{12-3}$$

where Z_N is the configuration integral

$$Z_N = \int \cdots \int e^{-U_N/kT} \, d\mathbf{r}_1 \, d\mathbf{r}_2 \cdots d\mathbf{r}_N \tag{12-4}$$

If we can neglect U_N in the configuration integral, then $Z_N = V^N$ and $Q = q^N/N!$ where $q(V, T) = (2\pi m k T/h^2)^{3/2} V$. The key point is that it is q being of the form $f(T)V$ that leads directly to the ideal gas equation of state. Although we have discussed only the case of monatomic gases here, the same result holds true for polyatomic gases. This is easily proved at the expense of introducing a number of angular integrations (*cf.* Chapters 6 and 8).

As the density of a gas is increased, the particles are closer on the average, and the intermolecular potential becomes nonnegligible. Thus the configuration integral is no longer simply V^N, and the ideal gas equation of state is not obtained as the equation of state of the gas. Of course, it is well known experimentally that real gases exhibit deviations from ideal gas behavior as the density is increased. A large number of empirical and semiempirical equations of state have been constructed to describe the deviations from the simple ideal gas law. The most fundamental of these, in the sense that it has the most sound theoretical foundation, is the so-called virial equation of state, originally proposed by Thiesen and developed by Kamerlingh-Onnes. The virial equation of state expresses the deviations from ideal behavior as an infinite power series in ρ:

$$\frac{p}{kT} = \rho + B_2(T)\rho^2 + B_3(T)\rho^3 + \cdots \tag{12-5}$$

The quantities $B_2(T)$, $B_3(T)$, ... are called the second, third, ... virial coefficients, respectively, and depend only on the temperature and on the particular gas under consideration, but are independent of density or pressure. The primary goal of this chapter is to derive expressions for B_2, B_3, and so on, in terms of intermolecular potentials.

We shall show in Section 12-1 that the jth virial coefficient can be calculated in terms of the interactions of j molecules in a volume V. This is proved most readily by means of the grand canonical partition function. Thus we shall show that the N-body problem of an imperfect gas can be reduced to a series of one-body, two-body, three-body problems, and so on. The initial deviations from ideality (up to 10 atm, say) rest in $B_2(T)$, which we shall see is easy to calculate since it involves only two-body interactions. The derivation presented in Section 12-1 is valid for any one-component gas, including polyatomic quantum-mechanical gases whose intermolecular forces are not pair-wise additive. Then in Section 12-2 we shall specialize these results to a classical monatomic gas whose intermolecular potential is pairwise additive. The next two sections are somewhat detailed discussions of the experimental and theoretical second and third virial coefficients. Section 12-5 is devoted to the calculation of virial coefficients higher than the third; Section 12-6 discusses quantum corrections to the second virial coefficient; Section 12-7 discusses the law of corresponding states; and lastly, Section 12-8 is a general discussion.

Before going on, however, it is helpful to consider Table 12-1, which gives the contribution of the first few terms of the virial expansion to $p/\rho kT$. The data are for argon at 25°C. The contributions of all the remaining terms are shown in the parentheses. It can be seen that the second and third virial coefficients alone give most of $p/\rho kT$ up to pressures approaching 100 atm.

Table 12-1. **The contribution of the first few terms in the virial expansion of** $p/\rho kT$ **for argon at 25°C**

p(atm)	$p/\rho kT$			
	$1 + B_2\rho$	$+ B_3\rho^2$	$+$	remainder
1	$1 - 0.00064$	$+ 0.00000$	$+ \cdots$	$(+0.00000)$
10	$1 - 0.00648$	$+ 0.00020$	$+ \cdots$	(-0.00007)
100	$1 - 0.06754$	$+ 0.02127$	$+ \cdots$	(-0.00036)
1000	$1 - 0.38404$	$+ 0.68788$	$+ \cdots$	$(+0.37232)$

Source: E. A. Mason and T. H. Spurling, *The Virial Equation of State* (New York: Pergamon, 1969).

12-1 THE VIRIAL EQUATION OF STATE FROM THE GRAND PARTITION FUNCTION

The grand partition function is [*cf*. Eq. (3–15)]

$$\Xi(V\ T, \mu) = \sum_{N=0}^{\infty} Q(N, V, T)\lambda^N \tag{12-6}$$

where $\lambda = \exp(\beta\mu)$. When $N = 0$, the system has only one state with $E = 0$, and so $Q(N = 0, V, T) = 1$. This allows us to write Eq. (12–6) as

$$\Xi(V, T, \mu) = 1 + \sum_{N=1}^{\infty} Q_N(V, T)\lambda^N \tag{12-7}$$

where we have written $Q_N(V, T)$ for $Q(N, V, T)$. The characteristic thermodynamic function associated with Ξ is pV according to the relation

$$pV = kT \ln \Xi \tag{12-8}$$

The average number of molecules in the system is given by [*cf*. Eq. (3–33)]

$$N = kT\left(\frac{\partial \ln \Xi}{\partial \mu}\right)_{V,T} = \lambda\left(\frac{\partial \ln \Xi}{\partial \lambda}\right)_{V,T} \tag{12-9}$$

Thus we have the pressure and essentially the density in terms of Ξ. The standard procedure to eliminate Ξ between these two quantities is to obtain a power series for $\ln \Xi$ in some convenient parameter and then to eliminate this parameter between Eqs. (12–8) and (12–9). The most obvious choice for this expansion parameter is λ, since we already have Ξ as a power series in λ in Eq. (12–7). It is more convenient, however, though not at all necessary, to define a new activity z, proportional to λ, such that $z \to \rho$ as $\rho \to 0$. By taking the limit $\lambda \to 0$ in Eq. (12–9), we find that

$$N = \lambda\left(\frac{\partial \ln \Xi}{\partial \lambda}\right)_{V,T} = \lambda Q_1 \qquad (\lambda \to 0)$$

Thus as $\lambda \to 0$, the density $\rho \to \lambda Q_1/V$, and so we set $z = \lambda Q_1/V$. In terms of this new activity, then

$$\Xi(V, T, \mu) = 1 + \sum_{N=1}^{\infty} \left(\frac{Q_N V^N}{Q_1^N}\right)z^N \tag{12-10}$$

It is convenient to define a quantity Z_N by

$$Z_N = N!\left(\frac{V}{Q_1}\right)^N Q_N \tag{12-11}$$

It will turn out that the classical limit of the Z_N defined here is just the configuration integral given in Eq. (12–4). With these definitions, Eq. (12–7) becomes

$$\Xi = 1 + \sum_{N=1}^{\infty} \frac{Z_N(V, T)}{N!} z^N \tag{12-12}$$

This gives us Ξ as a power series in z.

We now *assume* that the pressure can be expanded in powers of z according to

$$p = kT \sum_{j=1}^{\infty} b_j z^j \tag{12-13}$$

where we wish now to determine the unknown coefficients b_j in terms of the Z_N of Eq. (12–11). This can be done directly by substituting Eq. (12–13) into $\Xi = \exp(pV/kT)$, expanding the exponential, collecting like powers of z, equating the coefficients to those of Eq. (12–12), and finally solving for the b_j in terms of the Z_N. The result of this straightforward algebra is (see Problem 12–5)

$$b_1 = (1!\,V)^{-1}Z_1 = 1$$
$$b_2 = (2!\,V)^{-1}(Z_2 - Z_1{}^2)$$
$$b_3 = (3!\,V)^{-1}(Z_3 - 3Z_2 Z_1 + 2Z_1{}^3)$$
$$b_4 = (4!\,V)^{-1}(Z_4 - 4Z_3 Z_1 - 3Z_2{}^2 + 12Z_2 Z_1{}^2 - 6Z_1{}^4)$$
$$\cdots \tag{12–14}$$

It is possible to write down a general formula for b_j in terms of Z's, but we shall not need it. We now know the b_j in Eq. (12–13) in terms of the Z_N. Notice that the calculation of b_2, for example, involves the calculation of only Z_2 and Z_1, that is, essentially the partition functions of two and one particle, respectively. Similarly, b_3 involves the determination of a partition function for three particles at the most. Thus we see that we have reduced the original N-body problem to a series of few-body problems. This was accomplished by using the grand partition function.

We are not finished yet, however, since we really want an expansion of the pressure in terms of the density ρ and not some activity z. But we not only have the pressure in terms of z, we also have the density since the pressure and density are connected by $\ln \Xi$ through Eqs. (12–8) and (12–9). Thus we can write

$$\rho = \frac{N}{V} = \frac{\lambda}{V}\left(\frac{\partial \ln \Xi}{\partial \lambda}\right)_{V,T} = \frac{z}{V}\left(\frac{\partial \ln \Xi}{\partial z}\right)_{V,T}$$
$$= \frac{z}{kT}\left(\frac{\partial p}{\partial z}\right)_{V,T}$$

from which we have

$$\rho = \sum_{j=1}^{\infty} j b_j z^j \tag{12–15}$$

Now we have both p and ρ as power series in the activity z. The problem is to eliminate z between the two equations. There is a general mathematical technique to accomplish this (by means of complex variable theory), but we can determine the first few terms in a direct algebraic way (*cf.* Section 10–1). We simply write

$$z = a_1\rho + a_2\rho^2 + a_3\rho^3 + \cdots$$

Substitute this into Eq. (12–15) and equate like powers of ρ on the two sides of the equation to get

$$a_1 = 1$$
$$a_2 = -2b_2$$
$$a_3 = -3b_3 + 8b_2{}^2$$
$$\cdots$$

Now that we have z as a power series in ρ. We substitute this into the equation for p as a power series in z [Eq. (12–13)] to get the desired final result, namely, p as a power series in the density:

$$\frac{p}{kT} = \rho + B_2(T)\rho^2 + B_3(T)\rho^3 + \cdots$$

where

$$B_2(T) = -b_2 = -(2!\,V)^{-1}(Z_2 - Z_1^2) \tag{12–16}$$

$$B_3(T) = 4b_2{}^2 - 2b_3$$

$$= -\frac{1}{3V^2}\,[V(Z_3 - 3Z_2Z_1 + 2Z_1{}^3) - 3(Z_2 - Z_1{}^2)^2] \tag{12–17}$$

$\cdots$

The equations become increasingly complicated as one goes to higher virial coefficients, but Table 12–1 shows that the first few virial coefficients suffice to account for p–V–T data up to hundreds of atmospheres.

In the next section we shall apply these general equations to the important case of a classical monatomic gas. Before doing this, however, several comments are in order. We have expressed the virial coefficients in terms of the b_j's which, in turn, are given in terms of the Z_N. The b_j's then appear to be simply intermediate quantities with no physical significance. They do have an interesting physical significance, however, that is lost in the grand canonical ensemble derivation presented here. The first statistical mechanical development of the virial expansion is due to Mayer, who used the canonical ensemble as his starting point. In this approach, the b_j's turn out to be related in a way to clusters of j molecules and play a more central role in the development of the final equations. (See Mayer and Mayer in "Additional Reading.")

Another point involves the expansion of p or $\ln \Xi$ in powers of z. We have tacitly assumed that such an expansion exists in the first place. This assumption is correct for all the cases we shall discuss in this book, but it is not obvious that it is generally valid. It is possible that special cases of highly degenerate or strongly interacting systems do not allow such a power series expansion. For example, we shall see that the intermolecular potential between a pair of particles must go to zero more rapidly than r^{-3} in order that the second virial coefficient be finite. (See Problem 12–6.) Thus we see that a virial expansion for a fully ionized gas, that is, a plasma, is invalid.

In summary then, if the virial expansion exists, we have a method for calculating the virial coefficients from partition functions. Furthermore, this series can be shown to converge in some nonzero region for a large class of intermolecular potentials. At one time there was much effort to relate this radius of convergence to the point at which condensation sets in, but it is now felt unlikely that there is any such relation.

12–2 VIRIAL COEFFICIENTS IN THE CLASSICAL LIMIT

For simplicity we shall consider only monatomic gases in this section. The extension to include molecules with internal degrees of freedom is straightforward, but the equations are more complicated. (See Problems 12–29 through 12–32.) In the classical limit, $Q(N, V, T)$ is given by Eq. (12–2). In particular, for $N = 1$, in the absence of external forces we have $U = 0$, and so we find that

$$Q_1(V, T) = \left(\frac{2\pi m k T}{h^2}\right)^{3/2} V = \frac{V}{\Lambda^3} \tag{12–18}$$

For $N > 1$ we can still integrate over the momenta as we did to get Eq. (12–3), giving

$$Q_N = \frac{Z_N}{N! \Lambda^{3N}}$$

where Z_N is the configuration integral given by Eq. (12–4). But according to Eq. (12–18), we can write this as

$$Q_N = \frac{1}{N!} \left(\frac{Q_1}{V}\right)^N Z_N \tag{12–19}$$

Comparing this to Eq. (12–11), we see that the quantity Z_N defined there is just the configuration integral. This, of course, is why we used the symbol Z_N for both quantities.

In order to calculate the second and third virial coefficients then, we need

$$Z_1 = \int d\mathbf{r}_1 = V \tag{12–20}$$

$$Z_2 = \iint e^{-U_2/kT} \, d\mathbf{r}_1 \, d\mathbf{r}_2 \tag{12–21}$$

and

$$Z_3 = \iiint e^{-U_3/kT} \, d\mathbf{r}_1 \, d\mathbf{r}_2 \, d\mathbf{r}_3 \tag{12–22}$$

To calculate the second virial coefficient, we need U_2. For monatomic particles it is reasonable to assume that $U_2(\mathbf{r}_1, \mathbf{r}_2)$ depends only upon the separation of the two particles, so that $U_2 = u(r_{12})$, where $r_{12} = |\mathbf{r}_2 - \mathbf{r}_1|$. The intermolecular potential between two particles can be calculated, at least in principle, from quantum mechanics. We shall discuss this in Section 12–3. Figure 12–1(a) shows a typical intermolecular potential between two spherically symmetric molecules. We can derive an equation for $B_2(T)$ in terms of $u(r_{12})$ by substituting Z_2 and Z_1 into Eq. (12–16) to get

$$B_2(T) = -\frac{1}{2V}(Z_2 - Z_1^2)$$

$$= -\frac{1}{2V} \iint [e^{-\beta u(r_{12})} - 1] \, d\mathbf{r}_1 \, d\mathbf{r}_2 \tag{12–23}$$

For neutral molecules in their ground electronic states, $u(r_{12})$ goes to zero fairly rapidly, say in a few molecular diameters, and so the integrand in B_2 is zero unless the volume elements $d\mathbf{r}_1$ and $d\mathbf{r}_2$ are near each other. Thus it is possible to change the variables of integration to $\mathbf{r}_1$ and the relative coordinates $\mathbf{r}_{12} = \mathbf{r}_2 - \mathbf{r}_1$ and write

$$B_2(T) = -\frac{1}{2V} \int d\mathbf{r}_1 \int [e^{-\beta u(r_{12})} - 1] \, d\mathbf{r}_{12} \tag{12–24}$$

Now the integration over the relative separation of the two particles is independent of where the pair is in the volume V, except for when the pair of particles is near the

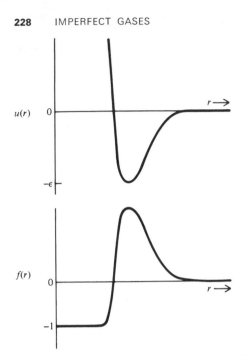

Figure 12–1. **(a) A typical intermolecular potential $u(r)$ and (b) a Mayer f-function $e^{-\beta u(r)} - 1$ plotted versus the intermolecular separation r.**

walls of the container. But thermodynamically we are interested only in the case that $V \to \infty$, and so this "surface effect" becomes entirely negligible. Thus we can carry out the integration over $d\mathbf{r}_1$ separately to give a factor of V and also write $4\pi r^2 \, dr$ for $d\mathbf{r}_{12}$ to obtain the final result

$$B_2(T) = -2\pi \int_0^\infty [e^{-\beta u(r)} - 1]r^2 \, dr \tag{12–25}$$

This gives the second virial coefficient as a simple quadrature of $u(r)$. We shall give a number of applications of this formula in Section 12–3. The limit of integration has been formally extended to infinity rather than just the walls of the container, since the integrand falls essentially to zero for r larger than a few molecular diameters. Note that B_2 is independent of V.

The term in brackets in the integrand of Eq. (12–25) appears throughout the equations of imperfect gas theory and is commonly denoted by $f(r)$,

$$f_{ij} = f(r_{ij}) = e^{-u(r_{ij})/kT} - 1 \tag{12–26}$$

Since $u(r) \to 0$ as r increases, we see that $f(r) \to 0$ as r increases. The importance of the function $f(r)$ was exploited by Mayer and is now called a Mayer f-function. Figure 12–1(b) shows a typical $f(r)$ plotted versus r.

To obtain the third virial coefficient, we need the potential $U_3(\mathbf{r}_1, \mathbf{r}_2, \mathbf{r}_3)$. Here is where the question of the pairwise additivity of intermolecular forces first arises. For many years it was common to assume that the intermolecular potential of a group of three molecules was the sum of the potentials of the three pairs taken one at a time, that is, to assume that

$$U_3(\mathbf{r}_1, \mathbf{r}_2, \mathbf{r}_3) \approx u(r_{12}) + u(r_{13}) + u(r_{23}) \tag{12–27}$$

This apparently can be used as a good approximation, but eventually suffers from close scrutiny. In general, we should write

$$U_3 = u(r_{12}) + u(r_{13}) + u(r_{23}) + \Delta(r_{12}, r_{13}, r_{23})$$

or in an obvious notation which we shall use often

$$U_3 = u_{12} + u_{13} + u_{23} + \Delta_3 \tag{12-28}$$

where Δ represents the deviation from pairwise additivity. We shall neglect Δ in what we do below, but Problem 12–26 involves the generalization to include a non-additive contribution to U_3. We shall now derive an expression for $B_3(T)$ under the assumption (apparently mild) of pairwise additivity. This derivation involves manipulations that may at first be unfamiliar, but are actually quite simple and, furthermore, are common in the statistical mechanical theories of gases and liquids.

In order to calculate B_3, we must calculate b_3, which is given by $6Vb_3 = Z_3 - 3Z_2 Z_1 + 2Z_1{}^3$ [*cf.* Eq. (12–14)]. For Z_3 we have

$$Z_3 = \iiint (1 + f_{12})(1 + f_{13})(1 + f_{23}) \, d\mathbf{r}_1 \, d\mathbf{r}_2 \, d\mathbf{r}_3$$

$$= \iiint [f_{12}f_{13}f_{23} + f_{12}f_{13} + f_{12}f_{23} + f_{13}f_{23} + f_{12} + f_{13} + f_{23} + 1] d\mathbf{r}_1 \, d\mathbf{r}_2 \, d\mathbf{r}_3 \tag{12-29}$$

The next step is to subtract $3Z_2 Z_1$ from this. Since $Z_1 = V$, we write

$$Z_1 Z_2 = V \iint (f_{12} + 1) \, d\mathbf{r}_1 \, d\mathbf{r}_2 = \iiint (f_{12} + 1) \, d\mathbf{r}_1 \, d\mathbf{r}_2 \, d\mathbf{r}_3 \tag{12-30}$$

where we have taken the volume in under the integral sign in order to obtain $d\mathbf{r}_1 \, d\mathbf{r}_2 \, d\mathbf{r}_3$ as in Z_3 above. Rather than to just subtract three times Eq. (12–30) from Z_3, however, we recognize that $Z_1 Z_2$ can also be written as

$$Z_1 Z_2 = V \iint (f_{13} + 1) \, d\mathbf{r}_1 \, d\mathbf{r}_3 = \iiint (f_{13} + 1) \, d\mathbf{r}_1 \, d\mathbf{r}_2 \, d\mathbf{r}_3$$

or equivalently

$$Z_1 Z_2 = V \iint (f_{23} + 1) \, d\mathbf{r}_2 \, d\mathbf{r}_3 = \iiint (f_{23} + 1) \, d\mathbf{r}_1 \, d\mathbf{r}_2 \, d\mathbf{r}_3$$

So instead of subtracting three times any one of these expressions for $Z_1 Z_2$ from Z_3, we subtract each once and get

$$Z_3 - 3Z_1 Z_2 = \iiint [f_{12}f_{23}f_{13} + f_{12}f_{13} + f_{12}f_{23} + f_{13}f_{23} - 2] \, d\mathbf{r}_1 \, d\mathbf{r}_2 \, d\mathbf{r}_3$$

For the $2Z_1{}^3$ that must be added to this to get $6Vb_3$, we add

$$2 \iiint d\mathbf{r}_1 \, d\mathbf{r}_2 \, d\mathbf{r}_3$$

to get

$$6Vb_3 = \iiint [f_{12}f_{13}f_{23} + f_{12}f_{13} + f_{12}f_{23} + f_{13}f_{23}] \, d\mathbf{r}_1 \, d\mathbf{r}_2 \, d\mathbf{r}_3 \tag{12-31}$$

Recall now that $B_3(T) = 4b_2{}^2 - 2b_3$, or more conveniently (in order to keep the factor of $6V$ with b_3)

$$B_3(T) = -\frac{1}{3V}(6Vb_3 - 12Vb_2{}^2)$$

The term $6Vb_3$ is given by Eq. (12–31), and so we need only write $12Vb_2{}^2$ in some convenient form and subtract it from $6Vb_3$. Since

$$b_2 = \frac{1}{2}\int f_{12}\,d\mathbf{r}_{12}$$

we can write

$$4b_2{}^2 = \left[\int f_{12}\,d\mathbf{r}_{12}\right]^2 = \left[\int f_{12}\,d\mathbf{r}_{12}\right]\left[\int f_{13}\,d\mathbf{r}_{13}\right]$$

or

$$4Vb_2{}^2 = \int d\mathbf{r}_1 \int f_{12}\,d\mathbf{r}_{12} \int f_{13}\,d\mathbf{r}_{13}$$

$$= \iiint f_{12}f_{13}\,d\mathbf{r}_1\,d\mathbf{r}_2\,d\mathbf{r}_3$$

where the last line is just the inverse of the procedure of converting to relative coordinates. Clearly the subscripts on the f-functions are arbitrary, and we can readily derive two alternative expressions for $4Vb_2{}^2$, one with $f_{13}f_{23}$ and one with $f_{12}f_{23}$ in the integrands. The three of these make $12Vb_2{}^2$, the quantity to be subtracted from $6Vb_3$ to give B_3. Thus we have finally

$$B_3(T) = -\frac{1}{3V}\iiint f_{12}f_{13}f_{23}\,d\mathbf{r}_1\,d\mathbf{r}_2\,d\mathbf{r}_3 \tag{12–32}$$

Let us examine the integrand of B_3 more closely. It involves three particles, and since $f_{ij} \to 0$ as particles i and j are separated, the product $f_{12}f_{13}f_{23}$ will vanish unless all the three particles are simultaneously close to one another. We can represent the integrand of B_3 pictorially in the following way. We draw a numbered circle for each different subscript appearing in the product and a line between each pair of particles connected by an f-function. For example, the integrands in the second and third virial coefficients can be represented schematically as shown in Fig. 12–2(a). Thus we see that all three particles in the integrand of B_3 are connected by f-functions. Since this represents a cluster of particles, we call diagrams as in Fig. 12–2 cluster diagrams. The only other cluster diagrams for three particles are shown in Fig. 12–2(b).

There is a general result of imperfect gas theory whose proof is too complicated to present here, but can be readily appreciated in terms of cluster diagrams. This result states that the virial coefficients are given by

$$B_{j+1} = \frac{-j}{j+1}\beta_j \tag{12–33}$$

where

$$\beta_j = \frac{1}{j!\,V}\int \cdots \int S'_{1,2,\ldots,j+1}\,d\mathbf{r}_1\,d\mathbf{r}_2 \cdots d\mathbf{r}_{j+1} \tag{12–34}$$

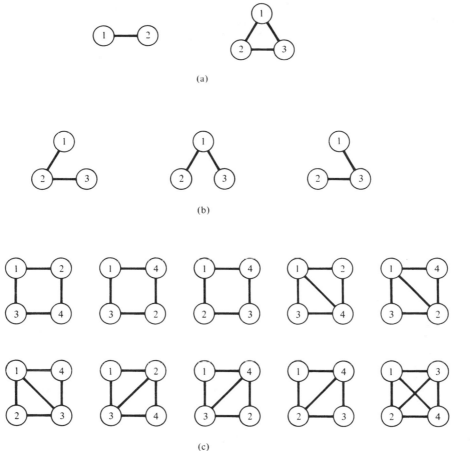

Figure 12–2. **Some examples of cluster diagrams. (a) The integrands of the second and third virial coefficients. (b) The three topologically equivalent diagrams of three particles that are singly connected. (c) The three topologically different stars or doubly connected diagrams for four particles. Compare the degeneracies of these with Eq. (12–35) and Table 12–2.**

where $S'_{1, 2, \ldots, j+1}$ is the sum of all products of f-functions that connect molecules 1, 2, ..., $j + 1$ such that the clusters are connected in such a way that the removal of *any* point, together with all of the lines associated with that point, still results in a connected graph, that is, all the particles connected to one another. Notice that all the graphs appearing in Fig. 12–2(a) are connected in this way, while those in Fig. 12–2(b) are not. Such diagrams are also called *doubly connected* since each particle is connected to any other by two independent paths. Doubly connected diagrams are also called *stars*.

All the stars of up to seven particles are listed in Appendix 1 of an article by Hoover and DeRocco.* Table 12–2 lists the number of topologically distinct connected and

Table 12–2. **The number of topologically different connected graphs $C(n)$ and star graphs $S(n)$ for $n \leq 7$**

n:	2	3	4	5	6	7
$C(n)$:	1	2	6	21	112	853
$S(n)$:	1	1	3	10	56	468

Source: W. G. Hoover and A. G. DeRocco, *J. Chem. Phys.*, **36**, p. 3141, 1962.

* W. G. Hoover and A. G. DeRocco, *J. Chem. Phys.*, **36**, p. 3141, 1962.

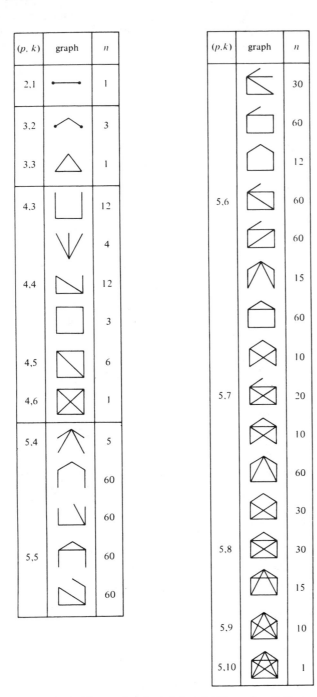

Figure 12–3. **The graphs of *p* points and *k* lines of up to five particles. *n* is the number of topologically equivalent graphs of that type.** (From G. E. Uhlenbeck and G. W. Ford, in *Studies in Statistical Mechanics* ed. by J. DeBoer and G. E. Uhlenbeck. New York: North-Holland Publishing Co., 1962.)

star graphs for $n \leq 7$. For example, for $n = 3$ there are two topologically different connected graphs, namely, $\angle$ and $\triangle$, and only one star graph, namely, $\triangle$. Figure 12–2(c) shows the three different types of star diagrams that occur for $n = 4$. The three connected diagrams that are not stars for $n = 4$ are of the form $\sqcup$, $\nwarrow$, and $\searrow$. Note that clusters like M and $\bowtie$ are really equivalent (topologically) to $\sqcup$. Figure 12–3 gives all the distinct connected graphs and star graphs of up to five particles. Note how Fig. 12–3 is consistent with Table 12–2.

In summary, then, the virial coefficients are integrals over sums of stars, and for $n = 2$, 3, and 4 these sums are given by

$$S'_{1,2} = -$$
$$S'_{1,2,3} = \triangle$$
$$S'_{1,2,3,4} = 3\square + 6\boxtimes + \boxtimes \tag{12–35}$$

The coefficients in $S'_{1,2,3,4}$ represent the number of equivalent stars of that form [see Fig. 12–2(c)]. We shall now discuss the calculation of $B_2(T)$ and $B_3(T)$ for a number of intermolecular potential functions.

12–3   SECOND VIRIAL COEFFICIENT

Equation (12–25) shows that once the intermolecular potential $u(r)$ is known, the second virial coefficient can be calculated as a function of temperature. Second virial coefficients can be measured experimentally over a large temperature range to within a few percent. Figure 12–4 shows some experimental second virial coefficients versus

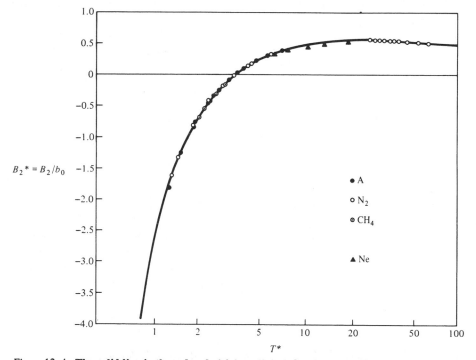

Figure 12–4. **The solid line is the reduced virial coefficient for the Lennard-Jones 6–12 potential as a function of the reduced temperature T^*. Experimental data of a number of substances are also given.** (From J. O. Hirschfelder, C. F. Curtiss, and R. B. Bird, *Molecular Theory of Gases and Liquids*. New York: Wiley, 1954.)

temperature. It is the goal of this section to understand this curve from a theoretical point of view.

In principle, $u(r)$ can be obtained from quantum mechanics, but this is a very difficult numerical problem and has not been done for anything more complicated than H_2.* One can show from perturbation theory, however, that asymptotically

$$u(r) \rightarrow -C_6 r^{-6} \tag{12-36}$$

Usually one uses simple analytical expressions with adjustable parameters for $u(r)$ that go asymptotically as r^{-6}. These adjustable parameters can then be varied to fit experimental data. Perhaps the most well-used form is

$$u(r) = \frac{n\varepsilon}{n-6} \left(\frac{n}{6}\right)^{6/(n-6)} \left\{ \left(\frac{\sigma}{r}\right)^n - \left(\frac{\sigma}{r}\right)^6 \right\} \tag{12-37}$$

where σ is the distance at which $u(r) = 0$, and ε is the depth of the well. The exponent n is usually taken to be an integer between 9 and 15, but for historical reasons, 12 is still the most popular value. The r^{-6} is included in Eq. (12–37) so that $u(r)$ has a correct asymptotic form. For $n = 12$, $u(r)$ is called the Lennard-Jones 6-12 potential:

$$u(r) = 4\varepsilon \left\{ \left(\frac{\sigma}{r}\right)^{12} - \left(\frac{\sigma}{r}\right)^6 \right\} \tag{12-38}$$

When plotted versus r, this potential is similar to that in Fig. 12–1(a). There are many other potentials in use nowadays, but they are all similar to this. We shall discuss some of these shortly.

It is not possible to integrate $B_2(T)$ analytically if Eq. (12–38) or any other realistic potential is used, and so before discussing this, however, let us consider some simpler but less realistic forms that have the advantage of allowing $B_2(T)$ to be integrated analytically.

A. HARD-SPHERE POTENTIAL

The hard-sphere potential has the form

$$u(r) = \begin{cases} \infty & r < \sigma \\ 0 & r > \sigma \end{cases} \tag{12-39}$$

This potential has no attractive part, but does simulate the steep repulsive part of realistic potentials. This is the simplest potential used and is the only potential for which the first seven virial coefficients have been calculated. (See, however, Problem 12–39.) It is the potential often used by theorists to try to understand things in a general way (we shall come back to this later). A system of particles with this potential is called a hard-sphere gas or fluid. With the hard-sphere potential, $B_2(T)$ is

$$B_2(T) = -\frac{1}{2} \int_0^\sigma (-)4\pi r^2 \, dr = \frac{2\pi\sigma^3}{3} \tag{12-40}$$

Note that this is four times the volume of a sphere and also is independent of temperature.

---

* W. Kolos and L. Wolniewicz, *J. Chem. Phys.*, **41**, p. 3663, 1964.

B. SQUARE-WELL POTENTIAL

An extension of the hard-sphere potential that includes an attractive term and yet is simple enough to handle analytically is the square-well potential:

$$u(r) = \begin{array}{ll} \infty & r < \sigma \\ -\varepsilon & \sigma < r < \lambda\sigma \\ 0 & r > \lambda\sigma \end{array} \qquad (12\text{–}41)$$

λ, the range of the attractive well, is usually taken to be between 1.5 and 2.0. If this potential is substituted into Eq. (12–25), one gets

$$B_2(T) = b_0\{1 - (\lambda^3 - 1)(e^{\beta\varepsilon} - 1)\} \qquad (12\text{–}42)$$

The quantity b_0 is the hard-sphere second virial coefficient $2\pi\sigma^3/3$. Note that as $\lambda \to 1$ or $\varepsilon \to 0$, Eq. (12–42) reduces to the hard-sphere result. Equation (12–42) can be used to fit experimental data very well, at least at ordinary temperatures. It does not pass through a maximum, however. (Why not?) Figure 12–5 shows the calculated and experimental second virial coefficient for argon, and Table 12–3 gives the square-well parameters for a number of molecules.

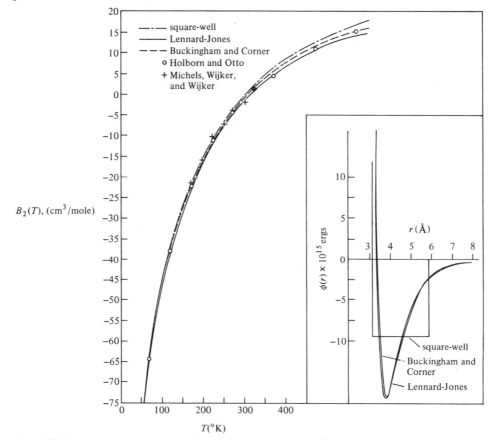

Figure 12–5. **Second virial coefficients for argon calculated for several molecular models. The potential functions obtained from the experimental $B_2(T)$ data are also shown.** (The experimental data are those of L. Holborn and J. Otto, *Z Physik*, **33**, 1, 1925, and A. Michels, Hub. Wijker, and Hk. Wijker, *Physica*, **15** 627, 1949, from J. O. Hirschfelder, C. F. Curtiss, and R. B. Bird, *Molecular Theory of Gases and Liquids*, New York: Wiley 1954.)

Table 12–3. **Potential parameters determined from second virial coefficient data**

substance	potential	λ	$\sigma(\text{Å})$	$\varepsilon/k(°\text{K})$
argon	sw	1.70	3.067	93.3
	LJ		3.504	117.7
krypton	sw	1.68	3.278	136.5
	LJ		3.827	164.0
methane	sw	1.60	3.355	142.5
	LJ		3.783	148.9
xenon	sw	1.64	3.593	198.5
	LJ		4.099	222.3
tetrafluoromethane	sw	1.48	4.103	191.1
	LJ		4.744	151.5
neopentane	sw	1.45	5.422	382.6
	LJ		7.445	232.5
nitrogen	sw	1.58	3.277	95.2
	LJ		3.745	95.2
carbon dioxide	sw	1.44	3.571	283.6
	LJ		4.328	198.2
n-pentane	sw	1.36	4.668	612.3
	LJ		8.497	219.5
benzene	sw	1.38	4.830	620.4
	LJ		8.569	242.7

Source: A. E. Sherwood and J. M. Prausnitz, *J. Chem. Phys.*, **41**, p. 429, 1964.

C. LENNARD-JONES POTENTIAL*

The hard-sphere and square-well potentials are simple enough to yield analytic expressions for virial coefficients, but presumably do not represent an actual inter-molecular potential function. As we mentioned above, the Lennard-Jones potential [Eq. (12–38)] is the most commonly used form that does qualitatively represent the behavior in Fig. 12–1(a). The second virial coefficient for this potential is

$$B_2(T) = -\frac{1}{2} \int_0^\infty \left[\exp\left\{ -\frac{4\varepsilon}{kT} \left[\left(\frac{\sigma}{r}\right)^{12} - \left(\frac{\sigma}{r}\right)^6 \right] \right\} - 1 \right] 4\pi r^2 \, dr \qquad (12\text{–}43)$$

This must be evaluated numerically. Before doing this, however, it is convenient to write Eq. (12–43) in a reduced form by defining a reduced distance, $x = r/\sigma$, and a reduced temperature $T^* = kT/\varepsilon$. If we define a reduced second virial coefficient $B_2^* = B_2/b_0$, then

$$B_2^*(T^*) = -3 \int_0^\infty \left[\exp\left\{ -\frac{4}{T^*}(x^{-12} - x^{-6}) \right\} - 1 \right] x^2 \, dx \qquad (12\text{–}44)$$

The reduced second virial coefficient B_2^* is a well-tabulated function. (See Hirschfelder, Curtiss, and Bird in " Additional Reading.")

There are a number of procedures for determining the "best" values of σ and ε from experimental second virial coefficient data. One way is to choose any two temperatures, say T_1 and T_2, and calculate the ratio

$$k_B = \frac{B_2(T_2)}{B_2(T_1)} \bigg|_{\text{exptl}}$$

and set this equal to

$$k_B = \frac{B_2^*(kT_2/\varepsilon)}{B_2^*(kT_1/\varepsilon)}$$

---

* Section 3–6 of Hirschfelder, Curtiss, and Bird (in "Additional Reading") has an excellent discussion of virial coefficients and the Lennard-Jones potential.

Then hunt for an ε by trial and error in the table of $B_2^*(T^*)$ until the calculated ratio agrees with the experimental ratio. This gives ε. Then calculate σ from

$$\frac{B_2(T_1)}{B_2^*(T_1^*)} = b_0 = \frac{2\pi\sigma^3}{3}$$

Hopefully, σ and ε will be independent of the choice of T_1 and T_2 and would be if the Lennard-Jones potential were really the "exact" potential. Table 12–3 gives the Lennard-Jones parameters for a number of molecules, and Fig. 12–4 shows the reduced second virial coefficient as a function of T^* for several gases. Figure 12–5 shows B_2 for the Lennard-Jones potential and the square-well potential. Notice that the two curves for B_2 are almost indistinguishable, although the insert shows that the two potentials themselves are quite different. It has often been said that the second virial coefficient is extremely insensitive to the potential used, but there is some indication that, if taken over a sufficiently large temperature range, it can be used to select potential functions.*

D. OTHER POTENTIALS

There are a number of other realistic potentials besides the Lennard-Jones, and B_2 has been tabulated and used for all of them. Fitts' review article† discusses several of these and their comparison with experimental data. One of the most successful of these simply includes an r^{-8} term in the potential given in Eq. (12–37) to give

$$u(r) = \frac{A}{r^n} - \frac{C_6}{r^6} - \frac{C_8}{r^8} \tag{12–45}$$

In terms of the usual parameters of ε, the depth of the potential well, and of σ, the distance at which the potential equals zero, $u(r)$ becomes

$$\frac{u(r)}{\varepsilon} = \frac{(6 + 2\gamma)}{n - 6}\left(\frac{d\sigma}{r}\right)^n - \frac{[n - \gamma(n - 8)]}{n - 6}\left(\frac{d\sigma}{r}\right)^6 - \gamma\left(\frac{d\sigma}{r}\right)^8 \tag{12–46}$$

where $d = r_m/\sigma$ and $\gamma = C_8/\varepsilon r_m^8$ and r_m is the distance at which the potential is a minimum. Since this potential has two more parameters than the Lennard-Jones 6-12 potential, it is able to give much better agreement with experimental data. Nevertheless, it has been subjected to a fairly critical test and has performed quite well.‡ Mason and Spurling discuss a large number of other intermolecular potentials. (See Mason and Spurling under "Additional Reading.")

12–4 THIRD VIRIAL COEFFICIENT

The third virial coefficient can also be measured experimentally, although not as accurately as the second virial coefficient. Equation (12–32) for $B_3(T)$ is

$$B_3(T) = -\frac{1}{3V} \iiint_V f_{12} f_{13} f_{23} \, d\mathbf{r}_1 \, d\mathbf{r}_2 \, d\mathbf{r}_3$$

$$= -\frac{1}{3} \iint f_{12} f_{13} f_{23} \, d\mathbf{r}_{12} \, d\mathbf{r}_{13} \tag{12–47}$$

We shall show how to calculate B_3 analytically for the hard-sphere and square-well potentials in the next section. For the Lennard-Jones potential, we can measure all

---

* A. E. Kingston, *J. Chem. Phys.*, **42**, p. 719, 1965.
† D. D. Fitts, *Ann. Rev. Phys. Chem.*, **17**, p. 59, 1966.
‡ H. J. M. Hanley and M. Klein, *J. Phys. Chem.*, **76**, p. 1743, 1972.

distances in Eq. (12–47) in terms of σ and define a reduced temperature T^*, and so on, to give

$$B_3^*(T^*) = \frac{B_3(T)}{b_0^2} = \frac{-3}{4\pi^2} \iint f_{12} f_{13} f_{23} \, d\mathbf{r}_{12}^* \, d\mathbf{r}_{13}^* \tag{12–48}$$

which, of course, must be evaluated numerically. This gives B_3/b_0^2 as a universal function of kT/ε, and this is shown as a function of T^* in Fig. 12–6. Although the figure caption suggests reasons for the discrepancy between the experimental points and the theoretical curve, the principal reason is probably due to the nonadditivity of the three-body potential.

Third virial coefficients, including a nonadditive contribution, have been calculated for several other potentials by Sherwood and Prausnitz.* They use the square-well, Kihara, and exp-6 potentials and make a very thorough comparison to experimental data. They find the inclusion of the nonadditive part of $U_3(\mathbf{r}_1, \mathbf{r}_2, \mathbf{r}_3)$ gives much better agreement with experimental data. This is shown in Fig. 12–7. Their article provides a readable and extensive study of virial coefficients and intermolecular potentials.

It becomes increasingly difficult to either determine higher virial coefficients experimentally or to calculate them theoretically for any realistic potential. A great deal of work has been done, however, in calculating higher virial coefficients for simple potentials such as the hard-sphere potential, not to compare such results to experiment, but to investigate the structure of the statistical mechanical equations themselves. For this reason we shall discuss some of the hard-sphere calculations of B_3 through B_7. Most of the results in this area are due to Ree and Hoover.†

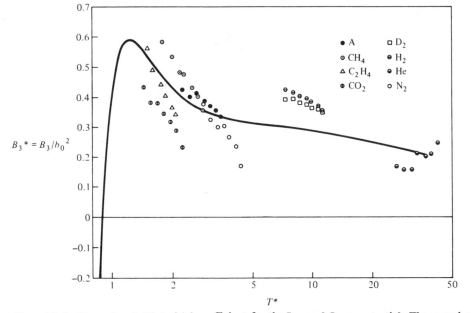

Figure 12–6. **The reduced third virial coefficient for the Lennard-Jones potential. The experimental values have been reduced using values of ε and σ determined by fitting the second virial coefficient to experimental data. The nonspherical molecules (carbon dioxide and ethylene) deviate markedly from the calculated curve. Also the light gases (hydrogen, deuterium and helium) exhibit different behavior because of quantum effects. (From R. B. Bird, E. L. Spotz, and J. O. Hirschfelder, *J. Chem. Phys.* 18, 1395, 1950.)**

* A. E. Sherwood and J. M. Prausnitz, *J. Chem. Phys.*, **41**, p. 429, 1964.
† F. H. Ree and W. G. Hoover, *J. Chem. Phys.*, **40**, p. 939, 1964.

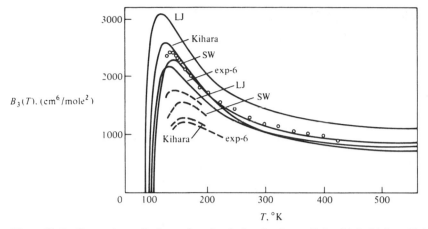

Figure 12–7. **Comparison of observed and calculated values of the third virial coefficient of argon. Solid lines include a nonadditivity correction; dashed lines show a portion of the additive third virial coefficient curve.** (From Sherwood and Prausnitz, *J. Chem. Phys.* **41,** 429, 1964.)

12–5   HIGHER VIRIAL COEFFICIENTS FOR THE HARD-SPHERE POTENTIAL

We have already seen that $B_2 = 2\pi\sigma^3/3$ for the hard-sphere potential. The third virial coefficient can be calculated by a geometrical consideration of the overlapping of three spheres, but this is a laborious and demanding route. We shall calculate B_3 here by a method due to Katsura* using Fourier transforms. Let $\boldsymbol{\rho}_j = \mathbf{r}_j - \mathbf{r}_1$ in Eq. (12–32). Then

$$B_3 = -\frac{1}{3} \iint f(|\boldsymbol{\rho}_2|) f(|\boldsymbol{\rho}_3|) f(|\boldsymbol{\rho}_3 - \boldsymbol{\rho}_2|) \, d\boldsymbol{\rho}_2 \, d\boldsymbol{\rho}_3 \tag{12–49}$$

Now let the Fourier transform of $f(|\boldsymbol{\rho}|)$ be (*cf.* Appendix B)

$$\gamma(t) = \gamma(|\mathbf{t}|) = (2\pi)^{-3/2} \int f(|\boldsymbol{\rho}|) e^{-i\mathbf{t}\cdot\boldsymbol{\rho}} \, d\boldsymbol{\rho} \tag{12–50}$$

$$= \left(\frac{2}{\pi}\right)^{1/2} \int_0^\infty \rho f(\rho) \frac{\sin t\rho}{t} \, d\rho \tag{12–51}$$

In Eq. (12–51), ρ is the scalar magnitude of $\boldsymbol{\rho}$. The second expression for $\gamma(t)$ is obtained by calling $\mathbf{t}$ the z-axis of a spherical coordinate system and measuring $\boldsymbol{\rho}$ relative to $\mathbf{t}$. Then we can set $\mathbf{t} \cdot \boldsymbol{\rho} = t\rho \cos\theta$, $d\boldsymbol{\rho} = \rho^2 \, d\rho \sin\theta \, d\theta \, d\phi$, and integrate over θ and ϕ (see Problem 12–41). Equation (12–50) can be inverted to give

$$f(\rho) = (2\pi)^{-3/2} \int \gamma(|\mathbf{t}|) e^{i\mathbf{t}\cdot\boldsymbol{\rho}} \, d\mathbf{t}$$

$$= \left(\frac{2}{\pi}\right)^{1/2} \int_0^\infty t\gamma(t) \frac{\sin \rho t}{\rho} \, dt \tag{12–52}$$

For hard spheres,

$$\begin{aligned} f(\rho) &= -1 \qquad 0 < \rho < 1 \\ &= 0 \qquad \rho > 1 \end{aligned} \tag{12–53}$$

---

* S. Katsura, *Phys. Rev.*, **115,** p. 1417, 1959.

and so according to Eq. (12–51)

$$\gamma(t) = \sigma^3 \left(\frac{2}{\pi}\right)^{1/2} \left\{\frac{\cos(\sigma t)}{(\sigma t)^2} - \frac{\sin(\sigma t)}{(\sigma t)^3}\right\}$$

$$= -\sigma^3 \frac{J_{3/2}(\sigma t)}{(\sigma t)^{3/2}} \tag{12–54}$$

where $J_{3/2}(t)$ is a Bessel function.

The trick now is to realize that Eq. (12–52) shows that

$$f(|\mathbf{\rho}_3 - \mathbf{\rho}_2|) = (2\pi)^{-3/2} \int \gamma(t) e^{it \cdot \mathbf{\rho}_3 - it \cdot \mathbf{\rho}_2} \, dt \tag{12–55}$$

and substitute this into Eq. (12–49) for B_3 to get

$$B_3 = -\frac{(2\pi)^{-3/2}}{3} \iiint \gamma(t) f(\rho_2) f(\rho_3) e^{it \cdot \mathbf{\rho}_3 - it \cdot \mathbf{\rho}_2} \, d\mathbf{\rho}_2 \, d\mathbf{\rho}_3 \, dt$$

$$= -\frac{(2\pi)^{3/2}}{3} \int dt \gamma^3(t)$$

$$= \frac{(2\pi)^{3/2}}{3} \sigma^9 \int_0^\infty 4\pi t^2 \frac{[J_{3/2}(\sigma t)]^3}{(\sigma t)^{9/2}} \, dt$$

$$= \frac{4\pi(2\pi)^{3/2}\sigma^6}{3} \int_0^\infty dx [J_{3/2}(x)]^3 x^{-5/2} \tag{12–56}$$

where we have converted the variable of integration from t to $x = \sigma t$.

The integral in Eq. (12–56) is not standard, but Katsura* has evaluated many integrals of the form

$$\int_0^\infty x^\lambda J_\alpha(x) J_\beta(x) J_\gamma(x) \, dx$$

of which Eq. (12–56) is a special case. Using Katsura's paper, Eq. (12–56) becomes (see Problem 12–40)

$$B_3 = \frac{4\pi(2\pi)^{3/2}\sigma^6}{3} \cdot \frac{5}{48(2\pi)^{1/2}} = \frac{5\pi^2\sigma^6}{18} = \frac{5}{8} b_0^2 \tag{12–57}$$

where b_0 is the second hard-sphere virial coefficient.

This is a technique which can be used in many applications involving r_{ij}-type integrands. The integration of the most highly connected stars such as ⊠ cannot be done in this way and are usually done numerically by Monte Carlo methods.†

The fourth virial coefficient B_4 involves the sum of three types of cluster integrals and is given by

$$B_4 = -\frac{1}{8V} \iiiint \{3\square + 6\boxslash + \boxtimes\} \, d\mathbf{r}_1 \, d\mathbf{r}_2 \, d\mathbf{r}_3 \, d\mathbf{r}_4$$

$$= (-0.9714 + 1.4167 - 0.1584)b_0^3$$

$$= 0.2869 b_0^3 \tag{12–58}$$

Ree and Hoover‡ have evaluated B_5 through B_7 by a variety of numerical methods.

* S. Katsura, *ibid.*
† G. W. Bern in *Modern Mathematics for the Engineer*, ed. by E. F. Bechenbach (New York: McGraw-Hill, 1956). B. J. Alder and W. G. Hoover, in *Physics of simple liquids*, ed. by H. N. V. Temperly, J. S. Rowlinson, and G. S. Rushbrooke. Amsterdam: North-Holland Publishing Co., 1968.
‡ F. H. Ree and W. G. Hoover, *J. Chem. Phys.*, **46**, p. 4181, 1967.

The number of integrals becomes quite large, there being 468 in the calculation of B_7. They find that

$$\frac{B_5}{b_0{}^4} = 0.1097 \pm 0.0003$$

$$\frac{B_6}{b_0{}^5} = 0.0386 \pm 0.0004$$

$$\frac{B_7}{b_0{}^6} = 0.0138 \pm 0.0004 \tag{12–59}$$

Not even the diagrams have ever been generated for B_8. These calculations were made to study virial coefficient equations themselves and also to determine how much one can learn about the liquid state from the first few virial coefficients. They found, for instance, that using just the truncated seventh degree polynomial in the density, they obtained 10 percent agreement with certain "computer experiments" on hard-sphere liquids. (We shall discuss these "computer experiments" later.) A virial expansion cannot really be expected to describe a liquid though, since the series itself diverges at certain densities and temperatures. Much work has been done examining this series and its convergence. This involves quite a sophisticated mathematical discussion of the β_j for large j, particularly their volume dependence. One of the questions that at one time was investigated is whether the phenomenon of condensation occurs at the density and temperature at which the virial expansion diverges. It is presently believed that these two are not related to each other.

Up to now we have concentrated on a one-component, classical, monatomic gas. It is straightforward to extend these results to two components (see Problems 12–17 and 12–18) and to polyatomic molecules (see Problems 12–29 through 12–32.) The main difference in the polyatomic-molecule case is that the intermolecular potential now depends upon the relative orientation of the molecules. This is discussed in Sections 3–4 and 3–8 in Hirschfelder, Curtiss, and Bird (see "Additional Reading") We can also extend these classical mechanical formulas to include quantum effects. To calculate the second virial coefficient in a complete quantum-mechanical scheme requires the detailed scattering state wave functions and phase shifts for $u(r)$, and this is very difficult. Chapter 6 of Hirschfelder, Curtiss, and Bird discusses such calculations for several potentials, such as the square well and Lennard-Jones. It is possible, however, to make an expansion in powers of h, and the first few terms not only give a measure of the degree of importance of quantum effects, but are also quite tractable. In the next section we calculate the first quantum correction to our classical $B_2(T)$.

12–6 QUANTUM CORRECTIONS TO $B_2(T)$

Recall Eq. (10–116) for Q_N, namely,

$$Q_N = \sum_m \int \psi_m{}^* e^{-\beta \mathcal{H}} \psi_m \, d\mathbf{r}_1 \cdots d\mathbf{r}_N$$

In Section 10–7 we showed that Q_N could be expanded in a power series in h, with the leading term being the classical partition function (without the $N!$ since we did not consider the symmetry requirements on the wave functions). We found that

$$Q_N = \frac{1}{h^{3N}} \int \cdots \int \exp(-\beta H) w(\mathbf{p}_1, \ldots, \mathbf{r}_N, \beta) \, d\mathbf{p}_1 \cdots d\mathbf{r}_N$$

where

$$w(\mathbf{p}_1, \ldots, \mathbf{r}_N, \beta) = \sum_{l=0}^{\infty} \hbar^l w_l(\mathbf{p}_1, \ldots, \mathbf{r}_N, \beta)$$

with

$$w_0 = 1,$$

$$w_1 = -\frac{i\beta^2}{2m} \sum_{j=1}^{N} \mathbf{p}_j \cdot \nabla_j U$$

and

$$w_2 = -\frac{1}{2m} \left\{ \frac{\beta^2}{2} \sum_{k=1}^{N} \nabla_k^2 U - \frac{\beta^3}{3} \left[\sum_{k=1}^{N} (\nabla_k U)^2 + \frac{1}{m} \left(\sum_{k=1}^{N} \mathbf{p}_k \cdot \nabla_k \right)^2 U \right] \right.$$
$$\left. + \frac{\beta^4}{4m} \left(\sum_{k=1}^{N} \mathbf{p}_k \cdot \nabla_k U \right)^2 \right\} \cdots$$

We can substitute this series into Q_N and integrate over the momenta to get

$$Q_N = \frac{(2\pi m k T)^{3N/2}}{h^{3N}} \iint e^{-\beta U} \left\{ 1 - \frac{\hbar^2 \beta^2}{12m} \sum_{k=1}^{N} \left(\nabla_k^2 U - \frac{\beta}{2} (\nabla_k U)^2 \right) + \cdots \right\} d\mathbf{r}_1 \cdots d\mathbf{r}_N$$

$$(12\text{--}60)$$

The contribution from w_1 vanishes upon integration, since it is an odd function of the momenta. We now divide this by $N!$ and substitute it into Eq. (12–11) and then that into Eq. (12–16) to get

$$B_2 = -b_2 = \frac{-1}{2V} (Z_2 - Z_1^2)$$

$$= -2\pi \int_0^{\infty} [e^{-\beta u(r)} - 1] r^2 \, dr + \frac{h^2}{24\pi m (kT)^3} \int_0^{\infty} e^{-\beta u(r)} \left(\frac{du}{dr} \right)^2 r^2 \, dr + O(h^3)$$

$$(12\text{--}61)$$

Table 12–4 shows the magnitude of the various quantum contributions to $B_2(T)$ for several gases at various temperatures. It lists not only the first correction which we have derived above, but also the second, which goes as h^4, and an h^3-term that arises

Table 12–4. **Contribution of various quantum terms to the second virial coefficient**

gas	$T(°K)$	B_{class}	h^2-term	h^4-term	ideal quantum gas
He⁴	27.3	−4.87	9.16	−4.05	0.50
	83.5	8.87	1.82	−0.19	0.093
	256.0	11.13	0.48	−0.01	0.017
H₂	49.2	−47.1	20.68	−8.63	0.57
	182.8	7.55	2.26	−0.19	0.080
	592.0	15.7	0.49	−0.01	0.014
D₂	37.0	78.94	19.60	−10.21	0.31
	182.8	7.55	1.13	−0.05	0.029
	592.0	15.7	0.25	0	0.004
Ne	35.6	−66.2	3.80	−0.47	0.03
	95.0	−6.23	0.55	−0.01	0.007
	392.0	12.1	0.07	0	0.0008

Source: J. O. Hirschfelder, C. F. Curtiss, and R. B. Bird, *Molecular Theory of Gases and Liquids* (New York: Wiley, 1954).

when the symmetry requirement of the wave functions is taken into account. This h^3-term is just the quantum-mechanical ideal gas second virial coefficient, Eqs. (10–15) and (10–48). You can see from Table 12–4 that this term is not numerically important. Clearly the quantum corrections become less important as the temperature increases. Table 12–4 indicates that the expansion may not be used below about 40°K for helium and 75°K for hydrogen. Of course, it is in the case of light molecules like He and H_2 that quantum effects are most important.

12-7   THE LAW OF CORRESPONDING STATES

In Sections 12–3 and 12–4 we discussed the second and third virial coefficients for several intermolecular potentials. For the Lennard-Jones potential, for example, we found that the virial coefficients could be written in a reduced form such that $B_2(T)/b_0$ or $B_3(T)/b_0{}^2$ is a function of only the reduced temperature $T^* = kT/\varepsilon$. Furthermore, assuming that all molecules interact through a Lennard-Jones potential (but with different parameters σ and ε), this function of T^* would be the same for all systems. This is illustrated in Fig. 12–4 and is an example of the law of corresponding states. We have seen examples of this law previously in Chapter 11, where we discussed the Debye theory of crystals. In this section we shall discuss a more general version of the law of corresponding states for classical monatomic systems.

We shall assume that the total intermolecular potential can be written in the form

$$U = \sum_{i,j} u(r_{ij}) = \sum_{i,j} \varepsilon\phi\left(\frac{r_{ij}}{\sigma}\right) \tag{12–62}$$

In particular, we are assuming pairwise additivity and that the pair potential can be written as an energy parameter ε times a function of only the reduced distance r/σ. This pair potential is quite general and need not be a Lennard-Jones potential. However, we do assume that the pair potential ϕ is the same function for all substances. With these assumptions, then, the configuration integral is

$$Z_N = \int \cdots \int e^{-U/kT} d\mathbf{r}_1 \cdots d\mathbf{r}_N$$

$$= \sigma^{3N} \int \cdots \int \exp\left\{-\frac{\varepsilon}{kT} \sum_{i,j} \phi\left(\frac{r_{ij}}{\sigma}\right)\right\} d\left(\frac{\mathbf{r}_1}{\sigma^3}\right) \cdots d\left(\frac{\mathbf{r}_N}{\sigma^3}\right)$$

$$= \sigma^{3N} f\left(T^*, \frac{V}{\sigma^3}, N\right)$$

where the function f is the same for all molecules. We have written V/σ^3 as one of the variables of f since all distances have been reduced by σ.

The N dependence of f is restricted by the fact that the Helmholtz free energy A is an extensive thermodynamic quantity. To see this, we use the fact that since A is extensive, A/N is intensive and, consequently, must be a function of only $v = V/N$ and T. In terms of Z_N, A/N is

$$\frac{A}{NkT} = -\frac{1}{N} \ln Q = -\frac{1}{N} \ln \frac{Z_N}{N!\Lambda^{3N}}$$

$$= -\frac{1}{N} \ln \frac{Z_N}{N!} + 3 \ln \Lambda$$

Since A/NkT is a function of v and T only, $N^{-1} \ln (Z_N/N!)$ is a function of v and T only, which says that

$$\frac{Z_N}{N!} = \frac{\sigma^{3N}}{N!} f\left(T^*, \frac{V}{\sigma^3}, N\right)$$

$$= \sigma^{3N}\left[g\left(T^*, \frac{v}{\sigma^3}\right)\right]^N \tag{12–63}$$

where g is the same function for all molecules (since f is).

Thus the partition function is of the form

$$Q(N, V, T) = \left[\frac{\sigma^3 g(kT/\varepsilon, v/\sigma^3)}{\Lambda^3}\right]^N \tag{12–64}$$

The pressure is

$$p = kT\left(\frac{\partial \ln Q}{\partial V}\right)_{N, T}$$

$$= \frac{kT}{N\sigma^3}\left(\frac{\partial \ln Q}{\partial (v/\sigma^3)}\right)_{N, T}$$

$$= \frac{kT}{\sigma^3}\left(\frac{\partial \ln g}{\partial (v/\sigma^3)}\right)_{T^*}$$

which shows that

$$\frac{pv}{kT} = \left(\frac{v}{\sigma^3}\right)\left(\frac{\partial \ln g}{\partial (v/\sigma^3)}\right)_{T^*} \tag{12–65}$$

must be the same function of $T^* = kT/\varepsilon$ and v/σ^3 for all substances. Table 12–5 shows the reduced critical parameters for a number of substances.

A look at Table 12–5 shows that the reduced critical constants of Ne, Ar, Kr, Xe, N_2, and CH_4 are quite similar, and hence these substances obey the law of corresponding states as we have formulated it. The other substances do not conform to it for various obvious reasons. For example, we assumed that classical statistics was applicable, and this is not valid for He at $5°K$. We also assumed that the intermolecular interactions were spherically symmetric or at least effectively spherically symmetric.

Table 12–5. **The critical constants and reduced critical constants, reduced by means of the Lennard-Jones parameters**

gas	$T_c(°K)$	$v_c(cm^3/mole)$	$p_c(atm)$	T_c^*	$V_c/N\sigma^3$	$p_c v_c/kT_c$
He	5.3	57.8	2.26	0.52	5.75	0.300
Ne	44.5	41.7	25.9	1.25	3.33	0.296
Ar	151	75.3	48.0	1.28	2.90	0.292
Kr	209	91.3	54.3	1.27	2.71	0.289
Xe	290	118.7	58.0	1.30	2.86	0.289
N_2	126	90.0	33.5	1.32	2.84	0.292
CO_2	304	94.0	72.8	1.53	1.93	0.274
CH_4	191	100.0	45.8	1.28	3.07	0.292
n-pentane	470	310.3	33.3	2.14	0.84	0.268
neopentane	434	302.5	31.6	1.87	1.21	0.268
benzene	563	260.0	48.6	2.32	0.68	0.274

Source: J. O. Hirschfelder, C. F. Curtiss, and R. B. Bird, *Molecular Theory of Gases and Liquids* (New York: Wiley, 1954).

Table 12–5 shows that this assumption is not good for molecules like pentane and benzene. Similar disagreement would have been observed if we had included polar molecules. The law of corresponding states can be extended to include polar substances and quantum effects, but we shall not do so here.*

Note that the law of corresponding states provides a useful method for estimating Lennard-Jones parameters. For example, for those substances that do obey a law of corresponding states, T_c^* is about 1.3, and v_c/σ^3 is 2.7. Thus we have

$$\frac{\varepsilon}{k} \approx \tfrac{3}{4} T_c$$

and

$$Nb_0 = \frac{2\pi N\sigma^3}{3} \approx \tfrac{3}{4} v_c \tag{12–66}$$

12–8   CONCLUSION

This almost concludes our discussion of gases. We have been able to derive many both useful and exact results (something which is quite rare) by appealing to the grand canonical ensemble. The area of imperfect gas theory is quite well understood.

Before leaving this, however, let us go back to B_5 for hard spheres and look at its calculation a little more closely. B_5 is a sum of ten types of cluster integrals, whose values are†

$$B_5 = -\frac{1}{30V}\int\cdots\int\{12\,\bigcirc + 60\,\bigcirc + 10\,\bigcirc + 60\,\bigcirc + 30\,\bigcirc + 10\,\bigcirc$$

$$+ 15\,\bigcirc + 30\,\bigcirc + 10\,\bigcirc + \bigcirc\}\,d\mathbf{r}_1\cdots d\mathbf{r}_5 \tag{12–67}$$

$$= -\frac{1}{30}(-45.70 + 152.72 + 23.43 - 114.28 - 47.55 - 20.55$$

$$+ 17.15 + 39.93 - 9.17 + 0.73)b_0{}^4$$

$$= -\frac{1}{30}(-237.25 + 233.96)b_0{}^4 = \frac{3.29}{30}b_0{}^4 = 0.11b_0{}^4 \tag{12–68}$$

It would appear that if one must add ten difficult-to-calculate terms, each of the order of ten or a hundred, to get 0.11, something is wrong somewhere. Is it possible to calculate 0.11 directly? Maybe our whole approach to gas theory is the wrong way to go about things, in spite of its rigor and physical appeal.

Along this same line, the determination of virial coefficients from experimental p–V–T data is not at all trivial. The usual method is to curve-fit the data to a polynomial, but it is well known that the coefficients depend upon the degree of the polynomial used. One usually uses a greater and greater degree polynomial and waits for the lower coefficients to settle down.‡ A much more satisfactory method is to expand p/kT not in a power series, but in terms of orthogonal polynomials in the density, since then the coefficients are independent of the number of terms used. To

---

* R. W. Hakala, *J. Chem. Phys.*, **71**, p. 1880, 1967. K. S. Pitzer, *J. Am. Chem. Soc.*, **77**, p. 3427, 1955.
† J. S. Rowlinson, *Proc. Roy. Soc.*, **A279**, p. 147, 1964.
‡ See, for example, K. R. Hall and F. B. Canfield, *Physica*, **33**, p. 481, 1967.

see this, consider the set of orthonormal polynomials $\{\phi_j(\rho)\}$ with weighting function $w(\rho)$.

Then

$$\frac{p}{kT} = \sum_{j=1} C_j(T)\phi_j(\rho)$$

and so

$$C_j(T) = \int \left(\frac{p}{kT}\right)\phi_j(\rho)w(\rho)\,d\rho$$

which is independent of the other $C_j(T)$ and any truncation of the series for the pressure. Is it possible to find a $\{\phi_j(\rho)\}$ and $w(\rho)$ such that the $C_j(T)$ can be written as integrals over physically appealing molecular aggregates and not involve the delicate cancellation that appears in the Mayer theory?

ADDITIONAL READING

General

FOWLER, R. H., and GUGGENHEIM, E. A. 1956. *Statistical thermodynamics.* Cambridge: Cambridge University Press. Chapter 7.

HILL, T. L. 1956. *Statistical mechanics.* New York: McGraw-Hill. Chapter 5.

HIRSCHFELDER, J. O., CURTISS, C. F., and BIRD, R. B. 1954. *Molecular theory of gases and liquids.* New York: Wiley. Chapters 3 and 6.

KAHN, B. 1965. In *Studies in statistical mechanics*, Vol. III, ed. by J. DeBoer and G. E. Uhlenbeck. Amsterdam: North-Holland Publishing Co.

KESTIN, J., and DORFMAN, J. R. 1971. *A course in statistical thermodynamics.* New York: Academic. Chapter 7.

KIHARA, T. *Rev. Mod. Phys.*, **25**, p. 831, 1953.

KILPATRICK, J. E. *Adv. Chem. Phys.*, **20**, p. 39, 1971.

MASON, E. A., and SPURLING, T. H. 1969. *The Virial equation of state.* New York: Pergamon.

MAYER, J. E. 1958. In *Encyclopedia of physics*, Vol. XII, ed. by S. Flügge. Berlin: Springer-Verlag.

MAYER, J. E., and MAYER, M. G. 1940. *Statistical mechanics.* New York: Wiley. Chapter 13.

MÜNSTER, A. 1969. *Statistical thermodynamics*, Vol. I. Berlin: Springer-Verlag. Chapters 8 and 9.

ROWLINSON, J. 1958. In *Encyclopedia of physics*, Vol. XII, ed. by S. Flügge. Berlin: Springer-Verlag.

TER HAAR, D. 1954. *Elements of statistical mechanics.* New York: Rinehard. Chapter 8.

UHLENBECK, G. E., and FORD, G. W. 1962. In *Studies in statistical mechanics*, Vol. I, ed. by J. DeBoer and G. E. Uhlenbeck. Amsterdam: North-Holland Publishing Co.

PROBLEMS

12–1. The usual form of a virial expansion is

$$\frac{pV}{RT} = 1 + \frac{B(T)}{V} + \frac{C(T)}{V^2} + \cdots$$

Some workers, however, prefer to express their data by expanding the compressibility factor in a power series in the pressure

$$\frac{pV}{RT} = 1 + B'(T)p + C'(T)p^2 + \cdots$$

Find the relations between the two sets of virial coefficients.

12–2. Find the second virial coefficient from the Dieterici equation:

$$p(v - b) = kTe^{-a/kTv}$$

12–3. Sketch the cluster diagrams corresponding to the following products of f-functions:

(a) $f_{12}f_{23}f_{34}f_{45}f_{51}f_{14}f_{25}$

(b) $f_{12}f_{23}f_{13}f_{34}f_{45}f_{46}f_{56}$

12–4. Find the second virial coefficient and the Boyle temperature for the Berthelot equation of state

$$\left(P + \frac{N^2 A}{V^2 T}\right)(V - NB) = NkT$$

where A and B are constants.

12–5. Derive Eqs. (12–14).

12–6. Show that the intermolecular potential must vanish more rapidly than r^{-3} in order for $B_2(T)$ to exist. Do this by breaking up the integral in B_2 into two regions, say 0 to L and L to ∞. Choose L large enough so that the exponential in B_2 can be expanded, and investigate this convergence.

12–7. Show that the second virial coefficient for a Lennard-Jones potential can be written in the form

$$\frac{B_2(T)}{b_0} = \sum_{j=0}^{\infty} b^{(j)} T^{*-(2j+1)/4}$$

with

$$b^{(j)} = -\frac{2^{j+1/2}}{4j!} \Gamma\left(\frac{2j-1}{4}\right)$$

The quantity $\Gamma(x)$ is the gamma function.

12–8. Show that the second virial coefficient for the Sutherland potential, defined by

$$u(r) = \infty \qquad\qquad r < \sigma$$
$$= -cr^{-\gamma} \qquad r > \sigma$$

is

$$B_2(T) = -\frac{2\pi\sigma^3}{3} \sum_{j=0}^{\infty} \frac{1}{j!} \left(\frac{3}{j\gamma - 3}\right)\left(\frac{c}{\sigma^\gamma kT}\right)^j$$

The parameter γ is usually taken to be 6.

12–9. Calculate the second virial coefficient for the triangle potential*

$$u(r) = \infty \qquad\qquad\qquad r < \sigma$$
$$= \frac{\varepsilon}{\sigma(\lambda - 1)} \{r - \lambda\sigma\} \qquad \sigma < r < \lambda\sigma$$
$$= 0 \qquad\qquad\qquad r > \lambda\sigma$$

12–10. Show that

$$B_2 = -\frac{1}{6kT} \int_0^{\infty} r \frac{du(r)}{dr} e^{-u(r)/kT} 4\pi r^2 \, dr$$

is equivalent to

$$B_2 = -\frac{1}{2} \int_0^{\infty} (e^{-\beta u(r)} - 1) 4\pi r^2 \, dr$$

State the condition on $u(r)$ that is necessary.

12–11. Find an exact B_2 for the potential $u(r) = \alpha/r^n$ for all r with $n > 3$, using the results of Problem 12–10. Carry the integration to the point of having a well-known function in your result for B_2.

* See M. J. Feinberg and A. G. DeRocco, *J. Chem. Phys.*, **41**, p. 3439, 1964.

12–12. Using the tables in Hirschfelder, Curtiss, and Bird (see "Additional Reading"), plot the second virial coefficient of argon versus temperature from its boiling point, 84°K, to 500°K, assuming a Lennard-Jones potential with parameters given in Table 12–3. Compare your result to the experimental values.

12–13. Using the data in F. Whalley and W. G. Schneider, *J. Chem. Phys.*, **23**, p. 1644, 1955; A. E. Kingston, *J. Chem. Phys.*, **42**, p. 719, 1965; B. E. F. Fender and G. D. Halsey, *J. Chem. Phys.*, **36**, p. 1881, 1962; and R. D. Weir, I. Wynn Jones, J. S. Rowlinson, and G. Saville, *Trans. Far. Soc.*, **63**, p. 1320, 1967, show that the second virial coefficient satisfies the law of corresponding states for argon, krypton, and xenon.

12–14. Using the experimental data in any of the references in Problem 12–13, determine ε and σ for the Lennard-Jones potential.

12–15. Calculate the compressibility factor for argon at 10 atm and 0°C.

12–16. Show that the virial expansion for the thermodynamic energy is

$$\frac{E}{NkT} = \frac{3}{2} - T \sum_{j=1}^{\infty} \frac{1}{j} \frac{dB_{j+1}}{dT} \rho^j$$

and that for entropy is

$$\frac{S}{Nk} = \frac{S_{\text{ideal}}}{Nk} - \sum_{j=1}^{\infty} \frac{1}{j} \frac{\partial}{\partial T} (TB_{j+1}) \rho^j$$

12–17. Show that for a binary mixture,

$$\frac{p}{kT} = \rho_1 + \rho_2 + B_{20}(T)\rho_1^2 + B_{11}(T)\rho_1\rho_2 + B_{02}(T)\rho_2^2 + \cdots$$

and derive expressions for the second virial coefficients. Show that if this virial expansion is written in the form

$$\frac{p}{kT} = \rho + B_2(T)\rho^2 + \cdots$$

with $\rho = \rho_1 + \rho_2$, then

$$B_2(x_1, T) = x_1^2 B_{20}(T) + x_1(1 - x_1)B_{11}(T) + (1 - x_1)^2 B_{02}(T)$$

where x_1 is the mole fraction of component 1.

12–18. Show that for a multicomponent mixture,

$$\frac{p}{kT} = \rho + B_2\rho^2 + B_3\rho^3 + \cdots$$

where $\rho = \rho_1 + \rho_2 + \cdots + \rho_n$ and

$$B_2(T) = \sum_{i=1}^{n} \sum_{j=1}^{n} B_{ij}(T)x_i x_j$$

$$B_3(T) = \sum_{i=1}^{n} \sum_{j=1}^{n} \sum_{k=1}^{n} B_{ijk} x_i x_j x_k$$

where the x's are mole fractions.

12–19. Using tables for the Lennard-Jones potential, calculate the volume change when 5 liters of N_2 and 2 liters of Ar are mixed at 10 atm pressure and 300°K.

12–20. Derive the first few terms of a virial expansion for the fugacity f.

12–21. For an ideal gas,

$$C_p - C_V = R$$

Derive a virial expansion for this difference for a real gas.

12–22. The Joule-Thomson coefficient μ is defined by

$$\mu = \left(\frac{\partial T}{\partial p}\right)_H = C_p^{-1}\left[T\left(\frac{\partial V}{\partial T}\right)_p - V\right]$$

Derive a density expansion for μ. At high temperatures μ is negative, and for sufficiently low temperatures, it is positive. The temperature at which μ is zero is called the inversion temperature. Show that for not too dense gases, the inversion temperature is given by $d(B_2/T)/dT = 0$.

12–23. Show that

$$\mu_0 C_p^0 = -2\pi N\beta \int_0^\infty e^{-\beta u}\left(u - \frac{r}{3}\frac{du}{dr}\right)r^2\,dr$$

where μ_0 and C_p^0 are the zero pressure limits of the Joule-Thomson coefficient and C_p.

12–24. Using the result of Problem 1–35, derive an expression for the first nonideal correction to the speed of sound in a gas.

12–25. Calculate the third virial coefficient for the potential*

$$u(r) = \begin{cases} \infty & r < \sigma \\ -\varepsilon & \sigma < r < 2\sigma \\ 0 & r > 2\sigma \end{cases}$$

12–26. Derive an expression for the third virial coefficient without assuming that the intermolecular potential is pairwise additive. Let $U_3(\mathbf{r}_1, \mathbf{r}_2, \mathbf{r}_3)$ be

$$U_3(\mathbf{r}_1, \mathbf{r}_2, \mathbf{r}_3) = u(r_{12}) + u(r_{13}) + u(r_{23}) + \Delta(r_{12}, r_{13}, r_{23})$$

12–27. Derive Eq. (12–51) from Eq. (12–50).

12–28. Evaluate the ring contribution to B_4 by the method of Fourier transformations in Section 12–5.

12–29. The Hamiltonian for a gas composed of N rigid rotors is

$$H = \sum_j \frac{1}{2m}p_j^2 + \sum_j \frac{1}{2I}\left(p_{\theta_i}^2 + \frac{p_{\phi_i}^2}{\sin^2\theta_i}\right) + U(\mathbf{q}_1, \ldots, \mathbf{q}_N)$$

where $\mathbf{q}_j$ is the set of five coordinates x_j, y_j, z_j, θ_j, and ϕ_j needed to specify the location and orientation of a molecule. Show that the second virial coefficient for such a system is given by

$$B_2(T) = -\frac{1}{32\pi^2 V}\iint f_{12}\,d\mathbf{q}_1\,d\mathbf{q}_2$$

where

$$d\mathbf{q}_j = dx_i\,dy_i\,dz_i\,\sin\theta_i\,d\theta_i\,d\phi_i$$

12–30. The potential between two dipolar molecules (see Fig. 12–8) can be approximated by

$$u(r, \theta_1, \theta_2, \phi_1 - \phi_2) = \infty \qquad\qquad r < \sigma$$

$$= -\frac{\mu^2}{r^3}g(\theta_1, \theta_2, \phi_2 - \phi_1) \qquad r > \sigma$$

where

$$g(\theta_1, \theta_2, \phi_1 - \phi_2) = 2\cos\theta_1\cos\theta_2 - \sin\theta_1\sin\theta_2\cos(\phi_2 - \phi_1)$$

Show that the second virial coefficient for this potential can be expressed as

$$B_2(T) = b_0\left[1 - \frac{1}{3}\left(\frac{\mu^2}{\sigma^3 kT}\right)^2 - \frac{1}{75}\left(\frac{\mu^2}{\sigma^3 kT}\right)^4 + \cdots\right]$$

* See S. Katsura, *Phys. Rev.*, **115**, p. 1417, 1959.

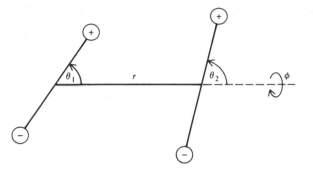

Figure 12–8. **The coordinates describing the mutual orientation of two polar molecules.** (From Hirschfelder, Curtiss, and Bird, *Molecular Theory of Gases and Liquids*. New York: Wiley, 1954.)

12–31. Show that

$$B_2(T) = -2\pi \int_0^\infty \left[\frac{kT}{u_2} e^{-u_1/kT} \left(\frac{\pi^{1/2}}{2} \operatorname{erf} \frac{\sqrt{u_2}}{kT} \right)^2 - 1 \right] r^2 \, dr$$

for a potential

$$\phi(r, \theta_1, \theta_2, \phi_1 - \phi_2) = u_1(r) + u_2(r)[\cos^2 \theta_1 + \cos^2 \theta_2]$$

12–32. Consider the angle dependent potential

$$u(r, \cos \theta) = u_0(r)[1 + aP_m(\cos \theta)]$$

where $P_m(\cos \theta)$ is a Legendre polynomial of degree m, and $u_0(r) = -\alpha/r^n$. Find an expression for the angular correction factor $f(a)$ in the equation

$$B_2[u(r, \cos \theta)] = f(a)b_2[u_0(r)]$$

Carry out the integration to obtain a "simple" expression for $f(a)$ for a "dipole" interaction $m = 1$ and long-range dispersion forces $n = 6$. As a check of your result, show

$$\lim_{a \to \infty} f(a) = 1$$

For a typical value of $a \sim \frac{1}{8}$, calculate $f(a)$.

12–33. Derive the quantum correction to $B_2(T)$ up through the term in h^2; that is, derive Eq. (12–61).

12–34. The hard-sphere virial equation can be written in terms of the variable*

$$y = \sqrt{2} \, \frac{\pi v_0}{6v} = \frac{b_0}{4v}$$

where v_0 is the closest packing volume, and b_0 is the hard-sphere second virial coefficient. In terms of y

$$\frac{pv}{kT} = 1 + 4y + 10y^2 + 18.36y^3 + 28.2y^4 + 39.5y^5 + \cdots$$

We can approximate the terms in this expansion by

$$\frac{pv}{kT} = 1 + 4y + 10y^2 + 18y^3 + 28y^4 + 40y^5 + \cdots$$

Now if you are clever enough to notice that B_n is given by $n^2 + n - 2$, you can write

$$\frac{pv}{kT} = 1 + \sum_{n=2}^{\infty} (n^2 + n - 2)y^{n-1}$$

* See N. F. Carnahan and K. E. Starling, *J. Chem. Phys.*, **51**, p. 635, 1969.

Sum this series exactly, and compare with the result of Thiele, *J. Chem. Phys.*, **39**, p. 474, 1963, as well as the exact molecular dynamics calculation of Alder and Wainwright, *J. Chem. Phys.*, **33**, p. 1439, 1960. In order to make this comparison, calculate the compressibility factor for $v/v_0 = 1.5, 2.0, 10.0$. The Alder and Wainwright values are 12.5, 5.89, 1.36, respectively. Furthermore, calculate the seventh hard-sphere virial coefficient from this scheme and compare to the result $0.0138 \pm 0.0004 b_0^6$ calculated from cluster integrals. Would you expect a phase transition for this infinite series virial expansion? Why or why not?

12–35. The virial expansion as originally developed by Ursell* was quite different from the method presented in this chapter. Consider the Boltzmann factor $W_N(\mathbf{r}_1, \mathbf{r}_2, \ldots, \mathbf{r}_N)$, which appears in the configuration integral. Ursell expressed W_N as a sum of products of functions U_l defined by

$$U_1(\mathbf{r}_i) = W_1(\mathbf{r}_i)$$
$$U_2(\mathbf{r}_i, \mathbf{r}_j) = W_2(\mathbf{r}_i, \mathbf{r}_j) - W_1(\mathbf{r}_i)W_1(\mathbf{r}_j)$$
$$U_3(\mathbf{r}_i, \mathbf{r}_j, \mathbf{r}_k) = W_3(\mathbf{r}_i, \mathbf{r}_j, \mathbf{r}_k) - W_2(\mathbf{r}_i \mathbf{r}_j), W_1(\mathbf{r}_k) - W_2(\mathbf{r}_j, \mathbf{r}_k)W_1(\mathbf{r}_i)$$
$$- W_2(\mathbf{r}_k, \mathbf{r}_i)W_1(\mathbf{r}_j) + 2W_1(\mathbf{r}_i)W_1(\mathbf{r}_j)W_1(\mathbf{r}_k)$$

Show that these U-functions, now called Ursell functions, are short-ranged functions in the sense that they vanish unless all of the l molecules in the argument of U_l are at least singly connected.

Invert these equations to write

$$W_1(\mathbf{r}_i) = U_1(\mathbf{r}_i) = 1$$
$$W_2(\mathbf{r}_i, \mathbf{r}_j) = U_2(\mathbf{r}_i, \mathbf{r}_j) + U_1(\mathbf{r}_i)U_1(\mathbf{r}_j)$$
$$W_3(\mathbf{r}_i, \mathbf{r}_j, \mathbf{r}_k) = U_3(\mathbf{r}_i, \mathbf{r}_j, \mathbf{r}_k) + U_2(\mathbf{r}_i, \mathbf{r}_j)U_1(\mathbf{r}_k)$$
$$+ U_2(\mathbf{r}_j, \mathbf{r}_k)U_1(\mathbf{r}_i) + U_2(\mathbf{r}_k, \mathbf{r}_i)U_1(\mathbf{r}_j) + U_1(\mathbf{r}_i)U_1(\mathbf{r}_j)U_1(\mathbf{r}_k)$$

This expresses the integrand of the configuration integral into a sum of products of short-ranged functions. This method can be used to write any property of N-molecules in terms of short-ranged functions (*cf.* Problem 12–36). It is easy to show (although we shall not) that the b_j of Section 12–1 are given by

$$b_j = (Vj!)^{-1} \int \cdots \int U_j(\mathbf{r}_1, \mathbf{r}_2, \ldots, \mathbf{r}_j) \, d\mathbf{r}_1 \ldots d\mathbf{r}_j$$

and that the configuration integral can be written as

$$\frac{Z_N}{N!} = \sum_{(m_j)}^{*} \prod_{j=1}^{N} \frac{(Vb_j)^{m_j}}{m_j}$$

where the asterisk denotes the condition

$$\sum jm_j = N$$

The Ursell development has the advantage of not being limited to pair-wise additive potentials.

12–36. Consider the polarizability of a system of N polarizable molecules $\alpha_N(\mathbf{r}_1, \ldots, \mathbf{r}_N)$. In general, this quantity cannot be rigorously written as the sum of N isolated-molecule polarizabilities, but can be expressed *exactly* by

$$\alpha_N(\mathbf{r}_1, \ldots, \mathbf{r}_N) = \sum_{i=1}^{N} \alpha_i(\mathbf{r}_i) + \sum\sum_{1 \le i < j \le N} \alpha_{ij}(\mathbf{r}_i, \mathbf{r}_j)$$
$$+ \sum\sum\sum_{1 \le i < j < k \le N} \alpha_{ijk}(\mathbf{r}_i, \mathbf{r}_j, \mathbf{r}_k) + \cdots$$

* See Section 3–2 of Hirschfelder, Curtiss, and Bird in "Additional Reading."

Show that the α_i are polarizabilities of isolated molecules; the α_{ij} are incremental polariz-abilities of pairs of molecules, and so on, or, in other words, by inversion determine the α_i, α_{ij}, α_{ijk}, and so on, in terms of the polarizabilities of groups of molecules.*

12–37. Find the set $\{m_j\}$ which maximizes $\ln Z_N/N!$, where

$$\frac{Z_N}{N!} = \sum_{(m_j)}^{*} \prod_{j=1}^{N} \frac{(Vb_j)^{m_j}}{m_j!}$$

where the asterisk signifies $\Sigma\, jm_j = N$ as a constraint. Call the undetermined multiplier z. The $m_j{}^*$ (those m_j that yield the maximum term) will come out in terms of z. So will $Z_N/N!$, and hence $Q(N, V, T)$ and p. Solve the equation

$$\sum_{j=1}^{N} jm_j{}^* = N$$

for z (at least the first few terms of z as a power series in ρ) and derive the first few terms of the virial expansion for p.†

12–38. Find the first few terms of $Z_N/N!$, written in the form (see Problem 12–37)

$$V^N\left(1 + \frac{a_1}{V} + \frac{a^2}{V^2} + \cdots + \frac{a_N}{V^N}\right)$$

expand the logarithm and use

$$p = kT\left(\frac{\partial \ln Q}{\partial V}\right)_{T,\,N}$$

to find the virial expansion.

12–39. A model "potential" that has been used in the theoretical study of imperfect gases is the so-called Gaussian gas, in which it is assumed that the Mayer f-function $f(r)$ can be approximated by a negative Gaussian, that is, by $-e^{-\alpha r^2}$, where α is a constant. Show that this "potential" simulates a soft repulsive potential. The great advantage of this "potential" is that the many-center integrals involved in the evaluation of the B_n can be done analytically, since it can be readily shown by matrix algebra** that

$$\int_{-\infty}^{\infty} \cdots \int e^{t\cdot x - (1/2)x\cdot A\cdot x}\, dx_1\, dx_2 \ldots dx_n = \frac{(2\pi)^{n/2}}{|A|^{1/2}}\, e^{(1/2)t\cdot A^{-1}\cdot t}$$

where $t = (t_1, t_2, \ldots, t_n)$, $x = (x_1, x_2, \ldots, x_n)$, A is a matrix, and where $|A|$ is the determinant of A. Use this to show that for a Gaussian gas,

$$B_2 = b \equiv \frac{1}{2}\left(\frac{\pi}{\alpha}\right)^{3/2}$$

$$B_3 = \frac{4}{9\sqrt{3}}\,b^2 = 0.257b^2$$

$$B_4 = -0.125b^3$$

and with patience

$$B_5 = 0.013b^4$$

---

* See H. B. Levine and D. A. McQuarrie, *J. Chem. Phys.*, **49**, p. 4181, 1968.
† See Chapter 13 of Mayer and Mayer in "Additional Reading."
‡ See T. L. Hill, *J. Chem. Phys.*, **28**, p. 61, 1958.
** See H. Cramer, *Mathematical Methods of Statistics* (Princeton, N.J.: Princeton University Press, 1946).

The sixth and seventh virial coefficients are also known for this potential, $B_6 = 0.038b^5$ and $B_7 = -0.030b^6$.*

12–40. Go to the tables in Katsura's paper† to verify the result for B_3 for hard spheres given in Eq. (12–57).

12–41. Derive Eq. (12–52).

* See *Studies in Statistical Mechanics*, vol. I, ed. by J. DeBoer and G. E. Uhlenbeck (Amsterdam: North-Holland Publishing Co., 1962, p. 182).
† In *Phys. Rev.*, **115**, p. 1417, 1959.

CHAPTER 13

DISTRIBUTION FUNCTIONS IN CLASSICAL MONATOMIC LIQUIDS

In the last chapter we have shown how one can derive a rigorous expansion of the pressure in terms of the density and thus reduce the many-body problem of a non-ideal gas to a two-body problem, a three-body problem, etc. Such a decomposition is not applicable to a liquid since each molecule in a liquid is in constant interaction with a large number of its neighbors. We shall, therefore, have to use the techniques that are more suitable for dense systems The central idea in most theories of liquids is the radial distribution function, and this chapter is devoted to a discussion of this function and a number of its close relatives.

The radial distribution function is defined in Section 13–2. Once having shown how to calculate thermodynamic functions in terms of the radial distribution function (in Section 13–3), we then derive an equation, the so-called Kirkwood equation, for the radial distribution function itself (Section 13–4). This equation is not exact, and so the rest of the chapter (except for the last sections) discusses the derivation of several other equations for the radial distribution function. These equations are derived quite differently from the Kirkwood equation, and Sections 13–5 and 13–6 are needed to provide the background for Section 13–7, in which two of the other equations are derived. Section 13–8 discusses the low-density limit of the various liquid theory expressions, and the last section is a comparison of the various results to experimental data.

It should be possible for someone who wants only a brief introduction to the theory of liquids to skip Sections 13–5 through 13–8. He simply must accept that two new equations, the hypernetted chain and the Percus-Yevick, are derived in these sections. These equations are somewhat similar to the Kirkwood equation, which is derived in Section 13-4.

13–1 INTRODUCTION

The virial expansion has a radius of convergence beyond which the series no longer represents the pressure. This point possibly has something to do with the onset of a liquid state in which the interactions can no longer be treated as a sequence of two-body, three-body interactions, etc. A theoretical treatment of liquids, therefore, requires us to seek new methods that are more directed to the many-body nature of liquids. Before going on to study some of these methods, however, let us see how much one can learn about dense gases and liquids from a knowledge of just the first few virial coefficients. Figure 13–1 shows a set of pressure-volume isotherms both above and below the critical temperature. To what extent can we reproduce curves like this with only a few virial coefficients? This question has been studied by Ree and Hoover.* They have used the first five or six hard-sphere virial coefficients to calculate equations of state and have compared these to " experimental data " on hard-sphere fluids. These " data " are actually the results of solving the equations of motion of several hundred hard spheres numerically on a large computer and then calculating observable macroscopic quantities by time averaging the appropriate microscopic equations. Such calculations are called *molecular dynamics* calculations and have been pioneered for hard spheres by Alder and Wainwright† and later extended to a fluid of molecules obeying a Lennard-Jones 6–12 potential by Rahman‡ and Verlet.§

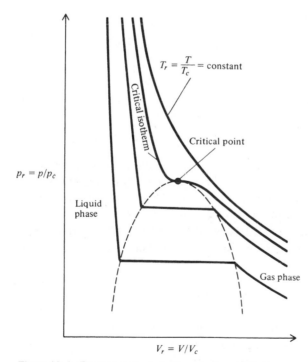

$$T_r = \frac{T}{T_c} = \text{constant}$$

Critical isotherm

Critical point

$p_r = p/p_c$

Liquid phase

Gas phase

$$V_r = V/V_c$$

Figure 13–1. **Pressure-volume isotherms of a real fluid.**

* F. H. Ree and W. G. Hoover, *J. Chem. Phys.*, **40**, p. 939, 1964; *ibid.* **46**, 4181, 1967.

† See, for example, B. J. Alder and T. Wainwright, in *Transport Processes in Statistical Mechanics*, edited by I. Prigogine (New York: Academic, 1958).

‡ A. Rahman, *Phys. Rev.*, **136**, p. 405, 1964.

§ L. Verlet, *Phys. Rev.*, **165**, p. 201, 1968.

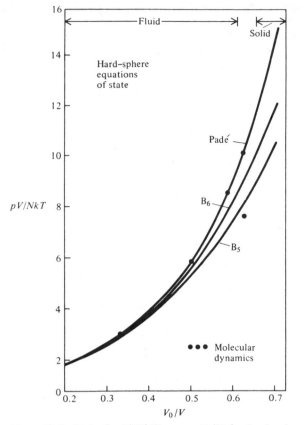

Figure 13–2. **Plot of** $pV/(NkT)$ **versus** V_0/V **for hard spheres.** V_0 **is the volume at closest packing,** $N\sigma^3/\sqrt{2}$. **The curves are: (1) virial series including** B_5 , **(2) virial series including** B_6 , **and (3) a Padé approximant. Molecular dynamics results of Alder and Wainwright are indicated by dots.** (From F. H. Ree and W. G. Hoover, *J. Chem. Phys.*, **40**, p. 939, 1964.)

Figure 13–2 shows the equation of state of a hard-sphere fluid calculated by Alder and Wainwright. It is interesting to note that the molecular dynamics calculations seem to indicate that a system of hard spheres exhibits both a fluid and a solid branch, with the transition occurring at around a volume $v = 1.6v_0$, where v_0, the closest-packed volume of a system of hard spheres, is equal to $N\sigma^3/\sqrt{2}$. On the same graph is plotted the equations of state obtained by truncating the virial series at the fifth and sixth virial coefficient.

The agreement of these truncated virial expansions is satisfactory up to a reduced density of about 0.5. It is well recognized in numerical analysis, however, that taking just the truncated power series of a function is an unsatisfactory method of approximating it and more sophisticated techniques are available. One such method is that of Padé approximants,* which simply represents the function by a ratio of polynomials, whose coefficients are found by expanding this ratio in a power series and then requiring the first n terms to correctly give the first n Taylor expansion coefficients of the function itself. For example, suppose we know the first n terms in the expansion of some function $f(x)$, i.e., we are given

$$f(x) = a_0 + a_1 x + \cdots + a_n x^n + O(x^{n+1}) \tag{13-1}$$

---

* A. Ralston, *A First Course in Numerical Analysis* (New York: McGraw-Hill, 1965).

We now construct the so-called N, M Padé approximant to this function by writing

$$f(x) \approx \frac{c_0 + c_1 x + \cdots + c_{N-1} x^{N-1}}{1 + d_1 x + \cdots + d_{N-1} x^{M-1}} \tag{13-2}$$

and then finding the c's and d's by expanding this in x and requiring the first $n + 1$ terms to be the same as those in Eq. (13-1). Clearly, the number of Padé coefficients to be determined, $M + N - 1$, must equal the number of known Taylor expansion coefficients, $n + 1$. This direct expansion of the denominator in Eq. (13-2) can become quite laborious, and fortunately the evaluation of the c's and d's can be written as a matrix problem. The appendix of the paper by Ree and Hoover* gives the matrix recipe.

Ree and Hoover* used the first six hard-sphere virial coefficients to construct the 3, 3 Padé approximant:

$$\left(\frac{p}{\rho kT} - 1\right) = b_0 \rho \frac{(1 + 0.063507 b_0 \rho + 0.017329 b_0{}^2 \rho^2)}{(1 - 0.561493 b_0 \rho + 0.239465 b_0{}^2 \rho^2)} \tag{13-3}$$

Figure 13-2 shows that this gives excellent agreement with the molecular dynamics calculations all along the fluid branch.

Fourth and fifth virial coefficients have been calculated for the Lennard-Jones potential by Barker and Monaghan.† They showed that the truncated virial series gives critical constants in fairly good agreement with experimental data. The results are shown in Table 13-1.

Table 13-1. **The reduced critical parameters obtained from a truncated Lennard-Jones virial series***

	$T_c{}^*$	$\rho_c{}^*$	$p_c{}^*/\rho_c{}^* T_c{}^*$
virial series through B_3	1.449	0.771	0.333
virial series through B_4	1.300	0.561	0.347
virial series through B_5	1.291	0.547	0.352
experimental values	1.29	0.66	0.29

* The temperature is reduced by ε/k and the volume by $2\pi N\sigma^3/3$.

All of this is fine, but it really avoids the true nature of liquids, namely, that each molecule is in constant interaction with all of its neighbors. This makes the virial expansion technique not applicable, and we must abandon it and treat the problem by methods that are more suitable for dense systems. The central idea in most theories of liquids is that of the radial distribution function, which we shall now discuss.

13-2 DISTRIBUTION FUNCTIONS

Consider a system of N particles in a volume V and at a temperature T. The probability that molecule 1 is in $d\mathbf{r}_1$ at $\mathbf{r}_1$, molecule 2 in $d\mathbf{r}_2$ at $\mathbf{r}_2$, etc., is given by

$$P^{(N)}(\mathbf{r}_1, \ldots, \mathbf{r}_N) \, d\mathbf{r}_1 \cdots d\mathbf{r}_N = \frac{e^{-\beta U_N} \, d\mathbf{r}_1 \cdots d\mathbf{r}_N}{Z_N} \tag{13-4}$$

where Z_N is the configuration integral. The probability that molecule 1 is in $d\mathbf{r}_1$ at $\mathbf{r}_1$, $\ldots$, molecule n in $d\mathbf{r}_n$ at $\mathbf{r}_n$, irrespective of the configuration of the remaining

* F. H. Ree and W. G. Hoover, *J. Chem. Phys.*, **40**, p. 939, 1964.
† J. A. Barker and J. J. Monaghan, *J. Chem. Phys.*, **36**, p. 2564, 1962.

$N - n$ molecules is obtained by integrating Eq. (13–4) over the coordinates of molecules $n + 1$ through N:

$$P^{(n)}(\mathbf{r}_1, \ldots, \mathbf{r}_n) = \frac{\int \cdots \int e^{-\beta U_N} \, d\mathbf{r}_{n+1} \cdots d\mathbf{r}_N}{Z_N} \tag{13–5}$$

Now the probability that *any* molecule is in $d\mathbf{r}_1$ at $\mathbf{r}_1$, $\ldots$, and *any* molecule is in $d\mathbf{r}_n$ at $\mathbf{r}_n$, irrespective of the configuration of the rest of the molecules, is

$$\rho^{(n)}(\mathbf{r}_1, \ldots, \mathbf{r}_n) = \frac{N!}{(N-n)!} \cdot P^{(n)}(\mathbf{r}_1, \ldots, \mathbf{r}_n) \tag{13–6}$$

This comes about since we have N choices for the first molecule, $N - 1$ for the second, etc.

The simplest distribution function is $\rho^{(1)}(\mathbf{r}_1)$. The quantity $\rho^{(1)}(\mathbf{r}_1) \, d\mathbf{r}_1$ is the probability that any one molecule will be found in $d\mathbf{r}_1$. For a crystal this is a periodic function of $\mathbf{r}_1$ with sharp maxima at the lattice sites, but in a fluid all points within V are equivalent and so $\rho^{(1)}(\mathbf{r}_1)$ is independent of $\mathbf{r}_1$. For a fluid, therefore, we can write

$$\frac{1}{V} \int \rho^{(1)}(\mathbf{r}_1) \, d\mathbf{r}_1 = \rho^{(1)} = \frac{N}{V} = \rho \qquad \text{(fluid)} \tag{13–7}$$

where we have used Eqs. (13–4) and (13–6) to equate the integral of $\rho^{(1)}(\mathbf{r}_1)$ to N.

We now define a correlation function $g^{(n)}(\mathbf{r}_1, \ldots, \mathbf{r}_n)$ by

$$\rho^{(n)}(\mathbf{r}_1, \ldots, \mathbf{r}_n) = \rho^n g^{(n)}(\mathbf{r}_1, \ldots, \mathbf{r}_n) \tag{13–8}$$

$g^{(n)}$ is called a correlation function since if the molecules were independent of each other, $\rho^{(n)}$ would equal simply ρ^n, and so the factor $g^{(n)}$ in Eq. (13–8) corrects for the "nonindependence" or, i.e., the correlation between the molecules.

Using Eq. (13–6) we see that (Problem 13–7)

$$g^{(n)}(\mathbf{r}_1, \ldots, \mathbf{r}_n) = \frac{V^n N!}{N^n (N-n)!} \cdot \frac{\int \cdots \int e^{-\beta U_N} \, d\mathbf{r}_{n+1} \cdots d\mathbf{r}_N}{Z_N}$$

$$= V^n (1 + O(N^{-1})) \frac{\int \cdots \int e^{-\beta U_N} \, d\mathbf{r}_{n+1} \cdots d\mathbf{r}_N}{Z_N} \tag{13–9}$$

We shall see shortly that $g^{(2)}(\mathbf{r}_1, \mathbf{r}_2)$ is particularly important since it can be determined experimentally. In a liquid of spherically symmetric molecules, $g^{(2)}(\mathbf{r}_1, \mathbf{r}_2)$ depends only upon the relative distance between molecules 1 and 2, i.e., upon r_{12}. We shall usually denote r_{12} simply by r, and since we shall concentrate mostly on $g^{(2)}(r)$, we shall therefore drop the superscript and write $g^{(2)}(r_{12}) = g(r)$. This is a standard notation.

Now $\rho g(r) \, d\mathbf{r}$ is the "probability" of observing a second molecule in $d\mathbf{r}$ given that there is a molecule at the origin of $\mathbf{r}$. Note that this "probability" is not normalized to unity, but we have instead

$$\int_0^\infty \rho g(r) 4\pi r^2 \, dr = N - 1 \approx N \tag{13–10}$$

In fact, Eq. (13–10) shows that $\rho g(r) 4\pi r^2 \, dr$ is really the number of molecules between r and $r + dr$ about a central molecule. The function $g(r)$ can also be thought of as the

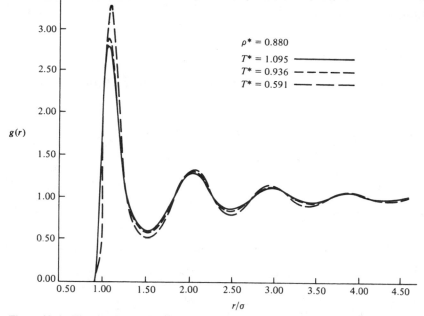

$\rho^* = 0.880$

$T^* = 1.095$ ———————

$T^* = 0.936$ — — — — —

$T^* = 0.591$ —— — —— —

Figure 13–3. **The radial distribution function of a fluid of molecules obeying a Lennard-Jones 6–12 potential from molecular dynamics calculations.** $T^* = kT/\varepsilon$ **and** $\rho^* = \sigma^3\rho$.

factor that multiplies the bulk density ρ to give a local density $\rho(r) = \rho g(r)$ about some fixed molecule. Clearly $g \to 0$ as $r \to 0$ since molecules become effectively "hard" as $r \to 0$. Also since the influence of the molecule at the origin diminishes as r becomes large, $g \to 1$ as $r \to \infty$. $g(r)$ is called the *radial distribution function* of the fluid. Figure 13–3 show the radial distribution function for a Lennard-Jones 6–12 fluid as determined from the molecular dynamics calculations of Verlet.*

The radial distribution function turns out to be of central importance in the theory of liquids for two reasons. First, we shall show in the next section that if we assume that the total potential energy of the N-body system is pair-wise additive, i.e., if we write

$$U_N(\mathbf{r}_1, \dots, \mathbf{r}_N) = \sum_{i<j} u(r_{ij}) \tag{13–11}$$

where the summation goes over all pairs of molecules, then all the thermodynamic functions of the system can be written in terms of $g(r)$. In addition, the radial distribution function can be determined by X-ray diffraction studies on liquids. In a solid the molecules are arranged in a regular repeating order, and this leads to a sharp X-ray diffraction pattern from which the order of the molecules in the solid can be obtained. The X-ray diffraction pattern of a liquid is more diffuse, but nevertheless can be used to determine the local short-range order in a fluid, i.e., the order exhibited in Fig. 13–3. The peaks in the curve represent smeared-out shells of first nearest neighbors, second nearest neighbors, etc., which in a sense are remnants of the ordering found in the solid.

We can give an outline of the relation between the radial distribution function and

* L. Verlet, *Phys. Rev.*, **165**, p. 201, 1968.

the scattering of electromagnetic radiation. Problem 13–47 shows that the scattering through an angle θ is given by

$$P(\theta) \propto \sum_i \sum_j \frac{\sin(sr_{ij})}{sr_{ij}} \tag{13–12}$$

where $s = (4\pi/\lambda)\sin(\theta/2)$. For a liquid the r_{ij} are continuously distributed, and so Eq. (13–12) can be written in the form

$$P(\theta) \propto \int_0^\infty 4\pi r^2 g(r) \frac{\sin(sr)}{sr} \, dr$$

It is convenient to write this as

$$P(\theta) \propto \int_0^\infty 4\pi r^2 (g(r) - 1) \frac{\sin sr}{sr} \, dr + \int_0^\infty 4\pi r^2 \frac{\sin sr}{sr} \, dr \tag{13–13}$$

Now it can be shown that this last integral vanishes for $s \neq 0$ (Problem 13–5), and so

$$P(\theta) \propto \int_0^\infty 4\pi r^2 (g(r) - 1) \frac{\sin sr}{sr} \, dr$$

or

$$P(\theta) \propto \int [g(r) - 1] e^{i\mathbf{s} \cdot \mathbf{r}} \, d\mathbf{r} \tag{13–14}$$

which shows that the Fourier transform of $g(r) - 1$ is proportional to the scattering through the angle θ, or really $\theta/2$ since $|\mathbf{s}|$ depends upon $\theta/2$ through $s = (4\pi/\lambda)\sin(\theta/2)$.

The function $g(r) - 1$, which goes to zero as $r \to \infty$, is usually denoted by $h(r)$, and the Fourier transform of $h(r)$ multiplied by the density is called the structure factor. Thus we can write

$$h(r) = g(r) - 1 \tag{13–15}$$

and

$$\hat{h}(s) = \rho \int h(r) e^{i\mathbf{s} \cdot \mathbf{r}} \, d\mathbf{r} \tag{13–16}$$

where we have denoted the structure factor by $\hat{h}(s)$.

The above derivation of the relation between X-ray scattering and the radial distribution function is incomplete since Eqs. (13–12) through (13–14) omit factors of proportionality, and the reader is referred to excellent discussions by Warren,[*] Gingrich,[†] and Kruh[‡] for more detail.

Since we can write thermodynamic functions in terms of $g(r)$ (next section), it is therefore possible to determine thermodynamic functions of a fluid from X-ray diffraction studies. To complete a statistical mechanical theory of liquids, however, we shall derive equations that give $g(r)$ from first principles in Sections 13–4 to 13–8.

[*] B. E. Warren, *J. Appl. Phys.*, **8**, p. 645, 1937.

[†] N. S. Gingrich, *Rev. Mod. Phys.*, **15**, p. 90, 1943.

[‡] R. E. Kruh, *Chem. Rev.*, **62**, p. 319, 1962.

13–3 RELATION OF THERMODYNAMIC FUNCTIONS TO $g(r)$

We first derive an equation for the energy E in terms of $g(r)$. Using the equation $Q_N = Z_N/N! \Lambda^{3N}$, we can write

$$E = \tfrac{3}{2} NkT + kT^2 \left(\frac{\partial \ln Z_N}{\partial T} \right)_{N,V}$$

$$= \tfrac{3}{2} NkT + \bar{U} \tag{13–17}$$

where

$$\bar{U} = \frac{\int \cdots \int U e^{-\beta U} \, d\mathbf{r}_1 \cdots d\mathbf{r}_N}{Z_N}$$

The first term in Eq. (13–17) is the mean kinetic energy and the second term is the mean potential energy. Now under the assumption of pair-wise additivity, Eq. (13–11), U is the sum of $N(N-1)/2$ terms, all of which give the same result upon carrying out the integration with respect to $\mathbf{r}_1$ through $\mathbf{r}_N$. Using $u(r_{12})$ as typical of these $N(N-1)/2$ terms in U, we have

$$\bar{U} = \frac{N(N-1)}{2Z_N} \int \cdots \int e^{-\beta U} u(r_{12}) \, d\mathbf{r}_1 \cdots d\mathbf{r}_N$$

$$= \frac{N(N-1)}{2} \iint u(r_{12}) \left\{ \frac{\int \cdots \int e^{-\beta U} \, d\mathbf{r}_3 \cdots d\mathbf{r}_N}{Z_N} \right\} d\mathbf{r}_1 \, d\mathbf{r}_2$$

$$= \frac{1}{2} \iint u(r_{12}) \rho^{(2)}(\mathbf{r}_1, \mathbf{r}_2) \, d\mathbf{r}_1 \, d\mathbf{r}_2$$

$$= \frac{N^2}{2V} \int_0^\infty u(r) g(r) 4\pi r^2 \, dr \tag{13–18}$$

The total energy E is then

$$\frac{E}{NkT} = \frac{3}{2} + \frac{\rho}{2kT} \int_0^\infty u(r) g(r, \rho, T) 4\pi r^2 \, dr \tag{13–19}$$

We have written $g(r, \rho, T)$ explicitly as a function of ρ and T as well as r to emphasize its dependence on these variables. Equation (13–19) could have been derived more physically by first choosing any molecule as a "central" molecule. The intermolecular potential energy between this central molecule and other molecules in the fluid at distances between r and $r + dr$ is $u(r) \cdot \rho g(r) \cdot 4\pi r^2 \, dr$. The total potential energy of the fluid is obtained by integrating over all values of r and multiplying by $N/2$ since any of the N molecules might be "central." The factor of two is inserted so that each pair interaction is counted only once. This gives Eq. (13–18) for $\bar{U}$.

Next we shall consider the pressure. For large V the pressure is independent of the shape of the container, and so for convenience, then, we assume that the container is a cube. The pressure p is given by

$$p = kT \left(\frac{\partial \ln Q}{\partial V} \right)_{N,T} = kT \left(\frac{\partial \ln Z_N}{\partial V} \right)_{N,T} \tag{13–20}$$

where

$$Z_N = \int_0^{V^{1/3}} \cdots \int e^{-\beta U} \, dx_1 \, dy_1 \, dz_1 \cdots dx_N \, dy_N \, dz_N$$

Before differentiating Z_N with respect to V, we change variables of integration in such a way that the limits become constants and U becomes an explicit function of V. Let the new variables be x_1', y_1', etc., where

$$x_K = V^{1/3} x_K', \text{ etc.}$$

Then

$$Z_N = V^N \int_0^1 \cdots \int_0^1 e^{-\beta U} \, dx_1' \cdots dz_N'$$

$$U = \sum_{1 \le i < j \le N} u(r_{ij}) \tag{13-21}$$

and

$$
\begin{aligned}
r_{ij} &= [(x_i - x_j)^2 + (y_i - y_j)^2 + (z_i - z_j)^2]^{1/2} \\
&= V^{1/3}[(x_i' - x_j')^2 + (y_i' - y_j')^2 + (z_i' - z_j')^2]^{1/2}
\end{aligned} \tag{13-22}
$$

Therefore,

$$
\left(\frac{\partial Z_N}{\partial V}\right)_{N,T} = N V^{N-1} \int_0^1 \cdots \int_0^1 e^{-\beta U} \, dx_1' \cdots dz_N'
$$
$$
- \frac{V^N}{kT} \int_0^1 \cdots \int_0^1 e^{-\beta U} \left(\frac{\partial U}{\partial V}\right) dx_1' \cdots dz_N'
$$

where

$$
\left(\frac{\partial U}{\partial V}\right) = \sum_{1 \le i < j \le N} \frac{du(r_{ij})}{dr_{ij}} \frac{dr_{ij}}{dV} = \sum_{1 \le i < j \le N} \frac{r_{ij}}{3V} \frac{du(r_{ij})}{dr_{ij}}
$$

Having carried out the differentiation with respect to V, we now transform back to the original variables $x_1, \ldots, z_N$. Also note that on integrating over the sum, $N(N-1)/2$ identical terms result. Therefore we finally get

$$
\left(\frac{\partial \ln Z_N}{\partial V}\right)_{N,T} = \frac{N}{V} - \frac{1}{6VkT} \iint_V r_{12} \frac{du(r_{12})}{dr_{12}} \rho^{(2)}(\mathbf{r}_1, \mathbf{r}_2) \, d\mathbf{r}_1 \, d\mathbf{r}_2
$$

which, when substituted into Eq. (13–20), gives

$$
\frac{p}{kT} = \rho - \frac{\rho^2}{6kT} \int_0^\infty r u'(r) g(r) 4\pi r^2 \, dr \tag{13-23}
$$

This equation is often called the *pressure equation*.

We have derived Eq. (13–23) with a liquid in mind, but it is valid for a gas as well since a liquid becomes a gas in the limit of low density. In fact, one could call Eq. (13–23) a *fluid* equation of state. In the previous chapter we have derived the equation of state of a gas as an expansion in the density ρ. In order to recover the virial expansion from Eq. (13–23), we write $g(r, \rho, T)$ in the form

$$
g(r, \rho, T) = g_0(r, T) + \rho g_1(r, T) + \rho^2 g_2(r, T) + \cdots \tag{13-24}
$$

and substitute this into Eq. (13–23) to get

$$
\frac{p}{kT} = \rho - \frac{\rho^2}{6kT} \sum_{j=0}^\infty \rho^j \int_0^\infty r u'(r) g_j(r, T) 4\pi r^2 \, dr
$$

By comparing this to Eq. (12–5), we see that

$$B_{j+2}(T) = -\frac{1}{6kT}\int_0^\infty ru'(r)g_j(r,\,T)4\pi r^2\,dr \qquad (13\text{–}25)$$

In particular, if we compare this to the first equation in Problem 12–10, we see that

$$g_0(r,\,T) = e^{-\beta u(r)} \qquad (13\text{–}26)$$

i.e., a simple Boltzmann factor type of expression. Note that this is obtained only in the limit of low density. We shall see in Section 13–4 that the energy factor in a Boltzmann expression is not just a potential energy when the system is dense.

Equation (13–25) has been employed a great deal in the literature to attempt to assess the relative merits of various approximate radial distribution functions. The procedure is to develop a density expansion of the approximate $g(r)$, substitute this into Eq. (13–25), and compare the resultant virial coefficients to the exact ones determined by the methods of Chapter 12. We shall consider this in some detail in Section 13–8.

We must calculate one last thermodynamic function in order to be able to calculate all the others. Notice that the two that we have dealt with up to now are both mechanical thermodynamic properties. We must calculate one nonmechanical thermodynamic property, and this is customarily taken to be the chemical potential. This is not as straightforward as the calculation of the two mechanical properties, however. We could use the thermodynamic equation

$$\left(\frac{\partial A/T}{\partial 1/T}\right)_{N,\,V} = E \qquad (13\text{–}27)$$

along with Eq. (13–19) for E, but this would require $g(r,\,\rho,\,T)$ as a function of T, something which is not available at present (Problem 13–6).

An alternate method to Eq. (13–27) is to introduce a so-called coupling parameter ξ, which varies from 0 to 1 and which has the effect of replacing the interaction of some central molecule, say 1, with the jth molecule of the system by $\xi u(r_{1j})$. In terms of this coupling parameter, then,

$$U(\mathbf{r}_1,\,\ldots,\,\mathbf{r}_N,\,\xi) = \sum_{j=2}^N \xi u(r_{1j}) + \sum_{2 \le i < j \le N} u(r_{ij}) \qquad (13\text{–}28)$$

Note that we can take molecule 1 in and out of the system by varying ξ from 0 to 1. This is a useful thing to be able to do since we can write the chemical potential μ in terms of Z_N and Z_{N-1}. We say that molecule 1 is "coupled" to the system to the degree ξ. All the other molecules of the system interact normally. The radial distribution function of the system now depends upon ξ and is written $g(r,\,\rho,\,T;\,\xi)$. Of course, $g(r,\,\rho,\,T;\,\xi)$ cannot be determined experimentally, but we shall derive theoretical expression for it.

Since N is very large, we can write essentially rigorously that

$$\mu = \left(\frac{\partial A}{\partial N}\right)_{V,\,T} = A(N,\,V,\,T) - A(N-1,\,V,\,T) \qquad (13\text{–}29)$$

and since

$$-\frac{A}{kT} = \ln Z_N - \ln N! - 3N \ln \Lambda$$

we get

$$-\frac{\mu}{kT} = \ln \frac{Z_N}{Z_{N-1}} - \ln N - \ln \Lambda^3 \tag{13-30}$$

The ratio of configuration integrals can easily be written in terms of the coupling parameter, as we shall now show. Clearly

$$Z_N(\xi = 1) = Z_N$$

and

$$Z_N(\xi = 0) = VZ_{N-1} \tag{13-31}$$

The factor of V in $Z_N(\xi = 0)$ comes from the integration over $d\mathbf{r}_1$. Therefore

$$\ln \frac{Z_N}{Z_{N-1}} = \ln \frac{Z_N(\xi = 1)}{Z_N(\xi = 0)} + \ln V$$

$$= \ln V + \int_0^1 \left(\frac{\partial \ln Z_N}{\partial \xi} \right) d\xi \tag{13-32}$$

where

$$Z_N(\xi) = \int \cdots \int e^{-\beta U_N(\xi)} \, d\mathbf{r}_1 \cdots d\mathbf{r}_N \tag{13-33}$$

Now, from Eqs. (13–28) and (13–32),

$$\frac{\partial Z_N}{\partial \xi} = -\frac{1}{kT} \int \cdots \int e^{-\beta U_N(\xi)} \left[\sum_{j=2}^N u(r_{1j}) \right] d\mathbf{r}_1 \cdots d\mathbf{r}_N$$

By dividing this by Z_N and collecting the $N - 1$ identical integrals, we can get

$$\frac{\partial \ln Z_N}{\partial \xi} = -\frac{1}{NkT} \iint_V u(r_{12}) \rho^{(2)}(\mathbf{r}_1, \mathbf{r}_2) \, d\mathbf{r}_1 \, d\mathbf{r}_2$$

$$= -\frac{\rho}{kT} \int_0^\infty u(r) g(r; \xi) 4\pi r^2 \, dr$$

Putting this back into Eq. (13–32) gives, then, the final result

$$\frac{\mu}{kT} = \ln \rho \Lambda^3 + \frac{\rho}{kT} \int_0^1 \int_0^\infty u(r) g(r; \xi) 4\pi r^2 \, dr \, d\xi \tag{13-34}$$

From E, p, and μ we can get all of the other thermodynamic functions. We now need only some equation for $g(r)$ [or $g(r; \xi)$] itself to have a complete theory of the liquid state. It is at this point that we run into difficulty. There have been a number of approximate equations derived, and we shall now look at the derivation of some of these and see the basis of each one.

13–4 THE KIRKWOOD INTEGRAL EQUATION FOR $g(r)$

One of the first and still one of the most important equations for $g(r)$ was derived in the 1930s by Kirkwood and is called the Kirkwood equation. To derive the Kirkwood equation we start with Eq. (13–6):

$$\rho^{(n)}(1, 2, \ldots, n; \xi) = \frac{N!}{(N-n)!} \cdot \frac{\int \cdots \int e^{-\beta U(\xi)} \, d\mathbf{r}_{n+1} \cdots d\mathbf{r}_N}{Z_N(\xi)} \tag{13-35}$$

where ξ is a coupling parameter for particle 1 and where we have written the argument of $\rho^{(n)}$ as 1, 2, ..., n instead of $r_1, r_2, \ldots, r_n$. We first differentiate this with respect to ξ to get

$$kT \frac{\partial \rho^{(n)}}{\partial \xi} = \frac{\rho^{(n)}}{Z_N(\xi)} \sum_{j=2}^{N} \int \cdots \int e^{-\beta U(\xi)} u(r_{1j}) \, d\mathbf{r}_1 \cdots d\mathbf{r}_N$$

$$- \frac{N!}{(N-n)!} \frac{1}{Z_N(\xi)} \sum_{j=2}^{N} \int \cdots \int e^{-\beta U(\xi)} u(r_{1j}) \, d\mathbf{r}_{n+1} \cdots d\mathbf{r}_N \qquad (13\text{-}36)$$

The first term here comes from differentiating the denominator of Eq. (13–35), and the second term comes from differentiating the numerator. The first integral in Eq. (13–36) can be written as

$$\frac{Z_N(\xi)}{N(N-1)} \iint_V u(r_{1j}) \rho^{(2)}(\mathbf{r}_1, \mathbf{r}_j) \, d\mathbf{r}_1 \, d\mathbf{r}_j$$

The second integral in Eq. (13–36) involves two separate cases. For $j = 2, \ldots, n$, $u(r_{1j})$ can be taken outside the integral, giving

$$\sum_{j=2}^{n} u(r_{1j}) \cdot \frac{N!}{(N-n)!} \frac{\int \cdots \int e^{-\beta U(\xi)} \, d\mathbf{r}_{n+1} \cdots d\mathbf{r}_N}{Z_N(\xi)} = \rho^{(n)}(1, 2, \ldots, n) \cdot \sum_{j=2}^{n} u(r_{1j})$$

For $j = n + 1, \ldots, N$, we have

$$\int_V u(r_{1j}) \left(\int \cdots \int e^{-\beta U} \, d\mathbf{r}_{n+1} \cdots d\mathbf{r}_{j-1} \, d\mathbf{r}_{j+1} \cdots d\mathbf{r}_N \right) d\mathbf{r}_j$$

$$= \frac{Z_N(\xi)(N-n-1)!}{N!} \int_V u(r_{1j}) \rho^{(n+1)}(1, 2, \ldots, n, j, \xi) \, d\mathbf{r}_j$$

Putting all this together and dividing by $\rho^{(n)}(1, 2, \ldots, n)$ gives

$$kT \frac{\partial \ln \rho^{(n)}}{\partial \xi} = - \sum_{j=2}^{n} u(r_{1j}) + \frac{1}{N} \iint_V u(r_{12}) \rho^{(2)}(\mathbf{r}_1, \mathbf{r}_2, \xi) \, d\mathbf{r}_1 \, d\mathbf{r}_2$$

$$- \int_V u(r_{1,n+1}) \frac{\rho^{(n+1)}(1, \ldots, n, n+1, \xi)}{\rho^{(n)}(1, \ldots, n, \xi)} \, d\mathbf{r}_{n+1} \qquad (13\text{-}37)$$

where we have taken account of the equivalence of the terms in the two summations. We now simply integrate Eq. (13–37) from 0 to ξ, noting that [see Eq. (13–35)]

$$\rho^{(n)}(1, \ldots, n, 0) = \frac{N!}{(N-n)! \, V} \frac{\int \cdots \int \exp(-\beta U_{N-1}) \, d\mathbf{r}_{n+1} \cdots d\mathbf{r}_N}{\int \cdots \int \exp(-\beta U_{N-1}) \, d\mathbf{r}_2 \cdots d\mathbf{r}_N}$$

$$= \rho \rho_{N-1}^{(n-1)}(\mathbf{r}_2, \ldots, \mathbf{r}_n) \qquad (13\text{-}38)$$

where $\rho_{N-1}^{(n-1)}$ here denotes a distribution function in a system containing $N - 1$ molecules. The integration gives us

$$kT \ln \rho^{(n)}(1, \ldots, n, \xi) = kT \ln \rho + kT \ln \rho_{N-1}^{(n-1)}(2, \ldots, n)$$

$$- \xi \sum_{j=2}^{n} u(r_{1j}) + \frac{1}{N} \int_0^{\xi} \iint_V u(r_{12}) \rho^{(2)}(\mathbf{r}_1, \mathbf{r}_2, \xi) \, d\mathbf{r}_1 \, d\mathbf{r}_2 \, d\xi$$

$$- \int_0^{\xi} \int u(r_{1,n+1}) \frac{\rho^{(n+1)}(1, \ldots, n, n+1, \xi)}{\rho^{(n)}(1, \ldots, n, \xi)} \, d\mathbf{r}_{n+1} \, d\xi$$

$$(13\text{-}39)$$

As it stands, this equation is applicable to any dense system. We now specialize this to the case of a fluid. Let $n = 2$ and introduce $g^{(2)}(r_{12})$ to get

$$-kT \ln g^{(2)}(1, 2, \xi) = \xi u(r_{12}) + \rho \int_0^\xi \int_V u(r_{13}) \left[\frac{g^{(3)}(1, 2, 3, \xi)}{g^{(2)}(1, 2, \xi)} \right.$$
$$\left. - g^{(2)}(1, 3, \xi) \right] d\mathbf{r}_3 \, d\xi \quad (13\text{–}40)$$

Notice that this gives $g^{(2)}$ in terms of $g^{(3)}$ or, in general, gives $g^{(n)}$ in terms of $g^{(n+1)}$. Such a set of coupled equations is called a hierarchy. It is exact, but unfortunately not very usable as it stands. If we could derive some other relation for $g^{(3)}$ in terms of $g^{(2)}$ we could "uncouple" this hierarchy and have a "closed" equation for $g^{(2)}$. This apparently cannot be done exactly. A further examination of $g^{(n)}$, however, will suggest an approximation which can be used to uncouple these equations.

Define a quantity $w^{(n)}(\mathbf{r}_1, \ldots, \mathbf{r}_n)$ by

$$g^{(n)}(\mathbf{r}_1, \ldots, \mathbf{r}_n) \equiv e^{-\beta w^{(n)}(\mathbf{r}_1, \ldots, \mathbf{r}_n)} \quad (13\text{–}41)$$

Substitute this into the defining equation for $g^{(n)}$ [Eq. (13–9)], take the logarithm of both sides, and then take the gradient with respect to the position of one of the n molecules, $1, \ldots, n$. This gives

$$-\nabla_j w^{(n)} = \frac{\int \cdots \int e^{-\beta U}(-\nabla_j U) \, d\mathbf{r}_{n+1} \cdots d\mathbf{r}_N}{\int \cdots \int e^{-\beta U} \, d\mathbf{r}_{n+1} \cdots d\mathbf{r}_N} \quad j = 1, 2, \ldots, n \quad (13\text{–}42)$$

Now $-\nabla_j U$ is the force acting on molecule j for any fixed configuration $\mathbf{r}_1, \ldots, \mathbf{r}_N$, and so the right-hand side is the mean force $f_j^{(n)}$ acting on particle j, averaged over the configurations of all the $n + 1, \ldots, N$ molecules not in the fixed set $1, \ldots, n$. Thus

$$f_j^{(n)} = -\nabla_j w^{(n)} \quad (13\text{–}43)$$

This says that $w^{(n)}$ is the potential that gives the mean force acting on particle j, or, i.e., $w^{(n)}$ is the *potential of mean force*. In particular, $w^{(2)}(r_{12})$ is the interaction between two molecules held a fixed distance r apart when the remaining $N - 2$ molecules of the fluid are canonically averaged over all configurations. Figure 13–4 shows the typical behavior of $w^{(2)}(r) \equiv w(r)$ for a dense fluid. When the density becomes very small, the two molecules fixed a distance r apart are not affected by the remaining $N - 2$ molecules, and so $w^{(2)}(r) \to u(r)$ as $\rho \to 0$, in agreement with Eq. (13–26). Note that in a dense system, however, the energy term in a Boltzmann factor is the potential of mean force rather than just the intermolecular potential. In a sense, one enters the realm of rigorous statistical mechanics when this distinction is clearly appreciated.

We shall see that the radial distribution function of a fluid of hard spheres is quite similar to that of a more realistic system. This means that the curves shown in Fig. 13–4 are also fairly representative of a hard-sphere system. Thus we see that although the hard-sphere potential itself has no attractive region, the corresponding potential of mean force does. The explanation for this can be seen by considering two hard spheres separated by a small distance and immersed in a bath of other hard spheres. The collisions that the right-hand sphere suffers with the other $N - 2$ spheres making up the bath will occur mostly from the right since the left-hand sphere of the pair is in the way of the bath spheres. There will be a net force to the left on the right-hand sphere of the given pair. Similarly, the left-hand sphere will be collided mostly from the left and so will experience a net force to the right. Thus the two spheres will be

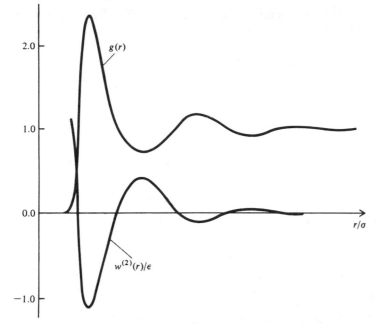

Figure 13–4. **The radial distribution function $g(r)$ and the corresponding potential of mean force $w^{(2)}(r)$ for a dense fluid. Note that $w^{(2)}(r)$ has minima where $g(r)$ has maxima and vice versa.**

driven toward each other due to the angular asymmetry of their collisions with the $N - 2$ bath spheres and so experience an effective attraction which manifests itself by the negative region in the potential of mean force.

Suppose now that we *assume* that the potential of mean force for a triplet of molecules is pair-wise additive, i.e., we *assume* that

$$w^{(3)}(1, 2, 3) \approx w^{(2)}(1, 2) + w^{(2)}(1, 3) + w^{(2)}(2, 3) \qquad (13\text{–}44)$$

Then this gives us that

$$g^{(3)}(1, 2, 3) \approx g^{(2)}(1, 2)g^{(2)}(1, 3)g^{(2)}(2, 3) \qquad (13\text{–}45)$$

Equation (13–45) has a probabilistic interpretation as well as a potential of mean force interpretation. If we assume that particles 1, 2, and 3 are completely independent of one another, then $g^{(3)}(1, 2, 3)$ would equal $g^{(1)}(1)g^{(1)}(2)g^{(1)}(3)$, essentially the product of the separate "probabilities." We see that Eq. (13–45) then assumes that the triplet correlation can be written as a product of pair-wise independent "probabilities." Substituting this approximation, called the *superposition approximation*, for $g^{(3)}$ into Eq. (13–40) gives finally the *Kirkwood equation* for $g(r)$, namely,

$$-kT \ln g(r_{12}, \xi) = \xi u(r_{12}) + \rho \int_0^\xi \int_V u(r_{13})g(r_{13}; \xi')[g(r_{23}) - 1] \, d\mathbf{r}_3 \, d\xi' \qquad (13\text{–}46)$$

Since the function to be determined in this equation occurs under an integral sign, an equation like Eq. (13–46) is called an integral equation. Thus Eq. (13–46) is an integral equation for $g(r_{12})$ and a nonlinear integral equation, in particular. The numerical solutions of this equation are difficult to obtain, but can be done on modern computers. We shall look at some of these solutions after we have derived a few other equations, but we can say here that the $g(r_{12})$ obtained from the Kirkwood and all the other equations, in fact, are qualitatively satisfactory. Note that Eq. (13–46) gives that $g(r_{12}) = \exp[-\xi u(r_{12})/kT]$ as $\rho \to 0$.

There is another integral equation for $g(r)$ that is derived somewhat similarly to the Kirkwood equation. Instead of differentiating with respect to a coupling parameter, we differentiate with respect to the coordinates of some particular molecule. This gives a coupled hierarchy similar to the Kirkwood equations and must also be uncoupled by means of the superposition approximation. We shall not go through the details, but the result is called the Born-Green-Yvon equation, which has the form*

$$-\frac{\partial}{\partial r}[kT \ln g(r, \xi) + \xi u(r)]$$

$$= \pi \xi \rho \int_0^\infty u'(s)g(s, \xi)\, ds \int_{|r-s|}^{r+s} \frac{(s^2 + r^2 - R^2)}{r^2} Rg(R)\, dR \quad (13\text{–}47)$$

The derivation of this equation is given through Problem 13–13. If one wishes to accept that the next few sections derive two alternative integral equations for $g(r)$, the hypernetted-chain equation [Eq. (13–84)] and the Percus-Yevick equation [Eq. (13–81)], then he can go directly to Section 13–9 for a comparison with experimental data.

13–5 THE DIRECT CORRELATION FUNCTION

In the late 1950s a new class of integral equations was derived by methods quite different from those used to derive the Kirkwood and BGY equations. These new equations came not out of a hierarchy, but rather through consideration of another type of correlation function, the direct correlation function, which we introduce and discuss in this section.

The distribution functions that we considered before were defined in a closed system. Since we are about to generalize these ideas to open systems, let us label these as $\rho_N^{(n)}(\mathbf{r}_1, \ldots, \mathbf{r}_n)$. Then the probability of observing n molecules in $d\mathbf{r}_1 \cdots d\mathbf{r}_n$ at $(\mathbf{r}_1, \ldots, \mathbf{r}_n)$, irrespective of N, is

$$\rho^{(n)} = \sum_{N \geq n} \rho_N^{(n)} P_N \tag{13–48}$$

where P_N is the probability that an open system contains N molecules, namely,

$$P_N = \frac{e^{\beta N \mu} Q(N, V, T)}{\Xi(\mu, V, T)} = \frac{z^N Z_N}{N! \, \Xi}$$

Substituting Eqs. (13–5) and (13–6) into (13–48) gives then

$$\rho^{(n)}(\mathbf{r}_1, \ldots, \mathbf{r}_n) = \frac{1}{\Xi}\left\{ z^n e^{-\beta U_n} + \sum_{N=n+1}^{\infty} \frac{z^N}{(N-n)!} \int \cdots \int e^{-\beta U_N} \, d\mathbf{r}_{n+1} \cdots d\mathbf{r}_N \right\}$$

$$\tag{13–49}$$

where U_j denotes the total intermolecular potential of a system of j particles. Note that

$$\int \cdots \int \rho^{(n)}(1, \ldots, n)\, d\mathbf{r}_1 \cdots d\mathbf{r}_n = \frac{1}{\Xi} \sum_{N \geq n} \frac{z^N Z_N}{(N-n)!} = \sum_{N \geq n} P_N \cdot \frac{N!}{(N-n)!}$$

$$= \left\langle \frac{N!}{(N-n)!} \right\rangle \tag{13–50}$$

---

* T. L. Hill, *Statistical Mechanics* (New York: McGraw-Hill, 1956), Chapter 6.

If we put $n = 2$, we get

$$\iint \rho^{(2)}(1, 2)\, d\mathbf{r}_1\, d\mathbf{r}_2 = \left\langle \frac{N!}{(N-2)!} \right\rangle = \langle N(N-1) \rangle = \overline{N^2} - \overline{N} \tag{13–51}$$

Furthermore,

$$\iint \rho^{(1)}(\mathbf{r}_1)\rho^{(1)}(\mathbf{r}_2)\, d\mathbf{r}_1\, d\mathbf{r}_2 = (\overline{N})^2 \tag{13–52}$$

Subtracting Eq. (13–52) from (13–51) gives

$$\frac{1}{\rho V} \iint_V [\rho^{(2)}(\mathbf{r}_1, \mathbf{r}_2) - \rho^{(1)}(\mathbf{r}_1)\rho^{(1)}(\mathbf{r}_2)]\, d\mathbf{r}_1\, d\mathbf{r}_N = \frac{\overline{N^2} - \overline{N}^2}{\overline{N}} - 1$$

$$= \rho kT\kappa - 1 \tag{13–53}$$

where κ is isothermal compressibility [cf. Eq. (3–51)]. For a fluid, Eq. (13–53) becomes

$$kT\left(\frac{\partial \rho}{\partial p}\right) = 1 + \rho \int [g(r) - 1]\, d\mathbf{r} \tag{13–54}$$

This is the equation we set out to derive. This is essentially another equation for p in terms of $g(r)$, and it often is referred to as the *compressibility equation*. Note that Eq. (13–54) shows that the isothermal compressibility is related to the structure factor at $s = 0$ [cf. Eq. (13–16)]. Also note that the derivation of this equation does not require the assumption of pair-wise additivity. This is not so for the pressure equation, Eq. (13–23).

We now introduce one final new function. Clearly $h(r_{12}) = g(r_{12}) - 1$ is a measure of the total influence of molecule 1 on molecule 2 at a distance r_{12}. Many years ago Ornstein and Zernike (1914)* proposed a division of $h(r_{12})$ into two parts, a direct part and an indirect part. The direct part is given by a function $c(r_{12})$ called the *direct correlation function*. The indirect part is the influence propagated directly from molecule 1 to a third molecule, 3, which in turn exerts its influence on 2, directly or indirectly through other particles. This effect is weighted by the density and averaged over all positions of molecules 3. With this decomposition of $h(r_{12})$, we can write

$$h(r_{12}) = c(r_{12}) + \rho \int c(r_{13})h(r_{23})\, d\mathbf{r}_3 \tag{13–55}$$

This equation is called the Ornstein-Zernike equation and can be considered to be the defining equation of the direct correlation function.

The direct correlation function does not seem to be as physically intuitive as $g(r)$, but it does have a simpler structure. (See Fig. 13–5.) The direct correlation function is important since, as we shall see, it is of much shorter range than $h(r)$, whose "tail" is almost entirely accounted for by the indirect term above. Equation (13–55) has the property that if we multiply through by $e^{i\mathbf{k}\cdot(\mathbf{r}_2 - \mathbf{r}_1)}$ and then integrate with respect to $d\mathbf{r}_1$ and $d\mathbf{r}_2$, we get

$$\int h(r_{12})e^{i\mathbf{k}\cdot\mathbf{r}_{12}}\, d\mathbf{r}_1\, d\mathbf{r}_2 = \int c(r_{12})e^{i\mathbf{k}\cdot\mathbf{r}_{12}}\, d\mathbf{r}_1\, d\mathbf{r}_2$$

$$+ \rho \iiint c(r_{13})e^{i\mathbf{k}\cdot(\mathbf{r}_2 - \mathbf{r}_1)}h(r_{23})\, d\mathbf{r}_1\, d\mathbf{r}_2\, d\mathbf{r}_3 \tag{13–56}$$

---

* This article is reproduced in H. Frisch and J. L. Lebowitz, *The Equilibrium Theory of Classical Fluids* (New York: Benjamin, 1964).

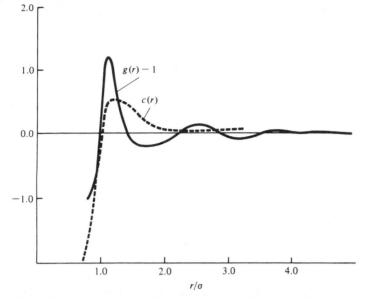

Figure 13–5. **A comparison of the behavior of the direct correlation function and the radial distribution function. Note that the direct correlation function has a much simpler structure and a much shorter range.**

If we denote the Fourier transforms of $h(r)$ and $c(r)$ by $\hat{H}(\mathbf{k})$ and $\hat{C}(\mathbf{k})$, we have (Problem 13–14)

$$\hat{H}(\mathbf{k}) = \hat{C}(\mathbf{k}) + \rho\hat{H}(\mathbf{k})\hat{C}(\mathbf{k}) \tag{13–57}$$

[Anyone familiar with the properties of Fourier transforms would write Eq. (13–57) immediately from Eq. (13–55).]

The compressibility equation [Eq. (13–54)] can be rewritten in the form

$$\frac{1}{kT}\left(\frac{\partial p}{\partial \rho}\right)_T = \frac{1}{1 + \rho \int h(r)\,d\mathbf{r}} = \frac{1}{1 + \rho\hat{H}(0)} = 1 - \rho\hat{C}(0)$$

$$= 1 - \rho \int c(r)\,d\mathbf{r} \tag{13–58}$$

where we have used Eq. (13–57) with $\mathbf{k} = 0$. The above manipulations probably served as the motivation for Ornstein and Zernike to invent Eq. (13–57), from which the defining equations for the somewhat physically obscure direct correlation function follows immediately.

Before using the results of this section to derive the two other integral equations, it is useful and interesting to study the density expansions of $g(r)$ and $c(r)$.

13–6 DENSITY EXPANSIONS OF THE VARIOUS DISTRIBUTION FUNCTIONS

In Chapter 12 we derived a density expansion of the pressure by means of the grand canonical ensemble. We first found the pressure as a power series in the activity z and then were able to convert that to a power series in the density. We can use the same procedure to calculate density expansions of the distribution functions. Equation (13–49) already gives $\rho^{(2)}(\mathbf{r}_1, \mathbf{r}_2)$ as an expansion in z:

$$\rho^{(2)}(\mathbf{r}_1, \mathbf{r}_2) = \frac{1}{\Xi} \left\{ z^2 e^{-\beta u(r_{12})} + \sum_{N=3}^{\infty} \frac{z^N}{(N-2)!} \int \cdots \int e^{-\beta U_N} \, d\mathbf{r}_3 \cdots d\mathbf{r}_N \right\} \qquad (13\text{–}59)$$

with

$$\Xi = 1 + \sum_{N=1}^{\infty} \frac{z^N}{N!} Z_N$$

It is convenient at this point to factor a $z^2 e^{-\beta u(r_{12})}$ out of Eq. (13–59) and write

$$\rho^{(2)}(\mathbf{r}_1, \mathbf{r}_2) = \frac{z^2 e^{-\beta u(r_{12})}}{\Xi} \left\{ 1 + \sum_{N=3}^{\infty} \frac{z^{N-2}}{(N-2)!} \int \cdots \int e^{-\beta U_N'} \, d\mathbf{r}_3 \cdots d\mathbf{r}_N \right\}$$

where the prime on U_N indicates that the $u(r_{12})$ term has been omitted. Straightforward expansion of this equation yields

$$\rho^{(2)}(\mathbf{r}_1, \mathbf{r}_2) = z^2 e^{-\beta u(r_{12})} \left[1 + z \left\{ \int e^{-\beta U_3'} \, d\mathbf{r}_3 - Z_1 \right\} \right.$$
$$\left. + \frac{z^2}{2} \left\{ \iint e^{-\beta U_4'} \, d\mathbf{r}_3 \, d\mathbf{r}_4 + 2Z_1^2 - Z_2 - Z_1 \int e^{-\beta U_3'} \, d\mathbf{r}_3 \right\} + \cdots \right]$$

$$(13\text{–}60)$$

If we assume pair-wise additivity and write $\exp(-\beta U_3')$ as $(1 + f_{13})(1 + f_{23})$ and recall that $Z_1 = V$, we find that

$$\int e^{-\beta U_3'} \, d\mathbf{r}_3 - Z_1 = \int \{ f_{13} + f_{23} + f_{13} f_{23} \} \, d\mathbf{r}_3$$

Note that the coordinates of particles 1 and 2 are fixed in this equation; i.e., we do not integrate over these coordinates. Such particles are indicated by open circles in the graphs corresponding to the integrands in the above equation, which becomes in a graph notation

$$\int e^{-\beta U_3'} \, d\mathbf{r}_3 - Z_1 = \int \{ \; \diagup_\circ + \; _\circ \diagdown + \; \wedge \; \} \, d\mathbf{r}_3 \qquad (13\text{–}61)$$

The next term in the expansion of $\rho^{(2)}(\mathbf{r}_1, \mathbf{r}_2)$ can be evaluated in a similar manner (Problem 13–8), and one eventually gets

$$\rho^{(2)}(\mathbf{r}_1, \mathbf{r}_2) = z^2 e^{-\beta u(r_{12})} \left[1 + z \int \{ \; \diagup_\circ + \; _\circ \diagdown + \; \wedge \; \} \, d\mathbf{r}_3 \right.$$
$$+ \frac{z^2}{2} \iint \{ 2 \; \vert\vert + 2 \; \ulcorner_\circ + 2 \; \lrcorner + \; \triangledown_\circ + \; \diagdown\!\!\!\triangle + 2 \; \square\!\!\!\diagup + 2 \; \boxtimes$$
$$\left. + 2 \; \boxslash + \; \boxtimes + \; \boxtimes \; \} \, d\mathbf{r}_3 \, d\mathbf{r}_4 + \cdots \right]$$

$$(13\text{–}62)$$

Note that the graphs that appear in Eq. (13–62) are such that there is no f-bond between the two root points, but the graph would become at least singly connected if there were an f-bond there. This leads us to define *modified cluster integrals* by

$$b_1^*(r_{12}) = e^{-\beta u(r_{12})}$$

$$b_l^*(r_{12}) = \frac{e^{-\beta u(r_{12})}}{l!} \int \cdots \int d\mathbf{r}_3 \cdots d\mathbf{r}_{l+1} \sum_i C_i(l+1) \qquad l \geq 2 \qquad (13\text{–}63)$$

where the summation is over all graphs of $l + 1$ points, $C_i(l + 1)$, which do not have a link between the fixed particles 1 and 2 and which would become at least singly connected if particles 1 and 2 were linked. The two fixed particles, 1 and 2, are called *root points*; the other $l - 1$ particles, which are integrated over, are called *field points*. With this definition, then, Eq. (13–62) takes on the simple looking form,

$$\rho^{(2)}(r_{12}) = \sum_{l=1}^{N-2} l b_l^* z^{l+1} \tag{13-64}$$

In the last chapter we derived an expression for z in powers of ρ, namely,

$$z = \rho - 2b_2 \rho^2 + (8b_2^2 - 3b_3)\rho^3 + \cdots$$

Substituting this into Eq. (13–64) gives

$$\rho^{(2)}(r_{12}) = b_1^* \rho^2 + (2b_2^* - 4b_1^* b_2)\rho^3$$
$$+ (3b_3^* - 12b_2^* b_2 - 6b_1^* b_3 + 20b_1^* b_2^2)\rho^4 + \cdots \tag{13-65}$$

The coefficients of each power of ρ can be considerably simplified, and one eventually can get (Problem 13–15):

$$\rho^{(2)}(r_{12}) = \exp\left[-\frac{u(r_{12})}{kT}\right]\left[\rho^2 + \rho^3 \int \wedge \, d\mathbf{r}_3\right.$$

$$+ \left(\frac{\rho^4}{2}\right)\int (2\,\square + 4\,\boxtimes + \boxtimes + \boxtimes)\, d\mathbf{r}_3 \, d\mathbf{r}_4$$

$$+ \left(\frac{\rho^5}{6}\right)\int (6\,\varhexagon + 6\,\varhexagon + 12\,\varhexagon + 12\,\varhexagon + 6\,\varhexagon + 6\,\varhexagon + \varhexagon + 12\,\varhexagon$$

$$+ 3\,\varhexagon + 12\,\varhexagon + 12\,\varhexagon + 12\,\varhexagon + 6\,\varhexagon + 6\,\varhexagon + 6\,\varhexagon + 3\,\varhexagon + 3\,\varhexagon$$

$$\left. + 12\,\varhexagon + 6\,\varhexagon + 6\,\varhexagon + 6\,\varhexagon + 3\,\varhexagon + 6\,\varhexagon + \varhexagon)\, d\mathbf{r}_3 \, d\mathbf{r}_4 \, d\mathbf{r}_5 + \cdots\right] \tag{13-66}$$

The integrals occurring in Eq. (13–66) are closely related to the integrals for the virial coefficients. Equation (13–66) suggests, and it can be proved in general, that all graphs that become stars when the line corresponding to f_{12} is added appear in the expansion of $\rho^{(2)}(r_{12})$, and, of course, $g(r)$ as well. Note that $\rho^{(2)}(r_{12}) \to \rho^2 e^{-\beta u(r_{12})}$ as $\rho \to 0$. This is the same as $w^{(2)}(r_{12}) \to u(r_{12})$.

The simpler integrals above can be calculated by the same method that we introduced in Section 12–5 to calculate the third virial coefficient. It we write

$$g(r_{12}, \rho, T) = g_0(r_{12}, T) + \rho g_1(r_{12}, T) + \rho^2 g_2(r_{12}, T) + \cdots$$

then Eq. (13–66) shows that

$$g_0(r_{12}, T) = e^{-\beta u(r_{12})}$$

and

$$g_1(r_{12}, T) = e^{-\beta u(r_{12})} \int f(r_{13})f(r_{23})\, d\mathbf{r}_3 \tag{13-67}$$

It is easy to evaluate $g_1(r_{12}, T)$ by the method of Fourier transforms (cf. Section 12–5). If we let $\gamma(t)$ be the Fourier transform of $f(r)$; i.e., let

$$\gamma(t) = (2\pi)^{-3/2} \int f(r) e^{-i t \cdot \mathbf{r}} \, d\mathbf{r}$$

$$= \left(\frac{2}{\pi}\right)^{1/2} \int_0^\infty r f(r) \frac{\sin tr}{t} \, dr$$

then one can show that (Problem 13–16)

$$\int f(r_{13}) f(r_{23}) \, d\mathbf{r}_3 = \frac{(2\pi)^{3/2}}{r_{12}^{1/2}} \int_0^\infty \{\gamma(t)\}^2 J_{1/2}(r_{12} t) t^{3/2} \, dt \tag{13–68}$$

where $J_{1/2}(x)$ is a Bessel function of $\frac{1}{2}$ order.

It is shown in Section 12–5 that for hard spheres

$$\gamma(t) = \sigma^3 \left(\frac{2}{\pi}\right)^{1/2} \left\{ \frac{\cos(\sigma t)}{(\sigma t)^2} - \frac{\sin(\sigma t)}{(\sigma t)^3} \right\}$$

$$= -\sigma^3 \frac{J_{3/2}(\sigma t)}{(\sigma t)^{3/2}}$$

Substituting this into Eq. (13–68), we get

$$\int f(r_{13}) f(r_{23}) \, d\mathbf{r}_3 = \frac{(2\pi)^{3/2} \sigma^{1/2}}{r_{12}^{1/2}} \int_0^\infty J_{3/2}(x) J_{3/2}(x) J_{1/2}(x) x^{-3/2} \, dx$$

Katsura* has evaluated integrals of this type, and we find that (Problem 13–17)

$$g_1(r_{12}, T) = 0 \qquad\qquad\qquad 0 < r_{12} < \sigma$$

$$= \pi \sigma^3 \left\{ \frac{4}{3} - \frac{r_{12}}{\sigma} + \frac{1}{12} \left(\frac{r_{12}}{\sigma}\right)^3 \right\} \qquad \sigma < r_{12} < 2\sigma$$

$$= 0 \qquad\qquad\qquad\qquad r_{12} > 2\sigma \tag{13–69}$$

It is readily shown that this form for $g_1(r, T)$ yields the correct third virial coefficient. The integrals in $g_2(r_{12}, T)$ can also be evaluated for a hard-sphere potential.† In addition, McQuarrie‡ has evaluated $g_1(r, T)$ and some of the terms in $g_2(r, T)$ for the square-well potential.

Equation (13–66) gives us a density expansion for $\rho^{(2)}(r_{12})$ in terms of integrals over graphs. Since $\rho^{(2)}(r_{12})$ is closely related to $g(r)$ and $h(r)$, Eq. (13–66) is essentially the expansion for $g(r)$ and $h(r)$ as well. For example, by dividing by ρ^2 and subtracting unity, we get

$$h(r_{12}) = f(r_{12}) + e^{-\beta u(r_{12})} \left[\rho \int \wedge \, d\mathbf{r}_3 + \frac{\rho^2}{2} \int (2 \, \square + 4 \, \boxtimes + \boxtimes + \boxtimes) \, d\mathbf{r}_3 \, d\mathbf{r}_4 + \cdots \right] \tag{13–70}$$

We can use this to derive a density expansion for the direct correlation function. Equation (13–55), the defining equation for $c(r)$, is

$$c(r_{12}) = h(r_{12}) - \rho \int c(r_{13}) h(r_{23}) \, d\mathbf{r}_3 \tag{13–71}$$

* S. Katsura, *Phys. Rev.*, **115**, p. 1417, 1959; **118**, p. 1667, 1960.
† B. R. A. Nijboer and L. van Hove, *Phys. Rev.*, **85**, p. 777, 1952.
‡ D. A. McQuarrie, *J. Chem. Phys.*, **40**, p. 3455, 1964.

Now a repeated iteration of this equation, obtained by successively substituting $c(r_{ij})$ from the left-hand side into the right-hand side, gives (Problem 13–22)

$$c(r_{12}) = h(r_{12}) - \rho \int h(r_{13})h(r_{23}) \, d\mathbf{r}_3 + \rho^2 \iint h(r_{13})h(r_{34})h(r_{42}) \, d\mathbf{r}_3 \, d\mathbf{r}_4 + \cdots$$

$$(13–72)$$

If we represent an h-bond by a wavy line, then Eq. (13–72) can be written graphically as

$$c(r_{12}) = \quad \sim\!\!\sim \; - \rho \int \wedge d\mathbf{r}_3 + \rho^2 \iint \sqcap d\mathbf{r}_3 \, d\mathbf{r}_4 - \rho^3 \iiint \Diamond \, d\mathbf{r}_3 \, d\mathbf{r}_4 \, d\mathbf{r}_5 + \cdots$$

$$(13–73)$$

If we now substitute Eq. (13–70) for $h(r_{ij})$ into Eq. (13–72), we get a density expansion for the direct correlation function, $c(r)$:

$$c(r_{12}) = \sim\!\!\sim + \rho \int \triangle \; d\mathbf{r}_3$$

$$+ \rho^2 \iint \left[\square + \tfrac{1}{2} \boxtimes + 2 \boxslash + \tfrac{1}{2} \boxtimes + \tfrac{1}{2} \boxtimes + \tfrac{1}{2} \boxtimes \right] d\mathbf{r}_3 \, d\mathbf{r}_4 + \cdots \qquad (13–74)$$

Note that $c(r_{12})$ is indeed a much shorter range function than $h(r_{12})$ since the graphs appearing in $c(r_{12})$ are much more tightly bound. All diagrams in $c(r_{12})$ are without *nodes*, i.e., a field point at which the graph can be separated into two or more parts by cutting at that point.

Substitution of Eq. (13–66) for $\rho^{(2)}(r_{12})$ into the pressure equation [Eq. (13–23)] yields the virial expansion, although this is not so easy to show. The substitution of the expansion for $c(r_{12})$ into the compressibility equation [Eq. (13–58)] yields the virial expansion in a direct way (Problem 13–23).

It is possible to extend the results of this section to higher-order correlation functions as well.* For example, one can show that (Problem 13–24)

$$g^{(3)}(\mathbf{r}_1, \mathbf{r}_2, \mathbf{r}_3) = \exp\{-\beta[u(r_{12}) + u(r_{13}) + u(r_{23})]\}\left[1 + \sum_{n=1}^{\infty} \rho^n \tau_n(1, 2, 3)\right]$$

$$(13–75)$$

where

$$\tau_1(1, 2, 3) = g^{(2)}(r_{12}) + g^{(2)}(r_{13}) + g^{(2)}(r_{23}) + \int \wedge\!\!\!\wedge \, d\mathbf{r}_4 \qquad (13–76)$$

Uhlenbeck and Ford (in "Additional Reading") give expressions for the higher-order terms.

13–7 DERIVATION OF TWO ADDITIONAL INTEGRAL EQUATIONS

The direct correlation function and the Ornstein-Zernike equation can be used to derive two additional integral equations which appear to be more satisfactory than either the Kirkwood or the Born-Green-Yvon equations. They were both derived in the late 1950s by methods that are quite different from these we shall use below. One of the equations, called the Percus-Yevick equation,† was originally derived by means

---

* E. E. Salpeter, *Ann. Phys.*, **5**, p. 183, 1958; see also D. Henderson, *J. Chem. Phys.*, **46**, p. 4306, 1967, for for a readable discussion of these expansions.

† J. K. Percus and G. J. Yevick, *Phys. Rev.*, **110**, p. 1, 1958.

of field theoretic techniques and later derived by Stell* by a detailed analysis of the graphical expansions of $g(r)$ and $c(r)$. He showed that the Percus-Yevick equation can be obtained by ignoring certain types of clusters that occur in $c(r)$. The other equation, called the hypernetted-chain equation, results from the work of a number of authors, but the initial impetus can be attributed to Rushbrooke and Scoins,† who derived a simpler equation (called the netted-chain equation) by a graphical analysis of $c(r)$ and the retention of only certain types of graphs. The names of the resulting equations are a reflection of the types of graphs retained.

The graphical analysis involved in the derivation of the Percus-Yevick and hypernetted-chain equations is quite involved and will not be presented here,* but it played an important role in assessing the success of the two integral equations. Later, in the 1960s, these equations were derived by an entirely different technique involving the theory of functionals.‡ A functional differs from a function in that whereas a function is a mapping from one number or set of numbers into another, a functional is a mapping of a function into a number. More precisely, we say that z is a functional of the function $x(t)$ in the interval (a, b) when it depends on all the values taken by $x(t)$ throughout the interval. A simple example is

$$I[x(t)] = \int_a^b x(t)\, dt$$

One that is relevant to statistical mechanics is $A[u(r)]$; i.e., the Helmholtz free energy is a functional of the intermolecular potential $u(r)$. It is possible to develop a calculus of functionals, § including such operations as functional differentiation ("differentiating with respect to a function") and functional Taylor expansions. This is a beautiful technique and can be used to derive not only the Percus-Yevick and hypernetted-chain equations but most of the other integral equations of liquid theory as well. It is not too difficult a technique, but it would take too much space here to present the necessary calculus of functionals. Several books dealing exclusively with the statistical mechanical theory of liquids present this powerful technique.|| In addition, the reader is referred to a paper by Verlet,¶ in which he reviews the basic ideas and derives an extended Percus-Yevick equation, known in the literature as the PY II equation.

We shall use neither the theory of graphs nor the theory of functionals, but for pedagogical reasons we shall derive the PY and HNC equations by appealing to a physical interpretation of the direct correlation function. The Ornstein-Zernike equation

$$h(r_{12}) = c(r_{12}) + \rho \int c(r_{13}) h(r_{23})\, d\mathbf{r}_3$$

can be considered to be a definition of the direct correlation function or, alternatively, can be considered to be a relation between $h(r)$ and $c(r)$. If $c(r)$ were given in terms of $g(r)$ or $h(r)$, say, its substitution into the Ornstein-Zernike equation would give a closed integral equation for $h(r)$. This can be done in the following approximate way.

---

* G. Stell, *Physica*, **29**, p. 517, 1963; see also Stell's chapter in H. Frisch and J. Lebowitz, ed., *Classical Fluids* (New York: Benjamin, 1964).

† G. S. Rushbrooke and H. I. Scoins, *Proc. Roy. Soc.*, **216A**, p. 203, 1953.

‡ J. K. Percus, *Phys. Rev. Lett.*, **8**, p. 462, 1962.

§ V. Volterra, *Theory of Functionals* (New York: Dover, 1959).

|| G. H. A. Cole, *Statistical Theory of Classical Simple Dense Fluids* (New York: Pergamon, 1967); S. A. Rice and P. Gray, *Statistical Mechanics of Simple Liquids* (New York: Interscience, 1965).

¶ L. Verlet, *Physica*, **30**, p. 95, 1964.

The direct correlation function was introduced to represent in a sense the direct correlation between two particles in a system containing $N - 2$ other particles. It is reasonable to represent this direct correlation function by

$$c(r) = g_{total}(r) - g_{indirect}(r) \tag{13–77}$$

where $g_{total}(r)$ is just the radial distribution function itself; i.e., $g(r) = \exp[-\beta w(r)]$, and $g_{indirect}(r)$ is the radial distribution function without the direct interaction $u(r)$ included; i.e., we write $g_{indirect}(r) = \exp\{-\beta[w(r) - u(r)]\}$. Thus we *approximate* $c(r)$ by

$$c(r) = e^{-\beta w(r)} - e^{-\beta[w(r) - u(r)]} \tag{13–78}$$

It is convenient to introduce one last function at this point. We define $y(r)$ by

$$y(r) = e^{\beta u(r)}g(r) \tag{13–79}$$

By comparing this definition to Eq. (13–66), we see that $y(r)$ is defined to go to unity as the density goes to zero. It simply eliminates the factor $\exp(-\beta u)$ that occurs in Eq. (13–66). In terms of $y(r)$ then, the direct correlation function is

$$\begin{aligned}
c(r) &= g(r) - y(r) \\
&= e^{-\beta u}y(r) - y(r) \\
&= f(r)y(r) \quad \text{(PY)}
\end{aligned} \tag{13–80}$$

This approximation for $c(r)$ is labeled PY since if it is substituted into the Ornstein-Zernike equation, one obtains the so-called *Percus-Yevick* equation:

$$y(r_{12}) = 1 + \rho \int f(r_{13})y(r_{13})h(r_{23})\,d\mathbf{r}_3 \tag{13–81}$$

Note that the direct correlation function given by Eq. (13–80) is a short-range function, namely, the same range as $f(r)$ or $u(r)$.

The hypernetted-chain equation is derived by expanding the $g_{indirect}(r)$ term in Eq. (13–78) and writing

$$\begin{aligned}
c(r) &= e^{-\beta w} - 1 + \beta[w(r) - u(r)] \\
&= g(r) - 1 - \ln y(r) \\
&= f(r)y(r) + [y(r) - 1 - \ln y(r)] \quad \text{(HNC)}
\end{aligned} \tag{13–82} \tag{13–83}$$

As the label indicates, if this approximation for $c(r)$ is substituted into the Ornstein-Zernike equation, one obtains *the hypernetted-chain equation*:

$$\ln y(r_{12}) = \rho \int \left[h(r_{13}) - \ln g(r_{13}) - \frac{u(r_{13})}{kT} \right][g(r_{23}) - 1]\,d\mathbf{r}_3 \tag{13–84}$$

Although it may not appear so at first sight, this form of $c(r)$ is short ranged since $f(r)$ is short ranged and the bracketed term becomes quite small as r increases much beyond the range of the intermolecular potential (Problem 13–26).

This gives us two more integral equations for the radial distribution function, giving a total of four. There are actually a number of others that have been presented in the literature, but we shall consider only the four that we have derived so far. We shall see in Section 13–9 that it is difficult to compare these equations. They are all qualitatively satisfactory, and no one seems to be quantitatively best for all systems under all conditions. We shall come back to this in Section 13–9.

It is easy and interesting to see what graphs occur in the Percus-Yevick approximation to $y(r)$. Write the $h(r_{23})$ appearing in the Percus-Yevick equation as

$$h(r_{23}) = g(r_{23}) - 1 = e^{-\beta u(r_{23})}y(r_{23}) - 1 = f(r_{23})y(r_{23}) + y(r_{23}) - 1$$

and substitute this into Eq. (13–81) to get

$$y(r_{12}) = 1 + \rho \int f(r_{13})y(r_{13})f(r_{23})y(r_{23})\, d\mathbf{r}_3 + \rho \int f(r_{13})y(r_{13})[y(r_{23}) - 1]\, d\mathbf{r}_3$$

$$(13\text{–}85)$$

This equation may be solved by iteration to get a series in ρ, whose first few terms are (Problem 13–27):

$$y(r) = 1 + \rho \int \triangle\, d\mathbf{r}_3 + \rho^2 \int \int \left\{ \square + \boxtimes + \boxtimes \right\} d\mathbf{r}_3\, d\mathbf{r}_4 + \cdots \qquad (13\text{–}86)$$

Percus and Yevick proved by simple induction that the diagrams that occur can be obtained from the following recipe. Draw a simple chain from base point 1 to base point 2 as a plane polygon of ordered vertices (omitting the f_{12} bond). Include in the sum this simple chain and all diagrams that can be formed from it by adding interior lines that do not cross. This expansion turns out to include fewer graphs than the $y(r)$ from the hypernetted-chain equation,* but nevertheless we shall see that the two equations compare favorably.

Up to now we have derived the four most popular integral equations for $g(r)$. Notice that the Kirkwood equation and the Born-Green-Yvon equation both arise from a hierarchy which is broken by the superposition approximation. Presumably both of these equations are as good as the superposition approximation. The HNC and the PY equations, on the other hand, can be obtained by a variety of methods, all of which involve the direct correlation function and the Ornstein-Zernike equation. In the next sections we shall compare these four equations to some known exact results and to experimental data.

13–8 DENSITY EXPANSIONS OF THE VARIOUS INTEGRAL EQUATIONS

All four of these equations are nonlinear integral equations and so are rather difficult to solve even numerically. Only in the past few years have numerical solutions begun to appear.

There is another way to try to assess these equations that is much easier, however. These are equations for liquids. As we let the density ρ become small, these equations should be applicable to dense gases. In the last chapter we treated dense gases exactly, and so we can compare the results of a density expansion of our approximate liquid theory equations with the exact equations of imperfect gas theory. Let us take the Percus-Yevick equation as an example. Write $y(r)$ in a density expansion,

$$y(r) = e^{\beta u}g(r) = 1 + \rho y_1(r) + \rho^2 y_2(r) + \cdots \qquad (13\text{–}87)$$

where the $y_j(r)$ are as yet undetermined. If we substitute this into both sides of the

---

* Rowlinson (in "Additional Reading") has an excellent, readable discussion of the graphical analysis of the PY and HNC equations.

Percus-Yevick equation and compare like powers of ρ on both sides, we get a set of simple equations for the $g_j(r)$:

$$\rho g_1(r_{12}) + \rho^2 g_2(r_{12}) + \cdots$$

$$= \rho \int f_{13}[1 + \rho g_1(r_{13}) + O(\rho^2)][e^{-\beta u_{23}} + \rho e^{-\beta u_{23}} g_1(r_{23}) + O(\rho^2) - 1]\, d\mathbf{r}_3$$

which gives

$$g_1(r_{12}) = \int f_{13} f_{23}\, d\mathbf{r}_3 = \int \wedge\, d\mathbf{r}_3 \tag{13–88}$$

$$g_2(r_{12}) = \int f_{13} g_1(r_{13}) f_{23}\, d\mathbf{r}_3 + \int f_{13}\, e^{-\beta u(r_{23})} g_1(r_{23})\, d\mathbf{r}_3$$

$$= \iint f_{13} f_{14} f_{34} f_{23}\, d\mathbf{r}_3\, d\mathbf{r}_4 + \iint f_{13}(1 + f_{23}) f_{24} f_{34}\, d\mathbf{r}_3\, d\mathbf{r}_4$$

$$= \iint \square\, d\mathbf{r}_3\, d\mathbf{r}_4 + 2 \iint \square\, d\mathbf{r}_3\, d\mathbf{r}_4 \tag{13–89}$$

The middle line of Eq. (13–89) is obtained by using

$$g_1(r_{13}) = \int f_{14} f_{34}\, d\mathbf{r}_4$$

in the first integral and

$$g_1(r_{23}) = \int f_{24} f_{34}\, d\mathbf{r}_4$$

in the second. Comparing these results to the exact expansion of $g(r)$ shows that $g_1(r)$ is correct and $g_2(r)$ is incorrect (as in all four equations). (Note how the graphs appearing in $g_2(r)$ above satisfy the recipe described earlier and how those that do not appear do not satisfy the PY recipe.)

A look at the pressure equation

$$\frac{p}{kT} = \rho - \frac{\rho^2}{6kT} \int_0^\infty ru'(r)g(r)4\pi r^2\, dr$$

shows that the jth virial coefficient is given by

$$B_j(T) = -\frac{1}{6kT} \int_0^\infty ru'(r)e^{-\beta u(r)} g_{j-2}(r, T)4\pi r^2\, dr \tag{13–90}$$

Since $g_1(r)$ is given correctly in all four equations, all four equations yield the correct third virial coefficient. Since $g_2(r)$ is given incorrectly by all four integral equations, they all yield incorrect and different fourth virial coefficients. Also, because of the approximate nature of $g(r)$, the fourth and higher virial coefficients depend upon whether one uses the pressure equation [Eq. (13–23)] or the compressibility equation [Eq. (13–58)] to calculate them. This type of discrepancy has been used extensively to compare and assess these four equations.* The results of such calculations for hard spheres are shown in Table 13–2.

Much work has been done in trying to devise schemes in which consistency of all of the equations up to the fifth virial coefficient is insured, but the fifth or at least higher virial coefficients remain both inconsistent and incorrect.† It is not clear that

* G. Stell, *J. Chem. Phys.*, **36**, p. 1817, 1962; G. S. Rushbrooke and P. Hutchinson, *Physica*, **27**, p. 647, 1961, and **29**, p. 675, 1963; D. A. McQuarrie, *J. Chem. Phys.*, **40**, p. 3455, 1964.
† G. H. A. Cole, *Repts. Prog. Phys.*, **31**, p. 419, 1968.

Table 13–2. **Virial coefficients for the rigid-sphere fluid from the several approximate theories***

	$B_2{}^p$	$B_2{}^c$	$B_3{}^p$	$B_3{}^c$	$B_4{}^p$	$B_4{}^c$	$B_5{}^p$	$B_5{}^c$
exact	1	1	$\frac{5}{8}$	$\frac{5}{8}$	0.2869	0.2869	0.1103	0.1103
BGY	1	1	$\frac{5}{8}$	$\frac{5}{8}$	0.2252	0.3424	0.0475	0.1335
K	1	1	$\frac{5}{8}$	$\frac{5}{8}$	0.1400	0.4418		
HNC	1	1	$\frac{5}{8}$	$\frac{5}{8}$	0.4453	0.2092	0.1147	0.0493
PY	1	1	$\frac{5}{8}$	$\frac{5}{8}$	0.2500	0.2969	0.0859	0.121

* These values are reduced by the hard-sphere second virial coefficient $b_0 = 2\pi\sigma^3/3$.

such attempts are fruitful, and we quote Rice and Gray (in "Additional Reading") to this effect: "It is important to emphasize that the failure of the theories to reproduce the virial coefficients does not necessarily imply that the theories are useless in the liquid region. In each case, contributions from all orders of the density are included in the integral equation representation. We may, therefore, expect the equation of state to be superior to a four- or five-term virial expansion. Indeed, if the contributions of each order of the density are those most important in the liquid range (highly connected diagrams), the approximate theories might be quite good at high densities even if the virial coefficients are not exact. It is, therefore, important to examine other predictions of the various theories before deciding on their relative merits."

13–9 COMPARISONS OF THE INTEGRAL EQUATIONS TO EXPERIMENTAL DATA

The four integral equations that we have been discussing are all nonlinear integral equations. The numerical solution of such equations is quite difficult, and it has been only in the past few years that such solutions have been produced. Almost all work has focused on two potentials, the hard-sphere potential and the 6–12 potential.

It turns out that the PY equation can be solved analytically for hard spheres (although apparently not for disks).* The two equations of state corresponding to the pressure equation and compressibility equation are

$$\frac{p}{\rho kT} = \frac{1 + 2y + 3y^2}{(1-y)^2} \qquad \text{[from the pressure equation, Eq. (13–23)]} \qquad (13\text{–}91)$$

$$\frac{p}{\rho kT} = \frac{1 + y + y^2}{(1-y)^3} \qquad \text{[from the compressibility equation, Eq. (13–54)]} \qquad (13\text{–}92)$$

respectively, where $y = \pi\rho\sigma^3/6 = \sqrt{2}\,\pi v_0/6v$. Of course, these would be the same if the Percus-Yevick equation were exact. The other integral equations must be solved numerically. The results of these calculations are plotted in Fig. 13–6. It is not clear from the figure, but both the Kirkwood and the Born-Green-Yvon equations do not have solutions above certain densities, presumably indicating some kind of a phase transition. On the other hand, the Percus-Yevick equation allows the density to exceed the closest-packed volume since the equations of state [Eqs. (13–91) or (13–92)] predict finite pressures until $y \to 1$. This corresponds to $v = (\sqrt{2}\,\pi/6)v_0$, which is physically impossible since it is less than the closest-packed volume. Yet it does describe the equation of state best over the fluid region.

* E. Thiele, *J. Chem. Phys.*, **39**, p. 474, 1963; M. S. Wertheim, *Phys. Rev. Lett.*, **10**, p. 321, 1963.

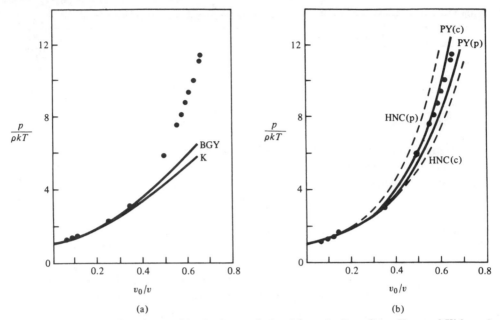

Figure 13–6. (a) **Equation of state of hard spheres calculated from the Born-Green-Yvon and Kirkwood integral equations compared with the results of molecular dynamics calculations.** v_0 **is the closest-packing volume,** $N\sigma^3/\sqrt{2}$. (b) **Equation of state of hard spheres calculated from the HNC and Percus-Yevick integral equations compared with the results of molecular dynamics calculations.** (From D. Henderson, *Ann. Rev. Phys. Chem.*, **15**, p. 31, 1964.)

The pressure equation can be integrated analytically for hard spheres. Recall that the pressure equation is [Eq.(13–23)]

$$\frac{p}{kT} = \rho - \frac{\rho^2}{6kT} \int_0^\infty ru'(r)g(r)4\pi r^2 \, dr$$

Now for a hard-sphere potential, $u'(r)$ is zero everywhere except at $r = \sigma$, where it is essentially a delta function, and so Eq. (13–23) becomes for hard spheres (Problem 13–33)

$$\frac{p}{\rho kT} = 1 + \frac{2\pi\sigma^3}{3} \rho g(\sigma+) \tag{13–93}$$

where $g(\sigma+)$ is the value of $g(r)$ at $\sigma + \varepsilon$, where $\varepsilon \to 0$ from the positive direction. Remember that for hard spheres $g(r)$ is discontinuous at $r = \sigma$, and so we must specify whether $g(\sigma)$, the value of $g(r)$ at contact, is evaluated from the right or from the left. Equation (13–93), in conjunction with any of the excellent hard-sphere equations of state such as the Ree–Hoover Padé approximant [Eq. (13–3)] or the Carnahan-Starling equation (Problem 12–34), gives g at contact. This quantity is sometimes useful in transport theories (Section 19–5).

Remember that both the Kirkwood and BGY equations are derived through the superposition approximation, which states that

$$g^{(3)}(\mathbf{r}_1, \mathbf{r}_2, \mathbf{r}_3) = g^{(2)}(r_{12})g^{(2)}(r_{13})g^{(2)}(r_{23})$$

Alder,* using molecular dynamics, has made an interesting study of the adequacy of

---

* B. J. Alder, *Phys. Rev. Lett.*, **12**, p. 317, 1964.

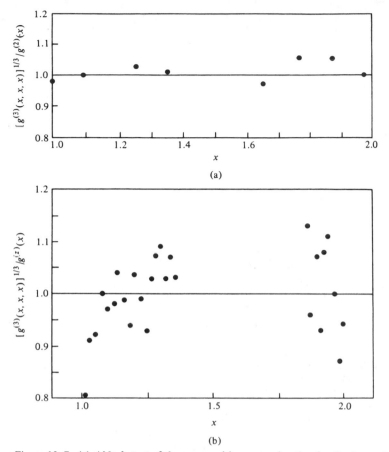

Figure 13–7. (a) **Alder's test of the superposition approximation for the dense rigid sphere fluid.** (From B. J. Alder, *Phys. Rev. Lett.*, **12**, p. 317, 1964. (b) **Rahman's test of the superposition approximation for a Lennard-Jones 6–12 fluid.** (From A. Rahman, *Phys. Rev. Lett.*, **12**, p. 575, 1964.) (From S. A. Rice and P. Gray, *Statistical Mechanics of Simple Liquids.* New York: Interscience, 1965).

this approximation by computing the function $g^{(3)}$ and comparing to its superposition approximation. Alder shows that the superposition approximation may be satisfactory to within a few percent at high densities in spite of the fact that it is much worse at low densities (i.e., giving fourth virial coefficients that are incorrect by 20 percent or so). Figure 13–7(a) shows a typical result from Alder's paper. This seems to indicate that the superposition approximation is capable of yielding satisfactory results at high densities. Rahman* has found similar results for the Lennard-Jones 6–12 potential [Fig. 13–7(b)], although Wang and Krumhansl† have shown that the superposition approximation suffers under a more extensive study.

In an interesting paper, Reiss‡ has derived the Kirkwood superposition approximation by means of a variational principle. His method is applicable to higher-order distribution functions as well.

Figure 13–8 shows the radial distribution function for hard spheres as calculated

* A. Rahman, *Phys. Rev. Lett.*, **12**, p. 575, 1964.
† S. Wang and J. A. Krumhansl, *J. Chem. Phys.*, **56**, p. 4287, 1972.
‡ H. Reiss, *J. Stat. Phys.*, **6**, p. 39, 1972.

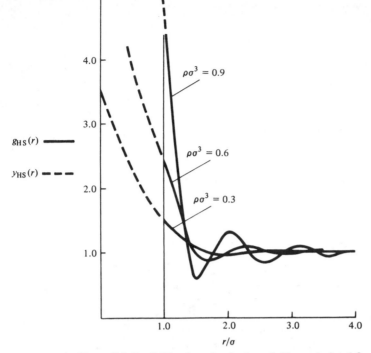

$g_{HS}(r)$ ——

$y_{HS}(r)$ - - -

$\rho\sigma^3 = 0.9$

$\rho\sigma^3 = 0.6$

$\rho\sigma^3 = 0.3$

r/σ

Figure 13–8. **The radial distribution for a hard-sphere fluid as calculated from the Percus-Yevick equation by Throop and Bearman.** (From *J. Chem. Phys.*, **42**, p. 2408, 1965). **The dash lines represent the function** $y(r) = \exp(u(r))g(r)$.

from the Percus-Yevick equation.* Although the equation of state can be written in a simple analytic form, namely, Eqs. (13–91) and (13–92), this is not so for the radial distribution function. Very accurate radial distribution functions for hard spheres have recently been tabulated by Barker and Henderson† by Monte Carlo methods,‡ and it is found that the Percus-Yevick results are quite similar to the accurate results. One important feature to notice in Fig. 13–8 is the discontinuity in $g(r)$ at the hard-sphere diameter σ. This is due to the factor of $\exp[-\beta u(r)]$ that occurs in $g(r)$ [cf. Eq. (13–66)]. This is exactly why it is convenient to introduce the function $y(r) = \exp[\beta u(r)]g(r)$. This function is continuous and smooth at $r = \sigma$ since the factor that produces the discontinuity in $g(r)$ has been taken out by multiplying by $\exp[\beta u(r)]$. A plot of $y_{HS}(r)$ is given by the dashed lines in Fig. 13–8. It turns out that the form of $y(r)$ is fairly insensitive to the potential, so that a plot of $y(r)$ versus r for a more realistic potential would appear much the same.

Although the radial distribution function cannot be determined analytically, the direct correlation function can be and is given in papers by Thiele,§ Wertheim,‖, and Baxter.¶

* G. J. Throop and R. J. Bearman, *J. Chem. Phys.*, **42**, p. 2408, 1965.

† J. A. Barker and D. Henderson, *Mol. Phys.*, **21**, p. 187, 1971.

‡ The Monte Carlo method is a numerical technique to evaluate multifold integrals. The application of this method to the evaluation of the configuration integral of a system involving a few hundred particles was pioneered by Metropolis *et al.* (*J. Chem. Phys.*, **21**, p. 1087, 1953) and well reviewed by Wood in Temperley, Rowlinson, and Rushbrooke (in "Additional Reading").

§ E. Thiele, *J. Chem. Phys.*, **39**, p. 474, 1963.

‖ M. S. Wertheim, *Phys. Rev. Lett.*, **10**, p. 321, 1963.

¶ R. J. Baxter, *Phys. Rev.*, **154**, p. 170, 1968.

Table 13–3. **Comparative values of the compressibility factor** z **and the entropy** $\Delta S = -(S - S^0)/R$ **for a hard-sphere fluid***

density		Kirkwood		BGY		PY(c)		PY(p)		Padé (3, 3)		CS	
y	v/v_0	z	ΔS	z	ΔS	z	ΔS	z	ΔS	z	ΔS	z	ΔS
0.0884	8.38	1.44	0.03	1.44	0.03	1.447	0.028	1.444	0.029	1.446	0.028	1.446	0.028
0.1562	4.74	1.91	0.12	1.93	0.11	1.965	0.101	1.946	0.105	1.960	0.102	1.959	0.102
0.2128	3.48	2.39	0.24	2.46	0.23	2.579	0.212	2.520	0.219	2.563	0.214	2.559	0.215
0.2616	2.83	2.89	0.39	3.04	0.37	3.304	0.360	3.170	0.365	3.266	0.362	3.260	0.361
0.3060	2.42	3.40	0.55	3.65	0.56	4.187	0.548	3.930	0.546	4.110	0.548	4.101	0.547
0.3444	2.15	3.91	0.73	4.21	0.76	5.192	0.765	4.757	0.747	5.057	0.761	5.047	0.759
0.3817	1.94	4.43	0.92	4.75	1.00	6.461	1.04	5.757	0.992	6.234	1.03	6.226	1.00
0.4150	1.78	4.93	1.14	5.33	1.23	7.978	1.36	6.857	1.25	7.619	1.33	7.617	1.33
0.4515	1.64	5.44	1.37	5.96	1.49	10.03	1.78	8.358	1.61	9.464	1.73	9.474	1.72
0.4840	1.53	5.95	1.60	6.54	1.77	12.50	2.27	10.031	2.00	11.64	2.18	11.68	2.17
0.5142	1.44	6.46	1.84			15.52	2.84	11.596	2.46	14.26	2.70	14.33	2.69
0.5405	1.37	6.99	2.07			18.89	3.44	14.007	2.86	17.14	3.24	17.26	3.24
0.5700	1.30	7.50	2.32			23.76	4.27	16.845	3.87	21.24	3.98	21.44	3.98
0.5972	1.24	7.93	2.60			29.89	5.26	20.119	4.07	26.32	4.83	26.63	4.85

* S^0 is the entropy of an ideal gas at the same pressure and temperature.
Source: N. F. Carnahan and K. E. Starling, *J. Chem. Phys.*, **53**, p. 600, 1970.

Table 13–3 compares the values of the compressibility factor $z = pv/kT$ and the so-called excess entropy calculated by several methods for a hard-sphere fluid. The excess entropy is the difference between the entropy of the system of interest and that of an ideal gas at the same pressure and temperature. The excess entropy can be calculated in terms of the compressibility factor by means of the thermodynamic relation for hard spheres (Problem 13–35):

$$\frac{S^E}{R} = \ln z - \int_0^y \frac{(z - 1)}{y} \, dy \tag{13–94}$$

where $y = \pi\sigma^3\rho/6$. Since there are two equations of state obtained from the Percus-Yevick equation (the pressure equation and the compressibility equation), there are two entries for this equation in Table 13–3. The column labeled CS is calculated using an equation of state derived by Carnahan and Starling* (Problem 12–34). It can be seen from this table that the Percus-Yevick equation gives the best agreement with the values calculated from the Padé approximant equation of state of Hoover and Ree† or the CS equation of state.

Before we conclude this discussion of a hard-sphere fluid, we point out an interesting paper by Rice and Lekner‡ in which they improve the superposition approximation and then use this improved version to truncate the Born-Green-Yvon hierarchy. The exact formal expression for the triplet correlation function is given by Eq. (13–75) namely,

$$g^{(3)}(1, 2, 3) = \exp[-\beta(u_{12} + u_{13} + u_{23})]\left[1 + \sum_{n=1}^{\infty} \rho^n \tau_n(1, 2, 3)\right]$$

Rice and Lekner evaluate the first few terms in the above summation over n and then use these to construct a Padé approximant for the summation. The BGY equation was then solved using this approximation for $g^{(3)}$, and the pressure obtained from the

* N. F. Carnahan and K. E. Starling, *J. Chem. Phys.*, **51**, p. 635, 1969.
† F. H. Ree and W. G. Hoover, *J. Chem. Phys.*, **40**, p. 939, 1964.
‡ S. A. Rice and J. Lekner, *J. Chem. Phys.*, **42**, p. 3559, 1965.

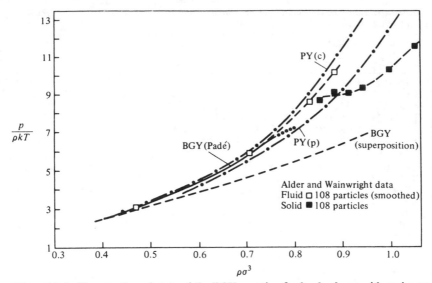

Figure 13–9. **The equation of state of the BGY equation for hard spheres with an improved version of the superposition approximation compared to several other hard-sphere equations of state.** (From S. A. Rice and J. Lekner, *J. Chem. Phys.*, **42**, p. 3559, 1965.)

contact value of $g^{(2)}$ is in almost perfect agreement with the molecular dynamics equations of state data of Alder and Wainwright up to $p/\rho kT = 6.80$, which is the Kirkwood upper limit of the stability of a fluid of hard spheres, i.e., the point beyond which no fluid-type solution exists to the BGY equation. Figure 13–9 compares the equation of state obtained in this way with several other approximate theories. The improvement over the ordinary BGY solution is seen to be quite substantial.

A comparison of the various integral equations for a more realistic potential such as the Lennard-Jones potential is more difficult since in this case temperature enters in as another variable. All four of these equations have been solved for the 6–12 potential.

Figure 13–10 shows the compressibility and virial equations of state for the Lennard-Jones 6–12 potential with $T^* = 2.74$. It can be seen from this that the Percus-Yevick equation is the most satisfactory at this reduced temperature. Figure 13–11 shows the radial distribution function from the Percus-Yevick equation for the Lennard-Jones 6–12 potential. Again it can be seen that the comparison to molecular dynamics results is quite good. This agreement deteriorates as the temperature gets lower and the density gets higher.

Figure 13–12 shows the Percus-Yevick equation of state for a 6–12 fluid. This figure has an entry that we have not yet mentioned. It has been shown by Barker and Henderson and colleagues* that one can obtain good results by integrating the energy with respect to the temperature to obtain the Helmholtz free energy and then differentiating with respect to the density to obtain the pressure. The energy equation appears to be less sensitive to errors in $g(r)$, and this is seen in Fig. 13–12.

Table 13–4 compares the compressibility factor and the average potential energy from the Percus-Yevick, hypernetted-chain, and Born-Green-Yvon equations to molecular dynamics calculations. The calculations were done at a temperature well

---

* J. A. Barker, D. Henderson, and R. O. Watts, *Phys. Lett.*, **A31**, p. 48, 1970; M. Chen, D. Henderson, and J. A. Barker, *Can. J. Phys.*, **47**, p. 2009, 1969; D. Henderson and M. Chen, *Can. J. Phys.*, **48**, p. 634, 1970.

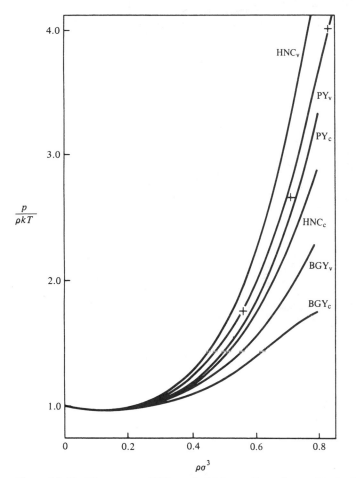

Figure 13–10. **The compressibility and virial equations of state for the Lennard-Jones 6–12 potential. The BGY, HNC, and PY results are those of D. Levesque** (*Physica*, **32**, p. 1985, 1966) **and A. A. Broyles** *et al.* (*J. Chem. Phys.*, **37**, p. 2462, 1962) **and the Monte Carlo results are due to Wood and Parker** (*J. Chem. Phys.*, **27**, p. 720, 1957).

Table 13–4. **Comparison of thermodynamic functions from the BGY, HNC, PY, and molecular dynamics calculations at** $T^* = 2.74$

$\rho\sigma^3$	molecular dynamics	$(p/\rho kT)^p$ PY	HNC	BGY
0.400	1.2–1.5	1.24	1.28	1.26
0.833	4.01	4.01	5.11	2.3
1.000	7.0	6.8	9.1	3.1
1.111	7.8	9.2	13.2	3.8

$\rho\sigma^3$	molecular dynamics	$\bar{U}/NkT$ PY	HNC	BGY
0.400	−0.86	−0.865	−0.859	−0.85
0.833	−1.58	−1.61	−1.40	−1.8
1.000	−1.60	−1.67	−1.19	−2.2
1.111	−1.90	−1.59	−0.78	−2.6

Source: S. A. Rice and P. Gray, *The Statistical Mechanics of Simple Liquids* (New York: Interscience, 1965).

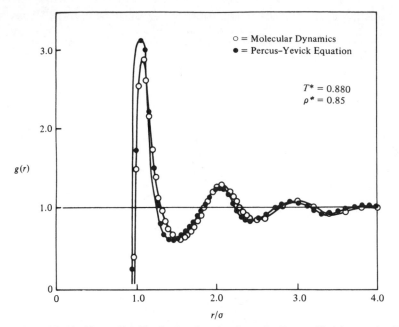

○ = Molecular Dynamics
● = Percus–Yevick Equation

$T^* = 0.880$
$\rho^* = 0.85$

Figure 13–11. **The radial distribution function from the Percus-Yevick equation for the Lennard-Jones 6–12 potential.** (From F. Mandel, R. J. Bearman, and M. Y. Bearman, *J. Chem. Phys.*, **52**, p. 3315, 1970.)

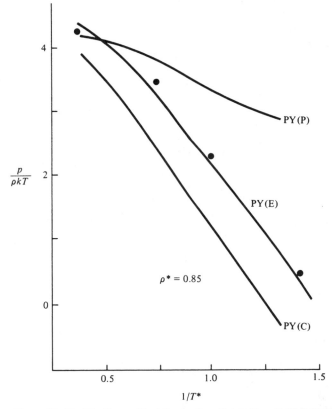

$\rho^* = 0.85$

PY(P)

PY(E)

PY(C)

Figure 13–12. **The Percus-Yevick equation of state for a 6–12 fluid. The three curves marked** P, C, **and** E **are the Percus-Yevick results from the pressure, compressibility, and energy equations, respectively.** (From J. A. Barker and D. Henderson, *Ann. Rev. Phys. Chem.*, **23**, p. 439 1972.)

above the critical temperature ($T_c^* \approx 1.29$). It can be seen that the agreement is quite good at the lower densities, but becomes poorer at the higher densities. The Percus-Yevick equation gives the best agreement, however.

Table 13–5 shows a more extensive comparison for the Percus-Yevick equation. Again the agreement, in this case with actual experimental data, is quite good. Table 13–6 shows that such agreement is not found for the pressure, at least, if the Percus-Yevick equation is pushed to very low temperatures and high densities. Nevertheless, the Percus-Yevick equation seems to be a satisfactory equation up to $\sim 1\frac{1}{2}$ times the critical density and near the critical temperatures.

Table 13–5. **A comparison of the pressure, the internal energy, and the excess entropy as calculated from the Percus-Yevick equation with experimental data for fluid argon**

Experimental and theoretical argon pressure, p^*/kT^*

ρ^*	$T^* = 1.3$		$T^* = 1.4$		$T^* = 1.5$		$T^* = 2.0$	
	Exp.	Calc.	Exp.	Calc.	Exp.	Calc.	Exp.	Calc.
0.1	0.070	0.071	0.075	0.075	0.078	0.079	0.089	0.090
0.2	0.097	0.098	0.113	0.114	0.125	0.127	0.167	0.169
0.3	0.106	0.105	0.136	0.137	0.161	0.164	0.250	0.254
0.4	0.115	0.118	0.164	0.170	0.207	0.215	0.362	0.369
0.5	0.153	0.172	0.233	0.249	0.303	0.317	0.546	0.552
0.6	0.311	0.326	0.431	0.437	0.535	0.532	0.881	0.865

Experimental and calculated internal energy, E^*

ρ^*	$T^* = 1.3$		$T^* = 1.4$		$T^* = 1.5$		$T^* = 2.0$	
	Exp.	Calc.	Exp.	Calc.	Exp.	Calc.	Exp.	Calc.
0.1	−0.825	−0.781	−0.785	−0.748	−0.755	−0.723	−0.665	−0.647
0.2	−1.581	−1.534	−1.499	−1.451	−1.442	−1.400	−1.287	−1.266
0.3	−2.233	−2.181	−2.129	−2.081	−2.063	−2.024	−1.879	−1.865
0.4	−2.785	−2.734	−2.705	−2.640	−2.647	−2.622	−2.457	−2.455
0.5	−3.337	−3.322	−3.278	−3.271	−3.229	−3.226	−3.030	−3.042
0.6	−3.934	−3.938	−3.876	−3.885	−3.826	−3.835	−3.593	−3.613

Experimental and calculated excess entropy, S^E/Nk

ρ^*	$T^* = 1.3$		$T^* = 1.4$		$T^* = 1.5$		$T^* = 2.0$	
	Exp.	Calc.	Exp.	Calc.	Exp.	Calc.	Exp.	Calc.
0.1	−0.317	−0.298	−0.288	−0.274	−0.266	−0.256	−0.213	−0.212
0.2	−0.621	−0.604	−0.560	−0.544	−0.521	−0.510	−0.429	−0.431
0.3	−0.887	−0.866	−0.810	−0.795	−0.764	−0.757	−0.656	−0.665
0.4	−1.116	−1.096	−1.056	−1.053	−1.015	−1.022	−0.904	−0.925
0.5	−1.380	−1.396	−1.336	−1.362	−1.302	−1.331	−1.188	−1.220
0.6	−1.728	−1.772	−1.684	−1.733	−1.649	−1.694	−1.516	−1.556

Source: R. O. Watts, *J. Chem. Phys.*, **50**, p. 984, 1969.

Table 13–6. **Comparison of the Percus-Yevick and molecular dynamics thermodynamic properties for a Lennard-Jones fluid**

T^*	ρ^*	pressure ($\beta P/\rho$)			energy ($\beta E/N$)	
		Percus-Yevick(c)	molecular dynamics	Percus-Yevick(p)	molecular dynamics	Percus-Yevick
1.128	0.85	1.71	2.78	3.57	−5.05	−4.98
0.880	0.85	0.54	1.64	3.17	−6.75	−6.61
0.786	0.85	(−0.10)	0.99	2.97	−7.70	−7.51
0.719	0.85	(−0.60)	0.36	2.82	−8.51	−8.28

Source: F. Mandel, R. J. Bearman, and M. Y. Bearman, *J. Chem. Phys.*, **52**, p. 3315, 1970.

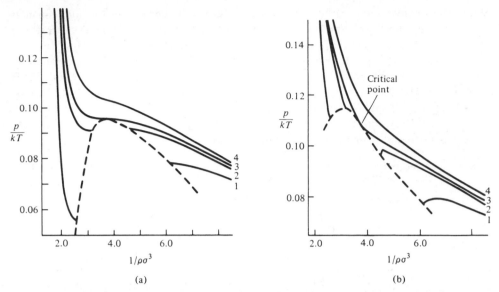

Figure 13–13. (a) **The Percus-Yevick equation of state for the Lennard-Jones 6–12 potential derived from the compressibility equation for several isotherms. The curves are (1)** $T^* = 1.2$; **(2)** $T^* = 1.263$; **(3)** $T^* = 1.275$; **(4)** $T^* = 1.3$. - - - **indicates the region where the P–Y equation has no solutions. (b) The Percus-Yevick equation of state calculated from the pressure equation.** (From R. O. Watts, *J. Chem. Phys.*, **48**, p. 50, 1968.)

Table 13–7. **The reduced critical parameters obtained from various theories of the liquid state***

	T_c^*	ρ_c^*	$p_c^*/\rho_c^*T_c^*$	reference
experimental	1.28	0.66	0.29	
3-term virial series	1.45	0.75	0.34	(a)
4-term virial series	1.30	0.56	0.35	(a)
BGY(p)	1.68	1.08	0.36	(b)
HNC(p)	1.25	0.55	0.30	(c)
HNC(c)	1.39	0.59	0.38	(c)
PY(p)	1.25	0.61	0.30	(c)
PY(c)	1.32	0.59	0.36	(c)
van der Waals	0.30	0.24	0.38	(d)
Lennard–Jones Devonshire	1.30	1.19	0.59	(d)

* The temperature is reduced by ε/k, and the density is reduced by $2\pi N\sigma^3/3$.

Sources: (a) J. A. Barker and J. J. Monaghan, *J. Chem. Phys.*, **36**, p. 2564, 1962; (b) J. G. Kirkwood, V. A. Lewison, and B. J. Alder, *J. Chem. Phys.*, **20**, p. 929, 1952; (c) L. Verlet and D. Levesque, *Physica*, **36** p. 254, 1967; (d) Chapter 12.

Figure 13–13 displays a set of p–V isotherms for the PY equation and 6–12 potential for both the pressure and compressibility equations. It is interesting to see that below $T^* = 1.275$ no solutions were found, possibly indicating a phase transition.

Table 13–7 lists the critical constants determined from a number of equations of state. The van der Waals equation will be discussed in the next chapter.

Lastly, we mention that the most reliable calculations of the radial distribution are molecular dynamics or Monte Carlo calculations. One advantage of this method is that the intermolecular potential is known, and so these calculations serve as "experimental" data on which to test various theories. The molecular dynamics radial distribution function for hard spheres was first given by Alder and Hecht* and Barker,

* B. J. Alder and C. E. Hecht, *J. Chem. Phys.*, **50**, p. 2032, 1969.

Watts, and Henderson;* the radial distribution function for the Lennard-Jones potential has been calculated by Verlet;† and the radial distribution function for an inverse 12 potential has been calculated by Hansen and Weis.‡ A program that generates the hard-sphere radial distribution function in excellent agreement with the molecular dynamics values is given in Appendix D. This program is due to Grundke and Henderson.

In addition, Barker, Fisher, and Watts§ have presented extensive molecular dynamics and Monte Carlo calculations, including three-body forces. Molecular dynamics and Monte Carlo calculations are particularly useful in the perturbation theories to be studied in the next chapter and in the transport theories to be studied in later chapters.

ADDITIONAL READING

COLE, G. H. A. 1967. *The statistical theory of classical simple dense fluids.* Oxford: Pergamon.
DE BOER, J. *Repts. Prog. Phys.,* **12**, p. 305, 1949.
FRISCH, H. L. *Adv. Chem. Phys.,* **6**, p. 229, 1964.
———, and LEBOWITZ, J. L., eds. 1964. *The equilibrium theory of classical fluids.* New York: Benjamin.
———, and SALSBURG, Z. W., eds. 1968. *Simple dense fluids.* New York: Academic.
HENDERSON, D. *Ann. Rev. Phys. Chem.,* **15**, p. 31, 1964.
———, and LEONARD, P. J. 1971. In *Physical chemistry, an advanced treatise,* Vol. 8B, ed. by H. Eyring, D. Henderson, and W. Jost. New York: Academic.
HILL, T. L. 1956. *Statistical mechanics.* New York: McGraw-Hill.
KAHN, B. *Studies in Stat. Mech.,* **3**, p. 281, 1965.
KIRKWOOD, J. G. 1968. *Theory of liquids,* ed. by B. J. Alder. New York: Gordon and Breach.
LEVELT, J. M. H., and COHEN, E. G. D. *Studies in Stat. Mech.,* **2**, p. 107, 1964.
McDONALD, I. R., and SINGER, K. *Quart. Rev.,* **24**, p. 238, 1970.
PRIGOGINE, I. 1957. *The molecular theory of solutions.* Amsterdam: North-Holland Publ.
RICE, S. A., and GRAY, P. 1965. *The statistical mechanics of simple liquids.* New York: Interscience.
ROWLINSON, J. S. *Repts. Prog. in Phys.,* **28**, p. 169, 1965.
———. 1969. *Liquids and liquid mixtures,* 2nd ed. London: Butterworths.
SALPETER, E. E. *Ann. Phys.,* **5**, p. 183, 1958.
TEMPERLEY, H. N. V., ROWLINSON, J. S., and RUSHBROOKE, G. S., eds. 1968. *Physics of simple liquids.* Amsterdam: North-Holland Publ.
UHLENBECK, G. E., and FORD, B. W. *Studies in Stat. Mech.,* **1**, p. 119, 1962.
WATTS, R. O. *Specialist Periodical Reports of the Chemical Society, Statistical mechanics, Vol. 1.* 1973. London: The Chemical Society.

PROBLEMS

13–1. Construct a (2, 2) Padé approximant for the pressure of a hard-sphere system and compare to the molecular dynamics values. Use the Appendix of the paper by Ree and Hoover‖ to construct the Padé approximant and the paper of Carnahan and Starling¶ for the molecular dynamics values as well as some other theoretical expressions.

13–2. Derive an equation for the thermodynamic energy E in terms of $g^{(2)}$ and $g^{(3)}$ for an intermolecular potential that can be written in the form

$$U(\mathbf{r}_1, \mathbf{r}_2, \ldots, \mathbf{r}_N) = \sum_{i<j} u^{(2)}(r_{ij}) + \sum_{i<j<k} u^{(3)}(r_{ij}, r_{jk}, r_{ik})$$

13–3. Use the pressure equation to derive an equation for the Helmholtz free energy in terms of $g(r, \rho, T)$.

* J. A. Barker, R. O. Watts, and D. Henderson, *Mol. Phys.,* **21**, p. 187, 1971.
† L. Verlet, *Phys. Rev.,* **165**, 201, 1968.
‡ J. P. Hansen and J. J. Weis, *Mol. Phys.,* **23**, p. 853, 1972.
§ J. A. Barker, R. A. Fisher, and R. O. Watts, *Mol. Phys.,* **21**, p. 657, 1971.
‖ *J. Chem. Phys.,* **40**, p. 939, 1964.
¶ *J. Chem. Phys.,* **51**, p. 635, 1969.

13–4. Show that the fugacity is related to the intermolecular potential by

$$\ln f = \ln(\rho kT) + \frac{\rho}{kT} \int_0^\infty \int_0^1 u(r)g(r, \rho, \xi)4\pi r^2 \, dr \, d\xi$$

Using this formula, derive a virial expansion for f.

13–5. Show that the last integral in Eq. (13–13) vanishes for $s \neq 0$.

13–6. Derive an equation for the Helmholtz free energy in terms of $g(r, T)$ by using the Gibbs-Helmholtz equation [Eq. (13–27)]. Use this to derive an equation for the chemical potential.

13–7. Show that $V^n N!/N^n(N-n)!$ can be written as $V^n[1 + O(N^{-1})]$. This factor occurs in Eq. (13–9) for $g^{(n)}(\mathbf{r}_1, \ldots, \mathbf{r}_n)$. Show that this implies that $g^{(2)} \to 1 + O(N^{-1})$ as $r \to \infty$.

13–8. Derive Eq. (13–62).

13–9. Define the potential of mean force in words. What does it have to do with the superposition approximation?

13–10. Show that the expression for the chemical potential as an integral over a coupling parameter is actually equivalent to the use of the Gibbs-Helmholtz equation and an integration with respect to $1/T$.

13–11. Derive a superposition approximation for $g^{(4)}$ by assuming that the potential of mean force $w^{(4)}$ is pair-wise additive.

13–12. Derive the second member of the Kirkwood hierarchy; i.e., using the concept of a coupling parameter, derive an integral equation for $g^{(3)}$ in terms of $g^{(4)}$. What superposition approximation would you make on $g^{(4)}$?*

13–13. Derive the Born-Green-Yvon integral equation [Eq. (13–47)] by taking the gradient with respect to the position of particle 1 in Eq. (13–9) for $\rho^{(n)}$ and then manipulating much in the same manner as in the derivation of the Kirkwood equation.†

13–14. If $\mathscr{H}(\mathbf{k})$ is the Fourier transform of $h(r_{12})$ and $C(\mathbf{k})$ is the Fourier transform of $c(r_{12})$, then show that

$$C(\mathbf{k}) = \frac{\mathscr{H}(\mathbf{k})}{1 + \rho\mathscr{H}(\mathbf{k})}$$

by taking the Fourier transform of the Ornstein-Zernike equation.

13–15. Show that

$$\rho^{(2)}(r_{12}) = e^{-\beta u(r_{12})}\left[\rho^2 + \rho^3 \int \wedge d\mathbf{r}_3 + \frac{\rho^4}{2} \int\int [2\,\square + 4\,\boxtimes + \boxtimes + \boxtimes]\, d\mathbf{r}_3 \, d\mathbf{r}_4 + \cdots\right]$$

13–16. Derive Eq. (13–68).

13–17. Show that $g_1(r, T)$ for hard spheres [Eq. (13–69)] is given by

$$
\begin{aligned}
g_1(r, T) &= 0 & 0 &< r < \sigma \\
&= \pi\sigma^3\left\{\frac{4}{3} - \frac{r}{\sigma} + \frac{1}{12}\frac{r^3}{\sigma^3}\right\} & \sigma &< r < 2\sigma \\
&= 0 & r &> 2\sigma
\end{aligned}
$$

13–18. Show that the density derivative of $\rho^n g^{(n)}$ is given by

$$\left[\frac{\partial(\rho^n g^{(n)})}{\partial\rho}\right]_\beta = \frac{n\rho^{n-1}g^{(n)} + \rho^n \int \{g^{(n+1)}(\mathbf{r}_1, \ldots, \mathbf{r}_{n+1}) - g^{(n)}(\mathbf{r}_1, \ldots, \mathbf{r}_n)\}\, d\mathbf{r}_n}{1 + \rho \int \{g(r) - 1\}\, d\mathbf{r}}$$

---

* See H. Reiss, *J. Stat. Phys.*, **6**, p. 39, 1972, for an excellent discussion of the superposition approximation applied to $g^{(4)}$.

† See T. L. Hill, *Statistical Mechanics*, Section 33, in "Additional Reading."

Hint: Start with Eq. (13–49) for $\rho^{(n)}$ and differentiate this with respect to $\bar{N}$ at constant volume, using the key fact that the dependence of $\rho^{(n)}$ and $\bar{N}$ on density at constant temperature and volume is only in Z. Thus * we can use

$$\left(\frac{\partial \rho^{(n)}}{\partial \rho}\right)_\beta = V\left(\frac{\partial \rho^{(n)}}{\partial z}\right)_\beta \left(\frac{\partial \bar{N}}{\partial z}\right)_\beta$$

13–19. Show that the temperature derivative of $g(r)$ involves $g^{(3)}$ and $g^{(4)}$.†

13–20. Derive an expression for the first two terms in the density expansion of the structure factor for a system of hard spheres.

13–21. For a nonadditive intermolecular potential, $y(r)$ has the density expansion

$$y(r) = 1 + y_1^{(A)}(r)\rho + y_1^{(NA)}(r)\rho + \cdots$$

where $y_1^{(A)}(r)$ and $y_1^{(NA)}(r)$ denote the additive and nonadditive contributions to $y_1(r)$, respectively. Show that

$$y_1^{(A)}(r) = \int \wedge \, d\mathbf{r}_3$$

and that‡

$$y_1^{(NA)}(r) = \int \exp[-\beta(u_{23} + u_{13})][\exp(-\beta u_{123}) - 1]\, d\mathbf{r}_3$$

13–22. Solve the Ornstein-Zernike equation for $c(r)$ by iteration to derive Eq. (13–72).

13–23. Substitute Eq. (13–74) into Eq. (13–58) to derive the virial expansion for the pressure.

13–24. Derive Eq. (13–75) for $g^{(3)}$.

13–25. Substitute Eq. (13–66) into Eq. (13–23) to derive the virial expansion for the pressure.

13–26. Show that the second term in the direct correlation function of the hypernetted-chain approximation $[y(r) - 1 - \ln y(r)]$ is small as r increases beyond the first maximum in $g(r)$, say.

13–27. Show that $y(r)$ from the Percus-Yevick equation is given by Eq. (13–86).

13–28. Show that all four of the integral equations discussed in this chapter yield the correct third virial coefficient.

13–29. Show that the hypernetted-chain equation leads to a density expansion for $y(r)$ in the form

$$y(r) = 1 + y_1(r)\rho + y_2(r)\rho^2 + \cdots$$

where§

$$y_1(r) = \int \wedge \, d\mathbf{r}_3$$

$$y_2(r) = \iint [\tfrac{1}{2} \bowtie + \sqcap + 2 \boxtimes]\, d\mathbf{r}_3\, d\mathbf{r}_4$$

13–30. Make the simple approximation that $c(r) = f(r)$ and derive an integral equation for $h(r)$ by substituting this into the Ornstein-Zernike equation. Solve this equation by iteration and derive a density expansion for $h(r)$. Can you identify the general term graphically? Solve this equation for the Fourier transform of $h(r)$. Can you invert this? Does it give a correct third virial coefficient? Calculate the structure function $h(k)$ and plot this versus $k\sigma$.

* See P. Schofield, *Proc. Phys. Soc.*, **88**, p. 149, 1968.

† See P. Schofield, *Proc. Phys. Soc.*, **88**, p. 149, 1968.

‡ See C. T. Chen and R. D. Present, *J. Chem. Phys.*, **53**, p. 1585, 1970.

§ See D. A. McQuarrie, *J. Chem. Phys.*, **40**, p. 3455, 1964.

13–31. Invent a reasonable approximate form for $c(r)$ and derive your own integral equation for $h(r)$. Does it reproduce any virial coefficients? Can the equation be solved by iteration to derive a density expansion for $h(r)$?

13–32. Prove generally that the superposition approximation yields the correct third virial coefficient.

13–33. Show that the equation of state of hard spheres can be written

$$\frac{p}{\rho kT} = 1 + \frac{2\pi\rho\sigma^3}{3} g(\sigma+)$$

where $g(\sigma+)$ is the value of $g(r)$ at $\sigma + \varepsilon$ as $\varepsilon \to 0$ from the positive direction.*

13–34. The entropy per molecule of a monatomic ideal gas is given by

$$T\tilde{s} = kT \ln v - kT \ln \Lambda^3 + \tfrac{5}{2}kT$$

Define an excess entropy per molecule s_v^E at any volume v by

$$s = k \ln v - k \ln \Lambda^3 + \tfrac{5}{2}k + s_v^E$$

Show that

$$s_v^E = -k \int_v^\infty \left[\left(\frac{\partial p}{\partial T}\right)_v - \frac{1}{v} \right] dv$$

by integrating the thermodynamic equation

$$ds = \left(\frac{\partial p}{\partial T}\right) dv \qquad T \text{ constant}$$

between v and v^0 where $v^0 \to \infty$.†

13–35. Define excess thermodynamic quantities at any pressure p by

$$\mu = kT \ln p + kT \ln\left(\frac{\Lambda^3}{kT}\right) + \mu_p^E$$

$$E = \tfrac{3}{2}NkT + E_p^E$$

$$S = -Nk \ln p - Nk \ln\left(\frac{\Lambda^3}{kT}\right) + \tfrac{5}{2}Nk + S_p^E$$

First derive the equation

$$TS_p^E = E_p^E - N\mu_p^E + (pV - NkT)$$

Using the fact that $E^E = 0$ for hard spheres and that

$$A^0 - A(V) = -\int_V^{v^0} p\, dV = (N\mu^0 - NkT) - (N\mu - pV)$$

derive the equations

$$\frac{\mu_p^E}{kT} = -\ln\left(\frac{pv}{kT}\right) + \left(\frac{pv}{kT} - 1\right) + \int_v^\infty \left(\frac{p}{kT} - \frac{1}{v}\right) dv$$

and

$$\frac{S_p^E}{Nk} = \ln\left(\frac{pv}{kT}\right) - \int_v^\infty \left(\frac{p}{kT} - \frac{1}{v}\right) dv$$

---

* See T. L. Hill, *Statistical Mechanics*, Section 35, in "Additional Reading."
† See T. L. Hill, *Statistical Mechanics*, Section 36, in "Additional Reading," for an application of this equation.

13–36. Using the Carnahan-Starling equation of state (Problem 12–34)

$$\frac{p}{\rho kT} = \frac{1 + y + y^2 - y^3}{(1-y)^3}$$

where $y = \pi \rho \sigma^3/6$, derive analytical expressions for the excess entropy, excess energy, excess Helmholtz free energy, excess enthalpy, and excess Gibbs free energy for a hard-sphere fluid.*

13–37. We shall see in the next chapter that one can obtain an excellent approximation to the correlation function $y(r)$ associated with the intermolecular potential $u(r)$ by writing

$$y(r) = e^{-\beta u(r)} y_d(r)$$

where $y_d(r)$ is the correlation function of a system of hard spheres of diameter d. In the next chapter we shall give a prescription for the choice of the value of d, but for now simply test this approximation for an inverse-12 potential† $u(r) = 4\varepsilon(\sigma/r)^{12}$ and the Lennard-Jones 6–12 potential‡ by varying d around σ. For the hard sphere $y(r)$, use either the molecular dynamics data of Alder and Hecht§ or the program listed in Appendix D.

13–38. Derive expressions for the pressure and thermodynamic energy of a multicomponent system, namely,

$$\frac{p}{\rho kT} = 1 - \frac{2\pi}{3} \frac{\rho}{kT} \sum_{i,j} x_i x_j \int_0^\infty g_{ij}(r) u_{ij}'(r) r^3 \, dr$$

$$E = \tfrac{3}{2} NkT + 2\pi N\rho \sum_{i,j} x_i x_j \int_0^\infty g_{ij}(r) u_{ij}(r) r^2 \, dr$$

where x_i is the mole fraction of component i.

13–39. The generalization of the Ornstein-Zernike equation for mixtures is

$$h_{ij}(r_{12}) = c_{ij}(r_{12}) + \rho \sum_i x_i \int h_{il}(r_{13}) c_{lj}(r_{23}) \, d\mathbf{r}_3$$

Use this to derive the Percus-Yevick equation for a multicomponent system. This equation has been solved analytically for additive hard spheres by Lebowitz‖ and Baxter.¶ Lebowitz's paper gives an analytical expression for the Laplace transform of $rg_{ij}(r)$.

13–40. A difficulty with the radial distribution function approach to solutions is that the radial distribution function of a mixture depends upon so many variables, such as the densities of each component, the temperature, and the intermolecular potential parameters ε_{ij} and σ_{ij}, say for a Lennard-Jones potential. There is a set of approximate mixture theories that attempts to treat the solution as an ideal mixture of pure fluids with an *effective* intermolecular potential. These theories are called *n-fluid theories*, the simplest of which is the *random mixture theory*. In this theory, we start with

$$E = \tfrac{3}{2} NkT + 2\pi N\rho \sum_{i,j} x_i x_j \int_0^\infty u_{ij}(r) g_{ij}(r) r^2 \, dr$$

and let

$$g_{11}(r) = g_{22}(r) = g_{12}(r) = \cdots = g(r)$$

---

* See Problems 13–34, 13–35 and N. F. Carnahan and K. E. Starling, *J. Chem. Phys.*, **53**, p. 600, 1970, for the definition of excess thermodynamic quantities.

† J. P. Hansen and J. J. Weis, *Mol. Phys.*, **23**, p. 853, 1972.

‡ L. Verlet, *Phys. Rev.*, **165**, p. 201, 1968.

§ *J. Chem. Phys.*, **50**, p. 2032, 1969.

‖ *Phys. Rev.*, **133**, p. A895, 1964.

¶ *J. Chem. Phys.*, **52**, p. 4559, 1970.

to get

$$E = \tfrac{3}{2}NkT + 2\pi N\rho \int_0^\infty u(r)g(r)r^2\, dr$$

where

$$u(r) = \sum_{i,j} x_i x_j u_{ij}(r)$$

Thus we see that the random mixture theory treats the mixture as a single pure fluid with an effective intermolecular potential given above. Hence it is often called *a one-fluid theory*. Show that if $u_{ij}(r) = 4\varepsilon_{ij}\{(\sigma_{ij}/r)^{12} - (\sigma_{ij}/r)^6\}$, then the random mixture theory uses an effective Lennard-Jones potential with

$$\varepsilon = \frac{(\sum_{i,j} x_i x_j \varepsilon_{ij}\sigma_{ij}^6)^2}{\sum_{i,j} x_i x_j \varepsilon_{ij}\sigma_{ij}^{12}}$$

$$\sigma = \left\{\frac{\sum_{i,j} x_i x_j \varepsilon_{ij}\sigma_{ij}^{12}}{\sum_{i,j} x_i x_j \varepsilon_{ij}\sigma_{ij}^6}\right\}^{1/6}$$

13–41. A more sophisticated approximation than the random mixture theory is the *average potential model*, in which one assumes that

$$g_{12}(r) = \tfrac{1}{2}[g_{11}(r) + g_{22}(r)]$$

Show that is the case that the effective intermolecular potential is

$$u_i(r) = x_i u_{ii}(r) + x_j u_{ij}(r)$$

which for a Lennard-Jones potential gives potential parameters

$$\varepsilon_i = \frac{(\sum_j x_j \varepsilon_{ij}\sigma_{ij}^6)^2}{\sum_j x_j \varepsilon_{ij}\sigma_{ij}^{12}}$$

$$\sigma_i = \left\{\frac{\sum_j x_i \varepsilon_{ij}\sigma_{ij}^{12}}{\sum_j x_j \varepsilon_{ij}\sigma_{ij}^6}\right\}^{1/6}$$

Thus a two-component solution is treated as a mixture of two pure pseudocomponents, and hence this approximation is often called a *two-fluid theory*.

13–42. If we assume that the intermolecular potential is of the form

$$u_{ij}(r) = \varepsilon_{ij} F\left(\frac{r}{\sigma_{ij}}\right)$$

where F is some universal function, then the energy can be written as

$$E = \tfrac{3}{2}NkT + 2\pi N\rho \sum_{i,j} x_i x_j \varepsilon_{ij} \int_0^\infty F\left(\frac{r}{\sigma_{ij}}\right) g_{ij}(r)r^2\, dr$$

Now the $g_{ij}(r)$ are functions that scale as

$$g_{ij} = g_{ij}\left(\frac{r}{\sigma_{ij}}\right)$$

Show that

$$E = \tfrac{3}{2}NkT + 2\pi N\rho \sum_{i,j} x_i x_j \varepsilon_{ij}\sigma_{ij}^3 \int_0^\infty F(r_{ij}^*)g_{ij}(r_{ij}^*)r_{ij}^{*2}\, dr_{ij}^*$$

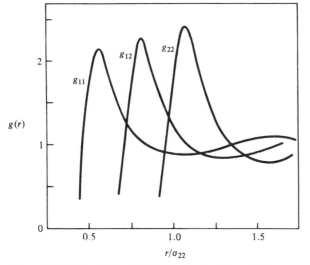

$g(r)$

Figure 13–14. **Typical radial distribution functions of a two-component mixture.**

Recall that the g_{ij} are functions of the x_i, ρ, σ_{22}/σ_{11}, kT/ε_{ij} as well as $r_{ij}{}^* = r/\sigma_{ij}$. Some typical values of g_{ij} are shown in Fig. 13–14. From this figure we can see that the approximations made in the random mixture theory and the average potential model are quite poor. A much better approximation is to let

$$g_{11}(r_{11}{}^*) = g_{12}(r_{12}{}^*) = g_{22}(r_{22}{}^*) = g(r^*)$$

Show that this leads to

$$E = \tfrac{3}{2}NkT + 2\pi N\rho \sum_{i,j} x_i x_j \varepsilon_{ij} \sigma_{ij}{}^3 \int_0^\infty F(x)g(x)x^2 \, dx$$

This equation is of the one-fluid form if we interpret $g(r)$ as the distribution function corresponding to a system in which

$$\varepsilon\sigma^3 = \sum_{i,j} x_i x_j \varepsilon_{ij} \sigma_{ij}{}^3$$

This is called *the van der Waals 1 theory* and is discussed in some detail by Henderson.

13–43. In Problem 13–42 we developed the so-called van der Waals 1 theory. We can derive a van der Waals 2 theory by assuming that

$$g_{12}(r_{12}{}^*) = \tfrac{1}{2}[g_{11}(r_{11}{}^*) + g_{22}(r_{22}{}^*)]$$

Show that this leads to

$$E = \tfrac{3}{2}NkT + 2\pi N\rho \sum_i x_i \varepsilon_i \sigma_i{}^3 \int_0^\infty F(r_{ii}{}^*)g_{ii}(r_{ii}{}^*)r_{ii}^{*2} \, dr_{ii}{}^*$$

where

$$\varepsilon_i \sigma_i{}^3 = x_i \varepsilon_{ii} \sigma_{ii}{}^3 + x_j \varepsilon_{ij} \sigma_{ij}{}^3$$

Figures 13–15 through 13–17 show some excess thermodynamic properties for the van der Waals 1 theory. Excess thermodynamic functions are defined as the difference between the thermodynamic function of mixing and the value corresponding to an ideal mixture at the same temperature, pressure, and composition.

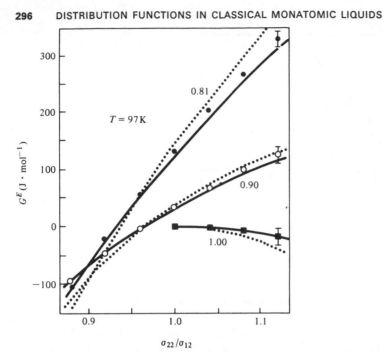

Figure 13–15. **Excess free energy of mixing as a function of σ_{22}/σ_{12} for an equimolar mixture at 97°K and** $p = 0$ $(\varepsilon_{12}/k = 133.5°K,$ $\sigma_{12} = 3.596$ Å. **The points give Monte Carlo data of Singer and Singer** (*Mol. Phys.*, **24**, p. 357, 1972), **the dotted curve gives the van der Waals 1 theory, and the solid curve gives the results of perturbation theory** (see Chapter 14). **The curves are labeled with appropriate values of** $\varepsilon_{22}/\varepsilon_{12}$. (From E. W. Grundke, D. Henderson, J. A. Barker, and P. J. Leonard, *Mol. Phys.*, **25**, p. 883, 1973.)

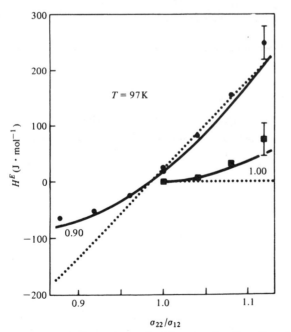

Figure 13–16. **Excess heat of mixing as a function of σ_{22}/σ_{12} for an equimolar mixture at 97°K and** $p = 0$ $(\varepsilon_{12}/k = 133.5°K,$ $\sigma_{12} = 3.596$ Å. **The points and curves have the same interpretation as in Figure 13–15.**

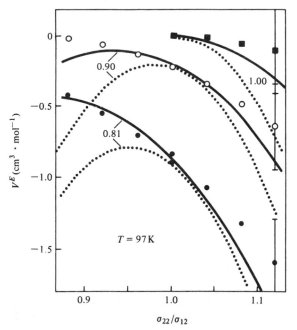

Figure 13-17. **Excess volume of mixing as a function of** σ_{22}/σ_{12} **for an equimolar mixture at 97°K and** $p = 0$ ($\varepsilon_{12}/k = 133.5°K$, $\sigma_{12} = 3.596$ Å. The points and curves have the same interpretation as in Figure 13-15.

13-44. Show that the Carnahan-Starling hard-sphere equation of state (Problem 12-34) can be written in the form

$$Z_{CS} = \tfrac{1}{3}(2Z_{PY}{}^C + Z_{PY}{}^P)$$

where Z is the compressibility factor $p/\rho kT$, and $Z_{PY}{}^C$ and $Z_{PY}{}^P$ are given by Eqs. (13-92) and (13-91), respectively. Mansoori *et al.*[*] have used this observation to derive the corresponding result for mixtures. Using the equations of state from the Percus-Yevick equation for a mixture of hard spheres,[†] derive an equation of state for a mixture of hard spheres. Mansoori *et al.* have used this equation of state to calculate the thermodynamic properties of a mixture of hard spheres.

13-45. Molecular dynamics calculations have proved to be quite useful in generating "experimental data" to which one can unambiguously compare various theories. The data are the results of solving the equations of motion of N particles under a specified known potential. The observable macroscopic properties are then calculated by time averaging the appropriate microscopic quantities. Write down the equations of motion of a small system, say five particles, under a hard-sphere potential and indicate how the observed macroscopic properties, such as the energy or the pressure, would be calculated.

13-46. Using Eq. (13-74), show that for hard spheres the coefficient of ρ is the linear term in the density and is of the form of a cubic polynomial in r with the quadratic term missing. It so happens that this same functional form is found in the ρ^2 term as well. This observation played a central role in Wertheim's analytic solution of the Percus-Yevick equation for hard spheres.[‡]

[*] *J. Chem. Phys.*, **54**, p. 1523, 1971.
[†] J. L. Lebowitz, *Phys. Rev.*, **133**, p. A895, 1964.
[‡] (See *Phys. Rev. Lett.*, **10**, p. 231, 1963.)

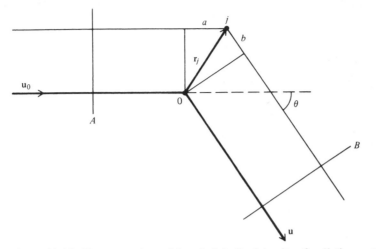

Figure 13–18. **The geometry used to calculate the intensity of radiation scattered through an angle** θ. **The plane** A **represents the incident radiation, and** B **represents the plane of an observer.**

13–47. Consider the light scattered from a system of N scatterers. Let the incident oscillating electric field be propagated in the direction of the unit vector $\mathbf{u}_0$ and let the scattered field be in the direction of the unit vector $\mathbf{u}$. Figure 13–18 shows these unit vectors and the position of the jth scatterer with respect to some arbitrary origin. The location of this origin can have no effect on the final result. Let the incident radiation be assumed to have the same phase over the plane A, and let plane B represent the position of an observer of the scattered radiation. The radiation arriving at B will, in general, no longer be in phase because of the extra distance $a + b$ traveled by the upper beam in the figure. Hence the intensity of the radiation arriving at B will be less than the incident intensity. Define a quantity $P(\theta)$ by

$$P(\theta) = \frac{\text{scattered intensity from a collection of scatters}}{\text{initial intensity}}$$

Note that as $\theta \rightarrow 0$ in Fig. 13–18, the distance $a + b \rightarrow 0$, and hence the intensity of scattered radiation equals the intensity of incident radiation. Thus

$$P(\theta) = \frac{\text{intensity of radiation scattered through an angle } \theta}{\text{intensity of radiation scattered through an angle } \theta \rightarrow 0}$$

$$= \frac{I(\theta)}{I(0)} \qquad\qquad (13\text{–}95)$$

where $I(\theta)$ is the intensity of radiation scattered through θ. The quantity $P(\theta)$, which can be determined experimentally, yields much information concerning the configuration of the collection of scatterers. First show that the extra distance traveled by the radiation scattered by the jth element can be written as

$$a + b = \mathbf{r}_j \cdot (\mathbf{u}_0 - \mathbf{u})$$

Now let $\mathbf{n}$ be a unit vector in the direction of $\mathbf{u}_0 - \mathbf{u}$ and show that the vector $\mathbf{u}_0 - \mathbf{u}$ equals $2\mathbf{n} \sin \theta/2$ and

$$a + b = 2\mathbf{r}_j \cdot \mathbf{n} \sin \frac{\theta}{2}$$

The phase difference ϕ_j due to this extra difference is $(a + b)/\lambda$, where λ is the wavelength of the radiation, and so

$$\phi_j = \mathbf{r}_j \cdot \mathbf{n}\left(\frac{2}{\lambda} \sin \frac{\theta}{2}\right) \tag{13-96}$$

The electric field at B due to the scattering by the element at j is

$$\mathscr{E}_j = A \cos[2\pi(\nu t - \phi_j)]$$

where ν is the frequency of the radiation, and A can be considered to be simply a constant. The electric field from the collection of N scatterers is

$$\mathscr{E}_s = \sum_{j=1}^{N} A \cos[2\pi(\nu t - \phi_j)]$$

The intensity is the energy that falls on 1 cm² of area per second and can be obtained by averaging $\mathscr{E}_s{}^2$ over one period:

$$I(\theta) = \frac{\int_0^{1/\nu} dt\, \mathscr{E}_s{}^2}{\int_0^{1/\nu} dt} = \nu A^2 \int_0^{1/\nu} dt \left(\sum_{j=1}^{N} \cos \alpha_j\right)^2$$

where $\alpha_j = 2\pi(\nu t - \phi_j)$. Show that this is equal to

$$I(\theta) = \frac{A^2}{2} \sum_{i=1}^{N} \sum_{j=1}^{N} \cos 2\pi(\phi_i - \phi_j)$$

As $\theta \to 0$, each $\phi_i \to 0$, and so $I(\theta) \to A^2 N^2/2$. The ratio of $I(\theta)$ to $I(0)$ then is

$$P(\theta) = \frac{1}{N^2} \sum_{i=1}^{N} \sum_{j=1}^{N} \cos[2\pi(\phi_i - \phi_j)]$$

$$= \frac{1}{N^2} \sum_{i=1}^{N} \sum_{j=1}^{N} \cos[s\mathbf{n} \cdot (\mathbf{r}_i - \mathbf{r}_j)] \tag{13-97}$$

where

$$s = \frac{4\pi}{\lambda} \sin \frac{\theta}{2} \tag{13-98}$$

Note that only the difference in the positions of particles appears here and that the arbitrarily placed origin in Fig. 13–18 drops out of the equations.

Equation (13–97) is for a collection of scatterers in a fixed orientation with respect to the incident radiation. We must now average Eq. (13–97) over all orientations. This can be done by averaging over all orientations of $\mathbf{n}$. To do this, take a spherical coordinate system with the z-axis to be along $\mathbf{r}_i - \mathbf{r}_j$ with the angle between $\mathbf{n}$ and this z-axis denoted by α and the other angle in the spherical coordinate system denoted by β. Show then that $d\mathbf{n} = \sin \alpha \, d\alpha \, d\beta$ and

$$\cos[s\mathbf{n} \cdot \mathbf{r}_{ij}] = \frac{1}{4\pi} \int_0^{2\pi} \int_0^{\pi} \cos(sr_{ij} \cos \alpha) \sin \alpha \, d\alpha \, d\beta$$

$$= \frac{\sin sr_{ij}}{sr_{ij}}$$

The scattering function $P(\theta)$ is given by

$$P(\theta) = \frac{1}{N^2} \sum_{i=1}^{N} \sum_{j=1}^{N} \frac{\sin sr_{ij}}{sr_{ij}} \tag{13-99}$$

CHAPTER 14

PERTURBATION THEORIES OF LIQUIDS

In the previous chapter we have studied the radial distribution function theory of liquids. This constitutes a rigorous statistical mechanical theory. Although we eventually had to introduce some sort of approximations into the formalism, we at least started from general principles. Because of this, however, the results are not particularly simple to use in practical calculations. There is an interesting and important result from the developments in Chapter 13 that we shall exploit in this chapter. Much evidence points to the fact that the structure of a liquid is primarily determined by the short-range repulsive forces and that the relatively longer-range attractive part of the potential provides a net force that gives a somewhat uniform attractive potential. Thus, in a sense we picture the repulsive part of the potential as determining the structure of the liquid and the attractive part as holding the molecules together at some specified density.

For example, except for the discontinuity at $r = \sigma$, the radial distribution function of a hard-sphere fluid is fairly similar to that of a real fluid. In particular, if this discontinuity is removed by multiplying $g(r)$ by $\exp[u(r)]$ to give $y(r)$, we find that this function is quite similar for hard spheres and real fluids. The function $y(r)$ is shown in Fig. 13–8 for hard spheres, but a similar result would be found for a Lennard-Jones potential.

Molecular dynamics calculations have also been used to show that the structure of a real fluid is similar to a hard-sphere fluid. Verlet* has carried out extensive molecular dynamics calculations on a system of molecules interacting with a Lennard-Jones 6–12 potential and has calculated the so-called structure factor $\hat{h}(k)$ of the fluid. The structure factor is related to the Fourier transform of $h(r) = g(r) - 1$, i.e., by

$$\hat{h}(k) = \rho \int e^{-i\mathbf{k}\cdot\mathbf{r}} h(r)\, d\mathbf{r} \qquad (14\text{–}1)$$

Verlet calculated $\hat{h}(k)$ since it is directly measured by X-ray scattering from the liquid.

* L. Verlet, *Phys. Rev.*, **165**, p. 201, 1968.

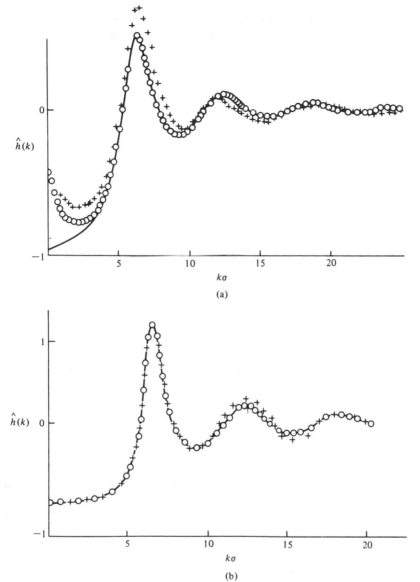

Figure 14–1. **The structure factor, $\hat{h}(k)$, plotted as a function of the wave vector $k\sigma$. The circles are the results from molecular dynamics calculations; the solid line is a hard-sphere model with an effective hard-sphere diameter; and the crosses represent neutron diffraction data. In (a) the reduced temperature is 1.326 and the reduced density is 0.5426. In (b), these values are 0.827 and 0.75, respectively.** (From L. Verlet, *Phys. Rev.*, **165**, p. 201, 1968.)

Figure 14–1 shows two graphs taken from Verlet's paper. The dots are the molecular dynamics results, and the solid lines in both Fig. 14–1(a) and (b) are the results for a hard-sphere fluid with an effective hard-sphere diameter that is a function of the temperature and the density. The important point is that at high densities the structure factor of the real (Lennard-Jones 6–12) fluid can be well represented by a hard-sphere structure factor.

This physical picture suggests that we attempt to treat a fluid as a system of molecules governed by a repulsive potential with an attractive potential that is treated as a small perturbation. The unperturbed system, i.e., the system of repulsive molecules, is usually taken to be a system of hard spheres since this system is fairly well understood from Chapter 13. In Section 14–1 we shall introduce the basic statistical mechanical perturbation theory due to Zwanzig. Then in Section 14–2 we shall make a simple application of this perturbation theory formalism to derive the van der Waals equation. Remember that van der Waals in 1873 treated a liquid as a system of hard-sphere molecules with an attractive potential that gave a uniform background energy. Thus he was one of the earliest to use the perturbation theory ideas presented here. Van der Waals did not have the benefit of modern developments in the statistical mechanics of hard-sphere fluids and statistical mechanical theory of fluids in general, and we shall see how it is possible to improve upon the van der Waals equation in a very obvious and organized way. There were a number of developments along these lines in the late 1960s, and Section 14–3 will discuss some of these. We shall see in that section that the perturbation approach to liquids can be brought to the point of providing a simple, convenient theory of liquids that is numerically reliable well past the critical point and into the temperatures and densities of dense liquids. Recall that the integral equations of Chapter 13 gave poor numerical agreement near and below the critical region.

14–1   STATISTICAL MECHANICAL PERTURBATION THEORY

In this section we describe the statistical mechanical perturbation theory developed by Zwanzig in 1954.* The results are obtained by performing a perturbation expansion on the canonical partition function of the system and on the configuration integral in particular.

Let the total potential energy be separated into two parts,

$$U_N = U_N^{(0)} + U_N^{(1)} \tag{14–2}$$

where $U_N^{(0)}$ is the potential energy of an unperturbed (reference) system and $U_N^{(1)}$ is the perturbation. The reference system is usually taken to be a hard-sphere system, but this is, of course, not necessary. The perturbation potential is usually *defined* by the difference between the potential of the real system and the potential of the reference system, i.e., by $U_N^{(1)} = U_N - U_N^{(0)}$.

The configurational integral for this potential is

$$Z_N = \int \cdots \int e^{-\beta[U_N^{(0)} + U_N^{(1)}]} \, d\mathbf{r}_1 \cdots d\mathbf{r}_N$$

We now multiply and divide by

$$\int \cdots \int \exp(-\beta U_N^{(0)}) \, d\mathbf{r}_1 \, d\mathbf{r}_2 \cdots d\mathbf{r}_N$$

to get

$$Z_N = \int \cdots \int e^{-\beta U_N^{(0)}} \, d\mathbf{r}_1 \cdots d\mathbf{r}_N \frac{\int \cdots \int e^{-\beta[U_N^{(0)} + U_N^{(1)}]} d\mathbf{r}_1 \cdots d\mathbf{r}_N}{\int \cdots \int e^{-\beta U_N^{(0)}} d\mathbf{r}_1 \cdots d\mathbf{r}_N}$$

Note that the second factor here can be considered to be the average of $\exp(-\beta U_N^{(1)})$ over the *unperturbed* or *reference* system. Thus we can write

$$Z_N = Z_N^{(0)} \langle \exp(-\beta U_N^{(1)}) \rangle_0 \tag{14–3}$$

* R. W. Zwanzig, *J. Chem. Phys.*, **22**, p. 1420, 1954.

where $Z_N^{(0)}$ is the configurational integral of the unperturbed system and $\langle \ \rangle_0$ indicates a canonical average in the *unperturbed* system. We assume that we know $Z_N^{(0)}$, and we shall consider $U_N^{(1)}$ to be small enough to expand the exponential in $U_N^{(1)}$. This gives an expansion in powers of β or $1/T$, according to

$$\langle \exp(-\beta U_N^{(1)}) \rangle_0 = 1 - \beta \langle U_N^{(1)} \rangle_0 + \frac{\beta^2}{2!} \langle (U^{(1)2}) \rangle_0 + \cdots \tag{14-4}$$

We wish, however, to express the free energy and related thermodynamic functions as a power series in β. Since $A = -kT \ln Q$ and $Q = Z_N/N! \Lambda^{3N}$, we can write

$$-\beta A = \ln\left(\frac{Z_N^{(0)}}{N! \Lambda^{3N}}\right) + \ln\langle \exp(-\beta U_N^{(1)}) \rangle_0 \tag{14-5}$$

$$= -\beta A_0 - \beta A^{(1)}$$

where clearly A_0 is the free energy of the reference system and $A^{(1)}$ is the perturbation free energy:

$$A^{(1)} = -kT \ln\langle \exp(-\beta U_N^{(1)}) \rangle_0 \tag{14-6}$$

It is this quantity that we wish to express as a power series in β. We write

$$A^{(1)} = \sum_{n=1}^{\infty} \frac{\omega_n}{n!} (-\beta)^{n-1} \tag{14-7}$$

where the ω_n are to be determined in terms of the $\langle (U_N^{(1)})^n \rangle_0$ of Eq. (14-4) by writing

$$\exp(-\beta A^{(1)}) = \exp\left(\sum_{n=1}^{\infty} \frac{\omega_n}{n!} (-\beta)^n\right)$$

We now expand the right-hand side in powers of β and compare like coefficients of β^n to

$$\exp(-\beta A^{(1)}) = \langle \exp(-\beta U_N^{(1)}) \rangle_0$$

$$= \sum_{k=0}^{\infty} \frac{(-\beta)^k}{k!} \langle (U_N^{(1)})^k \rangle_0$$

Straightforward algebra gives the first few ω_n:

$$\begin{aligned} \omega_1 &= \langle U_N^{(1)} \rangle_0 \\ \omega_2 &= \langle (U_N^{(1)})^2 \rangle_0 - \langle U_N^{(1)} \rangle_0^2 \\ \omega_3 &= \langle (U_N^{(1)})^3 \rangle_0 - 3\langle (U_N^{(1)})^2 \rangle_0 \langle U_N^{(1)} \rangle_0 + 2\langle U_N^{(1)} \rangle_0^3 \end{aligned} \tag{14-8}$$

It is possible to derive a general formula for this, but it is not necessary here.[*]
Thus we have shown that at high temperatures, the free energy is

$$A = A_0 + \omega_1 - \frac{\omega_2}{2kT} + O(\beta^2) \tag{14-9}$$

The first term in the expansion, $\omega_1 = \langle U_N^{(1)} \rangle_0$, simplifies considerably when $U_N^{(1)}$ can be written as a sum of pair potentials,

$$U_N^{(1)} = \sum_{i<j} u^{(1)}(r_{ij})$$

[*] R. W. Zwanzig, *J. Chem. Phys.*, **22**, p. 1420, 1954.

since in this case

$$
\begin{aligned}
\langle U_N{}^{(1)}\rangle_0 &= \left\langle \sum_{i<j} u^{(1)}(r_{ij}) \right\rangle_0 = \frac{N(N-1)}{2}\langle u^{(1)}(r_{12})\rangle_0 \\
&= \frac{N(N-1)/2}{Z_N{}^{(0)}} \int \cdots \int e^{-\beta U_N{}^{(0)}} u^{(1)}(r_{12})\, d\mathbf{r}_1 \cdots d\mathbf{r}_N \\
&= \frac{1}{2}\iint_V u^{(1)}(r_{12})\rho_0{}^{(2)}(r_{12})\, d\mathbf{r}_1\, d\mathbf{r}_2 = \frac{\rho^2 V}{2}\int u^{(1)}(r_{12})g_0{}^{(2)}(r_{12})\, d\mathbf{r}_{12} \quad (14\text{--}10)
\end{aligned}
$$

The zero subscript in the last two integrals indicates that the distribution function is that of the reference system.

On the other hand, terms like $\langle (U_N{}^{(1)})^2\rangle_0$ are awkward since they introduce higher-order density functions, namely, $P^{(3)}$ and $P^{(4)}$, which are due to terms like $u(r_{ij})u(r_{jk})$ and $u(r_{ij})u(r_{kl})$ which arise when $U_N{}^{(1)}$ is squared. We shall not require this second-order term in most of what we do here, and so we do not write it out.

Equations (14–3), (14–4), and (14–8) are the basic results of this section. In order to apply them, we must choose some reference fluid and hence $U_N{}^{(0)}$ and $U_N{}^{(1)}$. This is where the various applications, which we shall discuss in Section 14–3, differ from one another. Before going on to discuss these, however, we shall derive the van der Waals equation from a simple application of the equations of this section.

14–2   THE VAN DER WAALS EQUATION

We start with Eq. (14–3),

$$
Z_N = Z_N{}^{(0)}\langle \exp(-\beta U_N{}^{(1)})\rangle_0
$$

and take the reference system to be a hard-sphere fluid. We furthermore assume that U_N is pair-wise additive and to be of the form $u(r) = u_{HS}(r) + u^{(1)}(r)$, where $u^{(1)}(r)$ is some arbitrary attractive part (and so is negative). We then assume that $\beta U_N{}^{(1)}$ is small enough that we can write

$$
\begin{aligned}
\langle \exp(-\beta U_N{}^{(1)})\rangle_0 &\approx 1 - \beta\langle U_N{}^{(1)}\rangle_0 \\
&\approx \exp\langle -\beta U_N{}^{(1)}\rangle_0 \quad (14\text{--}11)
\end{aligned}
$$

Equation (14–11) follows since we are assuming that terms in β^2 and higher can be neglected. From Eq. (14–10),

$$
\langle U_N{}^{(1)}\rangle_0 = \frac{\rho^2 V}{2}\int_0^\infty u^{(1)}(r)g_{HS}(r)4\pi r^2\, dr
$$

Certainly $g_{HS}(r)$ was not available to van der Waals, and he effectively approximated $g_{HS}(r)$ by

$$
g_{HS}(r) = \begin{cases} 0 & r < \sigma \\ 1 & r > \sigma \end{cases} \quad (14\text{--}12)
$$

This form is correct for hard spheres only in the limit $\rho \to 0$. Using this $g_{HS}(r)$, we have

$$
\begin{aligned}
\langle U_N{}^{(1)}\rangle_0 &= 2\pi\rho^2 V \int_\sigma^\infty u^{(1)}(r)r^2\, dr \\
&= -aN\rho \quad (14\text{--}13)
\end{aligned}
$$

where

$$
a = -2\pi \int_\sigma^\infty u^{(1)}(r)r^2\, dr \quad (14\text{--}14)
$$

The minus sign in the definition of a has been included to make a a positive number. Remember that $u^{(1)}(r)$ is negative. The specific form of $u^{(1)}(r)$ is not important here.

So far now, Eq. (14–3) has been reduced to

$$Z_N = Z_N^{(0)} e^{\beta a \rho N} \tag{14–15}$$

and so we can write

$$\frac{p}{kT} = \left(\frac{\partial \ln Z_N^{(0)}}{\partial V} \right)_{N,T} - \frac{a\rho^2}{kT}$$

$$= \frac{p^{(0)}}{kT} - \frac{a\rho^2}{kT} \tag{14–16}$$

where $p^{(0)}$ is the pressure of the unperturbed system.

The final approximation of the van der Waals theory is to assume that the hard-sphere configuration integral is of the form V_{eff}^N, where the effective volume is determined by assuming that the volume available to a molecule in the fluid has a volume $4\pi\sigma^3/3$ excluded to it by each other molecule of the system. However, we have to divide this quantity by 2 since this factor of $4\pi\sigma^3/3$ arises from a pair of molecules interacting, and only half the effect can be assigned to a given molecule. Therefore $V_{\text{eff}} = V - 2\pi N\sigma^3/3$, and we can write

$$Z_N^{(0)} = (V - Nb)^N \qquad b = \frac{2\pi\sigma^3}{3} \tag{14–17}$$

Substituting Eq. (14–17) into Eq. (14–16), then, we get the famous van der Waals equation,

$$\frac{p}{kT} = \frac{\rho}{1 - b\rho} - \frac{a\rho^2}{kT} \tag{14–18}$$

Equation (14–18) is the van der Waals equation of state. Thus the constants a and b, defined in Eqs. (14–14) and (14–17), are the usual van der Waals constants, but here they are given in terms of the intermolecular potential function. We can calculate a and b for the square-well potential, for example, and compare the results to the empirically determined van der Waals parameters. The agreement is not particularly satisfactory, indicating that the van der Waals equation is not a good quantitative equation of state. Nevertheless, it does indicate a critical point and condensation by way of a simple model.

The perturbation theoretic derivation of the van der Waals equation suggests a number of obvious improvements. The most obvious ones are:

(1) Use a better form for $g(r)$.
(2) Use a better expression for $Z_N^{(0)}$.
(3) Use a more realistic unperturbed system.
(4) Consider higher terms in β.

The results of Chapter 13 allow us easily to consider points (1) and (2). We could use the analytic Percus-Yevick equations of state [Eqs. (13–91) and (13–92)], the Ree and Hoover Padé approximant [Eq. (13–3)], or the Carnahan-Starling equation of state (Problem 12–34). Furthermore, we could use the $g_{\text{HS}}(r)$ obtained from the Percus-Yevick or hypernetted-chain equations or even that calculated from molecular dynamics or Monte Carlo calculations. Points (3) and (4) are more difficult to incorporate since for point (3) we still must rely upon the results from the radial distribution theories of the previous chapter to give the unperturbed part and for (4) because terms beyond the linear term in β require three- and four-body distribution functions. The recent advances in the statistical mechanical perturbation theory of fluids have considered points (3) and (4) above, and we discuss these in the next section.

14-3 SEVERAL PERTURBATION THEORIES OF LIQUIDS

In the late 1960s and early 1970s there was much activity in applying the statistical mechanical perturbation theory to dense fluids. In this section we shall discuss three of these theories. Each of these theories uses a fairly different approach and illustrates a particular manner in which the simple van der Waals theory of the preceding section can be improved.

THE BARKER-HENDERSON THEORY

The first generally successful approach was that due to Barker and Henderson.* For simplicity, we shall first discuss their theory for a square-well fluid and then extend their approach to real (6–12) fluids. Barker and Henderson were the first to consider the higher-order terms in the $1/T$ expansion in Eq. (14–9). The first-order term, i.e. ω_1 in Eq. (14–9), presents no difficulty [cf. Eq. (14–10)]. The second-order term involves three- and four-body distribution functions:

$$\omega_2 = \frac{N^4}{4} \iiiint [P_0^{(4)}(1, 2, 3, 4) - P_0^{(2)}(1, 2,)P_0^{(2)}(3, 4)]$$

$$\times u^{(1)}(1, 2)u^{(1)}(3, 4) \, d\mathbf{r}_1 \, d\mathbf{r}_2 \, d\mathbf{r}_3 \, d\mathbf{r}_4$$

$$+ N^3 \iiint P_0^{(3)}(1, 2, 3)u^{(1)}(1, 2)u^{(1)}(2, 3) \, d\mathbf{r}_1 \, d\mathbf{r}_2 \, d\mathbf{r}_3$$

$$+ \frac{N^2}{2} \iint P_0^{(2)}(1, 2)[u^{(1)}(1, 2)]^2 \, d\mathbf{r}_1 \, d\mathbf{r}_2 \tag{14-19}$$

where

$$P_0^{(n)}(1, 2, \ldots, n) = \frac{1}{Z_N^{(0)}} \int \cdots \int e^{-\beta U_N^{(0)}} \, d\mathbf{r}_{n+1} \cdots d\mathbf{r}_N \tag{14-20}$$

We have left ω_2 in terms of the $P_0^{(n)}$ rather than the more commonly used $g_0^{(n)}$ (1, 2, ..., n) because the conversion of Eq. (14–19) from integrals over the $P_0^{(n)}$ to integrals over $g_0^{(n)}$'s is somewhat subtle. This subtlety is due to the fact that $g(r) \to 1 + O(1/N)$ for large r,† and so the integrand in Eq. (14–19) differs from zero by terms of order 1 in a region of molecular dimension and by terms of order $1/N$ in the whole volume; so the integral of these terms contribute the same order of magnitude to the final result. We shall not discuss this fine point any further, but refer the reader to a discussion by Henderson and Barker.‡

The important point about Eq. (14–19) is that it involves three- and four-body distribution functions, quantities that are not generally known. Barker and Henderson have approximated ω_2 in two different ways. The first method goes as follows.

* J. A. Barker and D. Henderson, *J. Chem. Phys.*, **47**, p. 2856, 1967; *Accounts of Chem. Res.*, **4**, p. 303, 1971; *Adv. Treatise of Phys. Chem.*, **8A** (in "Additional Reading"); *Ann. Rev. Phys. Chem.*, **23**, p. 439, 1972.
† See T. Hill, *Statistical Mechanics*, Appendix 7 (New York: McGraw-Hill, 1956), in "Additional Reading."
‡ See D. Henderson and J. A. Barker, in *Physical Chemistry*, Vol. 8A, edited by H. Eyring, D. Henderson, and W. Jost (New York: Academic Press, 1971) in "Additional Reading."

Imagine the range of intermolecular distances divided into intervals (r_0, r_1), $(r_1, r_2), \ldots, (r_j, r_{j+1}), \ldots$. Let $p(N_0, N_1, N_2, \ldots) \equiv p(\{N_j\})$ be the probability that N_j molecules lie in the interval (r_j, r_{j+1}) in the *unperturbed* system. Also consider that the interval (r_j, r_{j+1}) is small enough that we can treat the perturbing potential $u^{(1)}(r)$ to have a constant value $u_j^{(1)}$ in the interval (r_j, r_{j+1}). The configuration integral, Eq. (14–3), can then be written in the form

$$Z_N = Z_N^{(0)} \left\langle \exp\left(-\beta \sum_j N_j u_j^{(1)}\right) \right\rangle_0$$

$$= Z_N^{(0)} \sum_{\{N_j\}} p(\{N_j\}) \exp\left(-\beta \sum_j N_j u_j^{(1)}\right) \tag{14–21}$$

The quantity $p(\{N_j\})$ is as difficult to get as Z_N, but only the first two moments are necessary in order to calculate through terms of $O(\beta^2)$. To see this, expand the exponential in Eq. (14–21) to get

$$\frac{Z_N}{Z_N^{(0)}} = \sum_{\{N_j\}} p(\{N_j\}) \exp\left(-\beta \sum_j N_j u_j^{(1)}\right)$$

$$= \sum_{\{N_j\}} p(\{N_j\}) - \beta \sum_j u_j^{(1)} \sum_{\{N_j\}} N_j p(\{N_j\})$$

$$+ \frac{\beta^2}{2} \sum_i \sum_j u_i^{(1)} u_j^{(1)} \sum_{\{N_j\}} N_i N_j p(\{N_j\}) + \cdots$$

$$= 1 - \beta \sum_j u_j^{(1)} \langle N_j \rangle + \frac{\beta^2}{2} \sum_i \sum_j u_i^{(1)} u_j^{(1)} \langle N_i N_j \rangle + \cdots$$

To get the expansion for the Helmholtz free energy, we take logarithms to get

$$\frac{A}{kT} = \frac{A_0}{kT} + \beta \sum_j \langle N_j \rangle u_j^{(1)} - \tfrac{1}{2}\beta^2 \sum_i \sum_j (\langle N_i N_j \rangle - \langle N_i \rangle \langle N_j \rangle) u_i^{(1)} u_j^{(1)} + O(\beta^3) \tag{14–22}$$

The first moment is easy and is

$$\langle N_j \rangle = 2\pi N\rho \int_{r_j}^{r_{j+1}} r^2 g_0(r)\, dr$$

$$= 2\pi N\rho r_j^2 g_0(r_j)(r_{j+1} - r_j) \tag{14–23}$$

The evaluation of the second-moment term in Eq. (14–22) must be done approximately. The N_j may be regarded as representing the numbers of molecules in spherical shells surrounding some central molecule. If these shells were large macroscopic volumes, the numbers of molecules in different shells would be uncorrelated, i.e.,

$$\langle N_i N_j \rangle - \langle N_i \rangle \langle N_j \rangle = 0 \qquad i \neq j \tag{14–24}$$

and the fluctuation of the number in a given shell would be given by [Eq. (3–53)]:

$$\langle N_j^2 \rangle - \langle N_j \rangle^2 = \langle N_j \rangle kT \left(\frac{\partial \rho}{\partial p}\right)_T \tag{14–25}$$

Thus the β^2 term in the expansion in Eq. (14–22) can be approximated by

$$- \tfrac{1}{2}\beta^2 \sum_j \langle N_j \rangle kT [u_j^{(1)}]^2 \left(\frac{\partial \rho}{\partial p}\right)_0$$

If we substitute Eq. (14–23) for $\langle N_j \rangle$ into this and then pass to the continuum limit, we get

$$\frac{A}{NkT} = \frac{A_0}{NkT} + \frac{\rho\beta}{2} \int u^{(1)}(r) g_0(r) 4\pi r^2 \, dr$$

$$- \frac{\rho\beta^2}{4} \int [u^{(1)}(r)]^2 kT \left(\frac{\partial\rho}{\partial p}\right)_0 g_0(r) 4\pi r^2 \, dr + O(\beta^3) \qquad (14\text{–}26)$$

This approximation for the β^2 term is called the *macroscopic compressibility approximation*. A more satisfactory approximation is obtained by arguing that one should use a "local" compressibility in Eq. (14–26), related to the pressure derivative of the density at a distance r from a given molecule; i.e., replace $(\partial\rho/\partial p)_0 \, g_0(r)$ in Eq. (14–26) by $(\partial/\partial p)[\rho g(r)]_0$. This so-called *local compressibility approximation* gives

$$\frac{A}{NkT} = \frac{A_0}{NkT} + \frac{\rho\beta}{2} \int u^{(1)}(r) g_0(r) 4\pi r^2 \, dr$$

$$- \frac{\rho\beta^2}{4} \int [u^{(1)}(r)]^2 kT \left[\frac{\partial}{\partial p}(\rho g_0)\right] 4\pi r^2 \, dr + O(\beta^3) \qquad (14\text{–}27)$$

Note that the Helmholtz free energy A is expressed in terms of hard-sphere quantities and the perturbing potential $u^{(1)}(r)$.

In applying this equation to a square-well system, Barker and Henderson took the hard-sphere core as their unperturbed potential and the well as the perturbing potential. The comparison of the numerical results of this equation with the computer calculations of Alder and Wainwright[*] and Rotenberg[†] is shown in Fig. 14–2. In

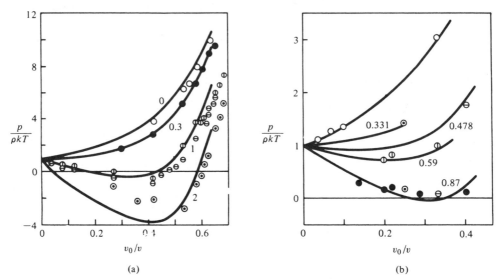

Figure 14–2. **Equation of state for the square-well potential. The points are the results of various molecular dynamics and Monte Carlo calculations, and the curves are isotherms calculated from Eq. (14–27). The numbers labeling the curves give the values of ε/kT. (From J. A. Barker and D. Henderson,** *J. Chem. Phys.*, **47, p. 2856, 1967.)**

* B. J. Alder and T. E. Wainwright, *J. Chem. Phys.* 33, p. 1439, 1960.
† A. Rotenberg, *J. Chem. Phys.*, 43, p. 1198, 1965.

obtaining these results, the Padé approximant of Hoover and Ree was used for the hard-sphere pressure $p^{(0)}$.

Smith *et al.** have also approximated ω_2 by using the superposition approximation on the three- and four-body distribution functions. This, of course, allows A to be written in terms of the two-body hard-sphere radial distribution function. The equations are too long to be given here, but the results are similar to the solid curves in Fig. 14–2. The agreement with the molecular dynamics results of Alder [Fig. 14–2(a)] is excellent. The agreement with the Monte Carlo results of Rotenberg is good except at the highest densities. Table 14–1 shows the first- and second-order terms for the Helmholtz free energy and the pressure. It can be seen that the perturbation series seems to converge rapidly. Table 14–2 compares the critical constants from the Barker-Henderson theory to the square-well molecular dynamics results of Alder. Again there is very good agreement.

Table 14–1. **The first- and second-order terms in the Barker-Henderson perturbation series for the Helmholtz free energy and the pressure for a square-well fluid**

		A_2/NkT			$p_2 v_0/kT$	
$\rho\sigma^3$	A_1/NkT	l.c.*	s.a.†	$p_1 v_0/kT$	l.c.*	s.a.†
0.00	0.0000	0.000	0.000	0.0000	0.000	0.000
0.10	−0.5339	−0.188	−0.188	−0.0403	−0.010	−0.010
0.20	−1.1405	−0.282	−0.282	−0.1816	−0.017	−0.016
0.30	−1.8159	−0.310	−0.312	−0.4502	−0.001	−0.005
0.40	−2.5511	−0.295	−0.306	−0.8601	+0.035	+0.018
0.50	−3.3292	−0.253	−0.289	−1.3993	0.088	0.026
0.60	−4.1234	−0.198	−0.281	−2.0120	0.147	−0.001
0.70	−4.8950	−0.140	−0.228	−2.5787	0.196	−0.034
0.80	−5.5933	−0.087	−0.291	−2.9072	0.216	+0.036
0.90	−6.1606	−0.045	−0.261	−2.7701	0.194	0.328
1.00	−6.5484	−0.018	—	−2.0513	0.137	—

* Local compressibility approximation.
† Superposition approximation.
Source: J. A. Barker and D. Henderson, *J. Chem. Phys.*, **47**, p. 2856, 1969; Symp. Thermophys. Properties, Papers, 4th, p. 30, 1968.

Table 14–2. **Critical constants for the square-well potential** ($\lambda = 1.5$)

	Barker-Henderson theory	molecular dynamics
kT_c/ε	1.32	1.28
V_0/V_c	∼0.22	0.235
$P_c V_0/NkT_c$	0.076	0.072

Barker and Henderson have applied their perturbation theory to a Lennard-Jones potential as well.† They actually develop the Helmholtz free energy as a double perturbation series. One variable, γ, is a measure of the depth of the attractive well, and the other variable, α, is a measure of the inverse steepness of the repulsive part of the

* W. R. Smith, D. Henderson, and J. A. Barker, *J. Chem. Phys.*, **53**, p. 508, 1970.
† J. A. Barker and D. Henderson, *J. Chem. Phys.*, **47**, p. 4714, 1967.

potential, i.e., the smaller α, the steeper the repulsive part. They do this by *defining* a modified potential function $v(\alpha, \gamma, d; r)$ corresponding to $u(r)$ by

$$v(\alpha, \gamma, d; r) = \begin{cases} u\left(d + \dfrac{r - d}{\alpha}\right) & d + \dfrac{r - d}{\alpha} < \sigma \\[2mm] 0 & \sigma < d + \dfrac{r - d}{\alpha} < d + \dfrac{\sigma - d}{\alpha} \\[2mm] \gamma u(r) & \sigma < r \end{cases} \tag{14–28}$$

The quantity d is a distance parameter which is as yet unspecified, and σ is customarily taken to be that point at which the potential $u(r)$ passes through zero. This appears to be complicated at first, but notice that $v(\alpha, \gamma, d; r)$ is independent of d and reduces to $u(r)$ when $\alpha = \gamma = 1$. When $\alpha = \gamma = 0$, on the other hand, v becomes a hard-sphere potential of diameter d, which is still arbitrary (Problem 14–10). Thus by varying γ and α, we can go from our original potential $u(r)$ to a hard-sphere potential.

The idea now is to express the Helmholtz free energy A in terms of this modified potential as a double power series about $\alpha = \gamma = 0$ according to

$$A = A_0 + \alpha \left(\frac{\partial A}{\partial \alpha}\right)_{\alpha=\gamma=0} + \gamma \left(\frac{\partial A}{\partial \gamma}\right)_{\alpha=\gamma=0} + \frac{\alpha^2}{2}\left(\frac{\partial^2 A}{\partial \alpha^2}\right)_{\alpha=\gamma=0} + \cdots \tag{14–29}$$

Note that the coefficients of α, γ, α^2, etc., here are evaluated at $\alpha = \gamma = 0$ or, in other words, are hard-sphere quantities. The differentiations indicated in Eq. (14–29) involve lengthy algebra,* but the final result is quite simple:

$$A = A_0 + \alpha 2\pi NkT\rho\, d^2 g_0(d)\left[d + \int_0^\sigma f(z)\,dz\right] + \gamma 2\pi N\rho \int_\sigma^\infty g_0(r)u(r)r^2\,dr$$

$$- \gamma^2 \pi N\rho \left(\frac{\partial \rho}{\partial p}\right)_0 \frac{\partial}{\partial \rho}\left[\rho \int_\sigma^\infty g_0(r)u^2(r)r^2\,dr\right] + O(\alpha^2) + O(\alpha\gamma) + \cdots \tag{14–30}$$

In this result A_0, g_0, and $(\partial \rho/\partial p)_0$ are the free energy, radial distribution function, and compressibility of a system of hard spheres of *diameter d* (as yet unspecified). The first-order terms in Eq. (14–30) are exact, but the γ^2 term has been approximated by the local compressibility approximation.

The value of d is still at our disposal. We choose for d the value

$$d = -\int_0^\sigma f(z)\,dz \tag{14–31}$$

so that the linear term in α vanishes. This gives d as a well-defined temperature-dependent effective hard-sphere diameter. Barker and Henderson argue that with this choice of d, the terms in α^2 and $\alpha\gamma$ in Eq. (14–30) are considerably smaller than the γ^2 term.

For $\alpha = \gamma = 1$, Eq. (14–30) is just the Helmholtz free energy for a system with potential $u(r)$, and so we finally write

$$A = A_0 + 2\pi N\rho \int_\sigma^\infty g_0(r)u(r)r^2\,dr - \pi\rho\beta\left(\frac{\partial}{\partial p}\right)_0\left[\rho \int_\sigma^\infty g_0(r)u^2(r)r^2\,dr\right] \tag{14–32}$$

where A_0, $g_0(r)$, and $(\partial \rho/\partial p)_0$ are the free energy, radial distribution function, and compressibility of a system of hard spheres of *diameter d*, given by Eq. (14–31).

---

* See the reference to Barker and Henderson in "Additional Reading" for complete details.

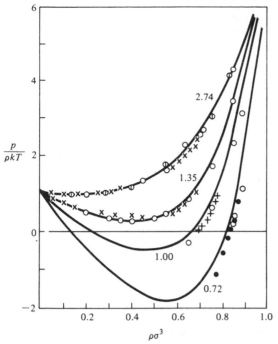

Figure 14–3. **The equation of state for the 6–12 potential according to Barker and Henderson. The curves are labeled by the value of** T^***. The points are a mixture of machine calculations and actual experimental data.** (From J. A. Barker and D. Henderson, *J. Chem. Phys.*, **47**, p. 4714, 1967.)

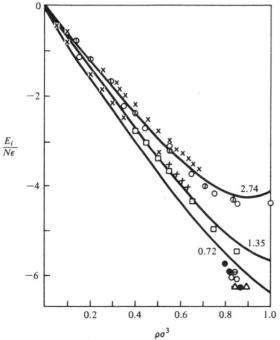

Figure 14–4. **Internal energy for the 6–12 potential according to Barker and Henderson. The curves are labeled with the values of** T^***. The points are a mixture of machine calculations and actual experimental data.** (From J. A. Barker and D. Henderson, *J. Chem. Phys.*, **47**, p. 4714, 1967.)

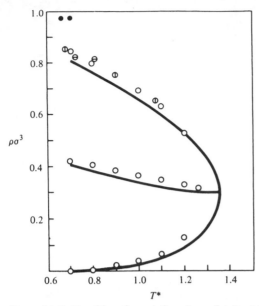

Figure 14–5. **Densities of coexisting phases for the 6–12 potential according to Barker and Henderson. The points are a mixture of machine calculations and actual experimental data.** (From J. A. Barker and D. Henderson, *J. Chem. Phys.*, **47**, p. 4714, 1967.)

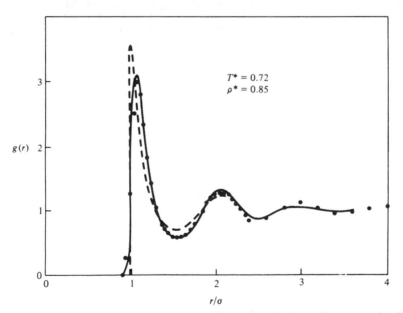

Figure 14–6. **Radial distribution function of the 6–12 liquid near its triple point. The points give the results of molecular dynamics calculations, and the curves give the results of the Barker-Henderson perturbation theory. The quantities ρ^* and T^* are $\rho\sigma^3$ and kT/ε, respectively.** (From J. A. Barker and D. Henderson, *Ann. Rev. Phys. Chem.*, **23**, p. 439, 1972.)

Figures 14–3 through 14–5 show the results from Eq. (14–32) and other thermo-dynamic functions. These comparisons indicate that the Barker-Henderson perturbation theory gives excellent results at temperatures that are not too low. Figures 14–6 and 14–7 compare the radial distribution function of the Barker-Henderson theory

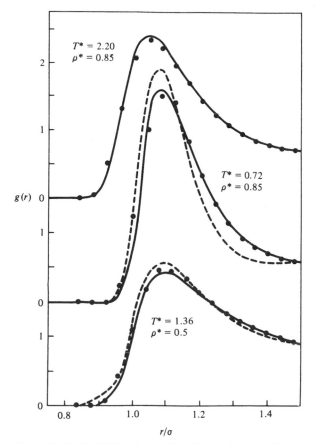

Figure 14–7. **Radial distribution function of the 6–12 fluid. The points give the results of molecular dynamics calculations, and the broken and solid curves give the result of the Percus-Yevick and Barker-Henderson theories, respectively. The quantities ρ^* and T^* are $\rho\sigma^3$ and kT/ε, respectively.** (From J. A. Barker and D. Henderson, *Ann. Rev. Phys. Chem.*, **23**, p. 439, 1972.)

to that from the Percus-Yevick equation and molecular dynamics. Note that the conditions of Fig. 14–6 are near the critical point and so this serves as a fairly severe test. Figure 14–7 compares the radial distribution function around the first maximum. Again one can see that the agreement is excellent.

Table 14–3 compares values of A/NkT for a 6–12 fluid for a number of theories. The column headed by $PY(E)$ indicates the Percus-Yevick results starting from the energy equation. We saw in Fig. 13–12 and the discussion surrounding it that this seems to be the most reliable treatment of the Percus-Yevick theory. The columns headed BH1(PY), BH1, and BH2 are the results of the Barker-Henderson theory calculated from first-order perturbation theory using the PY hard sphere $g(r)$, from first-order perturbation theory using the Monte Carlo values of the hard sphere $g(r)$, and from second-order perturbation theory using the Monte Carlo $g_0(r)$. The agreement of the BH2 results with those of the molecular dynamics is excellent. The other columns in Table 14–3 will be discussed shortly. Table 14–4 compares values of $p/\rho kT$ in a similar way. Table 14–5 compares the critical constants obtained from this theory to the molecular dynamics calculations of Verlet and experimental data of argon.

Table 14–3. **Values of** A/NkT **for the 6–12 potential**

kT/ε	$\rho\sigma^3$	molecular dynamics*	PY(E)	BH1-(PY)	BH1	BH2	CWA (PY)	CWA
2.74	0.60	−0.34	−0.33	−0.31	−0.31	−0.33	−0.32	−0.33
	0.70	+0.01	+0.01	+0.04	+0.02	+0.01	+0.02	+0.01
	0.80	0.43	0.43	0.45	0.46	0.42	0.43	0.41
	0.90	0.93	0.95	0.97	0.99	0.95	0.97	0.92
	1.00	1.59	1.61	1.65	1.66	1.62	1.66	1.56
1.35	0.60	−1.77	−1.75	−1.67	−1.65	−1.75	−1.73	−1.74
	0.70	−1.65	−1.62	−1.54	−1.51	−1.63	−1.62	−1.63
	0.80	−1.41	−1.37	−1.30	−1.26	−1.41	−1.39	−1.41
	0.90	−1.02	−0.99	−0.90	−0.84	−1.01	−0.98	−1.02
	0.95	−0.72	−0.72	−0.62	−0.55	−0.72	(−0.67)	(−0.74)
1.15	0.60	−2.29	−2.28	−2.16	−2.15	−2.30	−2.25	−2.26
	0.70	−2.25	−2.23	−2.11	−2.10	−2.26	−2.23	−2.24
	0.80	−2.06	−2.06	−1.95	−1.92	−2.10	−2.08	−2.09
	0.90	−1.79	−1.74	−1.61	−1.56	−1.76	−1.73	−1.77
0.75	0.60	−4.24		−3.99	−3.99	−4.29	−4.17	−4.18
	0.70	−4.53	−4.50	−4.26	−4.26	−4.28	−4.51	−4.51
	0.80	−4.69	−4.63	−4.38	−4.37	−4.74	−4.69	−4.69
	0.90		−4.55	−4.29	−4.26	−4.67	(−4.60)	(−4.62)

* Verlet *et al.*, *Phys. Rev.*, **182**, p. 307, 1969; **159**, p. 98, 1967.
Source: J. A. Barker and D. Henderson, *Ann. Rev. Phys. Chem.*, **23**, p. 439, 1972.

Table 14–4. **Values of** $p/\rho kT$ **for the 6–12 potential**

kT/ε	$\rho\sigma^3$	molecular dynamics*	Monte Carlo†	PY(E)	BH1-(PY)	BH1	BH2	CWA (PY)	CWA
2.74	0.65	2.22		2.23	2.23	2.24	2.22	2.21	2.18
	0.75	3.05		3.11	3.11	3.14	3.10	3.11	3.04
	0.85	4.38		4.42	4.42	4.48	4.44	4.50	4.30
	0.95	6.15		6.31	6.37	6.41	6.40	6.57	6.10
1.35	0.10	0.72		0.72	0.77	0.77	0.74	0.77	0.77
	0.20	0.50		0.51	0.54	0.55	0.52	0.53	0.53
	0.30	0.35		0.36	0.35	0.39	0.36	0.32	0.31
	0.40	0.27		0.29	0.25	0.26	0.26	0.17	0.17
	0.50	0.30		0.33	0.29	0.31	0.27	0.18	0.18
	0.55	0.41		0.43	0.40	0.43	0.35	0.27	0.27
	0.65	0.80		0.85	0.85	0.91	0.74	0.72	0.71
	0.75	1.73		1.72	1.77	1.87	1.64	1.70	1.64
	0.85	3.37		3.24	3.36	3.54	3.36	3.51	3.28
	0.95	6.32		5.65	5.96	6.21	6.32	(6.58)	(5.90)
1.00	0.65	−0.25		−0.22	−0.25	−0.21	−0.36	−0.51	−0.50
	0.75	+0.58	0.48	+0.57	+0.62	+0.71	+0.53	+0.43	+0.40
	0.85	2.27	2.23	2.14	2.30	2.48	2.25	2.41	2.20
	0.90	∼3.50		3.33	3.57	3.79	3.53	(3.96)	(3.55)
0.72	0.85	0.40	0.25	0.33	0.50	0.70	0.25	0.43	0.26
	0.90		∼1.60	1.59	1.90	2.15	1.63	(2.24)	(1.83)

* Verlet *et al.*, *Phys. Rev.*, **182**, p. 307, 1969; **159**, p. 98, 1967.
† McDonald and Singer, *J. Chem. Phys.*, **50**, p. 2308, 1969.
Source: J. A. Barker and D. Henderson, *Ann. Rev. Phys. Chem.*, **23**, p. 439, 1972.

Table 14–5. **Critical constants for the 6–12 potential**

	Barker-Henderson	Verlet*	Exp. (argon)†
T_c^*	1.35	1.32–1.36	1.26
ρ_c^*	0.30	0.32–0.36	0.316
p_c^*	0.14	0.13–0.17	0.117
$p_c V_c / NkT_c$	0.34	0.30–0.36	0.293

* L. Verlet, *Phys. Rev.*, **159**, p. 98, 1967.
† J. S. Rowlinson, *Liquids and Liquid Mixtures* (London: Butterworths, 1959).

The Barker-Henderson theory was the first generally successful perturbation theory and showed that perturbation theory is probably the most appealing approach to the liquid state. Barker and Henderson have extended and summarized their approach in a number of later papers.* Two important extensions are the generation of tables that allow one to use Monte Carlo values of A_2 and the inclusion of three-body forces. These are discussed in the most recent reference given below.

THE CHANDLER-WEEKS-ANDERSEN THEORY

In 1971 a new perturbation theory of liquids was formulated by Chandler, Weeks, and Andersen.† The primary difference between their approach and the Barker-Henderson approach is the way in which the intermolecular potential is divided into an unperturbed and a perturbed part. Chandler, Weeks, and Andersen argue that it is physically significant to separate the potential into a part containing all the repulsive *forces* and a part containing all the attractive *forces*. As they point out, "As far as the motion of a molecule in a liquid is concerned, the molecules do not 'know' the sign of their mutual potential energy, but do 'know' the sign of the derivative of the potential, i.e., the sign of the force." Notice that this separation is distinctly different from separating the potential into positive and negative parts as in the Barker-Henderson theory.

Thus the CWA theory separates the intermolecular potential according to

$$u(r) = u_0(r) + u^{(1)}(r)$$

where for a Lennard-Jones 6–12 potential:

$$
\begin{aligned}
u_0(r) &= u(r) + \varepsilon & r &< 2^{1/6}\sigma \\
&= 0 & r &\geq 2^{1/6}\sigma \\
u^{(1)}(r) &= -\varepsilon & r &< 2^{1/6}\sigma \\
&= u(r) & r &\geq 2^{1/6}\sigma
\end{aligned}
\tag{14–33}
$$

In Eq. (14–33), ε represents the depth of the potential well of $u(r)$, and the term $2^{1/6}\sigma$ is the same as r_{min} for a 6–12 potential. This division is shown in Fig. 14–8.

The separation utilized in the CWA theory has two advantages over other separations. Equation (14–8) shows that the higher-order terms ω_2, ω_3, ... are like central moments about some distribution, and so are smaller the more smoothly $u^{(1)}(r)$ varies. As Fig. 14–8 shows, the perturbing part of the potential chosen in the CWA theory

* J. A. Barker and D. Henderson, *J. Chem. Educ.*, **45**, p. 2, 1968; W. R. Smith, D. Henderson, and J. A. Barker, *J. Chem. Phys.*, **53**, p. 508, 1970; *ibid.*, **55**, p. 4027, 1971; J. A. Barker and D. Henderson, *Accts. Chem. Res.* **4**, p. 303, 1971; *Ann. Rev. Phys. Chem.*, **23**, p. 439, 1972.

† D. Chandler and J. D. Weeks, *Phys. Rev. Lett.*, **25**, p. 149, 1970; J. D. Weeks, D. Chandler, and H. C. Andersen, *J. Chem. Phys.*, **54**, p. 5237, 1971.

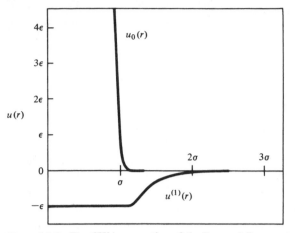

Figure 14–8. **The CWA separation of the Lennard-Jones potential,** $u(r)$, **into a part containing all the repulsive forces,** $u_0(r)$, **and a part containing all the attractive forces,** $u^{(1)}(r)$. (From J. D. Weeks, D. Chandler, and H. C. Andersen, *J. Chem. Phys.*, **54**, p. 5237, 1971.)

does indeed vary slowly for all r. This is to be contrasted with Barker and Henderson's separation of $u(r)$, in which $u_0(r)$ is that part of $u(r)$ greater than zero and $u^{(1)}(r)$ is that part less than zero. Their $u^{(1)}(r)$ varies quite rapidly with r, particularly between $r = \sigma$ and $r = 2^{1/6}\sigma$. Secondly, the reference system chosen by Chandler, Weeks, and Andersen is more realistic than a hard-sphere system. In fact, we shall see below that at high densities the radial distribution function of the Lennard-Jones systems can be very well approximated by the radial distribution function of a system with the potential $u_0(r)$ in Fig. 14–8, i.e., the reference system of CWA.

Using Eq. (14–33) in Eqs. (14–9) and (14–10), then we have

$$A = A_0 + \tfrac{1}{2}\rho^2 V \int g_0(r)u^{(1)}(r)\,dr + \cdots \tag{14–34}$$

where, of course, the zero subscripts indicate the reference system with potential $u_0(r)$ given in Eq. (14–33). For the purpose of comparison with experimental data and other theories, it is convenient to calculate not A itself, but the *excess* free energy with respect to the ideal gas at the same volume, temperature, and density. If we denote this excess free energy by ΔA, we can write Eq. (14–34) in the form

$$\frac{\beta \Delta A}{N} = \frac{\beta \Delta A_0}{N} + \tfrac{1}{2}\beta\rho \int d\mathbf{r}\, g_0(r)u^{(1)}(r) + \cdots \tag{14–35}$$

Chandler, Weeks, and Andersen argue on physical grounds and then show numerically (see below) that for *low* as well as high temperatures this equation will be approximately valid, i.e., errors 10 percent when the density is low, but will become much more accurate as the density increases.

The principal disadvantage of the CWA theory is that the properties of the reference state are not well known as they are for a hard-sphere system. In order to avoid having to perform expensive machine calculations to obtain this information, they present an approximate treatment of the reference system. The essence of their approximation goes back to Fig. 13–8, which shows the function $y(r) = g(r)\exp[\beta u(r)]$ versus r. It was stated there that the shape of $y(r)$ is not too sensitive to the intermolecular potential. Chandler, Weeks, and Andersen approximate $y(r)$ for the reference fluid, i.e.,

$y_0(r)$, by the similar function appropriate to a hard-sphere system of diameter d, $y_d(r)$. For $g_0(r)$, then, we have*

$$g_0(r) \simeq y_d(r)\exp[-\beta u_0(r)] \tag{14-36}$$

They choose the value of d by requiring the thermodynamic properties of the reference system to equal those of a hard-sphere system through the equation [Eq. (13-54)]

$$\frac{1}{\beta}\left(\frac{\partial \rho}{\partial p}\right)_\beta = 1 + \rho \int [g(r) - 1]\, d\mathbf{r}$$

Thus if we equate the compressibility of the reference system to the compressibility of a hard-sphere system, we have

$$\int d\mathbf{r}(y_d\, e^{-\beta u_0} - 1) = \int d\mathbf{r}(y_d\, e^{-\beta u_d} - 1) \tag{14-37}$$

where u_d denotes as hard-core repulsion of diameter d. This equation gives a unique value of d as a function of both temperature and density. Verlet and Weis† have presented a simple algorithm for $d(\beta, \rho)$. Once a value of d is obtained for a particular ρ and β, the free energy of the reference system is taken to be the free energy of a hard-sphere system with diameter d and density ρ. Figure 14-9 shows the behavior

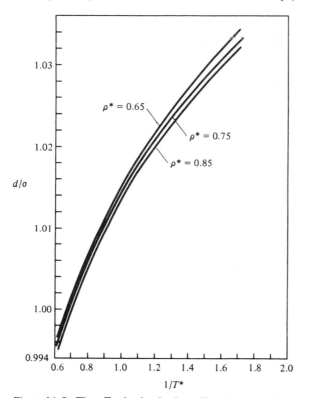

Figure 14-9. **The effective hard-sphere diameter according to the CWA theory [Eq. (14–37)].** (From J. D. Weeks, D. Chandler, and H. C. Andersen, *J. Chem. Phys.*, **54**, p. 5237, 1971.)

---

* Such an approximation had been used previously, e.g., By M. Orentlicher and J. M. Prausnitz, *Can. J. Chem.*, **45**, p. 595, 1967.

† L. Verlet and J. Weis, *Phys. Rev.*, **5A**, p. 939, 1972.

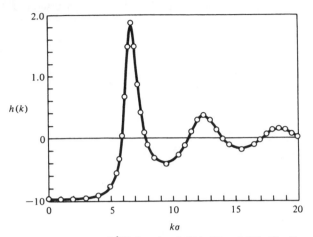

Figure 14–10. **Plot of $\hat{h}(k)$ for $\rho^* = 0.844$, $T^* = 0.723$. The line represents the use of Eq. (14–36); the circles are the molecular dynamics results of Verlet.** (From J. D. Weeks, D. Chandler, and H. C. Andersen, *J Chem. Phys.* **54**,, p. 5237, 1971.)

Table 14–6. **Perturbation theory and Monte Carlo results for the excess free energy on the $T^* = 0.75$, 1.15, and 1.35 isotherms**

			$-\beta \Delta A/N$	
T^*	ρ^*	CWA	Monte Carlo	Barker-Henderson
0.75	0.1	0.55	0.81	0.57
	0.2^a	1.15	1.48	1.16
	0.3^a	1.78	2.11	1.77
	0.4^a	2.42	2.68	2.38
	0.5^a	3.06	3.23	2.96
	0.6^a	3.65	3.74	3.48
	0.7^a	4.14	4.17	3.90
	0.8^a	4.46	4.47	4.16
	0.84	4.51	4.54	4.20
1.15	0.1	0.29	0.39	0.30
	0.2	0.60	0.73	0.61
	0.3	0.92	1.05	0.92
	0.4^a	1.23	1.33	1.20
	0.5^a	1.51	1.59	1.46
	0.55^a	1.63	1.69	1.56
	0.6^a	1.74	1.79	1.65
	0.65	1.82	1.84	1.72
	0.75	1.88	1.88	1.76
	0.85	1.77	1.78	1.63
1.35	0.1	0.22	0.30	0.23
	0.2	0.45	0.56	0.46
	0.3	0.68	0.80	0.69
	0.4	0.90	1.00	0.89
	0.5	1.09	1.16	1.05
	0.6	1.22	1.26	1.16
	0.7	1.26	1.29	1.18
	0.8	1.16	1.19	1.07

^{a} Corresponds to one-phase metastable states in the liquid-gas two-phase region.
Source: J. D. Weeks, D. Chandler, and H. C. Andersen, *J. Chem. Phys.*, **54**, p. 5237, 1971.

of d obtained numerically from Eq. (14–37) versus ρ and β. It is interesting to note that the curves of $d(\rho, \beta)$ obtained here are very similar to the empirical curves for an effective hard-sphere diameter determined empirically by Verlet* in the discussion surrounding Fig. 14–1. Figure 14–10 shows $\hat{h}(k)$ [Eq. (14–1)] obtained from the ap-

Table 14–7. **Approximate results for the excess free energy on the** $\rho^* = 0.88$, 0.85, 0.75, **and** 0.65 **isochores**

| ρ^* | T^* | $-\beta \Delta A/N$ | | |
		CWA	computer results	Barker-Henderson
0.88	1.095	1.94	1.99	1.80
	0.94	2.84	2.88	2.65
	0.591	6.80	6.86	6.30
0.85	2.889	−0.93	−0.86	−0.94
	2.202	−0.44	−0.37	−0.47
	1.214	1.51	1.55	1.39
	1.128	1.86	1.91	1.73
	0.880	3.32	3.35	3.10
	0.782	4.18	4.21	3.90
	0.786	4.14	4.17	3.86
	0.760	4.41	4.44	4.10
	0.719	4.87	4.90	4.53
	0.658	5.66	5.71	5.26
	0.591	6.74	6.81	6.25
0.75	2.849	−0.56	−0.50	−0.58
	1.304	1.36	1.41	1.27
	1.069	2.22	2.26	2.08
	1.071	2.21	2.25	2.07
	0.881	3.27	3.30	3.06
	0.827	3.66	3.69	3.43
0.65	2.557	−0.18	−0.20	−0.11
	1.585	0.79	0.85	0.73
	1.036	2.24	2.30	2.12
	0.900	2.90	2.96	2.74

Source: D. Chandler, and H. C. Andersen, *J. Chem. Phys.*, **54**, p. 5237, 1971.

Table 14–8. **Approximate and molecular dynamics results for the pressure and internal energy as obtained from Eqs. (14–38) and (14–39), respectively***

| ρ^* | T^* | $\beta p/\rho$ | | | $-\beta \Delta E/N$ | | |
		CWA	MD	PY	CWA	MD	PY
0.85	1.128	2.82	2.78	3.57	5.08	5.05	4.98
	0.88	1.82	1.64	3.17	6.77	6.75	6.61
	0.786	1.23	0.99	2.97	7.70	7.70	7.51
	0.719	0.69	0.36	2.82	8.52	8.51	8.28

* Columns 3 and 6 give the values obtained by applying the approximation $g(r) \simeq g_0(r)$. Columns 4 and 7 give the molecular dynamics results. The results obtained by using the solution of the Percus-Yevick equation for $g(r)$ are given in columns 5 and 8.
Source: J. D. Weeks, D. Chandler, and H. C. Andersen, *J. Chem. Phys.*, **54**, p. 5237, 1971.

proximate radial distribution function in Eq. (14–36) for $\rho^* = \rho \sigma^3 = 0.844$ and $T^* = 0.723$. It can be seen from this just how much the repulsive part of the potential determines the structure of a liquid at high densities. A similar result is found for the radial distribution function itself.

* L. Verlet, *Phys. Rev.*, **165**, p. 201, 1968.

Tables 14–6 through 14–8 show some thermodynamic results calculated from the CWA theory. Table 14–6 shows the excess free energy calculated according to Eq. (14–35). One sees that the excess free energy does converge at high densities ($\rho^* \geq 0.65$) to the values predicted by computer calculations. Table 14–7 shows some more detailed results for $\beta \Delta A / N$. The CWA theory does seem to improve as the density increases. Table 14–8 gives some results for the pressure and the excess internal energy. Although these quantities could be calculated from A, they were actually calculated from the equations

$$\beta p = \rho - \tfrac{1}{6}\beta\rho^2 \int ru'(r)g(r)\, d\mathbf{r} \tag{14-38}$$

and

$$\frac{\Delta E}{N} = \tfrac{1}{2}\rho \int u(r)g(r)\, d\mathbf{r} \tag{14-39}$$

with $g(r)$ replaced by simply $g_0(r)$. It can be seen that the results are far superior to a solution of the Percus-Yevick equation for a 6–12 potential. The poorer results for the pressure found at the lowest temperatures in Table 14–8 are attributed to approximating the reference system by an effective hard-sphere system rather than in any inadequacy of the CWA approach.

Tables 14–3 and 14–4 include values of A/NkT and $p/\rho kT$ calculated from the CWA theory. The column headed CWA(PY) indicates that the Percus-Yevick $g_0(r)$ was used, and the column headed by CWA indicates that Monte Carlo values of $g_0(r)$ were used. The numbers in parentheses were calculated using a hard-sphere system which was so dense that these results are of uncertain accuracy. As one would expect from a satisfactory theory, the results calculated from the more accurate $g_0(r)$ are in better agreement with the molecular dynamics and Monte Carlo results. The agreement is seen to be excellent.

There is at present some discussion* concerning the relative merits of the BH and CWA theories, but it can be said that the two theories are complementary rather than competitive. They are based on differing compromises, and each has its merits and drawbacks.† The Barker-Henderson theory uses a well-known reference state, but must include terms second order in β. The Chandler-Weeks-Andersen theory, on the other hand, is rapidly enough convergent that it does not require second-order terms, but the reference state is not universal, as in the case with perturbation theories based on hard spheres. Regardless, both theories are very successful and show that the perturbation theory approach to the liquid state is not only physically appealing but numerically very satisfactory.

ADDITIONAL READING

BARKER, J. A., and HENDERSON, D., *J. Chem. Educ.*, **45**, p. 2, 1968.
———, *Accts. Chem. Res.*, **4**, p. 303, 1971.
———, *Ann. Rev. Phys. Chem.*, **23**, p. 439, 1972.
HENDERSON, D., and BARKER, J. A. 1971. In *Physical chemistry, an advanced treatise*, Vol. 8A, ed. by H. Eyring, D. Henderson, and W. Jost. New York: Academic.
———, and LEONARD, P. J. 1971. In *Physical chemistry, an advanced treatise*, Vol. 8B, ed. by H. Eyring, D. Henderson, and W. Jost. New York: Academic.

* See e.g., J. A. Barker and D. Henderson, *Phys. Rev.*, **A4**, p. 806, 1971.

† D. Henderson, *Mol. Phys.*, **21**, p. 841, 1971; see also J. A. Barker and D. Henderson, *Ann. Rev. Phys. Chem.*, **23**, p. 477, 1972.

MANSOORI, G. A., and CANFIELD, F. B. *Ind. Eng. Chem.*, **62**, p. 12, 1970.
NEECE, G. A., and WIDOM, B. *Ann. Rev. Phys. Chem.*, **20**, p. 167, 1969.
RIGBY, M. *Quart. Rev.*, **24**, p. 416, 1970.
SMITH, W. R. *Specialist Periodical Reports of the Chemical Society, Statistical Mechanics, Vol. 1.* 1973. London: The Chemical Society.
VERA, J. H., and PRAUSNITZ, J. M. *Chem. Eng. J.*, **3**, p. 1, 1972.
WATTS, R. O. *Specialist Periodical Reports of the Chemical Society, Statistical Mechanics, Vol. 1.* 1973. London: The Chemical Society.

PROBLEMS

14–1. By using the van der Waals equation and the fact that $(\partial p/\partial V)_T$ and $(\partial^2 p/\partial V^2)_T$ equal zero at the critical point, show that

$$T_c = \frac{8a}{27bR}, \qquad V_c = 3b, \qquad p_c = \frac{a}{27b^2}$$

and

$$\frac{p_c V_c}{T_c} = \frac{3R}{8}$$

Show also that the van der Waals equation can be written in reduced form as

$$\left(p^* + \frac{3}{V^{*2}}\right)\left(V^* - \frac{1}{3}\right) = \frac{8}{3} T^*$$

where $p^* = 27b^2 p/a$, $V^* = V/3b$, and $T^* = 27RbT/8a$.

14–2. Determine the virial coefficients associated with the van der Waals equation. How do they compare to the exact ones?

14–3. The critical constants of nitrogen are $p_c = 33.6$ atm, $V_c = 0.090$ liters/mole, and $T_c = 126°$K. Calculate the van der Waals constants and determine the molar volume of nitrogen at 300°K and 1 atm; 10 atm; 100 atm. Compare these values to those calculated from the ideal gas equation of state.

14–4. Show that the van der Waals equation is a special case of the more general equation

$$p = \frac{NRT}{V} \beta(\rho) - \alpha(\rho)$$

where α and β are functions of the density ρ, but independent of temperature. Rigby* calls this a generalized van der Waals equation. Show that two thermodynamic consequences of this equation are that $(\partial p/\partial T)_V$ is a function only of the molar volume and that the entropy is determined solely by the $\beta(\rho)$ term.

14–5. Using Eqs. (14–14) and (14–17), calculate the van der Waals constants a and b for nitrogen. For $u(r)$, assume a Lennard-Jones 6–12 potential with ε and σ given in Table 12–3. Compare these calculated values to the experimentally determined values, $a = 1.39 \times 10^6$ cm$^6 \cdot$ atm/mole^2 and $b = 39.1$ cm^3/mole. Such poor agreement is quite typical, simply indicating the inadequacy of the van der Waals equation.

14–6. Show how terms in $(1/T)^2$ involve $g^{(3)}$ and $g^{(4)}$ as well as $g^{(2)}$ when part of the intermolecular potential is treated as a perturbation. Write out this term.

14–7. Show that the expression for the Helmholtz free energy of a square-well fluid to second order in the Barker-Henderson perturbation theory is

$$\frac{A - A_0}{NkT} = -2\pi\rho\left(\frac{\varepsilon}{kT}\right) \int_\sigma^{\lambda\sigma} r^2 g_0(r)\, dr - \pi\rho\left(\frac{\varepsilon}{kT}\right)^2 \frac{(1-\eta)^4}{1 + 4\eta + 4\eta^2} \frac{\partial}{\partial\rho}\left[\rho \int_\sigma^{\lambda\sigma} r^2 g_0(r)\, dr\right]$$

* *Quart. Rev.*, **24**, p. 416, 1970.

where $\eta = \pi \rho \sigma^3/6$ and the Percus-Yevick compressibility equation of state has been used for $(\partial \rho/\partial p)_0$.*

14–8. Evaluate the square-well second and third virial coefficients from the Barker-Henderson theory and compare them to the exact expression as a function of $T^* = kT/\varepsilon$.

14–9. Using the Grundke-Henderson program for the hard-sphere radial distribution function given in Appendix D, evaluate the integrals in Problem 14–7 and compare your final results for A to those of Barker and Henderson in their original paper.†

14–10. Show that Eq. (14–28) reduces to the hard-sphere case when $\alpha = \gamma = 1$.

14–11. In the second paper in the Barker-Henderson series,‡ the double Taylor expansion of the Helmholtz free energy, Eq. (14–30), is derived. Following the first few pages of their paper, derive Eq. (14–30).

14–12. Show that the Barker-Henderson effective hard-sphere diameter is always less than or equal to σ.

14–13. Evaluate the Barker-Henderson effective hard-sphere diameter $d(T)$ for a hard-sphere, square-well, and r^{-12} potential.

14–14. Evaluate the Barker-Henderson effective hard-sphere diameter $d(T)$ for the Lennard-Jones 6–12 potential and compare your result to Fig. 1 of Barker and Henderson's second paper.§

14–15. Show that if three-body forces are included in the intermolecular potential, then the Barker-Henderson theory gives

$$A = A_0 + 2\pi N\rho \int_\sigma^\infty g_0(r)u(r)r^2 \, dr$$

$$- \pi N\beta\rho \frac{(1-\eta)^4}{1+4\eta+4\eta^2} \frac{\partial}{\partial\rho} \left[\rho \int_\sigma^\infty g_0(r)u(r)r^2 \, dr \right]$$

$$+ \frac{N\rho^2}{6} \iint g_0(123)u(123) \, d\mathbf{r}_2 \, d\mathbf{r}_3$$

where the Percus-Yevick compressibility equation has been used for $(\partial \rho/\partial p)_0$.‖

14–16. Using the Grundke-Henderson program for the radial distribution function given in Appendix D, evaluate Eq. (14–32) for the Lennard-Jones 6–12 potential and compare the results to those given in Table 14–3.

14–17. The first step in the Chandler-Weeks-Andersen perturbation theory is the calculation of the effective hard-sphere diameter $d(\rho, T)$. This is defined through Eq. (14–37), but Verlet and Weis¶ have presented a simple algorithm from this defining equation. Define a quantity d_R^* by

$$d_R^*(T) = \int_0^\infty [1 - e^{-\beta v_0(x)}] \, dx \tag{14–40}$$

where $x = r/\sigma$ and $v_0(x)$ is the *CWA reference potential*. Verlet and Weis show that this is well represented by the empirical expression:

$$d_R^* = \frac{0.3837 + 1.068\beta^*}{0.4293 + \beta^*} \tag{14–41}$$

---

* See J. A. Barker and D. Henderson, *J. Chem. Phys.*, **47**, p. 2856, 1967.

† *J. Chem. Phys.*, **47**, p. 2856, 1967.

‡ *J. Chem. Phys.*, **47**, p. 4714, 1967.

§ *J. Chem. Phys.*, **47**, p. 4714, 1967.

‖ See J. A. Barker, D. Henderson, and W. R. Smith, *Phys. Rev. Lett.*, **21**, p. 134, 1968, for a discussion of the effect of the inclusion of three-body forces into perturbation theory.

¶ *Phys. Rev.*, **5A**, p. 939, 1972.

where $\beta^* = 1/T^* = \varepsilon/kT$. For $0.7 < T^* < 1.6$, the error here is less than 2×10^{-4}, and for $1.6 < T^* < 4.5$, the error may reach 8×10^{-4}. The error also decreases rapidly with density. In addition, Verlet and Weis show that the density dependence of the CWA value of d^*, to first order, is given by

$$d^* = d_R^*(1 + A\delta) \tag{14--42}$$

where

$$\delta = \frac{1}{210.21 + 404.6\beta^*} \tag{14--43}$$

is a function of temperature only, and

$$A = \frac{1 - 4.25\eta_w + 1.362\eta_w^2 - 0.8751\eta_w^3}{(1 - \eta_w)^2} \tag{14--44}$$

with

$$\eta_w = \eta - \tfrac{1}{16}\eta^2 \tag{14--45}$$

where $\eta = (\pi/6)\rho d^3$ is the packing fraction. This procedure gives the CWA values of d^* to within 1 percent. Before Eq. (14--45) can be used, however, d must be known. An iterative procedure will, however, suffice. As a first choice, set $d^* = d_R^*$, then evaluate the quantities appearing in Eqs. (14--43), (14--44), and (14--45). Thus d^* can be found from (14--42) and will yield the second choice for d^*. Three iterations will yield ample convergence. Use this Verlet-Weis algorithm to reproduce the curves in Fig. 14--9.

14--18. Using the Verlet-Weis algorithm represented in Problem 14--17 and the Grundke-Henderson program for the hard-sphere radial distribution function given in Appendix D, calculate the CWA reference system radial distribution function according to Eq. (14--36), namely,

$$g_0(r) \approx y_d(r)\exp[-\beta u_0(r)]$$

and compare this to the molecular dynamics results of Hansen and Weis,[*] and to the molecular dynamics calculations of Verlet.[†]

14--19. The CWA expression for the excess free energy is given by Eq. (14--35):

$$\frac{\beta \Delta A}{N} = \frac{\beta \Delta A_0}{N} + \tfrac{1}{2}\beta\rho \int d\mathbf{r}\, g_0(r)u(r)$$

Using the results from Problems 14--17 and 14--18, reproduce Tables I and II from the paper of Weeks, Chandler, and Andersen.[‡]

14--20. Using Eqs. (14--38) and (14--39) with $g(r)$ replaced by

$$g_0(r) \approx y_d(r)\exp[-\beta u_0(r)]$$

calculate the pressure and excess internal energy and reproduce Table III in the paper of Weeks, Chandler, and Andersen.[§]

14--21. According to the approximation discussed in Problem 14--18 show that the structure factor of the reference fluid may be written as

$$h_0(k) = h_d(k) + \rho \int d\mathbf{r}\, y_d(r)[\exp(-\beta u_0) - \exp(-\beta u_d)]\exp(-i\mathbf{k}\cdot\mathbf{r})$$

Using this, reproduce Figs. 3 and 4 in the paper of Weeks, Chandler, and Andersen.[¶]

[*] *Mol. Phys.*, **23**, p. 853, 1972.

[†] *Phys. Rev.*, **165**, p. 201, 1968.

[‡] *J. Chem. Phys.*, **54**, p. 5237, 1971. See also *J. Chem. Phys.*, **55**, p. 5422, 1971

[§] *J. Chem. Phys.*, **54**, p. 5237, 1971. See also *J. Chem. Phys.*, **55**, p. 5422, 1971.

[¶] *J. Chem. Phys.*, **54**, p. 5237, 1971.

14–22. Can the Barker-Henderson theory or the Chandler-Weeks-Andersen theory be applied to fused salts? Why or why not? Does the requirement of electroneutrality help?

14–23. The Barker-Henderson theory has been applied to mixtures of real fluids.* Assume that the pair potential is of the form

$$u_{\lambda\mu}(r) = 4\varepsilon_{\lambda\mu}\left\{\left(\frac{\sigma_{\lambda\mu}}{r}\right)^{12} - \left(\frac{\sigma_{\lambda\mu}}{r}\right)^{6}\right\}$$

along with the combining rules

$$\varepsilon_{12} = \xi_{12}(\varepsilon_{11}\varepsilon_{22})^{1/2}$$

$$\sigma_{12} = \tfrac{1}{2}(\sigma_{11} + \sigma_{22})$$

Show that to first order

$$\frac{A - A_0}{NkT} = -4\pi\rho x_1 x_2 d_{12}{}^2 g_0{}^{12}(d_{12})[d_{12} - \delta_{12}] + 2\pi\beta\rho \sum_{\lambda,\mu} x_\lambda x_\mu \int_{\sigma_{\mu\lambda}}^{\infty} u_{\lambda\mu}(r)g_0{}^{\lambda\mu}(r)r^2 \, dr$$

where A_0 and $g_0{}^{\lambda\mu}(r)$ are the Helmholtz free energy and radial distribution functions of a hard-sphere mixture with

$$d_{\lambda\mu} = \tfrac{1}{2}(d_{\lambda\lambda} + d_{\mu\mu})$$

$$d_{\lambda\lambda} = \delta_{\lambda\lambda}$$

and

$$\delta_{\lambda\mu} = \int_0^{\sigma_{\lambda\mu}} [1 - e^{-\beta u_{\lambda\mu}(z)}] \, dz$$

Grundke *et al.* show that very good agreement with various experimental data can be obtained with this equation.

14–24. Neff and McQuarrie† have applied perturbation theory to the calculation of the Henry's law constant. Recall that this is defined by

$$K_H = \lim_{x_2 \to 0} \left(\frac{p_2}{x_2}\right)$$

where p_2 is the partial pressure of the solute above the solution and x_2 is the mole fraction of solute in solution.

From K_H, the molar heat of solution ΔH_s and the partial molar volume solute $\overline{V}_2$ can be obtained from

$$\frac{\Delta H_s}{RT} = \left(\frac{1}{T}\right)\left[\frac{\partial \ln K_H}{\partial T^{-1}}\right]_p$$

$$\overline{V}_2 = RT\left(\frac{\partial \ln K_H}{\partial p}\right)_T$$

Consider a liquid solution of some solute (denoted by a subscript 2) in some solvent (subscript 1) which is in equilibrium with gaseous solute at partial pressure p_2.

$$\mu_2{}^{\text{gas}} = \mu_2{}^{\text{soln}}$$

If we assume the gas phase to be ideal, $\mu_2{}^{\text{gas}} = \mu_2^*(T) + kT \ln(p_2/kT)$. The chemical potential of solute in the solution can be obtained from

$$\mu_2{}^{\text{soln}} = \left(\frac{\partial A}{\partial N_2}\right)_{V,T,N_1}$$

---

* Cf. e.g., E. W. Grundke, D. Henderson, J. A. Barker, and P. J. Leonard, *Mol. Phys.*, **25**, p. 883, 1973.
† *J. Phys. Chem.*, **77**, p. 413, 1973.

From Problem 14–23, we have that

$$\beta A = \beta A_0 - 4\pi \rho_1 N_2 \, d_{12}{}^2 g_0(d_{12})[d_{12} - \delta_{12}] + 2\pi\beta \sum_{i,j} \rho_i N_j \int_{\sigma_{ij}}^{\infty} dr \, r^2 u_{ij}(r) g_0{}^{ij}(r)$$

where all the symbols are defined in Problem 14–23.

Following Neff and McQuarrie, show that $\ln K_H$ can be written in the form

$$\ln K_H = \ln \frac{RT}{\bar{V}_1} + \left(\frac{\mu_2{}^{HS}}{kT}\right) + \left(\frac{\mu_2{}^{corr}}{kT}\right)$$

where

$$\frac{\mu_2{}^{HS}}{kT} = -\ln(1-y) + y\chi(y)R^3 + \left(\frac{1}{2}\right)\left[\frac{3y}{(1-y)}\right]^2 R^2 + \left[\frac{3y}{(1-y)}\right](R^2 + R) \cdot$$

$$y = \frac{\pi \rho_1 d_{11}{}^3}{6}$$

$$R = \frac{d_{22}}{d_{11}}$$

$$\chi(y) = \frac{(1 + y + y^2)}{(1-y)^3}$$

$$\frac{\mu_2{}^{corr}}{kT} = \left(\frac{-2\rho_1 a_{12}}{kT}\right) + \left(\frac{2\rho_1 I_{11}}{kT}\right) - 4\pi\rho_1 \, d_{12}{}^2(d_{12} - \delta_{12})g_0{}^{12}(d_{12})$$

$$a_{12} = -2\pi \int_{\sigma_{12}}^{\infty} dr \, r^2 u_{12}(r) g_0{}^{12}(r)$$

$$I_{11} = \pi N_1 \int_{\sigma_{11}}^{\infty} dr \, r^2 u_{11}(r) \left[\frac{\partial g_0{}^{11}(r)}{\partial N_2}\right]_{V,T,N_1}$$

Neff and McQuarrie compare this equation and its derivatives to experimental data.

CHAPTER 15

SOLUTIONS OF STRONG ELECTROLYTES

Up to this point we have not discussed solutions or mixtures to any extent. Several of the problems in Chapters 12 and 13 discuss the extension of certain one-component equations to mixtures, but we have not treated mixtures in any detailed way. It so happens that many of the concepts and techniques introduced in Chapters 12 and 13 can be applied to mixtures of nonelectrolytes, but we shall not do this here since not many really new ideas are needed. For example, the perturbation theories of Chapter 14 have been applied quite successfully to mixtures, and this subject is reviewed by Henderson and Leonard.* Their article should be quite easy to read since it is based upon the material discussed in Chapters 13 and 14. Other specialized treatments of mixtures are *Molecular Theory of Solutions*† by Prigogine, "Liquids and Liquid Mixtures"‡ by Rowlinson, and *Regular and Related Solutions* by Hildebrand, Prausnitz, and Scott.§

In this chapter we shall consider solutions of strong electrolytes. We do this in some detail not only because ionic solutions are of such great importance in physical chemistry and biophysical systems, but, as we shall see shortly, the coulombic potential between charged particles requires special techniques. To see why this is so, consider a dilute aqueous solution of some strong electrolyte, say NaCl. For simplicity, let us represent the ions as hard spheres of diameters σ_+ and σ_- with the charge of the ion located at the center of the sphere. Furthermore, assume that the solvent, water in this case, can be represented by a continuous medium of uniform dielectric constant ε. We also assume that the ions are each made of a material with the same dielectric constant as the solvent. This is a commonly used model for solutions of strong electrolytes and is referred to as the *primitive model* of electrolyte solutions. This primitive

---

* D. Henderson and P. J. Leonard, in *Physical Chemistry, An Advanced Treatise*, Vol. 8B (New York: Academic, 1971).

† I. Prigogine, *Molecular Theory of Solutions* (Amsterdam: North-Holland Publ., 1957).

‡ J. S. Rowlinson, *Liquids and Liquid Mixtures*, 2nd ed. (New York: Plenum, 1969).

§ J. H. Hildebrand, J. M. Prausnitz, and R. L. Scott, *Regular and Related Solutions* (New York: Van Nostrand, 1970).

model, then, suggests that we attempt to treat the solution as a gaseous mixture of positively and negatively charged hard spheres contained in some volume V and at a temperature T. The first deviation from ideality lies in the second virial coefficient, which, for a two-component system, is given by (Problem 12–17)

$$\frac{p}{kT} = \rho_1 + \rho_2 + B_{11}(T)\rho_1^2 + B_{12}(T)\rho_1\rho_2 + B_{22}(T)\rho_2^2 + \cdots \tag{15–1}$$

where in the thermodynamic limit (N and $V \to \infty$ with N/V fixed)

$$B_{jj}(T) = -2\pi \int_0^\infty [e^{-\beta u_{jj}(r)} - 1]r^2 \, dr \tag{15–2}$$

and

$$B_{12}(T) = -4\pi \int_0^\infty [e^{-\beta u_{12}(r)} - 1]r^2 \, dr \tag{15–3}$$

For the primitive model of an ionic solution, $u_{12}(r)$, is

$$u_{12}(r) = \infty \qquad r < \frac{(\sigma_+ + \sigma_-)}{2}$$

$$= \frac{q_+ q_-}{\varepsilon r} \qquad r > \frac{(\sigma_+ + \sigma_-)}{2} \tag{15–4}$$

where q_+ and q_- are the charges on the positive and negative ion, respectively. If we substitute this into Eq. (15–3), we find that the integral diverges. This is most readily seen by expanding the exponential and then integrating. Physically this divergence is due to the fact that the coulomb potential is so long ranged that the two-body interactions implied by a second virial coefficient do not exist. Thus it is not possible to decompose the many-body problem into a series of two-body, three-body problems, etc., and a straightforward virial expansion is not applicable. The long-range nature of the coulomb potential requires special consideration, and this is the topic of the present chapter.

Before going on to discuss solutions of strong electrolytes, however, we shall briefly mention a virial-type expansion of solutions of nonelectrolytes developed by McMillan and Mayer.* The essential results of the McMillan-Mayer theory show that there is a rigorous one-to-one correspondence between the equations of imperfect gas theory and dilute solutions of nonelectrolytes. They showed that the pressure of the gas maps into the osmotic pressure of the solution and obeys a virial expansion of the form

$$\frac{\Pi}{kT} = c_2 + \sum_{j=2}^\infty B_j^*(\mu_1, T)c_2^n \tag{15–5}$$

where c_2 is the concentration of solute and μ_1 is the chemical potential of the solvent. The virial coefficients in this case are formally identical to those of imperfect gas theory, but instead of the potential U_N, we use the potential of mean force of N solute molecules in the pure solvent. The McMillan-Mayer theory can be extended to the development of distribution functions as well, and although in this chapter we shall represent the ionic solutions by an idealized model, we should keep in mind that its statistical mechanical basis lies ultimately in the McMillan-Mayer theory.

* For a simplified version of the McMillan-Mayer theory, see T. L. Hill, *J. Chem. Phys.*, **30**, p. 93, 1959.

In the remainder of this chapter we shall treat only the primitive model, and often the so-called *restricted* primitive model, in which the ions all have the same diameter. This means that we shall not be able to discuss pressure and temperature effects since this would require that we treat the solvent not as a continuum but to recognize explicitly the molecular nature of the solvent. In particular, it is the short-range potential of mean force that would have to be evaluated, and in the case of water as a solvent, this would involve a detailed study of orientation and interaction of water molecules around two ions. Such a calculation is at present still very difficult and represents a substantially undeveloped area.* The gross oversimplification of replacing the complicated short-range interaction (potential of mean force) by a hard-sphere potential is not too crucial for fairly dilute solutions, however, since the ions on the average will be far apart.

In Section 15–1 we discuss the famous Debye-Hückel theory, which we shall see is the exact *limiting* theory of ionic solutions, i.e., exact in the limit of small concentrations.

In Section 15–2 we shall discuss the connection between the Debye-Hückel theory and statistical mechanics. First, we shall show in this section that the Debye-Hückel theory is an exact limiting law, and then discuss the application of the integral equation theories of Chapter 13 to ionic solutions. Lastly, we shall introduce briefly some recent theories of ionic solutions that have proved to be very successful.

15–1   THE DEBYE-HÜCKEL THEORY

Consider a system of positive and negative point charges in a continuum medium of dielectric constant ε. If we have a set of point charges with charges $\{q_j{}^+\}$ and $\{q_k{}^-\}$ located at the points $\{\mathbf{r}_j\}$ and $\{\mathbf{r}_k\}$, then the potential at the arbitrary point $\mathbf{r}$ is

$$\phi(\mathbf{r}) = \sum_j \frac{q_j{}^+}{\varepsilon|\mathbf{r} - \mathbf{r}_j|} + \sum_k \frac{q_k{}^-}{\varepsilon|\mathbf{r} - \mathbf{r}_k|} \tag{15–6}$$

If instead of point charges we have some continuous charge density $\rho^+(\mathbf{r})$ and $\rho^-(\mathbf{r})$, then

$$\phi(\mathbf{r}) = \int \frac{\rho^+(\mathbf{r}') \, d\mathbf{r}'}{\varepsilon|\mathbf{r} - \mathbf{r}'|} + \int \frac{\rho^-(\mathbf{r}') \, d\mathbf{r}'}{\varepsilon|\mathbf{r} - \mathbf{r}'|}$$

$$= \int \frac{\rho(\mathbf{r}') \, d\mathbf{r}'}{\varepsilon|\mathbf{r} - \mathbf{r}'|} \tag{15–7}$$

where $\rho(\mathbf{r}) = \rho^+(\mathbf{r}) + \rho^-(\mathbf{r})$ is simply the total charge density at $\mathbf{r}$ (more properly, $\rho(\mathbf{r}) \, d\mathbf{r}$ is the total charge density in $d\mathbf{r}$ at $\mathbf{r}$).

Another fundamental equation of electrostatics that relates the potential to the charge density is Poisson's equation,

$$\nabla^2 \phi = -\frac{4\pi\rho}{\varepsilon} \tag{15–8}$$

This equation can be derived in a number of ways, the most straightforward being to take the Laplacian of both sides of Eq. (15–7). (See Problems 15–2 and 15–3.) Equations (15–7) and (15–8) are equivalent, and Eq. (15–7) can be considered to be the

---

* See, for example, *Chemical Physics of Ionic Solutions*, ed. by B. E. Conway and R. G. Barradas (New York: Wiley, 1966).

general solution to Poisson's equation. Either equation can be used in developing the Debye-Hückel theory, but it is customary to use the differential form, namely, Eq. (15–8).

One other fundamental law of electrostatics is Gauss' law, which states that

$$\int \mathbf{E} \cdot d\mathbf{S} = \frac{4\pi q}{\varepsilon} \tag{15–9}$$

where the integral is taken over some arbitrary closed surface S and where q is the total charge enclosed by the surface. The integrand $\mathbf{E} \cdot d\mathbf{S}$ is the magnitude of the component of the electric field perpendicular to the small surface area dS times the magnitude of dS itself. We shall use Gauss' law only for the simple special case of spherical symmetry. Suppose that there is a charge q located at some point. If we draw a sphere of radius r around this point, the electric field due to this charge will be automatically perpendicular to S and also constant at all points on S. Thus Eq. (15–9) becomes

$$E4\pi r^2 = \frac{4\pi q}{\varepsilon}$$

which gives Coulomb's law, namely, that

$$E = \frac{q}{\varepsilon r^2} \tag{15–10}$$

With a little bit of vector analysis, Poisson's equation can be derived from Gauss' law; so the two equations are really two formulations of the same physical law (Problem 15–3). Gauss' law will be used to derive Eq. (15–37).

With these preliminary electrostatic considerations out of the way, we go on to consider the electrostatic Helmholtz free energy of the primitive model of a system of N cations and N anions in a volume V and temperature T. By electrostatic Helmholtz free energy we mean $A - A_0$, where A is the Helmholtz free energy of the charged system and A_0 is the free energy of the same system but with all the charges set equal to zero. If U_N is the total potential energy of the "charged-up" system and $U_N{}^0$ is that of the uncharged system of rigid spheres, then

$$e^{-\beta(A-A_0)} = \frac{Z_N}{Z_N{}^0} = \frac{\int \cdots \int e^{-\beta U_N} d\mathbf{r}_1 \cdots d\mathbf{r}_{2N}}{\int \cdots \int e^{-\beta U_N{}^0} d\mathbf{r}_1 \cdots d\mathbf{r}_{2N}} \tag{15–11}$$

where

$$U_N = \frac{1}{2} \sum_{i,j}' \frac{q_i q_j}{\varepsilon |\mathbf{r}_i - \mathbf{r}_j|} + \tfrac{1}{2} \sum_{i,j}' u^{(s)}(r_{ij}) \tag{15–12}$$

As usual, the prime on the summations indicates that the $i = j$ term is not included in the summation, and q_k represents the charge on the kth ion. The quantity $u_{ij}{}^{(s)}$ is the short-range interaction between ions i and j. We shall omit from consideration the self-energy of the ions since this self-energy is unaffected by changes in concentration, etc. In the primitive model, $u^{(s)}(r_{ij})$ is taken to be that of a rigid sphere.

The quantity

$$\psi_j(\mathbf{r}_j) = \sum_{i \neq j} \frac{q_i}{\varepsilon |\mathbf{r}_i - \mathbf{r}_j|} \tag{15–13}$$

is the electrostatic potential acting upon the jth ion whose center is located at the point $\mathbf{r}_j$. This potential is due to all the other ions in the solution. The total electrostatic potential of the entire system, then, can be written in the form

$$U_{N,\text{elec}} = \tfrac{1}{2} \sum_j q_j \psi_j(\mathbf{r}_j) \tag{15-14}$$

We now define the canonical ensemble average of $\psi_j(\mathbf{r}_j)$ by

$$\langle \psi_j \rangle = \frac{\int \cdots \int \psi_j(\mathbf{r}_j) e^{-\beta U_N} \, d\mathbf{r}_1 \cdots d\mathbf{r}_{2N}}{\int \cdots \int e^{-\beta U_N} \, d\mathbf{r}_1 \cdots d\mathbf{r}_{2N}} \tag{15-15}$$

Given this definition, then clearly Eq. (15–11) shows that (Problem 15–8)

$$\left(\frac{\partial A}{\partial q_j}\right)_{V,T} = \langle \psi_j \rangle \qquad j = 1, 2, \ldots, 2N \tag{15-16}$$

The total differential of A at constant V and T is

$$dA = \sum_j \left(\frac{\partial A}{\partial q_j}\right) dq_j = \sum_j \langle \psi_j \rangle \, dq_j \tag{15-17}$$

Thus we see that if we can calculate $\langle \psi_j \rangle$, we can calculate the Helmholtz free energy of the system. We can vary each charge in the system simultaneously from 0 to its full value q_j by writing $dq_j = q_j \, d\lambda$ for each j where $0 \leq \lambda \leq 1$. Then Eq. (15–17) for the electrostatic free energy can be written as

$$A - A_0 = \sum_j q_j \int_0^1 \langle \psi_j(\lambda) \rangle \, d\lambda \tag{15-18}$$

where $\langle \psi_j(\lambda) \rangle$ is the average electrostatic potential acting upon the jth ion when each of the ions of the system has the charge λq_j. The Debye-Hückel theory involves an elegant method for calculating $\langle \psi_j(\lambda) \rangle$ and hence the electrostatic Helmholtz free energy, or essentially the canonical ensemble partition function.

To proceed with the Debye-Hückel theory, first consider the potential at some point $\mathbf{r}$

$$\psi(\mathbf{r}) = \sum_i \frac{q_i}{\varepsilon |\mathbf{r} - \mathbf{r}_i|} \tag{15-19}$$

where the summation runs over all the ions in the system. We now canonically average $\psi(\mathbf{r})$, keeping any one particle, say 1, fixed at $\mathbf{r}_1$ to give

$$^1\langle \psi(\mathbf{r}, \mathbf{r}_1) \rangle = \frac{\int \cdots \int \psi(\mathbf{r}) e^{-\beta U_N} \, d\mathbf{r}_2 \cdots d\mathbf{r}_{2N}}{\int \cdots \int e^{-\beta U_N} \, d\mathbf{r}_2 \cdots d\mathbf{r}_{2N}} \tag{15-20}$$

Remember that the left superscript here means that particle 1 is fixed at $\mathbf{r}_1$. It is easy to show that $^1\langle \psi(\mathbf{r}, \mathbf{r}_1) \rangle$ satisfies Poisson's equation by taking the Laplacian of both sides of Eq. (15–20) with respect to the variable $\mathbf{r}$:

$$\nabla^2(^1\langle \psi(\mathbf{r}, \mathbf{r}_1) \rangle) = \frac{\int \cdots \int \nabla^2 \psi(\mathbf{r}) e^{-\beta U_N} \, d\mathbf{r}_2 \cdots d\mathbf{r}_{2N}}{\int \cdots \int e^{-\beta U_N} \, d\mathbf{r}_2 \cdots d\mathbf{r}_{2N}}$$

$$= \frac{\int \cdots \int -\dfrac{4\pi \rho(\mathbf{r})}{\varepsilon} e^{-\beta U_N} \, d\mathbf{r}_2 \cdots d\mathbf{r}_{2N}}{\int \cdots \int e^{-\beta U_N} \, d\mathbf{r}_2 \cdots d\mathbf{r}_{2N}}$$

$$= -\frac{4\pi}{\varepsilon} {}^1\langle \rho(\mathbf{r}, \mathbf{r}_1) \rangle \tag{15-21}$$

We can relate $^1\langle\psi(\mathbf{r}, \mathbf{r}_1)\rangle$ to the quantity $\langle\psi_j\rangle$ that appears in the expression for the electrostatic Helmholtz free energy [Eq. (15–18)]. The quantity $^1\langle\psi(\mathbf{r}, \mathbf{r}_1)\rangle$ is the electrostatic potential energy at the point $\mathbf{r}$ in a system in which particle 1 is fixed at $\mathbf{r}_1$, but the other ions are canonically averaged over the volume V. The quantity

$$^1\langle\psi_1(\mathbf{r})\rangle \equiv {}^1\langle\psi(\mathbf{r}, \mathbf{r}_1)\rangle - \frac{q_1}{\varepsilon|\mathbf{r} - \mathbf{r}_1|} \tag{15–22}$$

is the average electrostatic potential at $\mathbf{r}$ due to all ions *except* 1 fixed at $\mathbf{r}_1$. Finally, we can write that

$$\langle\psi_1\rangle = {}^1\langle\psi_1(\mathbf{r}_1)\rangle \tag{15–23}$$

since $^1\langle\psi_1(\mathbf{r}_1)\rangle$ is the average electrostatic potential acting upon particle 1 fixed at $\mathbf{r}_1$, and this is just $\langle\psi_1\rangle$ if the system is an isotropic fluid. Thus the $^1\langle\psi(\mathbf{r}, \mathbf{r}_1)\rangle$ governed by Eq. (15–21) directly gives the terms needed to calculate the electrostatic Helmholtz free energy according to Eq. (15–18).

The total charge density that appears on the right-hand side of Eq. (15–21) can be written in terms of radial distribution functions according to

$$^1\langle\rho(\mathbf{r}, \mathbf{r}_1)\rangle = \sum_{s=1}^{2} c_s q_s g_{1s}(\mathbf{r}, \mathbf{r}_1) \tag{15–24}$$

where c_s is the bulk number concentration of the s-type ions and $g_{1s}(\mathbf{r}, \mathbf{r}_1)$ is the radial distribution function of s-type ions about the central ion located at $\mathbf{r}_1$. Except for a negligibly small region near the walls of the container we can set $\mathbf{r}_1 = 0$ or, alternatively, simply recognize that $^1\langle\psi(\mathbf{r}, \mathbf{r}_1)\rangle$ and $g_{1s}(\mathbf{r}, \mathbf{r}_1)$ depend only upon $|\mathbf{r} - \mathbf{r}_1|$. We can furthermore express the radial distribution functions in Eq. (15–24) in terms of potentials of mean force to give

$$\nabla^2({}^1\langle\psi(\mathbf{r})\rangle) = -\frac{4\pi}{\varepsilon} \sum_s c_s q_s e^{-\beta w_{1s}(\mathbf{r})} \tag{15–25}$$

This equation is exact up to this point and forms the starting point for the Debye-Hückel theory.

The first approximation of Debye and Hückel and is to set the potential of mean force equal to

$$w_{1s}(\mathbf{r}) = q_s \,{}^1\langle\psi(\mathbf{r})\rangle \equiv q_s \phi_1(\mathbf{r}) \qquad r > a \tag{15–26}$$

where for simplicity of notation we have written $\phi_1(\mathbf{r})$ for $^1\langle\psi(\mathbf{r})\rangle$ and are letting the radii of all the ions be $a/2$. This last restriction is easily relaxed, but the notation is simpler if we use it. In the primitive model, the right-hand side of Eq. (15–25) vanishes for $r < a$. If Eq. (15–26) is substituted into Eq. (15–25), we get the Poisson-Boltzmann equation,

$$\nabla^2\phi_1(\mathbf{r}) = -\frac{4\pi}{\varepsilon} \sum_s c_s q_s e^{-\beta q_s \phi_1(\mathbf{r})} \qquad r > a \tag{15–27}$$

which is a nonlinear differential equation for $\phi_1(\mathbf{r})$, which in turn is used to calculate the electrostatic Helmholtz free energy of the system. Being nonlinear, this equation is generally difficult to solve, and the second approximation of the Debye-Hückel theory is to linearize the right-hand side by expanding the exponentials. Thus we write

$$\sum_s c_s q_s e^{-\beta q_s \phi_1(\mathbf{r})} \approx \sum_s c_s q_s - \beta \sum_s c_s q_s^2 \phi_1(\mathbf{r}) \tag{15–28}$$

Remember that c_s is the number concentration of species s, i.e., that $c_s = N_s/V$, and so the first summation here vanishes because of electroneutrality; the total positive charge must equal the total negative charge in the system.

This linearization procedure gives the linear Poisson-Boltzmann equation:

$$\nabla^2 \phi_1(\mathbf{r}) = \kappa^2 \phi_1(\mathbf{r}) \qquad r > a \tag{15-29}$$

where

$$\kappa^2 = \frac{4\pi\beta}{\varepsilon} \sum_s q_s^2 c_s \tag{15-30}$$

Notice that κ^2 is closely related to the ionic strength,

$$I = \tfrac{1}{2} \sum_s q_s^2 c_s$$

This is the basic equation of the Debye-Hückel theory. We have introduced two approximations in going from the exact Eq. (15-25) to the linear Poisson-Boltzmann equation. We shall see that both of these approximations become justified (in fact are not approximations) in the limit of small concentrations. Thus although it is not obvious from the above derivation, Eq. (15-29) is exact as $\kappa \to 0$.

Equation (15-29) is appropriate only for the region in which $r > a$. For $r < a$, the centers of the surrounding ions are excluded (see Fig. 15-1) and hence the appropriate equation is Laplace's equation, namely,

$$\nabla^2 \phi_1(\mathbf{r}) = 0 \qquad 0 < r \le a \tag{15-31}$$

We now wish to solve Eqs. (15-29) and (15-31) for $\phi_1(r)$. The boundary conditions that $\phi_1(r)$ must satisfy are that $\phi_1(r)$ must vanish at infinity, that $\phi_1(r)$ be continuous across the surface specified by $r = a$, and that ε times the normal derivative of $\phi_1(r)$ be continuous across the surface $r = a$. This last condition comes from the fact that the normal component of the displacement vector $\mathbf{D}$ (Problem 15-6) is continuous across any charge free surface. These three conditions will specify $\phi_1(r)$ uniquely.

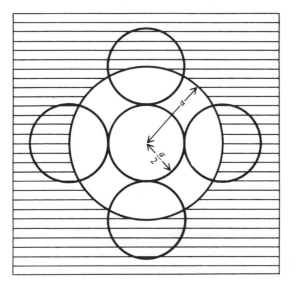

Figure 15-1. **An illustration of the boundaries involved in solving the linear Poisson-Boltzmann equation for the restricted primitive model.**

Since particle 1 is fixed at the origin and $\phi_1(r)$ is spherically symmetric, we naturally use spherical coordinates, which gives

$$\frac{1}{r^2}\frac{d}{dr}\left(r^2\frac{d\phi_1}{dr}\right) = \kappa^2\phi_1(r) \qquad r > a \tag{15-32}$$

The general solution of this equation is (Problem 15-9)

$$\phi_1(r) = \frac{A_1 e^{-\kappa r}}{r} + \frac{B_1 e^{\kappa r}}{r} \qquad r > a \tag{15-33}$$

where A_1 and B_1 are constants of integration. Since ϕ_1 must vanish as $r \to \infty$, we set $B_1 = 0$. So we have

$$\phi_1(r) = \frac{A_1}{r}e^{-\kappa r} \qquad r > a \tag{15-34}$$

We can determine A_1 by applying the boundary conditions at the surface $r = a$.

For the region between $r = 0$ and $r = a$ (cf. Fig. 15-1), we have

$$\frac{1}{r^2}\frac{d}{dr}\left(r^2\frac{d\phi_1}{dr}\right) = 0 \qquad 0 < r < a \tag{15-35}$$

which immediately gives

$$\phi_1(r) = \frac{A_2}{r} + B_2 \qquad 0 < r < a \tag{15-36}$$

We can use Gauss' law to evaluate the constant A_2. Applied to this problem, Gauss' law says that for $0 < r < a$,

$$4\pi\varepsilon r^2\left(-\frac{\partial\phi_1}{\partial r}\right) = 4\pi q_1 \qquad 0 < r < a$$

from which we find that $A_2 = q/\varepsilon$. Equation (15-36) then becomes

$$\phi_1(r) = \frac{q_1}{\varepsilon r} + B_2 \qquad 0 < r < a \tag{15-37}$$

The conditions that $\phi_1(r)$ and $\varepsilon d\phi_1/dr$ [ε times the normal derivative of $\phi_1(r)$] must be continuous at $r = a$ can be used to determine the two remaining constants A_1 and B_2. These two conditions give

$$\frac{A_1}{a}e^{-\kappa a} = \frac{q_1}{\varepsilon a} + B_2$$

$$\frac{A_1}{a^2}e^{-\kappa a} + \frac{\kappa A_1}{a}e^{-\kappa a} = \frac{q_1}{\varepsilon a^2}$$

from which we find that

$$A_1 = \frac{q_1 e^{\kappa a}}{\varepsilon(1 + \kappa a)} \tag{15-38}$$

$$B_2 = -\frac{q_1\kappa}{\varepsilon(1 + \kappa a)} \tag{15-39}$$

Putting all this together, then, we have

$$
\phi_1(r) = \frac{q_1}{\varepsilon r} - \frac{q_1 \kappa}{\varepsilon(1 + \kappa a)} \qquad 0 < r < a
$$

$$
= \frac{q_1 e^{-\kappa(r-a)}}{\varepsilon r(1 + \kappa a)} \qquad r > a
$$
(15–40)

Note that Eq. (15–40) is of the form $e^{-\kappa r}/r$, i.e., what is called a screened coulombic potential.

According to Eqs. (15–22) and (15–23)

$$
\langle \psi_1 \rangle = {}^1\langle \psi_1(\mathbf{r}) \rangle = \phi_1(r) - \frac{q_1}{\varepsilon r}
$$
(15–41)

which, when Eq. (15–40) is used, gives

$$
\langle \psi_1 \rangle = -\frac{q_1 \kappa}{\varepsilon(1 + \kappa a)}
$$
(15–42)

or more generally

$$
\langle \psi_j \rangle = -\frac{q_j \kappa}{\varepsilon(1 + \kappa a)}
$$
(15–43)

This, then, is the average electrostatic potential acting on the jth ion fixed at the origin. This potential is due to all the other ions in the solution.

Before going on to derive the thermodynamic expressions associated with the Debye-Hückel theory, we shall first discuss some general requirements that $\langle \psi_j \rangle$ must satisfy, and then we shall discuss the linearization approximation of the Debye-Hückel theory.

The quantity $\langle \psi_j \rangle$ must satisfy two conditions. According to Eq. (15–26), Debye and Hückel assume that

$$
w_{js}(r) = q_s{}^j \langle \psi(\mathbf{r}) \rangle
$$
(15–44)

where we have written j instead of 1. From the formal definition of the potential of mean force in a fluid, we see that

$$
w_{js}(r) = w_{sj}(r)
$$
(15–45)

which requires that ${}^k\langle \psi(r) \rangle$ be such that

$$
q_j{}^s \langle \psi(r) \rangle = q_s{}^j \langle \psi(r) \rangle
$$
(15–46)

Note that ${}^k\langle \psi(r) \rangle$ given by Eq. (15–40) does, in fact, satisfy this self-consistency condition. Another condition on ${}^k\langle \psi(r) \rangle$ follows directly from Eq. (15–17). Since A is an exact differential, we have the requirement that

$$
\frac{\partial \langle \psi_i \rangle}{\partial q_j} = \frac{\partial \langle \psi_j \rangle}{\partial q_i}
$$
(15–47)

for all i and j. It is easy to show that $\langle \psi_j \rangle$ given by Eq. (15–43) satisfies this condition (Problem 15–18). It is interesting to note that solutions to the nonlinear Poisson-Boltzmann equation do not satisfy these self-consistency conditions, and so the linearization employed by Debye and Hückel is a necessary step in deriving an exact limiting theory.

It is instructive to examine $\phi_j(r)$ further. For example, the total charge surrounding the central ion j is given by

$$\int_a^\infty {}^j\langle\rho(r)\rangle 4\pi r^2 \, dr \tag{15-48}$$

which, according to Poisson's equation, can be written as

$$-\frac{\varepsilon}{4\pi}\int_a^\infty \nabla^2\{{}^j\langle\psi(r)\rangle\}4\pi r^2 \, dr \tag{15-49}$$

The Debye-Hückel theory sets $\nabla^2\{{}^j\langle\psi(r)\rangle\}$ equal to $\kappa^{2\,j}\langle\psi(r)\rangle \equiv \kappa^2\phi_j(r)$, where $\phi_j(r)$ is given by Eq. (15–40). Substituting this into Eq. (15–49) gives

$$-\frac{\varepsilon}{4\pi}\int_a^\infty \frac{\kappa^2 q_j e^{-\kappa(r-a)}}{\varepsilon r(1+\kappa a)} 4\pi r^2 \, dr = -q_j \tag{15-50}$$

This shows that the total charge surrounding a central j-ion is equal and opposite to the charge q_j on the j ion.

Equation (15–50) shows that

$$p(r)\,dr = -\frac{\varepsilon\kappa^2\phi_j}{4\pi}\cdot 4\pi r^2 \, dr = -\frac{q_j\kappa^2}{1+\kappa a}e^{-\kappa(r-a)}r\,dr \qquad r>a \tag{15-51}$$

is the fraction of charge between the spherical shells of radius r and $r + dr$. The quantity $p(r)$ is plotted in Fig. 15–2, which shows that there is a tendency for ions having a charge opposite to that of the central ion to distribute themselves around the central ion. Thus it is said that Eq. (15–51) and Fig. 15–2 describe the *ionic atmosphere* around a central ion.

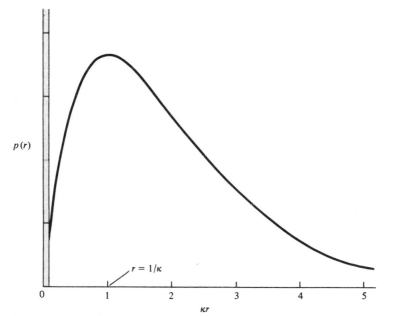

Figure 15–2. **The fraction of charge between the spherical shells of radius** r **and** $r + dr$**. This illustrates the idea of an ionic atmosphere.**

The maximum in $p(r)$ is given when $r = 1/\kappa$, and so values of r around this value are the most important in determining the thermodynamic properties of the solution. Equation (15–30) shows that $\kappa \propto c^{1/2}$, where c is the concentration of electrolyte, and so the important values of r become larger as c becomes smaller. For an aqueous solution of a 1–1 electrolyte at 25°C, $\kappa^{-1} = 3.04c^{-1/2}$ Å (angstroms) when c is the concentration of salt expressed in moles per liter, and so κ^{-1} is about 30 Å when $c = 0.01$ mole/liter. At these concentrations then, the short-range interaction between ions does not play an important role, and its precise nature is unimportant. Problem 15–20 shows that at concentrations around 0.01 mole/liter or less, the linearization of the Poisson-Boltzmann equation is valid as well.

The above paragraph shows that the Debye-Hückel theory is valid only for fairly small concentrations and becomes better as the concentration becomes smaller. We cannot show at this point that the approximation $w_{ij}(r) = q_i\phi_j(r)$ is valid for small κa or concentration, but we shall show in Section 15–2 that in the limit of small κa that

$$w_{ij}(r) \to \frac{q_i q_j e^{-\kappa r}}{\varepsilon r} \tag{15–52}$$

This is just what is obtained from Eq. (15–40) for small κa. Thus we say that the Debye-Hückel theory is an *exact* limiting law, and we expect that the theory correctly describes the thermodynamic properties of all solutions of strong electrolytes as the concentrations approach zero. It is a reliable theory for 1–1 electrolytes for concentrations around 0.005 mole/liter or less or, in other words, for quite dilute solutions. Nevertheless, the Debye-Hückel theory represented a great advance in the theory in ionic solutions since before it there was some amount of disagreement and confusion concerning the nature of such solutions.

We shall now derive the thermodynamic expressions of the Debye-Hückel theory. The electrostatic Helmholtz free energy is given by Eq. (15–18):

$$A - A_0 = \sum_j q_j \int_0^1 \langle \psi_j(\lambda) \rangle \, d\lambda \tag{15–53}$$

Remembering that κ depends upon the magnitude of the charge on the ions, we write

$$A - A_0 = -\sum_j \frac{q_j^2}{\varepsilon} \int_0^1 \frac{\lambda\kappa(\lambda) \, d\lambda}{1 + \kappa(\lambda)a} \tag{15–54}$$

Using Eq. (15–30) for κ, this integral is easy and gives (Problem 15–12)

$$\frac{\beta A^{el}}{V} \equiv \frac{\beta(A - A_0)}{V} = -\frac{\kappa^3}{12\pi} \tau(\kappa a) \tag{15–55}$$

where

$$\tau(\kappa a) = \frac{3}{\kappa^3 a^3} \left\{ \ln(1 + \kappa a) - \kappa a + \frac{\kappa^2 a^2}{2} \right\} \tag{15–56}$$

For small κ, $\tau(\kappa a)$ becomes

$$\tau(\kappa a) \approx 1 - \tfrac{3}{4}\kappa a + \tfrac{3}{5}(\kappa a)^2 + \cdots \tag{15–57}$$

The electrostatic chemical potential of the jth ionic species is given by

$$\mu_j^{el} = \left(\frac{\partial A^{el}}{\partial N_j} \right)_{V,T} = \frac{V}{\beta} \left[\frac{\partial}{\partial N_j} \left(\frac{\beta A^{el}}{V} \right) \right]_{T,V} \tag{15–58}$$

Using Eq. (15–55) for $\beta A^{\mathrm{el}}/V$, we find that

$$\mu_j^{\mathrm{el}} = -\frac{\kappa q_j^2}{2\varepsilon(1 + \kappa a)} \tag{15–59}$$

This equation can be derived in quite a different manner by using a simple generalization of Eq. (13–34). (See Problem 15–14.) We can calculate the electrostatic Gibbs free energy from this by using

$$G^{\mathrm{el}} = \sum_j N_j \mu_j^{\mathrm{el}}$$

This gives

$$\frac{\beta G^{\mathrm{el}}}{V} = -\frac{\kappa^3}{8\pi(1 + \kappa a)} \tag{15–60}$$

In addition, we can determine the electrostatic osmotic pressure from the equation

$$\frac{\beta G^{\mathrm{el}}}{V} = \frac{\beta A^{\mathrm{el}}}{V} + \beta p^{\mathrm{el}}$$

Using Eqs. (15–55) and (15–60), we get

$$\beta p^{\mathrm{el}} = -\frac{\kappa^3}{24\pi}\left\{\frac{3}{1 + \kappa a} - 2\tau(\kappa a)\right\} \tag{15–61}$$

Fowler and Guggenheim write this as (Problem 15–15)

$$\beta p^{\mathrm{el}} = -\frac{\kappa^3}{24\pi}\,\sigma(\kappa a) \tag{15–62}$$

where

$$\sigma(\kappa a) = \frac{3}{(\kappa a)^3}\left\{1 + \kappa a - \frac{1}{1 + \kappa a} - 2\ln(1 + \kappa a)\right\}$$

$$\approx 1 - \tfrac{3}{2}\kappa a + \tfrac{9}{5}(\kappa a)^2 + \cdots \tag{15–63}$$

The above expressions for the various thermodynamic functions, although frequently seen in the literature, are somewhat misleading since, as we have seen earlier, the entire theory should be valid only in the limit $\kappa a \to 0$. We should really use only the limiting forms of these equations. For example, Eq. (15–55) for $\beta A^{\mathrm{el}}/V$ is

$$\frac{\beta A^{\mathrm{el}}}{V} = -\frac{\kappa^3}{12\pi}\,\tau(\kappa a)$$

The limit of this as $\kappa a \to 0$ is

$$\frac{\beta A^{\mathrm{el}}}{V} = -\frac{\kappa^3}{12\pi} \tag{15–64}$$

This goes for the other thermodynamic functions as well:

$$\frac{\beta G^{\mathrm{el}}}{V} = -\frac{\kappa^3}{8\pi} \tag{15–65}$$

$$\beta p^{\mathrm{el}} = -\frac{\kappa^3}{24\pi} \tag{15–66}$$

and

$$\mu_j^{\mathrm{el}} = -\frac{\kappa q_j^2}{2\varepsilon} \tag{15–67}$$

Since these electrostatic quantities are taken as $\kappa a \to 0$, we can write

$$\frac{p}{kT} = c_1 + c_2 - \frac{\kappa^3}{24\pi} \qquad (15\text{-}68)$$

and

$$\mu_j = kT \ln \Lambda_j{}^3 + kT \ln c_j - \frac{\kappa q_j{}^2}{2\varepsilon} \qquad (15\text{-}69)$$

We can use these equations to present two other thermodynamic functions that are commonly used in the study of ionic solutions. We define the osmotic coefficient ϕ by

$$\phi = \frac{p/kT}{c_1 + c_2} \qquad (15\text{-}70)$$

which, in the Debye-Hückel theory, is

$$\phi = 1 - \frac{\kappa^3}{24\pi(c_1 + c_2)} \qquad (15\text{-}71)$$

For an ideal solution ϕ is unity.

The other thermodynamic function we wish to discuss is the activity coefficient γ_j, defined through the equation

$$\mu_j = kT \ln \Lambda_j{}^3 + kT \ln c_j + kT \ln \gamma_j \qquad (15\text{-}72)$$

By comparing this equation with Eq. (15-69), we get

$$\ln \gamma_j = -\frac{\kappa q_j{}^2}{2\varepsilon kT} \qquad (15\text{-}73)$$

If instead of Eq. (15-67) for $\mu_j{}^{el}$ we use Eq. (15-59), we find that

$$\ln \gamma_j = -\frac{\kappa q_j{}^2}{2kT\varepsilon(1 + \kappa a)} \qquad (15\text{-}74)$$

Although this equation is not, strictly speaking, valid, it is often used in the treatment of experimental data and will, in fact, be used in the next section. This and Eq. (15-40) are often referred to as the extended form of the Debye-Hückel theory.

Of course, individual activity coefficients cannot be measured experimentally. One measures the mean activity coefficient,* which for a binary salt is defined by

$$\gamma_\pm = (\gamma_+{}^{\nu_+} \gamma_-{}^{\nu_-})^{1/\nu} \qquad (15\text{-}75)$$

where $\nu = \nu_+ + \nu_-$ and ν_+ and ν_- are given by the chemical equation describing the dissociation of the electrolyte:

$$C_{\nu_+} A_{\nu_-} \to \nu_+ C^+ + \nu_- A^-$$

From Eq. (15-75) we write

$$\ln \gamma_\pm = \frac{\nu_+ \ln \gamma_+ + \nu_- \ln \gamma_-}{\nu_+ + \nu_-}$$

---

* W. J. Moore, *Physical Chemistry*, 4th ed. (Englewood Cliffs, N.J.: Prentice-Hall, 1971).

which, with the use of Eq. (15–73) for $\ln \gamma_j$, becomes

$$\ln \gamma_\pm = -\left(\frac{v_+ q_+{}^2 + v_- q_-{}^2}{v_+ + v_-}\right) \frac{\kappa}{2\varepsilon kT} \qquad (15\text{–}76)$$

Because of electroneutrality, $v_+ q_+ + v_- q_- = 0$, this can be rewritten as (Problem 15–17)

$$\ln \gamma_\pm = -|q_+ q_-| \frac{\kappa}{2\varepsilon kT} \qquad (15\text{–}77)$$

A plot of $\ln \gamma_\pm$ versus the square root of the ionic strength should yield a straight line for sufficiently low concentrations, and the slope of the line should depend upon the type of electrolyte through the factor $|q_+ q_-|$. Figure 15–3 shows such a plot for several different types of electrolytes. At higher concentrations we expect to find deviations from such limiting behavior, and this is shown in Fig. 15–4 for a number of 1–1 electrolytes. Note, however, that all of these salts yield the same limiting behavior. At the higher concentrations, the size and the nature of the short-range interactions of the various salts become important, and each salt shows individual behavior.

In this section, then, we have discussed in some detail the Debye-Hückel theory, which is the exact limiting law for electrolytic solutions. Deviations are expected and observed for concentrations greater than 0.01 mole/liter. In the next sections we shall discuss the statistical basis and extensions of the Debye-Hückel theory to higher concentrations.

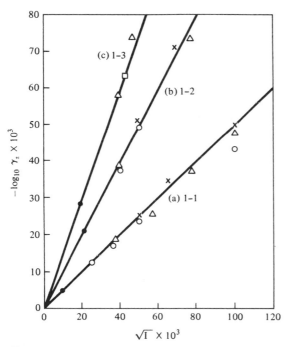

Figure 15–3. **Activity coefficients of sparingly soluble salts.** The quantity I, called the ionic strength, is given by $I = \frac{1}{2} \sum q_j^2 c_j$.

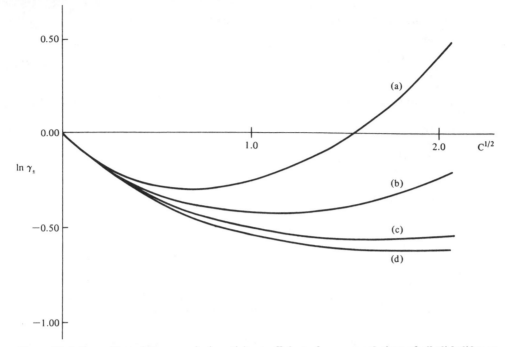

Figure 15–4. **Logarithm of the mean ionic activity coefficient of aqueous solutions of alkali halides at 25°C. (a) LiCl, (b) NaCl, (c) KCl, (d) RbCl.** (Data from R. A. Robinson and R. H. Stokes, "Electrolyte Solutions," 2nd ed., Butterworths: London, 1965.)

15–2 SOME STATISTICAL MECHANICAL THEORIES OF IONIC SOLUTIONS

In spite of the great success of the Debye-Hückel theory, when it was originally proposed its range of validity was not at all clear. Its success stimulated a great deal of work to place the Debye-Hückel theory within a statistical mechanical framework or, in simple terms, to derive it from a partition function. There are two important reasons for doing this. One is elucidate just what approximations are needed to derive the final Debye-Hückel equations and to investigate the range of concentrations over which these approximations are valid. The other reason is to formulate a method by which the approximate theory can be successively corrected. For example, it is not obvious how to improve the Debye-Hückel theory as we have presented in Section 15–1, but if we could derive it by starting from the configurational integral, for example, then we could formally write down correction terms and investigate these. This could lead to a series of systematic corrections.

An appealing approach is to use the virial expansion developed in Chapter 12 (or really the McMillan-Mayer theory since this is a solution rather than a gas). We saw earlier in this chapter, however, that the virial coefficients diverge for the coulombic potential. A great breakthrough along these lines was made by Mayer* in 1950. He showed that although individual cluster integrals diverge, it is possible to combine the infinite parts of all the virial coefficients B_n, for $n \geq 2$, so that they mutually cancel and yield a finite result. By a careful evaluation of the graphs from all the virial coefficients, he was able to show that the Debye-Hückel theory was obtained if only

---

* J. E. Mayer, *J. Chem. Phys.*, **18**, p. 1426, 1950; see also E. Haga, *J. Phys. Soc. Japan*, **8**, p. 714, 1953.

terms in the lowest order in the concentration were kept. A complete discussion of Mayer's approach is too involved to give here, but it is possible to present the general idea.

We start with the equations for an imperfect gas, namely,

$$\frac{p}{kT} = c + \sum_{k \geq 2} B_k c^k \qquad\qquad (15\text{--}78)$$

where the virial coefficients are given in terms of cluster integrals

$$B_k = -\frac{k-1}{k} \beta_{k-1}$$

and the cluster integrals are given by

$$\beta_{k-1} = \frac{1}{(k-1)! \, V} \int \cdots \int S'_{1, 2, \ldots, k} \, d\mathbf{r}_1 \cdots d\mathbf{r}_k$$

where $S'_{1, 2, \ldots, k}$ is the sum of all stars (doubly connected diagrams) of k particles. We wish to generalize these equations to more than one component, and to this end it is convenient to rewrite them in a slightly different form. Substituting Eq. (12–33) into Eq. (15–78) gives

$$\frac{p}{kT} = c - \sum_{k \geq 2} \frac{(k-1)}{k} \beta_{k-1} c^k$$

We now define B_k by

$$B_k \equiv \frac{\beta_{k-1}}{k} = \frac{1}{k! \, V} \int \cdots \int S'_{1, 2, \ldots, k} \, d\mathbf{r}_1 \cdots d\mathbf{r}_k \qquad\qquad (15\text{--}79)$$

and so we can write

$$\frac{p}{kT} = c - \sum (k-1) B_k c^k.$$

For two components this equation becomes simply[*]

$$\frac{p}{kT} = c_1 + c_2 - \sum_{\substack{k_1 \geq 1 \\ k_2 \geq 1}} (k_1 + k_2 - 1) B_{k_1 k_2} c_1^{k_1} c_2^{k_2} \qquad\qquad (15\text{--}80)$$

where

$$B_{k_1 k_2} = \frac{1}{k_1! \, k_2! \, V} \int \cdots \int S'_{1, 2, \ldots, k_1, k_1+1, \ldots, k_1+k_2} \, d\mathbf{r}_1 \cdots d\mathbf{r}_{k_1+k_2} \qquad\qquad (15\text{--}81)$$

The integrand of $B_{k_1 k_2}$ is a sum of products of f-functions. Now as Mayer indicated, Eq. (15–80) is a multiple sum in which the order of summation has been implicitly

[*] J. E. Mayer, "Theory of Real Gases," in *Handbuch der Physik*, Vol. 12, edited by S. Flügge (Berlin: Springer-Verlag, 1958).

specified, namely, that one sums over all products in the integrand for a specific $B_{k_1 k_2}$ and *then* over all values of k_1 and k_2. As long as convergent results are obtained, it is equally legitimate to sum first over all values of k_1 and k_2 for some specified type of product in the integrand, and then sum over all types of products. By doing this in a particular way, we will obtain convergent results for the coulomb potential.

Mayer defines a summation function S by

$$S = \sum_{\substack{k_1 \geq 1 \\ k_2 \geq 1}} B_{k_1 k_2} c_1^{k_1} c_2^{k_2} \tag{15-82}$$

In terms of S (Problem 15–27)

$$\frac{p}{kT} = c_1 + c_2 + S - c_1 \frac{\partial S}{\partial c_1} - c_2 \frac{\partial S}{\partial c_2} \tag{15-83}$$

and .

$$-\ln \gamma_j = \frac{\partial S}{\partial c_j} \tag{15-84}$$

If we classify all of the graphs generated in the $B_{k_1 k_2}$ into various mutually exclusive sets, say $\{j\}$, then

$$S = \sum_j S^{(j)} \tag{15-85}$$

where $S^{(j)}$ is Eq. (15–82) applied only this class of graph.

To see what types of graphs are generated, consider

$$u_{ij} = \frac{q_i q_j e^{-\alpha r}}{\varepsilon r} + u_{ij}^{(s)} \tag{15-86}$$

The factor $e^{-\alpha r}$ is a "convergence factor" to assure that the various integrals to arise converge. We are, of course, only interested in the limit $\alpha \to 0$. Now define

$$\lambda = \frac{4\pi}{\varepsilon kT}$$

$$g(r) = \frac{e^{-\alpha r}}{4\pi r} \tag{15-87}$$

$$k(r) = e^{-u^{(s)}/kT} - 1$$

The *f*-function that occurs in any $B_{k_1 k_2}$ can be written as

$$f_{ij} = k_{ij} + (k_{ij} + 1)(e^{-\lambda q_i q_j g_{ij}} - 1)$$

$$= k_{ij} + (k_{ij} + 1) \sum_{l \geq 1} \frac{(-1)^l}{l!} \lambda^l q_i^l q_j^l g_{ij}^l \tag{15-88}$$

This is simply an identity (Problem 15–29). If this is substituted into any particular cluster diagram, we get an infinite number of terms involving k_{ij}, powers of g_{ij}, and products of k_{ij} and powers of g_{ij}. These may be represented graphically by drawing a dotted line for a k-bond between two particles and a solid line for each g-bond. Some examples of the types of graphs that occur are

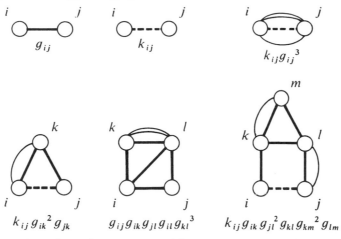

Certainly the types of graphs encountered here are more complex than those of imperfect gas theory, but they can, in fact, be classified and summed in an organized way. The delicate part of Mayer's analysis is to classify them according to the power of the concentration that they contribute to S. Mayer showed that if only the so-called cyclic graphs are included in S (cf. Fig. 15–5), then the Debye-Hückel expressions are obtained and that all other types of graphs contribute a higher order in the concentration. This is too involved to show here, but Problem 15–31 treats the evaluation of $S^{(\text{cyclic})}$. The result is

$$S^{(\text{cyclic})} = \frac{\kappa^3}{12\pi} \tag{15–89}$$

from which the Debye-Hückel expressions for p/kT and $\ln \gamma_j$ are obtained through Eqs. (15–83) and (15–84).

Equation (15–89) is the leading term in an expansion of S in the concentration

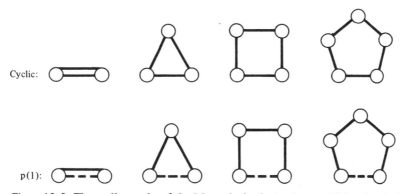

Figure 15–5. **The cyclic graphs of the Mayer ionic cluster theory which, when included in S, give the Debye-Hückel limiting law. Also included is another set of graphs which Problem 15–30 shows gives no contribution to S because of electroneutrality.**

For the primitive model, the expansion has been extended to higher order, and to terms in $c^2 \ln c$, we have

$$S = \frac{\kappa^3}{12\pi} - \frac{2\pi}{3} a^3 (c_1 + c_2)^2 - \frac{\kappa^4 a}{16\pi} + \frac{\kappa_1^4}{192\pi^2} [\gamma + \ln(3\kappa a) - 3\kappa a]$$

$$+ 2\pi a^2 \sum_{j=4}^{\infty} \frac{(-1)^j}{j!(j-3)} \frac{\kappa_{j-2}^4}{(4\pi a)^j} + O(\kappa^5) + O(\kappa^5 \ln \kappa) \qquad (15\text{-}90)$$

where a is the diameter of the hard-sphere core and

$$\kappa_{l-2}^2 = \lambda^{1/2}(c_1 q_1^l + c_2 q_2^l) \qquad (15\text{-}91)$$

This κ_{l-2}^2 reduces to the Debye-Hückel κ^2 when $l = 2$.

The various thermodynamic functions can be easily obtained from Eq. (15–90). Poirier* has made an extensive application of Mayer's theory to experimental data, and the "practical" aspects of the theory have been discussed by Frank and Tsao.† It seems to be useful up to concentrations of 0.4 molar for 1–1 electrolytes. We see then that Mayer's theory is an ionic solution analog to the cluster expansion of imperfect gas theory.

Another approach that suggests itself is the application of the integral equation methods of Chapter 13. The central function there is the radial distribution function of ions around some central one. Falkenhagen and Kelbg‡ review much of the work done prior to 1959, most of that being done by Falkenhagen and Kirkwood and Poirier.§ More recently, however, the Percus-Yevick and hypernetted-chain equations have found most favor.

Let us go back to the Ornstein-Zernike equation

$$h(r_{12}) = c(r_{12}) + \rho \int c(r_{13}) h(r_{23}) \, d\mathbf{r}_3$$

Both the Percus-Yevick and hypernetted-chain equations are obtained from this by assuming a particular relation between the direct correlation function $c(r)$ and the radial distribution function. According to Eq. (13–74), at low densities $c(r) = f(r)$, and the direct correlation function of both the Percus-Yevick and hypernetted-chain equations reduces to this correct limit at low densities. We shall see that if we substitute this expression for $c(r)$ into Eq. (13–55) and linearize the exponentials, we shall get the Debye-Hückel potential of mean force. Before doing this, however, we need the generalization of Eq. (13–55) for two components,|| which is

$$h_{ij}(\mathbf{r}, \mathbf{r}') = c_{ij}(\mathbf{r}, \mathbf{r}') + \sum_{l=1}^{2} \rho_l \int c_{il}(\mathbf{r}, \mathbf{x}) h_{lj}(\mathbf{x}, \mathbf{r}') \, d\mathbf{x} \qquad (15\text{-}92)$$

We can immediately obtain the Debye-Hückel result from this be letting

$$c_{ij} = f_{ij} = e^{-\beta u_{ij}} - 1 = -\beta u_{ij} \qquad (15\text{-}93)$$

where $u_{ij} = q_i q_j / \varepsilon r$, and writing $h_{ij} = g_{ij} - 1 = -\beta w_{ij}$ where w_{ij} is the potential of mean force. We are neglecting the hard-core cutoff since we are interested only in the

* J. C. Poirier, *J. Chem. Phys.*, **21**, p. 965, 1953; **21**, p. 972, 1953.

† H. S. Frank and M. Tsao, *Ann. Rev. Phys. Chem.*, **5**, p. 43, 1954.

‡ H. Falkenhagen and G. Kelbg, in *Modern Aspects of Electrochemistry*, Vol. II, edited by J. O. M. Bockris (New York: Academic, 1959).

§ J. G. Kirkwood and J. C. Poirier, *J. Phys. Chem.*, **58**, p. 591, 1954.

|| J. L. Lebowitz, *Phys. Rev.*, **133**, p. 895A, 1964.

limit of dilute solutions. This same condition also applies to the linearization of the exponentials in $f(r)$ and $h(r)$. This gives

$$w_{ij} = \frac{q_i q_j}{\varepsilon r_{ij}} - \sum_{l=1}^{2} \rho_l \int \frac{q_i q_l}{\varepsilon r_{il}} \frac{w_{lj}}{kT} \, d\mathbf{r}_l \tag{15-94}$$

It is convenient to write w_{ij} as $(q_i q_j / \varepsilon) w(r)$, in which case Eq. (15–94) becomes

$$w(r) = \frac{1}{r} - \frac{1}{\varepsilon kT} \sum_{l=1}^{2} \rho_l q_l^2 \int \frac{w(r_{lj})}{r_{il}} \, d\mathbf{r}_l$$

Now the integral on the right-hand side here is independent of l since we are integrating over $\mathbf{r}_l$, and so we can write more neatly

$$w(r) = \frac{1}{r} - \frac{\kappa^2}{4\pi} \int \frac{w(|\mathbf{r} - \mathbf{r}'|)}{\mathbf{r}'} \, d\mathbf{r}' \tag{15-95}$$

This is an integral equation for $w(r)$ and can be solved easily in several ways. The most straightforward is to introduce Fourier transforms as we did in going from Eq. (13–55) to Eq. (13–57). If we denote the Fourier transform of $w(r)$ by $\hat{W}$ and that of r^{-1} by $\hat{U}$, then Eq. (15–95) becomes (Problem 15–32)

$$\hat{W}(t) = \frac{\hat{U}(t)}{1 + \dfrac{(2\pi)^{3/2}}{4\pi} \kappa^2 \hat{U}(t)} \tag{15-96}$$

According to Eq. (12–51),

$$\hat{U}(t) = \left(\frac{2}{\pi}\right)^{1/2} \lim_{\alpha \to 0} \int \frac{e^{-\alpha r} \sin tr}{t} \, dr$$

$$= \left(\frac{2}{\pi}\right)^{1/2} \lim_{\alpha \to 0} \left(\frac{1}{t^2 + \alpha^2}\right) = \frac{(2/\pi)^{1/2}}{t^2}$$

Notice that we have introduced $e^{-\alpha r}$ as a convergence factor. If this is substituted into Eq. (15–96), we get

$$\hat{W}(t) = \frac{(2/\pi)^{1/2}}{t^2 + \kappa^2}$$

This can be inverted to give $w(r)$ by using Eq. (12–52):

$$w(r) = \left(\frac{2}{\pi}\right) \frac{1}{r} \int_0^\infty \frac{t \sin tr}{t^2 + \kappa^2} \, dt$$

This integral is elementary, and we find

$$w(r) = \frac{e^{-\kappa r}}{r}$$

which, if the factor $q_i q_j / \varepsilon$ is reinserted, gives the Debye-Hückel expression for the potential of mean force. Thus again we find that the Debye-Hückel theory is an exact statistical mechanical result for dilute conditions. This same result can be obtained from the Kirkwood and Born-Green-Yvon equations (Problem 15–33).

We can readily extend the above argument to include a hard-sphere repulsion of the ions. In this case, $w(r) = \infty$ for $r < a$ and for $r > a$ is given by Eq. (15–95) but with the restriction that $|\mathbf{r} - \mathbf{r}'|$ cannot be less than a:

$$w(r) = \frac{1}{r} - \frac{\kappa^2}{4\pi} \int\limits_{|\mathbf{r}-\mathbf{r}'|>a} \frac{w(|\mathbf{r} - \mathbf{r}'|)}{\mathbf{r}'} \, d\mathbf{r}' \qquad r > a \qquad (15\text{–}97)$$

Kirkwood and Poirier* have derived this equation in quite a different manner. This equation is much more difficult to solve than Eq. (15–95), but some general features of the solution were extracted by Kirkwood and Poirier. An intriguing result is that at a certain concentration, given by the condition $\kappa a > 1.032$, $w(r)$ assumes an oscillatory form, indicating zones of alternating positive and negative charge densities. This oscillatory behavior has been found in other formulations of the theory of ionic solutions as well.†

Recently Rasaiah and Friedman‡ have solved the Percus-Yevick and hypernetted-chain equations numerically. Since these equations have found much success in the theory of liquids, one expects that this should be a powerful method for concentrated solutions. Most of the results have been tabulated for the restrictive primitive model. It has been found that the hypernetted-chain equation‡ is superior to the Percus-Yevick equation§ at least for a 1–1 aqueous electrolyte at concentrations up to 1.0 M. Fortunately, Monte Carlo calculations have been performed by Card and Valleau‖ to which the integral equation results can be compared. Table 15–1 shows a comparison of the osmotic coefficient for an aqueous solution of a 1–1 electrolyte calculated from the PY equation and HNC equations to the Monte Carlo calculations for an aqueous solution of a 1–1 electrolyte. Remember that the value of the pressure calculated from the integral equations depends upon whether one uses the so-called pressure equation [Eq. (13–23)] or the compressibility equation [Eq. (13–54)]. It can be seen from the table that the results from the hypernetted-chain equation are more internally consistent than those from the Percus-Yevick equation and that the general agreement is quite good even up to almost 2 M.

Figures 15–6 and 15–7 show the osmotic coefficient and the logarithm of the mean activity coefficient versus the square root of the ionic strength. The curves labeled

Table 15–1. **A comparison of the solutions of the hypernetted-chain and Percus-Yevick equations to the Monte Carlo results for the osmotic coefficient for a restricted primitive model of a 1–1 electrolyte with $a = 4.25$ Å, $\varepsilon = 78.5$, $T = 298°$K**

C (moles/liter)	$\phi_v{}^{MC}$	$\phi_v{}^{HNC}$	$\phi_c{}^{HNC}$	$\phi_v{}^{PYA}$	$\phi_c{}^{PYA}$
0.00911	0.9701 ± 0.0008	0.9703	0.9705	0.9703	0.9705
0.10376	0.9445 ± 0.0012	0.9453	0.9458	0.9452	0.9461
0.42502	0.9774 ± 0.0046	0.9796	0.9800	0.9765	0.9844
1.0001	1.094 ± 0.005	1.0926	1.0906	1.0789	1.1076
1.9676	1.346 ± 0.009	1.3514	1.3404	1.3114	1.386

Source: J. C. Rasaiah, D. N. Card, and J. P. Valleau, *J. Chem. Phys.*, **56**, p. 248, 1972.

* J. G. Kirkwood and J. C. Poirier, *J. Phys. Chem.*, **58**, p. 591, 1964.

† C. W. Outhwaite, *J. Chem. Phys.*, **50**, p. 2277, 1969; F. H. Stillinger and R. Lovett, *J. Chem. Phys.*, **48**, p. 3858, 1968.

‡ J. C. Rasaiah and H. L. Friedman, *J. Chem. Phys.*, **48**, p. 2742, 1968; **50**, p. 3965, 1969; J. C. Rasaiah, *Chem. Phys. Lett.*, **7**, p. 260, 1970; *J. Chem. Phys.*, **52**, p. 704, 1970.

§ As modified by Allnatt: A. R. Allnatt, *Mol. Phys.*, **8**, p. 533, 1964.

‖ D. N. Card and J. P. Valleau, *J. Chem. Phys.*, **52**, p. 6232, 1970.

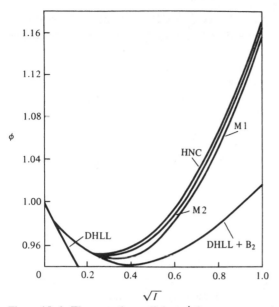

Figure 15–6. **The osmotic coefficient ϕ for an aqueous 1–1 restricted primitive model electrolyte with hard-sphere diameter 4.6 Å, obtained in various ways. DHLL denotes the Debye-Hückel limiting law (sum of ring diagrams in cluster theory). DHLL + B_2 denotes the sum of all ring diagrams and all two-particle cluster diagrams. HNC denotes the solution of the hypernetted-chain equation, and M1 and M2 denote the mode expansion results.** (From D. Chandler and H. C. Andersen, *J. Chem. Phys.*, **54**, p. 26, 1971.)

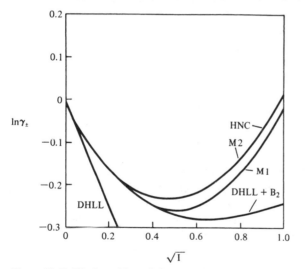

Figure 15–7. **The logarithm of the mean ionic activity coefficient, $\ln \gamma_{\pm}$, for an aqueous 1–1 restricted primitive model electrolyte with hard-sphere diameter 4.6 Å, obtained in various ways. See Fig. 15–6.** (From D. Chandler and H. C. Andersen, *J. Chem. Phys.*, **54**, p. 26, 1971.)

DHLL are the Debye-Hückel limiting law; those labeled DHLL + B2 are for the sum of all ring diagrams (DHLL) and all two-particle cluster diagrams in Mayer's ionic cluster theory; HNC is for the hypernetted chain as modified by Allnatt;* and the designations M1 and M2 will be discussed shortly. The application of integral equation

---

* A. R. Allnatt, *Mol. Phys.*, **8**, p. 533, 1964.

methods to ionic solutions has been very well developed by Rasaiah and Friedman, who discuss the numerical techniques involved and present a thorough analysis of the accuracy of their solutions.

Figures 15–6 and 15–7 contain two other curves, labeled M1 and M2, that we have not yet discussed. They are the results of a series of three papers by Andersen and Chandler* on a novel and powerful expansion for the Helmholtz free energy of a classical system. The potential energy of the system is assumed to be composed of two parts: a "reference system" potential and a perturbation potential, which is the sum of two-particle potentials. This two-particle potential must have a Fourier transform. They then introduce collective variables, which are Fourier transforms of the single-particle density, and the excess Helmholtz free energy is written as an infinite series

$$- \frac{\beta A^{ex}}{V} \equiv \mathscr{A} = \mathscr{A}_0 + \sum_{n=1}^{\infty} a_n \qquad (15\text{–}98)$$

where $\mathscr{A}_0$ is the excess free energy of the reference system and the a_n are determined by the perturbation potential and the distribution functions (two-, three-, and higher-particle) of the reference system, which are assumed to be known. The expansion [Eq. (15–98)] is called a *mode expansion* by Andersen and Chandler.

In the second paper of their mode expansion series, they apply their technique to the calculation of the thermodynamic properties of the restricted primitive model of a 1–1 electrolyte. In this model the reference potential is a hard core, and the perturbation potential is coulombic outside the hard core but arbitrary within. Chandler and Andersen found that the rate of convergence of the mode expansion depended upon what was chosen for the perturbation potential within the hard core (cf. their Fig. 2 for examples) but that very rapid convergence could be achieved from simple choices. The results of using a_1 and a_2 in Eq. (15–98) are shown in Figs. 15–6 and 15–7 as M1 and M2. The equations are not difficult to apply in this case, and the numerical results compare quite favorably with the integral equation methods of Rasaiah and Freidman. Tables 15–2 and 15–3 show a comparison of the mode expansion method with the hypernetted-chain results. Table 15–2 also shows the convergence of the method for the excess Helmholtz free energy. It should be pointed out that the results in Tables 15–2 and 15–3 agree quite well with the Monte Carlo calculations of Card and Valleau.

For high values of $z_1 z_2 e^2/\varepsilon k T R$, the convergence of the mode expansion is not fast, i.e., a_2 is not $\ll a_1$. In the third paper of their mode expansion series, Andersen and Chandler describe a criterion for optimizing the rate of convergence of the mode expansion. This criterion ensures that the distribution functions vanish in the physically inaccessible region within the hard core. The series [Eq. (15–98)] truncated at $n = 1$ is called the random-phase approximation (RPA), and when the Andersen-Chandler optimization is applied, it is called the optimized random-phase approximation (ORPA). A remarkable result of this optimization criterion is that ORPA is equivalent to another theory, the mean spherical model, if the hard-core reference system is approximated by the Percus-Yevick theory. The mean spherical model (MSM) was applied to ionic solution theory by Waisman and Lebowitz† and is discussed in Problems 15–37 through 15–39. The addition of the $n = 2$ term in Eq. (15–98) then represents an improvement over MSM and is now referred to as the MEX theory.

---

* H. C. Andersen and D. Chandler, *J. Chem. Phys.*, **53**, p. 547, 1970; **54**, p. 26, 1971; **55**, p. 1497, 1971.

† E. Waisman and J. L. Lebowitz, *J. Chem. Phys.*, **52**, p. 4307, 1970; **56**, p. 3086, 1972; **56**, p. 3093, 1972.

Table 15–4 compares the thermodynamic properties of the restricted primitive model of a 1–1 electrolyte calculated by the MEX theory to the Monte Carlo data of Card and Valleau.

Table 15–2. **Excess free energy of a model ionic solution**

I^*	$\mathscr{A}_0/c$	a_1/c	a_2/c	$\mathscr{A}/c$†	HNC $\mathscr{A}/c$‡	difference
0.05	−0.0247	0.2291	−0.0157	0.2587		
0.1	−0.0495	0.3990	−0.0249	0.3247	0.3262	−0.0015
0.2	−0.0998	0.5225	−0.0362	0.3866	0.3854	+0.0012
0.3	−0.1508	0.6055	−0.0431	0.4117	0.4078	+0.0039
0.5	−0.2553	0.7203	−0.0506	0.4143	0.408	+0.006
0.7	−0.3635	0.8011	−0.0541	0.3837	0.376	+0.008
0.8	−0.4185	0.8341	−0.0551	0.3605	0.352	+0.009
0.9	−0.4747	0.8636	−0.0556	0.3333	0.328	+0.005
1.0	−0.5318	0.8902	−0.0558	0.3025	0.298	+0.005

* Ionic strength or concentration in moles per liter.
† These numbers are the sum of reference system, one-mode, and two-mode contributions. They contain computational errors in the fourth decimal place.
‡ Calculated from osmotic coefficients and activity coefficients obtained from solutions of the hypernetted-chain equation.
Source: D. Chandler and H. C. Andersen, *J. Chem. Phys.*, **54**, p. 26, 1971.

Table 15–3. **Mean ionic activity coefficient of a model ionic solution**

I^*	$\ln \gamma_\pm$†	HNC $\ln \gamma_\pm$‡	difference
0.1	−0.2104	−0.2101	−0.0003
0.2	−0.2315	−0.2286	−0.0029
0.3	−0.2275	−0.2221	−0.0054
0.5	−0.1854	−0.1776	−0.0078
0.7	−0.1186	−0.1099	−0.0087
0.8	−0.0789	−0.0704	−0.0085
0.9	−0.0358	−0.0277	−0.0081
1.0	+0.0104	+0.0181	−0.0077

* Ionic strength or concentration in moles per liter.
† These results are the sum of reference system, one-mode, and two-mode contributions. They contain computational errors in the fourth decimal place.
‡ Obtained from solutions of the hypernetted-chain equation.
Source: D. Chandler and H. C. Andersen, *J. Chem. Phys.*, **54**, p. 26, 1971.

Table 15–4. **Thermodynamic properties of the restricted primitive model of a 1–1 electrolyte solution:** $T = 298.0°K$, $\varepsilon = 78.5$, **and** $d = 4.25$ Å

m	ϕ_2^*	ϕ_{MC}†	$(\ln \gamma_\pm)_2$‡	$(\ln \gamma_\pm)_{MC}$§	$(−E/2NkT)_2$‖	$(−E/2NkT)_{MC}$#
0.00911	0.9706	0.9701 ± 0.0008	−0.0957	−0.0973	0.0993	0.1029 ± 0.0013
0.10376	0.9451	0.9445 ± 0.0012	−0.2273	−0.2311	0.2678	0.2739 ± 0.0014
0.42502	0.9786	0.9774 ± 0.0046	−0.2587	−0.2643	0.4285	0.4341 ± 0.0017
1.0001	1.0906	1.094 ± 0.005	−0.1263	−0.1265	0.5472	0.5516 ± 0.0016
1.9676	1.3426	1.346 ± 0.009	+0.2587	+0.254	0.6519	0.6511 ± 0.0020

* The mode expansion osmotic coefficient, including the two-mode contribution.
† Monte Carlo osmotic coefficient.
‡ The mode expansion logarithm of the mean ionic activity coefficient, including the two-mode contribution.
§ The Monte Carlo result. The standard error for these results are only slightly larger than those for the osmotic coefficient at the same concentration.
‖ Mode expansion result, including the two-mode contribution, for the negative reduced excess internal energy per ion.
The Monte Carlo result, with standard errors.
Source: H. C. Andersen and D. Chandler, *J. Chem. Phys.*, **55**, p. 1497, 1971.

In a second series of three papers,* Andersen and Chandler further investigate the concept of the appropriate choice of perturbation potential. They start by transforming the usual Mayer diagrammatic expansion for the Helmholtz free energy and pair correlation function of a multicomponent fluid using certain topological reduction techniques that are beyond the level of this book. The result of this analysis is that the diagrams are expressed in terms of a so-called renormalized potential $\mathscr{C}(r)$ instead of $u(r)$ itself. This renormalized potential is defined as a sum of certain kinds of diagrams and, when evaluated for a 1–1 restricted primitive electrolyte, turns out to be

$$\mathscr{C}_{ij}(r) = \frac{(-1)^{i+j}}{(2\pi)^3} \int e^{-i\mathbf{k}\cdot\mathbf{r}}\, \hat{v}(k)\, d\mathbf{k} \qquad (15\text{–}99)$$

where $\hat{v}(k) = -\beta\hat{u}(k)/[1 + 2\beta c\hat{u}(k)]$ and $\hat{u}(k)$ is the Fourier transform of the optimized potential, which is coulombic outside the hard core and chosen such that $g(r) = 0$ within. The expansion of the excess Helmholtz free energy is

$$\mathscr{A} = \mathscr{A}_{\text{ORPA}} + B_2 + S \qquad (15\text{–}100)$$

where

$$B_2 = \tfrac{1}{2} \sum_{i,j} \rho_i \rho_j \int d\mathbf{r}\, \{(e^{\mathscr{C}_{ij}} - 1 - \mathscr{C}_{ij})g_{ij}{}^0 - \tfrac{1}{2}\mathscr{C}_{ij}{}^2\}$$

They also show that the radial distribution function can be written as

$$g_{ij}(r) = g_{ij}{}^{\text{HS}}(r)\exp[\mathscr{C}_{ij}(r)] + S' \qquad (15\text{–}101)$$

In Eqs. (15–100) and (15–101) S and S' are the sum of all the diagrams not included in the first terms.

When the functional optimization described before is applied, the result is that the contributions of S and S' are negligible. Equations (15–100) and (15–101) with S and S' equal to zero and the potential optimized are referred to as the ORPA $+ B_2$ and

Table 15–5. **Thermodynamic properties for the restricted primitive model for z_1–z_2 electrolyte solutions:** $T = 298.16°\text{K}$, $\varepsilon = 78.358$, and $d = 4.2$ Å

charge type	C_{st}	$-\Delta E/NkT$			ϕ		
		ORPA $+ B_2$	HNC	MSM	ORPA $+ B_2$	HNC	MSM
		(± 0.001)			(± 0.001)		
2–1	0.005	0.265	0.2631	0.2479	0.923	0.9239	0.9248
2–1	0.05	0.644	0.6324	0.6130	0.847	0.8502	0.8466
2–1	0.2	0.973	0.9492	0.9489	0.830	0.8358	0.8299
2–1	1.0	1.432	1.3870	1.4059	1.041	1.0475	1.0615
2–1	1.3333	1.523	1.4730	1.4899	1.167	1.1778	1.1994
3–1	0.005	0.585	0.5638	0.4981	0.849	0.8530	0.8520
3–1	0.01	0.759	0.7279	0.6568	0.812	0.8186	0.8130
3–1	0.1	1.486	1.4178	1.4233	0.708	0.7242	0.6960
3–1	0.5	2.124	2.0023	2.1088	0.788	0.8065	0.7981
3–1	1.0	2.433	2.2830	2.4113	1 011	1.0264	1.0544
2–2	0.005	1.055	0.8158	0.5607	0.785	0.8077	0.8294
2–2	0.0625	2.062	1.737	1.4554	0.643	0.6452	0.6303
2–2	0.5625	2.916	2.757	2.6438	9.595	0.6028	0.5732
2–2	2.0	3.579	3.396	3.3799	0.857	0.8865	0.9112

Source: S. Hudson and H. C. Andersen, *J. Chem. Phys.*, **60**, p. 2188, 1974.

* H. C. Andersen and D. Chandler, *J. Chem. Phys.*, **57**, p. 1918, 1972; **57**, p. 1930, 1972; **57**, p. 2626, 1972.

exponential approximation (EXP), respectively. These approximations have been applied to z_1–z_2 electrolytes by Hudson and Andersen,* and the results are shown in Table 15–5 where it can be seen that the agreement among the various theories is quite good. This is some indication, however, that the agreement does begin to break down as z_1 and z_2 increase. All of the recent advances described in this section are discussed very well by Rasaiah (in "Additional Reading").

ADDITIONAL READING

FALKENHAGEN, H. 1934. *Electrolytes*. London: Oxford University Press.
————, and KELBG, G. *Mod. Aspects Electrochem.*, **2**, p. 1, 1959.
FOWLER, R. H., and GUGGENHEIM, E. A. 1956. *Statistical thermodynamics*. Cambridge: Cambridge University Press.
FRIEDMAN, H. L. 1962. *Ionic solution theory*. New York: Interscience.
————. *Mod. Aspects Electrochem.*, **6**, p. 1, 1970.
————. *J. Solution Chem.*, **1**, p. 387, 1972.
GUGGENHEIM, E. A., and STOKES, R. H. 1969. *Equilibrium properties of aqueous solutions of single strong electrolytes*. New York: Pergamon.
OLIVARES, W., and McQUARRIE, D. A. *Biophys. J.*, **15**, p. 143, 1975.
RASAIAH, J. C. *J. Solution Chem.*, **2**, p. 301, 1973.
RESIBOIS, P. M. V. 1968. *Electrolyte theory*. New York: Harper & Row.
RICE, S. A., and NAGASAWA, M. 1961. *Polyelectrolyte solutions*. New York: Academic.
ROBINSON, R. A., and STOKES, R. H. 1965. *Electrolyte solutions*. 2nd. ed. New York: Academic.

PROBLEMS

15–1. Show that the second virial coefficient diverges for a coulombic potential.

15–2. Show that Eq. (15–7) is a solution to Poisson's equation. To do this you need to know that $\nabla^2(1/r) = \delta(r)$.

15–3. Using the divergence theorem from vector calculus, derive Poisson's equation from Gauss' law.

15–4. Use Gauss' law to show that the electric field due to an infinite uniformly charged plane is $4\pi\sigma$, where σ is the surface charge density.

15–5. Calculate the electric field within a sphere bounded by a uniformly charged surface.

15–6. Use Gauss' law to show that the normal component of $\mathbf{D} = \varepsilon\mathbf{E}$ on crossing a charged surface has a discontinuity equal to $4\pi\sigma$, where σ is the surface charge density.

15–7. Use Gauss' law to show that the electric field due to an infinite uniformly charged line of linear charge density e is $2e/r$. What does this say about the potential? Is there any difficulty in choosing the zero of energy?

15–8. Derive Eq. (15–16), i.e., that $\langle\psi_j\rangle = (\partial A/\partial q_j)$.

15–9. Show that the Debye-Hückel differential equation, Eq. (15–32),

$$\nabla^2\phi = \frac{1}{r^2}\frac{d}{dr}\left(r^2\frac{d\phi}{dr}\right) = \kappa^2\phi$$

can be written as

$$\frac{1}{r}\frac{d^2}{dr^2}(r\phi) = \kappa^2\phi$$

This is a standard substitution in handling a spherically symmetric Laplacian operator.

---

* S. Hudson and H. C. Andersen, *J. Chem. Phys.*, **60**, p. 2188, 1974.

15–10. The charge density around the jth central ion in the Debye-Hückel theory is

$$n_j(r) = -\frac{z_j e \kappa^2 e^{-\kappa(r-a)}}{4\pi r(1+\kappa a)}$$

Find the most probable value and the mean value of r in terms of κ.

15–11. Calculate the mean thickness of the Debye-Hückel ionic atmosphere and ionic strength of a 0.1 and 0.01 M solution of a 1–1 electrolyte in (a) water at 25°C ($\varepsilon = 78$), and (b) methanol at 25°C ($\varepsilon = 31.5$).

15–12. Derive Eq. (15–55).

15–13. Show that for small κa, $\tau(\kappa a) = 1 - \frac{3}{4}\kappa a + \cdots$.

15–14. Derive Eq. (15–59) from a generalization of Eq. (13–34).

15–15. Show that $\beta p^{el} = -\kappa^3 \sigma(\kappa a)/24\pi$, where $\sigma(\kappa a)$ is given by Eq. (15–63).

15–16. Show that $\sigma(\kappa a)$ in the previous problem becomes $1 - \frac{3}{2}\kappa a$ for small κa.

15–17. Show that the electroneutrality condition allows one to write

$$\frac{v_+ q_+^2 + v_- q_-^2}{v_+ + v_-} = |q_+ q_-|$$

Use this to derive Eq. (15–77) from Eq. (15–76).

15–18. Show that the Debye-Hückel expression for $\langle \psi_j \rangle$, i.e., Eq. (15–43), satisfies the symmetry condition of Eq. (15–47).

15–19. Show that $\kappa^{-1} = 3.04c^{-1/2}$ Å for a 1–1 aqueous electrolyte solution at 25°C when c is the concentration of salt expressed in moles per liter.

15–20. Show that the linearization of the exponents that leads to the Debye-Hückel equation is valid for concentrations less than around 0.01 M for a 1–1 aqueous electrolyte solution.

15–21. The mean molar activity coefficients of NaCl at 25°C are

0.001 M	0.005 M	0.01 M	0.05 M	0.1 M	0.5 M	1.0 M	2.0 M
0.966	0.929	0.904	0.823	0.778	0.682	0.685	0.671

Calculate $\gamma_\pm$ according to the Debye-Hückel theory and compare the result to the above data.

15–22. Using the Debye-Hückel theory, calculate the mean ionic activity coefficient of a 1–1 aqueous electrolyte solution at 25°C as a function of concentration. Do the same for a 2–2 electrolyte. Compare the results to those in Table 15–3.

15–23. Calculate the corresponding entries in Table 15–4 for the Debye-Hückel theory, with and without the κa terms.

15–24. Show that the solution to the Debye-Hückel differential equation for the case in which the charge q of the central ion is uniformly distributed over its surface is

$$\psi(r) = \frac{q}{\varepsilon a}\left(1 - \frac{\kappa a}{1+\kappa a}\right) \qquad 0 < r < a$$

Using the equation

$$W_{el} = \int_0^q \psi(q') \, dq'$$

for the electrostatic free energy of the central ion, show that

$$W_{el} = \frac{q^2}{2\varepsilon a}\left(1 - \frac{\kappa a}{1+\kappa a}\right)$$

Now show that if the fixed central ion has radius b and it is immersed in a solution of mobile ions whose centers can approach the center of the sphere to within a distance a (i.e., $a = b +$ the radius of the mobile ions), then

$$W_{el} = \frac{q^2}{2\varepsilon b}\left(1 - \frac{\kappa b}{1 + \kappa a}\right)$$

Show that this result is independent of the dielectric constant of the central ion (see Tanford in "Additional Reading").

15–25. Show that the radial distribution functions of the Debye-Hückel theory are given by

$$g_{ij}(r) = 1 - \frac{q_i q_j}{\varepsilon kT}\frac{e^{-\kappa r}}{r}$$

(in the limiting law limit of $\kappa a \to 0$). Interpret this equation. Show how this is consistent with the expression for the net charge density in the ionic atmosphere, Eq. (15–51).

15–26. Derive expressions for all the thermodynamic properties of an ionic solution from the Debye-Hückel radial distribution functions given in Problem 15–25 and

$$\beta E(c, \beta) = \beta \sum_{i,j=1}^{2} c_i c_j \int_{a_{ij}}^{\infty} g_{ij}(r)u_{ij}(r)4\pi r^2\, dr$$

where

$$u_{ij} = \infty \qquad r < a_{ij}$$

$$= \frac{q_i q_j}{\varepsilon r} \qquad r > a_{ij}$$

Hint: Recall the Gibbs-Helmholtz equation for A in terms of a temperature integration over β.

15–27. Show that in terms of the summation function S defined by Eq. (15–82), the equation state is given by

$$\frac{p}{kT} = c_1 + c_2 + S - c_1\frac{\partial S}{\partial c_1} - c_2\frac{\partial S}{\partial c_2}$$

15–28. Derive expressions for the osmotic coefficient ϕ and the logarithm of the activity coefficient $\ln \gamma_\pm$ in terms of $\mathscr{A} = -\beta A^{ex}/V$.

15–29. Show the Eq. (15–88) for f_{ij} is simply an identity.

15–30. Show that the summation over the set of graphs labeled by $p(1)$ in Fig. 15–5 gives no contribution to the thermodynamic properties because of electroneutrality.

15–31. Show that the contribution to S due to the cyclic graphs is

$$S^{(cyclic)} = \frac{\kappa^3}{12\pi}$$

15–32. Show that the Fourier transform of Eq. (15–95) is given by Eq. (15–96).

15–33. Use either the Kirkwood or Born-Green-Yvon equation to show that the Debye-Hückel theory is an exact limiting law.

15–34. An apparently excellent approximation for the interionic radial distribution functions is to write

$$g_{ij}(r) = g_{ij}^{HS}(r)e^{-\phi^{DH}(r)/kT}$$

where

$$\phi^{DH}(r) = \frac{q_i q_j\, e^{-\kappa(r-a)}}{\varepsilon r(1 + \kappa a)}$$

Using this approximation in the case in which all the ions have the same size, calculate βE and all the other thermodynamic properties of a 1–1 electrolyte by the method outlined in Problem 15–26. Compare your results to Tables 15–3 and 15–4. Use the Grundke-Henderson program for $g^{HS}(r)$ given in Appendix D.*

15–35. When an electrode or any charged surface is immersed in an ionic solution, an ionic atmosphere is formed next to the electrode. Since the net charge of the ionic atmosphere is opposite that of the electrode, one can picture the electrode and its ionic atmosphere as two oppositely charged planes separated by a distance around $1/\kappa$. Hence this is called a diffuse double layer and occurs whenever a large charged body is immersed in an ionic solution. The theory of the diffuse double layer was given by Gouy and Chapman and is now called the Gouy-Chapman theory. They start with the nonlinear Poisson-Boltzmann equation, which in this case depends only upon one variable x, the distance from the charged planar surface. If we let the potential be ϕ, then we have

$$\frac{d^2\phi}{dx^2} = -\frac{4\pi}{\varepsilon} \sum_s c_s q_s e^{-\beta q_s \phi}$$

Show that for the case of a symmetrical binary electrolyte, i.e., one in which $q_+ = -q_- = q$, this equation becomes

$$\frac{d^2\phi}{dx^2} = \frac{8\pi q c}{\varepsilon} \sinh(\beta q \phi)$$

Let $\phi^* = \beta q \phi$ and $\kappa^2 = 8\pi q^2 c/\varepsilon k T$, and write this equation in the reduced form

$$\frac{d^2\phi^*}{d\xi^2} = \sinh \phi^*$$

where $\xi = \kappa x$. Using the identity

$$\frac{1}{2}\frac{d}{dx}\left(\frac{dy}{dx}\right)^2 = \left(\frac{dy}{dx}\right)\left(\frac{d^2y}{dx^2}\right)$$

integrate this equation once to get

$$\frac{d\phi^*}{d\xi} = -(2\cosh\phi^* - 2)^{1/2} = -2\sinh\left(\frac{\phi^*}{2}\right)$$

where the fact that ϕ and $d\phi/dx \to 0$ as $x \to \infty$ has been used. If we let the electrode have a fixed potential ϕ_0, integrate once more to get

$$\exp(\phi^*/2) = \frac{e^{\phi_0^*/2} + 1 + (e^{\phi_0^*/2} - 1)e^{-\xi}}{e^{\phi_0^*/2} + 1 - (e^{\phi_0^*/2} - 1)e^{-\xi}}$$

This is called the Gouy-Chapman equation. Plot ϕ versus x and show that ϕ decreases roughly exponentially as a function of x. Show that if the Poisson-Boltzmann is linearized by replacing $\sinh(\beta q \phi)$ by $\beta q \phi$, then

$$\phi(x) = \phi_0 e^{-\kappa x}$$

Compare this to the Gouy-Chapman equation as a function of concentration. (The work by Verwey and Overbeek in "Additional Reading" is the classic reference for the Gouy-Chapman theory.)

15–36. By starting with the fact that the total charge surrounding a central ion must be equal in magnitude but of opposite sign to charge on the central ion, i.e., with

$$\sum_j q_j c_j \int_0^\infty g_{ij}(r) 4\pi r^2 \, dr = -q_i$$

---

* See W. Olivares and D. A. McQuarrie, *Biophys. J.*, **15**, p. 143, 1975.

show that

$$\sum_{i,j} q_i c_i q_j c_j \int_0^\infty g_{ij}(r) 4\pi r^2 \, dr = -\frac{\varepsilon k T \kappa^2}{4\pi}$$

is an expression of the condition of electroneutrality. This is an interesting way to write this equation since Stillinger and Lovett* have derived the additional condition

$$\sum_{i,j} q_i c_i q_j c_j \int_0^\infty g_{ij}(r) 4\pi r^4 \, dr = -\frac{3\varepsilon k T}{2\pi}$$

This is now called the second-moment condition. The condition of electroneutrality can be thought of as the zero-moment condition. Show that $g_{ij}^{DH}(r)$ satisfies both moment conditions.

15–37. In 1966, Lebowitz and Percus† derived an approximate integral equation, which applied to simple liquids or ionic solutions, is now called the mean spherical model and consists of the conditions

$$g_{ij}(r) = 0 \qquad\qquad r < R_{ij}$$

$$c_{ij}(r) = -\beta u_{ij}(r) \qquad r > R_{ij}$$

along with the Ornstein-Zernike equation. Note that the first condition is exact for a potential with a hard-core repulsive part (such as in the primitive model of electrolytes) and that the approximation consists of setting $c_{ij}(r) = -\beta u_{ij}(r)$ outside of the hard-core diameter. Using the Ornstein-Zernike equation, one can then solve for $g_{ij}(r)$ outside of R_{ij} and $c_{ij}(r)$ inside R_{ij}. For a strictly hard-sphere potential, $u_{ij} = 0$ for $r > R_{ij}$ and so $c_{ij}(r) = 0$ there also. Show that in this case, the mean spherical model is the same as the Percus-Yevick equation for hard spheres. Since the mean spherical model, then, is an extension of the hard-sphere Percus-Yevick equation, it is not surprising that it can be solved analytically for certain $u_{ij}(r)$, and Waisman and Lebowitz‡ have solved it for the restricted primitive model. They find that for a 1–1 electrolyte that

$$c_{ij}(r) = c_{ij}^0(r) - \left(\frac{\beta}{\varepsilon R}\right) q_i q_j \left(2B - \frac{B^2 r}{R}\right) \qquad r < R$$

$$= -\frac{\beta q_i q_j}{\varepsilon r} \qquad\qquad\qquad\qquad r > R$$

where $c_{ij}^0(r)$ is the direct correlation function of a system of uncharged hard spheres. R is the diameter of the ions (restricted primitive model), and

$$B = x^{-2}[x^2 + x - x(1 + 2x)^{1/2}]$$

where

$$x^2 = \kappa^2 R^2 = \left[\left(\frac{4\pi\beta}{\varepsilon}\right) \sum_{j=1}^m c_i q_i^2\right] R^2$$

Waisman and Lebowitz§ show that $c_{ij}(r)$ can be used directly to give

$$E^{ex}(x, \beta) = -\frac{x^2 + x - x(1 + 2x)^{1/2}}{4\pi\beta R^3}$$

* *J. Chem. Phys.*, **49**, p. 1991, 1968.
† *Phys. Rev.*, **144**, p. 251, 1966.
‡ *J. Chem. Phys.*, **52**, p. 4307, 1970; **56**, p. 3086, 1972; **56**, p. 3093, 1972.
§ *J. Chem. Phys.*, **56**, p. 3086, 1972.

Using the Gibbs-Helmholtz equation, show that

$$\beta \frac{(A - A^0)}{V} = \int_0^\beta E^{ex}(\rho, \beta') \, d\beta'$$

$$= -(12\pi R^3)^{-1}(6x + 3x^2 + 2 - 2(1 + 2x)^{3/2}]$$

Show from this that the osmotic coefficient is

$$\phi_E = \phi^0 + [4\pi(c_1 + c_2)R^3]^{-1}[x + x(1 + 2x)^{1/2} - \tfrac{2}{3}(1 + 2x)^{3/2} + \tfrac{2}{3}]$$

where the subscript E indicates that ϕ is calculated through the equation for $E^{ex}(x, \beta)$. Show also that the logarithm of the mean activity coefficient is

$$\ln \gamma_\pm^{el} = \left[\frac{x(1 + 2x)^{1/2} - x - x^2}{4\pi(c_1 + c_2)R^3} \right]$$

Table 15–6 gives a comparison of this equation to several other approaches.

Table 15–6. **Comparison of the osmotic coefficient calculated by various theories***

C (moles/liter)	ϕ_v^{MC}	ϕ_v^{HNC}	ϕ_c^{HNC}	ϕ_v^{PYA}	ϕ_c^{PYA}	ϕ_v^{MSM}	ϕ_E^{MSM}	ϕ^{MEX}
0.00911	0.9701 ± 0.0008	0.9703	0.9705	0.9703	0.9705	0.9687	0.9709	0.9707
0.10376	0.9445 ± 0.0012	0.9453	0.9458	0.9452	0.9461	0.9312	0.9454	0.9452
0.42502	0.9774 ± 0.0046	0.9796	0.9800	0.9765	0.9844	0.9446	0.9806	0.9787
1.0001	1.094 ± 0.005	1.0926	1.0906	1.0789	1.1076	1.039	1.097	1.091
1.9676	1.346 ± 0.009	1.3514	1.3404	1.3114	1.386†	1.2757	1.3595	1.342†

* For these calculations, $a = 4.25$ Å, $\varepsilon = 78.5$, and $T = 298°$K.
† There is some uncertainty in the last digit.
Source: J. C. Rasaiah, *J. Solution Chem.*, **2**, p. 301, 1973.

The mean spherical model has been solved analytically for fluids of hard spheres with permanent electric dipole moments by Wertheim* and for a mixture of such particles by Adelman and Deutch.†

15–38. Plot $\ln \gamma_\pm^{el}$ of the mean spherical model (see previous problem) as a function of $x = (8\pi e^2 c\beta R^2/\varepsilon)^{1/2}$ for an aqueous solution of 1–1 electrolyte at 25°C. Plot on the same graph $\ln \gamma_\pm^{DH}$ and $\ln \gamma_\pm^{DHLL}$, where DH denotes Debye-Hückel theory with the κa terms retained.

15–39. Using the analytic expression for $\ln \gamma_\pm^{el}$ in the mean spherical model (Problem 15–37), show that in the limit $x \to 0$ one obtains the Debye-Hückel limiting law. Note that $\ln \gamma_\pm^{el} = \ln \gamma_\pm - \ln \gamma^{HS}$ is the electrostatic contribution to the logarithm of the activity coefficient.

* *J. Chem. Phys.*, **55**, p. 4291, 1971.
† *J. Chem. Phys.*, **59**, p. 3971, 1973.

CHAPTER 16

KINETIC THEORY OF GASES AND MOLECULAR COLLISIONS

In the first 15 chapters we have discussed only systems that are in thermodynamic equilibrium. We started with simple systems such as ideal gases, where the intermolecular potential can be neglected, and progressed to more complicated systems where the intermolecular potential plays a key role. In the next seven chapters we shall study the molecular theory of transport through gases and liquids. This is now a very active area of research, and it has been only in the 1960s that this field has been set on a basis comparable to equilibrium statistical mechanics. There are still a number of tricky conceptual problems that rise up now and then, but on the whole and within the level of this book, nonequilibrium or irreversible statistical mechanics is only slightly more complicated than equilibrium statistical mechanics (statistical thermodynamics).

In this chapter we shall first review the elementary kinetic theory of gases and molecular collisions. In the next chapter we shall discuss the macroscopic equations of continuum mechanics or hydrodynamics, such as the continuity equation, the macroscopic momentum balance equation, and energy balance equation. This is essentially the "thermodynamic" background for nonequilibrium statistical mechanics. In the next chapter we shall review the derivation of the most fundamental equation of statistical mechanics, namely, the Liouville equation, and then introduce the various distribution functions and flux vectors. Then we shall derive the fundamental equation of change of dilute gases, the Boltzmann equation. We shall discuss some of its direct consequences, and then the standard method of solving this integrodifferential equation, the so-called Chapman-Enskog method. Then we shall present the main results of the Chapman-Enskog theory and make an extensive comparison to experimental data. In the remaining chapters we shall study several approaches to the transport theory of liquids ending with what is probably the most successful and well accepted to date, the time-correlation-function method. Before going on to study these various theories of transport in gases and liquids, however, let us review some elementary kinetic theory of gases, since that is what we are about to generalize and put into a general statistical mechanical framework.

16–1 ELEMENTARY KINETIC THEORY OF TRANSPORT IN GASES

In this section we shall discuss an elementary kinetic theory treatment of transport phenomena in dilute gases. We shall derive expressions for the principal thermal transport coefficients, namely, the coefficients of diffusion, viscosity, and thermal conductivity. The kinetic theory of transport in gases can be approached from a number of levels, varying from very simple treatments, in which all the molecules are assumed to have the same (average) velocity and move only in the x-, y-, and z-directions, to very sophisticated treatments which make no unnecessary assumptions. An interesting thing is that the results of these various theories differ only by constant factors of the order of unity. One can generate pages of algebra to get a more exact equation which may have the same form as the simple equation but introduce a missing factor like $\frac{3}{8}$. In order to introduce the basic ideas of transport in gases, we shall use a treatment intermediate in rigor. There are many texts dealing with the kinetic theory of gases. Several of these are given in the "Additional Readings."

Recall from Chapter 7 that the distribution of molecular velocities in a one-component gas of molecules with mass m is given by the Maxwell-Boltzmann distribution:

$$f(v_x, v_y, v_z)\, dv_x\, dv_y\, dv_z = \left(\frac{m}{2\pi kT}\right)^{3/2} \exp\left\{-\frac{m}{2kT}\left(v_x^{\,2} + v_y^{\,2} + v_z^{\,2}\right)\right\} dv_x\, dv_y\, dv_z$$

or

$$f(v)\, dv = 4\pi\left(\frac{m}{2\pi kT}\right)^{3/2} v^2 \exp\left\{-\frac{mv^2}{2kT}\right\} dv \tag{16–1}$$

where $v^2 = v_x^{\,2} + v_y^{\,2} + v_z^{\,2}$. Some average values derived from these distributions are (cf. Problems 7–15 through 7–24)

$$\bar{v}_x = \bar{v}_y = \bar{v}_z = 0$$

$$\overline{v_x^{\,2}} = \overline{v_y^{\,2}} = \overline{v_z^{\,2}} = \tfrac{1}{3}\overline{v^2} = \frac{kT}{m}$$

$$\bar{v} = \left(\frac{8kT}{m\pi}\right)^{1/2} \tag{16–2}$$

$$(\overline{v^2})^{1/2} = \text{root-mean-square velocity} = \left(\frac{3kT}{m}\right)^{1/2}$$

Note that the values of $\overline{v_x^{\,2}}$, $\overline{v_y^{\,2}}$, $\overline{v_z^{\,2}}$, and $\overline{v^2}$ are the results expected from equipartition (Chapter 7).

We now introduce the idea of the mean free path. In general, the average distance traveled by a molecule between collisions cannot be well defined since a collision itself cannot be unambiguously defined. We can get a good idea of this quantity, however, if we use rigid spheres as a guide. Consider a molecule with diameter σ moving in the x-direction. Since there will be a collision if the *center* of our one molecule comes within a distance σ of the center of one of the other molecules in the gas, each of these molecules presents a target of effective diameter 2σ, and hence of area $\pi\sigma^2$. This is shown in Fig. 16–1. The number of such targets in a plane of unit area perpendicular to the x-direction and of thickness dx is $\rho\, dx$, where ρ is the number density of molecules in the gas. Neglecting overlap, the total target area presented by these molecules

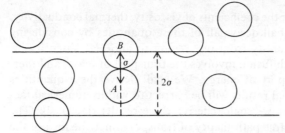

Figure 16–1. **An illustration of why the effective collision diameter is 2σ instead of just σ itself.**

is $\pi\sigma^2\rho\,dx$. The probability that our one molecule will suffer a collision is then the ratio of this area to the total area (1 cm^2):

$$\text{probability of a collision} = \pi\sigma^2\rho\,dx \qquad (16\text{--}3)$$

Now consider a beam of n_0 molecules with approximately equal velocities in the positive x-direction. Let them all start at $x = 0$. The number of molecules that undergo a collision between x and $x + dx$ is the number of molecules reaching x, $n(x)$, multiplied by the probability of a collision in dx, $\pi\sigma\rho\,dx$. We can write, then,

$$n(x + dx) - n(x) = -\pi\sigma^2\rho n(x)\,dx$$

or

$$\frac{dn}{dx} = -\pi\sigma^2\rho n(x)$$

whose solution is

$$n(x) = n_0\,e^{-\pi\sigma^2\rho x} \qquad (16\text{--}4)$$

The mean distance traveled is obtained by multiplying the distance x by $dn(x)$, the number of molecules colliding between x and $x + dx$, integrating over all values of x, and then dividing by the total number of molecules. If we denote the mean free path by l, then

$$l = \frac{1}{n_0}\int_0^{n_0} x\,dn = -\frac{1}{n_0}\int_0^{\infty} x\,\frac{dn}{dx}\,dx = \frac{1}{\pi\sigma^2\rho} \qquad (16\text{--}5)$$

which is the result we were after. This differs by a numerical factor of order unity from a more rigorous derivation, and we shall see in Section 16–3 that the correct expression for l is $1/2^{1/2}\pi\sigma^2\rho$.

We can get a feel for the order of magnitude of the mean free path by using the ideal gas equation to eliminate ρ, giving

$$l = \frac{kT}{\pi\sigma^2 p} \qquad (16\text{--}6)$$

At 300°K and for $\sigma = 5.0$ Å, $l = 5 \times 10^{-6}$ cm for a pressure of one atmosphere and 4×10^{-3} cm for a pressure of 1 mm Hg. The average time between collisions is obtained by dividing l by the average velocity. At standard conditions, this number is approximately 10^{-10} sec, i.e., one molecule of a gas at standard conditions undergoes about 10^{10} collisions per second.

We shall now derive expressions for the coefficients of viscosity, thermal conductivity, and diffusion for a dilute gas. We shall derive all of these quantities by considering the transport of some molecular quantity from one region to another. Viscosity involves the transport of momentum; diffusion involves the transport of mass; and thermal conductivity involves the transport of energy. We shall derive these quantities somewhat nonrigorously, but our final results will be correct to within numerical factors of order unity. In later chapters we shall rederive these results rigorously. The derivation we use is called the mean free path theory of transport since the idea of the mean free path will be of central importance. Let us consider viscosity first.

The usual mechanical setup in an experimental determination of the viscosity of gases is a pair of parallel plates a fixed distance a from each other. This is shown in Fig. 16–2. Let the z-axis be perpendicular to these plates, with one plate located at $z = 0$ and one at $z = a$. The lower plate is kept at rest, and the upper one is moving with a uniform velocity U in the x-direction. If the distance between the plates is large compared to the mean free path, the layer of gas next to the plates assumes the same velocity as the plates themselves. Because of this, the gas has $\bar{v}_x = U$ at $z = a$ and $\bar{v}_x = 0$ at $z = 0$. The average x-component of the velocity between the plates is denoted by $u(z)$. We now assume that $u(z)$ varies linearly with z so that $u(z) = Uz/a$. The average momentum at the height z will be

$$G(z) = \frac{mUz}{a} \tag{16–7}$$

where m is the mass of a molecule.

Although equally many molecules from above and below reach the height z per unit time, the ones coming from above will, on the average, carry more momentum than the ones from below. Consequently there will be a downward flow of momentum through any horizontal plane. Let $\psi(z)$ be this flow per square centimeter per second.

The number of molecules per unit time with velocity $\mathbf{v} = (v_x, v_y, v_z)$ passing through a unit horizontal plane located at z is equal to the number of molecules initially located within a parallelepiped with a 1 cm^2 base and a height $|v_z|$. The volume is then $|v_z|$. The average number of molecules in this parallelepiped is $|v_z|\rho f(v_x, v_y, v_z)$. If v_z is positive, the particles come from below, and if v_z is negative, they come from above. Since a $+v_z$ occurs as often as a $-v_z$, the net flow through the surface is zero, indicating a steady state.

The molecules passing through the unit horizontal plane at z have traveled in a straight path since undergoing their last collision and, on the average, have traveled a distance l. The last collision of a molecule with velocity $\mathbf{v}$, therefore, occurred at a height $z' = z - v_z l/v$. Note that if v_z is positive, the last collision occurs from below, and if v_z is negative, it occurs from above.

We now make the assumption that during the last collision the molecule has come

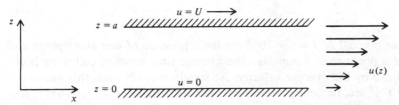

Figure 16–2. **The geometry involved in the experimental setup used to measure the viscosity of a gas.**

into *local equilibrium* with its environment and so carries the momentum associated with that height, i.e., z'. This momentum is

$$G(z') = G\left(z - \frac{v_z l}{v}\right) = G(z) - \frac{v_z l}{v}\frac{dG}{dz} + \cdots \tag{16-8}$$

where we have expanded $G(z')$ about $G(z)$. The net flow of momentum G is given by integrating the product of $G(z')$ and the number of molecules crossing the surface from the height z' over all velocities:

$$\psi(z) = \iiint\limits_{-\infty}^{\infty} G(z')v_z\,\rho f(v_x, v_y, v_z)\,dv_x\,dv_y\,dv_z \tag{16-9}$$

Substituting Eq. (16–8) for $G(z')$ into this gives the sum of two integrals, the first of which is zero and the second is

$$\psi(z) = -\rho l\frac{dG}{dz}\iiint\limits_{-\infty}^{\infty} \frac{v_z^2}{v}\, f(v_x, v_y, v_z)\,dv_x\,dv_y\,dv_z \tag{16-10}$$

Since there is nonzero average velocity in the x-direction, the distribution of molecular velocities in this integral is given by

$$f(v_x, v_y, v_z) = \left(\frac{m}{2\pi kT}\right)^{3/2} \exp\left\{-\frac{m}{2kT}\left[(v_x - u(z))^2 + v_y^2 + v_z^2\right]\right\}$$

If u is small compared to the molecular velocities, however, we can simply ignore the $u(z)$ in the exponent to get

$$\psi(z) = -\tfrac{1}{3}l\rho\bar{v}\frac{dG}{dz} \tag{16-11}$$

The minus sign indicates that the flow is in the direction from higher G to lower G.

Recall now that $G(z) = mUz/a$. Substituting this into Eq. (16–11) gives

$$\psi = -\tfrac{1}{3}\rho m l\bar{v}\,\frac{U}{a} \tag{16-12}$$

which we can write as

$$\psi = -\tfrac{1}{3}\rho m l\bar{v}\,\text{grad } u \tag{16-13}$$

Since ψ represents the rate of momentum transfer, it is really the viscous force acting upon the upper plate. The frictional force per unit area of surface acting on the upper plate is usually given by the macroscopic law (Newton's law of viscous flow)

$$\mathbf{F} = -\eta\,\text{grad } u \tag{16-14}$$

where η is just a proportionality constant called the coefficient of viscosity. Comparison of Eqs. (16–13) and (16–14) gives

$$\eta = \tfrac{1}{3}\rho m l\bar{v} \tag{16-15}$$

Using Eq. (16–5) for l and Eq. (16–2) for $\bar{v}$ gives

$$\eta = \frac{1}{3}\left(\frac{2}{\pi}\right)^{3/2}\frac{(mkT)^{1/2}}{\sigma^2} \tag{16-16}$$

This equation predicts that the coefficient of viscosity is independent of the density, or pressure, and is a function of temperature only. This result was first predicted by Maxwell and has since been demonstrated experimentally over a wide range of pressures.

Notice that we did not need to specify the form of $G(z)$ to derive Eq. (16–11). We can consider $\psi(z)$ to be the flow of some quantity $G(z)$ per second per unit area through a plane parallel to the xy plane located at a height z. When $G(z)$ was specified to the momentum, $\psi(z)$ became the momentum transfer or drag on the moving plate. Suppose then that both plates are stationary, but now there is a temperature gradient because the plates are fixed at different temperatures. The average energy per molecule $\bar{\varepsilon}$ will vary with height, i.e.,

$$G(z) = \bar{\varepsilon}(z) \tag{16–17}$$

$$\frac{dG}{dz} = \frac{d\bar{\varepsilon}}{dT}\frac{dT}{dz} = \frac{C_v}{N_0}\,\text{grad}\,T \tag{16–18}$$

where N_0 is Avogadro's number and C_v is the molar heat capacity. Substitute this into Eq. (16–11) to get the heat flow per unit area through a horizontal plane

$$\psi(z) = -\tfrac{1}{3}\rho l \frac{C_v}{N_0}\,\bar{v}\,\text{grad}\,T \tag{16–19}$$

The macroscopic law here is Fourier's heat law, which says that the heat flow is proportional to the temperature gradient. This law is usually written in the form

$$\psi(z) = -\lambda\,\text{grad}\,T \tag{16–20}$$

where λ is called the thermal conductivity. Comparing Eq. (16–19) with Eq. (16–20) gives

$$\lambda = \tfrac{1}{3}l\rho\bar{v}\frac{C_v}{N_0} \tag{16–21}$$

A comparison of this equation with Eq. (16–15) for η shows that

$$\lambda = \frac{\eta C_v}{M} \tag{16–22}$$

where M is the molecular weight of the gas. Let $C_v/M = c_v$, the specific heat of the substance. Equation (16–22) predicts that $\eta c_v/\lambda = 1$. A more refined theory still predicts that this ratio should be a constant, but somewhat different from unity. Table 16–1 gives this ratio for several gases.

So far we have studied the flow of momentum and the flow of energy. This led to expressions for the coefficient of viscosity and the thermal conductivity. Lastly, we shall study diffusion, which is the net transport of molecules through a surface. The macroscopic law in this case is Fick's law, which says that the flow of molecules of a certain type through a gas at uniform pressure is proportional to the concentration gradient of these molecules, or

$$\text{flow} = -D_1\,\text{grad}\,\rho_1 \tag{16–23}$$

The proportionality constant is called the diffusion coefficient. We shall now derive a kinetic theory expression for D_1. In order to do this we must consider a two-component system.

Let us set up an idealized experimental system consisting of a tube filled with a mixture of two gases of kinds 1 and 2. Let the axis of the tube be the z-direction. The densities of the two gases are functions of z only. In order to look at a flow that is due only to a concentration gradient, and not a pressure or temperature gradient, we must have the total density be the same for all z, i.e.,

$$\rho_1(z) + \rho_2(z) = \rho = \text{constant}$$

Each component has its own distribution of velocities, and so we can write

$$\rho_1(\mathbf{v}, z) = \rho_1(z)f_1(\mathbf{v}) \qquad \rho_2(\mathbf{v}, z) = \rho_2(z)f_2(\mathbf{v})$$

Let us now calculate the flow of molecules of species 1. A particle of species 1 arriving at z comes, on the average, from $z' = z - v_z l_1/v$, where l_1 is the mean free path of species 1 molecules. The density of these molecules at z' is

$$\rho_1(\mathbf{v}, z') = \rho_1(z')f_1(\mathbf{v})$$

$$= \rho_1(z)f_1(\mathbf{v}) - \frac{v_z l_1}{v}\frac{d\rho_1}{dz}\,f_1(\mathbf{v}) + \cdots$$

The number passing through a unit plane perpendicular to the z-axis in the $+z$-direction per second is then

$$v_z \rho_1(\mathbf{v}, z') = \rho_1(z)v_z f_1(\mathbf{v}) - \frac{v_z^2 l_1}{v}\frac{d\rho_1}{dz}\,f_1(\mathbf{v}) + \cdots$$

The net flow is obtained by integrating this over all velocities, giving

$$\psi_1(z) = -l_1\frac{d\rho_1}{dz}\iiint\limits_{-\infty}^{\infty}\frac{v_z^2}{v}\,f_1(v_x, v_y, v_z)\,dv_x\,dv_y\,dv_z$$

or

$$\psi_1(z) = -\tfrac{1}{3}l_1\bar{v}_1\,\text{grad}\,\rho_1 \tag{16-24}$$

Comparison of this to Fick's law gives

$$D_1 = \tfrac{1}{3}l_1\bar{v}_1 \tag{16-25}$$

There is a similar expression for the other component. The mean free path l_1 occurring in this equation is the mean free path of species 1 in a mixture. In a mixture of two gases, the mean free paths are, to the same degree of approximation as Eq. (16-5) (see Problem 16-16),

$$l_1 = \frac{4}{4\pi\sigma_1^2\rho_1 + \pi(\sigma_1 + \sigma_2)^2\rho_2}$$

$$l_2 = \frac{4}{\pi(\sigma_1 + \sigma_2)^2\rho_1 + 4\pi\sigma_2^2\rho_2} \tag{16-26}$$

Using these values for l_1 and l_2, D_1 and D_2 become

$$D_1 = \frac{8}{3}\left(\frac{2kT}{\pi m_1}\right)^{1/2}\frac{1}{4\pi\sigma_1^2\rho_1 + \pi(\sigma_1 + \sigma_2)^2\rho_2} \tag{16-27}$$

$$D_2 = \frac{8}{3}\left(\frac{2kT}{\pi m_2}\right)^{1/2}\frac{1}{\pi(\sigma_1 + \sigma_2)^2\rho_1 + 4\pi\sigma_2^2\rho_2} \tag{16-28}$$

This predicts that the diffusion coefficient of gases should vary inversely with the pressure at constant temperature, i.e., diffusion is proportional to the mean free path. This is observed experimentally.

A particularly interesting quantity for theoretical purposes is the coefficient of self-diffusion, which is the diffusion of some particular "marked" or "tagged" molecule through a gas of otherwise identical molecules. The self-diffusion coefficient is difficult to measure experimentally, but can be measured by studying the interdiffusion of isotopic species. The self-diffusion coefficient is obtained from Eq. (16–28) by letting $\sigma_1 = \sigma_2$ and dropping the subscripts to get

$$D = \frac{1}{3} \left(\frac{2}{\pi}\right)^{3/2} \left(\frac{kT}{m}\right)^{1/2} \frac{1}{\rho \sigma^2} \tag{16–29}$$

We can easily derive the relation

$$\frac{\rho m D}{\eta} = 1 \tag{16–30}$$

by dividing Eq. (16–29) by Eq. (16–15). Table 16–1 tests this ratio.

We said at the beginning of this chapter that the kinetic molecular theory can be approached from various levels of rigor and that the only difference in the results of these various treatments was numerical constants of the order of unity. Let us collect our results here and compare them to the rigorous expressions for hard spheres. We shall derive these rigorous expressions in a later chapter. Our approximate expressions are

$$\eta_{\text{approx}} = \frac{1}{3} \left(\frac{2}{\pi}\right)^{3/2} \frac{(mkT)^{1/2}}{\sigma^2} \tag{16–31}$$

$$\lambda_{\text{approx}} = \frac{1}{3} \left(\frac{2}{\pi}\right)^{3/2} \left(\frac{kT}{m}\right)^{1/2} \frac{C_v}{N_0 \sigma^2} \tag{16–32}$$

$$D_{\text{approx}} = \frac{1}{3} \left(\frac{2}{\pi}\right)^{3/2} \left(\frac{kT}{m}\right)^{1/2} \frac{1}{\sigma^2 \rho} \tag{16–33}$$

$$\frac{\rho m D_{\text{approx}}}{\eta_{\text{approx}}} = 1 \tag{16–34}$$

$$\frac{\eta_{\text{approx}} C_v}{\lambda_{\text{approx}}} = 1 \tag{16–35}$$

The rigorous results are

$$\eta_{\text{rigorous}} = \frac{5}{16\pi^{1/2}} \frac{(mkT)^{1/2}}{\sigma^2} \tag{16–36}$$

$$\lambda_{\text{rigorous}} = \frac{25}{32} \left(\frac{kT}{\pi m}\right)^{1/2} \frac{C_v}{N_0 \sigma^2} \tag{16–37}$$

$$D_{\text{rigorous}} = \frac{3}{8\pi^{1/2}} \left(\frac{kT}{m}\right)^{1/2} \frac{1}{\sigma^2 \rho} \tag{16–38}$$

$$\frac{\rho m D_{\text{rigorous}}}{\lambda_{\text{rigorous}}} = \frac{6}{5} \tag{16–39}$$

$$\frac{\eta_{\text{rigorous}} C_v}{\lambda_{\text{rigorous}}} = \frac{2}{5} \tag{16–40}$$

From all this we see that our approximate expressions differ from the exact expressions by only numerical factors. We see that $\eta_{approx}/\eta_{rigorous} = 32\sqrt{2}/15\pi = 0.96$; $\lambda_{approx}/\lambda_{rigorous} = 64\sqrt{2}/75\pi = 0.38$; $D_{approx}/D_{rigorous} = 16\sqrt{2}/9\pi = 0.80$.

Remember that these are hard-sphere expressions. Even the exact results do not give the observed temperature dependence of the various transport coefficients. Presumably this is because the *effective* hard-sphere diameter of real molecules is a function of temperature. An early attempt to rectify this deficiency replaced the fixed hard-sphere value of σ^2 by

$$\sigma^2 = \sigma_0^2 \left(1 + \frac{S}{T} \right) \tag{16-41}$$

where S is called the Sutherland constant of the particular molecule. This is often a useful way to save the rigorous mean free path expressions for the transport coefficients, but has little fundamental basis. The real problem is that the concept of a mean free path becomes less well defined when there is an attractive potential as in the Lennard-Jones potential. In later chapters we shall develop a molecular theory of transport in gases without appealing to a mean free path. This will be done through

Table 16–1. **Some results from the mean free path theory of transport**

gas	experimental* $\eta \times 10^5$ (g/cm · sec)	$\lambda \times 10^6$ (cal/cm · sec · deg)	D (cm²/sec)	calculated from the rigorous theory σ from η (Å)	σ from λ (Å)	σ from D (Å)	$\eta c_v/\lambda$	$\rho m D/\eta$	Suther-land con-stant (°K)
Ne	29.7	11.1	0.45	2.64	2.57	2.41	0.40	1.37	56
Ar	21.0	3.9	0.16	3.67	3.65	3.47	0.40	1.32	142
Kr	23.3	2.1	0.08	4.20	4.17	4.01	0.40	1.30	188
N₂	16.6	5.5	0.18	3.75	4.30	3.48	0.53	1.39	104
CO₂	13.7	3.5	0.10	4.64	5.76	4.30	0.60	1.40	254
CH₄	10.3	7.2	0.21	4.15	4.80	3.75	0.53	1.43	164
C₂H₆	8.5	4.4	—	5.34	7.00	—	0.69	—	252

* Values at 0°C and 1 atm pressure.

the so-called Boltzmann transport equation. The result of this development will be the derivation of equations that look much like Eqs. (16–36) through (16–38), but with an effective collision diameter which is given rigorously in terms of the intermolecular potential.

We can use Eqs. (16–36) through (16–38) to calculate σ from experimental data. All three equations should give the same value for σ. Table 16–1 gives σ calculated from each of them. It can be seen that not only are the values reasonable, i.e., of order of 10^{-8} cm, but that they are fairly close to one another. Table 16–1 also contains some experimental values of transport coefficients, Sutherland constants, and the ratios in Eqs. (16–39) and (16–40).

16–2 CLASSICAL MECHANICS AND MOLECULAR COLLISIONS

The rigorous molecular theory of transport that we are going to study depends upon the mechanics of the molecular collisions that occur in the gas. In this section then, we shall discuss the elementary classical mechanics of collisions. We assume that the interaction between the molecules depends only upon the distance between them.

Let us consider a collision between two molecules with masses m_1 and m_2. Let their velocities before the collision be $\mathbf{v}_1$ and $\mathbf{v}_2$, and their velocities after the collision be $\mathbf{v}_1{}'$ and $\mathbf{v}_2{}'$. We shall generally use primes to indicate quantities after the collision has occurred. If no chemical reaction takes place, we have the conservation of mass condition, namely, that $m_1 = m_1{}'$ and $m_2 = m_2{}'$. Note that the total mass before the collision is equal to the total mass after the collision,

$$m_1 + m_2 = m_1{}' + m_2{}' \tag{16-42}$$

We, furthermore, have conservation of momentum (as long as there are no external forces):

$$m_1\mathbf{v}_1 + m_2\mathbf{v}_2 = m_1\mathbf{v}_1{}' + m_2\mathbf{v}_2{}' \tag{16-43}$$

Assuming that the energy contained in the internal degrees of freedom is not changed because of the collision, the kinetic energy before the collision must equal the kinetic energy after

$$\tfrac{1}{2}m_1v_1{}^2 + \tfrac{1}{2}m_2v_2{}^2 = \tfrac{1}{2}m_1v_1'^2 + \tfrac{1}{2}m_2v_2'^2 \tag{16-44}$$

This defines what is called an elastic collision. Each of these conservation conditions is of the form

$$\psi_1 + \psi_2 = \psi_1{}' + \psi_2{}' \tag{16-45}$$

Quantities that obey this equation are called *summational invariants*.

The velocities $\mathbf{v}_1$, $\mathbf{v}_2$, $\mathbf{v}_1{}'$, and $\mathbf{v}_2{}'$ discussed above are absolute velocities. In terms of some arbitrary coordinate system, the positions of the colliding molecules are $\mathbf{r}_1$ and $\mathbf{r}_2$ and the velocities are the time derivatives of these position vectors. As we have seen in Chapter 1, it is natural to define

$$\mathbf{r}_c = \frac{m_1\mathbf{r}_1 + m_2\mathbf{r}_2}{m_1 + m_2} \tag{16-46}$$

and

$$\mathbf{r} = \mathbf{r}_1 - \mathbf{r}_2 \tag{16-47}$$

These are the center-of-mass coordinates and relative coordinates, respectively.

If we differentiate $\mathbf{r}_c$ twice with respect to time, we find

$$\ddot{\mathbf{r}}_c = \frac{m_1\ddot{\mathbf{r}}_1 + m_2\ddot{\mathbf{r}}_2}{m_1 + m_2} = \frac{\mathbf{F}_1(r) + \mathbf{F}_2(r)}{m_1 + m_2} = 0 \tag{16-48}$$

since $\mathbf{F}_1(r) = -\mathbf{F}_2(r)$ by Newton's law. Equation (16–48) shows that the center of mass of a colliding pair of atoms undergoes uniform rectilinear motion.

If we differentiate the relative coordinates in the same way, we find that

$$\ddot{\mathbf{r}} = \ddot{\mathbf{r}}_1 - \ddot{\mathbf{r}}_2 = \frac{\mathbf{F}_1(r)}{m_1} - \frac{\mathbf{F}_2(r)}{m_2} = \left(\frac{1}{m_1} + \frac{1}{m_2}\right)\mathbf{F}_1(r)$$

or that

$$\mu\ddot{\mathbf{r}} = \mathbf{F}(r) \tag{16-49}$$

where $\mu = m_1m_2/(m_1 + m_2)$ is the reduced mass of the colliding pair. Since $\mathbf{F}(r)$ is a central force, i.e., is in the same direction as $\mathbf{r}$, Eq. (16–49) is the same as the equation of motion of a single particle of mass μ in a central force field.

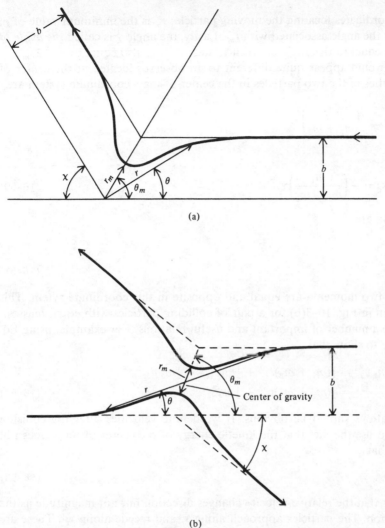

Figure 16–3. (a) **The trajectory of a collision in the relative coordinate system for a potential such as the Lennard-Jones potential. In this coordinate system, the particles approach with velocity v_r and recede with velocity v_r'. The moving particle is located by the coordinates r and θ; r_m is the minimum value of r; θ_m is the angle θ associated with r_m; b is the impact parameter; and χ is the angle of deflection. (b) The trajectory of a collision in the center-of-mass coordinate system for a pair of particles with equal masses. The symbols here are the same as in (a). Note that the particles approach each other with equal and opposite velocities in this coordinate system.**

The vectors $\mathbf{r}$ and $\dot{\mathbf{r}}$ define a plane that contains both particles and their center of mass. Since $\mathbf{F}(r)$ lies in this plane, the relative acceleration $\ddot{\mathbf{r}}$ also lies in this plane, and so the entire relative motion takes place in a plane. The entire plane itself is translated uniformly with velocity $\dot{\mathbf{r}}_c$.

In the relative coordinate system, the collision looks as though one of the particles is fixed and the other has mass μ, initial velocity $\mathbf{v}_r$, and final velocity $\mathbf{v}_r'$. This is shown in Fig. 16–3(a), where several other quantities are defined. If there were no interaction between the particles at all, b would be the distance of closest approach during the collision. This is called the *impact parameter* of the collision. The r and θ are nothing but the polar coordinates locating the moving particle; r_m is the minimum value of r; and θ_m is simply the angle associated with r_m. Lastly, the angle χ is called *the angle of deflection* and is equal to the angle between $\mathbf{v}_r$ and $\mathbf{v}_r'$. It is also equal to $\pi - 2\theta_m$.

The collision would appear quite different to an observer located at the center of mass. The velocities of the two particles in the center-of-mass coordinate system are

$$\mathbf{v}_{1c} = \mathbf{v}_1 - \mathbf{v}_c = \left(\frac{m_2}{m_1 + m_2}\right)\mathbf{v}_r$$

and

$$\mathbf{v}_{2c} = \mathbf{v}_2 - \mathbf{v}_c = -\left(\frac{m_1}{m_1 + m_2}\right)\mathbf{v}_r \tag{16-50}$$

and the momenta are

$$\mathbf{p}_{1c} = \mu\mathbf{v}_r$$

$$\mathbf{p}_{2c} = -\mu\mathbf{v}_r \tag{16-51}$$

We see that the two momenta are equal and opposite in this coordinate sytem. This collision is shown in Fig. 16-3(b) for a pair of colliding particles with equal masses.

We can derive a number of important and useful relations. For example, using Eq. (16-50) it is easy to show that

$$\tfrac{1}{2}m_1 v_1^2 + \tfrac{1}{2}m_2 v_2^2 = \tfrac{1}{2}(m_1 + m_2)v_c^2 + \tfrac{1}{2}m_1 v_{1c}^2 + \tfrac{1}{2}m_2 v_{2c}^2$$
$$= \tfrac{1}{2}(m_1 + m_2)v_c^2 + \tfrac{1}{2}\mu v_r^2 \tag{16-52}$$

whose interpretations should be obvious. If we denote velocities after the collision with a prime and use the fact that the kinetic energy of the center of mass does not change, we see that

$$\tfrac{1}{2}\mu v_r^2 = \tfrac{1}{2}\mu v_r'^2 \tag{16-53}$$

We see from this that the relative velocity changes direction but not magnitude as the result of a collision. The particles approach along $\mathbf{v}_r$ and recede along $\mathbf{v}_r'$. These are labeled in Fig. 16-3(a). One can also show (Problem 16-24) that the magnitudes of v_{1c} and v_{2c} do not change.

The equations of motion of the two molecules are best written in the polar coordinate system shown in Fig. 16-3(a). They are most easily obtained by simply writing down the conservation of energy and conservation of angular momentum conditions. If v_r is the initial relative velocity, then conservation of total energy gives

$$\tfrac{1}{2}\mu v_r^2 = \tfrac{1}{2}\mu(\dot{r}^2 + r^2\dot{\theta}^2) + u(r) \tag{16-54}$$

The left-hand side represents the kinetic energy before or after the collision, i.e., when $r \to \pm\infty$ in Fig. 16-3(a), and the right-hand side is the energy during the collision, when $u(r)$ is nonzero. The angular momentum before or after the collision is $\mu b v_r$, and the angular momentum at any time during the collision is $\mu r^2\dot{\theta}$, and so we have

$$\mu b v_r = \mu r^2\dot{\theta} \tag{16-55}$$

Equations (16-54) and (16-55) are the equations of motion and completely describe the collision in terms of v_r, $u(r)$, and b, which characterize the collision.

We can now eliminate θ between these two equations to get one closed equation for $r(t)$:

$$\tfrac{1}{2}\mu v_r{}^2 = \tfrac{1}{2}\mu\dot{r}^2 + \tfrac{1}{2}\mu v_r{}^2\left(\frac{b^2}{r^2}\right) + u(r) \tag{16-56}$$

This equation can be thought of as describing the one-dimensional motion of a particle of mass μ with total energy $\tfrac{1}{2}\mu v_r{}^2$ moving in an effective potential

$$u_{\text{eff}} = u(r) + \frac{\mu v_r{}^2 b^2}{2r^2} \tag{16-57}$$

The only quantity that will enter into our transport equations is the deflection angle χ. The most direct way to get at this is to solve Eq. (16–56) for dr/dt and then divide it by $d\theta/dt$ from Eq. (16–55):

$$\frac{dr}{d\theta} = \frac{dr/dt}{d\theta/dt} = -\frac{r^2}{b}\left[1 - \frac{u(r)}{\tfrac{1}{2}\mu v_r{}^2} - \frac{b^2}{r^2}\right]^{1/2} \tag{16-58}$$

Let r_m and θ_m be the distance and angle at the distance of closest approach. We may get θ_m by integrating Eq. (16–58) [cf. Fig. 16–3(a)]:

$$\theta_m = \int_0^{\theta_m} d\theta = -\int_\infty^{r_m} \frac{b\,dr}{r^2\left[1 - \dfrac{2u(r)}{\mu v_r{}^2} - \dfrac{b^2}{r^2}\right]^{1/2}} \tag{16-59}$$

and hence $\chi(b, v_r)$:

$$\chi(b, v_r) = \pi - 2b\int_{r_m}^\infty \frac{dr}{r^2\left[1 - \dfrac{2u(r)}{\mu v_r{}^2} - \dfrac{b^2}{r^2}\right]^{1/2}} \tag{16-60}$$

The lower limit here, the distance of closest approach, is found by setting $dr/d\theta$ equal to zero, i.e., by solving

$$\tfrac{1}{2}\mu v_r{}^2 - u(r) - \frac{\mu v_r{}^2 b^2}{2r^2} = 0 \tag{16-61}$$

for r.

The transport coefficients of dilute gases all may be expressed in terms of weighted integrals of the angle of deflection $\chi(b, v_r)$. For example, the coefficient of viscosity is given by (Chapter 19)

$$\eta(T) = \frac{5(mkT)^{1/2}}{16\pi^{1/2}\displaystyle\int_0^\infty e^{-\gamma^2}\gamma^7\left[\int_0^\infty \sin^2\chi\, b\, db\right] d\gamma} \tag{16-62}$$

where γ is a reduced relative velocity defined by $\gamma^2 = \mu v_r{}^2/2kT$. We shall see in Chapter 19 that the other transport coefficients have fairly similar forms.

The only intermolecular potential functions for which Eq. (16–60) can be integrated analytically are the hard-sphere potential and an r^{-4} repulsive potential. This latter potential function is not too realistic for most intermolecular interactions, but it is of historical interest since Maxwell realized over a hundred years ago that it was possible to carry through a rigorous kinetic theory of gases for this potential. Hypothetical molecules that have a repulsive r^{-4} potential are called Maxwellian molecules. For the rigid-sphere potential, the distance of the closest approach is σ if $b \leq \sigma$ and is b if $b \geq \sigma$. Consequently, Eq. (16–60) can be easily integrated (Problem 16–26):

$$\chi(b, g) = 2 \arccos\left(\frac{b}{\sigma}\right) \qquad b \leq \sigma$$

$$= 0 \qquad\qquad\qquad b \geq \sigma \tag{16–63}$$

Note that χ is independent of γ for a hard-sphere potential. For more realistic potentials, Eq. (16–60) must be integrated numerically. This has been done for a number of potentials, and we shall return to this in Chapter 19.

In the next and last section of this chapter, we shall calculate the average square of the momemtum change for a binary collision. We do this not only because we shall need this quantity in Section 21–9, but also because it serves to derive correct expressions for the mean free path, the total rate of collisions, and the rate of collisions with relative velocities in the range v_r and $v_r + dv_r$ and impact parameter in the range $b + db$.

16–3   MEAN-SQUARE MOMENTUM CHANGE DURING A COLLISION

The momentum change of some particular molecule upon a collision is

$$\begin{aligned}
\Delta(m_1\mathbf{v}_1) &= m_1\mathbf{v}_1' - m_1\mathbf{v}_1 \\
&= m_1(\mathbf{v}_{1c}' - \mathbf{v}_{1c}) \\
&= \mu(\mathbf{v}_r' - \mathbf{v}_r)
\end{aligned} \tag{16–64}$$

where we have used Eq. (16–50). The square of the momentum change is

$$\begin{aligned}
(\Delta m_1\mathbf{v}_1)^2 &= \mu^2(v_r'^2 - 2\mathbf{v}_r \cdot \mathbf{v}_r' + v_r^2) \\
&= 2\mu^2 v_r^2(1 - \cos\chi)
\end{aligned} \tag{16–65}$$

since $v_r' = v_r$ and the deflection angle χ is the angle between $\mathbf{v}_r$ and $\mathbf{v}_r'$. We wish to average this over all collisions, and so we need to calculate the collision rate in the gas.

For generality we consider collisions between two dissimilar groups of molecules, one having velocity distribution $f_1(\mathbf{v}_1) \, d\mathbf{v}_1$ and the other having the distribution $f_2(\mathbf{v}_2) \, d\mathbf{v}_2$. Let dn_1 and dn_2 be the number densities of these two groups so that

$$dn_j = \rho_j f(\mathbf{v}_j) \, d\mathbf{v}_j \qquad j = 1, 2$$

Now consider a collision between a molecule of the first group and one of the second and choose a coordinate system fixed on molecule 2 so that molecule 2 is at rest and molecule 1 has velocity between $\mathbf{v}_r$ and $\mathbf{v}_r + d\mathbf{v}_r$. The rate at which such collisions occur with impact parameter between b and $b + db$ is $v_r \, dn_1 \, 2\pi b \, db$. Since there are dn_2 of the second kind of molecule per unit volume, the number of such collisions per second per unit volume is

$$dZ = 2\pi b \, db \, v_r \, dn_1 \, dn_2$$

$$dZ = \rho_1 \rho_2 2\pi b \, db \, v_r f_1(\mathbf{v}_1) \, d\mathbf{v}_1 f_2(\mathbf{v}_2) \, d\mathbf{v}_2$$

If the molecules of the two groups are the same, we can use this equation as long as we divide it by 2 in order to avoid overcounting since the collision with $\mathbf{v}_1 = \mathbf{v}_\alpha$, $\mathbf{v}_2 = \mathbf{v}_\beta$ is the same as that with $\mathbf{v}_1 = \mathbf{v}_\beta$, $\mathbf{v}_2 = \mathbf{v}_\alpha$. Thus we write

$$dZ = \kappa \rho_1 \rho_2 \, 2\pi b \, db \, v_r f_1(v_1) \, d\mathbf{v}_1 f_2(v_2) \, d\mathbf{v}_2 \tag{16-66}$$

where $\kappa = \frac{1}{2}$ for collisions between like molecules and $\kappa = 1$ for collisions between unlike molecules.

If we now substitute the Maxwell-Boltzmann distribution for $f_1(v_1)$ and $f_2(v_2)$, Eq. (16–66) becomes

$$dZ = \kappa \rho_1 \rho_2 \frac{(m_1 m_2)^{3/2}}{(2\pi kT)^3} \, v_r \, 2\pi b \, db \, e^{-(m_1 v_1^2 + m_2 v_2^2)/2kT} \, d\mathbf{v}_1 \, d\mathbf{v}_2 \tag{16-67}$$

Transform from $\mathbf{v}_1$ and $\mathbf{v}_2$ to center of mass and relative coordinates. The Jacobian of this transformation is unity since

$$v_{1x} = v_{cx} + \left(\frac{m_2}{m_1 + m_2}\right) v_{rx}$$

$$v_{2x} = v_{cx} + \left(\frac{m_1}{m_1 + m_2}\right) v_{rx}$$

and

$$J\begin{pmatrix} v_{1x} & v_{2x} \\ v_{cx} & v_{rx} \end{pmatrix} = \begin{vmatrix} \partial v_{1x}/\partial v_{rx} & \partial v_{1x}/\partial v_{cx} \\ \partial v_{2x}/\partial v_{rx} & \partial v_{2x}/\partial v_{cx} \end{vmatrix} = \begin{vmatrix} m_2/M & 1 \\ -m_1/M & 1 \end{vmatrix} = 1$$

with a similar result for the y- and z-directions. Equation (16–52) says that

$$\tfrac{1}{2}m_1 v_1^2 + \tfrac{1}{2}m_2 v_2^2 = \tfrac{1}{2}(m_1 + m_2)v_c^2 + \tfrac{1}{2}\mu v_r^2$$

and so Eq. (16–67) becomes

$$dZ = \kappa \rho_1 \rho_2 \frac{(m_1 m_2)^{3/2}}{(2\pi kT)^3} \, 2\pi b \, db \, v_r e^{-(m_1 + m_2)v_c^2/2kT} \, d\mathbf{v}_c \, e^{-\mu v_r^2/2kT} \, d\mathbf{v}_r$$

We integrate this over the center-of-mass coordinates to get the desired result for the number of collisions per unit time per unit volume with relative velocity in the interval v_r and $v_r + dv_r$ and impact parameter in the interval b and $b + db$

$$dZ = \kappa \rho_1 \rho_2 \left(\frac{\mu}{kT}\right)^{3/2} \left(\frac{2}{\pi}\right)^{1/2} 2\pi b \, db \, v_r^3 e^{-\mu v_r^2/2kT} \, dv_r \tag{16-68}$$

Notice that this distribution has an extra factor of v_r, which essentially reflects the fact that molecules with a higher relative velocity collide more frequently. Problem 16–32 discusses a paper by Levine and Birnbaum in which they present an elegant application of this distribution.

To find the total rate of collisions, we integrate this over b and v_r. In the case of hard spheres, the integral over $2\pi b \, db$ gives just $\pi \sigma^2$, and we find that

$$Z = \kappa \rho_1 \rho_2 \, \pi \sigma^2 (\overline{v_1^2} + \overline{v_2^2})^{1/2} \tag{16-69}$$

It is shown in Problem 16–25 that $\bar{v}_r = (\overline{v_1^2} + \overline{v_2^2})^{1/2} = (8kT/\mu\pi)^{1/2}$, and so this formula for Z is

$$Z = \kappa \rho_1 \rho_2 \, \pi \sigma^2 \bar{v}_r \tag{16-70}$$

We can derive a number of interesting quantities from this one formula. If $\kappa = 1$ we get the rate of unlike-molecule collisions per unit volume

$$Z_{12} = \rho_1 \rho_2 \pi \sigma^2 \bar{v}_r \tag{16-71}$$

If $\kappa = \frac{1}{2}$, we get the rate of collisions per unit volume in a pure gas

$$Z_{11} = \frac{1}{\sqrt{2}} \rho^2 \pi \sigma^2 \bar{v} \tag{16-72}$$

where we have used the fact that $\bar{v}_r = 2^{1/2} \bar{v}$ where recall that $\bar{v} = (8kT/\pi m)^{1/2}$. Lastly, if we focus upon one particular tagged molecule, then the number of collisions that it undergoes per unit time per unit volume is

$$Z_1 = \rho_2 \pi \sigma^2 \bar{v}_r \tag{16-73}$$

We shall now calculate the mean-square change in momentum upon a collision. To be more precise, we wish to calculate the quantity

$$I = \int \langle (\Delta \mathbf{p}_1)^2 \rangle 2\pi b \, db \tag{16-74}$$

where the angular brackets denote an average over the relative velocity of the colliding particles. If we divide Eq. (16-68) by Eq. (16-70) (without the $2\pi b \, db$ or $\pi \sigma^2$), we get the fraction of collisions with relative velocity between v_r and $v_r + dv_r$

$$p(v_r) \, dv_r = \frac{1}{2} \left(\frac{\mu}{kT} \right)^2 v_r^3 e^{-\mu v_r^2 / 2kT} \, dv_r \tag{16-75}$$

We now multiply Eq. (16-65) by Eq. (16-75) and integrate over v_r to get

$$\langle (\Delta \mathbf{p}_1)^2 \rangle = \mu^2 \left(\frac{\mu}{kT} \right)^2 \int_0^\infty v_r^5 (1 - \cos \chi) e^{-\mu v_r^2 / 2kT} \, dv_r \tag{16-76}$$

If we define a collision cross section by

$$\sigma(v_r) = 2\pi \int_0^\infty (1 - \cos \chi) b \, db \tag{16-77}$$

then Eq. (16-74) becomes

$$I = \mu^2 \left(\frac{\mu}{kT} \right)^2 \int_0^\infty v_r^5 \sigma(v_r) e^{-\mu v_r^2 / 2kT} \, dv_r$$

$$= 8\mu kT \int_0^\infty \gamma^5 \sigma(\gamma) e^{-\gamma^2} \, d\gamma \tag{16-78}$$

where $\gamma^2 = \mu v_r^2 / 2kT$. We shall use this equation directly in Section 21-9.

In the next chapter we shall introduce a subject that is probably unfamiliar to most readers, but one that plays a very important role in any discussion of transport processes and liquids in motion in general. We shall derive some of the principal equations of hydrodynamics and then discuss their simpler solutions and applications.

ADDITIONAL READING

BOLTZMANN, L. 1964. *Lectures on gas theory* (trans. by S. G. Brush). Berkeley, Calif.: University of California Press.

BRUSH, S. G. 1965. *Kinetic theory*, Vol. 1. New York: Pergamon.

———. 1966. *Kinetic theory*, Vol. 2. New York: Pergamon.

GOLDEN, S. 1964. *Elements of the theory of gases*. Reading, Mass.: Addison-Wesley.

GUGGENHEIM, E. A. 1960. *Kinetic theory of gases*. New York: Pergamon.

JEANS, J. 1954. *The dynamical theory of gases*. New York: Dover.

———. 1959. *Kinetic theory of gases*. Cambridge: Cambridge University Press.

KAUZMAN, W. 1966. *Kinetic theory of gases*. New York: Benjamin.

KENNARD, E. H. 1938. *Kinetic theory of gases*. New York: McGraw-Hill.

LOEB, L. B. 1934. *Kinetic theory of gases*. New York: McGraw-Hill.

PRESENT, R. D. 1958. *Kinetic theory of gases*. New York: McGraw-Hill.

PROBLEMS

16–1. Estimate the mean free path in a gas at 300°K and 1 atm: 0.1 atm; 0.01 atm. Assume a size parameter $\sigma = 3$ Å.

16–2. Estimate the average time between collisions in a gas under the conditions of the previous problem.

16–3. Estimate the number of collisions per second experienced by one molecule in a gas under the conditions of the previous two problems.

16–4. Estimate the total number of collisions per second in a gas under the conditions of the previous three problems.

16–5. At 25°C, H_2 has a viscosity of 88 micropoise. Estimate the diameter of the hydrogen molecule.

16–6. At 0°C and 1 atm, helium has a viscosity of 186 micropoise. Estimate the thermal conductivity of helium under these conditions. The experimental value is 33.8×10^{-5} cal/sec · cm · deg.

16–7. Give order-of-magnitude values for the following quantities in a gas at normal conditions

(a) speed of sound
(b) mean free path
(c) mean time between collisions
(d) duration of a collision
(e) diffusion coefficient (include units)

16–8. The van der Waals constant, b, of argon is 29.5 cm³/mole. Estimate an atomic diameter from this and hence calculate the viscosity of Ar at 25°C.

16–9. Show that the escape velocity from a planet's surface is given by

$$v = (2gR)^{1/2}$$

where g is the gravitational constant and R the radius of the planet. Given that the gravitational constant of the earth is 980 cm/sec² and the radius is 6.4×10^8 cm, calculate the fraction of hydrogen, helium, nitrogen, and oxygen molecules having velocities exceeding the escape velocity.

16–10. Repeat the previous problem for the moon, for which $g = 167$ cm/sec² and $R = 1.8 \times 10^8$ cm.

16–11. Derive an expression for the mean free path in a two-dimensional gas.

16–12. Derive the analog of Eq. (16–11) for a two-dimensional gas.

16–13. Derive the analog of Eq. (16–31) for a two-dimensional gas.

16–14. Derive the analog of Eq. (16–32) for a two-dimensional gas.

16–15. Derive the analog of Eq. (16–33) for a two-dimensional gas.

16–16. Show that the mean free paths in a binary mixture are given by

$$l_1 = \frac{4}{4\pi\sigma_1^2\rho_1 + \pi(\sigma_1 + \sigma_2)^2\rho_2}$$

and

$$l_2 = \frac{4}{\pi(\sigma_1 + \sigma_2)^2\rho_1 + 4\pi\sigma_2^2\rho_2}$$

16–17. Prove that the distribution of free path lengths is given by

$$p(L)\, dL = e^{-L/l}\, d/l$$

Calculate the fraction of molecules with free paths greater than $0.5l$; $1.0l$; $2.0l$; $3.0l$. Note that this distribution is quite broad.

16–18. What is the probability that a nitrogen molecule at STP travels (a) 10^{-3} cm, (b) 1 cm without experiencing a collision.

16–19. In Section 16–1 we derived the formula $\lambda = \varepsilon \eta c_V$, where $\varepsilon = 1$ in the simple theory and $\varepsilon = \frac{5}{2}$ in the more rigorous theory. The underlying physical explanation of this discrepancy was first pointed out by Eucken in 1913 (cf. Loeb in "Additional Reading," pp. 247–250). He pointed out that for translational motion only, molecules with larger velocities have larger mean free paths, so that they travel farther and carry more kinetic energy than the more slowly moving molecules. Hence the transport of kinetic energy by these molecules is more effective, and this fact leads to a value of $\varepsilon > 1$. Eucken argued that this would not be so for the internal energy carried by a polyatomic molecule, and that in this case ε would be unity. Thus Eucken was able to present a plausible, though certainly not rigorous, extension of the formula $\lambda = \varepsilon \eta C_V$ for polyatomic molecules. He argued that the thermal transport coefficient is made up of two contributions:

$$\lambda = (\varepsilon_{\text{trans}} C_{V,\text{trans}} + \varepsilon_{\text{int}} C_{V,\text{int}})\eta$$

where $\varepsilon_{\text{trans}} = \frac{5}{2}$, $\varepsilon_{\text{int}} = 1$, $C_{V,\text{trans}} = 3R/2$, and $C_{V,\text{int}} = C_V - C_{V,\text{trans}}$. Show that this leads to the formula

$$\lambda = \tfrac{1}{4}(9\gamma - 5)\eta C_V$$

where $\gamma = C_p/C_V$. This formula is known as Eucken's correction. Show that this reduces to the rigorous result for a monatomic gas. Table 16–2 shows the agreement of Eucken's formula with experimental data.

Table 16–2. **A test of the Eucken correction formula**

	η	λ	c_V	γ	$\dfrac{\lambda}{\eta c_V}$	$\tfrac{1}{4}(9\gamma - 5)$
	$(10^{-7}$ cgs$)$	$\dfrac{10^{-3}\text{ cal}}{\text{cm}\cdot\text{sec}\cdot\text{deg}}$	$\dfrac{\text{cal}}{\text{g}\cdot\text{deg}}$			
Helium	1875	0.344	0.753	1.660	2.44	2.485
Neon	2986	0.1104	0.150	1.64	2.47	2.44
Argon	2100	0.0387	0.0763	1.67	2.42	2.51
H_2	840	0.416	2.40	1.410	2.06	1.92
N_2	1664	0.0566	0.178	1.406	1.91	1.91
O_2	1918	0.0573	0.156	1.395	1.92	1.89
H_2O at 100°C	1215	0.0551	0.366	1.32	1.24	1.72
CO_2	1377	0.0340	0.151	1.31	1.64	1.70
NH_3	915	0.0514	0.401	1.32	1.40	1.72
CH_4 methane	1027	0.0718	0.400	1.31	1.75	1.70
C_2H_4, ethene	948	0.0404	0.282	1.25	1.51	1.56
C_2H_6, ethane	854	0.0428	0.325	1.23	1.54	1.52

16–20. Estimate the fractions of the coefficient of thermal conductivity due to the translational and internal degrees of freedom for Ar; Cl_2; CCl_4; C_2H_6.

16–21. Prove that

$$m_1(\mathbf{v}_{1c}' - \mathbf{v}_{1c}) = \mu(\mathbf{v}_r' - \mathbf{v}_r)$$

16–22. Prove that the square of the momentum change during a collision is

$$(\Delta m_1 \mathbf{v}_1)^2 = 2\mu^2 v_r^2(1 - \cos \chi)$$

16–23. Prove that $\chi = \pi - 2\theta_m$.

16–24. Show that the magnitudes of $\mathbf{v}_{1c}$ and $\mathbf{v}_{2c}$ do not change during a collision.

16–25. Show that the average relative velocity is given by

$$\bar{v}_r = (\bar{v}_1{}^2 + \bar{v}_2{}^2)^{1/2}$$

$$= \left(\frac{8kT}{\mu\pi}\right)^{1/2}$$

where μ is the reduced mass of the colliding pair.

16–26. Show that the scattering angle $\chi(b, g)$ for hard spheres can be determined analytically to get

$$\chi(b, g) = 2 \arccos\left(\frac{b}{\sigma}\right) \qquad b \leq \sigma$$

$$= 0 \qquad b \geq \sigma$$

16–27. Derive an expression for the fraction of collisions in which the relative kinetic energy of the colliding pair of molecules is between E and $E + dE$. The integral of this from E^* to ∞ is an important quantity in many theories of gas-phase chemical kinetics. Why?

16–28. Derive an expression for the distribution of relative velocities in two dimensions. Use this to calculate the average relative velocity in a two-dimensional gas.

16–29. Derive an expression for $\chi(b, g)$ in terms of $u(r)$ for a two-dimensional system.

16–30. Derive the analog of Eq. (16–78) for a two-dimensional gas.

16–31. Equation (16–73) says that the number of collisions of some one particular tagged molecule per unit time per unit volume is

$$Z_1 = \rho\pi\sigma^2\bar{v}_r$$

Use this to show that the mean free path is given by

$$l = \frac{1}{2^{1/2}\pi\rho\sigma^2}$$

16–32. Levine and Birnbaum[*] have published an elegant treatment of the classical theory of collision-induced absorption in rare-gas mixtures which is based upon the kinetic theory of gases. First let us review what is meant by collision-induced absorption. We have all learned in physical chemistry that gases such as Ar, N_2, CO_2, etc., do not absorb in the microwave or far-infrared regions since the molecules do not possess permanent dipole moments, the necessary condition for absorption in that region. Consider, however, a mixture of two rare gases such as helium and argon. There will be many helium-argon collisions in this mixture, and during such a collision we can consider the helium-argon pair to be a transient heteronuclear diatomic molecule, albeit undergoing just one vibration. Regardless of the transient nature of this heteronuclear diatomic molecule, it does possess a (transient) dipole moment and hence leads to absorption in the microwave and far-infrared regions. Since this absorption occurs only as the result of a collision, it is called collision-induced absorption. Of course, the observed intensity is much less than in conventional microwave absorption, but nevertheless it was first observed in 1959 by Kish and Welsh[†] and later in more detail by Bosomworth and Gush.[‡]

There have been a number of theoretical analyses of such collision-induced absorption, but the one by Levine and Birnbaum referred to above is particularly simple and instructive.

[*] *Phys. Rev.*, 154, p. **86**, 1967.

[†] *Phys. Rev. Lett.*, 2, p. 166, 1959.

[‡] *Can. J. Phys.*, 43, p. 729, 1965.

Their starting point is based upon some electromagnetic theory with which the reader may not be familiar, but if he is willing to accept this, the rest of the problem is fairly straightforward. Levine and Birnbaum start with Kirchhoff's law, which relates the absorption coefficient $\alpha(\omega)$ to the power due to spontaneous emission per unit frequency interval per unit volume of sample $I(\omega)$ by

$$\alpha(\omega) = \frac{I(\omega)}{c\rho(\omega, T)}$$

where c is the speed of light and $\rho(\omega, T)$ is the energy density in the blackbody radiation field. This is given by Planck's law, Eq. (10–99),

$$\rho(\omega, T)\, d\omega = \frac{\hbar\omega^3\, d\omega}{\pi^2 c^3 [\exp(\hbar\omega/kT) - 1]}$$

The quantity $I(\omega)$ can be written as a statistical average over the power due to collisions of impact parameter b and relative speed v_r according to

$$I(\omega) = \iint Z(b, v_r)\hat{I}(b, v_r, \omega)\, db\, dv_r$$

where $Z(b, V_r)$ is the number of collisions with impact parameter b and relative speed v_r [Eq. (16–68)] and $\hat{I}(b, v_r, \omega)$ is the power emitted by collisions of this type.

The emission intensity $\hat{I}(b, v_r, \omega)$ is given by electromagnetic theory by the so-called Larmor power formula:

$$\hat{I}(b, v_r, \omega) = \frac{2\omega^4}{3c^3\pi} \left| \int_{-\infty}^{\infty} dt\, e^{i\omega t} \mu(b, v_r, t) \right|^2$$

where $\mu(b, v_r, t)$ is the electric dipole moment of the colliding pair as a function of time for a given b and v_r. This formula is essentially the power emitted by an oscillating dipole moment.

Levine and Birnbaum write the dipole moment as

$$\boldsymbol{\mu} = \boldsymbol{\mu}(\mathbf{r}) = \frac{\mu(r)\mathbf{r}}{r}$$

to denote that $\boldsymbol{\mu}$ depends only upon the separation r of the two colliding atoms, and they write

$$\mathbf{r} = \mathbf{r}(b, v_r, t)$$

to specify the collision trajectory. Clearly $\mathbf{r}$ depends on the intermolecular potential $u(r)$, but Levine and Birnbaum neglect the potential and assume straight-line paths, with

$$x = v_r t \qquad y = b \qquad z = 0$$

Although this is certainly not correct, particularly for small v values, nevertheless it has two virtues. The first is that it leads to simple mathematics; in particular, we do not have to compute trajectories, and the subsequent derivation of the line shape is easy. Second, the comparison of the results of such a calculation with experiment helps us to assess the importance of the potential in fixing the lineshape, i.e., helps us decide whether or not lineshape measurements can be expected to yield useful information about the potential function.

Although the dependence of the dipole moment function on the separation r is not yet known, it is probably *approximately* Gaussian looking, and Levine and Birnbaum assume that

$$\mu(r) = \mu_0 \gamma r e^{-\gamma^2 r^2}$$

where γ^{-1} is the range of the dipole moment and μ_0 characterizes its strength. The factor γr has the following significance: If one considers a realistic potential, then its repulsive part prevents two atoms from approaching too close to each other; i.e., the distance of closest approach is of the order of the range of the repulsive component of the potential. In a manner of speaking, the potential function "shuts off" the dipole moment for very small r values. Similarly, the factor γr "shuts off" the dipole moment when $r \lesssim \gamma^{-1}$. Thus, roughly speaking, the absence of a repulsive potential has been corrected for in the form used for $\mu(r)$.

Now the calculation of the complete lineshape $\alpha(\omega)$ for the collision-induced absorption in rare-gas mixtures follows in a fairly straightforward manner. Show that the Fourier transform of μ has two nonvanishing components:

$$\tilde{\mu}_x(b, v_r, \omega) = \frac{\mu_0 \omega \pi^{1/2}}{2\gamma^2 v_r^2} \exp\left\{-\gamma^2 b^2 - \frac{\omega^2}{4\gamma^2 v_r^2}\right\}$$

$$\tilde{\mu}_y(b, v_r, \omega) = \frac{\mu_0 b\pi^{1/2}}{v_r} \exp\left\{-\gamma^2 b^2 - \frac{\omega^2}{4\gamma^2 v_r^2}\right\}$$

where

$$\hat{I}(b, v_r, \omega) = \left(\frac{2\omega^4}{3c^3\pi}\right)(\tilde{\mu}_x^2 + \tilde{\mu}_y^2)$$

Now multiply $\hat{I}(b, v_r, \omega)$ by $Z(b, v_r)$ [Eq. (16-68)], integrate over b, and show that the remaining integrals are of the form

$$\int_0^\infty v_r^n \exp\left\{-\frac{\mu v_r^2}{2kT} - \frac{\omega^2}{2\gamma^2 v_r^2}\right\} dv_r$$

with $n = -1$ for the x-component and $n = 1$ for the y-component. Using the fact that

$$\int_0^\infty \exp\left(-\frac{a}{4x} - cx\right) dx = \left(\frac{a}{c}\right)^{1/2} K_1(\sqrt{ac})$$

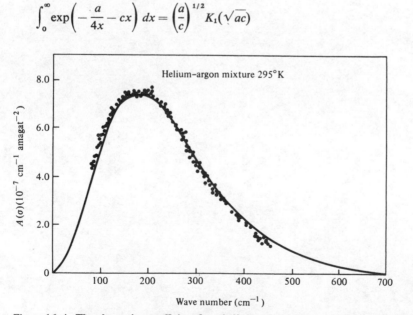

Figure 16–4. **The absorption coefficient for a helium-argon mixture at 295°K. The points are the experimental data of Bosomworth and Gush** (*Can. J. Phys.* **43**, 729, 1965), **and the solid line is calculated by Levine and Birnbaum for** $\gamma = 1.357$ Å^{-1} **and** $\mu_0 = 0.166$ **debyes. (From H. B. Levine and G. Birnbaum,** *Phys. Rev.,* **154**, p. 86, 1967.)**

where $K_n(z)$ is a modified Bessel function of the second kind and that

$$\frac{dK_n(z)}{dz} = \tfrac{1}{2}[K_{n-1}(z) - K_{n+1}(z)]$$

and

$$zK_{n-1}(z) + zK_{n+1}(z) = 2nK_n(z)$$

show that the line shape is given by

$$\alpha(\omega) = \frac{\mu_0^2 \pi^3 \rho_1 \rho_2}{12c\gamma^2} \left(\frac{2}{\pi\mu kT}\right)^{1/2} x^4 K_2(x)$$

where x is a reduced frequency

$$x = \frac{\omega}{\gamma}\left(\frac{\mu}{kT}\right)^{1/2} = \frac{2\pi c\sigma}{\gamma}\left(\frac{\mu}{kT}\right)^{1/2}$$

where σ is the frequency in wave numbers.

The reduced lineshape function $A(\sigma) = \alpha(\sigma)(\rho_1 \rho_2)^{-1}(N_0/V_0)^2$ ($N_0 =$ Avogadro's number, $V_0 = 22413.6 \text{ cm}^3$), which is the absorption coefficient measured by Bosomworth and Gush, is compared to their experimental data in Fig. 16–4. It can be seen that μ_0 and γ can be chosen so as to give excellent agreement; Levine and Birnbaum discuss the significance of this agreement in some detail.

CHAPTER 17

CONTINUUM MECHANICS

In this chapter we shall give an introduction to a field that is generally known as continuum mechanics. Continuum mechanics includes the areas of heat flow, fluid flow or hydrodynamics, and the mechanics of deformable bodies as special cases. The central idea in continuum mechanics is to define a local density, velocity, and energy that are continuous functions of space and time. Because of the molecular nature of matter, such concepts can be quite awkward to define rigorously, but what we mean by a local density, for example, can be defined by considering the mass contained in a small volume, Δv say, where this small volume is large enough to contain many molecules (so that the density is a continuous function of $\mathbf{r}$) but small enough that the mass divided by Δv describes a local density at the point $\mathbf{r}$. In short, we treat matter as continuous. We then derive a set of partial differential equations describing the spatial and time variations of these local density functions. The solution of these equations for given conditions and geometries becomes a boundary-value problem of mathematics. The partial differential equations we derive in Section 17–1 are simply a mass balance equation, a momentum balance equation, and an energy balance equation. Thus we see that in Section 17–1 we simply formulate the fundamental conservation laws in a continuum mechanics formalism or notation.

Since continuum mechanics in general and hydrodynamics in particular are somewhat neglected topics in most undergraduate curricula nowadays, the material of this section will probably be unfamiliar to most readers. Therefore, Sections 17–2 and 17–3 are devoted to solving the basic continuity equations for a number of special cases. For example, in Section 17–2 we shall derive and solve the diffusion equation, the heat flow equation, and Euler's equation; in Section 17–3 we shall derive and discuss the Navier-Stokes equation. These last two equations are fundamental equations in hydrodynamics, and from these we shall derive some familiar results such as the Poiseuille formula for the viscosity of fluid flow through a cylindrical capillary. This is a standard physical chemistry experiment for the determination of the viscosity of a liquid.

17–1 DERIVATION OF THE CONTINUITY EQUATIONS

For simplicity, we shall consider only a one-component system. The extension of the results of this section to multicomponent systems is straightforward. We shall consider the three basic continuity laws and derive a balance equation for each one.

MASS BALANCE

Let $\rho_m(\mathbf{r}, t)$ be the mass density of the system at the point $\mathbf{r}$ at time t. Let $\mathbf{u}(\mathbf{r}, t)$ be the velocity of the mass at $\mathbf{r}$ and t. Furthermore, let v be an arbitrary fixed volume located within our fluid and S be the surface of this volume. The total mass within v is

$$M = \int_v \rho_m(\mathbf{r}, t) \, d\mathbf{r} \tag{17–1}$$

The rate of change of mass within v is

$$\frac{dM}{dt} = \int_v \frac{\partial \rho_m}{\partial t} \, d\mathbf{r} \tag{17–2}$$

since v is fixed in space. Now since mass is neither created nor destroyed, the rate of change of mass within v must be given by the rate at which mass flows through the surface S.

We define a surface element vector $d\mathbf{S}$ such that the magnitude of $d\mathbf{S}$ is equal to the area dS and the direction of $d\mathbf{S}$ is directed normally outward from the volume that the entire surface encloses. If $\mathbf{n}$ is a unit vector directed outward from the enclosed volume, then $d\mathbf{S} = \mathbf{n} \, dS$. Figure 17–1 shows this surface area element and the vector $\mathbf{n}$.

The rate of flow of mass is $\rho_m \mathbf{u}$, and so the component of flow normal to a surface element $d\mathbf{S}$ is $\rho_m \mathbf{u} \cdot d\mathbf{S}$. Integrating this over the entire surface gives

$$\frac{dM}{dt} = - \int_S \rho_m \mathbf{u} \cdot d\mathbf{S} \tag{17–3}$$

as the total rate of flow of mass through the surface. The negative sign simply indicates that an outflow of mass yields a negative value for $\dot{M} \equiv dM/dt$. Recall that $\mathbf{u} \cdot d\mathbf{S}$ is positive if $\mathbf{u}$ is directed outward and negative if $\mathbf{u}$ is directed inward. The surface integral in Eq. (17–3) can be transformed to a volume integral by means of Gauss' divergence theorem:

$$\frac{dM}{dt} = - \int_v \nabla \cdot (\rho_m \mathbf{u}) \, d\mathbf{r} \tag{17–4}$$

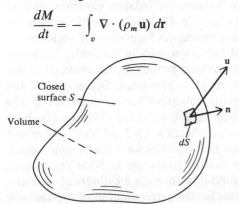

Closed surface S

Volume

$\mathbf{u}$

$\mathbf{n}$

$d\mathbf{S}$

Figure 17–1. **An elementary surface area dS, the unit vector n, and the mass velocity vector u.**

If we subtract Eq. (17–2) from Eq. (17–4) for $\dot{M}$, we get

$$\int_v \left[\frac{\partial \rho_m}{\partial t} + \nabla \cdot (\rho_m \mathbf{u}) \right] d\mathbf{r} = 0$$

but since the volume v is arbitrary, we must have

$$\frac{\partial \rho_m}{\partial t} + \nabla \cdot (\rho_m \mathbf{u}) = 0 \tag{17–5}$$

This is the continuity of mass equation.

By using the vector equation (Problem 17–1)

$$\nabla \cdot (a\mathbf{v}) = a\nabla \cdot \mathbf{u} + \mathbf{u} \cdot \nabla a \tag{17–6}$$

we can rewrite Eq. (17–5) in the form

$$\frac{D\rho_m}{Dt} + \rho_m \nabla \cdot \mathbf{u} = 0 \tag{17–7}$$

where

$$\frac{D}{Dt} = \frac{\partial}{\partial t} + \mathbf{u} \cdot \nabla \tag{17–8}$$

This derivative occurs frequently in continuum mechanics and is called the hydro-dynamic derivative, the substantial derivative, or the Stokes' operator. Physically, $Dh(\mathbf{r}, t)/Dt$ represents the change of $h(\mathbf{r}, t)$ along the flow of the system, i.e., along a streamline (Problem 17–2).

MOMENTUM BALANCE

We can now apply the same method to derive a momentum balance equation. The momentum density at any point is given by $\rho_m \mathbf{u}$, and the momentum within v is

$$\mathbf{G} = \int_v \rho_m \mathbf{u} \, d\mathbf{r} \tag{17–9}$$

Its rate of change is

$$\frac{d\mathbf{G}}{dt} = \int_v \frac{\partial}{\partial t} (\rho_m \mathbf{u}) \, d\mathbf{r} \tag{17–10}$$

There are two contributions to the rate of change of momentum One is the total force acting on the mass within v, and the other is the rate at which momentum flows in and out of v. There are two force terms to consider: (1) external forces such as gravitational or centrifugal forces, and (2) forces exerted by the fluid surrounding v, i.e., stress or pressure forces. If $\mathbf{K}$ is the external force per unit mass, then the first contribution $\mathbf{F}_1$ is given by

$$\mathbf{F}_1 = \int_v \rho_m \mathbf{K} \, d\mathbf{r} \tag{17–11}$$

In order to discuss the contribution due to the stress or pressure force, we must introduce the concept of a stress or pressure tensor.

Pressure is a concept that is familiar to all chemists as force per unit area. One is accustomed to a pressure reading of 10^6 dynes/cm^2, for example. A single number like

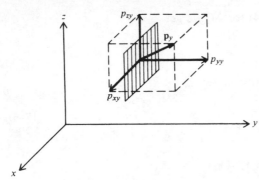

Figure 17–2. **Significance of the components of the pressure tensor.**

this implies that pressure is a scalar quantity, but this is not so in general. In fact, pressure is neither a scalar nor a vector quantity, but is a tensor. To understand what is meant by this, consider a small planar element of unit area located within a fluid. Let the normal of this surface area be the direction along the y-axis as in Fig. 17–2. We now investigate the forces acting upon this area. The force on this area has three components. Let the component of the force acting in the x-direction be denoted by p_{xy}, that in the y-direction be p_{yy}, and that in the z-direction by p_{zy}. In other words, $p_{\alpha y}$ is the force in the αth direction acting on a surface of unit area that is perpendicular to the y-axis. Now we can do the same thing for areas perpendicular to the x- and z-axis. Each of these will have three components of force acting upon it, and hence we will have six more components of pressure, namely, $p_{\alpha x}$ and $p_{\alpha z}$, $\alpha = x, y, z$. Since any arbitrary area can be resolved into three mutually perpendicular components, it takes nine pressure components to completely describe the pressure acting upon this area. These nine quantities are the components of the pressure tensor p, which can be displayed in the form

$$\mathsf{p} = \begin{pmatrix} p_{xx} & p_{xy} & p_{xz} \\ p_{yx} & p_{yy} & p_{yz} \\ p_{zx} & p_{zy} & p_{zz} \end{pmatrix} \qquad (17\text{--}12)$$

The only way a fluid can exert a component of pressure in the plane of a test area is by a viscous drag. Such a force is called a shear force and will be nonzero only for a viscous medium. In the special case of a nonviscous medium, i.e., one in which we can ignore viscosity effects, the off-diagonal terms of p are zero, and we have

$$\mathsf{p} = \begin{pmatrix} p_{xx} & 0 & 0 \\ 0 & p_{yy} & 0 \\ 0 & 0 & p_{zz} \end{pmatrix}$$

At equilibrium $p_{xx} = p_{yy} = p_{zz} = p$, and we have

$$\mathsf{p} = \begin{pmatrix} p & 0 & 0 \\ 0 & p & 0 \\ 0 & 0 & p \end{pmatrix} = p \begin{pmatrix} 1 & 0 & 0 \\ 0 & 1 & 0 \\ 0 & 0 & 1 \end{pmatrix} = p\mathsf{I} \qquad (17\text{--}13)$$

where I is the unit tensor, whose components are the Kroenecker delta, δ_{ij}. (The Kroenecker delta is a tensor.) The scalar quantity p is what is usually referred to as the pressure. We should see from this that pressure is a somewhat more complicated notion than many chemists are used to thinking.

The ith component of the force acting upon an arbitrary area S is then

$$F_i = \sum_j p_{ij} S_j \tag{17-14}$$

which we write in tensor notation as

$$\mathbf{F} = \mathsf{p} \cdot \mathbf{S}$$

Equation (17–14) defines this tensor operation. The total pressure force acting upon the mass enclosed within the volume v is

$$\mathbf{F}_2 = -\int_S \mathsf{p} \cdot d\mathbf{S} \tag{17-15}$$

where the negative sign signifies that a pressure acting upon the mass in v gives a positive force.

Lastly, we need to calculate the rate of change of momentum due to a flow of momentum in and out of v. Momentum is given by $\rho_m \mathbf{u}$, and so a flow of momentum is given by $\rho_m \mathbf{uu}$. The quantity $\mathbf{uu}$ has nine components, $u_i u_j$, which represent the ith component of momentum being carried in the jth direction. The quantity $\mathbf{uu}$ is the direct product of two vectors and is actually a special kind of tensor called a dyadic. The rate at which momentum flows across S is

$$\text{flow} = -\int_S \rho_m \mathbf{uu} \cdot d\mathbf{S} \tag{17-16}$$

The dot product of the tensor $\mathbf{uu}$ and the vector $d\mathbf{S}$ appearing here can be readily handled by writing it as $\mathbf{u}(\mathbf{u} \cdot d\mathbf{S})$. Sometimes it is most convenient to manipulate $\mathbf{uu}$ as a formal tensor quantity, however.

If we add Eqs. (17–10), (17–11), (17–15), and (17–16) and use the divergence theorem to transform all the surface integrals into volume integrals and then realize that v is arbitrary, we get the momentum balance equation (Problem 17–4)

$$\frac{\partial(\rho_m \mathbf{u})}{\partial t} + \nabla \cdot (\rho_m \mathbf{uu}) = \rho_m \mathbf{K} - \nabla \cdot \mathsf{p} \tag{17-17}$$

This equation involves taking the divergence of the tensor $\mathbf{uu}$. Generally, the jth component of $\nabla \cdot \mathbf{A}$ is

$$(\nabla \cdot \mathbf{A})_j = \sum_i \frac{\partial}{\partial x_i} A_{ij} \tag{17-18}$$

Equation (17–17) is the momentum balance equation. By using various vector and tensor relations, it can be written in several equivalent forms. For example, if we use the relations (Problem 17–6)

$$\nabla \cdot (\rho_m \mathbf{uu}) = \rho_m (\mathbf{u} \cdot \nabla)\mathbf{u} + \mathbf{u}(\nabla \cdot \rho_m \mathbf{u}) \tag{17-19}$$

and

$$\frac{\partial(\rho_m \mathbf{u})}{\partial t} = \rho_m \frac{\partial \mathbf{u}}{\partial t} + \mathbf{u}\frac{\partial \rho_m}{\partial t} \tag{17-20}$$

and the mass continuity equation, we get (Problem 17–5)

$$\rho_m \frac{D\mathbf{u}}{Dt} = \rho_m \mathbf{K} - \nabla \cdot \mathsf{p} \tag{17-21}$$

The form of Eq. (17–21) clearly shows that this is just the continuum mechanics expression for Newton's second law. It is is the equation of motion in a continuum notation.

ENERGY BALANCE

Lastly, we derive the energy balance equation. The total energy contained within v is

$$E = \int_v \rho_m[\tilde{E} + \tfrac{1}{2}u^2]\, d\mathbf{r} \tag{17–22}$$

where $\rho_m \tilde{E}$ is the total energy, including potential energy due to external forces but not including the kinetic energy due to the mass flow $\tfrac{1}{2}\rho_m u^2$. The rate of change of E is

$$\frac{dE}{dt} = \int_v \frac{\partial}{\partial t}\{\rho_m(\tilde{E} + \tfrac{1}{2}u^2)\}\, d\mathbf{r} \tag{17–23}$$

Now E can change with time for three reasons: (1) Work is done on the mass in v; (2) energy flows in or out by convection currents; and (3) energy flows in or out by conduction. These three contributions are given by

$$\frac{dE}{dt} = -\int_S \mathbf{u} \cdot \mathsf{p} \cdot d\mathbf{S} - \int_S \rho_m[\tilde{E} + \tfrac{1}{2}u^2]\mathbf{u} \cdot d\mathbf{S} - \int_S \mathbf{q} \cdot d\mathbf{S} \tag{17–24}$$

The vector $\mathbf{q}$ in the third term is the heat flux vector. The only term that should not be clear is the first. To see that this is, in fact, the work done on the mass in v, recall that $\mathsf{p} \cdot d\mathbf{S}$ is a force, and that the scalar product of force with velocity is rate of work, i.e., $\mathbf{u} \cdot (\mathsf{p} \cdot d\mathbf{S})$ is the rate of work through the area $d\mathbf{S}$.

Equations (17–23) and (17–24) give

$$\frac{\partial}{\partial t}[\rho_m(\tilde{E} + \tfrac{1}{2}u^2)] = -\nabla \cdot (\mathbf{u} \cdot \mathsf{p}) - \nabla \cdot [\rho_m \mathbf{u}(\tilde{E} + \tfrac{1}{2}u^2)] - \nabla \cdot \mathbf{q} \tag{17–25}$$

We can express this in a more convenient form by using the identity (Problem 17–6)

$$\nabla \cdot (\mathbf{u} \cdot \mathsf{p}) = \mathbf{u} \cdot (\nabla \cdot \mathsf{p}) + \mathsf{p} : \nabla \mathbf{u} \tag{17–26}$$

where the double dot notation denotes the standard tensorial manipulation

$$\mathsf{A} : \mathsf{B} = \sum_i \sum_j A_{ij} B_{ji}$$

which in the above case becomes

$$\mathsf{p} : \nabla \mathbf{u} = \sum_{i,j} p_{ij} \frac{\partial u_i}{\partial x_j} \tag{17–27}$$

Using this and Eq. (17–6), Eq. (17–25) becomes

$$\frac{D}{Dt}[\rho_m(\tilde{E} + \tfrac{1}{2}u^2)] = -\mathbf{u} \cdot (\nabla \cdot \mathsf{p}) - \mathsf{p} : \nabla \mathbf{u} - \rho_m(\tilde{E} + \tfrac{1}{2}u^2)\nabla \cdot \mathbf{u} - \nabla \cdot \mathbf{q} \tag{17–28}$$

The left-hand side can be written as

$$\frac{D}{Dt}[\rho_m(\tilde{E} + \tfrac{1}{2}u^2)] = \rho_m \frac{D}{Dt}(\tilde{E} + \tfrac{1}{2}u^2) + (\tilde{E} + \tfrac{1}{2}u^2)\frac{D\rho_m}{Dt}$$

Using the mass continuity equation in the form of Eq. (17–7), we see that the second term in this equation cancels the third term on the right-hand side of Eq. (17–28), giving

$$\rho_m \frac{D}{Dt}[\tilde{E} + \tfrac{1}{2}u^2] = -\mathbf{u} \cdot (\nabla \cdot \mathsf{p}) - \mathsf{p} : \nabla\mathbf{u} - \nabla \cdot \mathbf{q} \tag{17-29}$$

We can use the momentum balance equation, Eq. (17–21), to write

$$\mathbf{u} \cdot (\nabla \cdot \mathsf{p}) = \rho_m \mathbf{u} \cdot \mathbf{K} - \rho_m \mathbf{u} \cdot \frac{D\mathbf{u}}{Dt}$$

$$= \rho_m \mathbf{u} \cdot \mathbf{K} - \tfrac{1}{2}\rho_m \frac{Du^2}{Dt}$$

and so Eq. (17–29) can be written in the form (Problem 17–32)

$$\rho_m \frac{D\tilde{E}}{Dt} - \rho_m \mathbf{u} \cdot \mathbf{K} = -\mathsf{p} : \nabla\mathbf{u} - \nabla \cdot \mathbf{q} \tag{17-30}$$

Since $\tilde{E}$ is the total energy including potential energy due to external sources, but not including the kinetic energy due to mass flow, and $\rho_m \mathbf{u} \cdot \mathbf{K}$ is the rate of change in potential energy due to external forces, Eq. (17–30) can be written in the form (Problem 17–32)

$$\rho_m \frac{DE}{Dt} = -\mathsf{p} : \nabla\mathbf{u} - \nabla \cdot \mathbf{q} \tag{17-31}$$

where E is the total energy per unit mass within v *exclusive* of the potential energy due to external fields and the kinetic energy due to the mass flow.

Equations (17–7), (17–21), and (17–31) are the fundamental equations of continuum mechanics. They can hardly be solved as they stand since they contain the unknown quantities p and $\mathbf{q}$, which in some sense are properties of the medium itself. A common assumption about $\mathbf{q}$ is to assume that $\mathbf{q}$ is given by Fourier's law, i.e., that

$$\mathbf{q} = -\lambda \,\mathrm{grad}\, T \tag{17-32}$$

where λ is the thermal conductivity of the system. The empirical law for the pressure tensor is probably less familiar. The most general linear expression for p is to write

$$p_{ij} = -\sum_{k,l} \alpha_{ij}{}^{kl} \frac{\partial u_l}{\partial x_k} \tag{17-33}$$

Note that this is a generalization of Eq. (16–14). The $\{\alpha_{ij}{}^{kl}\}$ is a set of 81 coefficients, but fortunately not all of these are independent. If the body is not undergoing any translation or rotation, i.e., only being distorted, and if the medium is isotropic, e.g., a fluid, it can be shown (cf. Hirschfelder, Curtiss, and Bird in "Additional Reading") that p must be of the form

$$\mathsf{p} = [p + (\tfrac{2}{3}\eta - \kappa)\nabla \cdot \mathbf{u}]\mathsf{I} - 2\eta \,\mathrm{sym}(\nabla\mathbf{u}) \tag{17-34}$$

In Eq. (17–34), η is the coefficient of shear viscosity; κ is the coefficient of bulk viscosity, and $\mathrm{sym}(\nabla\mathbf{u})$ is a tensor with components

$$(\mathrm{sym}\,\nabla\mathbf{u})_{ij} = \frac{1}{2}\left\{\frac{\partial u_i}{\partial x_j} + \frac{\partial u_j}{\partial x_i}\right\} \tag{17-35}$$

The coefficient of shear viscosity is a measure of the resistance to a shearing force, such as in capillary flow. This is what is usually measured and what is referred to as "viscosity." The coefficient of bulk viscosity is a measure of the resistance to a compressive force and is fairly difficult to measure experimentally. Equation (17–34) is called the Newtonian pressure tensor. Note that if there are no viscous forces, i.e., if $\eta = \kappa = 0$, then p reduces to pI. Note also that at equilibrium, $\mathbf{u} = 0$, and p again reduces to pI.

17–2 SOME APPLICATIONS OF THE FUNDAMENTAL EQUATIONS OF CONTINUUM MECHANICS

In this section we shall reduce the general, exact equations of continuum mechanics to some simple, special cases. For example, suppose that we assume that the rate of flow of mass $\rho_m \mathbf{u}$ is proportional to the gradient of the density. This is known as Fick's law of diffusion. Mathematically, this says that

$$\rho_m \mathbf{u} = - D \text{ grad } \rho_m$$

where D is a constant called the diffusion constant. This equation is known to be empirically valid when the gradient of the density is small. Substituting this into the continuity equation [Eq. (17–5)] gives

$$\frac{\partial \rho_m}{\partial t} = D \text{ div grad } \rho_m = D\nabla^2 \rho_m \qquad (17\text{–}36)$$

This is the diffusion equation.

The diffusion equation is usually applied to the diffusion of one species through some medium, in which case it is written in the form

$$\frac{\partial c}{\partial t} = D \nabla^2 c \qquad (17\text{–}37)$$

where c is the concentration of the diffusing species. The solution to this equation depends upon the geometry of the system and the initial concentration distribution. A fundamental solution to this equation, however, is for diffusion in an infinite medium (no boundaries) with the initial condition that the diffusing substances is initially located at the origin. This solution is called the Green's function of the diffusion equation, and one can use this to construct solutions for other geometries and initial conditions.

Let us consider the case of diffusion in one dimension. In this case, Eq. (17–37) becomes

$$\frac{\partial c}{\partial t} = D \frac{\partial^2 c}{\partial x^2} \qquad -\infty < x < \infty \qquad (17\text{–}38)$$

with the initial condition

$$c(x, 0) = c_0 \,\delta(x) \qquad (17\text{–}39)$$

This equation is most easily solved by defining the Fourier transform of $c(x, t)$ by

$$\hat{C}(k, t) = (2\pi)^{-1/2} \int_{-\infty}^{\infty} e^{ikx} c(x, t)\, dx \qquad (17\text{–}40)$$

with

$$c(x, t) = (2\pi)^{-1/2} \int_{-\infty}^{\infty} e^{-ikx} \hat{C}(k, t)\, dk \qquad (17\text{–}41)$$

It is elementary to show by integration by parts that

$$\int_{-\infty}^{\infty} e^{ikx} \frac{\partial^2 c}{\partial x^2} \, dx = -k^2 \hat{C}(k, t)$$

If we take the Fourier transform of both sides of Eq. (17–38) and use the above expression, we get

$$\frac{\partial \hat{C}(k, t)}{\partial t} = -Dk^2 \hat{C}(k, t) \tag{17–42}$$

with the initial condition

$$\hat{C}(k, 0) = (2\pi)^{-1/2} c_0 = \hat{C}_0 \tag{17–43}$$

Note that the application of the Fourier transform to the partial differential equation has eliminated one of the independent variables x and given an ordinary differential equation, Eq. (17–42). This is readily solved to give

$$\hat{C}(k, t) = \hat{C}_0 e^{-k^2 Dt} \tag{17–44}$$

We can now invert this transform according to Eq. (17–41), giving

$$c(x, t) = \frac{\hat{C}_0}{(2\pi)^{1/2}} \int_{-\infty}^{\infty} e^{-ikx} e^{-k^2 Dt} dk$$

$$= \frac{c_0}{\pi} \int_0^{\infty} e^{-k^2 Dt} \cos kx \, dk$$

$$= \frac{c_0}{2(\pi Dt)^{1/2}} e^{-x^2/4Dt} \tag{17–45}$$

This is the well-known fundamental solution to the one-dimensional diffusion equation. This equation shows that the diffusing material, which is initially localized at the origin, will spread out as a function of time. Figure 17–3 shows this behavior, and Problem 17–12 discusses several interesting properties of this solution.

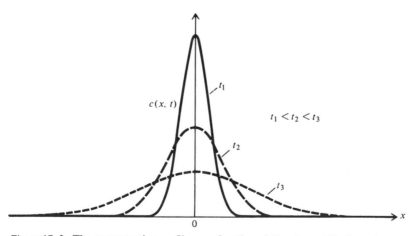

Figure 17–3. **The concentration profile as a function of time for a diffusing substance which is initially localized at the origin.**

It is instructive to solve this same problem in three-dimensional spherical coordinates assuming that the diffusion is isotropic. In this case, the diffusion equation is

$$\frac{\partial c}{\partial t} = D\nabla^2 c \tag{17-46}$$

with the initial condition

$$c(\mathbf{r}, 0) = c_0\, \delta(\mathbf{r}) \tag{17-47}$$

We define a three-dimensional Fourier transform of $c(\mathbf{r}, t)$ according to

$$\hat{C}(\mathbf{k}, t) = (2\pi)^{-3/2} \iiint_{-\infty}^{\infty} e^{i\mathbf{k}\cdot\mathbf{r}} c(\mathbf{r}, t)\, d\mathbf{r} \tag{17-48}$$

whose inverse is

$$c(\mathbf{r}, t) = (2\pi)^{-3/2} \iiint_{-\infty}^{\infty} e^{-i\mathbf{k}\cdot\mathbf{r}} \hat{C}(\mathbf{k}, t)\, d\mathbf{k} \tag{17-49}$$

From Eq. (17–49) we see that

$$\text{grad } c(\mathbf{r}, t) = -\frac{i\mathbf{k}}{(2\pi)^{3/2}} \iiint_{-\infty}^{\infty} e^{-i\mathbf{k}\cdot\mathbf{r}} \hat{C}(\mathbf{k}, t)\, d\mathbf{k}$$

and that

$$\text{div grad } c(\mathbf{r}, t) = \nabla^2 c(\mathbf{r}, t)$$

$$= -\frac{k^2}{(2\pi)^{3/2}} \iiint_{-\infty}^{\infty} e^{-i\mathbf{k}\cdot\mathbf{r}} \hat{C}(\mathbf{k}, t)\, d\mathbf{k}$$

$$= -k^2 c(\mathbf{r}, t)$$

Thus if we take the Fourier transform of Eq. (17–46), we get

$$\frac{\partial \hat{C}(\mathbf{k}, t)}{\partial t} = -Dk^2 \hat{C}(\mathbf{k}, t) \tag{17-50}$$

with the initial condition that

$$\hat{C}(\mathbf{k}, 0) = (2\pi)^{-3/2} c_0 = \hat{C}_0 \tag{17-51}$$

The solution to Eq. (17–50) is

$$\hat{C}(\mathbf{k}, t) = \hat{C}_0\, e^{-Dk^2 t} \tag{17-52}$$

Although this equation is similar to Eq. (17–44), their inverses are quite different since Eq. (17–44) describes a one-dimensional Fourier transform and Eq. (17–52) describes a three-dimensional Fourier transform. The inverse of Eq. (17–52) is given by (cf. Appendix B)

$$c(\mathbf{r}, t) = (2\pi)^{-3/2} \iiint_{-\infty}^{\infty} e^{-i\mathbf{k}\cdot\mathbf{r}} \hat{C}(\mathbf{k}, t)\, d\mathbf{k}$$

$$= c_0 (2\pi)^{-3} \iiint_{-\infty}^{\infty} e^{-i\mathbf{k}\cdot\mathbf{r}} e^{-k^2 Dt}\, d\mathbf{k}$$

$$= \frac{c_0}{8(\pi Dt)^{3/2}}\, e^{-r^2/4Dt} \tag{17-53}$$

One can show from this that

$$\int_0^\infty c(r, t) 4\pi r^2 \, dr = c_0 \tag{17-54}$$

The reader is referred to the problems and to the "Additional Readings" for further discussion of the diffusion equation.

As another simple example, if we set $\mathbf{u} = 0$ in Eq. (17–31), use Fourier's law, and assume that $E = E(T)$, we get

$$\rho_m c_v \frac{\partial T}{\partial t} = \lambda \nabla^2 T \tag{17-55}$$

where c_v is the specific heat. This equation is the heat conduction equation and has essentially the same form and the same solutions as the diffusion equation.

The mass balance and energy balance equations, then, are just generalizations of the diffusion equation and the heat flow equation, probably familiar to most physical chemistry students. The momentum balance equation, on the other hand, reduces to equations that are probably not familiar. These equations, however, are fundamental equations of hydrodynamics, and so it is necessary that we become familiar with these equations and some of their simpler applications.

Perhaps the simplest hydrodynamic equation is that describing nonviscous flow. This equation is called Euler's equation and is obtained immediately from Eqs. (17–21) and (17–34) by setting $\eta = \kappa = 0$ in the pressure tensor. Euler's equation is

$$\left(\frac{\partial}{\partial t} + \mathbf{u} \cdot \nabla \right) \mathbf{u} + \frac{1}{\rho_m} \nabla p = \mathbf{K} \tag{17-56}$$

and is a fundamental hydrodynamic equation describing nonviscous flow.

Let us now consider a monatomic ideal gas with no heat flow and no viscosity. We can cancel the factor of m that occurs in both terms of Eq. (17–7) to give

$$\frac{D\rho}{Dt} + \rho \nabla \cdot \mathbf{u} = 0 \tag{17-57}$$

This equation, with the number density ρ instead of the mass density ρ_m, is the equation for conservation of number of particles. If we use the fact that $1 : \nabla \mathbf{u} = \nabla \cdot \mathbf{u}$ (Problem 17–6), then Eq. (17–31) becomes

$$\frac{DT}{Dt} + \frac{p}{c_v \rho_m} \nabla \cdot \mathbf{u} = 0 \tag{17-58}$$

where $p/c_v \rho_m$ in Eq. (17–58) is equal to $2T/3$ for an ideal monatomic gas. The set of Eqs. (17–56) to (17–58) is called the ideal hydrodynamic equations. We can subtract Eqs. (17–57) and (17–58) to get

$$\frac{D\rho}{Dt} - \frac{3}{2} \frac{\rho}{T} \frac{DT}{Dt} = 0 \tag{17-59}$$

or

$$\frac{D}{Dt} (\rho T^{-3/2}) = 0$$

or

$$\rho T^{-3/2} = \text{constant} = C_1 \text{ (along a streamline)} \tag{17-60}$$

Using the equation of state of an ideal gas, $p = \rho kT$, Eq. (17–60) becomes

$$pV^{5/3} = \text{constant} = C_2 \qquad (17\text{–}61)$$

which says that the gas expands (or compresses) adiabatically along a streamline since Eq. (17–61) is the thermodynamic equation for the adiabatic change of an ideal gas, i.e., $pV^\gamma = \text{constant}$, where $\gamma = C_p/C_v = \frac{5}{3}$ for a monatomic ideal gas.

We can now use this equation and Euler's equation to derive the equation for the propagation of sound waves through a gas. To do this we assume that the gas is in equilibrium except for the passage of the sound waves. We can then assume that **u** and all of the space and time derivatives of **u**, ρ_m, and T are small first-order quantities. Neglecting second-order quantities then and setting $\mathbf{K} = 0$, Eqs. (17–56) and (17–57) become

$$\frac{\partial \rho_m}{\partial t} = \rho_m \cdot \nabla \mathbf{u} = 0 \qquad (17\text{–}62)$$

$$\rho_m \frac{\partial \mathbf{u}}{\partial t} + \nabla p = 0$$

Multiply the first of these by $\partial/\partial t$ and the second by div and again neglect second-order terms to get (Problem 17–9)

$$\frac{\partial^2 \rho_m}{\partial t^2} = \nabla^2 p \qquad (17\text{–}63)$$

Ordinarily the pressure is a function of the density and the temperature, but the density and temperature are related in this case by Eq. (17–59), which becomes

$$\frac{3}{2} \rho_m \frac{\partial T}{\partial t} = T \frac{\partial \rho_m}{\partial t}$$

when second-order quantities are neglected. We choose density as the independent variable and write

$$\nabla^2 p = \nabla \cdot [\text{grad } p] = \nabla \cdot \left[\left(\frac{\partial p}{\partial \rho_m} \right)_S \nabla \rho_m \right] \approx \left(\frac{\partial p}{\partial \rho_m} \right)_S \nabla^2 \rho_m \qquad (17\text{–}64)$$

where in the last step we have assumed that $(\partial p/\partial \rho)_S$ does not depend on the density. The subscript S refers to constant entropy since we have no heat flow. From the equation $pV^{5/3} = \text{constant}$, we find that $(\partial p/\partial \rho)_S = 5p/3\rho$, which is independent of ρ for an ideal monatomic gas.

Define an adiabatic compressibility by

$$\kappa_S = \frac{1}{\rho_m} \left(\frac{\partial \rho_m}{\partial p} \right)_S = \frac{1}{\rho} \left(\frac{\partial \rho}{\partial p} \right)_S \qquad (17\text{–}65)$$

Substitution of Eqs. (17–64) and (17–65) into (17–63) gives

$$\nabla^2 \rho = \rho_m \kappa_S \frac{\partial^2 \rho}{\partial t^2} \qquad (17\text{–}66)$$

for the spatial and time dependence of the density. This is a wave equation and describes the propagation of density waves (sound waves) through a gas with velocity

$$c = \frac{1}{(\rho_m \kappa_S)^{1/2}} \qquad (17\text{–}67)$$

For an ideal monatomic gas,

$$c = \left(\frac{5p}{3\rho_m}\right)^{1/2} = \left(\frac{5kT}{3m}\right)^{1/2} = \left(\frac{5\pi}{24}\right)^{1/2} \bar{v} = 0.81\bar{v} \tag{17-68}$$

This shows that the speed of sound is approximately equal to the mean molecular velocity of the molecules of the gas. This compares very favorably with experimental values.

We can almost immediately derive Bernoulli's equation from Euler's equation. Remember that Bernoulli's principle states that the pressure exerted on a surface varies as the inverse square of the velocity of the fluid passing over the surface. Start with Euler's equation (Eq. 17–56)

$$\left(\frac{\partial}{\partial t} + \mathbf{u} \cdot \nabla\right)\mathbf{u} = \mathbf{K} - \frac{1}{\rho_m}\nabla p \tag{17-69}$$

Assume we have a steady flow, i.e., that $\partial \mathbf{u}/\partial t = 0$, and that the external force is conservative, i.e., that $\mathbf{K} = -\nabla\psi$. We can use the vector identity (Problem 17–6)

$$\tfrac{1}{2}\nabla u^2 = (\mathbf{u} \cdot \nabla)\mathbf{u} + \mathbf{u} \times (\nabla \times \mathbf{u}) \tag{17-70}$$

to write Eq. (17–69) in the form

$$\nabla\left\{\tfrac{1}{2}u^2 + \frac{p}{\rho_m} + \psi\right\} = \mathbf{u} \times (\nabla \times \mathbf{u}) - \frac{kT\,\nabla\rho_m}{m\rho_m}$$

Now assume that the fluid is incompressible, which means that the density is constant throughout the volume of fluid and throughout its motion. From the equation of continuity, Eq. (17–7), we find that the condition of incompressibility can be written in the form $\nabla \cdot \mathbf{u} = 0$. Furthermore, assume that $\nabla \times \mathbf{u} = 0$. When this condition is satisfied, one says that the flow is irrotational. In the modern literature, $\nabla \times \mathbf{u}$ is denoted by curl $\mathbf{u}$, but in the older literature you will find rot $\mathbf{u}$. The quantity $\nabla \times \mathbf{u}$ is called the vorticity vector in hydrodynamics. With these two conditions, we have

$$\nabla\left\{\tfrac{1}{2}u^2 + \frac{p}{\rho_m} + \psi\right\} = 0$$

or

$$\tfrac{1}{2}u^2 + \frac{p}{\rho_m} + \psi = \text{constant} \tag{17-71}$$

This is Bernoulli's equation and shows that the pressure is inversely proportional to u^2. In the next section we shall discuss the solution of the Navier-Stokes equation for several simple systems.

17–3   THE NAVIER-STOKES EQUATION AND ITS SOLUTION

Euler's equation is the fundamental hydrodynamic equation of motion for non-viscous fluid flow. The fundamental hydrodynamic equation describing viscous flow is obtained by substitution of the pressure tensor given by Eq. (17–34),

$$\mathbf{p} = \{p + (\tfrac{2}{3}\eta - \kappa)\nabla \cdot \mathbf{u}\}\mathbf{I} - 2\eta\,\text{sym}(\nabla\mathbf{u})$$

into the momentum balance equation, Eq. (17–21),

$$\rho_m\left(\frac{\partial}{\partial t} + \mathbf{u} \cdot \nabla\right)\mathbf{u} = \rho_m\mathbf{K} - \nabla \cdot \mathbf{p}$$

to get the Navier-Stokes equation (Problem 17–7)

$$\rho_m\left(\frac{\partial}{\partial t} + \mathbf{u} \cdot \nabla\right)\mathbf{u} = \rho_m \mathbf{K} + \eta \nabla^2 \mathbf{u} + \left(\frac{\eta}{3} + \kappa\right)\text{grad div } \mathbf{u} - \nabla p \tag{17–72}$$

The Navier-Stokes equation is the hydrodynamic equation governing the flow of a viscous fluid.

The Navier-Stokes equation is often seen written in the form (Problem 17–10)

$$\frac{\partial}{\partial t}\rho_m\mathbf{u} = \nabla \cdot (\boldsymbol{\sigma} - \rho_m\mathbf{uu}) + \rho_m\mathbf{K} \tag{17–73}$$

where the tensor $\boldsymbol{\sigma}$ is called the *stress tensor* and is given by

$$\boldsymbol{\sigma} = -\mathsf{p} \tag{17–74}$$

The stress tensor itself is often written in the form

$$\boldsymbol{\sigma} = -p\mathsf{I} + \kappa(\nabla \cdot \mathbf{u})\mathsf{I} + 2\eta\boldsymbol{\varepsilon} \tag{17–75}$$

where $\boldsymbol{\varepsilon}$ is the divergenceless part of $\nabla\mathbf{u}$,

$$\boldsymbol{\varepsilon} = \text{sym}(\nabla\mathbf{u}) - \tfrac{1}{3}(\nabla \cdot \mathbf{u})\mathsf{I} \tag{17–76}$$

and is called the rate of shear tensor.

Since the Navier-Stokes equation plays an important role in the statistical mechanical theory of viscosity (cf. Section 21–8) and, in addition, is probably unfamiliar to most readers, we shall discuss here several of its simpler applications. This section is somewhat of a digression since the results here are not used later in the book, but nevertheless, many of the standard textbook problems of hydrodynamics are not only interesting in themselves, but also serve as excellent exercises in boundary-value problems in various coordinate systems. In fact, it is unfortunate that hydrodynamics has disappeared from the curriculum of theoretical chemistry and physics since it is a rich applied mathematical experience, as the perusal through the references listed in the "Additional Reading" would show.

Let us consider first the simple case of the steady flow of a viscous fluid between two parallel plates as shown in Fig. 17–4. Choose the flow to be in the x-direction and the perpendicular to the plates to be the y-direction. We consider the parallel plates to be infinite in extent, or, in other words, we shall ignore edge effects. In addition, we shall assume that the fluid is incompressible, so that we have that $\nabla \cdot \mathbf{u} = 0$. The Navier-Stokes equation becomes

$$\rho_m \frac{D\mathbf{u}}{Dt} = -\nabla p + \rho_m \mathbf{K} + \eta \nabla^2 \mathbf{u} \tag{17–77}$$

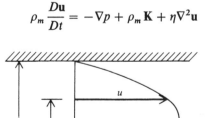

Figure 17–4. **The velocity profile of plane Poiseuille flow between two parallel infinite planes.**

Now due to the geometry of this simple system, $u_y = u_z = 0$ and $\partial u_x/\partial x = \partial u_x/\partial z = 0$ in the steady state. Thus, Eq. (17–77) becomes

$$\eta \frac{d^2 u_x}{dy^2} - \frac{\partial p}{\partial x} = 0$$

$$-\frac{\partial p}{\partial y} + \rho mg = 0 \tag{17–78}$$

where we have included the possibility of a gravitational force in the y-direction. That $\partial p/\partial x$ cannot depend upon y can be seen by differentiating the second of these equations with respect to x to get

$$\frac{\partial}{\partial x}\left(\frac{\partial p}{\partial y}\right) = \frac{\partial}{\partial y}\left(\frac{\partial p}{\partial x}\right) = 0$$

which shows that $\partial p/\partial x$ cannot depend upon y. If we let the gradient of the pressure, $\partial p/\partial x$, be a constant $-\gamma$, then the first of Eqs. (17–78) gives

$$u_x(y) = -\frac{\gamma}{2\eta} y^2 + c_1 y + c_2 \tag{17–79}$$

where c_1 and c_2 are two constants of integration. A standard boundary condition in viscous flow is to assume that the fluid sticks to the boundary surface, and so in this particular case we have $u_x(0) = u_x(b) = 0$ and Eq. (17–79) becomes

$$u_x(y) = \frac{\gamma}{2\eta} y(b - y) \tag{17–80}$$

This velocity profile is shown in Fig. 17–4. Flow of this type is called plane Poiseuille flow. Problem 17–27 illustrates plane Couette flow, which is the steady-state flow when one of the planes moves past the other with a constant velocity U, and Problem 17–29 illustrates the flow down an inclined plane.

The concept of Poiseuille flow should be familiar to most physical chemists since the timing of the flow of a viscous liquid down a cylindrical capillary is a standard physical chemistry experiment for the determination of viscosity. For this cylindrical case, Eqs. (17–78) take the form

$$\eta\left(\frac{d^2 u}{dr^2} + \frac{1}{r}\frac{du}{dr}\right) = \frac{\partial p}{\partial z} \tag{17–81}$$

where r is the radial distance and z is the distance along the cylinder. The radius of the capillary is R. We choose $\partial p/\partial z$ to be a constant $\gamma = (p_1 - p_2)/l$, where l is the length of the capillary and $(p_1 - p_2)$ is the difference in pressures at the two ends. Equation (17–81) is easily integrated if one uses the identity

$$\frac{d^2 u}{dr^2} + \frac{1}{r}\frac{du}{dr} = \frac{1}{r}\frac{d}{dr}\left(r\frac{du}{dr}\right) \tag{17–82}$$

Two simple integrations give

$$u(r) = \frac{\gamma r^2}{4\eta} + c_1 \ln r + c_2$$

Since $u(0)$ must be finite, c_1 must be zero, and since $u(R) = 0$ due to the no-slip boundary condition, we have

$$u(r) = \frac{\gamma}{4\eta}(r^2 - R^2)$$

The volume of fluid crossing any normal section of the capillary per unit time is

$$2\pi \int_0^R ru(r)\,dr = \frac{\pi R^4(p_2 - p_1)}{8\eta l} \tag{17–83}$$

which is the standard formula of Poiseuille flow.

The calculation of the velocity profile for the steady unidirectional flow through a rectangular conduit is an instructive problem. If we assume conditions similar to those in the previous examples, the Navier-Stokes equation is

$$\eta\left(\frac{\partial^2}{\partial x^2} + \frac{\partial^2}{\partial y^2}\right)u = \gamma \tag{17–84}$$

where γ is a constant pressure gradient and u is the velocity along the conduit. The geometry is shown in Fig. 17–5. Equation (17–84) is an inhomogeneous partial differential equation. Its solution can be found by first writing

$$u(x, y) = u_c(x, y) + u_p(x, y) \tag{17–85}$$

where $u_c(x, y)$ is the complementary solution, i.e., the solution to the homogeneous equation

$$\left(\frac{\partial^2}{\partial x^2} + \frac{\partial^2}{\partial y^2}\right)u_c = 0$$

and $u_p(x, y)$ is a particular solution to the inhomogeneous equation. Certainly

$$u_p(x, y) = \frac{\gamma}{2\eta}y^2 + c_1 y + c_2$$

is a particular solution to Eq. (17–84), and the no-slip boundary condition $u_p = 0$ for $y = \pm c/2$ gives

$$u_p = \frac{\gamma}{2\eta}\left(y^2 - \frac{c^2}{4}\right)$$

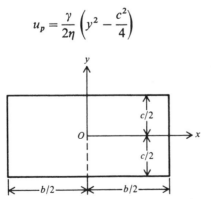

Figure 17–5. **The coordinate system for a rectangular cross section describing steady unidirectional flow through a rectangular conduit.**

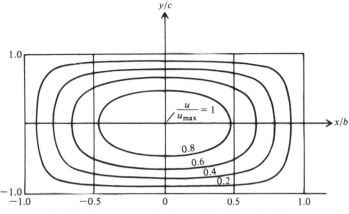

Figure 17-6. **Velocity contours for the steady unidirectional flow through a rectangular conduit.** (From H. Rouse, ed., *Advanced Fluid Mechanics.* New York: Wiley, 1959.)

The complementary solution can be found by separation of variables to be (Problem 17-31)

$$u_c(x, y) = \sum_{n=1}^{\infty} A_n \cosh \frac{(2n-1)\pi x}{2c} \cos \frac{(2n-1)\pi y}{2c} \tag{17-86}$$

where the A_n are given by

$$A_n \cosh \frac{(2n-1)\pi b}{4c} = \frac{2c^2\gamma}{\eta \alpha_n{}^3} \left(2 \sin \frac{\alpha_n}{2} - \alpha_n \cos \frac{\alpha_n}{2} \right)$$

where $\alpha_n = (2n-1)\pi/2$. The velocity contours, $u(x, y)$ are shown in Fig. 17-6 for the case $b = 2c$.

Although this section has been a rather lengthy digression, it is hoped that this brief introduction to fluid mechanics gives the reader some degree of comfort and familiarity with the Navier-Stokes equation.

This concludes our presentation of continuum mechanics. In the next chapter we shall derive these same equations from within a molecular framework. In this way, we shall obtain expressions for the various transport coefficients in terms of molecular properties. We now go on to a main goal of this book, that is, to determine the molecular basis of the theory of transport.

ADDITIONAL READING

BRODKEY, R. S. 1967. *The phenomena of fluid motions.* Reading, Mass.: Addison-Wesley.
CARSLAW, H. S., and JAEGER, J. C. 1941. *Operational methods in applied mathematics.* New York: Dover.
CHURCHILL, R. V. 1972. *Operational mathematics.* 3rd. ed. New York: McGraw-Hill.
CRANK, J. 1956. *The mathematics of diffusion.* London: Oxford University Press.
LANDAU, L. D., and LIFSHITZ, E. M. 1959. *Fluid mechanics.* Reading, Mass.: Addison-Wesley.
LASS, H. 1950. *Vector and tensor analysis.* New York: McGraw-Hill.
ROUSE, H., ed. 1959. *Advanced fluid mechanics.* New York: Wiley.
SNEDDON, I. N. 1951. *Fourier transforms.* New York: McGraw-Hill.
SOMMERFELD, A. 1950. *Mechanics of deformable bodies.* New York: Academic.
YIH, C-S. 1969. *Fluid mechanics.* New York: McGraw-Hill.

PROBLEMS

17-1. Prove the vector identities:

(a) div grad $a = \nabla^2 a$

(b) curl grad $a = 0$

(c) div curl $\mathbf{v} = 0$

(d) curl curl $\mathbf{v} = $ grad div $\mathbf{v} - \nabla^2 \mathbf{v}$

(e) $\nabla \cdot (a\mathbf{v}) = a\nabla \cdot \mathbf{u} + \mathbf{u} \cdot \nabla a$

(f) $\nabla \cdot (\mathbf{v} \times \mathbf{u}) = \mathbf{u} \cdot \nabla \times \mathbf{v} - \mathbf{v} \cdot \nabla \times \mathbf{u}$

(g) $\nabla(\mathbf{v} \cdot \mathbf{u}) = (\mathbf{v} \times \nabla) \times \mathbf{u} + (\mathbf{u} \times \nabla) \times \mathbf{v} + \mathbf{v}(\nabla \cdot \mathbf{u}) + \mathbf{u}(\nabla \cdot \mathbf{v})$

17-2. Convince yourself that

$$\frac{D\theta}{Dt} = \frac{\partial \theta}{\partial t} + \mathbf{u} \cdot \nabla \theta$$

represents the change in $\theta(\mathbf{r}, t)$ along the flow of the fluid, moving with velocity $\mathbf{u}$.

17-3. Show that the condition of incompressibility can be expressed by div $\mathbf{u} = 0$.

17-4. Derive the momentum balance equation, Eq. (17-17).

17-5. Convert Eq. (17-17) into Eq. (17-21).

17-6. Prove the tensorial identities

(a) $\nabla \cdot (\mathbf{uu}) = (\mathbf{u} \cdot \nabla)\mathbf{u} + \mathbf{u}(\nabla \cdot \mathbf{u})$

(b) $\nabla \cdot (a\mathbf{uu}) = a(\mathbf{u} \cdot \nabla)\mathbf{u} + \mathbf{u}(\nabla \cdot a\mathbf{u})$

(c) $\nabla \cdot (\mathbf{u} \cdot \mathbf{B}) = \mathbf{u} \cdot (\nabla \cdot \mathbf{B}) + \mathbf{B}:\nabla\mathbf{u}$

(d) $\mathbf{I}:\nabla\mathbf{u} = \nabla \cdot \mathbf{u}$

(e) $\nabla \cdot (\nabla \cdot \mathbf{u})\mathbf{I} = \nabla(\nabla \cdot \mathbf{u})$

(f) $\nabla \cdot (\text{sym } \nabla \mathbf{u}) = \frac{1}{2}\nabla^2\mathbf{u} + \frac{1}{2}\nabla(\nabla \cdot \mathbf{u})$

(g) $\mathbf{v} \cdot \nabla\mathbf{v} = \frac{1}{2}\nabla v^2 - \mathbf{v} \times (\nabla \times \mathbf{v})$

17-7. Substitute the Newtonian pressure tensor into the momentum balance equation to derive the Navier-Stokes equation,

$$\rho_m \frac{\partial \mathbf{u}}{\partial t} = -\nabla p - (\tfrac{2}{3}\eta - \kappa)\nabla \cdot (\nabla \cdot \mathbf{u})\mathbf{I} + 2\eta\nabla \cdot \text{sym } \nabla\mathbf{u})$$

Now use the tensorial identities proved in Problem 17-6 to derive Eq. (17-72).

17-8. For slow, smooth fluid flow, one often uses the so-called linearized Navier-Stokes equation

$$\rho_m \frac{\partial \mathbf{u}}{\partial t} = -\nabla p + \eta\nabla^2\mathbf{u} + (\tfrac{1}{3}\eta + \kappa)\nabla(\nabla \cdot \mathbf{u})$$

Discuss the physical assumptions involved in deriving the linearized Navier-Stokes equation from the complete equation. We shall use this linearized form in Section 21-8.

17-9. Derive Eq. (17-63).

17-10. Show that the Navier-Stokes equation can be written in the form

$$\frac{\partial}{\partial t}\rho_m\mathbf{u} = \nabla \cdot (\boldsymbol{\sigma} - \rho_m\mathbf{uu}) + \rho_m\mathbf{K}$$

where $\boldsymbol{\sigma} = -\mathbf{p}$ is called the stress tensor.

17-11. Show that the stress tensor can be written in the form

$$\boldsymbol{\sigma} = -p\mathbf{I} + \kappa(\nabla \cdot u)\mathbf{I} + 2\eta\varepsilon$$

where

$$\varepsilon = \text{sym}(\nabla\mathbf{u}) - \tfrac{1}{3}(\nabla \cdot \mathbf{u})\mathbf{I}$$

17–12. Solve the one-dimensional diffusion equation for an infinite region under the initial condition $c(x, t) = \delta(x - x_0)$. This solution is called the fundamental solution of the one-dimensional diffusion equation.

17–13. A two-dimensional Fourier transform pair has the form

$$\hat{c}(\mathbf{k}, t) = (2\pi)^{-1} \int_{-\infty}^{\infty} \int e^{i\mathbf{k}\cdot\mathbf{r}} c(\mathbf{r}, t)\, d\mathbf{r}$$

$$c(\mathbf{r}, t) = (2\pi)^{-1} \int_{-\infty}^{\infty} \int e^{-i\mathbf{k}\cdot\mathbf{r}} \hat{c}(\mathbf{k}, t)\, d\mathbf{k}$$

where $\mathbf{k} = (k_x, k_y)$, $\mathbf{r} = (x, y)$, $d\mathbf{r} = dx\, dy$, and $d\mathbf{k} = dk_x\, dk_y$. Show that if $c(\mathbf{r}, t)$ has circular symmetry, then

$$\hat{c}(\mathbf{k}, t) = (2\pi)^{-1} \int_{0}^{\infty} dr\, r\, c(r, t) \int_{0}^{2\pi} e^{ikr\cos\theta}\, d\theta$$

The integral over $d\theta$ here is one of the defining equations for the zero-order Bessel function J_0,

$$J_0(x) = \frac{1}{2\pi} \int_{0}^{2\pi} e^{ix\cos\theta}\, d\theta$$

Using this and the fact that $J_0(x)$ is an even function, show that

$$\hat{c}(\mathbf{k}, t) = \int_{0}^{\infty} r J_0(kr) c(r, t)\, dr$$

and

$$c(\mathbf{r}, t) = \int_{0}^{\infty} k J_0(kr) \hat{c}(k, t)\, dk$$

Compare this to the three-dimensional case given by Eqs. (12–51) and (12–52).

17–14. Using the result of the previous problem, find the fundamental solution of the two-dimensional diffusion equation. (The necessary integral is standard and can be found, for example, in *Handbook of Mathematical Functions*, edited by M. Abramawitz and I. Stegun, New York: Dover.) Show that the resulting $c(r, t)$ is normalized.

17–15. Show that the solution to the one-dimensional diffusion equation for an unbounded region with initial condition $c(x, 0) = f(x)$ is

$$c(x, t) = (4\pi Dt)^{-1/2} \int_{-\infty}^{\infty} f(u) e^{-(x-u)^2/4Dt}\, du$$

Note that this is simply the convolution of the fundamental solution (the Green's function) and the initial condition.

17–16. Use the fundamental solution to the three-dimensional diffusion equation [Eq. (17–53) with $c_0 = 1$] to show that the mean-square displacement of a diffusing particle is $6Dt$ in three dimensions. Derive the same result directly from the Fourier transform of $c(\mathbf{r}, t)$ by showing that

$$\left(\frac{\partial^2 \hat{c}(k, t)}{\partial k^2}\right)_{k=0} = -2Dt = -\tfrac{1}{3} \int_{0}^{\infty} 4\pi r^4 c(r, t)\, dr$$

$$= -\tfrac{1}{3}\langle r^2(t)\rangle$$

Discuss the implications of this concerning reversibility and irreversibility.

17–17. Show that

$$\frac{\partial c}{\partial t} = D_x \frac{\partial^2 c}{\partial x^2} + D_y \frac{\partial^2 c}{\partial y^2} + D_z \frac{\partial^2 c}{\partial z^2}$$

represents anisotropic diffusion and that its fundamental solution is

$$c(x, y, z, t) = \frac{1}{8(\pi t)^{3/2}(D_x D_y D_z)^{1/2}} \exp\left[-\frac{(x - x_0)^2}{4D_x t} - \frac{(y - y_0)^2}{4D_y t} - \frac{(z - z_0)^2}{4D_z t} \right]$$

17–18. Show that

$$\frac{\partial c}{\partial t} = \beta_x \frac{\partial c}{\partial x} + \beta_y \frac{\partial c}{\partial y} + \beta_z \frac{\partial c}{\partial z} + D_x \frac{\partial^2 c}{\partial x^2} + D_y \frac{\partial^2 c}{\partial y^2} + D_z \frac{\partial^2 c}{\partial z^2}$$

represents anisotropic diffusion with an anisotropic drift and that its fundamental solution is

$$c(x, y, z, t) = \frac{1}{8(\pi t)^{3/2}(D_x D_y D_z)^{1/2}} \exp\left[-\frac{(x - x_0 + \beta_x t)^2}{4D_x t} - \frac{(y - y_0 + \beta_y t)^2}{4D_y t} \right.$$

$$\left. -\frac{(z - z_0 + \beta_z t)^2}{4D_z t} \right]$$

17–19. The Debye theory of rotational relaxation is based upon the rotational diffusion equation (see, e.g., A. Carrington and A. D. McLachlan, *Introduction to Magnetic Resonance*, Appendix H)

$$\frac{\partial p}{\partial t} = D_R \nabla^2 p$$

where $p = p(\theta, \phi, t)$ is the probability that a molecular axis points in the direction (θ, ϕ), D_R is the rotational diffusion coefficient, and ∇^2 is the angular part of the Laplacian operator in spherical coordinates, i.e.,

$$\nabla^2 = \frac{1}{\sin \theta} \frac{\partial}{\partial \theta}\left(\sin \theta \frac{\partial}{\partial \theta} \right) + \frac{1}{\sin^2 \theta} \frac{\partial^2}{\partial \phi^2}$$

Using the fact that the spherical harmonics $Y_{lm}(\theta, \phi)$ satisfy the equation

$$\nabla^2 Y_{lm} = -l(l + 1) Y_{lm}$$

and assuming that the initial probability distribution is a delta function concentrated at $\theta = \phi = 0$, i.e., that

$$p(\theta, \phi, 0) = \frac{1}{2\pi} \delta(\cos \theta - 1)$$

show that

$$p(\theta, \phi, t) = \sum_l \left(\frac{2l + 1}{4\pi} \right) P_l(\cos \theta) e^{-l(l+1)D_R t}$$

and that

$$\langle \cos \theta(t) \rangle = \int_0^{2\pi} d\phi \int_0^{\pi} d\theta \sin \theta \cos \theta \, p(\theta, \phi, t)$$

$$= e^{-2D_R t}$$

and that

$$\langle \tfrac{1}{2}[3 \cos^2 \theta(t) - 1] \rangle = e^{-6D_R t}$$

17–20. Consider the one-dimensional diffusion equation for the region $-\infty < x < a$, with an absorbing wall located at $x = a$. Let the initial condition be simply $c(x, 0) = \delta(x)$. Thus we must solve the partial differential equation

$$\frac{\partial c}{\partial t} = D \frac{\partial^2 c}{\partial x^2} \qquad x < a$$

subject to the conditions

$$c(x, 0) = \delta(x)$$

$$c(a, t) = 0$$

This last condition is the mathematical statement of an absorbing barrier.

First show that the solution to the unbounded problem with initial condition $c(x, 0) = \delta(x - x_0)$ is

$$c(x, t; x_0) = (4\pi Dt)^{-1/2} \exp\left\{\frac{-(x - x_0)^2}{4Dt}\right\}$$

Clearly $c(x, t; 0)$ satisfies the partial differential equation and the initial condition, but not the boundary condition. For $x_0 > a$, show that the linear combination

$$c(x, t) = c(x, t; 0) + Ac(x, t; x_0)$$

also satisfies both the partial differential equation and the initial condition. The procedure now is to look for an A and x_0 such that the boundary condition is also satisfied.

Notice that the two terms in this trial solution can be thought of as the concentration profiles resulting from initial sources located at $x = 0$ and $x = x_0$. Thus we shall refer to them as source terms. The first term is real in the sense that the diffusion does start from the point $x = 0$, but the existence of the absorbing barrier located at $x = a$ prevents this from being a solution to the problem. This first term alone predicts that the concentration at $x = a$ would be $c(a, t; 0) = (4\pi Dt)^{-1/2} \exp(-a^2/4Dt)$ instead of zero as required by the absorbing barrier. To overcome this, we can place a fictitious source beyond the barrier in such a way that the concentration at the absorbing barrier is zero. Such a fictitious source is called an *image source*, and the method being introduced here is called *the method of images*.

Clearly, by symmetry the image source should be placed at $x = 2a$ and should be done so with a negative weight so that it is actually an image sink rather than an image source. Thus we see on physical grounds that we should choose $x_0 = 2a$ and $A = -1$. Show that the complete solution to the problem (partial differential equation, initial condition, *and* boundary condition) is given by

$$c(x, t) = (4\pi Dt)^{-1/2}\left\{\exp\left(-\frac{x^2}{4Dt}\right) - \exp\left[-\frac{(x - 2a)^2}{4Dt}\right]\right\}$$

Note that the solution to the problem with a barrier can be written down in terms of the fundamental solution $c(x, t; x_0)$. Sketch this solution, showing how the source term and the image sink term contribute to the total solution $c(x, t)$.

17–21. Use the method of images introduced in the previous problem to solve the diffusion problem with two absorbing barriers, located at $\pm a$. The problem to be solved is

$$\frac{\partial c}{\partial t} = D \frac{\partial^2 c}{\partial x^2} \qquad -a < x < a$$

$$c(a, t) = c(-a, t) = 0$$

$$c(x, 0) = \delta(x)$$

Show that the solution to the two absorbing barrier problems requires a doubly infinite system of images, with a set of "unit sources" located at the points $x_n' = 4na$ ($n = 0, \pm 1, \pm 2, \ldots$) and a set of "unit sinks" located at the points $x_n'' = 2a(1 - 2n)$, and thus

$$c(x, t) = (4\pi Dt)^{-1/2} \sum_{n=-\infty}^{\infty} \left\{ \exp\left[-\frac{(x - x_n')^2}{4Dt} \right] - \exp\left[-\frac{(x - x_n'')^2}{4Dt} \right] \right\}$$

17–22. Solve the previous problem by the method of separation of variables. The solution is

$$c(x, t) = \sum_{n=1}^{\infty} \frac{1}{a} \sin\left(\frac{n\pi}{2}\right) e^{-n^2\pi^2 Dt} \sin\left[\frac{n\pi(x + a)}{2a}\right]$$

Note that this appears to be different from the one obtained from the method of images in the previous problem. Of course the two solutions are equivalent, and the two series are alternative representations of the same function.

17–23. Show that the equation

$$\frac{\partial c}{\partial t} = D \frac{\partial^2 c}{\partial x^2} - \mu \frac{\partial c}{\partial x}$$

represents a system both diffusing and moving with a constant drift velocity μ. Show how Fick's law is modified for this case. This equation could represent the diffusion of a large charged particle such as a protein in a uniform electric field (cf. Section 22–3).

17–24. Show that the fundamental solution to the equation in the previous problem is

$$c(x, t; x_0) = (4\pi Dt)^{-1/2} \exp\left[-\frac{(x - x_0 - \mu t)^2}{4Dt} \right]$$

17–25. Use the method of images introduced in Problem 17–20 and the fundamental solution given in the previous problem to solve the problem of diffusion with a bias or a drift with an absorbing barrier located at $x = a$. In other words, solve the equation

$$\frac{\partial c}{\partial t} = D \frac{\partial^2 c}{\partial x^2} - \mu \frac{\partial c}{\partial x}$$

along with the conditions

$$c(x, 0) = \delta(x)$$

$$c(a, t) = 0$$

17–26. Repeat the previous problem, but this time with two absorbing barriers, located at $\pm a$. This is also the solution to Problem 17–21 with a drift.

17–27. Consider the flow of a fluid between two parallel plates, a distance b apart, where one of the plates moves with velocity U relative to the other. Take the velocity to be in the x-direction and the perpendicular to the plates to be the y-direction. Following the derivation of Eq. (17–80), find $u_x(y)$ for this system. This type of flow is called plane Couette flow.

17–28. In plane Couette-Poiseuille flow, the geometry is similar to that in the above problem, but there is a constant pressure gradient $\partial p/\partial x$ in addition to the relative motion of the plates. In other words, this problem is a combination of the previous one and the plane Poiseuille flow discussed in the chapter. Find $u_x(y)$ for this type of flow.

17–29. If the plates bounding the flow in the previous two problems are parallel but not horizontal, the solution is found by the same procedure that led to Eq. (17–80), except that if the x-direction is still taken to be the direction of flow; then $\partial p/\partial x$ should be replaced by $\partial p/\partial x - \rho m g \sin \beta$ and $\partial p/\partial y$ by $-\rho m g \cos \beta$, where β is the angle that the x-axis makes with the horizontal (cf. Fig. 17–7). Find $u_x(y)$ for this case.

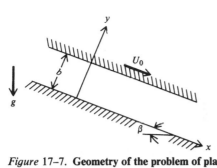

Figure 17-7. **Geometry of the problem of plane Couette-Poiseuille flow down an inclined plane.**

17-30. Show that Eq. (17-81) is the appropriate form of the Navier-Stokes equation for flow through a cylindrical capillary.

17-31. Show that Eq. (17-86) is the complementary solution to Eq. (17-84).

17-32. Derive Eq. (17-30).

CHAPTER 18

KINETIC THEORY OF GASES AND THE BOLTZMANN EQUATION

In Section 7–2 we introduced the concept of phase space and distribution functions in phase space. We also derived the Liouville equation, which is the equation of motion that the phase space distribution function must satisfy. Since we were interested only in equilibrium statistical mechanics at that time, we did not consider the Liouville equation in any detail. In this chapter we shall review the concept of phase space and derive the Liouville equation again. We shall then introduce reduced distribution functions and derive the Bogoliubov, Born, Green, Kirkwood, Yvon (BBGKY) hierarchy. This hierarchy is the nonequilibrium generalization of the Kirkwood integral equation hierarchy for the fluid distribution functions, $g^{(n)}(\mathbf{r}_1, \ldots, \mathbf{r}_n)$, of Chapter 13. Nobody has yet devised a successful way to uncouple the BBGKY hierarchy, and so in Section 18–4 we shall derive a physical, yet approximate, equation for the distribution function for gases. This equation, called the Boltzmann equation, is the central equation of the rigorous kinetic theory of gases. In Section 18–5, we shall derive some of the general consequences of the Boltzmann equation that can be determined without actually solving it completely. We shall discuss its solution in Chapter 19. The standard reference for most of this chapter is Hirschfelder, Curtiss, and Bird. Mazo (in "Additional Reading") also discusses these topics well.

18–1 PHASE SPACE AND THE LIOUVILLE EQUATION

Consider a system of N point particles. The classical dynamical state of this system is specified by the $3N$ momentum components $p_1, p_2, \ldots, p_{3N}$ and the $3N$ spatial coordinates $q_1, \ldots, q_{3N}$. We can construct a $6N$-dimensional space whose coordinates are $q_1, q_2, \ldots, p_1, \ldots, p_{3N}$. One point in this *phase space* completely specifies the microscopic dynamical state of our N-particle system. As the system evolves in time, this *phase point* moves through phase space in a manner completely dictated by the equations of motion of the system. Actually, one never knows (nor really cares to know) the $6N$ coordinates of a macroscopic system. Rather, one knows just a few macroscopic mechanical properties of the system, such as the energy, volume, velocity, etc. Clearly there are a great number of points in phase space that are compatible with the few variables that we know about the system. The set of all such phase points constitutes an *ensemble* of systems. The number of systems in an ensemble approaches infinity, and so the set of phase points that could possibly represent our system becomes quite dense. This allows us to define a *density of phase points* or *distribution function* as the fraction of phase points contained in the volume $dq_1\, dq_2 \cdots dp_{3N}$. We shall denote the phase space distribution function by $f_N(q_1, q_2, \ldots, p_{3N}, t)$, or more conveniently by $f_N(p, q, t)$. We shall often use this abbreviated notation. Similarly, we shall often denote $dq_1, dq_2 \cdots dp_{3N}$ by $dp\, dq$. The density $f_N(p, q, t)$ is normalized such that

$$\int f_N(p, q, t)\, dp\, dq = 1$$

Since each phase point moves in time according to the equations of motion of the system it describes, f_N itself must obey some sort of equation of motion. The equation that $f_N(p, q, t)$ satisfies can be readily determined by using the methods of the previous chapter, particularly, the argument associated with Eqs. (17–1) to (17–5). The number of phase points within some arbitrary volume v is

$$n = \mathcal{N} \int_v f_N(p, q, t)\, dp\, dq$$

where we are using the condensed notation of letting p and q denote all the spatial coordinates and momenta necessary to specify a system in the ensemble. The rate of change of the number of phase points within v is

$$\frac{dn}{dt} = \mathcal{N} \int_v \frac{\partial f_N}{\partial t}\, dp\, dq \qquad (18\text{–}1)$$

Since phase points are neither created nor destroyed, the rate of change of n must be given by the rate at which phase points flow through the surface enclosing v. The rate of flow of phase points is $\mathcal{N} f_N \mathbf{u}$, where $\mathbf{u}$ is not just the $3N$-dimensional vector $(\dot{q}_1, \dot{q}_2, \ldots, \dot{q}_{3N})$, but the $6N$-dimensional vector $(\dot{q}_1, \ldots, \dot{p}_1, \ldots, \dot{p}_{3N})$ since the spatial coordinates and momenta play an equivalent role in phase space. We integrate this flow over the surface to get

$$\frac{dn}{dt} = -\mathcal{N} \int_S f_N \mathbf{u} \cdot d\mathbf{S}$$

The negative sign here indicates that an outflow of phase points yields a negative value for dn/dt since $\mathbf{u} \cdot d\mathbf{S}$ is positive if $\mathbf{u}$ is directed outward from v and negative if $\mathbf{u}$ is directed inward.

The surface integral can be transformed to a volume integral by using Gauss' theorem to get

$$\frac{dn}{dt} = -\mathcal{N} \int_v \nabla \cdot (f_N \mathbf{u})\, dp\, dq \qquad (18\text{–}2)$$

If we subtract Eq. (18–1) from Eq. (18–2) and realize that this equation is valid for any choice of v, we have the equation for the conservation of phase points

$$\frac{\partial f_N}{\partial t} + \nabla \cdot (f_N \mathbf{u}) = 0 \qquad (18\text{–}3)$$

in which it should be clear that since we are dealing with phase space

$$\mathbf{u} = (\dot{q}_1, \ldots, \dot{q}_{3N}, \dot{p}_1, \ldots, \dot{p}_{3N})$$

and

$$\nabla \cdot f_N \mathbf{u} = \sum_{j=1}^{3N} \frac{\partial}{\partial q_j} (f_N \dot{q}_j) + \sum_{j=1}^{3N} \frac{\partial}{\partial p_j} (f_N \dot{p}_j)$$

$$= \sum_{j=1}^{3N} \left\{ \frac{\partial f_N}{\partial q_j} \dot{q}_j + \frac{\partial f_N}{\partial p_j} \dot{p}_j \right\} + \sum_{j=1}^{3N} \left\{ \frac{\partial \dot{q}_j}{\partial q_j} + \frac{\partial \dot{p}_j}{\partial p_j} \right\} f_N$$

But Eq. (7–27) shows that the summand of the second summation here is zero, and so Eq. (18–3) becomes

$$\frac{\partial f_N}{\partial t} + \sum_{j=1}^{3N} \frac{\partial f_N}{\partial q_j} \dot{q}_j + \sum_{j=1}^{3N} \frac{\partial f_N}{\partial p_j} \dot{p}_j = 0 \tag{18-4}$$

Using Hamilton's equations of motion,

$$\dot{p}_i = -\frac{\partial H}{\partial q_i} \quad \text{and} \quad \dot{q}_i = \frac{\partial H}{\partial p_i}$$

Eq. (18–4) can be written

$$\frac{\partial f_N}{\partial t} + \sum_{j=1}^{3N} \left(\frac{\partial H}{\partial p_j} \frac{\partial f_N}{\partial q_j} - \frac{\partial H}{\partial q_j} \frac{\partial f_N}{\partial p_j} \right) = 0 \tag{18-5}$$

The summation here is called a Poisson bracket and is commonly denoted by $\{H, f_N\}$; so Eq. (18–5) is often written as

$$\frac{\partial f_N}{\partial t} + \{H, f_N\} = 0 \tag{18-6}$$

This is the Liouville equation, the most fundamental equation of statistical mechanics. In fact, it can be shown that the Liouville equation is equivalent to the $6N$ Hamilton equations of motion of the N-body system.[*]
    In Cartesian coordinates, the Liouville equation reads

$$\frac{\partial f_N}{\partial t} + \sum_{j=1}^{N} \frac{\mathbf{p}_j}{m_j} \cdot \nabla_{\mathbf{r}_j} f_N + \sum_{j=1}^{N} \mathbf{F}_j \cdot \nabla_{\mathbf{p}_j} f_N = 0 \tag{18-6'}$$

In this equation $\nabla_{\mathbf{r}_j}$ denotes the gradient with respect to the spatial variables in f_N; $\nabla_{\mathbf{p}_j}$ denotes the gradient with respect to the momentum variables in f_N; and $\mathbf{F}_j$ is the total force on the jth particle.
    One often sees the Liouville equation written as

$$i \frac{\partial f_N}{\partial t} = L f_N \tag{18-7}$$

where L is the Liouville operator,

$$L = -i \left(\sum_{j=1}^{N} \frac{\mathbf{p}_j}{m_j} \cdot \nabla_{\mathbf{r}_j} + \sum_{j=1}^{N} \mathbf{F}_j \cdot \nabla_{\mathbf{p}_j} \right) \tag{18-8}$$

[*] Mazo, "Additional Reading," p. 23; M. Beran, *Amer. J. Phys.*, **35**, p. 242, 1967.

The Liouville operator has been defined in such a way as to bring the Liouville equation into the form of the Schrödinger equation. A formal, and sometimes useful, solution to Eq. (18–7) is

$$f_N(\mathbf{p}, \mathbf{r}, t) = e^{-iLt} f_N(\mathbf{p}, \mathbf{r}, 0) \tag{18–9}$$

Note that the operator $\exp(-iLt)$ displaces f_N ahead a distance t in time. This operator is called the *time displacement operator* of the system.

18–2   REDUCED DISTRIBUTION FUNCTIONS

Once we have the distribution function $f_N(p, q, t)$, we may compute the ensemble average of any dynamical variable, $A(p, q, t)$, from the equation

$$\langle A(t) \rangle = \int A(p, q, t) f_N(p, q, t) \, dp \, dq \tag{18–10}$$

It turns out that the dynamical variables of interest are functions of either the coordinates and momenta of just a few particles or can be written as a sum over such functions. A familiar example of this is the total intermolecular potential of the system. To a good approximation, this can be written as a sum over pair-wise potentials, and so

$$\langle U \rangle = \sum_{i,j} \int \cdots \int u(\mathbf{r}_i, \mathbf{r}_j) f_N(\mathbf{r}_1, \ldots, \mathbf{p}_N, t) \, d\mathbf{r}_1 \cdots d\mathbf{p}_N \tag{18–11}$$

We encountered similar integrands when we studied the equilibrium theory of liquids. There we integrated over the coordinates of all the particles except i and j and called the resulting function of $\mathbf{r}_i$ and $\mathbf{r}_j$ a radial distribution function. We do the same thing here. We define *reduced distribution functions* $f_N^{(n)}(\mathbf{r}_1, \ldots, \mathbf{r}_n, \mathbf{p}_1, \ldots, \mathbf{p}_n, t)$ by

$$f_N^{(n)}(\mathbf{r}_1, \ldots, \mathbf{r}_n, \mathbf{p}_1, \ldots, \mathbf{p}_n, t) = \frac{N!}{(N-n)!} \int \cdots \int f_N(\mathbf{r}_1, \ldots, \mathbf{p}_N, t)$$
$$d\mathbf{r}_{n+1} \cdots d\mathbf{r}_N \, \mathbf{p}_{n+1} \cdots d\mathbf{p}_N \tag{18–12}$$

We shall usually drop the N subscript and furthermore write this simply as $f^{(n)}(\mathbf{r}^n, \mathbf{p}^n, t)$. Usually only $f^{(1)}$ and $f^{(2)}$ are necessary, and therefore we want to derive an equation for $f^{(1)}$ and $f^{(2)}$. To do this, write the force $\mathbf{F}_j$ appearing in the Liouville equation as the sum of the forces due to the other molecules in the system $\sum_i \mathbf{F}_{ij}$ and an external force $\mathbf{X}_j$. Then multiply through by $N!/(N-n)!$ and integrate over $d\mathbf{r}_{n+1} \cdots d\mathbf{r}_N \, d\mathbf{p}_{n+1} \cdots d\mathbf{p}_N$ to get (Problem 18–2)

$$\frac{\partial f^{(n)}}{\partial t} + \sum_{j=1}^n \frac{\mathbf{p}_j}{m_j} \cdot \nabla_{\mathbf{r}_j} f^{(n)} + \sum_{j=1}^n \mathbf{X}_j \cdot \nabla_{\mathbf{p}_j} f^{(n)}$$

$$+ \frac{N!}{(N-n)!} \sum_{i,\,j=1}^N \int \cdots \int \mathbf{F}_{ij} \cdot \nabla_{\mathbf{p}_j} f \, d\mathbf{r}_{n+1} \cdots d\mathbf{r}_N \, d\mathbf{p}_{n+1} \cdots d\mathbf{p}_N = 0 \tag{18–13}$$

We have used the fact that f vanishes outside the walls of the container and when $\mathbf{p}_i = \pm\infty$. The last term in Eq. (18–13) can be broken up into two parts:

$$\sum_{i,\,j=1}^n \mathbf{F}_{ij} \cdot \nabla_{\mathbf{p}_j} f^{(n)}$$

$$+ \frac{N!}{(N-n)!} \sum_{j=1}^n \sum_{i=n+1}^N \int \cdots \int \mathbf{F}_{ij} \cdot \nabla_{\mathbf{p}_j} f \, d\mathbf{r}_{n+1} \cdots d\mathbf{r}_N \, d\mathbf{p}_{n+1} \cdots d\mathbf{p}_N$$

The second term here can be written as

$$\sum_{j=1}^{n} \iint \mathbf{F}_{j,\,n+1} \cdot \nabla_{\mathbf{p}_j} f^{(n+1)} \, d\mathbf{r}_{n+1} \, d\mathbf{p}_{n+1}$$

Putting all this together finally gives an exact equation for $f^{(n)}$, namely, (Problem 18–3),

$$\frac{\partial f^{(n)}}{\partial t} + \sum_{j=1}^{n} \frac{\mathbf{p}_j}{m_j} \cdot \nabla_{\mathbf{r}_j} f^{(n)} + \sum_{j=1}^{n} \mathbf{X}_j \cdot \nabla_{\mathbf{p}_j} f^{(n)}$$

$$+ \sum_{i,\,j=1}^{n} \mathbf{F}_{ij} \cdot \nabla_{\mathbf{p}_j} f^{(n)} + \sum_{j=1}^{n} \iint \mathbf{F}_{j,\,n+1} \cdot \nabla_{\mathbf{p}_j} f^{(n+1)} \, d\mathbf{r}_{n+1} \, d\mathbf{p}_{n+1} = 0 \qquad (18\text{–}14)$$

This is the so-called Bogoliubov, Born, Green, Kirkwood, Yvon (BBGKY) hierarchy. This is the time-dependent generalization of the hierarchy that we derived earlier in the equilibrium theory of fluids. In fact, if one assumes that

$$f^{(n)} = g^{(n)}(\mathbf{r}_1, \ldots, \mathbf{r}_n) \exp\left\{-\frac{1}{2mkT} \sum_{j=1}^{n} p_j^2\right\}$$

multiplies Eq. (18–14) through by $\mathbf{p}_i$, $1 \le i \le n$, and integrates over all momenta, one obtains the equilibrium hierarchy for $g^{(n)}(\mathbf{r}_1, \ldots, r_n)$ (Problem 18–4). It would seem natural at this point to truncate this hierarchy by some sort of a superposition approximation, but so far this approach has not been successful.* We shall end up deriving approximate equations for $f^{(1)}$ and $f^{(2)}$.

Everything we have done up to now has been independent of density; i.e., it has been applicable to any density. Now we shall specialize to systems of dilute gases.

18–3 FLUXES IN DILUTE GASES

In a dilute gas, most of the molecules are not interacting with any other molecule and are just traveling along between collisions. Because of this, the macroscopic properties of a gas depend upon only the singlet distribution function $f_j^{(1)}(\mathbf{r}, \mathbf{p}_j, t)$. The subscript j here denotes the singlet distribution function of species j. This is the central distribution function of any theory of transport in dilute gases. In this section we shall define a number of averages over $f_j^{(1)}$ and derive molecular expressions for the important flux quantities in terms of integrals over $f_j^{(1)}$. Since we shall be concerned only with gases in this and the following sections, we shall drop the superscript (1) from here on. We shall also write our equations in velocity space rather than momentum space, and so the distribution function of interest becomes $f_j(\mathbf{r}, \mathbf{v}_j, t)$. We shall renormalize f_j such that the integral of this distribution function over all velocities is the number density of j particles at the point $\mathbf{r}$ at time t, i.e.,

$$\rho_j(\mathbf{r}, t) = \int f_j(\mathbf{r}, \mathbf{v}_j, t) \, d\mathbf{v}_j \qquad (18\text{–}15)$$

Furthermore, if N_j is the total number of j molecules in our system, then

$$N_j = \iint f_j(\mathbf{r}, \mathbf{v}_j, t) \, d\mathbf{r} \, d\mathbf{v}_j \qquad (18\text{–}16)$$

We shall now define a number of important average velocities. $\mathbf{v}_j$ is the *linear*

* See, for example, R. G. Mortimer, *J. Chem. Phys.*, **48**, p. 1023, 1968.

velocity of a molecule of species j; i.e., it is the velocity with respect to a coordinate system fixed in space. The average velocity is given by

$$\mathbf{v}_j(\mathbf{r},\, t) = \frac{1}{\rho_j} \int \mathbf{v}_j f(\mathbf{r},\, \mathbf{v}_j,\, t)\, d\mathbf{v}_j \tag{18-17}$$

and represents the macroscopic flow of species j. The *mass average velocity* is defined by

$$\mathbf{v}_0(\mathbf{r},\, t) = \frac{\sum_j m_j \rho_j \mathbf{v}_j}{\sum_j m_j \rho_j} \tag{18-18}$$

Note that the denominator here is the mass density $\rho_m(\mathbf{r},\, t)$. This velocity is often called the flow velocity or stream velocity. The momentum density of the gas is the same as if all the molecules were moving with velocity $\mathbf{v}_0$. The *peculiar velocity* is the velocity of a molecule relative to the flow velocity. The peculiar velocity $\mathbf{V}_j$ is

$$\mathbf{V}_j = \mathbf{v}_j - \mathbf{v}_0 \tag{18-19}$$

The average of this peculiar velocity is the *diffusion velocity* (Problem 18–5). Clearly,

$$\overline{\mathbf{V}}_j = \frac{1}{\rho_j} \int (\mathbf{v}_j - \mathbf{v}_0) f_j(\mathbf{r},\, \mathbf{v}_j,\, t)\, d\mathbf{v}_j \tag{18-20}$$

It is easy to show that (Problem 18–6)

$$\sum_j \rho_j m_j \overline{\mathbf{V}}_j = 0 \tag{18-21}$$

When we studied the elementary kinetic theory of gases, we saw that the various transport coefficients were related to molecular transport of mass, momentum, and kinetic energy. Let these molecular properties be designated collectively by ψ_j, where j refers to the particular species. We now derive expressions for the fluxes of these properties. Figure 18–1 shows a surface dS moving with velocity $\mathbf{v}_0$. The quantity $\mathbf{n}$ is a unit vector normal to dS, and $d\mathbf{S} = \mathbf{n}\, dS$. All the molecules that have velocity $\mathbf{V}_j = \mathbf{v}_j - \mathbf{v}_0$ and that cross dS in the time interval $(t,\, t + dt)$ must have been in a cylinder of length $|\mathbf{V}_j|\, dt$ and base dS. This cylinder is shown in Fig. 18–1 and has a volume $(\mathbf{n} \cdot \mathbf{V}_j)\, dS\, dt$. Since there are $f_j\, d\mathbf{v}_j$ molecules per unit volume with relative velocity $\mathbf{V}_j$, the number of j molecules that cross dS in dt is given by

$$(f_j\, d\mathbf{v}_j)(\mathbf{n} \cdot \mathbf{V}_j)\, dS\, dt$$

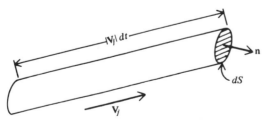

Figure 18–1. **The cylinder containing all those molecules of species j with velocity $\mathbf{V}_j$, which cross the surface dS during the time interval dt.** (From J. O. Hirschfelder, C. F. Curtiss, and R. B. Bird, *Molecular Theory of Gases and Liquids.* New York: Wiley, 1954.)

If each molecule carries with it a property ψ_j, then the flux of this property is

$$\psi_j f_j(\mathbf{n} \cdot \mathbf{V}_j) \, d\mathbf{v}_j$$

and the total flux across this surface is

$$\text{total flux} = \int \psi_j f_j(\mathbf{n} \cdot \mathbf{V}_j) \, d\mathbf{v}_j = \mathbf{n} \cdot \int \psi_j f_j \mathbf{V}_j \, d\mathbf{v}_j = \mathbf{n} \cdot \boldsymbol{\psi}_j \qquad (18\text{--}22)$$

The vector $\boldsymbol{\psi}_j$,

$$\boldsymbol{\psi}_j = \int \psi_j f_j \mathbf{V}_j \, d\mathbf{v}_j \qquad (18\text{--}23)$$

is called the flux vector associated with the property ψ_j. The component of this vector in any direction is the transport of the property ψ_j in that direction. Let us now consider the various examples of ψ_j.

TRANSPORT OF MASS

In this case, $\psi_j = m_j$, and

$$\boldsymbol{\psi}_j = m_j \int f_j \mathbf{V}_j \, d\mathbf{v}_j = \rho_j m_j \overline{\mathbf{V}_j} \equiv \mathbf{j}_j \qquad (18\text{--}24)$$

TRANSPORT OF MOMENTUM

Here $\psi_j = m_j V_{jx}$, and

$$\boldsymbol{\psi}_j = m_j \int V_{jx} f_j \mathbf{V}_j \, d\mathbf{v}_j = \rho_j m_j \overline{V_{jx} \mathbf{V}_j} \qquad (18\text{--}25)$$

which is the flux of the x-component of momentum relative to $\mathbf{v}_0$. The flux of momentum is a pressure, which has components

$$(\mathsf{p}_j)_{xx} = \rho_j m_j \overline{V_{jx} V_{jx}}$$

$$(\mathsf{p}_j)_{xy} = \rho_j m_j \overline{V_{jx} V_{jy}}, \text{ etc}$$

or, in general,

$$\mathsf{P}_j = \rho_j m_j \overline{\mathbf{V}_j \mathbf{V}_j} \qquad (18\text{--}26)$$

which is the partial pressure tensor of the jth species.

TRANSPORT OF KINETIC ENERGY

$$\psi_j = \tfrac{1}{2} m_j V_j^{\ 2}$$

and

$$\boldsymbol{\psi}_j = \frac{m_j}{2} \int v_j^{\ 2} \mathbf{V}_j f_j \, d\mathbf{v}_j = \tfrac{1}{2} \rho_j m_j \overline{V_j^{\ 2} \mathbf{V}_j} = \mathbf{q}_j \qquad (18\text{--}27)$$

the heat flux vector of the jth species.

It should be clear at this point that once we have an expression for $f_j(\mathbf{r}, \mathbf{v}_j, t)$, we can calculate all the fluxes and hence all the transport properties of a dilute gas. What we need now is f_j, or at least an equation that gives f_j as its solution. The only equation we have up to now is Eq. (18–14) with $n = 1$, and it can be seen that this also contains $f_j^{(2)}$. As we said earlier, nobody has found a successful way to uncouple this system. In the next section we shall derive an equation for f_j, the Boltzmann equation, which is the fundamental equation of the rigorous kinetic theory of gases.

18–4 THE BOLTZMANN EQUATION

In this section we shall derive the Boltzmann equation for f_j by a simple, physical argument. The gas is assumed to be dilute enough that only two-body interactions are ever important. The number of j-molecules in the phase space volume element $d\mathbf{r} \, d\mathbf{v}_j$ at the point $(\mathbf{r}, \mathbf{v}_j)$ is given by $f_j \, d\mathbf{r} \, d\mathbf{v}_j$. In the absence of collisions in the gas, the molecules at the point $(\mathbf{r}, \mathbf{v}_j)$ at time t move according to the equations of motion of the system and arrive at the point $(\mathbf{r} + \mathbf{v}_j \, dt, \mathbf{v}_j + (1/m_j)\mathbf{X}_j \, dt)$ at the time $(t + dt)$. The quantity $\mathbf{X}_j$ is an external force. Because all the points that start out end up at the same point (in the absence of collisions), we have that

$$f_j(\mathbf{r}, \mathbf{v}_j, t) = f_j\left(\mathbf{r} + \mathbf{v}_j \, dt, \mathbf{v}_j + \frac{\mathbf{X}_j}{m_j} \, dt, t + dt\right) \quad \text{(no collisions)} \tag{18–28}$$

But since collisions occur in the gas, not all those molecules that start out at the point $(\mathbf{r}, \mathbf{v}_j)$ at time t end up at $(\mathbf{r} + \mathbf{v}_j \, dt, \mathbf{v}_j + (1/m_j)\mathbf{X}_j \, dt)$ at the time $(t + dt)$. Some molecules leave this stream because of collisions, and furthermore some molecules enter this stream because of collisions. Let the number of j-molecules lost from the velocity range $(\mathbf{v}_j, \mathbf{v}_j + d\mathbf{v}_j)$ and the position range $(\mathbf{r}, \mathbf{r} + d\mathbf{r})$ because of collisions with i-molecules during the time interval $(t, t + dt)$ be $\Gamma_{ji}^{(-)} \, d\mathbf{r} \, d\mathbf{v}_j \, dt$. Similarly, let the number of j-molecules that join the group of molecules that starts at $(\mathbf{r}, \mathbf{v}_j)$ at time t because of collisions with i-molecules be $\Gamma_{ji}^{(+)} \, d\mathbf{r} \, d\mathbf{v}_j \, dt$. If we now include these collision terms in Eq. (18–28), we can write

$$f_j(\mathbf{r} + \mathbf{v}_j \, dt, \mathbf{v}_j + m_j^{-1}\mathbf{X}_j \, dt, t + dt) \, d\mathbf{r} \, d\mathbf{v}_j$$
$$= f_j(\mathbf{r}, \mathbf{v}_j, t) \, d\mathbf{r} \, d\mathbf{v}_j + \sum_i (\Gamma_{ji}^{(+)} - \Gamma_{ji}^{(-)}) \, d\mathbf{r} \, d\mathbf{v}_j \, dt \tag{18–29}$$

If we expand the left-hand side of the equation, we can get (Problem 18–8)

$$\frac{\partial f_j}{\partial t} + \mathbf{v}_j \cdot \nabla_\mathbf{r} f_j + \frac{\mathbf{X}_j}{m_j} \cdot \nabla_{\mathbf{v}_j} f_j = \sum_i (\Gamma_{ji}^{(+)} - \Gamma_{ji}^{(-)}) \tag{18–30}$$

The left-hand side of the equation represents the change in f_j due to the collisionless motion of the molecules, called streaming, and the right-hand side represents the change in f_j due to collisions. Notice that this equation looks very similar to the Liouville equation for $f_j^{(1)}$.

We now want to find an explicit expression for the collision terms in this equation. Let us look at $\Gamma_{ji}^{(-)} \, d\mathbf{r} \, d\mathbf{v}_j \, dt$ first. Consider a molecule of type j located at $\mathbf{r}$ with velocity $\mathbf{v}_j$. The probability that this molecule will collide with an i-molecule in the time interval dt and with impact parameter in a range db about b is given by the following argument. (See Fig. 18–2). If molecule j is considered to be fixed, the i-molecule approaches it with a relative velocity $(\mathbf{v}_i - \mathbf{v}_j) = \mathbf{g}_{ij}$. If A in Fig. 18–2 is the range of the intermolecular potential, any i-molecule within the cylindrical shell indicated in the figure will collide with the fixed j-molecule during the time interval dt. The probable number of i-molecules within this cylindrical shell is

$$2\pi f_i(\mathbf{r}, \mathbf{v}_i, t) g_{ij} b \, db \, dt$$

where $g_{ij} = |\mathbf{g}_{ij}|$. The total number of collisions that would occur with this one fixed j-molecule is

$$2\pi \, dt \iint f_i(\mathbf{r}, \mathbf{v}_i, t) g_{ij} b \, db \, d\mathbf{v}_i$$

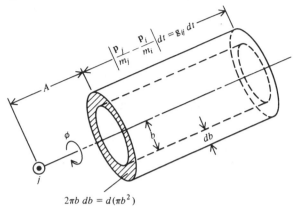

$$2\pi b\, db = d(\pi b^2)$$

Figure 18–2. **Collisions of molecules of type *i* with one molecule of type *j*. The distance *A* is essentially the intermolecular distance at which the potential begins to "take hold."** (From J. O. Hirschfelder, C. F. Curtiss, and R. B. Bird, *Molecular Theory of Gases and Liquids*. New York: Wiley, 1954.)

The probable number of molecules of type *j* located in the volume element $d\mathbf{r}$ about $\mathbf{r}$ with velocity between $\mathbf{v}_j$ and $\mathbf{v}_j + d\mathbf{v}_j$ is $f_j(\mathbf{r}, \mathbf{v}_j, t)\, d\mathbf{r}\, d\mathbf{v}_j$. Therefore we write

$$\Gamma_{ji}^{(-)}\, d\mathbf{r}\, d\mathbf{v}_j\, dt = 2\pi\, d\mathbf{r}\, d\mathbf{v}_j\, dt \iint f_j(\mathbf{r}, \mathbf{v}_j, t) f_i(\mathbf{r}, \mathbf{v}_i, t) g_{ij} b\, db\, d\mathbf{v}_i$$

and see that

$$\Gamma_{ji}^{(-)} = 2\pi \iint f_j f_i g_{ij} b\, db\, d\mathbf{v}_i \qquad (18\text{–}31)$$

We should notice at this point that we have assumed that the mean number of *i*-molecules about the fixed *j*-molecule is given by the product of f_i and f_j; i.e., we are assuming that their positions and velocities are uncorrelated. The assumption, known as the molecular chaos assumption or the stosszahlansatz, is not strictly true and is the weakest point in the entire derivation. This turns out to be a very important point, and we shall discuss this later on when we do the critique on the Boltzmann equation.

We can apply the very same argument that we used to derive Eq. (18–31) to the inverse collisions, i.e., those that scatter particles into the point $(\mathbf{r} + \mathbf{v}_j\, dt, \mathbf{v}_j + (1/m_j)\mathbf{X}_j\, dt)$. We can immediately write

$$\Gamma_{ji}^{(+)}\, d\mathbf{r}\, d\mathbf{v}_j\, dt = 2\pi\, d\mathbf{r}\, d\mathbf{v}_j'\, dt \iint f_i(\mathbf{r}, \mathbf{v}_i', t) f_j(\mathbf{r}, \mathbf{v}_j', t) g_{ij}' b'\, db'\, d\mathbf{v}_i' \qquad (18\text{–}32)$$

where we use primes to indicate those quantities before the collision which will go over into b, $\mathbf{v}_i$, and $\mathbf{v}_j$ after the collision. They can be computed from b, $\mathbf{v}_i$, and $\mathbf{v}_j$ by solving the collisional equations of motion. Liouville's theorem says that

$$d\mathbf{r}\, d\mathbf{v}_i\, d\mathbf{v}_j g_{ij}\, dt\, b\, db = d\mathbf{r}\, d\mathbf{v}_i'\, d\mathbf{v}_j' g_{ij}'\, dt\, b'\, db' \qquad (18\text{–}33)$$

Using this, we can rewrite Eq. (18–32) in the form

$$\Gamma_{ji}^{(+)} = 2\pi \iint f_i(\mathbf{r}, \mathbf{v}_i', t) f_j(\mathbf{r}, \mathbf{v}_j', t) g_{ij} b\, db\, d\mathbf{v}_i \qquad (18\text{–}34)$$

$$= 2\pi \iint f_i' f_j' g_{ij} b\, db\, d\mathbf{v}_i \qquad (18\text{–}35)$$

where the primes on the f_i and f_j indicate that the velocity arguments of these functions are primed.

Substituting Eqs. (18–31) and (18–35) into Eq. (18–30) finally gives *the Boltzmann equation*:

$$\frac{\partial f_j}{\partial t} + \mathbf{v}_j \cdot \nabla_\mathbf{r} f_j + \frac{1}{m_j} \mathbf{X}_j \cdot \nabla_{\mathbf{v}_j} f_j = 2\pi \sum_i \iint \{f_i' f_j' - f_i f_j\} g_{ij} b \, db \, d\mathbf{v}_i \tag{18–36}$$

This is an integrodifferential equation for $f_j^{(1)}$. We have one such equation for each component of the gas. Notice that the equations of motion, and hence the intermolecular potential, enter this equation implicitly in the integrand on the right-hand side. The functions f_i' and f_j' depend upon the postcollisional velocities $\mathbf{v}_i'$ and $\mathbf{v}_j'$, which depend upon the precollisional velocities $\mathbf{v}_i$ and $\mathbf{v}_j$ through the equations of motion governing the collision. Notice also that once the right-hand side is integrated over b and $\mathbf{v}_i$, the only variables left are $\mathbf{r}$, $\mathbf{v}_j$, and t, exactly those variables on which the left-hand side depends.

For several reasons which we shall discuss, the Boltzmann equation is a remarkable equation. The derivation we have presented here is the standard derivation and has the advantage of being simple and physical.

The Boltzmann equation is not only an integrodifferential equation, but a nonlinear one as well. Solving such an equation is not simple, but it turns out that there exists a very elegant and straightforward approximate scheme due to Hilbert, Chapman, and Enskog, which we shall present in the next chapter. Before doing this, however, we can extract some very interesting consequences from the Boltzmann equation without a great deal of work.

18–5   SOME GENERAL CONSEQUENCES OF THE BOLTZMANN EQUATION

First, we shall discuss the equations of change. The fundamental equations of continuum mechanics that we derived in Chapter 17 may be obtained from the Boltzmann equation without determining the form of f. If we multiply the Boltzmann equation for f_i by ψ_i and integrate over $\mathbf{v}_i$, we obtain

$$\int \psi_i \left\{ \frac{\partial f_i}{\partial t} + \mathbf{v}_i \cdot \nabla_\mathbf{r} f_i + \frac{1}{m_i} \mathbf{X}_i \cdot \nabla_{\mathbf{v}_i} f_i \right\} d\mathbf{v}_i$$
$$= 2\pi \sum_j \iiint \psi_i (f_i' f_j' - f_i f_j) g_{ij} b \, db \, d\mathbf{v}_i \, d\mathbf{v}_j \tag{18–37}$$

Each of the three terms on the left-hand side of Eq. (18–37) can be transformed into more convenient forms:

$$\int \psi_i \frac{\partial f_i}{\partial t} d\mathbf{v}_i = \frac{\partial}{\partial t} \int \psi_i f_i \, d\mathbf{v}_i - \int f_i \frac{\partial \psi_i}{\partial t} d\mathbf{v}_i$$
$$= \frac{\partial}{\partial t} (\rho_i \overline{\psi}_i) - \rho_i \overline{\frac{\partial \psi_i}{\partial t}} \tag{18–38}$$

$$\int \psi_i v_{ix} \frac{\partial f_i}{\partial x} d\mathbf{v}_i = \frac{\partial}{\partial x} \int \psi_i v_{ix} f_i \, d\mathbf{v}_i - \int f_i v_{ix} \frac{\partial \psi_i}{\partial x} d\mathbf{v}_i$$
$$= \frac{\partial}{\partial x} (\rho_i \overline{\psi_i v_{ix}}) - \rho_i \overline{v_{ix} \frac{\partial \psi_i}{\partial x}} \tag{18–39}$$

$$\int \psi_i \frac{\partial f_i}{\partial v_{ix}} d\mathbf{v}_i = \iint |\psi_i f_i|_{-\infty}^{\infty} dv_{iy} \, dv_{iz} - \int f_i \frac{\partial \psi_i}{\partial v_{ix}} d\mathbf{v}_i$$
$$= -\rho_i \overline{\frac{\partial \psi_i}{\partial v_{ix}}} \tag{18–40}$$

We have assumed that the external force X_i appearing in Eq. (18–37) is independent of v_i. Putting these results together, we can write Eq. (18–37) as

$$\frac{\partial(\rho_i\bar{\psi}_i)}{\partial t} + \nabla_{\mathbf{r}} \cdot \rho_i \overline{\psi_i \mathbf{v}_i} - \rho_i\left\{\overline{\frac{\partial\psi_i}{\partial t}} + \overline{\mathbf{v}_i \cdot \nabla_{\mathbf{r}}\psi_i} + \frac{X_i}{m_i} \cdot \overline{\nabla_{\mathbf{v}_i}\psi_i}\right\}$$

$$= 2\pi \sum_j \iiint \psi_i(f_i'f_j' - f_if_j)g_{ij}b \, db \, d\mathbf{v}_i \, d\mathbf{v}_j \tag{18–41}$$

This is known as Enskog's general equation of change for the property ψ_i. We shall now show that if ψ_i is any of the collisional invariants, i.e., m_i, $m_i\mathbf{v}_i$, or $\frac{1}{2}m_i v_i^2$, the right-hand side of Eq. (18–41) vanishes when summed over i. To prove this, first notice that the integral

$$\iiint \psi_i(f_i'f_j' - f_if_j)g_{ij}b \, db \, d\mathbf{v}_i \, d\mathbf{v}_j \tag{18–42}$$

is equal to the integral

$$\iiint \psi_i'(f_if_j - f_i'f_j')g_{ij}'b' \, db' \, d\mathbf{v}_i' \, d\mathbf{v}_j' \tag{18–43}$$

which is written in terms in inverse collisions (Problem 18–9). This is so because an integral over a collision in one direction must be the same as an integral over a collision in the opposite direction. This can be seen from Fig. 16–3. Furthermore, from Liouville's theorem and from the dynamics of molecular collisions, we know that

$$g_{ij} = g_{ij}', \qquad b = b', \qquad d\mathbf{v}_i \, d\mathbf{v}_j = d\mathbf{v}_i' \, d\mathbf{v}_j' \tag{18–44}$$

and so Eq. (18–43) may be written as

$$-\iiint \psi_i'(f_i'f_j' - f_if_j)g_{ij}b \, db \, d\mathbf{v}_i \, d\mathbf{v}_j \tag{18–45}$$

Since the integrals of Eqs. (18–42) and (18–45) are equal to each other, they must also each be equal to one-half the sum of the two. This gives us that (Problem 18–10)

$$\iiint \psi_i(f_i'f_j' - f_if_j)g_{ij}b \, db \, d\mathbf{v}_i \, d\mathbf{v}_j$$

$$= \frac{1}{2}\iiint (\psi_i - \psi_i')(f_i'f_j' - f_if_j)g_{ij}b \, db \, d\mathbf{v}_i \, d\mathbf{v}_j \tag{18–46}$$

Note that if $\psi_i = m_i$, then the integrand equals zero. This is simply because the mass of the ith species is conserved in an elastic collision.

Now if we sum Eq. (18–46) over both i and j and then interchange the dummy indices i and j, we get

$$\frac{1}{2}\sum_{i,j}\iiint (\psi_i - \psi_i')(f_i'f_j' - f_if_j)g_{ij}b \, db \, d\mathbf{v}_i \, d\mathbf{v}_j$$

$$= \frac{1}{2}\sum_{i,j}\iiint (\psi_j - \psi_j')(f_i'f_j' - f_if_j)g_{ij}b \, db \, d\mathbf{v}_i \, d\mathbf{v}_j \tag{18–47}$$

Since these two integrals are equal to each other, each equals one-half of the sum of the two, and so we can finally write (Problem 18–11)

$$\sum_{i,j}\iiint \psi_i(f_i'f_j' - f_if_j)g_{ij}b \, db \, d\mathbf{v}_i \, d\mathbf{v}_j$$

$$= \frac{1}{4}\sum_{i,j}\iiint (\psi_i + \psi_j - \psi_i' - \psi_j')(f_i'f_j' - f_if_j)g_{ij}b \, db \, d\mathbf{v}_i \, d\mathbf{v}_j \tag{18–48}$$

Now $\psi_i + \psi_j - \psi_i' - \psi_j'$ vanishes when $\psi_i = m_i$, $m_i \mathbf{v}_i$, or $\frac{1}{2} m_i v_i^2$, and so we have shown that the sum of the right-hand side of Eq. (18–41) over i equals zero. This leaves, then,

$$\frac{\partial}{\partial t} \sum_i (\rho_i \overline{\psi}_i) + \nabla_\mathbf{r} \cdot \sum_i (\rho_i \overline{\psi_i \mathbf{v}_i}) - \sum_i \rho_i \overline{\left\{ \frac{\partial \psi_i}{\partial t} + \mathbf{v}_i \cdot \nabla_\mathbf{r} \psi_i + \frac{\mathbf{X}_i}{m_i} \cdot \nabla_{\mathbf{v}_i} \psi_i \right\}} = 0 \qquad (18–49)$$

as the general equation of change.

If we let $\psi_i = m_i$, Eq. (18–41) becomes

$$\frac{\partial \rho_i}{\partial t} + \nabla_\mathbf{r} \cdot (\rho_i \mathbf{v}_i) = 0 \qquad (18–50)$$

This is the continuity equation for the ith species. If we sum this over i and use Eq. (18–18), we get the continuity equation for the entire system:

$$\frac{\partial \rho}{\partial t} + \nabla_\mathbf{r} \cdot (\rho \mathbf{v}_0) = 0 \qquad (18–51)$$

If we let $\psi_i = m_i \mathbf{v}_i$ and sum over i, Eq. (18–49) becomes (Problem 18–12)

$$\rho_m \frac{D\mathbf{v}_0}{Dt} = \sum_j \rho_j \mathbf{X}_j - \nabla \cdot \mathsf{p} \qquad (18–52)$$

where p is the pressure tensor defined in Eq. (18–26). Similarly, if we let $\psi_i = \frac{1}{2} m_i v_i^2$, for example, we get the energy balance equation, Eq. (17–31). (See Problem 18–13.) It is reassuring, although not surprising, that these equations should come out of the Boltzmann equation. The next general consequence we shall derive is not particularly obvious and, in fact, is a rather profound result.

Next we discuss the so-called Boltzmann H-theorem and the equilibrium solution to the Boltzmann equation. Consider a one-component system. The Boltzmann H-function is defined by

$$H(t) = \iint f(\mathbf{r}, \mathbf{v}, t) \ln f(\mathbf{r}, \mathbf{v}, t) \, d\mathbf{r} \, d\mathbf{v} \qquad (18–53)$$

Differentiate $H(t)$ with respect to t:

$$\frac{dH}{dt} = \iint \frac{\partial f}{\partial t} \ln f \, d\mathbf{r} \, d\mathbf{v} + \iint \frac{\partial f}{\partial t} \, d\mathbf{r} \, d\mathbf{v} \qquad (18–54)$$

The second term here vanishes since

$$\iint \frac{\partial f}{\partial t} \, d\mathbf{r} \, d\mathbf{v} = \frac{d}{dt} \iint f \, d\mathbf{r} \, d\mathbf{v} = \frac{dN}{dt} = 0$$

if the number of particles in the system is conserved. Equation (18–54) becomes then

$$\frac{dH}{dt} = \iint \frac{\partial f}{\partial t} \ln f \, d\mathbf{r} \, d\mathbf{v} \qquad (18–55)$$

To evaluate this integral, multiply the Boltzmann equation by $\ln f$ and integrate over $d\mathbf{r}$ and $d\mathbf{v}$. This gives

$$\iint \frac{\partial f}{\partial t} \ln f \, d\mathbf{r} \, d\mathbf{v} = -\iint (\ln f)\mathbf{v} \cdot \nabla_\mathbf{r} f \, d\mathbf{r} \, d\mathbf{v} - \iint (\ln f) \frac{\mathbf{X}}{m} \cdot \nabla_\mathbf{v} f \, d\mathbf{r} \, d\mathbf{v}$$
$$+ 2\pi \iiint \ln f \{ f' f_1' - f f_1 \} g b \, db \, d\mathbf{v} \, d\mathbf{v}_1 \qquad (18–56)$$

The subscript 1 in the collision integral here is to distinguish the two colliding molecules. The first two integrals on the right vanish since we assume as always that f vanishes at the walls of the container and as $\mathbf{v} \to \pm\infty$ (Problem 18–14). This leaves

$$\frac{dH}{dt} = 2\pi \iiint \ln f \{f'f_1' - ff_1\} gb \, db \, d\mathbf{v} \, d\mathbf{v}_1 \tag{18–57}$$

This integral may be symmetrized by the same method we used to derive Eq. (18–48) to give (Problem 18–15)

$$\frac{dH}{dt} = \frac{2\pi}{4} \iiint \ln\left(\frac{ff_1}{f'f_1'}\right) \{f'f_1' - ff_1\} gb \, db \, d\mathbf{v} \, d\mathbf{v}_1 \tag{18–58}$$

Now this integrand is of the form $-(x - y)\ln(x/y)$. If $x > y$, this function is negative; if $x < y$, it is also negative; and if $x = y$, it is equal to zero. Therefore, we get the result that

$$\frac{dH}{dt} \leq 0 \tag{18–59}$$

The definition of $H(t)$ shows that it is bounded, and hence $H(t)$ must approach a limit as $t \to \infty$. In this limit, $dH/dt = 0$, and so we have an equilibrium or steady-state situation with

$$f'f_1' = ff_1 \tag{18–60}$$

or

$$\ln f' + \ln f_1' = \ln f + \ln f_1 \tag{18–61}$$

This tells us that $\ln f$ is a summational invariant. However, we know that the only summational invariants of bimolecular collisions of spherically symmetric molecules are the mass, the momentum, and the kinetic energy. Therefore, $\ln f$ must be a linear combination of these three quantities, or

$$\ln f = \alpha m + \boldsymbol{\beta} \cdot (m\mathbf{v}) - \gamma \frac{mv^2}{2} = \alpha m + \frac{m}{2} \frac{\boldsymbol{\beta} \cdot \boldsymbol{\beta}}{\gamma} - \frac{m\gamma}{2}\left(\mathbf{v} - \frac{\boldsymbol{\beta}}{\gamma}\right)^2 \tag{18–62}$$

If we let $\alpha m + m\boldsymbol{\beta} \cdot \boldsymbol{\beta}/2\gamma$ be $\ln c$, then

$$\ln f = \ln c - \frac{m\gamma}{2}\left(\mathbf{v} - \frac{\boldsymbol{\beta}}{\gamma}\right)^2$$

or

$$f(\mathbf{r}, \mathbf{v}) = c \exp\left[-\frac{m\gamma}{2}\left(\mathbf{v} - \frac{\boldsymbol{\beta}}{\gamma}\right)^2\right] \tag{18–63}$$

Notice how closely this resembles a Maxwellian distribution. We can determine our unknown parameters c, $\boldsymbol{\beta}$, and γ from the following conditions. First, we have

$$\rho(\mathbf{r}, t) = \int f \, d\mathbf{v} = c \int \exp\left[-\frac{m\gamma}{2}\left(\mathbf{v} - \frac{\boldsymbol{\beta}}{\gamma}\right)^2\right] d\mathbf{v} \tag{18–64}$$

Evaluating the integral in Eq. (18–64) gives (Problem 18–16)

$$\rho(\mathbf{r}, t) = c \left(\frac{2\pi}{m\gamma}\right)^{1/2} \tag{18–65}$$

Similarly, we have (Problem 18-16)

$$\mathbf{v}_0 = \bar{\mathbf{v}} = \frac{1}{\rho} \int \mathbf{v} f \, d\mathbf{v} = \frac{\boldsymbol{\beta}}{\gamma} \tag{18-66}$$

Lastly, we define temperature by the usual relation

$$\tfrac{3}{2}kT = \frac{m}{2} \overline{(\mathbf{v} - \mathbf{v}_0)^2}$$

which gives (Problem 18-16)

$$\tfrac{3}{2}kT = \tfrac{3}{2}\gamma^{-1} \tag{18-67}$$

Equations (18-65), (18-66), and (18-67) then give

$$f(\mathbf{r}, \mathbf{v}) = \rho \left(\frac{m}{2\pi kT}\right)^{3/2} e^{-mV^2/2kT} \tag{18-68}$$

which is the classical expression for the Maxwellian distribution of velocities. Since ρ and T can be functions of position here, this is more generally a local Maxwellian distribution. This might be the distribution in a steady state rather than at equilibrium. We see then that the equilibrium solution to the Boltzmann equation is indeed the Maxwellian distribution, another reassuring, but nevertheless necessary, result.

The really interesting consequence of the Boltzmann equation is the H-theorem, whose significance we have glossed over. The H-theorem attributes a direction in time to the Boltzmann equation since it states that if we start out with some arbitrary distribution function, it will relax to the equilibrium (or steady-state) Maxwellian distribution. Of course, one can say that this is a necessary feature of any equation we derive to describe a gas since we know from the second law of thermodynamics that systems tend toward their equilibrium states. But the fact that it comes out of the Boltzmann equation was at first severely attacked for the following reasons.

The equations of motion of classical mechanics,

$$m \frac{d^2\mathbf{r}_j}{dt^2} = -\text{grad}_j \, U(\mathbf{r}_1, \ldots, \mathbf{r}_N) \qquad 1 \leq j \leq N \tag{18-69}$$

are symmetrical in time. If we let $t \to -t$ in Eq (18-69), one gets the same equations back again. This means that classical mechanical systems have no preferred direction in time; motion in one direction is no more preferred than motion in the opposite direction. On the other hand, the H-theorem shows that the Boltzmann equation does have a preferred direction. From a purely mechanical point of view, if all the molecules move in such a way to make H decrease, there is at least a possible mechanical motion where everything is reversed, and if everything is reversed, H must increase. This is part of the more general question of how the irreversible processes that we observe in nature can be reconciled with the basic reversibility of the underlying mechanical equations of motion. The H-theorem is a particular expression of this situation. This objection was first raised by Loschmidt. Boltzmann tried to answer this by claiming that Eqs. (18-31) and (18-35) should be interpreted as the probability of a collision rather than the actual number of collisions, and this means that $H(t)$ does, indeed, not always decrease, but that the probability that it decrease is far greater than it increase. The further the system is from equilibrium, the more likely $H(t)$ is to decrease. Boltzmann's arguments were not convincing to all of his critics, and it was left to the Ehrenfests and Smoluchowski some years later to clearly explain the statistical nature of the Boltzmann equation. They were able to show that on the average, the H-function decreases with time to its equilibrium value, but that fluctuations can and will always occur. Furthermore, H will almost always remain near its equilibrium value once it gets there. We have encountered a somewhat similar situation when we discussed entropy in Chapter 2. Notice, in fact, that the equilibrium value of the H-function is equal to $-S/k$ for an ideal gas. A decreasing H-function is in some sense an increasing entropy. The H-function can actually be computed as a function of time by molecular dynamics. In fact, one of the most powerful and interesting applications of molecular dynamics calculations is to nonequilibrium systems. Figure 18–3 shows $H(t)$ calculated by Alder and Wainwright in their pioneering review article (in "Additional Reading ").

If the objection by Loschmidt were not enough, there was also another paradox pointed out by Zermelo. This one was at the time more awkward than the time-reversal paradox. There is a general theorem in classical mechanics that says that any mechanical system enclosed in a finite volume will return arbitrarily close to its original state. This theorem is called the Poincaré recursion theorem, and the time it takes to essentially return to its original state is called the recurrence time. Such a cycle is called a Poincaré cycle. Zermelo vehemently pointed out that the H-theorem is at odds with the Poincaré recursion theorem, since how could H evolve toward an equilibrium value and remain there when classical mechanics dictates that the system must eventually retrace itself. Boltzmann was able to meet this objection as well by pointing out that for physically interesting systems, the recurrence times are ridiculously large. For example, Boltzmann estimated that a system consisting of 10^{18} atoms per cubic centimeter with an average velocity of 5×10^4 cm/sec would reproduce all of its

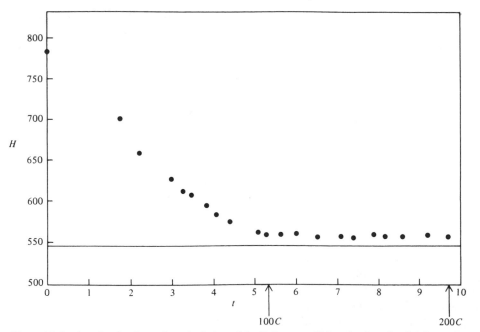

Figure 18–3. **A molecular dynamics calculation of the Boltzmann H-function for a hard-sphere system of 100 particles at a density $v/v_0 = 14.14$. The H-function is plotted versus the collision number C.** (From B. J. Alder and T. Wainwright, in *Transport Processes in Statistical Mechanics*, I. Prigogine, ed., New York: Interscience, 1958.)

coordinates within 10^{-7} cm and all of its velocities to within 100 cm/sec in a time of the order of $10^{10^{19}}$ years!

The scientific atmosphere of the time, however, was such that Boltzmann was not able to convince all of his critics, and it was finally Smoluchowski who clearly stated that the very concept of irreversibility is intimately involved with the length of the recurrence time. If one is initially in a state with a very long recurrence time, the process will appear to be irreversible. On the other hand, if the recurrence time is short, one does not speak about irreversibility. For any sensible physical system, the Poincaré recurrence times are extremely long.

The above results lead to the following picture of phase space. The overwhelming majority of phase space describes an equilibrium situation. Scattered throughout phase space are small regions that describe certain nonequilibrium configurations. Almost all of these nonequilibrium regions are completely surrounded by equilibrium regions, so that if we do happen to observe a system in a nonequilibrium state, it is almost always headed toward an equilibrium state. Once it reaches equilibrium, it will travel through various equilibrium states, only occasionally, rarely passing through other little regions of nonequilibrium. Such events are what we observe as fluctuations. This is so regardless of whether the phase point travels in a forward or backward direction, and so this picture is consistent with the classical mechanical requirement of time reversibility. Furthermore, the phase point must make its complete circuit through phase space before passing near its original position (since trajectories in phase space never cross). This length of time is the recurrence time. This would simply be observed as another fluctuation from equilibrium.

These little nonequilibrium regions in phase space vary in extent. The smaller the region, the less time the system spends away from equilibrium. It is conceivable, however, that some nonequilibrium region of phase space is such that a phase point can be trapped there. For example, consider a gas prepared such that the velocities of all of its molecules are parallel to each other and perpendicular to the walls of the container. If we assume that the walls are perfectly smooth, the molecules would then travel back and forth between the walls, remain parallel to each other, and hence never collide and come to equilibrium. The phase point describing this gas would be confined to some peculiar region of phase space and get hung up there. Clearly the H-theorem and the Boltzmann equation itself would not be applicable to such a pathological system. The Boltzmann equation tacitly requires that the system be sufficiently chaotic. This leaves open the question of how to decide when a system can be described by the Boltzmann equation (or statistical mechanics for that matter). All that can be safely said is that such a difficulty never seems to arise in practice. In our pathological system above, for example, any slight imperfection or roughness in the walls of the container would send the molecules off their parallel paths, and the system would become sufficiently chaotic in just a few series of collisions.

There are many long discussions about irreversibility, classical or quantum mechanics, and statistical mechanics in the literature, and the reader is referred to Tolman, Mazo, Huang, Uhlenbeck and Ford, and ter Haar for further discussions. (See "Additional Reading.")

It is interesting to note that Boltzmann (1844–1906) did his pioneering work at a time when atomic theories were not generally accepted. There were essentially two schools of scientific philosophy at that time: the atomists, led by Boltzmann himself, and the school of energeticists, led by Mach, Ostwald, Duhem, and others. Remember that this was a prequantum mechanical time, and the inadequacies of classical mechanics were becoming increasingly evident. For example, Boltzmann was never to understand why the internal degrees of freedom of molecules could not be successfully treated by classical mechanics. The extent of the controversy can be well summarized in Boltzmann's foreword to Part II of his great work, *Lectures on Gas Theory.* He says, "As the first part of *Gas Theory* was being printed, I had already almost completed the present second and last part, in which the more difficult parts of the subject were to have been treated. It was just at this time that attacks on the theory of gases began to increase. I am convinced that these attacks are merely based on a misunderstanding, and that the role of gas theory in science has not yet been played out. . . .

"In my opinion it would be a great tragedy for science if the theory of gases were temporarily thrown into oblivion because of a momentary hostile attitude toward it, as was for example the wave theory because of Newton's authority.

"I am conscious of being only an individual struggling weakly against the stream of time. But it still remains in my power to contribute in such a way that, when the theory of gases is again revived, not too much will have to be rediscovered. Thus in this book [this Part] I will now include the parts that are the most difficult and most subject to misunderstanding and give (at least in outline) the most easily understood exposition of them. . . ." Boltzmann was clearly pessimistic about the future of the kinetic theory. This led to severe fits of depression, ending with his suicide in 1906.

The Boltzmann equation is accepted today as giving a completely adequate description of the behavior of dilute gases. In the next chapter we shall study some of the numerical results of the Boltzmann equation.

ADDITIONAL READING

BOLTZMANN, L. 1964. *Lectures on gas theory* (trans. by S. G. Brush). Berkeley: University of California Press.

BRUSH, S. 1972. *Kinetic theory*, Vol. 3. Oxford: Pergamon.

CERCIGNANI, C. 1969. *Mathematical methods in kinetic theory.* New York: Plenum.

CHAPMAN, S., and COWLING, T. G. 1960. *The mathematical theory of non-uniform gases.* Cambridge: Cambridge University Press.

EHRENFEST, P., and EHRENFEST, T. 1959. *The conceptual foundations of the statistical approach in mechanics.* Ithaca, N.Y.: Cornell University Press.

FERZIGER, J. H., and KAPER, H. G. 1972. *Mathematical theory of transport processes in gases.* Amsterdam: North-Holland Publ.

GRAD, H. *Commum. on Pure and Appl. Math.*, **2**, p. 331, 1949.

———. *Handbuch der Physik*, Vol XII, p. 205, ed. by S. Flügge. Berlin: Springer-Verlag.

HANLEY, H. J. 1969. *Transport phenomena in fluids.* New York: Dekker.

HIRSCHFELDER, J. O., CURTISS, C. F., and BIRD, R. B. 1954. *Molecular theory of gases and liquids.* New York: Wiley.

HUANG, K. 1963. *Statistical mechanics.* New York: Wiley.

LEBOWITZ, J. L., and PENROSE, O. *Physics Today*, **26**, No. 2, p. 23, 1973.

LIBOFF, R. L. 1969. *Introduction to the theory of kinetic equations.* New York: Wiley.

MAZO, R. M. 1967. *Statistical mechanical theories of transport processes.* Oxford: Pergamon.

PRIGOGINE, I., ed. 1958. *Transport processes in statistical mechanics.* New York: Interscience.

REIF, F. 1965. *Fundamentals of statistical and thermal physics.* New York: McGraw-Hill.

TER HAAR, D. *Rev. Mod. Phys.*, **27**, p. 289, 1955.

———. 1954. *Elements of statistical mechanics.* New York: Rinehart.

THIRRING, H. *J. Chem. Educ.* **29**, p. 298, 1952.

TOLMAN, R. C. 1938. *Principles of statistical mechanics.* Oxford: Oxford University Press.

UHLENBECK, G. E., and FORD, G. W. 1963. *Lectures in statistical mechanics.* Providence, R.I.: American Mathematical Society.

WU, T-Y. 1966. *Kinetic equations of gases and plasmas.* Reading, Mass.: Addison-Wesley.

PROBLEMS

18–1. Why is the operator $\exp(-iLt)$ called the time displacement operator?

18–2 Derive Eq. (18–13).

18–3. Derive the BBGKY hierarchy for the reduced distribution function $f^{(n)}$.

18 4. Derive the Born-Green-Yvon-Kirkwood hierarchy for the equilibrium n-particle correlation function $g^{(n)}(\mathbf{r}_1, \mathbf{r}_2, \ldots, \mathbf{r}_n)$, Eq. (13–39), from the BBGKY hierachy for $f^{(n)}(\mathbf{p}_1, \ldots, \mathbf{p}_n, \mathbf{r}_1, \ldots, \mathbf{r}_n)$, Eq. (18–14).

18–5. Why is the average of the peculiar velocity called the diffusion velocity [cf. Eq. (18–20)]?

18–6. Prove that

$$\sum_i \rho_J m_J \overline{\mathbf{V}_J} = 0$$

18–7. Calculate both p and q given that $f(\mathbf{r}, \mathbf{v}, t)$ is a local Maxwell distribution

$$f = \rho(\mathbf{r}, t)\left[\frac{m}{2\pi kT(\mathbf{r}, t)}\right]^{3/2} \exp\left\{-\frac{m[\mathbf{v} - \mathbf{v}_0(\mathbf{r}, t)]^2}{2kT(\mathbf{r}, t)}\right\}$$

18–8. Derive Eq. (18–30).

18–9. Show that Eq. (18–42) is equal to Eq. (18–43).

18–10. Prove Eq. (18–46).

18–11. Prove Eq. (18–48).

18–12. Derive the momentum balance equation, Eq. (18–52), from Eq. (18–49).

18–13. Derive the energy balance equation from Eq. (18–49).

18–14. Show that the first two integrals on the right-hand side of Eq. (18–56) vanish.

18–15. Derive Eq. (18–58).

18–16. Prove that the Boltzmann H-theorem implies that the steady-state solution to the Boltzmann equation is a local Maxwellian distribution, i.e., prove Eqs. (18–65), (18–66), and (18–67).

18–17. Show that if a local Maxwellian distribution is used to calculate the averages in the equations of change, Eqs. (18–51) and (18–52), one gets the ideal hydrodynamic equations.

18–18. The collision term in the Boltzmann equation is often written approximately as

$$\left(\frac{\partial f}{\partial t}\right)_{\text{collisions}} \approx -\frac{(f - f_0)}{\tau}$$

where f_0 is a local Maxwellian distribution. Show that this implies that the difference of the distribution function from equilibrium decays as a simple exponential due to collisions. Interpret τ. Give an estimate of its magnitude. Should τ depend upon the velocity $\mathbf{v}$?

The standard method of solution to the Boltzmann equation, called the Chapman-Enskog method, is fairly long and involved. It is possible, however, to present an illuminating preview of the method by solving a simplified form of the Boltzmann equation in which the collision term is approximated by $-(f - f_0)/\tau$. This will be discussed in Problems 18–19 through 18–21.

18–19. We start with the approximate form of the Boltzmann equation (cf. Problem 18–18)

$$\frac{\partial f}{\partial t} + \mathbf{v} \cdot \nabla_{\mathbf{r}} f = -\frac{f - f_0}{\tau}$$

where the relaxation time τ may depend upon $\mathbf{v}$ and f_0 is a local Maxwellian distribution. We assume that the deviation of f from f_0 is linear in the velocity gradient and the temperature gradient. Show that under this assumption we can replace f by f_0 in the left-hand side of the equation to get

$$\frac{\partial f_0}{\partial t} + \mathbf{v} \cdot \nabla_{\mathbf{r}} f_0 = -\frac{f - f_0}{\tau}$$

Remember now that the temporal and spatial dependence of f_0 is given implicitly through the temporal and spatial behavior of $\rho(\mathbf{r}, t)$, $\mathbf{v}_0(\mathbf{r}, t)$, and $T(\mathbf{r}, t)$. Using the chain rule for $\partial f_0 / \partial t$, namely,

$$\frac{\partial f_0}{\partial t} = \frac{\partial f_0}{\partial \rho}\frac{\partial \rho}{\partial t} + \nabla_v f_0 \cdot \frac{\partial \mathbf{v}_0}{\partial t} + \frac{\partial f_0}{\partial T}\frac{\partial T}{\partial t}$$

and the ideal hydrodynamic equations (cf. Problem 18–17), show that the above Boltzmann equation becomes

$$f_0[(W^2 - \tfrac{5}{2})\mathbf{V} \cdot \nabla \ln T + \mathbf{b}:\nabla \mathbf{v}_0] = -\frac{f - f_0}{\tau}$$

where $\mathbf{W}$ is a reduced velocity

$$\mathbf{W} = \left(\frac{m}{2kT}\right)^{1/2}\mathbf{V}$$

and

$$\mathbf{b} = 2(\mathbf{WW} - \tfrac{1}{3}W^2\mathbf{I})$$

Except for the fact that we have approximated the collision term, this equation is the starting point for the second step in the Chapman-Enskog method.

18–20. Because of the simple approximation for the collision term used in the previous problem, we can solve the final equation immediately for f to get

$$f = f_0[1 - \tau(W^2 - \tfrac{5}{2})\mathbf{V} \cdot \nabla \ln T - \tau\mathbf{b}:\nabla \mathbf{v}_0]$$

We can now use this expression for f to calculate $\mathbf{p}$ and $\mathbf{q}$. We have already seen in Problem 18–7 that $\mathbf{p}^{(0)} = \rho kT\mathbf{I}$ and $\mathbf{q}^{(0)} = 0$. In this problem we shall calculate $\mathbf{q}$, and in the next we shall calculate $\mathbf{p}$. Show that

$$\mathbf{q} = -\frac{m}{2}\int V^2\mathbf{V}f_0\tau(W^2 - \tfrac{5}{2})\mathbf{V}\,d\mathbf{v} \cdot \nabla \ln T$$

and that

$$\lambda = \frac{2\pi m}{3T}\int_0^\infty f_0\,\tau V^6(W^2 - \tfrac{5}{2})\,dV$$

We wish to include the possibility that τ depends upon V. If we assume that

$$\tau = AV^s$$

show that

$$\lambda = Apk\left(\frac{2kT}{m}\right)^{1+(1/2)s}\frac{(s + 2)}{3\sqrt{\pi}}\Gamma[\tfrac{1}{2}(7 + s)]$$

Now let $\tau = l/V$. If the mean free path l is taken to be equal to $1/\sqrt{2}\pi\rho\sigma^2$, then show that

$$\lambda = \frac{2^{1/2}}{9}\left(\frac{2}{\pi}\right)^{3/2}\left(\frac{kT}{m}\right)^{1/2}\frac{C_v}{N_0\sigma^2}$$

where C_v is the molar heat capacity and N_0 is Avogadro's number. Compare this to Eqs. (16–32) and (16–37).

18–21. Using Eq. (18–25) and the expression for f given in the previous problem, show that

$$p_{ij} = p\delta_{ij} - 2m \sum_{k,\,l=x,\,y,\,z} \int f_0\,\tau V_i\,V_j(W_k\,W_l - \tfrac{1}{3}W^2\delta_{kl})\,d\mathbf{V}\,\frac{\partial v_{0k}}{\partial l}$$

or

$$\mathsf{p} = [p + (\tfrac{2}{3}\eta - \kappa)\nabla\cdot\mathbf{v}_0]\mathsf{l} - 2\eta\,\mathrm{sym}(\nabla\mathbf{v}_0)$$

where

$$\eta = 2m\int f_0\,\tau V_i\,V_j\,W_i\,W_j\,d\mathbf{V}$$

$$= \frac{4\pi m^2}{15kT}\int_0^\infty f_0\,\tau V^6\,dV$$

and

$$\kappa = 0$$

If we again assume that $\tau = A V^s$, show that

$$\eta = \frac{4\pi A m^2}{15kT}\,\rho\left(\frac{m}{2\pi kT}\right)^{3/2}\int_0^\infty v^{6+s}e^{-mv^2/2kT}\,dv$$

$$= A\rho kT\left(\frac{2kT}{m}\right)^{s/2}\frac{\Gamma[\tfrac{1}{2}(7+s)]}{\Gamma(7/2)}$$

Show that

$$\eta = \frac{4}{15\sqrt{2}}\left(\frac{2}{\pi}\right)^{3/2}\frac{(mkT)^{1/2}}{\sigma^2}$$

if $\tau = l/v$ where $l = 1/\sqrt{2}\,\pi\rho\sigma^2$. Compare this result to Eqs. (16–31) and (16–36).

18–22. Consider a uniform gas with no external forces, so that the distribution function will depend only upon $\mathbf{v}$ and t and not $\mathbf{r}$. Divide velocity space into equal finite cells ω_j and let $f_j(t)$ be the number of molecules whose velocity vectors end at time t in ω_j. Show that the Boltzmann equation can be written in the form

$$\frac{df_i}{dt} = \sum_{j,\,k,\,l}(a_{kl}{}^{ij}f_kf_l - a_{ij}{}^{kl}f_if_j)$$

where the number of collisions per unit time of molecules from cell ω_i with those from cell ω_j, which yield molecules in cells ω_k and ω_l, is $a_{ij}{}^{kl}\,f_if_j$. Show that the number of collisions that yields a molecule in ω_i is given by

$$\sum_{j,\,k,\,l}a_{kl}{}^{ij}f_kf_l$$

A common and fundamental assumption is that of microscopic reversibility, which says that

$$a_{kl}{}^{ij} = a_{ij}{}^{kl}$$

Interpret this physically.

18–23. The simplified Boltzmann equation of the preceding problem yields a more transparent proof of the H-theorem than the full Boltzmann equation. Define the H-function by

$$H = \sum_i f_i \ln f_i$$

and show that

$$\frac{dH}{dt} = \sum_i \frac{df_i}{dt} \ln f_i + \sum_i \frac{df_i}{dt}$$

$$= \sum_i \frac{df_i}{dt} \ln f_i$$

$$= \sum_{ij,\,kl} a_{ij}{}^{kl} \ln f_i (f_k f_l - f_i f_j)$$

$$= \tfrac{1}{2} \sum_{ijkl} a_{ij}{}^{kl} (\ln f_i + \ln f_j)(f_k f_l - f_i f_j)$$

$$= \tfrac{1}{4} \sum_{ijkl} a_{ij}{}^{kl} (\ln f_i f_j - \ln f_k f_l)(f_k f_l - f_i f_j)$$

$$\leq 0$$

Be sure to point out the symmetry property of the $a_{ij}{}^{kl}$ that is used in each line.

18–24. One often sees the Boltzmann equation written in the form

$$\frac{\partial f}{\partial t} + \mathbf{v} \cdot \nabla_r f + \frac{\mathbf{X}}{m} \cdot \nabla_v f = \int d\mathbf{v}_1 \int d\Omega \, gI(g, \, \theta)(f'f_1' - ff_1)$$

where $d\Omega = \sin\theta \, d\theta \, d\phi$ is the differential solid angle and $I(g, \, \theta)$ is the differential collision cross section for a collision, defined through

$$b\, db \, d\phi = I(g, \, \theta) \, d\Omega$$

Derive this form of the Boltzmann equation.

18–25. Show that the collision integral in the Boltzmann equation, i.e., the right-hand side of the Boltzmann equation, vanishes if $f(\mathbf{r}, \mathbf{v}, t)$ is a local Maxwellian distribution.

18–26. Show that the Maxwellian distribution of velocities satisfies the Boltzmann equation identically. Is this also true for a local Maxwellian distribution?

18–27. In Section 7–2 we proved the property of conservation of extension in phase space. We mentioned there that this could be proved directly from the equations of motion, which is the content of this problem.

Let the natural motion in phase space map the set of phase points (p_0, q_0) into (p_t, q_t). The property of conservation of extension in phase space can be written as

$$\int_{\Delta(0)} \cdots \int dp_1(0) \, dp_2(0) \cdots dq_n(0) = \int_{\Delta(t)} \cdots \int dp_1(t) \, dp_2(t) \cdots dq_n(t)$$

where $\Delta(0)$ is some volume in phase space taken at some initial time and $\Delta(t)$ is the result of $\Delta(0)$ under the mapping. Show that to prove this property or theorem, one must prove that the Jacobian of the (p_0, q_0) and (p_t, q_t) is independent of time, or, that

$$J = \frac{\partial(p_t, q_t)}{\partial(p_0, q_0)}$$

is independent of time. Now prove that this is so by proving that $dJ/dt = 0$ with the use of Hamilton's equations of motion. (See Mazo, in "Additional Reading," Section 2.4.)

A number of simple models have been introduced into the statistical mechanical literature that are meant to elucidate the concepts of irreversibility, recurrence times, and to try to probe the basic assumptions underlying the Boltzmann equation, particularly the so-called Stosszahlansatz assumption of molecular chaos, or Eq. (18–31). The next two problems will introduce two of these models.

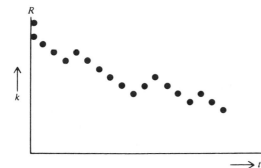

Figure 18–4. **A typical result of the Ehrenfest dog-flea experiment.**

18–28. This illuminating model is due to the Ehrenfests and is called the Ehrenfest dog-flea model. Suppose we have two dogs and $2R$ fleas, numbered from 1 to $2R$. Initially let all $2R$ fleas be on one dog. Then draw a number from 1 to $2R$ and move the flea with that number onto the other dog. Let this process be continued and let $N_A + N_B = 2R$ and $N_A - N_B = 2k$, where N_A and N_B are the number of fleas on the two dogs. If we draw a number and transfer a flea every τ seconds, then $t = s\tau$, where s is the number of draws. A typical result of such an experiment is shown in Fig. 18–4.

Show that at equilibrium, i.e., after a suitable number of draws, that the probability distribution of k is given by

$$W(k) = \frac{(2R)!}{N_A! N_B!} \, (\tfrac{1}{2})^{2R}$$

and that for large R and small k that

$$W(k) \approx (\pi R)^{-1/2} e^{-k^2/R}$$

This corresponds to the Maxwell distribution.

A valuable feature of this problem is that the "dynamics" can be solved exactly. Let the "state" of the system be designated by an integer varying from $-R$ to R. Let $P(n, m; s)$ be the probability that the system be in the state m given that it was in the state n s draws earlier, and let $Q(n, m) \equiv P(n, m; 1)$ be the transition probability. Show that

$$P(n, m; s) = \sum_k P(n, k; s - 1)Q(k, m)$$

$$= P(n, m; s - 1) + \sum_{k \ne m} P(n, k; s - 1)Q(k, m) - P(n, m; s - 1) \sum_{l \ne m} Q(m, l)$$

where of course $\sum_m Q(k, m) = 1$. Compare this equation with the Boltzmann equation. An equation like this one, which describes how a probability distribution changes with time, is called a *master equation*.

Show that the transition probability Q for the Ehrenfest dog-flea model is given by

$$Q(k, k') = \frac{R+k}{2R} \, \delta_{k-1, k'} + \frac{R-k}{2R} \, \delta_{k+1, k'}$$

and that the master equation becomes

$$P(n, m; s) = \frac{R+m+1}{2R} \, P(n, m+1; s-1) + \frac{R-m+1}{2R} \, P(n, m-1; s-1)$$

Now show that if one starts in the state n, i.e., if $P(n, m; 0) = \delta_{m, n}$, then

$$\overline{m(s)} = \sum_m mP(n, m; s) = \left(1 - \frac{1}{R}\right)\overline{m(s-1)} = n\left(1 - \frac{1}{R}\right)^s$$

which in the limit $R \to \infty$, $1/R\tau = \gamma$, $s\tau = t$ gives

$$\overline{m(t)} = ne^{-\gamma t}$$

Note that the average value of m decreases exponentially to its equilibrium value of zero. Thus on the average the system appears to decay irreversibly to its equilibrium state.

This model can be analyzed in great detail, particularly with respect to Poincare's theorem. Let $P'(n, m; s)$ be the probability that the system is in the state m for the first time after s draws given that it was in the state n initially. In particular, $P'(n, n; s)$ is the probability that the recurrence time of state n is $t = s\tau$. One can show that (cf. M. Kac, in *Selected Papers in Noise and Stochastic Processes*, edited by N. Wax, New York: Dover, 1954)

$$\sum_{s=1}^{\infty} P'(n, n; s) = 1$$

which says that each state of the system is bound to recur with probability 1. This is the analog of Poincaré's theorem. One can also show that the mean recurrence time

$$T_R = \sum_{s=1}^{\infty} s\tau P(n, n; s)$$

is equal to

$$T_R = \tau \frac{(R+n)!\,(R-n)!}{(2R)!}\, 2^{2R}$$

This is the analog of the Poincaré recurrence time. The proof of the last few results is somewhat involved, but is given in the article by Kac referred to above.

If $R + n$ and $R - n$ differ considerably, T_R is extremely large. For example, show that if $R = 10,000$, $n = 10,000$, and $\tau = 1$ sec, then

$$T_R = 2^{20,000} \text{ sec} \approx 10^{6,000} \text{ yr.}$$

Conversely, if $R + n$ and $R - n$ are comparable, T_R is quite short. Show that, for example, if we set $n = 0$ in the previous calculation

$$T_R \approx 175 \text{ sec}$$

It was Smoluchowski who clearly elucidated that if one starts in a state with a long recurrence time, the process will appear to be irreversible. For example, if one dog has 20,000 fleas and the other zero, then one would observe what would appear to be an irreversible flow of fleas from one dog to the other. On the other hand, if the recurrence is small, it makes no sense to speak about irreversibility. Thus we see that the concept of irreversibility itself is intimately tied up with the magnitude of the Poincaré recurrence time.

18–29. Another interesting model is discussed by M. Dresden in Vol. 1 of *Studies in Statistical Mechanics* ("Additional Reading") and is called by him the Kac Ring Model (Mark 1) after its originator, Mark Kac. Consider a circle with n equidistant points on the circumference. Let a set S of m points be permanently marked. We consider n and m to be large numbers, but that $m \ll n$. Now place a ball on each of the n points on the circle. These balls can be either black or white, there, of course, being n balls in all. This specifies what we may consider to be the initial state of the system. We now specify its " dynamics." During each elementary time interval, say 1 sec, all the balls move one step counterclockwise, with the rule that when a ball leaves one of the m marked points on the circle, it will change its color (white to black or black to white). We wish to determine the color distribution after a large number of steps given some initial distribution. Let $N_W(t)$ and $N_W(S, t)$ be the total number of white balls and the number of white balls in the set S (sitting on the m marked points), respectively, after t sec. Let $N_B(t)$ and $N_B(S, t)$ be the corresponding quantities for black balls. Clearly we have the relations

$$N_W(t) + N_B(t) \qquad = n$$
$$N_W(S, t) + N_B(S, t) = m$$

In addition, we define a quantity $\Gamma(t)$ through

$$\Gamma(t) = \frac{1}{n} [N_W(t) - N_B(t)]$$

Show that the "equations of motion" of the system are

$$N_W(t+1) = N_W(t) + N_B(S, t) - N_W(S, t)$$
$$N_B(t+1) = N_B(t) + N_W(S, t) - N_B(S, t)$$

These can be thought of as corresponding to the Liouville equation.

We now must resort to some approximation to proceed further, for otherwise these equations are simply formal statements. We make the apparently reasonable assumption that

$$N_W(S, t) = \frac{m}{n} N_W(t)$$

$$N_B(S, t) = \frac{m}{n} N_B(t)$$

These equations are analogous to the "stosszahlansatz." Substitute these into the above "Liouville equation" to derive the "Boltzmann equation" for this system. Show that the solution to these two equations is

$$\Gamma(t+1) = (1 - 2\mu)\Gamma(t)$$

where $\mu \ll 1$ is equal to m/n. Solve this by iteration to get

$$\Gamma(t) = (1 - 2\mu)^t \Gamma(0)$$

This represents a monotonic approach to the equilibrium state—the number of white balls being the same as the number of black balls, independent of the initial distribution. This corresponds to the "Maxwell distribution" of our system. Note that this result is quite independent of the nature of the set S.

This result is incorrect! Clearly the model is strictly periodic with period $2n$ (prove this). Hence this system has a Poincaré recurrence time of $2n$, so that $\Gamma(t) = \Gamma(t + 2n)$, which is a violent conflict with the above expression for $\Gamma(t)$. The only assumption that is made is the above "stosszahlansatz," and so this apparently reasonable assumption has somehow introduced irreversibility into the problem. Dresden also discusses how this model has an analog to the time reversibility objection as well. We refer the reader to Dresden's article for a thorough discussion of these apparent paradoxes.

CHAPTER 19

TRANSPORT PROCESSES IN DILUTE GASES

In the previous chapter we have derived and discussed the Boltzmann equation, which is the fundamental equation of the kinetic theory of dilute gases. The standard method for solving this integrodifferential equation is due to Chapman and Enskog and is now called the Chapman-Enskog method. This method is quite long and involved and so will be presented here only in outline. Chapman and Cowling as well as Hirschfelder, Curtiss, and Bird have detailed discussions of the Chapman-Enskog method (in "Additional Reading").

In Section 19–1 we outline the Chapman-Enskog method; in Section 19–2 we present and summarize the formulas for the transport-properties of dilute gases and then make a detailed comparison of these formulas to experimental data in Section 19–3. Lastly, in Section 19–4, we discuss some extensions of the Boltzmann equation. In particular, we discuss the Enskog theory of dense hard-sphere fluid, which results from a simple, physical modification of the Boltzmann equation.

19–1 OUTLINE OF THE CHAPMAN-ENSKOG METHOD

The types of solutions obtained by the Chapman-Enskog method are a very special class of solutions, called normal solutions, in which the spatial and time dependence of $f^{(1)}(\mathbf{r}, \mathbf{v}, t)$ appear *implicitly* through the local density, flow velocity, and temperature. These solutions describe the final stage of the relaxation of a dilute gas to its equilibrium state. This final stage is called the hydrodynamic stage. The Chapman-Enskog method expands f in a series of the form (dropping the superscript on $f^{(1)}$)

$$f = \frac{1}{\xi} f^{[0]} + f^{[1]} + \xi f^{[2]} + \cdots \tag{19–1}$$

where ξ is an ordering parameter and not a smallness parameter. It is set equal to unity later on.

If we substitute Eq. (19–1) into the Boltzmann equation, we get a set of equations for the $f^{[j]}$, which we then solve successively and uniquely (Problem 19–1). The equation for $f^{[0]}$ is easy to solve and gives (Problem 19–2)

$$f^{[0]}(\mathbf{r}, \mathbf{v}, t) = \rho(\mathbf{r}, t)\left(\frac{m}{2\pi k T(\mathbf{r}, t)}\right)^{3/2} \exp\left\{-\frac{m[\mathbf{v} - \mathbf{v}_0(\mathbf{r}, t)]^2}{2kT(\mathbf{r}, t)}\right\} \qquad (19\text{--}2)$$

The $f^{[j]}$ are chosen such that the equations of change that result from the Boltzmann equation keep the same form for increasing orders of f, but that the values of the parameters that appear in them, namely, p and **q**, do depend upon the order of f. In particular, since

$$\mathsf{p} = m \int \mathbf{VV} f \, d\mathbf{v}$$

and $\qquad\qquad\qquad\qquad\qquad\qquad\qquad\qquad\qquad\qquad\qquad\qquad\qquad (19\text{--}3)$

$$\mathbf{q} = \frac{m}{2} \int V^2 \mathbf{V} f \, d\mathbf{v}$$

Equation (19–2) gives $\mathsf{p}^{(0)} = \rho kT\mathsf{I}$ and $\mathbf{q}^{(0)} = 0$ (Problems 19–6 and 19–7). These expressions for $\mathsf{p}^{(0)}$ and $\mathbf{q}^{(0)}$ give the ideal hydrodynamic equations when substituted into the equations of change, i.e., Eqs. (18–51), (18–52), and (17–31). These equations specify ρ, $\mathbf{v}_0$, and T, and hence $f^{[0]}$ itself through Eq. (19–?)

The equation for $f^{[1]}$ is quite a bit more complicated. If we let $f^{[1]} = \phi f^{[0]}$, then ϕ is given by the linear inhomogeneous integral equation (Problem 19–3):

$$2\pi \iint f^{[0]} f_1^{[0]'}(\phi' + \phi_1' - \phi - \phi_1) gb \, db \, d\mathbf{v}_1 \equiv -\rho^2 I(\phi)$$

$$= \frac{\partial f^{[0]}}{\partial t} + \mathbf{v} \cdot \nabla_r f^{[0]} + \frac{1}{m}\mathbf{X} \cdot \nabla_\mathbf{v} f^{[0]}$$

$$(19\text{--}4)$$

It turns out that the existence condition for this equation for ϕ is just the ideal hydrodynamic equations. Since $f^{[0]}$ is a normal solution, given by Eq. (19–2), the time dependence is given implicitly through ρ, $\mathbf{v}_0$, and T. So we write

$$\frac{\partial f^{[0]}}{\partial t} = \frac{\partial f^{[0]}}{\partial \rho}\frac{\partial \rho}{\partial t} + \nabla_{\mathbf{v}_0} f^{[0]} \cdot \frac{\partial \mathbf{v}_0}{\partial t} + \frac{\partial f^{[0]}}{\partial T}\frac{\partial T}{\partial t}$$

and then use the equations of change for the time derivatives of ρ, $\mathbf{v}_0$, and T. This eventually gives (Problem 19–4)

$$\rho^2 I(\phi) = -f^{[0]}[(W^2 - \tfrac{5}{2})\mathbf{V} \cdot \nabla \ln T + \mathsf{b} : \nabla \mathbf{v}_0] \qquad (19\text{--}5)$$

where **W** is a reduced velocity,

$$\mathbf{W} = \left(\frac{m}{2kT}\right)^{1/2} \mathbf{V} \qquad (19\text{--}6)$$

and

$$\mathsf{b} = 2(\mathbf{WW} - \tfrac{1}{3}W^2\mathsf{I}) \qquad (19\text{--}7)$$

This suggests looking for a solution ϕ of the form

$$\phi = -\frac{1}{\rho}\left(\frac{2kT}{m}\right)^{1/2} A(W)\mathbf{W} \cdot \nabla \ln T - \frac{1}{\rho}B(W)(\mathbf{WW} - \tfrac{1}{3}W^2\mathsf{I}) : \nabla \mathbf{v}_0 \qquad (19\text{--}8)$$

where $A(W)$ and $B(W)$ are unknown scalar functions of W. Since $I(\phi)$ is a linear integral operator (Problem 19–5), we get two separate integral equations, one for $A(W)$ and one for $B(W)$.

Furthermore, ϕ can be used to calculate the next approximation to $\mathbf{p}$ and $\mathbf{q}$ by means of Eqs. (19–3). One eventually finds that (cf. Hirschfelder, Curtiss, and Bird, in "Additional Reading")

$$\mathbf{q}^{(1)} = -\frac{2}{3}\frac{k^2 T}{m\rho} \nabla T \int A(W)W^2(W^2 - \tfrac{5}{2})f^{[0]} \, d\mathbf{v} \tag{19–9}$$

and

$$\mathbf{p}^{(1)} = -\frac{2kT}{5\rho} [\text{sym } \nabla\mathbf{v}_0 - \tfrac{1}{3}(\nabla \cdot \mathbf{v}_0)\mathbf{I}] \int f^{[0]}B(W)(\mathbf{WW} - \tfrac{1}{3}W^2\mathbf{I}) : (\mathbf{WW} - \tfrac{1}{3}W^2\mathbf{I}) \, d\mathbf{v} \tag{19–10}$$

Thus we know $\mathbf{q}^{(1)}$ and $\mathbf{p}^{(1)}$ if we solve the integral equations for $A(W)$ and $B(W)$. By comparing $\mathbf{q}^{(0)} + \mathbf{q}^{(1)}$ to Fourier's law, we get

$$\lambda = \frac{2}{3}\frac{k^2 T}{m\rho} \int A(W)W^2(W^2 - \tfrac{5}{2})f^{[0]} \, d\mathbf{v} \tag{19–11}$$

for the thermal conductivity. Similarly, by comparing $\mathbf{p}^{(0)} + \mathbf{p}^{(1)}$ to the Newtonian pressure tensor [Eq. (17–34)], we find that

$$\eta = \frac{kT}{5\rho} \int f^{[0]}B(W)(\mathbf{WW} - \tfrac{1}{3}W^2\mathbf{I}) : (\mathbf{WW} - \tfrac{1}{3}W^2\mathbf{I}) \, d\mathbf{v} \tag{19–12}$$

Equations (19–9) and (19–10) show that $f^{[0]} + f^{[1]}$ gives the Navier-Stokes equations. The $A(W)$ and $B(W)$ are found by expanding them in truncated series of orthogonal polynomials (Sonine polynomials). This is a standard method to solve integral equations (Problems 19–46 and 19–47). The expressions we find for $A(W)$ and $B(W)$ depend upon the number of terms we include in their expansions in Sonine polynomials. Usually only one or two terms is needed. If we substitute two-term expansions into Eqs. (19–11) and (19–12), we get

$$\lambda = \frac{75k^2 T}{8m}\left[\frac{1}{a_{11}} + \frac{a_{12}^2/a_{11}}{a_{11}a_{22} - a_{12}^2} + \cdots\right] \tag{19–13}$$

$$\eta = \frac{5kT}{2}\left[\frac{1}{b_{11}} + \frac{b_{12}^2/b_{11}}{b_{11}b_{22} - b_{12}^2} + \cdots\right] \tag{19–14}$$

The a_{ij} and b_{ij} appearing in these equations are complicated integrals over the dynamics of bimolecular collisions and hence contain the information about the intermolecular potential. All of these quantities can be reduced to linear combinations of a set of collision integrals, $\Omega_{ij}^{(l,s)}$. For collisions between molecules of type i and type j, these integrals are defined by

$$\Omega_{ij}^{(l,s)} = \left(\frac{2\pi kT}{\mu_{ij}}\right)^{1/2} \int_0^\infty \int_0^\infty e^{-\gamma_{ij}^2}\gamma_{ij}^{2s+3}(1 - \cos^l \chi)b \, db \, d\gamma_{ij} \tag{19–15}$$

In these integrals, μ_{ij} is the reduced mass; g_{ij} is the relative velocity; $\chi = \chi(b, g_{ij})$ is the deflection angle; and γ_{ij} is a reduced relative velocity, $\gamma_{ij} = (\mu_{ij}/2kT)^{1/2}\mathbf{g}_{ij}$. These integrals must be evaluated numerically, and we shall discuss them in the next section. We simply list some results from Chapman and Cowling:

$$a_{11} = 4\Omega^{(2,2)}$$

$$a_{12} = 7\Omega^{(2,2)} - 2\Omega^{(2,3)}$$

$$a_{22} = \frac{77}{4}\Omega^{(2,2)} - 7\Omega^{(2,3)} + \Omega^{(2,4)}$$

$$b_{11} = 4\Omega^{(2,2)}$$

$$b_{12} = 7\Omega^{(2,2)} - 2\Omega^{(2,3)}$$

$$b_{22} = \frac{301}{12}\Omega^{(2,2)} - 7\Omega^{(2,3)} + \Omega^{(2,4)} \tag{19-16}$$

Equations (19–13) through (19–16) then give λ and η in terms of the intermolecular potential. In the next section we shall study λ and η as a function of intermolecular potential and temperature as we did for the second virial coefficient in Chapter 12.

In order to study diffusion, it is necessary to consider multicomponent systems. The main results are that the one-term expansion for the binary diffusion coefficient is given by

$$D_{12} = \frac{3}{16\rho\mu_{12}}\frac{kT}{\Omega_{12}^{(1,1)}} \tag{19-17}$$

Furthermore, there is the possibility of diffusion due to a temperature gradient as well. This phenomenon is known as thermal diffusion and was unobserved experimentally and unknown theoretically before being discovered by Enskog. The coefficient associated with this phenomenon is called the thermal diffusion coefficient D_1^T. Usually, however, it is the thermal diffusion ratio k_T,

$$k_T = \frac{\rho_m}{\rho^2 m_1 m_2}\frac{D_1^T}{D_{12}} \tag{19-18}$$

that is discussed, both experimentally and theoretically.

These results are all for the first two terms in Eq. (19–1). One could go on to calculate $f^{[2]}$, but this becomes even more tedious. It is interesting to note that if this is done, one gets hydrodynamic equations, which are more general than the Navier-Stokes equations. These more general equations are called the Burnett equations. They contain higher derivatives of the thermodynamic quantities and powers of lower derivatives. This makes these equations very difficult to use in practice. Nevertheless, it clearly shows that there is not one set of hydrodynamic equations but several, each set offering increasing detail and precision. The Burnett equations have been applied to shock waves in gases, but the agreement between experiment and theory is not good. This is probably because the entire Chapman-Enskog method is a development in stages away from a local Maxwellian velocity distribution. One must expect the results to be most applicable when the system is near equilibrium. Just how near is hard to say.

Grad has developed quite a different method to solve the Boltzmann equation. This method gives the ideal hydrodynamic equations in the zero approximation and gives the Navier-Stokes equations in the first approximation, but in the next approximation gives something quite different from the Burnett equations. Grad's method is discussed in Section 7–5 in Hirschfelder, Curtiss, and Bird and in the two books by Grad himself (in "Additional Reading").

We are now ready to discuss the results of the Chapman-Enskog solution.

19–2   SUMMARY OF FORMULAS

We have seen that all of our results can be expressed in terms of the collision integrals $\Omega_{ij}^{(l,s)}$. These are given by

$$\Omega_{ij}^{(l,s)} = \left(\frac{kT}{2\pi\mu_{ij}}\right)^{1/2} \int_0^\infty e^{-\gamma_{ij}^2} \gamma_{ij}^{2s+3} Q^{(l)}(\gamma_{ij})\, d\gamma_{ij} \tag{19–19}$$

where

$$Q^{(l)}(\gamma_{ij}) = 2\pi \int_0^\infty (1 - \cos^l \chi) b\, db \tag{19–20}$$

and

$$\chi(b, \gamma_{ij}) = \pi - 2b \int_{r_m}^\infty \frac{dr/r^2}{\left[1 - \dfrac{b^2}{r^2} - \dfrac{u_{ij}(r)}{kT\gamma_{ij}^2}\right]^{1/2}} \tag{19–21}$$

In these equations, $\gamma^2 = \mu g^2/2kT$, μ is the reduced mass, and r_m is the distance of closest approach of the colliding molecules.

For hard spheres of diameter σ, Eqs. (19–19) and (19–20) become (Problem 19–13)

$$\Omega_{\text{rig sph}}^{(l,s)} = \left(\frac{kT}{2\pi\mu}\right)^{1/2} \frac{(s+1)!}{2} Q_{\text{rig sph}}^{(l)} \tag{19–22}$$

with

$$Q_{\text{rig sph}}^{(l)} = \left[1 - \frac{1}{2}\frac{1 + (-)^l}{1 + l}\right]\pi\sigma^2 \tag{19–23}$$

Note that these quantities are independent of γ for hard spheres. Using Eqs. (19–13), (19–14), (19–16), and (19–17), we can write for hard spheres (Problem 19–14):

$$\lambda = \frac{75k^2T}{8ma_{11}} = \frac{25}{32}\frac{C_v}{N}\left(\frac{\pi kT}{m}\right)^{1/2}\frac{1}{\pi\sigma^2} \tag{19–24}$$

$$\eta = \frac{5kT}{2b_{11}} = \frac{5}{16}\frac{(\pi mkT)^{1/2}}{\pi\sigma^2} \tag{19–25}$$

$$D = \frac{3}{8m\rho}\frac{kT}{\Omega^{(1,1)}} = \frac{3}{8}\left(\frac{\pi kT}{m}\right)^{1/2}\frac{1}{\pi\sigma^2\rho} \tag{19–26}$$

To get these equations, remember that $\mu = m/2$ when the two colliding atoms are identical. These equations are exactly those presented in Chapter 16 as the rigorous kinetic theory expressions. The advantage here, however, is that these are special cases of a more general theory that includes any realistic intermolecular potential.

For most intermolecular potential functions, we can write the collision integrals given by Eqs. (19–19) to (19–21) in a reduced form by introducing reduced variables. If the intermolecular potential can be written in the form

$$u(r) = \varepsilon f\left(\frac{r}{\sigma}\right) \tag{19-27}$$

then we can introduce reduced variables:

$$r* = \frac{r}{\sigma}$$

$$b* = \frac{b}{\sigma}$$

$$u* = \frac{u}{\varepsilon}$$

$$T* = \frac{kT}{\varepsilon}$$

$$g*^2 = \frac{1}{2}\frac{\mu g^2}{\varepsilon} \tag{19-28}$$

Just as we reduced the second virial coefficient by its hard-sphere value, we reduce $Q^{(l)}(g)$ and $\Omega^{(l,s)}(T)$ by dividing them by their hard-sphere values. Thus

$$Q^{(l)*}(T*) = \frac{Q^{(l)}}{Q^{(l)}_{\text{rig sph}}} \tag{19-29}$$

$$\Omega^{(l,s)*}(T*) = \frac{\Omega^{(l,s)}}{\Omega^{(l,s)}_{\text{rig sph}}} \tag{19-30}$$

These reduced quantities physically represent the deviation of any particular molecular model from the hard-sphere model. In terms of these reduced quantities, we have (Problem 19–16)

$$\lambda = \frac{25}{32}\frac{C_v}{N}\left(\frac{\pi kT}{m}\right)^{1/2}\frac{1}{\pi\sigma^2\Omega^{(2,2)*}} \tag{19-31}$$

$$\eta = \frac{5}{16}\frac{(\pi mkT)^{1/2}}{\pi\sigma^2\Omega^{(2,2)*}} \tag{19-32}$$

$$D = \frac{3}{8}\left(\frac{\pi kT}{m}\right)^{1/2}\frac{1}{\rho\pi\sigma^2\Omega^{(1,1)*}} \tag{19-33}$$

These are the generalizations of Eqs. (19–24) to (19–26). They reduce to the hard-sphere results where $\Omega^{(l,s)*} = 1$.

These formulas are the result of using simply a one-term expansion for $A(W)$ and $B(W)$ in Eqs. (19–11) and (19–12). Equations (19–31) and (19–32) are, in fact, the first terms in Eqs. (19–13) and (19–14). Such one-term expansions turn out to be satisfactory for most transport coefficients (the thermal diffusion ratio is a notable exception). It is common to write the expressions for these transport coefficients in the form

$$\lambda = \frac{25}{32}\frac{C_v}{N}\left(\frac{\pi kT}{m}\right)^{1/2}\frac{f_\lambda}{\pi\sigma^2\Omega^{(2,2)*}} \tag{19-34}$$

for example, where f_λ is a factor close to unity that corrects λ to a more precise value. Equation (19–13) shows that f_λ is

$$f_\lambda = 1 + \frac{a_{12}^2}{a_{11}a_{22} - a_{12}^2} + \cdots \qquad (19\text{--}35)$$

The correction factors f_λ, f_η, and f_D, etc., are given in Appendix 8A of Hirschfelder, Curtiss, and Bird (in "Additional Reading").* Table 19–1 gives these factors as a function of temperature for the Lennard-Jones 6–12 potential. It clearly shows that these higher corrections are small. This substantiates our contention that the integral equations for $A(W)$ and $B(W)$ can be adequately solved by using just a one-term expansion in Sonine polynomials.

Table 19–1. **The effect of higher approximation of the thermal transport coefficients for a Lennard-Jones 6–12 potential***

$[\eta]_3 = [\eta]_1 f_\eta^{(3)}$ T^*	$[\lambda]_3 = [\lambda]_1 f_\lambda^{(3)}$ $f_\eta^{(3)}$	$f_\lambda^{(3)}$	$[D]_2 = [D]_1 f_D^{(2)}$ $f_D^{(2)}$
0.30	1.0014	1.0022	1.0001
0.50	1.0002	1.0003	1.0000
0.75	1.0000	1.0000	1.0000
1.00	1.0000	1.0001	1.0000
1.25	1.0001	1.0002	1.0002
1.5	1.0004	1.0006	1.0006
2.0	1.0014	1.0021	1.0016
2.5	1.0025	1.0038	1.0026
3.0	1.0034	1.0052	1.0037
4.0	1.0049	1.0076	1.0050
5.0	1.0058	1.0090	1.0059
10.0	1.0075	1.0116	1.0076
50.0	1.0079	1.0124	1.0080
100.0	1.0080	1.0125	1.0080
400.0	1.0080	1.0125	1.0080

* The superscript on the f's indicates the number of terms used in equations such as Eqs. (19–13) and (19–14).

Source: J. O. Hirschfelder, C. F. Curtiss, and R. B. Bird, *Molecular Theory of Gases and Liquids* (New York: Wiley, 1954).

Table 19–2. **The types of collision integrals that appear in the dilute gas expressions for the thermal transport coefficients**

property	1st approximation	2nd approximation
η	$\Omega^{(2,2)}$	$\Omega^{(2,s)}$ $s = 2, 3, 4$
λ	$\Omega^{(2,2)}$	$\Omega^{(2,s)}$ $s = 2, 3, 4$
D	$\Omega^{(1,1)}$	$\Omega^{(2,2)}$, $\Omega^{(1,s)}$ $s = 1, 2, 3$
k_r	—	$\Omega^{(2,t)}\Omega^{(1,s)}\Omega^{(3,3)}$ $s = 1, 2, 3, 4, 5$ $t = 2, 3, 4$

Table 19–2 gives the various types of collision integrals that appear in the first two approximations for λ, η, D, and k_T. Table 19–3 gives the fraction of the true value of the hard-sphere transport coefficients in the kth approximation. It can readily be seen that the thermal diffusion ratio is the most demanding quantity to calculate. A look at Hirschfelder, Curtiss, and Bird's Appendix 8A or Appendix I of an excellent summary

* See also E. A. Mason, *J. Chem. Phys.*, **22**, p. 169, 1954.

Table 19-3. **Fraction of the true value of the transport coefficients in the kth approximation***

P	$[P]_1/[P]$	$[P]_2/[P]$	$[P]_3/[P]$
η	0.984	0.999	0.999+
λ	0.976	0.998	0.999+
D	0.883	0.957	0.978
k_τ	0.77	0.88	—

* Based on the rigid sphere model. $[P_k]$ = kth approximation to the coefficient P; $[P]$ = true value of the coefficient P.

Source: J. O. Hirschfelder, C. F. Curtiss, and R. B. Bird, *Molecular Theory of Gases and Liquids* (New York: Wiley, 1954).

by Mason* will show that the formula for k_T can become quite long. You will also discover there another set of formulas for the f_λ, f_η, etc., due to Kihara, which seem to be more convenient than those obtained from the Chapman-Enskog procedure. They are all equal to unity to within just a few percent, however.

Equations (19-31) to (19-33), then, can be used to calculate λ, η, and D as a function of the intermolecular potential. All we need now are tables of the $\Omega^{(l,\,s)}$. We discuss these $\Omega^{(l,\,s)}$ in the next section.

19-3 TRANSPORT COEFFICIENTS FOR VARIOUS INTERMOLECULAR POTENTIALS

It is possible to evaluate the $\Omega^{(l,\,s)}$ analytically for the hard-sphere potential, but not for the other potential functions we shall discuss. We saw in Chapter 16 that the deflection angle $\chi(b, g)$ is

$$\chi(g, b) = 2 \arccos\left(\frac{b}{\sigma}\right) \qquad b \le \sigma$$

$$= 0 \qquad b \ge \sigma$$

for hard spheres of diameter σ. Substitution of this result into Eqs. (19-20) and (19-19) gives Eqs. (19-22) and (19-23), which in turn give Eqs. (19-24) to (19-26) for λ, η, and D. We have already discussed these rigid sphere results in Chapter 16.

Table 19-4. **Square-well parameters obtained from second virial coefficient data and viscosity data**

gas	from $B_2(T)$			from $\eta(T)$		
	$\varepsilon/k(^\circ K)$	$\sigma(\text{Å})$	λ	$\varepsilon/k(^\circ K)$	$\sigma(\text{Å})$	λ
Ar	93.3	3.07	1.70	167	2.98	1.96
N_2	95.2	3.28	1.58	80	3.36	2.08
CO_2	284	3.57	1.44	200	3.46	2.22
methane	142	3.35	1.60	244	3.81	2.29

Source: The second virial coefficient values are taken from Sherwood and Prausnitz, *J. Chem. Phys.*, **41**, p. 429, 1964, and the viscosity values are taken from J. O. Hirschfelder, C. F. Curtiss, and R. B. Bird, *Molecular Theory of Gases and Liquids* (New York: Wiley, 1954).

Contrary to the second virial coefficient, the collision integrals must be evaluated numerically for the square-well potential. Table 19-4 gives a comparison of the square-well potential parameters obtained from the second virial coefficient and the viscosity.

Of course, both of these theories are essentially exact, and so the discrepancy is due to the inadequacy of the square-well potential.

* E. A. Mason, *J. Chem. Phys.*, **22**, p. 169, 1954.

Table 19–5. **The integrals** $\Omega^{(l,\,s)*}$ **for calculating the transport coefficients for the Lennard-Jones 6–12 potential**

T^*	$\Omega^{(1,\,1)*}$	$\Omega^{(2,\,2)*}$	$\Omega^{(2,\,3)*}$	$\Omega^{(2,\,4)*}$
0.60	1.877	2.065	1.806	1.610
0.80	1.612	1.780	1.549	1.389
1.00	1.439	1.587	1.388	1.258
1.20	1.320	1.452	1.280	1.174
1.40	1.233	1.353	1.205	1.115
1.60	1.167	1.279	1.149	1.072
1.80	1.116	1.221	1.106	1.038
2.00	1.075	1.175	1.073	1.012
2.20	1.041	1.138	1.045	0.9895
2.40	1.012	1.107	1.022	0.9710
2.60	0.9878	1.081	1.002	0.9555
2.80	0.9672	1.058	0.9935	0.9485
3.00	0.9490	1.039	0.9708	0.9295
3.20	0.9328	1.022	0.9578	0.9185
3.60	0.9058	0.9932	0.9358	0.8995
4.00	0.8836	0.9700	0.9175	0.8840
5.00	0.8422	0.9269	0.8823	0.8530
6.00	0.8124	0.8963	0.8565	0.8295
8.00	0.7712	0.8538	0.8193	0.7945
10.00	0.7424	0.8242	0.7923	0.7690
20.00	0.6640	0.7432	0.7160	0.6950
40.00	0.5960	0.6718	0.6475	0.6285
50.00	0.5756	0.6504	0.6268	0.6085

Source: J. O. Hirschfelder, C. F. Curtiss, and R. B. Bird, *Molecular Theory of Gases and Liquids* (New York: Wiley, 1954).

Hirschfelder, Curtiss, and Bird discuss the transport coefficients of a Lennard-Jones 6–12 gas in great detail. Table 19–5 gives values of several $\Omega^{(l,\,s)*}$ as a function of the reduced temperature. Table 19–6 lists the Lennard-Jones 6–12 parameters obtained by fitting viscosity versus temperature data and second virial coefficient versus temperature data. Tables 19–5 and 19–6 allow one to calculate the viscosity at any temperature (Problem 19–20). Figure 19–1 shows the viscosity for a number of substances versus temperature plotted in reduced units. One can see from Eq. (19–32) that (Problem 19–28)

$$\eta^* = \frac{\eta\sigma^2}{(m\varepsilon)^{1/2}} \tag{19–36}$$

is a reduced viscosity. Similar agreement is found for the thermal conductivity. Figure 19–2 gives the reduced coefficient of self-diffusion versus reduced temperature.

Table 19–6. **Lennard-Jones 6–12 parameters determined for second virial coefficient and viscosity measurements**

gas	second virial coefficient		viscosity	
	$\varepsilon/k(°K)$	$\sigma(Å)$	$\varepsilon/k(°K)$	$\sigma(Å)$
Ar	117.7[a]	3.504[a]	124[b]	3.418[b]
Kr	164.0[a]	3.827[a]	190[b]	3.61[b]
Xe	222.3[a]	4.099[a]	229[b]	4.055[b]
N_2	95.2[a]	3.745[a]	91.5[b]	3.681[b]
CO_2	198.2[a]	4.328[a]	190[b]	3.996[b]
CH_4	148.9[a]	3.783[a]	137[b]	3.822[b]

Source: [a]A. E. Sherwood and J. M. Prausnitz, *J. Chem. Phys.*, **41**, p. 429, 1964; [b]J. O. Hirschfelder, C. F. Curtiss, and R. B. Bird, *Molecular Theory of Gases and Liquids* (New York: Wiley, 1954).

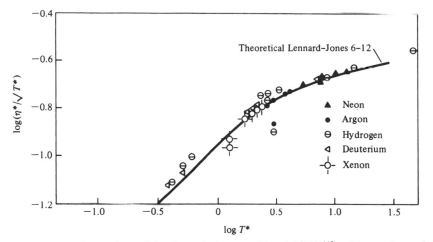

Figure 19–1. **Comparison of the theoretical curve of** log $(\eta^*/(T^*)^{1/2})$ **with experimental data. The experimental values are reduced according to the relation** $\eta^* = \eta\sigma^2/(m\varepsilon)^{1/2}$. (From J. O. Hirschfelder, C. F. Curtiss, and R. B. Bird, *Molecular Theory of Gases and Liquids.* New York: Wiley, 1954.)

The reduced coefficient of self-diffusion is obtained from Eq. (19–33) (Problem 19–30):

$$D^* = \frac{Dm^{1/2}}{\sigma\varepsilon^{1/2}} \tag{19–37}$$

Again it can be seen that the agreement is very good.

The thermal diffusion ratio is much more complicated than the three transport coefficients we have discussed above. It seems to be the most sensitive to the form of the intermolecular potential. Hanley and Klein* discuss the determination of the intermolecular potential from the thermal diffusion ratio in a mixture of isotopes.

It can be seen from Figs. 19–1 and 19–2 that the Lennard-Jones 6–12 potential, with its two parameters ε and σ, can be used to fit the thermal transport data of dilute gases quite well. The agreement is comparable to that found in Chapter 12, where we discussed a similar comparison for the second virial coefficient. A more detailed numerical study, however, would show that the Lennard-Jones 6–12 potential with fixed values of ε and σ is not able to yield satisfactory agreement over an extended temperature range. In addition, Table 19–6 shows that the Lennard-Jones parameters obtained by fitting second virial coefficient data do not agree with those obtained by fitting viscosity data. Of course, since the theoretical formulas expressing these quantities in terms of the intermolecular potential are essentially exact, this indicates that the Lennard-Jones 6–12 potential is inadequate.

A great deal of work has been done to devise a semiempirical intermolecular potential that is flexible enough to fit a number of experimental quantities simultaneously. These potentials usually involve more than two parameters and can become fairly complicated. The review articles by Fitts and Curtiss given in the "Additional Reading" discuss a number of these potentials and list references to numerical tabulations. Figure 19–3 shows the coefficient of viscosity for argon versus temperature for several widely different forms of the intermolecular potential. One can see from this figure that although $\eta(T)$ is somewhat insensitive to $u(r)$, there are, indeed, discrepancies.

* H. J. M. Hanley and M. Klein, *J. Chem. Phys.*, **50**, p. 4765, 1969.

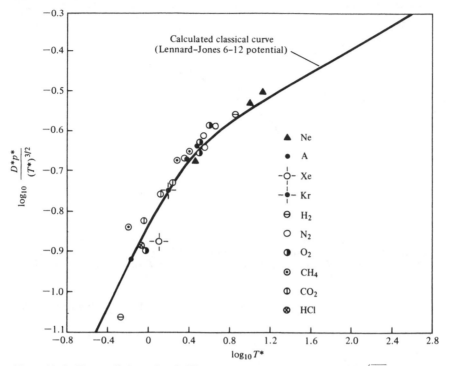

Figure 19–2. **The coefficient of self-diffusion in reduced units.** $D^* = (D/\sigma)\sqrt{m/\varepsilon}$. **The solid curve is the calculated curve:** $D^*p^*T^{*-3/2} = (3/8\sqrt{\pi})(1/\Omega^{(1,1)*})$. **The plotted points are the experimental values in reduced units.** (From J. O. Hirschfelder, C. F. Curtiss, and R. B. Bird, *Molecular Theory of Gases and Liquids*. New York: Wiley, 1954.)

These are clearly illustrated by means of a so-called deviation plot, in which a difference such as $(\eta_{exp} - \eta_{calc})/\eta_{exp}$ is plotted versus temperature. Figure 19–4 shows such a deviation plot for argon for the Lennard-Jones 6–12 potential, the exp-6 potential and the Kihara potential. The exp-6 potential is given by

$$u(r) = \frac{\varepsilon}{1 - 6/\alpha} \left\{ \frac{6}{\alpha} \exp\left[\alpha\left(\frac{1 - r}{r_{min}}\right)\right] - \left(\frac{r_{min}}{r}\right)^6 \right\} \qquad (19\text{--}38)$$

where r_{min} is the value of r for which u is a minimum, ε is the depth of the potential, and α is a parameter that governs the steepness of the repulsive part of the potential. This potential is discussed in Hirschfelder, Curtiss, and Bird, and the collision integrals have been tabulated by Mason.*

The Kihara potential is given by

$$u(r) = 4\varepsilon \left[\left(\frac{\sigma - a}{r - a}\right)^{12} - \left(\frac{\sigma - a}{r - a}\right)^6 \right] \qquad (19\text{--}39)$$

In this potential the finite size of the molecule is taken into consideration by including a core parameter a. The collision integrals for this potential have been tabulated by Barker, Fock, and Smith.† Figure 19–4 clearly shows marked deviations at temperatures below 200°K.

* E. A. Mason, *J. Chem. Phys.*, **22**, p. 169, 1954.
† J. A. Barker, W. Fock, and F. Smith, *Phys. Fluids*, **7**, p. 897, 1964.

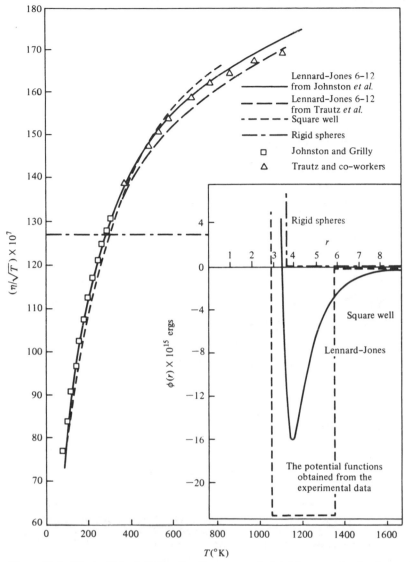

Figure 19–3. **The coefficient of viscosity of** *argon* **as calculated for several molecular models.** (From J. O. Hirschfelder, C. F. Curtiss, and R. B. Bird, *Molecular Theory of Gases and Liquids*, New York: Wiley, 1954.)

There has been much discussion in the literature concerning the relative merits of various intermolecular potentials. A simple potential that seems to correlate many experimental data over a large temperature range is the *m*–6–8 potential, which has the form

$$u(r) = \frac{A}{r^m} - \frac{c_6}{r^6} - \frac{c_8}{r^8} \qquad (19\text{--}40)$$

In terms of the usual parameters of ε, the depth of the well, and σ, the value of r at which the potential equals zero, $u(r)$ becomes

$$\frac{u(r)}{\varepsilon} = \frac{(6 + 2\gamma)}{n - 6}\left(\frac{d\sigma}{r}\right)^m - \frac{[n - \gamma(n - 8)]}{n - 6}\left(\frac{d\sigma}{r}\right)^6 - \gamma\left(\frac{d\sigma}{r}\right)^8 \qquad (19\text{--}41)$$

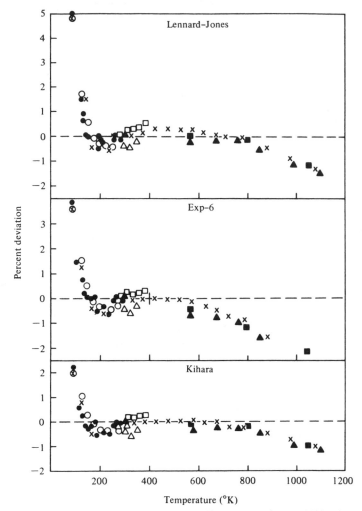

Figure 19–4. **Percentage deviation plots** $\{(\eta_{exp} - \eta_{calc})/\eta_{exp}\} \times 100$, **where** η_{exp} **are experimental values of the viscosity of argon and the** η_{calc} **are calculated for the Lennard-Jones 6–12 potential, the exp-6 potential, and the Kihara potential.** (From H. J. M. Hanley, *J. Chem. Phys.*, **44**, p. 4219, 1966.)

where $d = r_{min}/\sigma$ and $\gamma = c_8/\varepsilon(r_{min})^8$. Using this potential, Hanley and Klein* have performed extensive calculations and comparisons to experimental data. Figure 19–5 shows deviation plots for the second virial coefficient and viscosity coefficients of argon. According to Hanley and Klein, the m–6–8 potential is the most successful few-parameter potential to correlate both second virial coefficient and transport data. The m–6–8 parameters have not been determined for many substances yet, but Table 19–7 gives them for argon, krypton, and xenon. In addition, Appendix E gives the reduced second virial coefficient, several reduced collision integrals, and the Chapman-Enskog correction factors f for the 11–6–8 potential with $\gamma = 3$. Ely and Hanley† have

---

* H. J. M. Hanley and M. Klein, *J. Phys. Chem.*, **76**, p. 1743, 1972.
† J. F. Ely and H. J. M. Hanley, *Mol. Phys.*, **24**, p. 683, 1972.

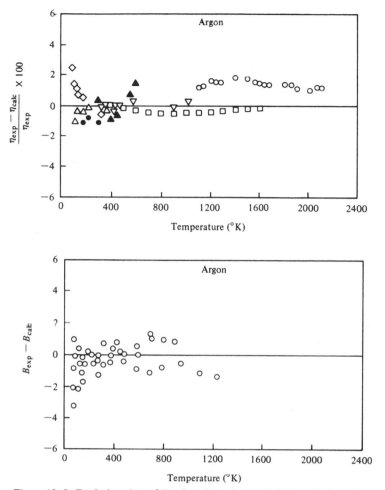

Figure 19-5. **Deviation plots of the viscosity and second virial coefficient of argon. The calculated values of B_2 and η are determined using the m-6-8 potential with $m = 11$, $\gamma = 3$, $\sigma = 3.292$ Å, and $\varepsilon/k = 153°K$).** (From H. J. M. Hanley and M. Klein, *J. Phys. Chem.*, **76**, p. 1743, 1972.)

Table 19-7. **The m-6-8 potential parameters for argon, krypton, and xenon**

gas	m	γ	$\sigma(\text{Å})$	$\varepsilon/k(°K)$
Ar	11	3	3.292	153
Kr	11	3	3.509	216
Xe	11	3	3.841	295

Source: H. J. M. Hanley and M. Klein, *J. Phys. Chem.*, **76**, p. 1743, 1972.

also applied the m-6-8 potential to the nonpolar polyatomic gases O_2, N_2, and CO_2. It is possible that this potential will emerge as the most useful potential for a serious calculation of the thermodynamic or transport properties of dilute gases, but this area is still a state of flux and discussion as an interesting recent paper by Hanley, Barker, Parson, Lee, and Klein shows.*

* H. J. M. Hanley, J. A. Barker, J. M. Parson, Y. T. Lee, and M. Klein, *Mol. Phys.*, **24**, p. 11, 1972.

19–4 EXTENSIONS OF THE BOLTZMANN EQUATION

Although the theory of transport phenomena in dilute gases made up of structureless spherical particles is well developed, all of the obvious extensions are still areas of research. For instance, there are still difficulties in trying to predict the transport properties of mixtures of monatomic gases. The equations rapidly become long and unwieldy, as one can see from Section 8.2 of Hirschfelder, Curtiss, and Bird. The difficulties in the case of polyatomic gases are even worse, however, since in this case there was still some question concerning the formulation of the Boltzmann equation itself until only recently. Furthermore, polyatomic molecules undergo inelastic as well as elastic collisions, and so the transport in polyatomic gases gets very involved with molecular scattering theory. We refer to the article by Curtiss in "Additional Reading" for a discussion of the basic formulation of this theory and to papers by Mason and co-workers* for practical applications.

The 1960s saw a great deal of research in trying to extend the Boltzmann equation to higher densities. We learned in Chapter 12 that one can expand equilibrium thermodynamic properties in a power series in the density, and it is natural to try to expand the transport coefficients in a similar way. One approach has been to derive a generalized Boltzmann equation from the BBGKY hierarchy by approximating the pair distribution function in the integral as a functional of the singlet distribution function. This was first suggested by Bogoliubov in 1946 and popularized by Uhlenbeck in this country. This development is very complicated, but expressions have been derived for the density corrections to the transport coefficients. An interesting result of all this is that terms in ρ^2 and higher do not exist. When the formal expressions derived for them are evaluated, they diverge. It appears now that logarithmic terms occur after the linear term. It has been found that a $\rho^2 \ln \rho$ term seems to occur in the expansion of the transport coefficients. If this is correct, then the transport coefficients are not analytic functions of the density, and it is therefore not surprising that formulas derived on this assumption give divergent results. The Bogoliubov theory is too involved to go into here, but is discussed in Chapter 6 of Mazo and in the review article by Ernst, Haines, and Dorfman.† Hanley, McCarty, and Sengers have presented a detailed analysis of experimental data of transport coefficients as a function of density.‡ The review article by Curtiss discusses several other approaches as well as the Bogoliubov theory.

There is an interesting and useful modification of the Boltzmann equation so that it yields an approximate theory of transport in dense hard-sphere fluids. This extension is due originally to Enskog§ and is now called the Enskog theory. The manipulations involved in deriving the final (simple!) results of the Enskog theory are fairly lengthy, and so we shall present only the end results here. The two references§ below, however, have excellent discussions on the Enskog theory.

The Boltzmann equation considers only binary collisions and so ordinarily is not applicable to dense systems, but in the case of hard spheres the collisions are instantaneous, and so the probability of multiple encounters is negligible. Enskog used this fact to graft a dense hard-sphere transport theory onto the Boltzmann equation.

---

* L. Monchick and E. A. Mason, *J. Chem. Phys.*, **35**, p. 1676, 1961; E. A. Mason and L. Monchick, *ibid.*, **36**, p. 1622, 1962; L. Monchick, K. S. Yun, and E. A. Mason, *ibid.*, **39**, p. 654, 1963.

† M. H. Ernst, L. K. Haines, and J. R. Dorfman, *Rev. Mod. Phys.*, **41**, p. 296, 1969.

‡ H. J. M. Hanley, R. D. McCarty, and J. V. Sengers, *J. Chem. Phys.*, **50**, p. 857, 1969.

§ Section 9.3 of Hirschfelder, Curtiss, and Bird ("Additional Reading") and Chapter 16 of Chapman and Cowling ("Additional Reading") are excellent references.

In the derivation of the Boltzmann equation, two assumptions that are made are that only binary collisions are important and that the molecules have zero size or, more exactly, that the molecular diameter σ is small compared to the mean free path of the gas. Both of these assumptions are sensible in a dilute gas but not in a dense system. As the density of a dilute gas is increased, two effects become important because the molecules how have a nonzero size. The first is a positional correlation, which will appear through the radial distribution function. Since $g(r)$ is greater than unity when the molecules are near each other, there is an increase in the rate of collisions. Secondly, collisional transfer of flux becomes important. In a dilute gas the only mechanism for the transport of flux is the movement of the molecule itself through a plane. In a dense system, on the other hand, it is possible for a molecule on one side of a plane to collide with another molecule on the other side of a plane and transfer some momentum or energy across the plane even though neither molecule itself crosses through the plane. This manner of transfer of flux is called collisional transfer. Of course, this mechanism is present in dilute gases also, but it does not become important until the density becomes large enough. In liquids, in fact, collisional transfer of flux is more important than molecular transfer. Two main factors, then, distinguish the Enskog theory from the Boltzmann equation; the frequency of collisions in a dense system is greater than in a dilute gas, and collisional transfer of flux is important.

Recognizing these modifications, one derives an equation similar to the Boltzmann equation, called the Enskog equation, which is then solved by the Chapman-Enskog method. We shall simply summarize the results of the Enskog theory. Following Hirschfelder, Curtiss, and Bird, we introduce a parameter $Y = b_0 \rho g(\sigma) = (p/\rho kT) - 1$. In terms of Y, we have

$$\frac{\eta}{\eta^0 b_0 \rho} = \frac{1}{Y} + 0.800 + 0.761\,Y \tag{19-42}$$

$$\frac{\kappa}{\eta^0 b_0 \rho} = 1.002\,Y \tag{19-43}$$

$$\frac{\lambda}{\lambda^0 b_0 \rho} = \frac{1}{Y} + 1.20 + 0.755\,Y \tag{19-44}$$

$$\frac{D}{D^0 b_0 \rho} = \frac{1}{Y} \tag{19-45}$$

The molecular dynamics rigid sphere transport coefficients have recently been calculated by Alder, Gass, and Wainwright.* Their results are given in Table 19-8. They identify the first terms in Eqs. (19-42) and (19-44) with the kinetic contribution, the third terms with the potential contribution, and the middle terms with a cross term. These are all presented separately in Table 19-8. One can see from there that the agreement with "experiment" is quite good except for densities approaching the solid phase ($v/v_0 \approx 1.5$). (See Fig. 13-2.) Alder, Gass, and Wainwright give a thorough discussion of the results of the Enskog theory.

The Enskog theory has been extended in a number of ways. For example, Thorne† has extended the Enskog theory to binary hard-sphere mixtures, and Tham and

---

* B. J. Alder, D. M. Gass, and T. E. Wainwright, *J. Chem. Phys.*, **53**, p. 3813, 1970.

† S. Chapman and T. G. Cowling, *The Mathematical Theory of Non-Uniform Gases* (Cambridge: Cambridge University Press, 1939); L. Barajas, L. S. Garcia-Colin, and E. Piña, *J. Stat. Phys.*, **7**, p. 161, 1973.

Table 19–8. A comparison of the Enskog theory with the molecular dynamics results of Alder, Gass, and Wainwright*

$1/v_0$	D/D_E	η/η_E	η^K/η_E^K	η^C/η_E^C	η^P/η_E^P	λ/λ_E	λ^K/λ_E^K	λ^C/λ_E^C	λ^P/λ_E^P	κ/κ_E
100	1.02	1.01 ± 0.02	1.01 ± 0.02	1.05 ± 0.06	1.07 ± 0.04	0.98 ± 0.02	0.98 ± 0.02	0.99 ± 0.03	1.00 ± 0.02	0.9
20	1.03	1.00 ± 0.02	1.00 ± 0.02	1.00 ± 0.03	1.01 ± 0.02	0.99 ± 0.02	1.00 ± 0.02	0.95 ± 0.03	1.03 ± 0.03	1.0
10	1.03	0.99 ± 0.04	1.00 ± 0.04	0.95 ± 0.04	0.97 ± 0.02					
5	1.09	0.99 ± 0.05	1.05 ± 0.05	0.90 ± 0.07	0.98 ± 0.07	0.97 ± 0.03	0.94 ± 0.02	0.96 ± 0.04	1.03 ± 0.04	0.9
3	1.22	1.02 ± 0.01	1.06 ± 0.01	1.01 ± 0.03	1.01 ± 0.01	1.00 ± 0.01	1.00 ± 0.01	0.96 ± 0.01	1.02 ± 0.01	0.98 ± 0.07
2	1.14	1.10 ± 0.04	1.13 ± 0.03	0.93 ± 0.05	1.13 ± 0.05	1.07 ± 0.03	1.11 ± 0.04	1.04 ± 0.05	1.07 ± 0.03	1.2
1.8	0.95	1.18 ± 0.03	1.08 ± 0.03	0.73 ± 0.04	1.16 ± 0.03	1.03 ± 0.02	0.99 ± 0.02	1.00 ± 0.03	1.03 ± 0.03	1.1
1.6	0.76	1.44 ± 0.07	1.08 ± 0.02	0.67 ± 0.08	1.54 ± 0.08	1.05 ± 0.02	1.00 ± 0.03	1.03 ± 0.03	1.05 ± 0.02	1.1
1.5	0.55	2.24 ± 0.09	1.16 ± 0.05	0.61 ± 0.13	2.40 ± 0.09	1.07 ± 0.03	1.04 ± 0.03	1.05 ± 0.04	1.07 ± 0.03	

* The columns show the ratio of the molecular dynamics results with the Enskog theory. The superscripts K, C, and P designate the kinetic, cross, and potential terms, respectively, in Eqs. (19–42) and (19–44).

Source: B. J. Alder, D. M. Gass, and T. E. Wainwright, J. Chem. Phys., **53**, p. 3813, 1970.

Gubbins* have published the multicomponent theory. In addition, Gass† has applied the Enskog theory to a fluid of rigid disks (i.e., a two-dimensional fluid), and Davis, Rice, and Sengers‡ have extended it to a square-well fluid.

In spite of the fact that the Enskog theory is strictly a rigid sphere theory, Enskog showed how the results could be applied in an ad hoc manner to real systems. He suggested that instead of relating Y to the actual pressure of the system, i.e.,

$$Y = \frac{p}{\rho kT} - 1 \tag{19-46}$$

that one should introduce the so-called thermal pressure, $T(\partial p/\partial T)_V$. The justification for this is that the pressure experienced by a single molecule is made up to two parts: the external pressure p due to the walls of the container and the "internal pressure" $(\partial E/\partial V)_T$, which represents the force of cohesion of the molecules. The sum of these is related to the thermal pressure by the thermodynamic relation

$$\left(\frac{\partial E}{\partial V}\right)_T + p = T\left(\frac{\partial p}{\partial T}\right)_V \tag{19-47}$$

Therefore, we write for Y:

$$Y = \frac{1}{\rho kT}\left[T\left(\frac{\partial p}{\partial T}\right)_V\right] - 1 \tag{19-48}$$

Furthermore, since $Y/\rho \to b_0$ as $\rho \to 0$, we must have

$$b_0 = B(T) + T\frac{dB}{dT}$$

where $B(T)$ is the second virial coefficient.

Equations (19–47) and (19–48), along with Eqs. (19–42) through (19–45), and the dilute gas expressions for η^0, λ^0, and D^0 given in Section 19–2 give us a complete approximate modification of the Enskog rigid sphere theory. One can calculate the transport coefficients of a dense real system from equation-of-state data alone. Tables 19–9 and 19–10 give an illustration of Enskog's semiempirical extension of his rigid

Table 19–9. **The viscosity of nitrogen at 50° C as a function of pressure according to the modified Enskog theory**

p (atm)	ρ (g/cm³)	$\left(\dfrac{\eta}{\rho}\right)_{exp} \times 10^7$ (cm²/sec)	$\eta \times 10^7$ (g · cm^{-1} · sec^{-1}) experimental	calculated
15.37	0.01623	117,900	1,913	1,810
57.60	0.06049	32,740	1,981	1,900
104.5	0.1083	19,280	2,088	2,050
212.4	0.2067	11,480	2,373	2,240
320.4	0.2875	9,520	2,737	2,660
430.2	0.3528	8,370	3,129	3,080
541.7	0.4053	8,660	3,509	3,480
630.4	0.4409	8,590	3,786	3,800
742.1	0.4786	8,700	4,163	4,180
854.1	0.5117	8,890	4,550	4,550
965.8	0.5404	9,090	4,913	4,920

Source: J. O. Hirschfelder, C. F. Curtiss, and R. B. Bird, *Molecular Theory of Gases and Liquids* (New York: Wiley, 1954).

* M. K. Tham and K. E. Gubbins, *J. Chem. Phys.*, **55**, p. 268, 1971.

† D. M. Gass, *J. Chem. Phys.*, **54**, p. 1898, 1971.

‡ H. T. Davis, S. A. Rice, and J. V. Sengers, *J. Chem. Phys.*, **35**, p. 2210, 1971. See also J. B. Schrodt, J. S. Ku, and K. D. Luks, *J. Chem. Phys.*, **57**, p. 4589, 1972.

Table 19–10. **The viscosity of carbon dioxide at 40.3°C as a function of pressure according to the modified Enskog theory**

p (atm)	ρ (g/cm³)	$\left(\dfrac{\eta}{\rho}\right)_{exp}$ × 10⁷ (cm²/sec)	$\eta \times 10^7$ (g · cm⁻¹ · sec⁻¹) experimental	calculated
45.3	0.100	18,000	1,800	1,910
64.3	0.170	11,500	1,960	1,990
75.9	0.240	9,080	2,180	2,160
82.7	0.310	7,840	2,430	2,400
86.8	0.380	7,240	2,750	2,730
89.2	0.450	7,020	3,160	3,150
91.7	0.520	7,040	3,660	3,670
94.9	0.590	7,220	4,260	4,270
101.6	0.660	7,560	4,990	4,980
114.6	0.730	7,950	5,800	5,780

Source: J. O. Hirschfelder, C. F. Curtiss, and R. B. Bird, *Molecular Theory of Gases and Liquids* (New York: Wiley, 1954).

sphere theory. One can see from the results that in spite of the approximate nature of the theory, the agreement with experiment is quite good when one compares the density variation of the transport coefficients at a fixed temperature.

Hanley, McCarty, and Cohen[*] have made a thorough comparison of the modified Enskog theory to experimental data. Specifically, experimental data and theoretical predictions for the first density corrections to the viscosity and thermal conductivity were examined, and the temperature and density dependences of the experimental and theoretical transport coefficients in the liquid were studied. Overall, the modified Enskog theory, with the exception of the critical region for the thermal conductivity, was found to give reasonable agreement with experiment (to within about 10 to 15 percent) for densities generally not exceeding twice the critical density. This reference is an excellent discussion of the modified Enskog theory and experimental transport data.

Ely and McQuarrie[†] have combined the modified Enskog theory with one of the statistical mechanical perturbation theory equations of state, thus eliminating the need for experimental p–V–T data. They arbitrarily chose the Barker-Henderson equation of state, which, under the macroscopic compressibility approximation, reads [Eq. (14–32)]

$$\frac{A}{NkT} = \frac{A_0}{NkT} + 2\pi\rho\beta \int_\sigma^\infty g_d(\eta, r)u(r)r^2\, dr - \pi\rho\beta\left(\frac{\partial\rho}{\partial p}\right)_0 \frac{\partial}{\partial\rho} \int_\sigma^\infty g_d(\eta, r)u^2(r)r^2\, dr$$

where A is the Helmholtz free energy, $g_d(\eta, r)$ is the radial distribution function of a system of hard spheres of diameter d, $\eta = \pi\rho d^3/6$,

$$d(T) = -\int_0^\sigma [e^{-u(r)/kT} - 1]\, dr$$

For the hard-sphere contribution to A, they use the Carnahan-Starling equation of state:

$$Z_0 = \frac{(1 + \eta + \eta^2 - \eta^3)}{(1 - \eta)^4}$$

[*] H. J. M. Hanley, R. D. McCarty, and E. G. D. Cohen, *Physica*, **60**, p. 322, 1972.
[†] J. F. Ely and D. A. McQuarrie, *J. Chem. Phys.*, **60**, p. 4105, 1974.

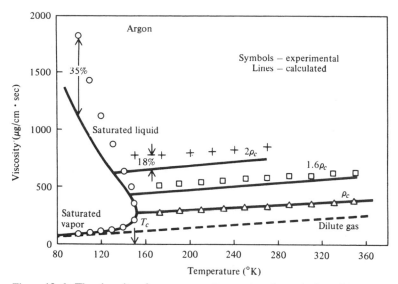

Figure 19–6. **The viscosity of argon versus temperature for a number of densities.** (From J. F. Ely and D. A. McQuarrie, *J Chem. Phys.*, **60**, 4105, 1974.)

It is then straightforward to use this equation of state to calculate Y from Eq. (19–48) and to use this value of Y to calculate the transport coefficients according to Eqs. (19–42) through (19–45). The results are shown in Fig. 19–6, where the viscosity is plotted versus temperature for argon at a number of densities. The results are quite good for temperatures above critical (150.86°K) and densities below approximately twice critical ($\rho_c \approx 0.52$ g/cm^3). Similar results are found for the thermal conductivity.

Regardless of the success of this approach, it is really an ad hoc theory and does not arise from a rigorous starting point. Consequently, it cannot be improved in an organized procedure of successively better approximations. It is still desirable, therefore, to develop a molecular theory of transport in dense fluids which begins with the Liouville equation and reduces it to a calculable form by means of a series of well-defined approximations. We shall present such an attempt in Chapters 21 and 22.

ADDITIONAL READING

BARKER, J. A., FOCK, W., and SMITH, F. *Phys. Fluids*, **7**, p. 897, 1964.

BOGOLIUBOV, N. N. *Stud. Stat. Mech.*, **1**, p. 1, 1962.

CHAPMAN, S., and COWLING, T. G. 1960. *The mathematical theory of non-uniform gases.* Cambridge: Cambridge University Press.

COHEN, E. G. D., ed. 1962. *Fundamental problems in statistical mechanics.* Amsterdam: North-Holland Publ.

CURTISS, C. F. *Ann. Rev. Phys. Chem.*, **18**, p. 125, 1967.

ERNST, M. H., HAINES, L. K., and DORFMAN, J. R. *Rev. Mod. Phys.*, **41**, p. 296, 1969.

FITTS, D. D. *Ann. Rev. Phys. Chem.*, **17**, p. 59, 1966.

HANLEY, H. J. 1969. *Transport phenomena in fluids*, New York: Dekker.

———, and KLEIN, M., *J. Chem. Phys.*, **50**, p. 4765, 1969.

HIRSCHFELDER, J. O., CURTISS, C. F., and BIRD, R. B. 1954. *Molecular theory of gases and liquids.* New York: Wiley.

HOFFMAN, D. K., and DAHLER, J. S. *J. Stat. Phys.*, **1**, p. 521, 1969.

MASON, E. A. *J. Chem. Phys.*, **22**, p. 169, 1954.

MONCHICK, L., and MASON, E. A. *J. Chem. Phys.*, **35**, p. 1676, 1961.

SMITH, F. J., MASON, E. A., and MUNN, R. J. *J. Chem. Phys.*, **42**, p. 1334, 1965.

WANG CHANG, C. S., and UHLENBECK, G. E. *Stud. Stat. Mech.*, **5**, p. 1, 1970.

———, UHLENBECK, G. E., and DE BOER, J. *Stud. Stat. Mech.*, **2**, p. 243, 1964.

PROBLEMS

19–1. Substitute the Chapman-Enskog expansion

$$f = \frac{1}{\xi}f^{[0]} + f^{[1]} + \xi f^{[2]} + \cdots$$

into the Boltzmann equation and derive the equations for the $f^{[J]}$.

19–2. Prove that

$$f^{[0]}(\mathbf{r}, \mathbf{v}, t) = \rho(\mathbf{r}, t)\left[\frac{m}{2\pi kT(\mathbf{r}, t)}\right]^{3/2} \exp\left\{-\frac{m[\mathbf{v} - \mathbf{v}_0(\mathbf{r}, t)]^2}{2kT(\mathbf{r}, t)}\right\}$$

19–3. Defining ϕ through $f^{[1]} = \phi f^{[0]}$, show that it is determined by Eq. (19–4).

19–4. Derive Eq. (19–5).

19–5. Prove that $I(\phi)$ is a linear integral operation.

19–6. Show that

$$\mathbf{p}^{[0]} = m \int \mathbf{V}\mathbf{V} f^{[0]} \, d\mathbf{v}$$

$$= \rho kT\mathbf{I}$$

19–7. Show that

$$\mathbf{q}^{[0]} = \frac{m}{2} \int V^2 \mathbf{V} f^{[0]} \, d\mathbf{v} = 0$$

19–8. Prove that when the above expression for $\mathbf{p}^{(0)}$ and $\mathbf{q}^{(0)}$ are substituted into the equations of change [Eqs. (18–51), (18–52), and (17–31)], one gets the ideal hydrodynamic equations.

19–9. Derive Eqs. (19–9) and (19–10).

19–10. By comparing $\mathbf{q}^{(0)} + \mathbf{q}^{(1)}$ to Fourier's law, show that

$$\lambda = -\frac{2}{3}\frac{k^2 T}{m\rho} \int A(W)W^2(W^2 - \tfrac{5}{2})f^{[0]} \, d\mathbf{v}$$

19–11. By comparing $\mathbf{p}^{(0)} + \mathbf{p}^{(1)}$ to the Newtonian stress tensor, show that

$$\eta = -\frac{kT}{5\rho} \int f^{[0]}B(W)(\mathbf{W}\mathbf{W} - \tfrac{1}{3}W^2\mathbf{I}):(\mathbf{W}\mathbf{W} - \tfrac{1}{3}W^2\mathbf{I}) \, d\mathbf{v}$$

19–12. Prove that

$$\chi(g, b) = 2\arccos\left(\frac{b}{\sigma}\right) \qquad b \leq \sigma$$

$$= 0 \qquad\qquad b \geq \sigma$$

for hard spheres.

19–13. Prove that

$$\Omega^{(l,s)}_{\text{hard spheres}} = \left(\frac{kT}{2\pi\mu}\right)^{1/2}\frac{(s+1)!}{2}Q^{(l)}_{\text{hard spheres}}$$

and

$$Q^{(l)}_{\text{hard spheres}} = \left[1 - \frac{1}{2}\frac{1 + (-)^l}{1 + l}\right]\pi\sigma^2$$

19–14. Derive Eqs. (19–24) through (19–26).

19–15. Show that

$$\Omega^{(l,s)*} = \frac{\Omega^{(l,s)}}{\Omega_{HS}^{(l,s)}}$$

is a function only of the reduced temperature T^*.

19–16. Derive Eqs. (19–31) through (19–33).

19–17. Given the equation

$$\lambda = \frac{75k^2T}{8m}\left[\frac{1}{a_{11}} + \frac{a_{12}^2/a_{11}}{a_{11}a_{22} - a_{12}^2} + \cdots\right]$$

Show that the second term here is much smaller than the first term. Use the 6–12 potential and several reduced temperatures, e.g., $T^* = 0.50$, 1.00, 2.00, and 4.00.

19–18. Given the equation

$$\eta = \frac{5kT}{2}\left[\frac{1}{b_{11}} + \frac{b_{12}^2/b_{11}}{b_{11}b_{22} - b_{12}^2} + \cdots\right]$$

Show that the second term here is much smaller than the first term. Use the 6–12 potential and several reduced temperatures, e.g., $T^* = 0.05$, 1.00, 2.00, and 4.00.

19–19. Hirschfelder, Curtiss, and Bird ("Additional Reading") give the formula [their Eq. (8.2–18)]

$$\eta \times 10^7 = 266.93 \, \frac{(MT)^{1/2}}{\sigma^2 \Omega^{(2,2)*}(T^*)}$$

where

$$\eta = \text{viscosity in poises}$$
$$T = \text{temperature in } {}^\circ K$$
$$T^* = \frac{kT}{\varepsilon}$$
$$M = \text{molecular weight}$$
$$\sigma = \text{collision diameter in Å}$$

Verify this result

19–20. Use the formula of the previous problem to calculate the viscosity of argon from 200°K to 500°K. Compare your results to experimental data (cf. Hirschfelder, Curtiss, and Bird, in "Additional Reading," Table 8.4–2).

19–21. Hirschfelder, Curtiss, and Bird ("Additional Reading") give the formula [their Eq. (8.2–31)]

$$\lambda \times 10^7 = 1989.1 \, \frac{(T/M)^{1/2}}{\sigma^2 \Omega^{(2,2)*}(T^*)}$$

where

$$\lambda = \text{thermal conductivity in cal/cm} \cdot \text{sec} \cdot {}^\circ K$$
$$M = \text{molecular weight}$$

Verify this formula.

19–22. Use the formula of the preceding problem to calculate the thermal conductivity of argon from 100°K to 600°K. Compare your results to experimental data (cf. Hirschefelder, Curtiss, and Bird in "Additional Reading," Table 8.4–9).

19–23. Hirschfelder, Curtiss, and Bird ("Additional Reading") give the formula [their Eq. (8.2–46)].

$$D = 0.002628 \, \frac{(T^3/M)^{1/2}}{p\sigma^2 \Omega^{(1,1)*}(T^*)}$$

where

D = coefficient of self-diffusion in cm²/sec

p = pressure in atmospheres

M = molecular weight

Verify this formula.

19–24. Use the formula in the preceding problem to calculate the coefficient of self-diffusion of argon at 1 atm and for $100°\text{K} \leq T \leq 350°\text{K}$. Compare your results to experimental data (cf. Hirschfelder, Curtiss, and Bird in "Additional Reading", Table 8.4–13).

19–25. Show that the collision integrals obey the recursion formula

$$\Omega^{(l,s+1)} = T \, \frac{\partial \Omega^{(l,s)}}{\partial T} + (s + \tfrac{3}{2}) \Omega^{(l,s)}$$

19–26. Show that the reduced collision integrals, $\Omega^{(l,s)*} = \Omega^{(l,s)}/\Omega_{\text{HS}}^{(l,s)}$ obey the recursion formula

$$\Omega^{(l,s+1)*} = \Omega^{(l,s)*} + \frac{1}{s+2} T^* \, \frac{\partial \Omega^{(l,s)*}}{\partial T^*}$$

19–27. Hattikudur and Thodos* give approximate, simple formulas for $\Omega^{(1,1)}(T^*)$ and $\Omega^{(2,2)}(T^*)$ for the 6–12 potential. Show that these formulas do, indeed, reproduce the values in Table 19–5. Now using the relation in Problem 19–25, derive approximate formulas for $\Omega^{(1,2)}$ and $\Omega^{(2,3)}$ and compare these to Tables in Hirschfelder, Curtiss, and Bird for $T^* = 0.3$, 1.00, 2.00, 10.00, and 100.0.

19–28. Prove that the reduced viscosity is given by

$$\eta^* = \frac{\eta \sigma^2}{(m\varepsilon)^{1/2}}$$

19–29. Prove that the reduced thermal conductivity is given by

$$\lambda^* = \frac{\lambda \sigma^2 m^{1/2}}{\varepsilon^{1/2}}$$

19–30. Prove that the reduced self-diffusion coefficient is given by

$$D^* = \frac{D m^{1/2}}{\sigma \varepsilon^{1/2}}$$

19–31. Using the m–6–8 parameters given in Table 19–7 and the collision integrals given in Appendix E, calculate the viscosity and thermal conductivity of argon from $100°\text{K}$ to $1000°\text{K}$. Compare your results to those of the Lennard-Jones 6–12 potential, using the parameters given in Table 19–6.

Calculate the second virial coefficient from both of these potentials and compare. (Use the *viscosity* Lennard-Jones parameters in Table 19–6.)

19–32. Calculate the viscosity, thermal conductivity, and self-diffusion coefficient of argon at STP and compare your results to experimental data. (See Hirschfelder, Curtiss, and Bird in "Additional Reading," pp. 561, 575, and 581.)

* *J. Chem. Phys.*, **52**, p. 4313, 1970.

19–33. Calculate the viscosity of methane at 300°K and compare your results to experimental data.

19–34. Calculate the viscosity of carbon dioxide at 500°K and compare your result to experimental data.

19–35. Calculate the thermal conductivity of krypton at 0°C and compare your result to experimental data.

19–36. Calculate the coefficient of self-diffusion of argon at 0°C and 1 atm and compare your result to the experimental value of 0.157 cm²/sec.

19–37. Calculate the diffusion coefficient for the gas-pair nitrogen-oxygen at 0°C and 1 atm. Compare your result to the experimental value of 0.181 cm²/sec. (For the mixed interaction parameters σ_{12} and ε_{12}, use the approximate combining rule,

$$\sigma_{12} = \tfrac{1}{2}(\sigma_1 + \sigma_2)$$

$$\varepsilon_{12} = (\varepsilon_1 \varepsilon_2)^{1/2}$$

19–38. In Problem 16–19, we introduced the Eucken correction factor to the thermal conductivity. This factor accounts for the fact that a polyatomic molecule carries energy in its internal modes as well as in its translational modes. The thermal conductivity is modified according to

$$\lambda = \frac{15}{4} \frac{R}{M} \eta \left(\frac{4}{15} \frac{C_v}{R} + \frac{3}{5} \right)$$

Note that the factor in parentheses is unity for a monatomic gas.

Use this expression to calculate the thermal conductivity of carbon dioxide and methane at 300°K and compare your results to experimental data (cf. Hirschfelder, Curtiss, and Bird in "Additional Reading," Table 8.4–10.)

19–39. Read Section 8.2b of Hirschfelder, Curtiss, and Bird ("Additional Reading") and calculate the viscosity of a binary mixture of argon and neon, verifying the calculated results in their Table 8.4–5. In order to do this, one needs to know the mixed interaction parameters ε_{12} and σ_{12}. For these, assume that

$$\sigma_{12} = \tfrac{1}{2}(\sigma_1 + \sigma_2)$$

$$\varepsilon_{12} = (\varepsilon_1 \varepsilon_2)^{1/2}$$

19–40. Using the collision integrals given in Table 19–5, calculate the quantities f_λ and f_η, through the second term and compare your results to the three-term expression for f_λ and f_η given in Table 19–1.

19–41. Using the collision integrals in Table 19–5 and the extended formulas for f_λ and f_η given either in Hirschfelder, Curtiss, and Bird ("Additional Reading") or Appendix I of E. A. Mason,* verify the entries in Table 19–1.

19–42. Verify the entries in Table 19–3 for the viscosity and thermal conductivity. Use the extended formulas for f_λ and f_η given in Appendix 8A of Hirschfelder, Curtiss, and Bird ("Additional Reading") or Appendix 1 of E. A. Mason,*

19–43. Kihara (cf. Hirschfelder, Curtiss, and Bird in "Additional Reading," Appendix 8A) has developed alternate expressions for the higher approximations to the transport coefficients. Unlike Chapman and Cowling, who express the transport coefficients in terms of ratios of infinite determinants which are then truncated, Kihara notes that the off-diagonal

---

* *J. Chem. Phys.*, **22**, p. 169, 1954.

elements in these determinants are small, and so he expands the infinite determinants in powers of the off-diagonal terms. For pure gases, Kihara obtains the results

$$f_\eta^{(\text{Kihara})} = 1 + \frac{3}{49} \left[\frac{4\Omega^{(2,3)*}}{\Omega^{(2,2)*}} - \frac{7}{2} \right]^2$$

$$f_\lambda^{(\text{Kihara})} = 1 + \frac{2}{21} \left[\frac{4\Omega^{(2,3)*}}{\Omega^{(2,2)*}} - \frac{7}{2} \right]^2$$

and

$$f_D^{(\text{Kihara})} = 1 + \frac{(6C^* - 5)^2}{16A^* + 40}$$

where

$$A^* = \frac{\Omega^{(2,2)*}}{\Omega^{(1,1)*}}$$

and

$$C^* = \frac{\Omega^{(1,2)*}}{\Omega^{(1,1)*}}$$

Evaluate these correction factors and compare your results to Table 19–1.

19–44. Joshi[*] has developed a procedure for evaluating the correction factors f_η and f_λ, in which the actual intermolecular potential energy function is considered to be a perturbation over the rigid spherical one. Use Eqs. 6 and 7 of Joshi's paper and calculate some of the entries in his Table 1.

19–45. Show that the exp-6 potential has a spurious maximum (at r_{max}) and then approaches $-\infty$ as $r \to 0$. How would you remedy this in doing calculations with this potential?[†]

19–46. Consider the integral equation

$$\int_a^b K(x, y) f(x) \, dx = h(y)$$

where $K(x, y)$ and $h(y)$ are known functions and $f(x)$ is to be determined. Now expand both $f(x)$ and $h(y)$ in a set of functions that is orthonormal in the interval (a, b):

$$f(x) = \sum_j c_j \phi_j(x)$$
$$h(y) = \sum_j h_j \phi_j(y)$$

Since $f(x)$ is not known, the c_j are as yet undetermined, but the h_j are given by

$$h_j = \int_a^b h(y) \phi_j(y) \, dy$$

Multiply the integral equation by $\phi_j(y)$ and integrate; substitute the expansion for $f(x)$ into the result; and derive the infinite set of algebraic equations

$$h_j = \sum_j c_i a_{ij}$$

where

$$a_{ij} = \int_a^b \int K(x, y) \phi_i(x) \phi_j(y) \, dx \, dy$$

---

[*] *Chem. Phys. Lett.*, **1**, 575, 1968.
[†] Compare E. A. Mason, *J. Chem. Phys.*, **22**, p. 169, 1954.

19–47. Repeat the previous problem, but this time assume that $f(x)$ and $h(y)$ can be represented by a *finite* number of terms

$$f(x) = \sum_{j=0}^{N} c_j \phi_j(x)$$

$$h(y) = \sum_{j=0}^{N} h_j \phi_j(y)$$

Show how this leads to a *finite* set of algebraic equations.

CHAPTER 20

THEORY OF BROWNIAN MOTION

The last few chapters have treated the transport theory of dilute gases. We have been able to derive essentially exact relations between the various transport coefficients and the intermolecular potential. A number of the more recent and successful theories of transport in dense fluids are based upon the ideas of Brownian motion, and so in this chapter we shall digress into the theory of Brownian motion. This is a rich and beautiful field in itself. It was developed many years ago by Einstein, Planck, Smoluchowski, and others, and much of this chapter is based on the classic review article by Chandrasekhar ("Additional Reading").

20-1 THE LANGEVIN EQUATION

In 1827, the botanist Robert Brown discovered that small pollen grains immersed in a fluid undergo a perpetual irregular motion. This irregular motion is now known to be due to incessant collisions of the pollen particle with the molecules of the surrounding fluid. To an observer, the trajectory of such a Brownian particle will appear as an irregular, random path. We assume that the force on such a particle can be split up into two parts. The first is a frictional force due to the drag exerted on the particle by the fluid. If $\mathbf{u}$ represents the velocity of the particle, then this force is assumed to be of the form given by Stokes' law, i.e., $-\gamma'\mathbf{u}$ where γ' is the friction constant. γ' is given by $6\pi a\eta$, where a is the radius of the particle and η is the viscosity of the medium. The second part of the total force acting upon a Brownian particle is assumed to be the fluctuating force, $\mathbf{A}'(t)$. This random force represents the constant molecular bombardment exerted by the surrounding fluid. We furthermore assume that $\mathbf{A}'(t)$ is independent of $\mathbf{u}$ and that $\mathbf{A}'(t)$ varies extremely rapidly compared to the variations in $\mathbf{u}$. This second assumption says that time intervals Δt exist such even though $\mathbf{u}(t + \Delta t) - \mathbf{u}(t)$ is expected to be very small, no correlation between $\mathbf{A}'(t)$ and $\mathbf{A}'(t + \Delta t)$ exists;

i.e., $\mathbf{A}'(t)$ will have undergone many fluctuations. The *Langevin equation* is simply the equation of motion for such a Brownian particle, namely,

$$m \frac{d\mathbf{u}}{dt} = -\gamma'\mathbf{u} + \mathbf{A}'(t)$$

or more conventionally,

$$\frac{d\mathbf{u}}{dt} = -\zeta\mathbf{u} + \mathbf{A}(t) \tag{20-1}$$

where $\zeta = \gamma'/m$ and $\mathbf{A}(t) = \mathbf{A}'(t)/m$. A differential equation such as Eq. (20–1), where one of the terms is a randomly varying function, is called a *stochastic differential equation*.

We must now solve this equation. What we mean by solving a stochastic equation is to find the *probability* that the solution is $\mathbf{u}$ at the time t, given that $\mathbf{u} = \mathbf{u}_0$ at $t = 0$. This is given by the probability density function $W(\mathbf{u}, t; \mathbf{u}_0)$. Clearly we require of W that

$$W(\mathbf{u}, t; \mathbf{u}_0) \rightarrow \delta(u_x - u_{x0})\delta(u_y - u_{y0})\delta(u_z - u_{z0}) \qquad t \rightarrow 0 \tag{20-2}$$

Furthermore, we expect that as $t \rightarrow \infty$, the particle must be in equilibrium at the temperature of the surrounding fluid, and so $W(\mathbf{u}, t; \mathbf{u}_0)$ must approach a Maxwellian distribution, *independently* of $\mathbf{u}_0$, as $t \rightarrow \infty$; i.e.,

$$W(\mathbf{u}, t; \mathbf{u}_0) \rightarrow \left(\frac{m}{2\pi kT}\right)^{3/2} \exp\left(-\frac{mu^2}{2kT}\right) \qquad t \rightarrow \infty \tag{20-3}$$

Equation (20–1) is a first-order linear differential equation and can be solved *formally* to give (Problem 20–1)

$$\mathbf{U} \equiv \mathbf{u}(t) - \mathbf{u}_0 e^{-\zeta t} = e^{-\zeta t} \int_0^t e^{\zeta\xi}\mathbf{A}(\xi)\, d\xi \tag{20-4}$$

This shows that the statistical properties of $\mathbf{u} - \mathbf{u}_0 e^{-\zeta t}$ must be the same as

$$e^{-\zeta t} \int_0^t e^{\zeta\xi}\mathbf{A}(\xi)\, d\xi$$

If we assume that the ensemble average of $\mathbf{A}(t)$ is zero, then taking the ensemble average of both sides of Eq. (20–4) gives

$$\langle\mathbf{u}\rangle = \mathbf{u}_0 e^{-\zeta t} \tag{20-5}$$

We can also readily find $\langle\mathbf{U}^2\rangle$ by squaring both sides of Eq. (20–4) and then ensemble averaging both sides to get (Problem 20–2)

$$\langle\mathbf{U}^2\rangle = \langle u^2\rangle - u_0^2 e^{-2\zeta t} = e^{-2\zeta t}\int_0^t\int_0^t e^{\zeta(t'+t'')}\langle\mathbf{A}(t') \cdot \mathbf{A}(t'')\rangle\, dt'\, dt'' \tag{20-6}$$

Now $\langle\mathbf{A}(t') \cdot \mathbf{A}(t'')\rangle$ represents the correlation of $\mathbf{A}$ at time t' with $\mathbf{A}$ at time t''. Since $\mathbf{A}$ is a rapidly varying function, we assume that $\langle\mathbf{A}(t') \cdot \mathbf{A}(t'')\rangle$ is a function only of $|t' - t''|$ and is nonzero only when $|t' - t''|$ is small; i.e., we assume that

$$\langle\mathbf{A}(t') \cdot \mathbf{A}(t'')\rangle = \phi_1(|t' - t''|) \tag{20-7}$$

where ϕ_1 ($|t' - t''|$) is very peaked at $t' = t''$. We then let $t' + t'' = x$ and $t' - t'' = y$ in the integral in Eq. (20–6); we get (Problem 20–3)

$$\langle U^2 \rangle = \tfrac{1}{2} e^{-2\zeta t} \int_0^{2t} e^{\zeta x} \, dx \int_{-\infty}^{\infty} \phi_1(y) \, dy$$

The limits $\pm \infty$ have been used since $\phi_1(y)$ is a very rapidly decreasing function. We can let the integral of $\phi_1(y)$ be a constant τ to get

$$\langle U^2 \rangle = \frac{\tau}{2\zeta} (1 - e^{-2\zeta t}) \tag{20–8}$$

We can evaluate τ by realizing that equipartition must apply when $t \to \infty$. This gives

$$\frac{3kT}{m} = \frac{\tau}{2\zeta} \tag{20–9}$$

or

$$\langle U^2 \rangle = \frac{3kT}{m} (1 - e^{-2\zeta t}) \tag{20–10}$$

Using Eq. (20–5), we may write this equation in the form

$$\langle u^2 \rangle = \frac{3kT}{m} + \left(u_0{}^2 - \frac{3kT}{m} \right) e^{-2\zeta t} \tag{20–11}$$

which shows how $\langle u^2 \rangle$ approaches its equipartition value. It is actually possible, although fairly tedious, to continue this process and calculate all the moments of **U**. The result is (Problem 20–4)

$$\langle U^{2n+1} \rangle = 0$$
$$\langle U^{2n} \rangle = 1 \cdot 3 \cdot 5 \cdots (2n - 1) \langle U^2 \rangle$$

which shows that the probability distribution of U is Gaussian (Problem 20–5), or that

$$W(\mathbf{u}, t; \mathbf{u}_0) = \left[\frac{m}{2\pi kT(1 - e^{-2\zeta t})} \right]^{3/2} \exp \left[- \frac{m |\mathbf{u} - \mathbf{u}_0 e^{-\zeta t}|^2}{2kT(1 - e^{-2\zeta t})} \right] \tag{20–12}$$

It can be readily seen that this becomes a Maxwellian distribution independently of $\mathbf{u}_0$ as $t \to \infty$.

In the very same way we can go on to consider the probability distribution function of the displacement **r** instead of the velocity **u**. Since

$$\mathbf{r} - \mathbf{r}_0 = \int_0^t \mathbf{u}(t') \, dt'$$

we have, according to Eq. (20–4), that

$$\mathbf{r} - \mathbf{r}_0 = \int_0^t dt' \left[\mathbf{u}_0 e^{-\zeta t'} + e^{-\zeta t'} \int_0^{t'} dt'' \, e^{\zeta t''} \mathbf{A}(t'') \right]$$

or

$$\mathbf{r} - \mathbf{r}_0 - \zeta^{-1} \mathbf{u}_0 (1 - e^{-\zeta t}) = \int_0^t dt' \, e^{-\zeta t'} \int_0^{t'} dt'' \, e^{\zeta t''} \mathbf{A}(t'')$$

We can simplify the right-hand side of this equation by integrating by parts to give

$$\mathbf{r} - \mathbf{r}_0 - \zeta^{-1}\mathbf{u}_0(1 - e^{-\zeta t}) = \int_0^t \zeta^{-1}[1 - e^{-\zeta(t'-t)}]\mathbf{A}(t')\, dt' \tag{20-13}$$

By taking the ensemble average of both sides, we get (Problem 20–6)

$$\langle \mathbf{r} - \mathbf{r}_0 \rangle = \mathbf{u}_0 \frac{(1 - e^{-\zeta t})}{\zeta}$$

By squaring and then averaging Eq. (20–13) in the same way that we did before (Problem 20–7)

$$\langle |\mathbf{r} - \mathbf{r}_0|^2 \rangle = \frac{u_0^2}{\zeta^2}(1 - e^{-\zeta t})^2 + \frac{3kT}{m\zeta^2}(2\zeta t - 3 + 4e^{-\zeta t} - e^{-2\zeta t}) \tag{20-14}$$

Just as for $\mathbf{u} - \mathbf{u}_0 e^{-\zeta t}$, it is possible to show that $\mathbf{r} - \mathbf{r}_0 - \zeta^{-1}\mathbf{u}_0(1 - e^{-\zeta t})$ is Gaussianly distributed, i.e., that (Problem 20–8)

$$W(\mathbf{r}, t; \mathbf{r}_0, \mathbf{u}_0) = \left[\frac{m\zeta^2}{2k\pi T(2\zeta t - 3 + 4e^{-\zeta t} - e^{-2\zeta t})}\right]^{3/2}$$

$$\times \exp - \left[\frac{m\zeta^2 |\mathbf{r} - \mathbf{r}_0 - \zeta^{-1}\mathbf{u}_0(1 - e^{-\zeta t})|^2}{2kT(2\zeta t - 3 + 4e^{-\zeta t} - e^{-2\zeta t})}\right] \tag{20-15}$$

Actually, Eqs. (20–12) and (20–15) are special cases of the general theorem, proved in Chandrasekhar's review article, which states that if

$$\mathbf{R} = \int_0^t \Psi(\xi)\mathbf{A}(\xi)\, d\xi \tag{20-16}$$

then the probability distribution of $\mathbf{R}$ is given by

$$W(\mathbf{R}) = \frac{1}{\left[\frac{4\pi\zeta kT}{m}\int_0^t \Psi^2(\xi)\, d\xi\right]^{3/2}} \exp\left[-\frac{R^2}{\frac{4\zeta kT}{m}\int_0^t \Psi^2(\xi)\, d\xi}\right] \tag{20-17}$$

Equation (20–14) has two interesting limiting conditions. When t is very small, we have

$$\langle |\mathbf{r} - \mathbf{r}_0|^2 \rangle \to |u_0^2|t^2 \qquad t \to 0 \tag{20-18}$$

which simply says that

$$\mathbf{r} - \mathbf{r}_0 = \int_0^t \mathbf{u}_0\, dt$$

for small t. On the other hand, for large times, we have

$$\langle |\mathbf{r} - \mathbf{r}_0|^2 \rangle = \frac{6kT}{m\zeta}t = 6Dt \tag{20-19}$$

where we have introduced the diffusion constant, $D = kT/m\zeta$. This result was first derived by Einstein and later verified by Perrin in a famous series of experiments on Brownian motion. It has more recently been illustrated in molecular dynamics calculations by Rahman* on a system of particles obeying a Lennard-Jones 6–12 potential. This is shown in Fig. 20–1.

* A. Rahman, *Phys. Rev.*, **136A**, p. 405, 1964.

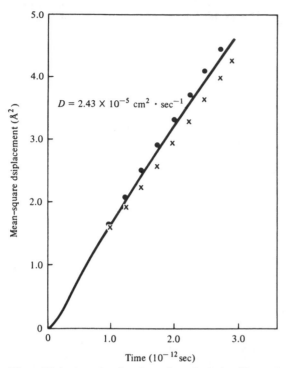

Figure 20–1. **A molecular dynamics calculation illustrating Eq. (20–19). Note now initially the mean-square displacement varies as t^2, but then goes into a linear dependence as t becomes larger.** (From A. Rahman, *Phys. Rev.*, **136A**, p. 405, 1964.)

So we see that for very short times the mean-square displacement goes as t^2, a result dictated by classical mechanics, but after a time long compared to ζ^{-1}, the statistical nature of the problem takes over, and the mean-square displacement goes linearly in time. Note the implications here concerning reversibility and irreversibility.

Equation (20–15) also has an interesting limit for $t \gg \zeta^{-1}$. In this limit we can ignore $\zeta^{-1}\mathbf{u}_0(1 - e^{-\zeta t})$ compared to $\mathbf{r} - \mathbf{r}_0$, which increases with increasing t, and get (Problem 20–10)

$$W(\mathbf{r}, t; \mathbf{r}_0, \mathbf{u}_0) \approx \frac{1}{(4\pi Dt)^{3/2}} \exp\left\{-\frac{|\mathbf{r} - \mathbf{r}_0|^2}{4Dt}\right\} \qquad t \gg \zeta^{-1} \tag{20–20}$$

This is the well-known solution to the diffusion equation, which goes to $\delta(\mathbf{r} - \mathbf{r}_0)$ as $t \to 0$ [cf. Eq. (17–53)]. In the next section we shall derive generalizations of the diffusion equation that give Eqs. (20–12) and (20–15) as their fundamental solutions. These "diffusion equations" will govern diffusion in velocity space and phase space itself rather than just diffusion in configuration space.

20–2 THE FOKKER-PLANCK EQUATION AND THE CHANDRASEKHAR EQUATION

We shall now show how to determine Eqs. (20–12) and (20–15) as solutions to appropriate boundary-value problems of certain partial differential equations. We are particularly interested in an equation governing $W(\mathbf{r}, \mathbf{u}, t)$, the density function in the phase space of a single particle since one of the more successful theories of transport in dense fluids approximates the reduced Liouville equations for $f^{(1)}$ and $f^{(2)}$ by diffusion equations in the six- and twelve-dimensional phase space of one and two particles.[*]

First, however, we shall derive a differential equation for just $W(\mathbf{u}, t)$. This differential equation, the diffusion equation in velocity space, is known as the Fokker-Planck equation.

Let Δt be a time interval long enough for $\mathbf{A}(t)$ to undergo many fluctuations but short enough so that $\mathbf{u}(t)$ does not change appreciably. As before, we assume that such a time interval exists. Under these circumstances, we should expect to derive $W(\mathbf{u}, t + \Delta t)$ from $W(\mathbf{u}, t)$ and a knowledge of the transition probability $\Psi(\mathbf{u}; \Delta \mathbf{u})$, the probability that $\mathbf{u}$ undergoes a change $\Delta \mathbf{u}$ in the interval Δt. In particular, we can write

$$W(\mathbf{u}, t + \Delta t) = \int W(\mathbf{u} - \Delta \mathbf{u}, t)\Psi(\mathbf{u} - \Delta \mathbf{u}; \Delta \mathbf{u})d(\Delta \mathbf{u}) \qquad (20\text{--}21)$$

Equation (20–21) is a fundamental equation in probability theory. It states that the course of the system at a time t depends only upon the instantaneous state of the system at the time t and is independent of its previous history. A probabilistic or stochastic process that has this characteristic is known as a *Markov process*. Equation (20–21) is the fundamental equation of Markov processes and is called the *Chapman-Kolmogorov equation*.

We now expand $W(\mathbf{u}, t + \Delta t)$, $W(\mathbf{u} - \Delta \mathbf{u}, t)$, and $\Psi(\mathbf{u} - \Delta \mathbf{u}; \Delta \mathbf{u})$ in Taylor series.

$$W(\mathbf{u}, t) + \frac{\partial W}{\partial t} \Delta t + O(\Delta t^2)$$

$$= \iiint_{-\infty}^{\infty} \left[W(\mathbf{u}, t) - \sum_j \frac{\partial W}{\partial u_j} \Delta u_j + \frac{1}{2} \sum_j \frac{\partial^2 W}{\partial u_j^2} \Delta u_j^2 + \sum_{i<j} \frac{\partial^2 W}{\partial u_i \, \partial u_j} \Delta u_i \, \Delta u_j + \cdots \right]$$

$$\times \left[\Psi(\mathbf{u}; \Delta \mathbf{u}) - \sum_i \frac{\partial \Psi}{\partial u_i} \Delta u_i + \frac{1}{2} \sum_i \frac{\partial^2 \Psi}{\partial u_i^2} \Delta u_i^2 + \sum_{i<j} \frac{\partial^2 \Psi}{\partial u_i \, \partial u_j} \Delta u_i \, \Delta u_j + \cdots \right]$$

$$\times d(\Delta u_1)d(\Delta u_2)d(\Delta u_3) \qquad (20\text{--}22)$$

In Eq. (20–22), $O(\Delta t)^2$ means terms such that $O(\Delta t^2)/\Delta t \to 0$ as $\Delta t \to 0$. Write

$$\langle \Delta u_i \rangle = \int_{-\infty}^{\infty} \Delta u_i \, \Psi(\mathbf{u}; \Delta \mathbf{u})d(\Delta \mathbf{u})$$

$$\langle \Delta u_i^2 \rangle = \int_{-\infty}^{\infty} \Delta u_i^2 \, \Psi(\mathbf{u}; \Delta \mathbf{u})d(\Delta \mathbf{u})$$

$$\langle \Delta u_i \, \Delta u_j \rangle = \int_{-\infty}^{\infty} \Delta u_i \, \Delta u_j \, \Psi(\mathbf{u}; \Delta \mathbf{u})d(\Delta \mathbf{u}) \qquad (20\text{--}23)$$

Using these definitions then, Eq. (20–22) becomes

$$\frac{\partial W}{\partial t} \Delta t + O(\Delta t^2) = -\sum_j \frac{\partial W}{\partial u_j} \langle \Delta u_j \rangle + \frac{1}{2} \sum_j \frac{\partial^2 W}{\partial u_j^2} \langle \Delta u_j^2 \rangle + \sum_{i<j} \frac{\partial^2 W}{\partial u_i \, \partial u_j} \langle \Delta u_i \, \Delta u_j \rangle$$

$$- \sum_j W \frac{\partial \langle \Delta u_j \rangle}{\partial u_j} + \sum_j \frac{\partial}{\partial u_j} \langle \Delta u_j^2 \rangle \frac{\partial W}{\partial u_j} + \sum_{i<j} \frac{\partial W}{\partial u_i} \frac{\partial \langle \Delta u_i \, \Delta u_j \rangle}{\partial u_j}$$

$$+ \frac{1}{2} \sum_j \frac{\partial^2}{\partial u_j^2} \langle \Delta u_j^2 \rangle W + \sum_{i<j} W \frac{\partial^2 \langle \Delta u_i \, \Delta u_j \rangle}{\partial u_i \, \partial u_j} + \cdots \qquad (20\text{--}24)$$

[*] S. A. Rice, and P. Gray, *The Statistical Mechanics of Simple Liquids* (New York: Interscience, 1965); R. M. Mazo, *Statistical Mechanical Theories of Transport Processes* (Oxford, Pergamon, 1967).

where the remaining terms involve quantities $\langle \Delta u_j{}^3 \rangle$, $\langle \Delta u_i \Delta u_j{}^2 \rangle$, and $\langle \Delta u_i \Delta u_j \Delta u_k \rangle$ and higher-order terms. Equation (20–24) can be written more conveniently in the form

$$\frac{\partial W}{\partial t} \Delta t + O(\Delta t^2) = -\sum_j \frac{\partial}{\partial u_j} (W \langle \Delta u_j \rangle) + \frac{1}{2} \sum_j \frac{\partial^2}{\partial u_j{}^2} (W \langle \Delta u_j{}^2 \rangle)$$

$$+ \sum_{i<j} \frac{\partial^2}{\partial u_i \, \partial u_j} (W \langle \Delta u_i \, \Delta u_j \rangle) + \cdots \quad (20\text{–}25)$$

This is the Fokker-Planck equation in its most general, albeit unusual, form. We must now evaluate the quantities in Eqs. (20–23). Equation (20–5) gives us $\langle \Delta u_i \rangle$. This may be readily seen by writing Eq. (20–5) in the form

$$\langle \Delta \mathbf{u} \rangle = \mathbf{u}(e^{-\zeta \Delta t} - 1)$$

to give (Problem 20–12)

$$\langle \Delta u_j \rangle = -\zeta u_j \, \Delta t + \cdots$$

Similarly, Eq. (20–11) gives (Problem 20–12)

$$\langle \Delta u_j{}^2 \rangle = \frac{2\zeta kT}{m} \Delta t + \cdots$$

and if we assume that different components of $\mathbf{A}(t)$ are uncorrelated, we have

$$\langle \Delta u_i \Delta u_j \rangle = 0 \qquad i \neq j$$

Equation (20–25) then can be written in the form

$$\frac{\partial W}{\partial t} \Delta t + O(\Delta t^2) = \left[\zeta \, \mathrm{div}_{\mathbf{u}}(W\mathbf{u}) + \frac{\zeta kT}{m} \nabla_{\mathbf{u}}{}^2 W \right] \Delta t + O(\Delta t^2)$$

which, in the limit $\Delta t \to 0$, becomes the Fokker-Planck equation:

$$\frac{\partial W}{\partial t} = \zeta \, \mathrm{div}_{\mathbf{u}}(W\mathbf{u}) + \frac{\zeta kT}{m} \nabla_{\mathbf{u}}{}^2 W \qquad (20\text{–}26)$$

It can be readily shown by substitution that Eq. (20–12) is the fundamental solution to this equation, i.e., the solution that tends to $\delta(\mathbf{u} - \mathbf{u}_0)$ as $t \to 0$ (Problem 20–13). This is the diffusion equation in velocity space. We are now ready to derive the diffusion equation in phase space.

Again let Δt be an interval of time long enough that $\mathbf{A}(t)$ undergoes many fluctuations but short enough that neither $\mathbf{u}$ nor $\mathbf{r}$ changes appreciably. Then we can write

$$\Delta \mathbf{r} = \mathbf{u} \, \Delta t$$

and

$$\Delta \mathbf{u} = -(\zeta \mathbf{u} - \mathbf{K}) \, \Delta t + \mathbf{B}(\Delta t) \qquad (20\text{–}27)$$

where $\mathbf{K}$ denotes an external force (per unit mass) and $\mathbf{B}(\Delta t)$ the net acceleration arising from fluctuations that occur in time Δt. Assuming that diffusion in phase space is a Markov process, we can write

$$W(\mathbf{r}, \mathbf{u}, t + \Delta t) = \iint W(\mathbf{r} - \Delta \mathbf{r}, \mathbf{u} - \Delta \mathbf{u}, t) \psi(\mathbf{r} - \Delta \mathbf{r}, \mathbf{u} - \Delta \mathbf{u}; \Delta \mathbf{r}, \Delta \mathbf{u}) d(\Delta \mathbf{r}) d(\Delta \mathbf{u})$$

$$(20\text{–}28)$$

Because of Eqs. (20–27) we write

$$\psi(\mathbf{r}, \mathbf{u}; \Delta\mathbf{r}, \Delta\mathbf{u}) = \Psi(\mathbf{r}, \mathbf{u}; \Delta\mathbf{u})\delta(\Delta\mathbf{r} - \mathbf{u}\,\Delta t) \tag{20–29}$$

where $\Psi(\mathbf{r}, \mathbf{u}; \Delta\mathbf{u})$ is the transition probability in velocity space. With this form for $\psi(\mathbf{r}, \mathbf{u}; \Delta\mathbf{r}, \Delta\mathbf{u})$, we can integrate over $\Delta\mathbf{r}$ in Eq. (20–28) to get

$$W(\mathbf{r}, \mathbf{u}, t) + \Delta t) = \int W(\mathbf{r} - \mathbf{u}\,\Delta t, \mathbf{u} - \Delta\mathbf{u}, t)\Psi(\mathbf{r} - \mathbf{u}\,\Delta t, \mathbf{u} - \Delta\mathbf{u}; \Delta\mathbf{u})d(\Delta\mathbf{u}) \tag{20–30}$$

or, equivalently,

$$W(\mathbf{r} + \mathbf{u}\,\Delta t, \mathbf{u}, t + \Delta t) = \int W(\mathbf{r}, \mathbf{u} - \Delta\mathbf{u}, \Delta t)\Psi(\mathbf{r}, \mathbf{u} - \Delta\mathbf{u}; \Delta\mathbf{u})d(\Delta\mathbf{u}) \tag{20–31}$$

We can now derive the phase space analog of Eq. (20–22) by expanding the various functions in Eq. (20–31) in Taylor series to obtain

$$\left(\frac{\partial W}{\partial t} + \mathbf{u}\cdot\nabla_\mathbf{r} W\right)\Delta t + O(\Delta t^2) = -\sum_j \frac{\partial}{\partial u_j}(W\langle\Delta u_j\rangle) + \frac{1}{2}\sum_j \frac{\partial^2}{\partial u_j{}^2}(W\langle\Delta u_j{}^2\rangle)$$

$$+ \sum_{i<j} \frac{\partial^2}{\partial u_i\,\partial u_j}(W\langle\Delta u_i\,\Delta u_j\rangle) + \cdots \tag{20–32}$$

This is the Chandrasekhar equation in its general form. We can calculate $\langle\Delta u_j\rangle$, $\langle\Delta u_j{}^2\rangle$, and $\langle\Delta u_i\,\Delta u_j\rangle$ from Eqs. (20–27).

The second of Eqs. (20–27) shows immediately that

$$\langle\Delta u_j\rangle = -(\zeta u_j - K_j)\,\Delta t + \cdots$$

To find $\langle\Delta u_j{}^2\rangle$, we proceed as we did in the derivation of the Fokker-Planck equation and get (Problem 20–20)

$$\langle\Delta u_j{}^2\rangle = \frac{2\zeta kT}{m}\,\Delta t + \cdots$$

and

$$\langle\Delta u_i\,\Delta u_j\rangle = 0 \qquad i \neq j$$

Therefore, Eq. (20–32) simplifies to in the limit $\Delta t \to 0$,

$$\frac{\partial W}{\partial t} + \mathbf{u}\cdot\nabla_\mathbf{r} W + \mathbf{K}\cdot\nabla_\mathbf{u} W = \zeta\,\mathrm{div}_\mathbf{u}(W\mathbf{u}) + \frac{\zeta kT}{m}\,\nabla_\mathbf{u}{}^2 W \tag{20–33}$$

This equation is the generalization of the Fokker-Planck equation to phase space. It is often referred to as the *Chandrasekhar equation*. Notice how the left-hand side of this equation is identical with the left-hand side of the reduced Liouville equation for $f^{(1)}$ [Eq. (18–14)]. The right-hand side of Eq. (20–33) is the Brownian motion representation of the collision terms in the Boltzmann equation. Both the Kirkwood and Rice-Allnatt theories* of transport in dense fluids derive Chandrasekhar-type equations for $f^{(1)}$ and $f^{(2)}$ or at least the parts of $f^{(1)}$ and $f^{(2)}$ due to the long-range part of the potential. They derive these equations directly from the reduced Liouville equations for $f^{(1)}$ and $f^{(2)}$.

We have derived the diffusion equation in velocity space and the diffusion equation in phase space. Before concluding this chapter let us derive the diffusion equation in

---

* S. A. Rice, and P. Gray, *The Statistical Mechanics of Simple Liquids* (New York: Interscience, 1965); R. M. Mazo, *Statistical Mechanical Theories of Transport Processes* (Oxford: Pergamon, 1967).

configuration space, the "ordinary" diffusion equation. Equation (20–20) says that for $\Delta t \gg \zeta^{-1}$, the transition probability that $\mathbf{r}$ changes by $\Delta \mathbf{r}$ in the time interval Δt is

$$\Psi(\Delta \mathbf{r}) = \frac{1}{(4\pi D\, \Delta t)^{3/2}} \exp\left(-\frac{|\Delta \mathbf{r}|^2}{4D\, \Delta t}\right) \tag{20–34}$$

Assuming then that the process is Markovian for $\Delta t \gg \zeta^{-1}$, we can write

$$W(\mathbf{r}, t + \Delta t) = \int W(\mathbf{r} - \Delta \mathbf{r}, t)\Psi(\Delta \mathbf{r})d(\Delta \mathbf{r}) \tag{20–35}$$

Applying the same procedure we have applied twice before, we get

$$\frac{\partial W}{\partial t} = D\, \nabla_{\mathbf{r}}^2 W \tag{20–36}$$

the ordinary diffusion equation. The fundamental solution of this equation is Eq. (20–20). (See Problem 20–15.)

ADDITIONAL READING

CHANDRASEKHAR, S. *Rev. Mod. Phys.*, **15**, p. 1, 1943.
CUKIER, R. I., LAKATOS-LINDENBERG, K., and SHULER, K. E. *J. Stat. Phys.*, **9**, p. 137, 1973.
JOHANNESMA, P. I. M. 1968. In *Neural networks*, ed. by E. R. Caianello. New York: Springer-Verlag.
KUBO, R. *Repts. Prog. Phys.*, **24**, p. 255, 1966.
LAX, M. *Rev. Mod. Phys.*, **38**, p. 541, 1966.
MAZUR, P. 1965. In *Statistical mechanics of equilibrium and non-equilibrium*, ed. by J. Meixner. Amsterdam: North-Holland Publ.
MCQUARRIE, D. A. 1975. In *Physical chemistry, an advanced treatise*, Vol. 11B, ed. by H. Eyring, D. Henderson, and W. Jost. New York: Academic.
OPPENHEIM, I., SHULER, K. E., and WEISS, G. H. *Adv. Mol. Relax. Proc.*, **1**, p. 12, 1967–1968.
PAPOULIS, A. 1965. *Probability, random variables, and stochastic processes*. New York: McGraw-Hill.
STEVENS, C. F. *Biophys. J.*. **4**, p. 417, 1964.
WAX, N., ed. 1954. *Selected papers on noise and stochastic processes*. New York: Dover.

PROBLEMS

20–1. Show that a formal solution to the Langevin equation

$$\frac{d\mathbf{u}}{dt} = -\zeta \mathbf{u} + \mathbf{A}(t)$$

is

$$\mathbf{u}(t) = \mathbf{u}_0 e^{-\zeta t} + e^{-\zeta t} \int_0^t e^{\zeta \xi} \mathbf{A}(\xi)\, d\xi$$

20–2. Derive Eq. (20–6).

20–3. Show that $\langle U^2 \rangle$ is given by

$$\langle U^2 \rangle \equiv \langle u^2 \rangle - u_0^2 e^{-2\zeta t} = \tfrac{1}{2} e^{-2\zeta t} \int_0^{2t} e^{\zeta x}\, dx \int_{-\infty}^{\infty} \phi_1(y)\, dy$$

where $\langle \mathbf{A}(t') \cdot \mathbf{A}(t'') \rangle = \phi_1(|t' - t''|)$.

20–4. Prove that the moments of $U \equiv u - u_0 e^{-\zeta t}$ are given by

$$\langle U^{2n} \rangle = 1 \cdot 3 \cdot 5 \cdots (2n - 1)\langle U^2 \rangle$$

$$\langle U^{2n+1} \rangle = 0$$

In order to evaluate these higher moments, one must make assumptions about the higher-order correlation functions $\langle A(t_1)A(t_2) \cdots A(t_j) \rangle$. It is a common assumption to assume that the random process $A(t)$ is Gaussian, which amounts to assuming that*

$$\langle A(t_1)A(t_2) \cdots A(t_{2n+1}) \rangle = 0$$
$$\langle A(t_1)A(t_2) \cdots A(t_{2n}) \rangle = \sum_{\text{all pairs}} \langle A(t_i)A(t_j) \rangle \langle A(t_k)A(t_l) \rangle \cdots$$

20–5. Prove that the result for the moments of **U** given in the previous problem implies that

$$W(\mathbf{u}, t; \mathbf{u}_0) = \left[\frac{m}{2\pi kT(1 - e^{-2\zeta t})} \right]^{3/2} \exp\left[-\frac{m|\mathbf{u} - \mathbf{u}_0 e^{-\zeta t}|^2}{2kT(1 - e^{-2\zeta t})} \right]$$

20–6. Show that

$$\langle (\mathbf{r} - \mathbf{r}_0) \rangle = \frac{\mathbf{u}_0(1 - e^{-\zeta t})}{\zeta}$$

20–7. Show that

$$\langle |\mathbf{r} - \mathbf{r}_0|^2 \rangle = \frac{u_0^2}{\zeta^2}(1 - e^{-\zeta t})^2 + \frac{3kT}{m\zeta^2}(2\zeta t - 3 + 4e^{-\zeta t} - e^{-2\zeta t})$$

20–8. Prove that the distribution of $\mathbf{r} - \mathbf{r}_0 - \zeta^{-1}\mathbf{u}_0(1 - e^{-\zeta t})$ is given by Eq. (20–15).

20–9. Using the theorem given in Eqs. (20–16) and (20–17), derive Eqs. (20–12) and (20–15).

20–10. Derive Eq. (20–20).

20–11. Chandrasekhar ("Additional Reading") proves the following lemma. Let

$$\mathbf{R} = \int_0^t \Psi(\xi)\mathbf{A}(\xi)\, d\xi$$

and

$$\mathbf{S} = \int_0^t \phi(\xi)\mathbf{A}(\xi)\, d\xi$$

Then the joint probability distribution of **R** and **S** is given by

$$W(\mathbf{R}, \mathbf{S}) = \frac{1}{8\pi^3(FG - H^2)^{3/2}} \exp\left[-\frac{(GR^2 - 2H\mathbf{R}\cdot\mathbf{S} + FS^2)}{2(FG - H^2)} \right]$$

where

$$F = 2q \int_0^t \Psi^2(\xi)\, d\xi$$

$$G = 2q \int_0^t \phi^2(\xi)\, d\xi$$

$$H = 2q \int_0^t \phi(\xi)\Psi(\xi)\, d\xi$$

where $q = \zeta kT = \gamma'kT/m$.

* See M. C. Wang and G. E. Uhlenbeck, *Rev. Mod. Phys.*, **17**, p. 323, 1945.

Letting

$$\mathbf{R} = \mathbf{r} - \mathbf{r}_0 - \zeta^{-1}\mathbf{u}_0(1 - e^{-\zeta t})$$
$$\mathbf{S} = \mathbf{u} - \mathbf{u}_0 e^{-\zeta t}$$

show that

$$\Psi(\xi) = \zeta^{-1}[1 - e^{\zeta((\xi - t)}]$$
$$\phi(\xi) = e^{\zeta(\xi - t)}$$

and that

$$F = q\zeta^{-3}(2\zeta t - 3 + 4e^{-\zeta t} - e^{-2\zeta t})$$
$$G = q\zeta^{-1}(1 - e^{-2\zeta t})$$
$$H = q\zeta^{-3}(1 - e^{-\zeta t})^2$$

20–12. Derive the equations

$$\langle \Delta u_j \rangle = - \zeta u_j \, \Delta t + \cdots$$

$$\langle \Delta u_j{}^2 \rangle = \frac{2\zeta kT}{m} \Delta t + \cdots$$

Hint: Remember that $\Delta u_j{}^2 = (u_j - u_{0j})^2$.

20–13. Show that Eq. (20–12) is the fundamental solution of the Fokker-Planck equation.

20–14. Derive Eq. (20–36) by starting with the Chapman-Kolmogorov equation and Eq. (20–34).

20–15. Show that Eq. (20–20) is the fundamental solution to Eq. (20–36).

20–16. Show that the distribution given in Problem 20–11 is the fundamental solution of the Chadrasekhar equation.

20–17. Show that if we set $u_0{}^2 = \langle u_0{}^2 \rangle$ in Eq. (20–14), then

$$\langle |\mathbf{r} - \mathbf{r}_0|^2 \rangle = \frac{6kT}{m\zeta^2} (\zeta t - 1 - e^{-\zeta t})$$

20–18. Assuming that $\mathbf{u}_0$ and $\mathbf{A}(\xi)$ are uncorrelated, show that the time correlation function of the velocity, defined as $\langle \mathbf{u}(t) \cdot \mathbf{u}(0) \rangle$, is given by

$$\langle \mathbf{u}(t) \cdot \mathbf{u}(0) \rangle = \langle u^2(0) \rangle e^{-\zeta t}$$

$$= \frac{3kT}{m} e^{-\zeta t}$$

We shall see in Chapter 21 that such time correlation functions play a central role in modern theories of transport.

Noting from Eq. (20–19) that the self-diffusion coefficient can be expressed as $D = kT/m\zeta$, show that

$$D = \frac{1}{3} \int_0^\infty \langle \mathbf{u}(t) \cdot \mathbf{u}(0) \rangle \, dt$$

We shall derive this result in Chapter 21.

20–19. The relation between the coefficient of self-diffusion and the velocity time correlation function suggested in the previous problem can be derived from the considerations given in this chapter. Consider Eq. (20–19):

$$D = \lim_{t \to \infty} \frac{1}{6t} \langle |\mathbf{r}(t) - \mathbf{r}(0)|^2 \rangle$$

Using the fact that

$$\mathbf{r}(t) = \mathbf{r}(0) + \int_0^t \mathbf{u}(t')\, dt'$$

show that

$$D = \lim_{t \to \infty} \frac{1}{6t} \int_0^t dt' \int_0^t dt'' \, \langle \mathbf{u}(t') \cdot \mathbf{u}(t'') \rangle$$

Using the stationarity of the equilibrium ensemble average and time reversibility, show that

$$\langle \mathbf{u}(t') \cdot \mathbf{u}(t'') \rangle = \langle \mathbf{u}(t' - t'') \cdot \mathbf{u}(0) \rangle = \langle \mathbf{u}(t'' - t') \cdot \mathbf{u}(0) \rangle$$

Now change the integration variable to $\tau = t'' - t'$ and interchange orders of integration to get

$$D = \frac{1}{3} \lim_{t \to \infty} \int_0^t \left(1 - \frac{\tau}{t} \right) \langle \mathbf{u}(0) \cdot \mathbf{u}(\tau) \rangle \, d\tau$$

If the velocity time correlation function decays sufficiently rapidly, we can finally write

$$D = \frac{1}{3} \int_0^\infty \langle \mathbf{u}(0) \cdot \mathbf{u}(\tau) \rangle \, d\tau$$

We shall discuss such expressions in great detail in Chapter 21.

20–20. In deriving the Fokker-Planck equation, an intermediate step was to prove that $\langle \Delta u_j \rangle = -\zeta u_j \Delta t + \cdots$ and that $\langle \Delta u_j{}^2 \rangle = (2\zeta kT/m) \Delta t + \cdots$ (Problem 20–12). An alternate derivation of this equation (cf. Chandrasekhar in "Additional Reading") argues that the transition probability $\Psi(\mathbf{u} - \Delta\mathbf{u}; \Delta u)$ in Eq. (20–21) has the form

$$\Psi(\mathbf{u}; \Delta\mathbf{u}) = (4\pi q \, \Delta t)^{-3/2} \exp\left\{ -\frac{|\Delta\mathbf{u} + \beta\mathbf{u} \, \Delta t|^2}{4q \, \Delta t} \right\}$$

where $q = \zeta kT/m$. Show that this leads to the same results as those given in the chapter.

20–21. Equation (20–9) is a very interesting equation, whose significance should be pointed out. Show that this equation can be written as

$$\zeta = \frac{m}{6kT} \int_{-\infty}^\infty \langle \mathbf{A}(0) \cdot \mathbf{A}(t) \rangle \, dt$$

where we have recognized the fact that the random process represented by $\mathbf{A}(t)$ is *stationary*.

This is a fundamental result, whose significance we shall discuss fully in Chapter 21. Nevertheless, note here that if the Langevin equation represents a Brownian particle randomly moving about in thermal equilibrium, then the friction constant ζ is related to the fluctuations of the random force. This is a simple example of a famous, fundamental theorem called the *fluctuation-dissipation theorem*.

20–22. The Langevin equation can be extended to include the case of Brownian motion in an external field. In this case the Langevin equation becomes (in one dimension for simplicity)

$$m \frac{du}{dt} = -\gamma' u - \frac{\partial V}{\partial x} + A'(t)$$

Show that the Fokker-Planck equation corresponding to this is

$$\left(\frac{\partial}{\partial t} + u \frac{\partial}{\partial x} - \frac{1}{m} \frac{\partial V}{\partial x} \frac{\partial}{\partial u} \right) W = \frac{\partial}{\partial u} \left(D_u \frac{\partial}{\partial u} + \zeta u \right) W$$

where $\zeta = \gamma'/m$ and $D_u = \zeta kT/m$.

20–23. Extend the analysis of the previous problem to three dimensions.

20–24. Show that the relation between the friction constant ζ and the force correlation function (Problem 20–21) is also valid when there is an external force.

20–25. Show that the fundamental solution to the multidimensional Fokker-Planck equation

$$\frac{\partial W}{\partial t} = -\sum_j \lambda_j \frac{\partial}{\partial x_j}(x_j W) + \frac{1}{2}\sum_{i,j} \sigma_{ij} \frac{\partial^2 W}{\partial x_i\,\partial x_j}$$

is a multivariable Gaussian distribution with average values

$$\langle x_j \rangle = x_{j0} e^{\lambda_j t}$$

and variances

$$\langle (x_i - \langle x_i \rangle)(x_j - \langle x_j \rangle)\rangle = -\frac{\sigma_{ij}}{\lambda_i + \lambda_j}[1 - e^{(\lambda_i + \lambda_j)t}]$$

20–26. Let $y(t)$ represent a random variable whose average value is zero. Furthermore, let $y(t)$ be such that its statistical properties are independent of the origin of time. Such a process is said to be stationary. Define the frequency spectrum of $y(t)$ to be

$$G(\omega) = \int_0^\infty d\tau \,\langle y(t)y(t+\tau)\rangle \cos \omega\tau$$

Convince yourself that this is independent of t.

Show that if $y(t)$ satifies the Langevin equation

$$\frac{dy}{dt} + \zeta y = F(t)$$

and

$$\langle F(t_1)F(t_2)\rangle = c\delta(t_1 - t_2)$$

then the frequency spectrum of $y(t)$ is

$$G(\omega) = \frac{c/2}{\zeta^2 + \omega^2}$$

20–27. In this and the next few problems we shall illustrate several methods of solving the Fokker-Planck equation. For simplicity consider the one-dimensional case

$$\frac{\partial W}{\partial t} = \zeta W + \zeta u\frac{\partial W}{\partial u} + \frac{\zeta kT}{m}\frac{\partial^2 W}{\partial u^2}$$

We wish to find the fundamental solution, i.e., the solution for no boundaries and for the initial condition

$$W(u, t; u_0) = \delta(u - u_0) \qquad t \to 0$$

Except for the last term on the right-hand side of the Fokker-Planck equation, the equation would be a linear first-order partial differential equation, which is fairly straightforward to solve. It is natural to expect that a study of the first order-equation will give insight into the solution of the full Fokker-Planck equation. To this end, let $\partial^2 W/\partial u^2 = 0$ for now to get

$$\frac{\partial W}{\partial t} - \zeta u\frac{\partial W}{\partial u} = \zeta W$$

Multiply this by an integrating factor $\mu(u, t)$ and compare the result to the perfect differential

$$dW = \frac{\partial W}{\partial t}dt + \frac{\partial W}{\partial u}du$$

Show from this comparison that we would have an exact differential if $dt = \mu$, $du = -\mu\zeta u$, and $dW = \mu\zeta W$, from which one writes

$$u = c_1 e^{-\zeta t} \quad \text{and} \quad W = c_2 e^{\zeta t}$$

This suggests that the variables $v = ue^{\zeta t}$ and $\chi = We^{-\zeta t}$ be used in the Fokker-Planck equation. Show that this transformation gives

$$\frac{\partial \chi}{\partial t} = \frac{\zeta kT}{m} e^{2\zeta t} \frac{\partial^2 \chi}{\partial v^2}$$

Show that this equation now becomes of the form of a simple diffusion equation if we introduce a new time variable τ such that $dt = e^{-2\zeta t} d\tau$, or

$$\tau = \frac{(e^{2\zeta t} - 1)}{2\zeta}$$

Show that the fundamental solution to this diffusion equation is

$$\chi(v, \tau) = (4\pi q\tau)^{-1/2} \exp\left[-\frac{(v - v_0)^2}{4q\tau}\right]$$

where $q = \zeta kT/m$.

Thus finally show that the fundamental solution to the Fokker-Planck equation is

$$W(u, t; u_0) = \left[\frac{m}{2\pi kT(1 - e^{-2\zeta t})}\right]^{1/2} \exp\left[-\frac{m(u - u_0 e^{-\zeta t})^2}{2kT(1 - e^{-2\zeta t})}\right]$$

20–28. Again we consider the one-dimensional Fokker-Planck equation for simplicity:

$$\frac{\partial W}{\partial t} = \zeta \frac{\partial}{\partial u}(uW) + q \frac{\partial^2 W}{\partial u^2}$$

where $q = \zeta kT/m$.

Define the Fourier transform of W by

$$\hat{W}(s, t) = (2\pi)^{-1/2} \int_{-\infty}^{\infty} W(u, t) e^{isu} \, du$$

Show that $\hat{W}(s, t)$ satifies the equation

$$\frac{\partial \hat{W}}{\partial t} = -qs^2 \hat{W} - \zeta s \frac{\partial \hat{W}}{\partial s}$$

Show that the term in $\hat{W}$ can be eliminated by the substitution (see the next problem)

$$\hat{V}(s, t) = \hat{W}(s, t) e^{qs^2/2\zeta}$$

to get

$$\frac{\partial \hat{V}}{\partial t} = -\zeta s \frac{\partial \hat{V}}{\partial s}$$

Show that this first-order linear partial differential equation has the general solution (I. Sneddon, *Partial Differential Equations*, New York: McGraw-Hill, 1957)

$$\hat{V}(s, t) = f\left(t - \frac{\ln s}{\zeta}\right)$$

where $f(x)$ is a differentiable function of x.

Using the initial condition that

$$W(u, 0) = \delta(u - u_0)$$

show that

$$\hat{W}(s, 0) = (2\pi)^{-1/2} e^{isu_0}$$

and that

$$f\left(-\frac{\ln s}{\zeta}\right) = (2\pi)^{-1/2} \exp\left(\frac{qs^2}{2\zeta} + isu_0\right)$$

Lastly show that

$$\hat{W}(s, t) = (2\pi)^{-1/2} \exp\left[-\frac{qs^2}{2\zeta}(1 - e^{-2\zeta t}) + isu_0 e^{-\zeta t}\right]$$

whose inverse is

$$W(u, t; u_0) = \left[\frac{\zeta}{2\pi q(1 - e^{-2\zeta t})}\right]^{1/2} \exp\left[-\frac{(u - u_0 e^{-\zeta t})^2}{2q(1 - e^{-2\zeta t})/\zeta}\right]$$

20–29. Show that the substitution

$$\theta = \phi e^{-ht}$$

reduces the equation

$$\frac{\partial\theta}{\partial t} = \kappa\frac{\partial^2\theta}{\partial x^2} - h\theta$$

to the one-dimensional diffusion equation.

20–30. Show that if

$$\frac{\partial\chi}{\partial t} = \phi(t)\nabla^2\chi$$

then the fundamental solution to this equation is

$$\chi = \left[4\pi\int_0^t \phi(s)\,ds\right]^{-3/2} \exp\left[-\frac{|\mathbf{r} - \mathbf{r}_0|^2}{4\int_0^t \phi(s)\,ds}\right]$$

Note how this was used in Problem 20–27.

CHAPTER 21

THE TIME-CORRELATION FUNCTION FORMALISM, I

The 1950s saw the beginning of the development of a new approach to transport processes that has grown into one of the most active and fruitful areas of nonequilibrium statistical mechanics. This work was initiated by Green and Kubo, who showed that the phenomenological coefficients describing many transport processes and time-dependent phenomena in general could be written as integrals over a certain type of function called a time-correlation function. We shall see shortly that these time-correlation functions play a similar role in nonequilibrium statistical mechanics that the partition function plays in equilibrium statistical mechanics. Up to now we have presented no rigorous general formalism of transport processes, but we shall finally do so in this and the next chapter.

Before considering any specific development, we shall first discuss the concept of a time-correlation function. We shall do this in the classical limit, but there is a quantum-statistical analog. Let $p(t)$ and $q(t)$ denote all the momenta and spatial coordinates necessary to describe our system, and let $p = p(0)$ and $q = q(0)$ denote the phase space coordinates at some initial time, $t = 0$. The $p(t)$ and $q(t)$ are related to the p and q through the equations of motion of the system. To emphasize this, we write

$$p(t) = p(p, q; t)$$
$$q(t) = q(p, q; t) \tag{21-1}$$

Let $A\{p(t), q(t)\}$ be some function of the phase space coordinates. By Eqs. (21–1) then, we can write

$$A\{p(t), q(t)\} = A(p, q; t) = A(t) \tag{21-2}$$

We define a classical time-correlation function of $A(t)$ by

$$C(t) = \langle A(0)A(t) \rangle = \int \cdots \int dp \, dq \, A(p, q; 0)A(p, q; t)f(p, q) \tag{21-3}$$

where $f(p, q)$ is the *equilibrium* phase space distribution function and where $dp \, dq$

stands for $d\mathbf{p}_1 \cdots d\mathbf{p}_N \, d\mathbf{q}_1 \cdots d\mathbf{q}_N$. If $A(t)$ is a vectorial function, such as velocity or momentum, then Eq. (21–3) becomes

$$C(t) = \langle \mathbf{A}(0) \cdot \mathbf{A}(t) \rangle = \int \cdots \int dp \, dq \, \mathbf{A}(p, q; 0) \cdot \mathbf{A}(p, q; t) f(p, q) \tag{21–4}$$

For simplicity, let us consider the simple case in which $A(t)$ is the velocity of some particular "tagged" molecule. Then the correlation function is

$$C(t) = \langle \mathbf{v}(0) \cdot \mathbf{v}(t) \rangle \tag{21–5}$$

We first point out that in order to evaluate $C(t)$ exactly even in this simple case, we would have to solve the equations of motion of one particle immersed in a system of others and then average the initial conditions over some equilibrium ensemble such as a canonical ensemble. Since $\mathbf{v}(t)$ depends upon the momenta and positions of many other particles in the system, this is clearly a very difficult calculation to do exactly. What, then, is the advantage of this time-correlation function formalism?

One advantage is that the resulting formulas for the transport coefficients are quite general in the sense that they do not depend upon the details of any particular model and are not limited to any particular density region. For example, we shall show in Section 21–8 that the coefficient of self-diffusion can be expressed in terms of the above velocity time-correlation function by

$$D = \tfrac{1}{3} \int_0^\infty \langle \mathbf{v}(0) \cdot \mathbf{v}(t) \rangle \, dt \tag{21–6}$$

This expression is valid for any density, for angle-dependent intermolecular forces, for polyatomic molecules, and generally for any classical system in which the diffusion is governed by the diffusion equation. Thus we see why we say above that the time-correlation function plays a somewhat similar role in nonequilibrium statistical mechanics that the partition function plays in equilibrium statistical mechanics. The analogy breaks down in one respect, however. Since the state of thermal equilibrium is unique, a single partition function gives all the thermodynamic properties, but since there are many different kinds of nonequilibrium states, we shall need a different time-correlation function for each type of transport process. For example, although the self-diffusion coefficient is given in terms of the velocity time-correlation function, other transport processes are given in terms of other time-correlation functions. Much of the discussion in this and the next chapter is concerned with determining what is the appropriate time-correlation function to use for a particular transport process of interest.

Another advantage of this formalism is that even though it may be prohibitively difficult to calculate a time-correlation function exactly, we at least have exact equations that we can start with to make sets of approximations. We use the same procedure in equilibrium statistical mechanics where we start with a partition function which is also prohibitively difficult to calculate exactly and introduce various well-defined approximations. The nature of a time-correlation function lends itself very nicely to this procedure. Consider the velocity correlation function [cf. Eq. (21–5)]. When $t = 0$, $C(t) = \langle \mathbf{v}(0) \cdot \mathbf{v}(0) \rangle$, which is simply the equilibrium average of v^2, which equals $3kT/m$ by equipartition. As time evolves, the particle will suffer collisions, and as we average over all these collisions, its velocity will change both in direction and magnitude and the velocity at time t, $\mathbf{v}(t)$, will be less and less correlated with its initial value

$\mathbf{v}(0)$. After a number of collisions, the velocity at that time will be completely uncorrelated with the initial velocity, and $C(t)$ will equal zero. Thus we expect the velocity
correlation here to start at its initial value $C(0) = 3kT/m$ and to decay to zero. It is
not an uncommon approximation to assume that $C(t)$ decays exponentially with some
time constant τ, i.e., to assume that $C(t) = (3kT/m)\exp(-t/\tau)$. We shall see in Section 21–9 that an approximation of this form can lead to the Enskog theory of transport in dense hard-sphere fluids. Although we have discussed the velocity correlation
function as an example here, the same argument holds for most any time-correlation
function.

It so happens that the time-correlation function formalism is more general than the
above discussion implies. We have discussed its application only to thermal transport
processes, but it is applicable to many other types of processes. A system that is
subject to a time-dependent perturbation will produce a time-dependent response to
this probe, which can be Fourier analyzed into what is called a frequency-dependent
susceptibility. For example, a time-dependent external electric field acting across a
conductor causes a time-dependent current to flow. The current depends upon the
nature of the medium and the frequency of the field and can be described by a frequency
dependent conductivity $\sigma(\omega)$. We shall show in Section 21–4 that $\sigma(\omega)$ is given by

$$\sigma(\omega) = \frac{1}{kT} \int_0^\infty dt\, e^{-i\omega t} \langle J(0)J(t) \rangle \tag{21-7}$$

where

$$J(t) = \sum_j q_j v_j$$

where q_j is the charge on the jth particle and v_j is its velocity in the direction of the
field.

Note that this formula differs from Eq. (21–6) in that it contains the additional
factor $e^{-i\omega t}$ in the integrand. [We call the integral in Eq. (21–7) a Fourier-Laplace
transform.] Equation (21–7) is more representative of the type of formula that results
from time-correlation function theory. Equation (21–6) for D implies that this expression is the zero-frequency limit of some more general quantity. Thus we say that an
ordinary thermal transport coefficient such as the self-diffusion coefficient is a zero-
frequency result, meaning that it is valid only for slow processes.

It is quite typical of the development to be presented in this chapter that we apply a
time-dependent perturbation to a system and assume that this perturbation induces a
time-dependent response which is linearly related to the perturbation. For this reason,
the time-correlation function approach is often called linear response theory.

We can also write time-correlation functions of two phase space functions,
$A\{p(t), q(t)\}$ and $B\{p(t), q(t)\}$, i.e., $\langle A(0)B(t) \rangle$. All of the results of this chapter will be
of the form

$$\psi(\omega) = \int_0^\infty dt\, e^{-i\omega t} \langle A(0)B(t) \rangle$$

where $\psi(\omega)$ is some generalized susceptibility. The familiar thermal transport coefficients will be the zero-frequency limit of such expressions. When $A = B$, the correlation function is often called an autocorrelation function.

In Section 21–1 we shall present the time-correlation function formalism of the
absorption of electromagnetic radiation. We choose this as the introductory application since the derivation of the key formulas is quite pedagogical, beginning with the

standard formulas of quantum-mechanical time-dependent perturbation theory (which is assumed to be known to the reader at this point). In this section we show that the lineshape of an infrared absorption band is given as the Fourier-Laplace transform of the dipole moment autocorrelation function of an absorbing molecule. In spite of the pedagogical nature of this section, it does require a certain degree of quantum-mechanical knowledge. If the reader is not familiar with the golden rule of time-dependent perturbation theory, the Dirac bra and ket notation or the Heisenberg picture of time-dependent quantum mechanics, he is referred to Appendixes F to H. Section 21–2 is somewhat of a digression on the classical theory of light scattering. This serves as a background for the next section in which we present the time-correlation function formulation of Raman scattering. The mathematical and quantum-mechanical apparatus of this third section is very similar to that of the first section. In Section 21–4 we take a more general approach and present an elementary derivation of the basic formulas of linear response theory. In this derivation, the linear response nature of the theory will be clearly evident. Then in Sections 21–5 and 21–6 we give two concrete examples of the general theory of Section 21–4. The first is the theory of dielectric relaxation, which leads naturally into the next section, which is a further discussion of molecular spectroscopy. Section 21–7 contains a derivation of the basic equations of linear response theory, this time starting from both the classical and quantum-mechanical Liouville equations. Although no new results are derived in this section, the method is fundamental and serves to introduce several important concepts. Then, in Section 21–8 we discuss the time-correlation function formulation of the thermal transport coefficients, such as self-diffusion, viscosity, and thermal conductivity. We shall find in this section that the derivation of the basic formulas for the thermal transport coefficients does not proceed as smoothly as for the previous applications. The reason for this is that it is not possible (at least not easy) to account for the thermal transport phenomena by means of a well-defined mechanical perturbation Hamiltonian. Because of this, the basic time-correlation function formulas do not have the general validity that the previous formulas have. This is why Eq. (21–6) does not have the factor $e^{-i\omega t}$ that appears in Eq. (21–7). Lastly in Section 21–9 we present a number of approximate calculations of the thermal transport coefficients starting from their time-correlation function expressions. We shall show that the time-correlation function formulas reduce to the kinetic theory of gases formulas in the low-density limit, but, in addition, we show how they can be used to approximately calculate the thermal transport coefficients of dense fluids. We shall also discuss the law of corresponding states of transport processes in this section.

21–1   ABSORPTION OF RADIATION*

Consider a system of N interacting molecules in the quantum state i (for initial). Let the Hamiltonian of this N-body system be $\mathscr{H}_0$, with $\mathscr{H}_0 \psi_j = E_j \psi_j$. If this system now interacts with an electric field of frequency ω, transitions into other quantum states f (for final) will occur if the frequency of the radiation is close to $(E_f - E_i)/\hbar$ according to the Bohr relation. If we let the field be monochromatic, then we can write

$$\mathbf{E}(t) = E_0 \boldsymbol{\varepsilon} \cos \omega t = \frac{E_0 \boldsymbol{\varepsilon}}{2} (e^{i\omega t} + e^{-i\omega t}) \tag{21–8}$$

* This section is based upon R. G. Gordon, *Adv. Mag. Resonance*, **3**, p. 1, 1968.

where E_0 is the amplitude of the field and ε is a unit vector along the electric field. We are assuming here here that the field is spatially uniform so that we can suppress the wave vector dependence that occurs in Eq. (10–79), for example. Since the field is uniform, or since the wavelength is large compared to molecular dimensions, the interaction between the field and the molecules can be written as

$$\mathscr{H}^{(1)}(t) = -\mathbf{M} \cdot \mathbf{E}(t) \tag{21-9}$$

where $\mathbf{M}$ is the total electric dipole moment operator of the N-body system.

According to the so-called Golden Rule of time-dependent quantum-mechanical perturbation theory (cf. Appendix F), the probability per unit time that a transition from the state i to the state f takes place is given by

$$P_{i \to f}(\omega) = \frac{\pi E_0^2}{2\hbar^2} |\langle f | \varepsilon \cdot \mathbf{M} | i \rangle|^2 [\delta(\omega_{fi} - \omega) + \delta(\omega_{fi} + \omega)] \tag{21-10}$$

where $\omega_{fi} = \omega_f - \omega_i$. If we multiply this by $\hbar\omega_{fi}$, this gives the rate of energy lost from the radiation in going from state i to f; if we sum this over all f, we get the rate of energy lost in going from the initial state i to any other state; and lastly, if we multiply this by ρ_i, the probability that the system was in the initial state, and then sum over all i, we finally get the rate of energy loss from the radiation to the system

$$-\dot{E}_{\text{rad}} = \sum_i \sum_f \rho_i \hbar \omega_{fi} P_{i \to f}$$

$$= \frac{\pi E_0^2}{2\hbar} \sum_f \sum_i \omega_{fi} \rho_i |\langle f | \varepsilon \cdot \mathbf{M} | i \rangle|^2 [\delta(\omega_{fi} - \omega) + \delta(\omega_{fi} + \omega)] \tag{21-11}$$

Since the summations i and f go over all the quantum states of the system, we may interchange these indices in the summation over the second delta function, giving (Problem 21–1)

$$-\dot{E}_{\text{rad}} = \frac{\pi E_0^2}{2\hbar} \sum_f \sum_i \omega_{fi} (\rho_i - \rho_f) |\langle f | \varepsilon \cdot \mathbf{M} | i \rangle|^2 \delta(\omega_{fi} - \omega) \tag{21-12}$$

If we assume that the system is initially in equilibrium, then

$$\rho_f = \rho_i e^{-\beta \hbar \omega_{fi}} \tag{21-13}$$

and so

$$\rho_i - \rho_f = \rho_i (1 - e^{-\beta \hbar \omega_{fi}}) \tag{21-14}$$

We can now substitute this into Eq. (21–12); we find that

$$-\dot{E}_{\text{rad}} = \frac{\pi E_0^2}{2\hbar} (1 - e^{-\beta \hbar \omega}) \omega \sum_i \sum_f \rho_i |\langle f | \varepsilon \cdot \mathbf{M} | i \rangle|^2 \delta(\omega_{fi} - \omega) \tag{21-15}$$

We have dropped the subscripts on the ω's since the delta function requires that $\omega_{fi} = \omega$. We shall define an absorption cross section $\alpha(\omega)$ such that this cross section, multiplied by the incident flux of radiation, is equal to the rate of energy lost from the field to the system. It is shown in Appendix I that the energy flux is given by the so-called Poynting vector $\mathbf{S}$, whose magnitude in this case is

$$S = \frac{v}{8\pi} \varepsilon E_0^2 = \frac{c}{8\pi} n E_0^2 \tag{21-16}$$

where v is the speed of light in the medium, ε is the dielectric constant, c is the speed of light in vacuo, and n is the index of the refraction of the medium. If we divide Eq. (21–15) by S, then we obtain

$$\alpha(\omega) = \frac{4\pi^2}{\hbar c n} \omega(1 - e^{-\beta\hbar\omega}) \sum_f \sum_i \rho_i |\langle f| \boldsymbol{\varepsilon} \cdot \mathbf{M} |i\rangle|^2 \, \delta(\omega_{fi} - \omega) \tag{21–17}$$

It is convenient to use this equation to define an absorption lineshape $I(\omega)$ by

$$I(\omega) \equiv \frac{3\hbar c n \alpha(\omega)}{4\pi^2 \omega(1 - e^{-\beta\hbar\omega})} = 3 \sum_i \sum_f \rho_i |\langle f| \boldsymbol{\varepsilon} \cdot \mathbf{M} |i\rangle|^2 \, \delta(\omega_{fi} - \omega) \tag{21–18}$$

As Gordon points out, this formula represents a Schrödinger representation of spectroscopy as transitions between Bohr stationary states since Eq. (21–10) is derived from first-order perturbation theory in which the operators are independent of time and the wave function of the system varies with time. In the Heisenberg representation of time-dependent quantum mechanics, the time evolution of the system is placed in the operators, and the states of the system are considered to be independent of time. These two different representations arise from the fact that observable quantities occur as inner products of the form $(\chi, A\psi)$. We can ascribe the time variation of such products to a time dependence of the wave functions or to the operators. The two representations agree for all times if

$$(\chi(0), A(t)\psi(0)) = (\chi(t), A(0)\psi(t)) \tag{21–19}$$

The correspondence between these two representations is shown in Appendix H, which briefly goes as follows. The time-dependent wave functions $\chi(t)$ and $\psi(t)$ obey the Schrödinger equation

$$i\hbar \frac{\partial \psi}{\partial t} = \mathscr{H} \psi \tag{21–20}$$

whose formal solution for a time-independent Hamiltonian operator is

$$\psi(t) = \exp\left(-\frac{i\mathscr{H}t}{\hbar}\right)\psi(0)$$

$$\equiv U(t)\psi(0) \tag{21–21}$$

with a similar equation for $\chi(t)$. If these are substituted into Eq. (21–19), we find

$$\begin{aligned}(\chi(0), A(t)\psi(0)) &= (\chi(t), A(0)\psi(t)) \\ &= (U(t)\chi(0), A(0)U(t)\psi(0)) \\ &= (\chi(0), U^*(t)A(0)U(t)\psi(0)) \end{aligned} \tag{21–22}$$

$U^*(t)$ is the adjoint operator of $U(t)$, which is simply $\exp(i\mathscr{H}t/\hbar)$ since $\mathscr{H}$ is Hermitian (Appendix H). Equation (21–22) suggests that we define the time-dependent operator $A(t)$ by

$$A(t) = e^{i\mathscr{H}t/\hbar}A(0)e^{-i\mathscr{H}t/\hbar} \tag{21–23}$$

Although we have explicitly displayed the zero-time argument in A in the right-hand side of the above equation, it is common practice to suppress it. Equation (21–23) is the so-called Heisenberg representation of an operator. We shall now see that the Heisenberg representation of Eq. (21–18) naturally leads to a time-correlation function of spectroscopy. In addition, we shall see that this has the advantage of having a clear correspondence with classical mechanics, which is not apparent in the Schrödinger picture.

We can convert Eq. (21–18) to the Heisenberg picture by introducing the Fourier transform of the Dirac delta function,

$$\delta(\omega) = \frac{1}{2\pi} \int_{-\infty}^{\infty} e^{i\omega t} \, dt \tag{21–24}$$

giving

$$I(\omega) = \frac{3}{2\pi} \sum_{i,f} \rho_i \langle i | \boldsymbol{\varepsilon} \cdot \mathbf{M} | f \rangle \langle f | \boldsymbol{\varepsilon} \cdot \mathbf{M} | i \rangle \int_{-\infty}^{\infty} dt \, \exp\left[\frac{(E_f - E_i)}{\hbar} - \omega \right] it \tag{21–25}$$

Now the states $|i\rangle$ and $|f\rangle$ are eigenstates of the system excluding the radiation, so that (Problem 21–2)

$$e^{-iE_i t/\hbar} | i \rangle = e^{-i\mathscr{H}_0 t/\hbar} | i \rangle \tag{21–26}$$

and

$$\langle f | e^{iE_f t/\hbar} = \langle f | e^{i\mathscr{H}_0 t/\hbar} \tag{21–27}$$

Thus Eq. (21–25) can be written as

$$I(\omega) = \frac{3}{2\pi} \int_{-\infty}^{\infty} dt \, e^{-i\omega t} \sum_{i,f} \rho_i \langle i | \boldsymbol{\varepsilon} \cdot \mathbf{M} | f \rangle \langle f | \boldsymbol{\varepsilon} \cdot \mathbf{M}(t) | i \rangle \tag{21–28}$$

where

$$\mathbf{M}(t) = e^{i\mathscr{H}_0 t/\hbar} \mathbf{M} e^{-i\mathscr{H}_0 t/\hbar} \tag{21–29}$$

The summation over the set of final states can be removed by using the closure relation (Appendix G)

$$\sum_f | f \rangle \langle f | = 1 \tag{21–30}$$

giving

$$I(\omega) = \frac{3}{2\pi} \int_{-\infty}^{\infty} dt \, e^{-i\omega t} \sum_i \rho_i \langle i | \boldsymbol{\varepsilon} \cdot \mathbf{M}(0) \boldsymbol{\varepsilon} \cdot \mathbf{M}(t) | i \rangle \tag{21–31}$$

This summation over i is nothing but an equilibrium ensemble average, which we shall denote by $\langle \ \rangle$, giving

$$I(\omega) = \frac{3}{2\pi} \int_{-\infty}^{\infty} dt \, e^{-i\omega t} \langle \boldsymbol{\varepsilon} \cdot \mathbf{M}(0) \boldsymbol{\varepsilon} \cdot \mathbf{M}(t) \rangle \tag{21–32}$$

Finally, for an isotropic fluid, one can show by averaging $\boldsymbol{\varepsilon}$ over all directions (Problem 21–3) that

$$I(\omega) = \frac{1}{2\pi} \int_{-\infty}^{\infty} dt \, e^{-i\omega t} \langle \mathbf{M}(0) \cdot \mathbf{M}(t) \rangle \tag{21–33}$$

This is the desired result, namely, the lineshape function $I(\omega)$ has been written as the Fourier transform of the time-correlation function of the dipole moment operator of the absorbing molecules in the absence of the field! Thus we see that a knowledge of the lineshape $I(\omega)$ can be used to study the motion of the dipole moments of the molecules by Fourier inversion of $I(\omega)$. As we pointed out in the introduction, a correlation function describes the average decay of a property of a system from some initial value. In the present case, the property is the dipole moment. In general, one must consider the dipole moment to be that of the entire system, i.e. that of the N interacting dipole moments. Thus in the quantity

$$\mathbf{M}(0) \cdot \mathbf{M}(t) = \left(\sum_{i=1}^{N} \boldsymbol{\mu}(0) \right) \cdot \left(\sum_{j=1}^{N} \boldsymbol{\mu}(t) \right)$$

there are, in addition to terms of the form $\boldsymbol{\mu}_j(0) \cdot \boldsymbol{\mu}_j(t)$, cross terms between the dipole moments on different molecules, $\boldsymbol{\mu}_i(0) \cdot \boldsymbol{\mu}_j(t)$. Because of these cross terms, one cannot simply interpret $\langle \mathbf{M}(0) \cdot \mathbf{M}(t) \rangle$ in terms of the reorientation of a single dipole when the concentration of dipolar molecules is large. If, on the other hand, the dipolar molecules are dilutely dissolved in some nonpolar solvent, the cross terms are negligible, and the correlation function

$$\langle \mathbf{M}(0) \cdot \mathbf{M}(t) \rangle = \left\langle \sum_{j=1}^{N} \boldsymbol{\mu}_j(0) \cdot \boldsymbol{\mu}_j(t) \right\rangle = N \langle \boldsymbol{\mu}(0) \cdot \boldsymbol{\mu}(t) \rangle$$

can be interpreted in terms of the orientation of the dipole moment of a single molecule.

We shall return to an application of these formulas, but it turns out for experimental reasons that the interpretation of the dipole correlation function is often simpler in the case of rotation-vibration bands in the near infrared, for it is usually a good approximation to treat the internal vibrational motion of the molecules as essentially separable from the rotational and translational motion. If we also assume that the vibrational Hamiltonian of the various molecules is additive, then the N-body vibrational wave function is a product of the individual molecular vibrational wave functions, and by orthogonality of the various excited vibrational states of the molecules, cross terms between dipoles on different molecules vanish. This means that a vibrational excitation is localized on a single molecule, and so the correlation function may be considered to be that of vibrational transition dipole moment of a typical molecule. By a transitional dipole moment, we mean in classical terms the dipole moment whose oscillation leads to the vibrational absorption. In a diatomic molecule, this is simply the dipole moment of the molecule itself. For more complicated molecules, however, we mean each infrared-active normal coordinate whose oscillation produces an oscillating dipole. For a linear molecule such as CO_2, the two infrared active modes are the antisymmetric stretch and the bending mode. Since in the first case the dipole moment oscillates along the axis of the molecule, the absorption band is called a parallel band, and since the dipole oscillates perpendicular to the axis in the bending mode, the absorption band due to this is called a perpendicular band. In this case the frequency ω in Eq. (21–33) is replaced by $\omega_0 + \omega$, where ω_0 is the vibrational band center and ω is the frequency displacement from ω_0. The distribution of absorption frequencies about the vibrational frequency ω_0, then, is the Fourier transform of the average motion of the particular transition dipole moment.

If experimental data are available to extend over the entire range of rotational-translational frequencies about a vibrational band, then the dipole moment correlation function can be determined from the inverse Fourier transform,

$$\langle \boldsymbol{\mu}(0) \cdot \boldsymbol{\mu}(t) \rangle = \int_{\text{band}} I(\omega) e^{i\omega t} \, d\omega \tag{21–34}$$

It is usually convenient to normalize the correlation function to its initial value by

$$\langle \boldsymbol{\mu}(0) \cdot \boldsymbol{\mu}(0) \rangle = \int_{\text{band}} I(\omega)\, d\omega \qquad (21\text{--}35)$$

By dividing Eq. (21–34) by Eq. (21–35), we get

$$\langle \mathbf{u}(0) \cdot \mathbf{u}(t) \rangle = \int_{\text{band}} \hat{I}(\omega) e^{i\omega t}\, d\omega \qquad (21\text{--}36)$$

where $\hat{I}(\omega)$ is a normalized lineshape function

$$\hat{I}(\omega) = \frac{I(\omega)}{\int_{\text{band}} I(\omega)\, d\omega} \qquad (21\text{--}37)$$

and $\mathbf{u}$ is a unit vector along the direction of the transition dipole moment. Figure 21–1 shows the normalized dipole correlation function of carbon monoxide in various environments. The negative region in some of these correlation functions indicates that it is probable that the molecule swing its dipole moment to a direction opposite that which it had at $t = 0$. Such correlation functions occur in gaseous systems, where

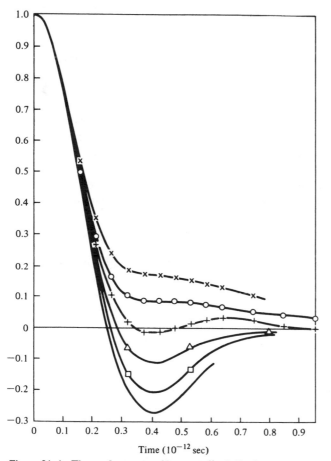

Figure 21–1. **The carbon monoxide normalized dipole moment correlation function in various environ ments.** $\times$: **CO in liquid** $CHCl_3$; $\bigcirc$: **CO in liquid** CCl_4; $+$: **CO in liquid** $n\text{-}C_2H_6$; $\triangle$: **CO in argon gas at 510 amagats;** $\square$: **CO in argon gas at 270 amagats;** $-$: **CO in argon gas at 6 amagats.** (From R. Gordon, *J. Chem. Phys.*, **43**, p. 1307, 1965.)

the molecule has some ability to rotate. In more dense systems, however, such easy reorientation is not permitted, and the correlation function simply decays monotonically from its initial value of unity to the long time value of zero.

Note that the procedure used in the derivation of the lineshape function in terms of the Fourier transform of a time-correlation function is to start with the quantum-mechanical perturbation theory expression [Eq. (21–10)] and then convert that into a Fourier transform by straightforward manipulation. The above analysis was concerned with infrared band shapes, but it should be clear that most any perturbation theory expression can be manipulated in the same way. In the next section we present somewhat of a digression on the classical theory of light scattering, and then in Section 21–3 we shall continue this analysis and discuss Raman light scattering.

21–2 CLASSICAL THEORY OF LIGHT SCATTERING

Since light scattering, in general, is beginning to play an increasingly important role in chemistry, physics, and biophysics, we shall digress a little here to study the classical theory of the Raman effect. This is not only instructive in itself, but will serve as background to Section 22–3, in which we shall discuss the general statistical mechanical theory of light scattering.

In the ordinary scattering of electromagnetic radiation by an interacting system of molecules, the frequency of the incident radiation and the scattered radiation is the same; i.e., there is no frequency change associated with the scattering. This type of scattering is known as Rayleigh scattering. In addition to Rayleigh scattered radiation, it is found that there exists, to a much smaller extent, scattered radiation in which there are definite frequency changes. This type of scattering, predicted theoretically in 1923 by Smekal and observed experimentally in 1928 by Raman, is called Raman scattering. The Raman scattered radiation has frequencies both higher and lower than that of the incident radiation; these differences, however, are independent of the frequency of the incident radiation (assumed to be monochromatic here for simplicity) and are characteristic of the scattering medium. It is common to define a Raman shift by

$$\Delta v = v_i - v_s \tag{21–38}$$

where v_i and v_s are the frequencies of the incident and scattered radiation. As pointed out above, Δv is characteristic of the scattering medium. The frequencies such that $\Delta v > 0$ are called Stokes lines and those for which $\Delta v < 0$ are called anti-Stokes lines. It turns out that the Stokes lines are often much more intense than the anti-Stokes lines. We shall now present a classical argument which shows that these Raman shifts are due to rotational and vibrational transitions occurring within the scattering medium.

If a system of molecules is placed in an electric field, the electrons and nuclei will be displaced in such a manner as to induce dipole moments. If we make the valid assumption that the dipole moments induced, μ, are proportional to the strength of the electric field E, then

$$\mu = \alpha E \tag{21–39}$$

where α is the polarizability of a molecule. If $E = E_0 \cos 2\pi v t$, then μ will vary with time according to

$$\mu(t) = \alpha E_0 \cos 2\pi v t \tag{21–40}$$

Thus we see that the field will induce a dipole moment that oscillates with frequency v. According to classical electromagnetic theory, this oscillating dipole will emit radiation of the same frequency, and hence the result will be Rayleigh scattering in which the incident and scattered frequencies are the same.

Now suppose that the molecules in the system are undergoing rotational and vibrational motions. For simplicity, let these be diatomic molecules. As the two nuclei vibrate along the internuclear axis, the polarizability of the molecule will vary. Since the amplitude of vibration is normally small, we can write the variation of the polarizability due to this vibration as

$$\alpha(t) = \alpha_0 + \alpha_1 \cos 2\pi v_v t \tag{21-41}$$

where v_v represents the fundamental vibrational frequency of the molecule and α_1 is a measure of how much the polarizability varies with vibration. If we substitute into Eq. (21–40), we see that

$$\mu(t) = \alpha_0 E_0 \cos 2\pi v t + \alpha_1 E_0 \cos 2\pi v t \cos 2\pi v_v t$$

$$= \alpha_0 E_0 \cos 2\pi v t + \frac{\alpha_1 E_0}{2} [\cos 2\pi(v - v_v)t + \cos 2\pi(v + v_v)t] \tag{21-42}$$

This equation shows us that the induced dipole oscillates not only with the incident frequency v, but also with frequencies $v - v_v$ and $v + v_v$. The Raman shifts here are $\Delta v = \pm v_v$ and so are independent of v itself but characteristic of the molecules.

Without going through the derivation, it should be clear that the rotational motion of the molecule can also lead to Raman scattering. As the molecule rotates, the orientation of the molecule with respect to the incident field changes, and if the molecule has different polarizabilities in different directions, its polarization, i.e., the induced dipole moment, will vary as a function of time. Note that for there to be a rotational Raman effect, the molecule must have different polarizabilities in different directions. Unless the molecule has a very high degree of symmetry, this will be the case. For example, if the molecule is diatomic, there will be a different polarizability along the internuclear axis than perpendicular to it. Generally, the polarizability of a molecule is a 2nd-rank tensor whose nine components are given by

$$\mu_X = \alpha_{XX} E_X + \alpha_{XY} E_Y + \alpha_{XZ} E_Z$$
$$\mu_Y = \alpha_{YX} E_X + \alpha_{YY} E_Y + \alpha_{YZ} E_Z \tag{21-43}$$
$$\mu_Z = \alpha_{ZX} E_X + \alpha_{ZY} E_Y + \alpha_{ZZ} E_Z$$

or in more compact notation

$$\boldsymbol{\mu} = \alpha \mathbf{E} \tag{21-44}$$

Just as in the case of the moment of inertia tensor discussed in Chapter 8, one can find a set of axes in the molecule such that

$$\mu_1 = \alpha_1 E_1$$
$$\mu_2 = \alpha_2 E_2 \tag{21-45}$$
$$\mu_3 = \alpha_3 E_3$$

Such axes are called principal axes of polarizability, and the associated polarizabilities are called the principal values of the polarizability. These principal axes are usually self-evident from the symmetry of the molecule.

Classical electromagnetic theory also tells us that the mean rate of radiation emitted over all directions by a dipole oscillating according to $\mu = \mu_0 \cos 2\pi vt$ is (Appendix J)

$$I = \frac{16\pi^4 v^4}{3c^3} \mu_0^2 \tag{21-46}$$

In cgs units, I has units of ergs per second. Referring to Fig. 21-2, the rate of radiation emitted per unit solid angle in the X-direction is given by* (Appendix J)

$$\left(\frac{dI}{d\Omega}\right)_X = \frac{2\pi^3 v^4}{c^3} (\mu_{0Y}^2 + \mu_{0Z}^2) \tag{21-47}$$

Note that this gives Eq. (21-46) back if we average μ_{0Y}^2 and μ_{0Z}^2 over angles to give $\mu_0^2/3$ for each of them and then multiply by the total solid angle 4π.

Light-scattering experiments are often performed with polarized incident radiation. Figure 21-2(a) shows incident radiation directed in the Y-direction and plane-polarized with the electric vector parallel to the direction of observation, i.e., along the X-direction. The radiation scattered in the X-direction will be depolarized, and the total scattered intensity in this case is designated by $I_T(\text{obs. } \parallel)$. Figure 21-2(b) shows the incident radiation to be plane-polarized along the Z-axis. The intensities in this case are $I_T(\text{obs. } \perp)$, $I_\parallel(\text{obs. } \perp)$, and $I_\perp(\text{obs. } \perp)$. In the first case, the incident electric field lies only along the X-axis and so $\mu_{0Y} = \alpha_{YX} E_0$ and $\mu_{0Z} = \alpha_{ZX} E_0$, so that Eq. (21-47) gives

$$I_T(\text{obs. } \parallel) = \frac{2\pi^3 v^4}{c^3} (\alpha_{YX}^2 + \alpha_{ZX}^2) E_0^2. \tag{21-48}$$

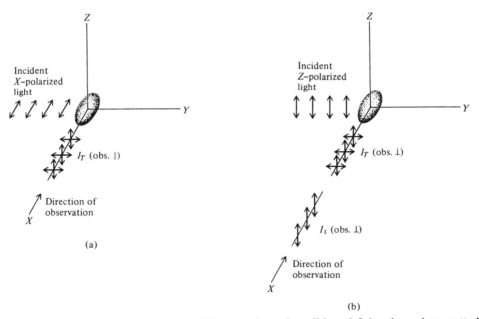

Figure 21-2. **Schematic representation of the experimental conditions defining the various scattering intensities. (a) The incident radiation is polarized along the X-axis. (b) The incident radiation is polarized along the Z-axis.** (From E. G. Wilson, Jr., J. C. Decius, and P. C. Cross, *Molecular Vibrations.* New York: McGraw-Hill, 1955.)

---

* E. B. Wilson, Jr., J. C. Decius, and P. C. Cross, *Molecular Vibrations* (New York: McGraw-Hill, 1955), p. 43. Much of the material of this section comes from this reference.

Similarly, in the second case, in which the incident electric field vector is plane-polarized along the Z-axis

$$I_T(\text{obs. } \perp) = \frac{2\pi^3 v^4}{c^3} (\alpha_{YZ}{}^2 + \alpha_{ZZ}{}^2)E_0{}^2 \tag{21-49}$$

$$I_{\parallel}(\text{obs. } \perp) = \frac{2\pi^3 v^4}{c^3} \alpha_{ZZ}{}^2 E_0{}^2 \tag{21-50}$$

$$I_{\perp}(\text{obs. } \perp) = \frac{2\pi^3 v^4}{c^3} \alpha_{YZ}{}^2 E_0{}^2 \tag{21-51}$$

Equations (21-48) through (21-51) are appropriate for a single molecule of fixed orientation. In order to apply these equations to a gas of N molecules, for example, we must average the α_{AB}'s over all angles. Note that the α_{AB} are the polarizability components in a spaced-fixed coordinate system. If we let α_{AB} be the components in a molecule-fixed system (see Fig. 21-3), then one set is given in terms of the other by (Problem 21-4)

$$\alpha_{AA'} = \sum_{a, a'} \alpha_{aa'} C_{Aa'} C_{A'a'} \tag{21-52}$$

where the C's are the direction cosines between the spaced-fixed and molecule-fixed axes. Recall that the direction cosines satisfy $C_{Xj}{}^2 + C_{Yj}{}^2 + C_{Zj}{}^2 = 1$ and $C_{Xi}C_{Xi} + C_{Yi}C_{Yj} + C_{Zi}C_{Zj} = 0$ (Problem 21-5). In case the three unprimed axes in Fig. 21-3 are principal axes, then (Problem 21-7)

$$\alpha_{AA'} = \sum_{j=1}^{3} \alpha_j C_{Aj} C_{A'j} \tag{21-53}$$

We wish to average $\alpha_{AA'}^2$ over all orientations, and so we write

$$\begin{aligned}
\overline{\alpha_{AA'}^2} &= \overline{\left(\sum_j \alpha_j C_{Aj} C_{A'j}\right)^2} \\
&= \sum_j \alpha_j{}^2 \overline{C_{Aj}{}^2 C_{A'j}^2} + 2\sum_{i<j} \alpha_i \alpha_j \overline{C_{Ai} C_{A'i} C_{Aj} C_{A'j}}
\end{aligned} \tag{21-54}$$

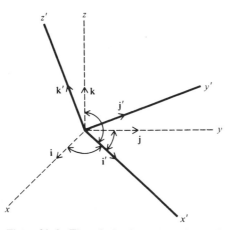

Figure 21-3. **The relation between molecule-fixed axes (primes) and space-fixed axes (no primes). The direction cosines of the X'-axis with respect to the space-fixed coordinate system are** $C_{X'X} = \cos(i', i) = i' \cdot i$, $C_{X'Y} = \cos(i', j) = i' \cdot j$, **and** $C_{X'Z} = \cos(i' \cdot k) = i' \cdot k$ **with similar relations for the six other direction cosines.**

To evaluate these averages of orientations, first note that

$$\overline{C_{Aj}{}^4} = \overline{\cos^4 \theta} = \frac{1}{4\pi} \int_0^{2\pi} d\phi \int_0^{\pi} d\theta \sin\theta \cos^4\theta = \tfrac{1}{5} \tag{21-55}$$

Now square the relation

$$C_{Xj}{}^2 + C_{Yj}{}^2 + C_{Zj}{}^2 = 1 \tag{21-56}$$

average, and then use symmetry to get (Problem 21–6)

$$3\overline{C_{Aj}{}^4} + 6\overline{C_{Aj}{}^2 C_{A'j}^2} = 1 \qquad A \neq A' \tag{21-57}$$

Using Eq. (21–55), we see that

$$\overline{C_{Aj}{}^2 C_{A'j}^2} = \tfrac{1}{15} \tag{21-58}$$

To evaluate the averages involved in the second summation of Eq. (21–54), start with

$$C_{Xi}C_{Xj} + C_{Yi}C_{Yj} + C_{Zi}C_{Zj} = 0 \qquad i \neq j \tag{21-59}$$

Squaring, averaging, and using symmetry give

$$3\overline{C_{Ai}{}^2 C_{Aj}{}^2} + 6\overline{C_{Ai}C_{A'i}C_{Aj}C_{A'j}} = 0 \tag{21-60}$$

Using Eq. (21–58), we find that

$$\overline{C_{Ai}C_{A'i}C_{Aj}C_{A'j}} = -\tfrac{1}{30} \qquad i \neq j \quad A \neq A' \tag{21-61}$$

We can now use all of these results to evaluate the average of Eqs. (21–48) through (21–51) over all orientations. If we multiply by the density ρ of molecules in the gas and introduce the incident flux of radiation $I_0 = cE_0{}^2/8\pi$, we find (Problem 21–8)

$$I_T(\text{obs. } \|) = \frac{1}{15} \frac{16\pi^4 v^4}{c^4} \rho I_0 \left(2\sum_j \alpha_j{}^2 - 2\sum_{i<j} \alpha_i \alpha_j \right) \tag{21-62}$$

$$I_T(\text{obs. } \perp) = \frac{1}{15} \frac{16\pi^4 v^4}{c^4} \rho I_0 \left(4\sum_j \alpha_j{}^2 + \sum_{i<j} \alpha_i \alpha_j \right) \tag{21-63}$$

$$I_\|(\text{obs. } \perp) = \frac{1}{15} \frac{16\pi^4 v^4}{c^4} \rho I_0 \left(3\sum_j \alpha_j{}^2 + 2\sum_{i<j} \alpha_i \alpha_j \right) \tag{21-64}$$

The summations in these equations can be written in more lucid form by introducing the spherical part of the polarizability $\boldsymbol{\alpha}$,

$$\bar{\alpha} = \tfrac{1}{3}(\alpha_1 + \alpha_2 + \alpha_3) = \tfrac{1}{3}\operatorname{Tr}\boldsymbol{\alpha} \tag{21-65}$$

and the anisotropic part $\boldsymbol{\beta}$,

$$\boldsymbol{\beta} = \boldsymbol{\alpha} - \bar{\alpha}\mathbf{I} \tag{21-66}$$

where $\mathbf{I}$ is the unit tensor. Note that (Problem 21–9)

$$\begin{aligned}
\beta^2 &\equiv \operatorname{Tr}\boldsymbol{\beta}^2 = \tfrac{1}{9}[(2\alpha_1 - \alpha_2 - \alpha_3)^2 + (2\alpha_2 - \alpha_1 - \alpha_3)^2 + (2\alpha_3 - \alpha_1 - \alpha_2)^2] \\
&= \tfrac{1}{3}[(\alpha_1 - \alpha_2)^2 + (\alpha_2 - \alpha_3)^2 + (\alpha_3 - \alpha_1)^2]
\end{aligned} \tag{21-67}$$

Note that for a linear molecule, the three principal values of the polarizability are $\alpha_\|, \alpha_\perp$, and $\alpha_\perp$, where $\alpha_\|$ is the polarizability along the internuclear axis and $\alpha_\perp$ is that perpendicular to the internuclear axis. If we let the internuclear axis be the Z-axis, the $\alpha_\parallel$ is α_{ZZ} and the $\alpha_\perp$ are α_{XX} and α_{YY}, which are equal due to the cylindrical symmetry. Table 21–1 gives some values of α_1, α_2, and α_3 for a number of molecules.

For linear molecules or, in fact, for symmetric top molecules as well (Problem 21–10)

$$\bar\alpha = \tfrac{1}{3}(\alpha_\parallel + 2\alpha_\perp) \tag{21–68}$$

$$\beta^2 = \tfrac{2}{3}(\alpha_\parallel - \alpha_\perp)^2 \tag{21–69}$$

Values of $\bar\alpha$ and β^2 can be calculated from Table 21–1. In terms of $\bar\alpha^2$ and β^2 then, Eqs. (21–62) through (21–64) become (Problem 21–11)

$$I_T(\text{obs. } \|) = \frac{16\pi^4 v^4}{c^4} \rho I_0 \frac{\beta^2}{5} \tag{21–70}$$

$$I_T(\text{obs. } \perp) = \frac{16\pi^4 v^4}{c^4} \rho I_0 \frac{45\bar\alpha^2 + \frac{21}{2}\beta^2}{45} \tag{21–71}$$

$$I_\|(\text{obs. } \perp) = \frac{16\pi^4 v^4}{c^4} \rho I_0 \frac{45\bar\alpha^2 + 6\beta^2}{45} \tag{21–72}$$

Table 21–1. **Polarizability of molecules**

	$\bar\alpha \cdot 10^{25}$ (cm^3)	$\alpha_1 \cdot 10^{25}$ (cm^3)	$\alpha_2 \cdot 10^{25}$ (cm^3)	$\alpha_3 \cdot 10^{25}$ (cm^3)	location of principal axes and direction of dipole moment; structure of molecule
H_2	7.9	9.3	7.1	7.1	α_1 symm. axis
N_2	17.6	23.8	14.5	14.5	α_1 symm. axis
O_2	16.0	23.5	12.1	12.1	α_1 symm. axis
Cl_2	46.1	66.0	36.2	36.2	α_1 symm. axis
HCl	26.3	31.3	23.9	23.9	α_1 symm. axis
HBr	36.1	42.2	33.1	33.1	α_1 symm. axis
HI	54.4	65.8	48.9	48.9	α_1 symm. axis
CO	19.5	26.0	16.25	16.25	α_1 symm. axis
CO_2	26.5	40.1–41.0	19.7–19.3	19.7–19.3	α_1 symm., linear
SO_2	37.2	54.9	27.2	24.9	$\mu = \mu_3$; $\alpha_2 \perp$ to OSO-plane
N_2O	30.0	48.6	20.7	20.7	α_1 symm. axis
HCN	25.9	39.2	19.2	19.2	α_1 symm. axis, linear
H_2S	37.8	40.4	34.4	40.1	$\mu = \mu_3$; $\alpha_2 \perp$ to HSH-plane
CS_2	87.4	151.4	55.4	55.4	$\alpha_1 =$ symm., axis; linear
NH_3	22.6	24.2	21.8	21.8	α_1 symm. axis; $\mu = \mu_1$; pyramidal
CH_4	26.0	26.0	26.0	26.0	Reg. tetrahedron
C_2H_6	44.7	54.8	39.7	39.7	α_1 symm. axis
$CH_2{=}CH_2$	42.6	(56.1)	(35.9)	(35.9)	α_1 symm. axis
$CH{\equiv}CH$	33.3	51.2	24.3	24.3	α_1 symm. axis, linear
C_3H_8	62.9	50.1	69.3	69.3	$\alpha_1 \perp$ CCC-plane
C_6H_6	103.2	123.1	63.5	123.1	$\alpha_1 \perp$ plane of ring
CH_3Cl	45.6	54.2	41.4	41.4	$\mu = \mu_1$; α_1 symm. axis
CH_2Cl_2	64.8	50.2	84.7	59.6	$\mu = \mu_3$; $\alpha_1 \perp$ ClCCl-plane
$CHCl_3$	82.3	66.8	90.1	90.1	α_1 symm. axis; $\mu = \mu_1$
CCl_4	105.0	105.0	105.0	105.0	Reg. tetrahedron

Source: J. O. Hirschfelder, C. F. Curtiss, and R. B. Bird, *Molecular Theory of Gases and Liquids* (New York: Wiley, 1954).

A quantity that is often experimentally observed and reported is the depolarization ratio, defined as

$$\frac{I_T(\text{obs. }\parallel)}{I_T(\text{obs. }\perp)} = \frac{9\beta^2}{45\bar{\alpha}^2 + \frac{21}{2}\beta^2} \tag{21–73}$$

For future reference, we note that

$$I_\perp(\text{obs. }\perp) = I_T(\text{obs. }\perp) - I_\parallel(\text{obs. }\perp) = \frac{16\pi^4 v^4}{c^4}\rho I_0 \frac{\beta^2}{10} \tag{21–74}$$

depends only upon the anisotropic part of the polarizability β^2 and that

$$I_\parallel(\text{obs. }\perp) - \tfrac{4}{3}I_\perp(\text{obs. }\perp) = \frac{16\pi^4 v^4}{c^4}\rho I_0 \bar{\alpha}^2 \tag{21–75}$$

depends only upon the spherical average $\bar{\alpha}$. We shall see in the next section that $I_\parallel - \frac{4}{3}I_\perp$ and $I_\perp$ can be expressed in terms of the time-correlation function of $\bar{\alpha}$ and $\boldsymbol{\beta}$, respectively.

Although this section is mainly a degression for preparation for the next section on Raman light scattering and later sections on the general theory of light scattering, we are in a position to discuss an interesting pioneering paper by McTague and Birnbaum* on the so-called collision-induced light scattering by dense rare gases. As Eq. (21–73) shows, for molecules that are optically isotropic, such as rare gas atoms, there should be no depolarization since the polarizability is isotropic, and in fact there should be no frequency shift since the scattering should be purely Rayleigh scattering. This is indeed so for single atomic scatterers, but consider the situation at high enough densities, e.g., 100 atm, where there are many binary collisions. As two rare gas atoms collide, they form what can be considered to be a transient homonuclear diatomic molecule. The collision process is essentially a transient vibrational motion, accompanied by a polarizability change due to the distortion of the electronic structure during collision. This transient change in polarizability should lead to a transient anisotropic polarizability which, in turn, should lead to a depolarization of the incident radiation according to Eq. (21–74). This effect was first seen by McTague and Birnbaum who analyzed their experimental results essentially by the equations that we have presented above.

They start with Eqs. (21–71) and (21–70), which say that for the case where the detector measures both the y- and z-polarized components of the scattered radiation, one has

$$I_T(\text{obs. }\perp) = I_0 \rho \frac{\omega^4}{c^4}(\bar{\alpha}^2 + \tfrac{7}{30}\beta^2) \tag{21–76}$$

and

$$I_T(\text{obs. }\parallel) = I_0 \rho \frac{\omega^4}{c^4}\frac{\beta^2}{5} \tag{21–77}$$

where $\omega = 2\pi v$.

Figure 21–4 shows the experimental data for this scattering. Note that the intensity of the scattering is proportional to ρ^2, suggesting that this effect is due to bimolecular

---

* From J. P. McTague and G. Birnbaum, *Phys. Rev. Lett.*, **21**, p. 661, 1968.

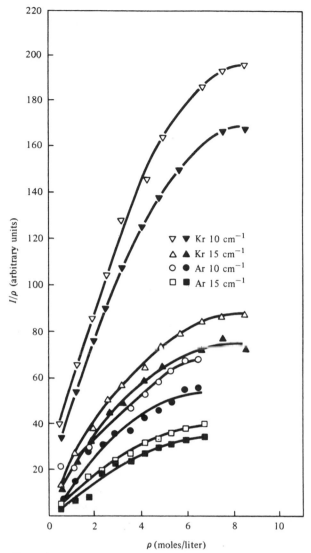

Figure 21–4. **Density dependence of scattering intensity at two frequencies.** z-polarized intensity is indicated by open symbols, x-polarized intensity by filled symbols. (From J. P. McTague and G. Birnbaum, *Phys. Rev., Lett.*, **21**, p. 661, 1968.)

collisions. These expressions can be readily modified for this collision-induced light scattering by noting that in this case both $\bar{\alpha}$ and β^2 are functions of the interatomic distance r. Realizing this, we can modify Eqs. (21–76) and (21–77) by considering $\bar{\alpha}^2$ and β^2 to be given by

$$\bar{\alpha}^2 = \frac{\rho}{2} \int_0^\infty g(r)\bar{\alpha}^2(r)4\pi r^2 \, dr \tag{21–78}$$

and

$$\beta^2 = \frac{\rho}{2} \int_0^\infty g(r)\beta^2(r)4\pi r^2 \, dr \tag{21–79}$$

where $g(r)$ is the radial distribution function, which at these densities is simply

$\exp[-u(r)/kT]$, where $u(r)$ is the intermolecular potential and the factor of $\frac{1}{2}$ in front of the integral is due to the fact that we are dealing with a collision between like atoms.

Equations (21–76) and (21–77) become

$$I_T(\text{obs.} \perp) = 2\pi I_0 \rho^2 \frac{\omega^4}{c^4} \int_0^\infty [\bar{\alpha}^2(r) + \tfrac{7}{30}\beta^2(r)]g(r)r^2 \, dr \tag{21–80}$$

$$I_T(\text{obs.} \parallel) = \frac{2\pi I_0 \rho^2 \omega^4}{5c^4} \int_0^\infty \beta^2(r)g(r)r^2 \, dr \tag{21–81}$$

where $\bar{\alpha}(r)$ and $\beta(r)$ refer to the polarizability increments relative to $r = \infty$.

McTague and Birnbaum find that the depolarization ratio $I_T(\text{obs.} \parallel)/I_T(\text{obs.} \perp)$ is 0.86 ± 0.04 for argon and krypton at frequencies of 10 and 15 cm^{-1}. Assuming that the lineshapes for the incident x- and z-polarizations are the same (Fig. 21–2), these results show that $\langle \bar{\alpha}^2(r) \rangle \ll \langle \beta^2(r) \rangle$, where $\langle \ \rangle$ indicates that average over r indicated in Eqs. (21–80) and (21–81). This shows that as two rare gas atoms collide, the anisotropic part of their transient polarizability change is much larger than the change of their isotropic or spherical part. These measurements serve as a very nice probe into the electronic structures of two colliding molecules. Levine and McQuarrie[*] and DuPré and McTague[†] have presented simple theoretical calculations of the transverse and longitudinal polarizabilities of a pair of colliding rare gas atoms.

21–3 RAMAN LIGHT SCATTERING[‡]

We begin with the quantum-mechanical expression for the differential scattering cross section for scattering into a frequency range $d\omega$ and a solid angle $d\Omega$:[‡]

$$\frac{d^2\sigma}{d\omega \, d\Omega} = \lambda_s^{-4} \sum_{i,f} \rho_i |\langle i| \boldsymbol{\varepsilon}^0 \cdot \boldsymbol{\alpha} \cdot \boldsymbol{\varepsilon}^s |f\rangle|^2 \delta(\omega - \omega_{fi}) \tag{21–82}$$

In this equation, $2\pi\lambda_s$ is the wavelength of the scattered radiation; $\boldsymbol{\varepsilon}^0$ and $\boldsymbol{\varepsilon}^s$ are unit vectors in the direction of the incident and scattered radiation, respectively; $\boldsymbol{\alpha}$ is the polarizability tensor of the group of interacting molecules; and as before, $\omega_{fi} = (E_f - E_i)/\hbar$. The polarizability tensor is given by

$$\boldsymbol{\alpha} \equiv 2e^2 \sum_v \frac{\langle 0|\mathbf{r}|v\rangle\langle v|\mathbf{r}|0\rangle}{E_v - E_0} \tag{21–83}$$

where $\mathbf{r} = \Sigma \mathbf{r}_j$, and the summation here is over the coordinates of all the electrons. Although Eq. (21–82) is very similar in appearance to Eq. (21–10), it is, in fact, not as generally valid. The derivation of Eq. (21–82) involves a number of assumptions: The frequency of the incident radiation must be far from the molecular absorption frequencies (i.e., nonresonant scattering); the frequency shifts must be small compared with both the incident and scattered frequencies; the ground electronic state must be nondegenerate; and $(E_{\min}/\hbar) \gg \omega_0 \gg |\omega_0 - \omega_s|$, where $E_{\min}$ is the electronic excitation energy of the lowest optical transition.

We can now manipulate Eq. (21–82) in much the manner that we used to derive Eq. (21–33). We introduce the Fourier representation for $\delta(\omega - \omega_{fi})$, use Eqs. (21–26)

[*] H. B. Levine and D. A. McQuarrie, *J. Chem. Phys.*, **49**, p. 4181, 1968.

[†] D. B. DuPré and J. P. McTague, *J. Chem. Phys.*, **50**, p. 2024, 1969.

[‡] R. G. Gordon, *Adv. Mag. Resonance*, **3**, p. 1, 1968; *J. Chem. Phys.*, **42**, p. 3658, 1965.

and (21–27) for $e^{-iE_f t/\hbar}|i\rangle$ and $\langle f|e^{iE_f t/\hbar}$, eliminate the summation over f by using closure, and finally get the analog of Eq. (21–31), namely,

$$\lambda^4 \frac{d^2\sigma}{d\omega\, d\Omega} = \frac{1}{2\pi} \int_{-\infty}^{\infty} dt\, e^{-i\omega t} \langle (\boldsymbol{\varepsilon}^0 \cdot \boldsymbol{\alpha}(0) \cdot \boldsymbol{\varepsilon}^s)(\boldsymbol{\varepsilon}^0 \cdot \boldsymbol{\alpha}(t) \cdot \boldsymbol{\varepsilon}^s) \rangle \tag{21–84}$$

As in Section 21–1, we have transformed the scattering cross section from the Schrödinger picture into the Heisenberg picture.

For a spatially isotropic system such as a liquid or a gas, we may average Eq. (21–84) over all (equivalent) directions $\boldsymbol{\varepsilon}^0$ and $\boldsymbol{\varepsilon}^s$. The integrand of Eq. (21–84) contains the factor

$$(\boldsymbol{\varepsilon}^0 \cdot \boldsymbol{\alpha}(0) \cdot \boldsymbol{\varepsilon}^s)(\boldsymbol{\varepsilon}^0 \cdot \boldsymbol{\alpha}(t) \cdot \boldsymbol{\varepsilon}^s) = \left(\sum_{i=1}^{3} \sum_{j=1}^{3} \varepsilon_i{}^0 \alpha_{ij}(0) \varepsilon_j{}^s \right) \left(\sum_{k=1}^{3} \sum_{l=1}^{3} \varepsilon_k{}^0 \alpha_{kl}(t) \varepsilon_l{}^s \right)$$

$$= \sum_{i=1}^{3} \sum_{j=1}^{3} \sum_{k=1}^{3} \sum_{l=1}^{3} \varepsilon_i{}^0 \varepsilon_j{}^s \varepsilon_k{}^0 \varepsilon_l{}^s \alpha_{ij}(0) \alpha_{kl}(t) \tag{21–85}$$

and so we must evaluate the spherical average of $\varepsilon_i{}^0 \varepsilon_j{}^s \varepsilon_k{}^0 \varepsilon_l{}^s$. If we denote this average by $\langle\ \rangle_{\text{sphere}}$, then Eq. (21–84) can be written as

$$\lambda^4 \frac{d^2\sigma}{d\omega\, d\Omega} = \frac{1}{2\pi} \int_{-\infty}^{\infty} dt\, e^{-i\omega t} \sum_{i=1}^{3} \sum_{j=1}^{3} \sum_{k=1}^{3} \sum_{l=1}^{3} \langle \varepsilon_i{}^0 \varepsilon_j{}^s \varepsilon_k{}^0 \varepsilon_l{}^s \rangle_{\text{sphere}} \langle \alpha_{ij}(0) \alpha_{kl}(t) \rangle \tag{21–86}$$

where it should be clear that the angular brackets without a subscript denotes an ensemble average. We shall consider separately the two cases in which $\boldsymbol{\varepsilon}^s$ is parallel to $\boldsymbol{\varepsilon}^0$ and $\boldsymbol{\varepsilon}^s$ is perpendicular to $\boldsymbol{\varepsilon}^0$. In order to do this average in the parallel case, we need to consider quantities $\langle \varepsilon_i{}^{\|} \varepsilon_j{}^{\|} \varepsilon_k{}^{\|} \varepsilon_l{}^{\|} \rangle_{\text{sphere}}$, where $\varepsilon_i{}^{\|}$ is the ith component ($\alpha = x, y, z$) of $\boldsymbol{\varepsilon}^0 = \boldsymbol{\varepsilon}^s$ and the notation $\langle\rangle_{\text{sphere}}$ denotes an average over the surface of a sphere. For example, the following special cases are elementary.

$$\langle \varepsilon_z{}^{\|} \varepsilon_z{}^{\|} \varepsilon_z{}^{\|} \varepsilon_z{}^{\|} \rangle = \langle \cos^4 \theta \rangle = \frac{1}{4\pi} \int_0^{2\pi} d\phi \int_0^{\pi} d\theta \sin\theta \cos^4\theta = \tfrac{1}{5}$$

$$= \langle \varepsilon_x{}^{\|} \varepsilon_x{}^{\|} \varepsilon_x{}^{\|} \varepsilon_x{}^{\|} \rangle = \langle \varepsilon_y{}^{\|} \varepsilon_y{}^{\|} \varepsilon_y{}^{\|} \varepsilon_y{}^{\|} \rangle \tag{21–87}$$

$$\langle \varepsilon_x{}^{\|} \varepsilon_x{}^{\|} \varepsilon_z{}^{\|} \varepsilon_z{}^{\|} \rangle = \langle \sin^2\theta \cos^2\phi \cos^2\theta \rangle = \tfrac{1}{15}$$

$$= \langle \varepsilon_y{}^{\|} \varepsilon_y{}^{\|} \varepsilon_z{}^{\|} \varepsilon_z{}^{\|} \rangle = \langle \varepsilon_x{}^{\|} \varepsilon_x{}^{\|} \varepsilon_y{}^{\|} \varepsilon_y{}^{\|} \rangle \tag{21–88}$$

$$\langle \varepsilon_x{}^{\|} \varepsilon_y{}^{\|} \varepsilon_z{}^{\|} \varepsilon_z{}^{\|} \rangle = 0, \text{ etc.} \tag{21–89}$$

These results can be written in the compact form

$$\langle \varepsilon_i{}^{\|} \varepsilon_j{}^{\|} \varepsilon_k{}^{\|} \varepsilon_l{}^{\|} \rangle_{\text{sphere}} = \tfrac{1}{15}(\delta_{ij}\delta_{kl} + \delta_{ik}\delta_{jl} + \delta_{il}\delta_{jk}) \tag{21–90}$$

The corresponding calculation for the perpendicular case is a little more complicated since one must evaluate $\langle \varepsilon_i{}^{\|} \varepsilon_j{}^{\|} \varepsilon_k{}^{\perp} \varepsilon_l{}^{\perp} \rangle$, where $\varepsilon_k{}^{\perp}$ represents to kth component of $\boldsymbol{\varepsilon}^s$, which in this case is perpendicular to $\boldsymbol{\varepsilon}^0$. One must first express the $\varepsilon_k{}^{\perp}$ in terms of $\varepsilon_i{}^{\|}$ and then perform the integration over the sphere (see Problem 21–12). The result is

$$\langle \varepsilon_\alpha{}^{\|} \varepsilon_\beta{}^{\|} \varepsilon_\gamma{}^{\perp} \varepsilon_\delta{}^{\perp} \rangle_{\text{sphere}} = \tfrac{1}{30}(4\delta_{\alpha\beta}\delta_{\gamma\delta} - \delta_{\alpha\gamma}\delta_{\beta\delta} - \delta_{\alpha\delta}\delta_{\beta\gamma}) \tag{21–91}$$

If we now substitute these results into Eq. (21–86), we get (Problem 21–14)

$$\lambda^4 \left(\frac{d^2\sigma}{d\omega\, d\Omega} \right)_{\|} = \frac{1}{2\pi} \int_{-\infty}^{\infty} dt\, e^{-i\omega t} \frac{1}{15} \sum_{i,j} \langle \alpha_{ii}(0)\alpha_{jj}(t) + 2\alpha_{ij}(0)\alpha_{ij}(t) \rangle \tag{21–92}$$

$$\lambda^4 \left(\frac{d^2\sigma}{d\omega\, d\Omega} \right)_{\perp} = \frac{1}{2\pi} \int_{-\infty}^{\infty} dt\, e^{-i\omega t} \frac{1}{30} \sum_{i,j} \langle 3\alpha_{ij}(0)\alpha_{ij}(t) - \alpha_{ii}(0)\alpha_{jj}(t) \rangle \tag{21–93}$$

These can be rewritten as

$$\lambda^4 \left(\frac{d^2\sigma}{d\omega \, d\Omega}\right)_{\parallel} = \frac{1}{30\pi} \int_{-\infty}^{\infty} dt \, e^{-i\omega t} \langle 2 \, \mathrm{Tr} \, \boldsymbol{\alpha}(0) \cdot \boldsymbol{\alpha}(t) + [\mathrm{Tr} \, \boldsymbol{\alpha}(0)][\mathrm{Tr} \, \boldsymbol{\alpha}(t)] \rangle \qquad (21\text{-}94)$$

$$\lambda^4 \left(\frac{d^2\sigma}{d\omega \, d\Omega}\right)_{\perp} = \frac{1}{60\pi} \int_{-\infty}^{\infty} dt \, e^{-i\omega t} \langle 3 \, \mathrm{Tr} \, \boldsymbol{\alpha}(0) \cdot \boldsymbol{\alpha}(t) - [\mathrm{Tr} \, \boldsymbol{\alpha}(0)][\mathrm{Tr} \, \boldsymbol{\alpha}(t)] \rangle \qquad (21\text{-}95)$$

At this point it is convenient to decompose the polarizability tensor into a spherical part $\bar{\alpha}\mathsf{I}$ and a traceless anisotropic part $\boldsymbol{\beta}$

$$\boldsymbol{\alpha} = \bar{\alpha}\mathsf{I} + \boldsymbol{\beta} \qquad (21\text{-}96)$$

where I is the unit tensor,

$$\bar{\alpha} = \tfrac{1}{3}(\alpha_1 + \alpha_2 + \alpha_3) \qquad (21\text{-}97)$$

and

$$\mathrm{Tr} \, \boldsymbol{\beta} = 0 \qquad (21\text{-}98)$$

Substituting this into Eqs. (21–94) and (21–95) gives (Problem 21–15)

$$\lambda^4 \left(\frac{d^2\sigma}{d\omega \, d\Omega}\right)_{\parallel} = \frac{1}{2\pi} \int_{-\infty}^{\infty} dt \, e^{-i\omega t} \, \tfrac{1}{15} \langle 15\bar{\alpha}(0)\bar{\alpha}(t) + 2 \, \mathrm{Tr} \, \boldsymbol{\beta}(0) \cdot \boldsymbol{\beta}(t) \rangle \qquad (21\text{-}99)$$

$$\lambda^4 \left(\frac{d^2\sigma}{d\omega \, d\Omega}\right)_{\perp} = \frac{1}{2\pi} \int_{-\infty}^{\infty} dt \, e^{-i\omega t} \tfrac{1}{10} \langle \mathrm{Tr} \, \boldsymbol{\beta}(0) \cdot \boldsymbol{\beta}(t) \rangle \qquad (21\text{-}100)$$

Note how these equations compare with Eqs. (21–72) and (21–74). Following through to Eq. (21–75), this leads us to define

$$\frac{d^2\sigma_{\mathrm{pol}}}{d\omega \, d\Omega} \equiv \left(\frac{d^2\sigma}{d\omega \, d\Omega}\right)_{\parallel} - \frac{4}{3} \left(\frac{d^2\sigma}{d\omega \, d\Omega}\right)_{\perp} \qquad (21\text{-}101)$$

and

$$\frac{d^2\sigma_{\mathrm{depol}}}{d\omega \, d\Omega} \equiv 10 \left(\frac{d^2\sigma}{d\omega \, d\Omega}\right)_{\perp} \qquad (21\text{-}102)$$

so that

$$\lambda_s^4 \frac{d^2\sigma_{\mathrm{pol}}}{d\omega \, d\Omega} = \frac{1}{2\pi} \int_{-\infty}^{\infty} dt \, e^{-i\omega t} \langle \bar{\alpha}(0)\bar{\alpha}(t) \rangle \qquad (21\text{-}103)$$

and

$$\lambda_s^4 \frac{d^2\sigma_{\mathrm{depol}}}{d\omega \, d\Omega} = \frac{1}{2\pi} \int_{-\infty}^{\infty} dt \, e^{-i\omega t} \langle \mathrm{Tr} \, \boldsymbol{\beta}(0) \cdot \boldsymbol{\beta}(t) \rangle \qquad (21\text{-}104)$$

If the scattering process produces no change in the vibrational state of the molecules, then the polarized component above corresponds to Rayleigh scattering, and the depolarized component corresponds to the conventional rotational Raman scattering. In the vibrational Raman effect, on the other hand, when the scattering process produces a vibrational transition, the polarized component corresponds to a polarized Q-branch of the vibrational band, while the depolarized component corresponds to the anisotropic part of the vibrational band. This becomes an important matter only when one is actually analyzing spectroscopic data and involves a greater degree of spectroscopic knowledge than is necessary here.

The depolarized component turns out to be particularly interesting. As in Section 21-1, it is convenient both theoretically and experimentally to define a normalized intensity distribution

$$\hat{I}(\omega) = \frac{\lambda^4 \left(\dfrac{d^2\sigma}{d\omega\,d\Omega}\right)_\perp}{\int \lambda^4 \left(\dfrac{d^2\sigma}{d\omega\,d\Omega}\right)_\perp d\omega} \tag{21–105}$$

and a normalized correlation function, which decays from unity at $t = 0$:

$$\hat{C}(t) = \frac{\langle \text{Tr } \boldsymbol{\beta}(0) \cdot \boldsymbol{\beta}(t)\rangle}{\langle \text{Tr } \boldsymbol{\beta}(0) \cdot \boldsymbol{\beta}(0)\rangle} \tag{21–106}$$

The normalized band shape and normalized correlation function are related through the Fourier transform:

$$\hat{I}(\omega) = \frac{1}{2\pi} \int_{-\infty}^{\infty} \hat{C}(t) e^{-i\omega t}\, dt \tag{21–107}$$

$$\hat{C}(t) = \int_{-\infty}^{\infty} \hat{I}(\omega) e^{i\omega t}\, d\omega \tag{21–108}$$

The form of $\hat{C}(t)$ becomes quite simple for a totally symmetric vibration. A totally symmetric vibration in a linear or symmetric top molecule maintains the symmetry of the molecule. For a totally symmetric vibration, the anisotropic part of the polarizability has the form

$$\beta_{ij} = \text{constant}(u_i u_j - \tfrac{1}{3}\delta_{ij}) \tag{21–109}$$

where u_i is the ith component of a unit vector fixed along the symmetry axis of the molecule. Thus

$$\begin{aligned}
\text{Tr } \boldsymbol{\beta}(0) \cdot \boldsymbol{\beta}(t) &= \sum_{i=1}^{3} \sum_{j=1}^{3} [u_i(0)u_j(0) - \tfrac{1}{3}\delta_{ij}][u_i(t)u_j(t) - \tfrac{1}{3}\delta_{ij}] \\
&= \sum_{i=1}^{3} \sum_{j=1}^{3} [u_i(0)u_j(0)u_i(t)u_j(t)] - \tfrac{1}{3}
\end{aligned} \tag{21–110}$$

and the normalized correlation function $\hat{C}(t)$ is

$$\hat{C}(t) = \tfrac{1}{2}\left\langle 3\sum_{i=1}^{3} \sum_{j=1}^{3} u_i(0)u_j(0)u_i(t)u_j(t) - 1 \right\rangle \tag{21–111}$$

In the classical limit, $u(0)$ and $u(t)$ commute, and so $\hat{C}(t)$ can then be written in the form

$$\hat{C}(t) = \langle P_2[\mathbf{u}(0) \cdot \mathbf{u}(t)]\rangle \tag{21–112}$$

where $P_2(x)$ is a Legendre polynomial, $P_2(x) = \tfrac{1}{2}(3x^2 - 1)$. Remember that $\mathbf{u}$ here is a unit vector pointing along the symmetry axis of the molecule. Thus we see that the depolarized component of a Raman spectrum can be Fourier-inverted to give the time-correlation function of the rotation of the molecule in terms of $P_2[\mathbf{u}(0) \cdot \mathbf{u}(t)]$. Figure 21–5 shows such a correlation for liquid nitrogen. Values of the correlation function at times longer than those indicated in the figure are unreliable because of the fairly low resolution in the experimental Raman data used.

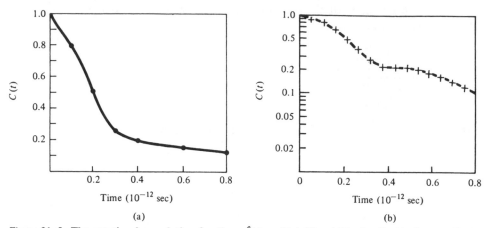

Figure 21–5. **The rotational correlation function, $\hat{C}(t) = \langle P_2(\mathbf{u}(0) \cdot \mathbf{u}(t))\rangle$, for liquid nitrogen. In (a),
$\hat{C}(t)$ is plotted on a linear scale, and in (b) it is plotted on a logarithmic scale.** (From R. G.
Gordon, *J. Chem. Phys.*, **42**, p. 3658, 1965.)

Note that in the case of infrared absorption treated in Section 21–1, the normalized
correlation function can be written in terms of a Legendre polynomial of first order,
$P_1(x) = x$, or

$$\hat{C}(t) = \langle P_1[\mathbf{u}(0) \cdot \mathbf{u}(t)]\rangle$$

We see that infrared and Raman spectra determine different measures of molecular
rotations. Figure 21–6 shows the two different time-correlation functions for liquid
methane at 95°K. Gordon* presents an interesting interpretation and discussion of
these two correlation functions.

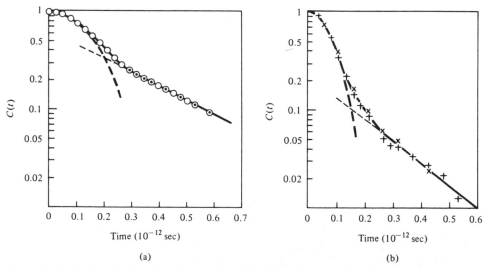

Figure 21–6. (a) **The dipole correlation function, $\langle P_1(\mathbf{u}(0) \cdot \mathbf{u}(t))\rangle = \langle \mathbf{u}(0) \cdot \mathbf{u}(t)\rangle$ for liquid methane at
98°K.** (b) **The rotational correlation function, $\langle P_2(\mathbf{u}(0) \cdot \mathbf{u}(t))\rangle$ for liquid methane at 95°K.**
(From R. G. Gordon, *J. Chem. Phys.*, **43**, p. 1307, 1965.)

* R. G. Gordon, *J. Chem. Phys.*, **43**, p. 1307, 1965.

The two effects that we have considered thus far, infrared and Raman spectra, are both developed in the same manner. One starts with the quantum-mechanical expression for the appropriate cross section in the so-called Schrödinger picture and then transforms it into the Heisenberg picture, which ends up in the form of a Fourier transform of a time-correlation function. The corresponding experimental spectrum can then be numerically inverted to give the time-correlation function itself as a function of time, and this can be used to give much insight into the molecular motion. From the two cases studied so far, we can learn about the molecular orientation as a function of time, both in the gaseous and liquid states.

We could proceed at this point to investigate other applications of this approach. For example, neutron scattering in liquids can be used to probe the time dependence of the translational motion of molecules in liquids. One starts with the appropriate cross section, which in this case is the standard Born approximation for the differential scattering cross section of a collision of a neutron with an atom and then manipulates this expression into the Heisenberg picture and hence a time-correlation function. We shall actually do this in Section 22–1, but now we shall present in the next few sections a more general and abstract formulation of the time-correlation function approach to molecular motion from which the results obtained so far are special cases.

21–4 AN ELEMENTARY DERIVATION OF THE BASIC FORMULAS

In this section we shall present an elementary derivation of the basic formulas of the time-correlation function formalism. This derivation is not only elementary (in the sense that it does not require any specialized mathematical techniques such as operator calculus), but also displays a simple one-to-one correspondence between the mathematical steps in the derivation and the steps in the performance of an experiment. This derivation is due to Zwanzig, whose 1965 review* still serves as an excellent introduction to this subject. For concreteness only, we shall consider the case of electrical conductivity, although the formalism itself is quite general.

In an experiment designed to measure the electrical conductivity of a substance, one applies an electric field across the system and measures the electric current that flows in response to this field. Usually this experiment is repeated a number of times, and the results, the measured currents, are averaged. If we denote the electric current by $\mathbf{J}$, the conductivity (tensor) by σ, and the applied field by $\mathbf{E}$, then the conductivity is defined by the relation

$$\langle \mathbf{J} \rangle = \boldsymbol{\sigma} \cdot \mathbf{E} + O(E^2) \tag{21–113}$$

We are assuming here that the field is small enough that nonlinear effects are of no importance; i.e., we are assuming that Ohm's law is valid. This is an excellent assumption under normal conditions.

Equation (21–113) does not indicate that the quantities involved, particularly the conductivity, may be functions of the frequency of the applied field. As it stands, Eq. (21–113) implies a steady or dc conductance. In order to determine the frequency dependence of the conductivity, one uses a time-dependent electric field. This is turned

* R. Zwanzig, *Ann. Rev. Phys. Chem.*, **16**, p. 67, 1965.

on at $t = 0$, and then one measures the average current $\langle \mathbf{J}(t) \rangle$, which will now, of course, be a function of time. In this case, Eq. (21–113) reads

$$\langle \mathbf{J}(t) \rangle = \int_0^t dt' \Phi(t - t') \cdot \mathbf{E}(t') + O(E^2) \tag{21–114}$$

The (tensor) function $\Phi(t - t')$ here is called the after-effect function and shows that, in general, there may be a delay between the field and the current induced by the field. Values of the field at times earlier than t affect the average current at t. The after-effect function represents a time lag of the system. If the system can respond instantaneously to $\mathbf{E}(t)$, then Φ is a delta function. Ordinarily, however, the system lags behind $\mathbf{E}(t)$, and $\Phi(t)$ is a monotonically decaying function of time.

The integral in Eq. (21–114) is of the form of a Laplace transform convolution. If we take the Fourier-Laplace transform of Eq. (21–114), then we get

$$\langle \mathbf{J}_\omega \rangle = \sigma(\omega) \mathbf{E}_\omega \tag{21–115}$$

where

$$\langle \mathbf{J}_\omega \rangle = \int_0^\infty dt \, e^{-i\omega t} \langle \mathbf{J}(t) \rangle \tag{21–116}$$

$$\mathbf{E}_\omega = \int_0^\infty dt \, e^{-i\omega t} \mathbf{E}(t) \tag{21–117}$$

and, most importantly for our purposes,

$$\sigma(\omega) = \int_0^\infty dt \, e^{-i\omega t} \Phi(t) \tag{21–118}$$

This gives us the frequency-dependent conductivity of the system under study.

Although we are focusing on the case of electrical conductivity, it should be clear at this point that the development presented thus far is general. If we let B be the quantity that responds to the external field F, then Eq. (21–114) becomes

$$\langle B(t) \rangle = \int_0^t \phi(t - t') F(t') \, dt' \tag{21–119}$$

where ϕ is the after-effect function for this particular process. Equation (21–115) becomes

$$\langle B_\omega \rangle = \chi(\omega) F_\omega \tag{21–120}$$

where $\chi(\omega)$ is called the (frequency-dependent) susceptibility of the system.

We shall now show how to express the frequency-dependent conductivity or susceptibility, in general, in terms of a Fourier-Laplace transform of a time-correlation function. Following Zwanzig, we note that there are three stages to an experiment designed to measure electrical conductivity. First, the system must be prepared; then the field is turned on; and then the current is measured. Each of these stages has an analog in the following derivation.

In the first stage the experimenter may select a length of wire from a spool, and then he repeats this process a number of times. In a statistical mechanical sense he is selecting samples from a thermal equilibrium (canonical) ensemble. Let $\mathcal{H}_0$ be the Hamiltonian operator for the entire N-body sample, with corresponding eigenfunctions and eigenvalues given by

$$\mathcal{H}_0 \psi_n = E_n \psi_n \tag{21–121}$$

The probability of observing the system in the state n is given by

$$\rho_n = \frac{e^{-\beta E_n}}{\sum_n e^{-\beta E_n}} = \frac{e^{-\beta E_n}}{Q} \tag{21-122}$$

The average current, or response in general, is found by calculating the response in some particular eigenstate n and then averaging over the canonical ensemble distribution of eigenstates.

In the second stage of the experiment, the field is turned on. This field will interact with the total dipole moment **M** of the system and add a term of the form

$$\mathscr{H}_1(t) = -\mathbf{M} \cdot \mathbf{E}(t) \tag{21-123}$$

to the unperturbed Hamiltonian $\mathscr{H}_0$. As the notation $\mathbf{E}(t)$ suggests, we are assuming for simplicity that the field is uniform over the system, i.e., does not depend upon **r**. Recall that the total dipole moment operator is

$$\mathbf{M} = \sum_j e_j \mathbf{r}_j \tag{21-124}$$

where the summation runs over all charges in the system. If we let the field be directed in the a-direction, then

$$\mathscr{H}_1(t) = -M_a E_a(t) \tag{21-125}$$

The presence of the applied time-dependent field will alter the state of the system, which now must be governed by the time-dependent Schrödinger equation

$$i\hbar \frac{\partial \Psi}{\partial t} = [\mathscr{H}_0 + \mathscr{H}_1(t)]\Psi(t) \tag{21-126}$$

with the initial condition that $\Psi(0) = \psi_n$ with probability ρ_n. If $\mathscr{H}_1(t)$ is small, which is so here, it is a standard problem in time-dependent perturbation theory to show that to first order in the applied field strength (cf. Appendix F)

$$\Psi(t) = \psi_n e^{-i\omega_n t} + \sum_{m \neq n} c_m(t) \psi_m e^{-i\omega_m t} \tag{21-127}$$

where

$$c_m(t) = -i \int_0^t dt' \, e^{i(\omega_m - \omega_n)t'} \langle m | \mathscr{H}_1(t') | n \rangle \tag{21-128}$$

$\omega_n = E_n/\hbar$, and $\hbar$ is set equal to unity.

In the third and final stage of the experiment, one measures the current. In the statistical mechanical theory, this corresponds to calculating the expectation value of the current operator, first with respect to $\Psi(t)$ and then with respect to the canonical distribution of initially prepared states n. The component of the current in the bth direction is

$$J_b = \sum_j e_j v_{jb} = \sum_j e_j \dot{r}_{jb} = \dot{M}_b \tag{21-129}$$

where the notation v_{jb} represents the bth component of the velocity of the jth charged particle. The expectation value of J_b over $\Psi(t)$ is

$$\int \Psi^*(t) J_b \Psi(t)\, d\mathbf{r}^N = \langle n|J_b|n\rangle$$

$$+ \sum_{m \neq n} i^{-1} \int_0^t dt' \langle n|J_b|m\rangle e^{-i(\omega_m - \omega_n)(t - t')} \langle m|\mathcal{H}_1(t')|n\rangle$$

$$- \sum_{m \neq n} i^{-1} \int_0^t dt' \langle m|J_b|n\rangle e^{-i(\omega_n - \omega_m)(t - t')} \langle n|\mathcal{H}_1(t')|m\rangle$$

$$+ O(E^2) \tag{21–130}$$

$$= (J_b)_{nn}$$

$$+ \sum_{m \neq n} i \int_0^t dt' (J_b)_{nm} (M_a)_{mn} e^{-i(\omega_m - \omega_n)(t - t')} E_a(t')$$

$$- \sum_{m \neq n} i \int_0^t dt' (M_a)_{nm} (J_b)_{mn} e^{-i(\omega_n - \omega_m)(t - t')} E_a(t')$$

$$+ O(E^2) \tag{21–131}$$

where we have introduced the more compact notation $(A)_{nm} \equiv \langle n|A|m\rangle$.

This is the expectation value of J_a for a given initial state n. We now canonically average over n by multiplying by ρ_n and summing over n to get

$$\langle J_b(t)\rangle = \langle J_b\rangle_{eq} + \int_0^t dt'\, \phi_{ba}(t - t') E_a(t') + O(E^2) \tag{21–132}$$

where $\langle J_b\rangle_{eq}$ is the average current in the bth direction at equilibrium

$$\langle J_b\rangle_{eq} = \sum_n \rho_n (J_b)_{nn} \tag{21–133}$$

Since there is no current before the field is turned on, $\langle J_b\rangle_{eq} = 0$. The after-effect function indicated in Eq. (21–132) is given by

$$\phi_{ba}(t) = \sum_n \sum_{m \neq n} i\rho_n [(J_b)_{nm}(M_a)_{mn} e^{-i(\omega_m - \omega_n)t} - (M_a)_{nm}(J_b)_{mn} e^{-i(\omega_n - \omega_m)t}] \tag{21–134}$$

We can use the Heisenberg equation of motion to write the matrix elements of J_b in terms of those of M_a. The Heisenberg equation of motion for the operator $\mathbf{M}$ is

$$i\hbar \frac{d\mathbf{M}}{dt} = i\hbar \mathbf{J} = [\mathbf{M}, \mathcal{H}] \tag{21–135}$$

Since we are considering only a linear relation between the current and the field, ϕ_{ba} must be independent of the field, and so (Problem 21–16)

$$i\hbar(\mathbf{J})_{mn} = ([\mathbf{M}, \mathcal{H}_o])_{mn} + O(E)$$
$$= (\mathbf{M}\mathcal{H}_o)_{mn} - (\mathcal{H}_o \mathbf{M})_{mn} + O(E)$$
$$= E_n(\mathbf{M})_{mn} - E_m(\mathbf{M})_{mn} + O(E)$$
$$= (E_n - E_m)(\mathbf{M})_{mn} + O(E) \tag{21–136}$$

In addition, it is easy to show that (Problem 21–17)

$$[\mathbf{J}(t)]_{mn} = e^{-i(\omega_n - \omega_m)t} (\mathbf{J})_{mn} + O(E) \tag{21–137}$$

We can use these last few relations to derive several equivalent formulas for $\phi_{ba}(t)$. If we substitute Eq. (21–137) directly into Eq. (21–134), we see that

$$\phi_{ba}(t) = \frac{1}{i\hbar} \sum_n \sum_{m \neq n} \rho_n \{(M_a)_{nm}[J_b(t)]_{mn} - [J_b(t)]_{nm}(M_a)_{mn}\}$$

$$= \frac{1}{i\hbar} \sum_n \rho_n \{[M_a J_b(t)]_{nn} - [J_b(t)M_a]_{nn}\}$$

$$= \frac{1}{i\hbar} \mathrm{Tr}\{\rho[M_a, J_b(t)]\} \tag{21–138}$$

We have used the fact that $[J_b(t)]_{nn} = (J_b)_{nn} = 0$, according to Eq. (21–136).

Another standard alternative expression for $\phi_{ba}(t)$ can be derived from Eq. (21–138) by noting that

$$\mathrm{Tr}(ABC) = \mathrm{Tr}(BCA) = \mathrm{Tr}(CAB) \tag{21–139}$$

which says that the trace is invariant to a cyclic permutation of the operators or matrices involved. This allows us to write

$$\phi_{ba}(t) = \frac{1}{i\hbar} \mathrm{Tr}\{\rho[M_a, J_b(t)]\} \tag{21–140}$$

$$= \frac{1}{i\hbar} \{\mathrm{Tr}[\rho M_a J_b(t)] - \mathrm{Tr}[\rho J_b(t) M_a]\} \tag{21–141}$$

$$= \frac{1}{i\hbar} \{\mathrm{Tr}[\rho M_a J_b(t)] - \mathrm{Tr}[M_a \rho J_b(t)]\} \tag{21–142}$$

$$= \frac{1}{i\hbar} \{\mathrm{Tr}[\rho, M_a] J_b(t)\} \tag{21–143}$$

Equations (21–140) and (21–143) are two standard expressions for $\phi_{ba}(t)$.

We can use Eqs. (21–138) and (21–136) to derive one other expression for $\phi_{ba}(t)$ (Problem 21–26)

$$\phi_{ba}(t) = \sum_n \sum_{m \neq n} \rho_n (J_a)_{nm}[J_b(t)]_{mn} \frac{e^{-\beta(E_m - E_n)} - 1}{E_n - E_m} \tag{21–144}$$

This can be written in a compact form by defining a so-called Kubo transform by the relation

$$\tilde{J}_a = \frac{1}{\beta} \int_0^\beta d\lambda \, e^{\lambda \mathscr{H}_0} J_a e^{-\lambda \mathscr{H}_0} \tag{21–145}$$

The matrix elements of $\tilde{J}_a$ are (Problem 21–28)

$$(\tilde{J}_a)_{mn} = \frac{1}{\beta} \int_0^\beta d\lambda \, e^{\lambda(E_m - E_n)}(J_a)_{mn}$$

$$= \frac{1}{\beta} \left[\frac{e^{-\beta(E_n - E_m)} - 1}{E_m - E_n} \right] (J_a)_{mn} \tag{21–146}$$

and so Eq. (21–144) can be written as

$$\phi_{ba}(t) = \beta \sum_n \sum_{m \neq n} \rho_n (\tilde{J}_a)_{nm} [J_b(t)]_{mn}$$
$$= \beta \, \mathrm{Tr}[\rho \tilde{J}_a J_b(t)]$$
$$= \beta \langle \tilde{J}_a J_b(t) \rangle \tag{21–147}$$

This last form is most convenient for taking the classical limit since in the classical limit the Kubo transform of an operator A becomes simply the classical expression for A (Problem 21–20). The classical limit of the other two expressions for $\phi_{ba}(t)$ is more difficult to obtain since both the numerators and denominators go to zero as $\hbar \to 0$ (Problem 21–19). Since $\mathrm{Tr}\,\rho A$ corresponds to an integral over phase space in the classical limit, we see that

$$\mathrm{Tr}\,[\rho \dot{A}B(t)] \leftrightarrow \int \cdots \int dp \, dq \, f_0(p, q) \dot{A}(0) B(t) \tag{21–148}$$

where

$$\rho = \frac{e^{-\beta \mathscr{H}}}{\mathrm{Tr}\,(e^{-\beta \mathscr{H}})} \leftrightarrow f_0 = \frac{e^{-\beta H(p, q)}}{\int \cdots \int dp \, dq \, e^{-\beta H(p, q)}} \tag{21–149}$$

Similarly, since the quantum-mechanical commutator corresponds to the classical mechanical Poisson bracket, then we also have

$$\frac{1}{i\hbar} \mathrm{Tr}[\{[\rho, A]B(t)\} \leftrightarrow \int \cdots \int dp \, dq \, (f_0, A)B(t) \tag{21–150}$$

$$\frac{1}{i\hbar} \mathrm{Tr}\{\rho[A, B(t)]\} \leftrightarrow \int \cdots \int dp \, dq \, f_0(A, B(t)) \tag{21–151}$$

where $(A, B(t))$ here denotes the Poisson bracket notation

$$(A, B) = \sum_i \left(\frac{\partial A}{\partial q_i} \frac{\partial B}{\partial p_i} - \frac{\partial A}{\partial p_i} \frac{\partial B}{\partial q_i} \right) \tag{21–152}$$

Although we have obtained these classical limits essentially by inspection, they can be derived directly from the Liouville equation. We shall do this in Section 21–7 after we have discussed a few applications of the general formalism derived here.

In summary, then, we have derived an expression for the frequency-dependent conductivity in terms of a Laplace-Fourier transform of a time-correlation function:

$$\sigma_{ba}(\omega) = \beta \int_0^\infty dt \, e^{-i\omega t} \langle \tilde{J}_a J_b(t) \rangle \tag{21–153}$$

Although we have used the example of electrical conductivity for concreteness, the above derivation is quite general. It is straightforward to show (Problem 21–29) that if

$$\mathscr{H}_1(t) = -AF(t) \tag{21–154}$$

and if the response is denoted by B, then

$$\langle B(t) \rangle = \int_0^t dt' \, \phi(t - t')F(t') + O(F^2) \tag{21–155}$$

and

$$\langle B_\omega \rangle = \chi(\omega)F_\omega \tag{21–156}$$

where

$$\chi(\omega) = \int_0^\infty dt \, e^{-i\omega t} \phi(t) \qquad (21\text{-}157)$$

The quantity $\chi(\omega)$ is the frequency-dependent susceptibility of the system. The equivalent forms of $\phi(t)$ corresponding to Eqs. (21-143), (21-138), and (21-147) are

$$\phi(t) = \frac{1}{i\hbar} \text{Tr}\{[\rho, A]B(t)\} \qquad (21\text{-}158)$$

$$= \frac{1}{i\hbar} \text{Tr}\{\rho[A, B(t)]\} \qquad (21\text{-}159)$$

$$= \beta \, \text{Tr}[\rho \tilde{A} B(t)] \qquad (21\text{-}160)$$

where the tilde over $\dot{A}$ denotes the Kubo transform

$$\tilde{A} = \frac{1}{\beta} \int_0^\beta d\lambda \, e^{\lambda \mathscr{H}} \dot{A} e^{-\lambda \mathscr{H}} \qquad (21\text{-}161)$$

Problem 21-22 investigates some of the properties of $\phi(t)$. For example, it is shown there that $\phi(t)$ is real and odd if $A = B$.

It may not be obvious at this point that the spectroscopic applications of Sections 21-1 and 21-3 are readily derived from this formalism. In the next two sections we shall show that this is so.

21-5 DIELECTRIC RELAXATION

In this section we shall discuss the behavior of a set of noninteracting dipoles in a time-dependent external field. At low enough frequencies, the orientation of the dipoles can keep up with the instantaneous value of the field, but as the frequency increases, the dipoles start to lag behind the field. This time lag is commonly referred to as dielectric relaxation, and we shall see below that it is responsible for an absorption of energy from the external field.

Consider a nonconducting isotropic polarizable gas and an externally applied electric field $\mathbf{E}(t)$, which we take to be in the z-direction. Since the system is nonconducting, this electric field induces a net dipole moment in the system. The function $B(t)$ of the previous section, then, is the total dipole moment $M_z(t)$. The Hamiltonian of the system is

$$\mathscr{H} = \mathscr{H}_0 - M_z E_z(t) \qquad (21\text{-}162)$$

where $\mathscr{H}_0$ is the Hamiltonian of the entire system in the absence of the electric field, and so the function A of the previous section is M_z. The susceptibility $\chi(\omega)$ is given by

$$\chi(\omega) = \int_0^\infty e^{-i\omega t} \phi(t) \, dt \qquad (21\text{-}163)$$

with

$$\phi(t) = \frac{1}{i\hbar} \text{Tr}\{\rho[M_z, M_z(t)]\} \qquad (21\text{-}164)$$

$$= \beta \, \text{Tr}[\rho \tilde{M}_z M_z(t)] \qquad (21\text{-}165)$$

The susceptibility can be expressed in terms of the frequency-dependent dielectric constant. If **P** is the polarization of the gas, i.e., the total dipole moment per unit volume, then

$$\mathbf{P} = \chi(\omega)\mathbf{E} \tag{21-166}$$

Since $\mathbf{D} = \mathbf{E} + 4\pi\mathbf{P}$ and $\mathbf{D} = \varepsilon\mathbf{E}$, we have the relation between the susceptibility and the dielectric constant:

$$\varepsilon(\omega) = 1 + 4\pi\chi(\omega) \tag{21-167}$$

Notice that since $\phi(t)$ is real (Problem 21–22), $\chi(\omega)$ defined by Eq. (21–163) is a complex quantity. This means that $\varepsilon(\omega)$ is formally complex also. Of course, physical quantities are real and not complex, but it is often convenient to define certain functions to be complex and then identify the real and imaginary parts with measurable quantities. For example, if one introduces a harmonic electric field $e^{i\omega t}$ into the equations of electromagnetism, one is led quite naturally to define a complex dielectric constant whose imaginary part is directly related to the absorption of electromagnetic radiation by the system. This is discussed in some detail in Appendix K. It is shown there that if we write $\varepsilon(\omega)$ as

$$\varepsilon(\omega) = \varepsilon'(\omega) - i\varepsilon''(\omega) \tag{21-168}$$

then the imaginary part $\varepsilon''(\omega)$ is given by

$$\varepsilon''(\omega) = \frac{nc\alpha(\omega)}{\omega} \tag{21-169}$$

where n is the index of refraction, c is the speed of light, and $\alpha(\omega)$ is the absorption coefficient of Lambert's law, $I = I_0 \exp(-\alpha x)$. Since $\varepsilon''(\omega)$ represents the absorption by the system it is often called the dielectric loss. Substances with large values of $\varepsilon''(\omega)$ are said to be "lossy" in that frequency region.

Now let us consider the classical limit of Eq. (21–164) for a dilute gas. We can ignore the tilde denoting the Kubo transform in this limit and write

$$\phi(t) = \beta\langle \dot{M}_z(0)M_z(t)\rangle$$
$$= \beta\sum_{i,j}\langle \dot{\mu}_{zi}(0)\mu_{zj}(t)\rangle \tag{21-170}$$

In a dilute gas, we can neglect those terms here in which $i \neq j$; i.e., we can assume that the dipole moments on different molecules are uncorrelated. Thus we can write $\phi(t)$ as

$$\phi(t) = \beta N\langle \dot{\mu}_z(0)\mu_z(t)\rangle \tag{21-171}$$

We can write $\langle \dot{\mu}_z(0)\mu_z(t)\rangle$ as the time derivative of $\langle \mu_z(0)\mu_z(t)\rangle$ by means of the following set of manipulations. First, we must recognize that since the angular brackets in the expressions for time-correlation functions denote an *equilibrium* ensemble average, we can write that

$$\langle A(0)B(t)\rangle = \langle A(s)B(t+s)\rangle \tag{21-172}$$

An average that satisfies this condition is said to be *stationary*. Physically, it means that there is no preferred origin of time in a system that is in equilibrium. In particular, if we let $s = -t$,

$$\langle A(0)B(t)\rangle = \langle A(-t)B(0)\rangle$$

Using this, we can write

$$\langle \dot\mu_z(0)\mu_z(t)\rangle = \langle \dot\mu_z(-t)\mu_z(0)\rangle$$

$$= -\frac{d}{dt}\langle \mu_z(-t)\mu_z(0)\rangle$$

$$= -\frac{d}{dt}\langle \mu_z(0)\mu_z(t)\rangle \tag{21–173}$$

The reader should get used to the kind of manipulations that lead to Eq. (21–173).

In order to calculate the time-correlation function, we would have to solve the equations of motion of the gas to find $\mu_z(t)$, which we would then canonically average. This is certainly possible in the case of a dilute gas, but becomes a formidable task for dense systems. Fortunately, time-correlation functions have a physical interpretation which allows one to devise sensible approximations. The dipole moment time-correlation function is easy to picture physically. It is a measure of the correlation of $\mu_z(t)$ with $\mu_z(0)$. Initially it is $\langle \mu_{z0}^2\rangle$, but as the molecule suffers collisions with its neighbors, this correlation decays to zero. The molecule no longer remembers the initial direction of its dipole moment. The Debye theory of dielectric behavior is obtained by simply assuming that $\langle \mu_z(0)\mu_z(t)\rangle$ decays exponentially from its initial value, i.e.,

$$\langle \mu_z(0)\mu_z(t)\rangle = \mu_{z0}^2 e^{-|t|/\tau} \tag{21–174}$$

where τ is a measure of the duration of the correlation. The time τ is called a relaxation time. If we substitute Eq. (21–174) into Eq. (21–173), this result into Eq. (21–170), and that into Eq. (21–163), we get

$$\chi(\omega) = \frac{\beta\mu_0^2}{\tau}\left(\int_0^\infty e^{-t/\tau}\cos\omega t\,dt - i\int_0^\infty e^{-t/\tau}\sin\omega t\,dt\right) \tag{21–175}$$

$$= \beta\mu_0^2\left(\frac{1}{1+\omega^2\tau^2} - \frac{i\omega\tau}{1+\omega^2\tau^2}\right) \tag{21–176}$$

This gives

$$\varepsilon'(\omega) - 1 = 4\pi\beta\mu_0^2\left(\frac{1}{1+\omega^2\tau^2}\right) \tag{21–177}$$

$$\varepsilon''(\omega) = 4\pi\beta\mu_0^2\left(\frac{\omega\tau}{1+\omega^2\tau^2}\right) \tag{21–178}$$

These are the famous Debye equations. They show that $\varepsilon'(\omega)$ decreases from $\varepsilon'(0)$ to unity with increasing frequency. Most of this decrease occurs within a fairly small frequency range centered about $\omega = 1/\tau$. Similarly, $\varepsilon''(\omega)$ changes from a small value through a maximum to a small value again. This maximum occurs at $\omega = 1/\tau$. Physically, as the frequency of the external field approaches $1/\tau$, the dipoles start to lag behind the field, and eventually the frequency becomes high enough that they essentially do not respond to the field at all, and hence $\varepsilon'(\omega) \to 1$. Equation (21–178) shows that as the molecules begin to lag behind the field, they absorb energy from the field, and continue to do so until the frequency becomes so high that the dipoles no longer respond at all.

Figure 21–7 shows experimental data for iso-butyl bromide. The agreement between theory and experiment is seen to be very good in this case. It should be pointed out,

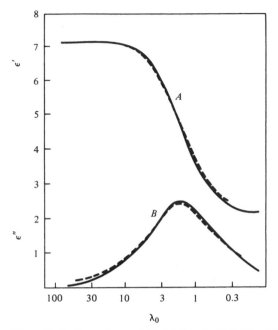

Figure 21–7. **Dependence of $\varepsilon'(\omega)$ (A) and $\varepsilon''(\omega)$ (B) of i-butyl bromide upon the logarithm of the wavelength (cm). The solid lines are the Debye curves, and the dashed lines are drawn through the experimental data.** (From E. J. Hennelly, W. M. Heston, Jr., and C. P. Smyth, *J. Amer. Chem. Soc.*, **70**, p. 4102, 1948.)

however, that this is somewhat exceptional. In many cases the dispersion occurs over a wider frequency range, with a maximum value of $\varepsilon''(\omega)$ lower than that predicted by Eq. (21–178). This effect has been attributed to the fact that there is actually a number of, or a distribution of, relaxation times, rather than just the one as implied in Eq. (21–174). The reader may notice in Fig. 21–7 that $\varepsilon'(\omega)$ levels off between 2 and 3 rather than unity. This is simply due to the fact that there are other mechanisms that contribute to the dielectric constant at the frequencies given in the figure.

Before leaving this section we wish to discuss an important relation between $\varepsilon'(\omega)$ and $\varepsilon''(\omega)$ or, more generally, between $\chi'(\omega)$ and $\chi''(\omega)$. This connection, called the Kramers-Kronig relation, is

$$\chi'(\omega) = \frac{2}{\pi} \int_0^\infty \chi''(\omega') \frac{\omega' \, d\omega'}{\omega'^2 - \omega^2} \tag{21–179}$$

$$\chi''(\omega) = \frac{2}{\pi} \int_0^\infty \chi'(\omega') \frac{\omega \, d\omega'}{\omega^2 - \omega'^2} \tag{21–180}$$

The Kramers-Kronig relations show that $\chi'(\omega)$ and $\chi''(\omega)$ are not independent quantities; a knowledge of one implies a knowledge of the other.

The proof of the Kramers-Kronig relations goes as follows. Since

$$\chi(\omega) = \chi'(\omega) - i\chi''(\omega) = \int_0^\infty e^{-i\omega t} \phi(t) \, dt \tag{21–181}$$

we have

$$\chi'(\omega) = \int_0^\infty \phi(t) \cos \omega t \, dt \tag{21–182}$$

and

$$\chi''(\omega) = \int_0^\infty \phi(t) \sin \omega t \, dt \tag{21-183}$$

Equations (21–182) and (21–183) can both be inverted to give

$$\phi(t) = \frac{2}{\pi} \int_0^\infty \chi'(\omega) \cos \omega t \, d\omega \tag{21-184}$$

$$= \frac{2}{\pi} \int_0^\infty \chi''(\omega) \sin \omega t \, d\omega \tag{21-185}$$

To prove Eq. (21–179), we substitute Eq. (21–185) into Eq. (21–182):

$$\chi'(\omega) = \frac{2}{\pi} \int_0^\infty dt \cos \omega t \int_0^\infty \chi''(\omega') \sin \omega' t \, d\omega'$$

$$= \frac{2}{\pi} \lim_{R \to \infty} \int_0^\infty d\omega' \, \chi''(\omega') \int_0^R \cos \omega t \sin \omega' t \, dt$$

$$= \frac{2}{\pi} \lim_{R \to \infty} \int_0^\infty d\omega' \, \chi''(\omega') \frac{1}{2} \left[\frac{1 - \cos(\omega' + \omega)R}{\omega' + \omega} + \frac{1 - \cos(\omega' - \omega)R}{\omega' - \omega} \right]$$

Now the integrals containing the cosine terms vanish as $R \to \infty$ because the integrands oscillate infinitely rapidly. Therefore, we finally have

$$\chi'(\omega) = \frac{2}{\pi} \int_0^\infty \chi''(\omega') \frac{\omega' \, d\omega'}{\omega'^2 - \omega^2} \tag{21-186}$$

which proves Eq. (21–179). The proof of Eq. (21–180) is very similar (Problem 21–23).

Kittel* gives an interesting application of the Kramers-Kronig relations. Problem 21–24 asks the reader to show that the Debye equations, Eqs. (21–177) and (21–178), satisfy the Kramers-Kronig relations.

21–6 TIME-CORRELATION FUNCTION FORMALISM OF MOLECULAR SPECTROSCOPY

We saw in the previous section that the imaginary part of the dielectric constant was directly related to the absorption of electromagnetic radiation. This is what provides a way for us to investigate the shape of spectral lines in terms of the dipole moment correlation function. As indicated in the first few sections, this approach was developed by Gordon and has been applied to infrared absorption, Raman scattering, fluorescence, nuclear magnetic resonance, and many other phenomena.† We shall study some of these applications in this section within the more general formalism developed in Section 21–4.

According to Eqs. (21–163) and (21–167), the imaginary part of the dielectric constant is given by

$$\varepsilon''(\omega) = 4\pi \int_0^\infty \phi(t) \sin \omega t \, dt \tag{21-187}$$

---

* *Statistical Physics*, New York: Wiley, 1958.

† See, e.g., R. G. Gordon, *J. Chem. Phys.*, **42**, p. 3658, 1965; **43**, p. 1307, 1965; **44**, p. 3083, 1966; **45**, p. 1643, 1966; **45**, p. 1649, 1966; *Adv. Mag. Resonance*, **3**, p. 1, 1968.

Problem 21–61 shows that $\phi(t)$ is an odd function of t, and so we can write Eq. (21–187) in the form

$$\varepsilon''(\omega) = -\frac{4\pi}{2i} \int_{-\infty}^{\infty} \phi(t) e^{-i\omega t} \, dt$$

$$= -\frac{2\pi}{\hbar} \int_{-\infty}^{\infty} \text{Tr}[\mu_z(t)\mu_z \rho_0 - \mu_z \mu_z(t)\rho_0] e^{-i\omega t} \, dt \qquad (21\text{–}188)$$

where we have used Eq. (21–164) for $\phi(t)$.

This integral can be written in a more convenient form by using the relation

$$\int_{-\infty}^{\infty} \text{Tr}[\rho A B(t)] e^{-i\omega t} \, dt = e^{\beta \hbar \omega} \int_{-\infty}^{\infty} \text{Tr}[\rho B(t) A] e^{-i\omega t} \, dt \qquad (21\text{–}189)$$

The proof of this equation goes as follows. Substituting in the expressions for ρ and $B(t)$, we have

$$\int_{-\infty}^{\infty} \text{Tr}[\rho A B(t)] e^{-i\omega t} \, dt = \frac{1}{Q} \int_{-\infty}^{\infty} \text{Tr}[e^{-\beta \mathscr{H}} A e^{i\mathscr{H}t/\hbar} B e^{-i\mathscr{H}t/\hbar}] e^{-i\omega t} \, dt \qquad (21\text{–}190)$$

We can change the integration variables from t to $t - i\beta\hbar$ by bringing the factor $\exp(-\beta\mathscr{H})$ from the first to the last position in the expression in brackets (the trace is unaffected by this cyclic permutation), inserting $\exp(\beta\mathscr{H}) \exp(-\beta\mathscr{H})$ directly before B, and multiplying the front by $\exp(\beta\hbar\omega)$. Doing all this gives

$$\frac{1}{Q} \int_{-\infty}^{\infty} \text{Tr}[e^{-\beta \mathscr{H}} A e^{i\mathscr{H}t/\hbar} B e^{-i\mathscr{H}t/\hbar}] e^{-i\omega t} \, dt$$

$$= \frac{e^{\beta \hbar \omega}}{Q} \int_{-\infty}^{\infty} \text{Tr}[A e^{i\mathscr{H}(t - i\beta\hbar)/\hbar} e^{-\beta\mathscr{H}} B e^{-i\mathscr{H}(t - i\beta\hbar)/\hbar}] e^{-i\omega(t - i\beta\hbar)} \, dt \qquad (21\text{–}191)$$

If we now bring A from the first to the last position, bring $\exp(-\beta\mathscr{H})$ through to the first position (this term commutes with $\exp[i\mathscr{H}(t - i\beta\hbar)/\hbar]$ since $\mathscr{H}$ commutes with itself), let $t - i\beta\hbar = z$, we get that

$$\int_{-\infty}^{\infty} \text{Tr}[\rho A B(t)] e^{-i\omega t} \, dt = e^{\beta \hbar \omega} \int_{-\infty - i\beta\hbar}^{\infty + i\beta\hbar} \text{Tr}[\rho B(z) A] e^{-i\omega z} \, dz$$

$$= e^{\beta \hbar \omega} \int_{-\infty}^{\infty} \text{Tr}[\rho B(t) A] e^{-i\omega t} \, dt \qquad (21\text{–}192)$$

where we have assumed that $\text{Tr}[\rho B(t) A] \to 0$ rapidly enough at $t \pm \to \infty$ and that it is an analytic function of z for $0 \geq \text{Im } z \geq -\beta\hbar$. This equation says that a factor of $\exp(\beta\hbar\omega)$ results when the order of the operators in the Fourier transform of a quantum-mechanical correlation function is reversed. Of course, this factor is unity in the classical limit, where the operators become ordinary dynamical functions, which commute. This is just one of many relationships among various forms of correlation functions.

If we use this result on Eq. (21–188), we have

$$\varepsilon''(\omega) = \frac{2\pi(1 - e^{-\beta\hbar\omega})}{\hbar} \int_{-\infty}^{\infty} \langle \mu_z \mu_z(t) \rangle e^{-i\omega t} \, dt \qquad (21\text{–}193)$$

We define an absorption lineshape $I(\omega)$ by

$$I(\omega) \equiv \frac{\hbar \varepsilon''(\omega)}{4\pi^2(1 - e^{-\beta\hbar\omega})} = \frac{1}{2\pi} \int_{-\infty}^{\infty} \langle \mu_z \mu_z(t) \rangle e^{-i\omega t} \, dt \qquad (21\text{--}194)$$

Using the reaction between $\varepsilon''(\omega)$ and $\alpha(\omega)$ given by Eq. (21–169), we see that this is Eq. (21–18), which we derived in Section 21–1 by starting from quantum-mechanical time-dependent perturbation theory.

Note carefully that the time-dependent operator $\mu_z(t)$ occurs on the right-hand side in the above correlation function. Such a correlation function is said to be *one sided*. Several important properties of the one-sided correlation function $\langle \mu_z \mu_z(t) \rangle$ follow from the fact that the frequency spectrum $I(\omega)$ is a real quantity, as is clear from its definition in Eq. (21–194). Thus letting $C(t) \equiv \langle \mu_z \mu_z(t) \rangle$ we have

$$I(\omega) = I^*(\omega) = \frac{1}{2\pi} \int_{-\infty}^{\infty} dt \, C(t)^* e^{i\omega t}$$

$$= \frac{1}{2\pi} \int_{-\infty}^{\infty} dt \, C(-t)^* e^{-i\omega t} \qquad (21\text{--}195)$$

where we have changed the integration variable from t to $-t$. Comparing Eq. (21–194) and Eq. (21–195), we have

$$C(t) = C(-t)^* \qquad (21\text{--}196)$$

If we separate this into its real and imaginary parts, we get

$$\text{Re } C(t) = \text{Re } C(-t) \qquad (21\text{--}197)$$

and

$$\text{Im } C(t) = -\text{Im } C(-t) \qquad (21\text{--}198)$$

Thus the real part of a one-sided correlation function is an even function of time, and the imaginary part is an odd function of time. Since classical mechanical correlation functions are real functions, we find that they must be even functions of time. Quantum-mechanical correlation functions, on the other hand, are generally complex quantities and have both even and odd parts as a function of time. The imaginary part, being an odd function of time according to Eq. (21–198), represents a strictly quantum-mechanical contribution, which becomes negligible as the temperature increases. A classical correlation function, being real and even in time, leads to a lineshape function $I(\omega)$ that is even in ω, or, i.e., symmetric about ω_0, and so the observed asymmetry in $I(\omega)$ can be used to determine the strictly quantum-mechanical part of $C(t)$. In practice, for most molecular gases at most temperatures, classical mechanics provides a sufficiently good approximation to $C(t)$.

Even though a calculation of time-correlation function for all times is a difficult dynamical calculation, it is possible to write $C(t)$ as a power series expansion about the initial time. By Taylor's theorem, the expansion is

$$\langle A(0)A(t) \rangle = \sum_{n=0}^{\infty} \frac{t^n}{n!} \left| \frac{d^n}{dt^n} \langle A(0)A(t) \rangle \right|_{t=0} \qquad (21\text{--}199)$$

The time derivatives may be evaluated by a repeated application of Heisenberg's equation of motion for $A(t)$:

$$\frac{dA}{dt} = \frac{i}{\hbar} [\mathcal{H}, A] \qquad (21\text{--}200)$$

This gives the general expansion

$$\langle A(0)A(t)\rangle = \sum_{n=0}^{\infty} \frac{(it)^n}{\hbar^n n!} \langle A(0), [\mathcal{H}, [\mathcal{H}, [\mathcal{H}, \ldots, [\mathcal{H}, A(0)]\ldots]]]\rangle \qquad (21\text{--}201)$$

in which the commutator is repeated until the Hamiltonian operator appears n times in the nth term.

An important point to notice here is that the coefficients of t^n involve no dynamical calculation; they are simply averages of equilibrium quantities. This can be readily seen by considering the classical limit of Eq. (21–199). In that case, we have

$$\langle A(0)A(t)\rangle = \left\langle A(0)\left[A(0) + t\dot{A}(0) + \frac{t^2}{2!}\ddot{A}(0) + \cdots\right]\right\rangle$$

$$= \langle A(0)A(0)\rangle + t\langle A(0)\dot{A}(0)\rangle + \frac{t^2}{2!}\langle A(0)\ddot{A}(0)\rangle + \cdots \qquad (21\text{--}202)$$

The coefficient of t^2 can be written in a different form by using the relation

$$\langle A(0)\ddot{A}(0)\rangle = \left[\frac{d^2}{dt^2}\langle A(0)A(t)\rangle\right]_{t=0}$$

$$= \left[\frac{d}{dt}\langle A(0)\dot{A}(t)\rangle\right]_{t=0}$$

$$= \left[\frac{d}{dt}\langle A(-t)\dot{A}(0)\rangle\right]_{t=0}$$

$$= -\langle \dot{A}(0)\dot{A}(0)\rangle \qquad (21\text{--}203)$$

In addition, we can show that the coefficient of t vanishes:

$$\langle A(0)\dot{A}(0)\rangle = \frac{1}{2}\left\langle \frac{d}{dt}A^2(t)\right\rangle_{t=0} = 0 \qquad (21\text{--}204)$$

since the rate of change of a quantity must be zero at equilibrium. Equation (21–202) then becomes

$$\langle A(0)A(t)\rangle = \langle A(0)A(0)\rangle - \frac{t^2}{2!}\langle \dot{A}(0)\dot{A}(0)\rangle + \cdots \qquad (21\text{--}205)$$

Notice that this result is consistent with the fact that classical time-correlation functions must be even in time. Furthermore, notice from Eq. (21–201) that the even powers of t are real and that it is the odd powers of t that are imaginary. Not only do the odd powers of t not vanish in the quantum-mechanical case, but these terms are responsible for the imaginary part of a one-sided quantum-mechanical time-correlation function.

The coefficients in the time power series may be given a physical identification as moments of the appropriate frequency spectrum. We can invert Eq. (21–194) and expand the exponential in the integrand, giving

$$\langle \mu_z \mu_z(t) \rangle = \int_{-\infty}^{\infty} e^{i\omega t} I(\omega) \, d\omega$$

$$= \sum_{n=0}^{\infty} \frac{(it)^n}{n!} \int_{-\infty}^{\infty} \omega^n I(\omega) \, d\omega \tag{21–206}$$

Comparing this with Eq. (21–201), we see that the moments of $I(\omega)$ may be calculated by an equilibrium average. Often the first few moments give a satisfactory parametrization of the lineshape.

Perhaps we should say a word at this point about the limits of ω in Eq. (21–206). We ordinarily do not think of a frequency as taking on negative values, but it is often very useful to do so. For example, if $I(\omega)$ represents the lineshape of some particular vibrational absorption band, it is convenient to take the zero of frequency at the band peak. In Fig. 21–10, which shows the profile of a particular vibrational absorption of CH_4 broadened by He, we would consider the curves to be centered about $\omega = 0$. Similarly, in Fig. 21–9, which shows the far infrared absorption spectrum of an argon-neon mixture, we would extend the curve symmetrically through $\omega = 0$. This is strictly a matter of convenience, however.

Let us now return to Eq. (21–194):

$$I(\omega) = \frac{1}{2\pi} \int_{-\infty}^{\infty} \langle \mu_z \mu_z(t) \rangle e^{-i\omega t} \, dt \tag{21–207}$$

We can derive approximate lineshapes by assuming simple, physical approximations for $\langle \mu_z \mu_z(t) \rangle$. We shall consider only the classical limit; so our approximations will be real, and hence $I(\omega)$ will be a symmetric function. We treat three examples.

COLLISIONLESS MICROWAVE ABSORPTION

Consider a collisionless gas of permanent dipolar molecules. Since there are no collisions, the molecules simply rotate undisturbed with some angular frequency ω_0, and the dipole moment time-correlation function can be written as

$$\langle \mu_z \mu_z(t) \rangle = \mu_{0z}^2 \cos \omega_0 t \tag{21–208}$$

If we substitute this into Eq. (21–207), we get

$$I(\omega) = \frac{\mu_{0z}^2}{2} [\delta(\omega_0 - \omega) + \delta(\omega_0 + \omega)] \tag{21–209}$$

This result for $I(\omega)$ says that the system will absorb radiation only at $\omega = \omega_0$ and, in fact, all the incident radiation at that frequency. The lineshape for this system, then, is an infinitely narrow line at $\omega = \omega_0$. This is the correct classical electromagnetic result. Of course, the molecules in a gas experience collisions, and this has the effect of destroying the long-time correlation in Eq. (21–208).

PRESSURE BROADENING OF MICROWAVE LINES

In order to include the effect of molecular collisions in the correlation function, we say that the probability that a molecule will not experience a collision by the time t is given by $e^{-t/\tau}$, where τ represents the mean time between collisions. We can write a simple approximation for $\langle \mu_z(0)\mu_z(t)\rangle$ by using this factor to modulate the $\cos \omega_0 t$ in Eq. (21–208) to write

$$\langle \mu_z \mu_z(t)\rangle = \mu_{0z}^{\ 2}e^{-|t|/\tau}\cos \omega_0 t \qquad (21\text{–}210)$$

The lineshape corresponding to this correlation function is

$$I(\omega) = \frac{\mu_{0z}^{\ 2}}{4\pi\tau}\left[\frac{1}{\dfrac{1}{\tau^2} + (\omega_0 + \omega)^2} + \frac{1}{\dfrac{1}{\tau^2} + (\omega_0 - \omega)^2}\right]$$

$$\approx \frac{\mu_{0z}^{\ 2}}{4\pi}\left[\frac{1/\tau}{\dfrac{1}{\tau^2} + (\omega - \omega_0)^2}\right] \qquad \text{for } \omega > 0 \qquad (21\text{–}211)$$

Note that the term in brackets here becomes $\pi\delta(\omega - \omega_0)$ as $\tau \to \infty$ (Problem 21–25), and hence this expression for $I(\omega)$ reduces to the collisionless case as the mean time between collisions approaches infinity. Equation (21–211) for $I(\omega)$ is known as a Lorentzian lineshape and is shown in Fig. 21–8, for several values of $1/\tau$. It can be seen there that the effect of collisions is to broaden the delta function of Eq. (21–209). A Lorentzian curve does not have finite moments beyond the first. This is because the $\exp(-|t|/\tau)$ in Eq. (21–210) is not analytic at $t = 0$, and hence $\langle \mu_z \mu_z(t)\rangle$ does not possess a Taylor expansion in time about $t = 0$. It is customary to express the width of a Lorentzian curve as the width at half the maximum value. The width, then, is $2/\tau$. The mean time between collisions is approximately the mean free path divided by the average velocity, and so is $O(10^{-10} \text{ sec})$ at standard temperature and pressure. Consequently, pressure broadening of microwaves lines is expected to be of the order of less than 1 cm^{-1}. This is in accord with experimental results. Of course, the correlation function we have used here is just an approximation, but it does seem to capture

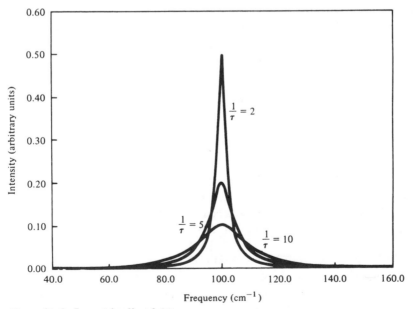

Figure 21–8. **Lorentzian line shape.**

much of the "physics" of the problem. There is an entire research field of pressure broadening, whose primary goal, in a sense, is to calculate the broadening of spectral lines in terms of the intermolecular interaction between the colliding molecules.

COLLISION-INDUCED ABSORPTION

You learn in physical chemistry that a molecule must possess a permanent dipole moment in order to absorb radiation in the microwave or far infrared region. Although this is true for an isolated molecule, it is not entirely correct for a collection of molecules. Equation (21–194) simply requires that there be a dipole moment time-correlation function, and this does not require a permanent dipole moment. To see this, consider a gaseous mixture of two rare gases, say neon and argon. As a neon and argon atom collide, they form a transient heteronuclear diatomic molecule, and so possess a transient dipole moment. This dipole moment lasts only during the collision, and its correlation function might be approximated by

$$\langle \mu_z \mu_z(t) \rangle = \mu_0^2 e^{-t^2/\tau^2} \qquad (21\text{–}212)$$

where now τ is the duration of a collision. Since this correlation function is due only to collisions, the absorption associated with this process is called collision-induced absorption. It is possible to observe collision-induced absorption. The spectrum associated with Eq. (21–212) is

$$I(\omega) = \frac{\mu_0^2 \tau}{2\pi^{1/2}} \, e^{-\omega^2\tau^2/4} \qquad (21\text{–}213)$$

This equation predicts that the spectrum is Gaussian, with a variance given by $2/\tau^2$ or a standard variation of $2^{1/2}/\tau$. The mean duration of a collision is the range of the potential divided by the average speed, or $O(10^{-13}$ sec). This predicts that the absorption predicted by Eq. (21–213) should be several hundred wave numbers wide. The experimental data for neon-argon absorption are shown in Fig. 21–9. Although Eq. (21–213) is approximate, it nevertheless is probably a fair representation of the true correlation

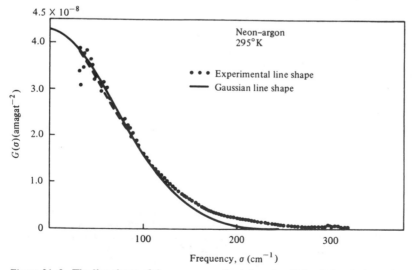

Figure 21–9. **The line shape of the neon-argon far infrared collision-induced absorption spectrum.** (From D. R. Bosomworth and H. P. Gush, *Can. J. Phys.*, **43**, pp. 751–759, 1965. Reproduced by permission of the National Research Council of Canada.)

function. From Eqs. (21–206) and (21–199) one can show that the nth moment of $I(\omega)$ is given by

$$\int_{-\infty}^{\infty} \omega^n I(\omega)\, d\omega = (-i)^n \langle \boldsymbol{\mu}(0) \cdot \boldsymbol{\mu}^{(n)}(0) \rangle \tag{21–214}$$

where $\boldsymbol{\mu}^{(n)}(0)$ represents the nth time derivative of $\boldsymbol{\mu}(t)$ *evaluated at $t = 0$*. The important point here is that the above average is an equilibrium ensemble average and does not involve any time dependence. It is straightforward to show that in the classical limit (Problem 21–62) that

$$\langle \boldsymbol{\mu}(0) \cdot \boldsymbol{\mu}^{(n)}(t) \rangle = 0 \qquad\qquad n \text{ odd}$$
$$= \langle \boldsymbol{\mu}^{(n/2)}(0) \cdot \boldsymbol{\mu}^{(n/2)}(0) \rangle \qquad n \text{ even} \tag{21–215}$$

The $\boldsymbol{\mu}^{(n/2)}$ can be determined from a repeated application of the equation of motion of $\boldsymbol{\mu}(t)$ in the Poisson bracket notation

$$\frac{d\boldsymbol{\mu}}{dt} = (\boldsymbol{\mu}, H) \tag{21–216}$$

This approach has been used by Brenner and McQuarrie* to derive a lineshape theory of collision-induced absorption in rare gas mixtures.

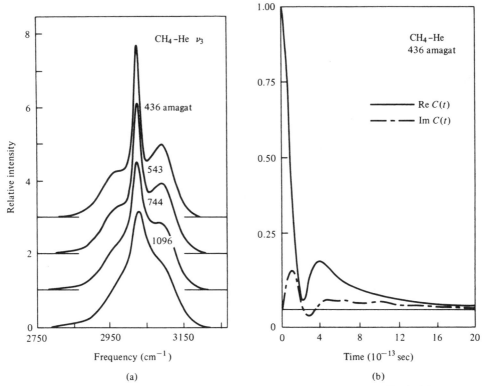

Figure 21–10. (a) **The effect of various densities of helium on the ν_3 band of CH_4. The individual profiles have been normalized to show equal areas; the relative intensity scales have been displaced for clarity of presentation. (b) The real and imaginary parts of the correlation function as obtained from a CH_4–He ν_3 band profile at 436 amagats.** (From R. L. Armstrong, S. M. Blumenfeld, and C. G. Gray, *Can. J. Phys.*, **46**, pp. 1331–1340, 1968. Reproduced by permission of the National Council of Canada.)

* S. L. Brenner and D. A. McQuarrie, *Can. J. Phys.*, **49**, p. 837, 1971.

So far we have calculated the lineshapes of certain processes by approximating the time-correlation function by simple functions. As we saw in Section 21–1, it is also possible to invert experimental data and determine an experimental time-correlation function. This is shown in Fig. 21–1. Another example is shown in Fig. 21–10.

Figure 21–10(a) shows the effect of various densities of helium on the v_3 band of methane. Since a classical correlation function gives a symmetric lineshape function, the asymmetry in these bands allows the imaginary part of the correlation function to be determined also. Both the real and imaginary parts of $C(t)$ are shown in Fig. 21–10 (b). Note that the imaginary part is, in fact, much smaller than the real part. This is in agreement with the approximating $C(t)$ by a classical correlation function.

21–7 DERIVATION OF THE BASIC FORMULAS FROM THE LIOUVILLE EQUATION

In this section we shall derive our previous results by starting from the Liouville equation. Although we shall not derive any new equations, the method is often used and also introduces several useful new concepts. The first half of this section will deal with the classical mechanical derivation, and the second half will treat the quantum-mechanical one.

Consider a classical system in equilibrium. At $t = 0$ we apply a force $F(t)$. We assume that the force may be represented by an additional term $\mathscr{H}_{ext}$ of the Hamiltonian, so that we may write

$$\mathscr{H} = \mathscr{H}_0 + \mathscr{H}_{ext} = \mathscr{H}_0 - AF(t) \tag{21–217}$$

where $A = A(p, q)$ is a function defined in phase space. Let the response be denoted by $B(p, q)$; then according to Eq. (21–119), the average of $B(t)$ is related to $F(t)$ by

$$\langle B(t) \rangle = \int_0^t dt' \, \phi(t - t')F(t') \tag{21–218}$$

We wish to find the response function $\phi(t)$ or the susceptibility $\chi(\omega)$ in terms of the phase space functions A and B. To do this, we simply must determine the phase space distribution function $f(p, q, t)$ to terms linear in $F(t)$ and then use it to determine $\langle B(t) \rangle$ according to the standard formula for the evaluation of a phase space average, namely,

$$\langle B(t) \rangle = \int dp \, dq \, B(p, q)f(p, q, t) \tag{21–219}$$

Assuming that the external field is small, we can write the distribution function as an unperturbed part plus a small perturbed part

$$f = f_0 + \Delta f \tag{21–220}$$

By substituting this and Eq. (21–217) into the Liouville equation and using

$$\frac{\partial f_0}{\partial t} = (\mathscr{H}_0, f_0) \tag{21–221}$$

we get

$$\frac{\partial \, \Delta f}{\partial t} = (\mathscr{H}_0, \Delta f) - F(t)(A, f_0) \tag{21–222}$$

to terms in first order of the perturbation.

It is convenient to introduce the Liouville operator,

$$\mathcal{L}\psi = i(\mathcal{H}, \psi) \tag{21-223}$$

where ψ here is just an arbitrary phase space function on which $\mathcal{L}$ operates. In terms of this Liouville operator, the Liouville equation is

$$\frac{\partial f}{\partial t} = -i\mathcal{L}f \tag{21-224}$$

whose formal solution is

$$f(p, q, t) = e^{-i\mathcal{L}t}f(p, q, 0) \tag{21-225}$$

The equation of motion of any dynamical quantity $\alpha(p, q)$, not depending explicitly on time, can be written in terms of the Poisson brackets or the Liouville operator:

$$\frac{d\alpha}{dt} = -(\mathcal{H}, \alpha) = i\mathcal{L}\alpha \tag{21-226}$$

The formal solution to this equation is

$$\alpha(t) = e^{it\mathcal{L}}\alpha(0) \tag{21-227}$$

This is to be compared to the formal solution of the Liouville equation, Eq. (21–225).

The expectation value of $\alpha(p, q)$, weighted according to the ensemble density $f(p, q, t)$, is

$$\begin{aligned}
\langle \alpha | f(t) \rangle &= \int \cdots \int dp\, dq\, \alpha(p, q)f(p, q, t) \\
&= \langle \alpha | e^{-it\mathcal{L}}f(0) \rangle \tag{21-228} \\
&= \langle e^{i\mathcal{L}t}\alpha | f(0) \rangle \\
&= \langle \alpha(t) | f(0) \rangle \tag{21-229}
\end{aligned}$$

In going from the second to the third line above, we have used the fact that $\exp(\pm it\mathcal{L})$ is unitary. This follows from the fact that we have defined $\mathcal{L}$ such that it is Hermitian, i.e., such that

$$\int \cdots \int g^*(p, q)\mathcal{L}h(p, q)\, dp\, dq = \int \cdots \int (\mathcal{L}g)^*h(p, q)\, dp\, dq \tag{21-230}$$

for functions $g(p, q)$ and $h(p, q)$ that vanish sufficiently rapidly at infinity in momentum space and outside of some volume in coordinate space. We shall use the fact that $\exp(\pm it\mathcal{L})$ is unitary below. Equations (21–228) and (21–229) show that expectation value of α at time t can be calculated in two ways: by following the evolution of the ensemble density or the evolution of the dynamical function.

In terms of the Liouville operator, Eq. (21–222) becomes

$$\frac{\partial\, \Delta f}{\partial t} = -i\mathcal{L}_0\, \Delta f - F(t)(A, f_0) \tag{21-231}$$

This equation can be solved formally as a linear first-order differential equation to give (Problem 21–33)

$$\Delta f(t) = -\int_0^t e^{i(t'-t)\mathcal{L}_0}(A, f_0)F(t')\, dt' \tag{21-232}$$

We have assumed that $\Delta f = 0$ at $t = 0$, i.e., that the system is in equilibrium at $t = 0$.

The expectation value of $B(p, q)$ is given by

$$\langle B| f(t)\rangle = \langle B| f_0\rangle + \langle B| \Delta f(t)\rangle \qquad (21\text{--}233)$$

Without loss of generality, we may assume that the expectation value of B is zero for the system before the external field is turned on. Then we can write

$$
\begin{aligned}
\langle B(t)\rangle &= \langle B| \Delta f(t)\rangle \\
&= -\int_0^t dt' \int \cdots \int dp\, dq\, B(p, q) e^{i(t'-t)\mathscr{L}_0}[A(t'), f_0(t')]F(t') \\
&= -\int_0^t dt' \int \cdots \int dp\, dq [e^{-i(t'-t)\mathscr{L}_0}B]\,[A(t'), f_0(t')]F(t') \\
&= -\int_0^t dt' F(t') \int \cdots \int dp\, dq\, [A(t'), f_0(t')]B(t-t') \qquad (21\text{--}234)
\end{aligned}
$$

In going from the first integral to the second, we have used the unitary property of $\exp(i\mathscr{L}t)$, and to write the last line we have used the fact that $e^{i\mathscr{L}t}B(0) = B(t)$ according to Eq. (21–227). This may be written as

$$\langle B(t)\rangle = \int_0^t dt'\, F(t')\phi(t-t') \qquad (21\text{--}235)$$

where

$$
\begin{aligned}
\phi(t) &= \int \cdots \int dp\, dq\, [f_0, A(0)]B(t) \\
&= \int \cdots \int dp\, dq\, f_0[A(0), B(t)] \\
&= \langle [A(0), B(t)]\rangle \qquad (21\text{--}236)
\end{aligned}
$$

The second line follows from the first by an integration by parts.

If the distribution f_0 is assumed to be canonical,

$$f_0 = e^{-\beta \mathscr{H}_0} \Big/ \int \cdots \int e^{-\beta \mathscr{H}_0}\, dp\, dq$$

then Eq. (21–226) can be used to show that (Problem 21–21)

$$(f_0, A) = \beta \dot{A} f_0 \qquad (21\text{--}237)$$

If we substitute this into Eq. (21–236), we get another form for $\phi(t)$, namely,

$$\phi(t) = \beta\langle \dot{A}(0)B(t)\rangle \qquad (21\text{--}238)$$

Furthermore, since the angular brackets denote an *equilibrium* ensemble average, we have the stationarity condition

$$\langle A(0)B(t)\rangle = \langle A(s)B(t+s)\rangle \qquad (21\text{--}239)$$

which gives

$$
\begin{aligned}
\frac{d}{ds}\langle A(s)B(t+s)\rangle &= \langle \dot{A}(s)B(t+s)\rangle + \langle A(s)\dot{B}(t+s)\rangle \\
&= \langle \dot{A}(0)B(t)\rangle + \langle A(0)\dot{B}(t)\rangle = 0 \qquad (21\text{--}240)
\end{aligned}
$$

We see then that another form for $\phi(t)$ is

$$\phi(t) = -\beta\langle A(0)\dot{B}(t)\rangle \qquad (21\text{--}241)$$

We had presented most of these expressions for $\phi(t)$ in Section 21–4, but here they were derived directly from the classical Liouville equation. This is an important exercise in itself since the Liouville is the most fundamental equation of statistical mechanics, and many statistical mechanical transport theories start with this equation.

We shall now derive the corresponding quantum-mechanical time-correlation function expressions. In making the correspondence between classical mechanical and quantum-mechanical expressions, the distribution function is replaced by the density matrix ρ, the Poisson bracket by a commutator, and the integration over phase space by the trace operation. The quantum-mechanical Liouville equation is (Problem 21–34)

$$\frac{\partial \rho}{\partial t} = \frac{1}{i\hbar}[\mathcal{H}, \rho] = -i\mathcal{L}\rho \tag{21–242}$$

where the square brackets denote the commutator of $\mathcal{H}$ and ρ, i.e., $\mathcal{H}\rho - \rho\mathcal{H}$, and we have introduced the quantum-mechanical Liouville operator

$$\mathcal{L}\rho = \frac{1}{\hbar}[\mathcal{H}, \rho] = \frac{1}{\hbar}[\mathcal{H}\rho - \rho\mathcal{H}] \tag{21–243}$$

The quantum-mechanical analog of the classical equation of motion is the Heisenberg equation of motion. In the quantum-mechanical case, the $\alpha(p, q)$ of Eq. (21–226) is an operator, whose time evolution is given by

$$\frac{d\alpha}{dt} = -\frac{1}{i\hbar}[\mathcal{H}, \alpha] = i\mathcal{L}\alpha \tag{21–244}$$

The formal solutions of Eqs. (21–242) and (21–244) are

$$\rho(t) = e^{-it\mathcal{L}}\rho(0) \tag{21–245}$$

$$\alpha(t) = e^{it\mathcal{L}}\alpha(0) \tag{21–246}$$

The more conventional form of these equations is the Heisenberg representation for the time dependence of a quantum-mechanical operator:

$$\rho(t) = e^{-i\mathcal{H}t/\hbar}\rho(0)e^{i\mathcal{H}t/\hbar} \tag{21–247}$$

$$\alpha(t) = e^{i\mathcal{H}t/\hbar}\alpha(0)e^{-i\mathcal{H}t/\hbar} \tag{21–248}$$

The expectation value of α is given by

$$\langle \alpha \,|\, \rho(t) \rangle = \text{Tr}[\alpha\rho(t)] \tag{21–249}$$

with analogous forms for Eq. (21–229). (See Problem 21–35.)

The equation for $\Delta\rho$ is

$$\frac{\partial \Delta\rho}{\partial t} = -i\mathcal{L}_0 \Delta\rho - \frac{1}{i\hbar}[A, \rho_0]F(t) \tag{21–250}$$

whose solution is

$$\Delta\rho = -\int_0^t e^{i(t'-t)\mathcal{L}_0}\frac{1}{i\hbar}[A, \rho_0]F(t')\,dt' \tag{21–251}$$

The expectation of B is

$$\langle B(t) \rangle = \langle B | \Delta \rho \rangle = \text{Tr}[B \, \Delta \rho(t)]$$

$$= -\frac{1}{i\hbar} \int_0^t dt' \, F(t') \, \text{Tr}\{Be^{i(t'-t)\mathcal{L}_0}[A, \rho_0]\}$$

$$= -\frac{1}{i\hbar} \int_0^t dt' \, F(t') \, \text{Tr}\{B(t-t')[A(t'), \rho_0(t')]\} \tag{21-252}$$

The response function is

$$\phi(t) = \frac{1}{i\hbar} \text{Tr}\{[\rho_0, A]B(t)\} \tag{21-253}$$

Thus we have derived Eq. (21–158) by starting with the quantum-mechanical Liouville equation. By using the fact that the trace of a product is invariant with respect to a cyclic permutation, i.e., that (Problem 21–18) $\text{Tr}(ABC) = \text{Tr}(CAB) = \text{Tr}(BCA)$, it is easy to show that

$$\phi(t) = \frac{1}{i\hbar} \text{Tr}\{\rho_0[A, B(t)]\} \tag{21-254}$$

which was derived before [cf. Eq. (21–159)].

Lastly, we can derive Eq. (21–160) by assuming that ρ is canonical, i.e., that $\rho_0 = e^{-\beta \mathcal{H}_0}/\text{Tr} \, e^{-\beta \mathcal{H}_0}$, and making use of the identity

$$[\rho_0, A] \equiv \int_0^\beta \rho_0 e^{\lambda \mathcal{H}_0}[A, \mathcal{H}_0]e^{-\lambda \mathcal{H}_0} \, d\lambda$$

$$= i\hbar \int_0^\beta \rho_0 \dot{A}(-i\hbar\lambda) \, d\lambda$$

$$= i\hbar\beta\rho_0\tilde{A} \tag{21-255}$$

where as before we have defined the Kubo transform of $\dot{A}$ by

$$\tilde{A} = \frac{1}{\beta} \int_0^\beta \dot{A}(-i\hbar\lambda) \, d\lambda$$

$$= \frac{1}{\beta} \int_0^\beta e^{\lambda \mathcal{H}_0}\dot{A}e^{-\lambda \mathcal{H}_0} \, d\lambda \tag{21-256}$$

Equation (21–253) for $\phi(t)$ can then be written as

$$\phi(t) = \beta \, \text{Tr}[\rho_0 B\tilde{A}(t)] \tag{21-257}$$

which is Eq. (21–160).

We are now in a position to summarize the results that we have derived so far. For an external force $F(t)$ that adds a term $-AF(t)$ to the Hamiltonian of the system, the response $B(t)$ is given by

$$B(t) = \int_0^t \phi(t-t')F(t') \, dt' \tag{21-258}$$

where we have the following correspondences between the various expressions for the response function $\phi(t)$:

classical	*quantum*

$$\int \cdots \int dp\, dq\, [f_0, A(0)]B(t) \qquad \frac{1}{i\hbar} \, \mathrm{Tr}\{[\rho_0, A\,]B(t)\}$$

$$\int \cdots \int dp\, dq\, f_0[A(0), B(t)] \qquad \frac{1}{i\hbar} \, \mathrm{Tr}\{\rho_0[A, B(t)]\}$$

$$\beta \int \cdots \int dp\, dq\, f_0\, \dot{A}(0)B(t) \qquad \beta\, \mathrm{Tr}\{\rho_0\, \tilde{A}B(t)\}$$

We have explicitly subscripted the ensemble distribution functions in these expressions to emphasize that they are equilibrium distribution functions.

21–8   TIME-CORRELATION FUNCTION EXPRESSIONS FOR THE THERMAL TRANSPORT COEFFICIENTS

In the previous sections we have presented several derivations of the basic formulas expressing a susceptibility $\chi(\omega)$ as the Fourier-Laplace transform of some appropriate time-correlation function. Since all of these alternative derivations are exact to within certain general assumptions (such as linear response), they must all be rigorously equivalent to one another, which, in fact, they are. They are all based upon the fact that the externally applied field adds a perturbation term to the Hamiltonian of the system. For the applications we have discussed so far, this perturbation term arises naturally since it is of a mechanical or electromagnetic nature, but what if we wish to discuss thermal transport phenomena such as diffusion, viscosity, and thermal conductivity? There is no simple or obvious way that a concentration gradient or a temperature gradient can be incorporated into a well-defined perturbation Hamiltonian since these are of a thermal or nonmechanical nature, and so the previous derivations do not lend themselves to these processes in any straightforward manner. Nevertheless, it is possible to derive expressions for the standard transport coefficients such as the coefficient of self-diffusion D, the viscosity η, and the thermal conductivity λ in terms of integrals over certain time-correlation functions. Because of the lack of a well-defined mechanical perturbation Hamiltonian, however, these derivations are more subtle than those in the previous sections. The reason for this is that a number of assumptions must be made in the derivations, and it is often difficult to assess the quantitative limitations of the resulting equations. The first such approach was presented by Green in his important pioneering work in the 1950s.* His work contains a number of approximations and assumptions, principally that the dynamical process underlying the transport process of interest is a Markov process, and the departure from thermal equilibrium is small. His derivation is expressed in terms of a type of Fokker-Planck equation in a space of certain macroscopic variables (called "gross variables" by Green). He was the first to obtain expressions involving time-correlation functions for the coefficients of shear and bulk viscosity, thermal conductivity, diffusion, and thermal diffusion.

Following Green's work, a number of alternative derivations of these basic equations appeared in the 1950s and 1960s. Some of these approaches differed greatly from one

---

* M. S. Green, *J. Chem. Phys.*, **20**, p. 1281, 1952; **22**, p. 398, 1954.

another, although they all gave essentially the same final result. In Zwanzig's 1965 review article,* he discusses six different derivations of the equations for the thermal transport coefficients! This was a very active area of research in the 1960s, and the limitations and generalizations of these equations became fairly well understood by the end of that decade. This work involves more advanced concepts and techniques than the level of this introductory chapter, but in the next chapter we shall derive exact equations that are somewhat complicated and abstract but which reduce to the simpler equations to be derived below when appropriate well-defined limits are taken. These more rigorous derivations show that the simpler equations are exact in the limit of small disturbances and long times, i.e., in the hydrodynamic limit.

We wish to present a more elementary and heuristic derivation, however, in this section in order to introduce and discuss the basic time-correlation function equations in the hydrodynamic limit. Out of all the possibilities, we have chosen the method presented by Helfand in 1961† since it is not only pedagogically one of the most appropriate but the one we shall generalize in Section 22–5.

As an introduction to Helfand's approach, we shall consider the case of self-diffusion. We start with the diffusion equation

$$\frac{\partial G(\mathbf{r}, t)}{\partial t} = D\nabla^2 G(\mathbf{r}, t) \tag{21–259}$$

with the initial condition

$$G(\mathbf{r}, 0) = \delta(\mathbf{r}) \tag{21–260}$$

This is the macroscopic equation governing the probability $G(\mathbf{r}, t)\, d\mathbf{r}$ that a particle is located in the volume element $d\mathbf{r}$ at $\mathbf{r}$ at time t given that it was initially located at the origin. We are going to consider only the case of isotropic diffusion, so that $G(\mathbf{r}, t) = G(r, t)$, but we shall continue to write $G(\mathbf{r}, t)$ to remind ourselves that G is a three-dimensional function. Since the time-correlation function expression for D that we are in the process of deriving is based upon the validity of this equation, the formula itself must be subject to the same limitations, namely, that spatial inhomogeneities must be smoothly varying and the time must be long compared to microscopic times (cf. Section 20–1).

If we write the Fourier transform of $G(\mathbf{r}, t)$ as

$$F_s(\mathbf{k}, t) = \int d\mathbf{r}\, e^{i\mathbf{k} \cdot \mathbf{r}} G(\mathbf{r}, t) \tag{21–261}$$

then clearly

$$F_s(\mathbf{k}, 0) = 1 \tag{21–262}$$

It is easy to show by taking the Fourier transform of both sides of Eq. (21–259) that (Problem 21–36)

$$\frac{\partial F_s(\mathbf{k}, t)}{\partial t} = -k^2 D F_s(\mathbf{k}, t) \tag{21–263}$$

whose solution is

$$F_s(\mathbf{k}, t) = \exp(-k^2 Dt) \tag{21–264}$$

* R. Zwanzig, *Ann, Rev. Phys. Chem.*, **16**, p. 67, 1965.
† E. Helfand, *Phys. Fluids*, **4**, p. 681, 1961.

According to this equation for $F_s(\mathbf{k}, t)$,

$$\left(\frac{\partial^2 F_s}{\partial k^2}\right)_{k=0} = -2Dt \tag{21-265}$$

We can derive another expression for this derivative by differentiating Eq. (21–261) directly, giving

$$\left(\frac{\partial^2 F_s}{\partial k^2}\right)_{k=0} = -\frac{1}{3}\int_0^\infty 4\pi r^4 G(\mathbf{r}, t)\, dr$$

$$= -\tfrac{1}{3}\langle r^2(t)\rangle \tag{21-266}$$

giving the average square displacement as a function of time. Thus we have that

$$\langle r^2(t)\rangle = 6Dt \tag{21-267}$$

an equation that we derived in Chapter 20 [cf. Eq. (20–19)]. This equation, in fact, was first derived by Einstein in his early work on Brownian motion. If the diffusing particle had originated at the point $\mathbf{r}_0$ instead of the origin, then Eq. (21–267) would be simply

$$\langle|\mathbf{r}(t) - \mathbf{r}_0|^2\rangle = 6Dt \tag{21-268}$$

It is instructive to derive Eq. (21–268) in another way. We have interpreted $G(\mathbf{r}, t)$ as the conditional probability that some particular, for example tagged, molecule is located in $d\mathbf{r}$ at $\mathbf{r}$ at time t given that it was at the origin at $t = 0$. We can also interpret it as the fraction of particles in $d\mathbf{r}$ at $\mathbf{r}$ at time t given that they were located somewhere else at $t = 0$. Thus

$$G(\mathbf{r}, t) = \frac{1}{N}\left\langle \sum_{j=1}^N \delta\{\mathbf{r} - [\mathbf{r}_j(t) - \mathbf{r}_j(0)]\}\right\rangle \tag{21-269}$$

The Fourier transform of this is

$$F_s(\mathbf{k}, t) = \frac{1}{N}\left\langle \sum_{j=1}^N e^{i\mathbf{k}\cdot[\mathbf{r}_j(t) - \mathbf{r}_j(0)]}\right\rangle \tag{21-270}$$

Note that this is the time-correlation function of the dynamical quantity $\exp[i\mathbf{k}\cdot\mathbf{r}(t)]$ since it can be written in the form (if the dynamical quantity is complex, the time correlation function is defined as $\langle A^*(t)A(0)\rangle$)

$$F_s(\mathbf{k}, t) = \frac{1}{N}\left\langle \sum_{j=1}^N e^{-i\mathbf{k}\cdot\mathbf{r}_j(0)}e^{i\mathbf{k}\cdot\mathbf{r}_j(t)}\right\rangle \tag{21-271}$$

Differentiation of Eq. (21–270) gives Eq. (21–266).

We can now express this equation as the integral of a time-correlation function. Clearly

$$\mathbf{r}(t) - \mathbf{r}_0 = \int_0^t dt'\, \mathbf{v}(t') \tag{21-272}$$

and

$$[\mathbf{r}(t) - \mathbf{r}_0]^2 = \int_0^t dt' \int_0^t dt''\, \mathbf{v}(t')\cdot\mathbf{v}(t'') \tag{21-273}$$

where $\mathbf{v}(t)$ is the velocity of the particle. If we ensemble average both sides of this equation, we get

$$\langle[\mathbf{r}(t) - \mathbf{r}_0]^2\rangle = \int_0^t dt' \int_0^t dt'' \langle\mathbf{v}(t') \cdot \mathbf{v}(t'')\rangle \tag{21-274}$$

Due to the stationarity of the equilibrium ensemble average and the time reversibility of the classical mechanical equations of motion, we have that

$$\langle\mathbf{v}(t') \cdot \mathbf{v}(t'')\rangle = \langle\mathbf{v}(t' - t'') \cdot \mathbf{v}(0)\rangle = \langle\mathbf{v}(t'' - t') \cdot \mathbf{v}(0)\rangle \tag{21-275}$$

If we now change integration variables in Eq. (21–274) to $\tau = t'' - t'$, change the order of integration, and carry out the one obvious integration, we get (Problem 2i–37)

$$\langle[\mathbf{r}(t) - \mathbf{r}_0]^2\rangle = 6Dt = 2t \int_0^t \left(1 - \frac{\tau}{t}\right) \langle\mathbf{v}(0) \cdot \mathbf{v}(\tau)\rangle \, d\tau \tag{21-276}$$

If the time-correlation function $\langle\mathbf{v}(0) \cdot \mathbf{v}(t)\rangle$ decays sufficiently rapidly and t is chosen sufficiently large, then one can neglect the τ/t in the integrand and write

$$6Dt = 2t \int_0^t \langle\mathbf{v}(0) \cdot \mathbf{v}(\tau)\rangle \, d\tau \qquad t \to \infty \tag{21-277}$$

Since it is assumed that $\langle\mathbf{v}(0) \cdot \mathbf{v}(\tau)\rangle$ has fallen to zero long before t, this integral is insensitive to the value of t for large values of t and so it is permissible to write finally

$$D = \frac{1}{3} \int_0^\infty \langle\mathbf{v}(0) \cdot \mathbf{v}(\tau)\rangle \, d\tau \tag{21-278}$$

which is our desired expression for D. It should be clear from the derivation that this equation is limited to processes for which Eq. (21–259) is a suitable description, which means for "normal" diffusion processes, i.e., ones that occur slowly with respect to microscopic times and for which spatial variations are smooth.

Equation (21–278) is often written in terms of the normalized velocity autocorrelation,

$$C(\tau) = \frac{\langle\mathbf{v}(0) \cdot \mathbf{v}(\tau)\rangle}{\langle\mathbf{v}(0) \cdot \mathbf{v}(0)\rangle} \tag{21-279}$$

Using the fact that $\langle\mathbf{v}(0) \cdot \mathbf{v}(0)\rangle = 3kT/m$, we have

$$D = \frac{kT}{m} \int_0^\infty C(\tau) \, d\tau \tag{21-280}$$

Clearly it is no simple matter to calculate $C(\tau)$ exactly, but several molecular dynamics calculations of $C(\tau)$ have been made. Figure 21–11(a) shows $C(\tau)$ for a system of hard spheres for several densities, and Fig. 21–11(b) shows $C(\tau)$ for a system of Lennard-Jones molecules. The negative region that occurs at high densities was first observed by Alder and Wainwright and is interpreted as the back scattering that occurs when a molecule collides with one of the neighboring molecules forming a cage around the central one. We shall return to Eq. (21–278) for D after we derive the corresponding expressions for the other transport coefficients.

The calculation of the other coefficients proceeds in a similar manner. In order to treat the shear viscosity, one starts with the Navier-Stokes equation with no external field [Eq. (17–73)].

$$\frac{\partial \rho_m \mathbf{u}}{\partial t} = \nabla \cdot (\boldsymbol{\sigma} - \rho_m \mathbf{u}\mathbf{u}) \tag{21-281}$$

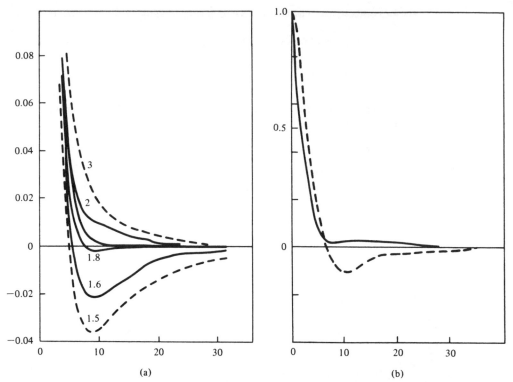

Figure 21–11. **Velocity autocorrelation functions calculated from molecular dynamics. (a) Calculated on a hard-sphere system for various densities, indicated by the values of v/v_0. The abscissa is given in multiples of the mean collision time. (From B. J. Alder, D. M. Gass, and T. E. Wainwright, *J. Chem. Phys.*, 53, p. 3813, 1970.) (b) Calculated on a system of Lennard-Jones molecules at a reduced density of $\rho^* = \rho\sigma^3 = 0.85$. The solid line is for a reduced temperature of 4.70, and the dashed line is for a reduced temperature of 0.76. As in (a), the abscissa is given in units of mean collision times. (From D. Levesque and L. Verlet, *Phys. Rev.*, A2, p. 2514, 1970.)**

where ρ_m is the mass density $m\rho$. The stress tensor has the form

$$\boldsymbol{\sigma} = -[p + (\tfrac{2}{3}\eta - \kappa)\nabla \cdot \mathbf{u}]\mathsf{I} + 2\eta \,\text{sym}(\nabla \mathbf{u}) \tag{21-282}$$

where η and κ are the shear and bulk viscosities, respectively. We shall work with the linearized form of the Navier-Stokes equation in which $\mathbf{u}$ and $\partial\rho/\partial t$ are small, so that we have (Problem 21–38)

$$m\rho \,\frac{\partial \mathbf{u}}{\partial t} = -\nabla p - (\tfrac{2}{3}\eta - \kappa)\nabla \cdot (\nabla \cdot \mathbf{u})\mathsf{I} + 2\eta\nabla \cdot (\text{sym } \nabla \mathbf{u}) \tag{21-283}$$

It is a straightforward exercise to show that (Problem 21–39)

$$\nabla \cdot (\text{sym } \nabla \cdot \mathbf{u}) = \tfrac{1}{2}\nabla^2 \mathbf{u} + \tfrac{1}{2}\nabla(\nabla \cdot \mathbf{u}) \tag{21-284}$$

and

$$\nabla \cdot (\nabla \cdot \mathbf{u})\mathsf{I} = \nabla(\nabla \cdot \mathbf{u}) \tag{21-285}$$

and so we get the linearized Navier-Stokes equation (Problem 21–40)

$$m\rho \,\frac{\partial \mathbf{u}}{\partial t} = -\nabla p + \eta\nabla^2 \mathbf{u} + (\tfrac{1}{3}\eta + \kappa)\nabla(\nabla \cdot \mathbf{u}) \tag{21-286}$$

This equation plays the same role as the diffusion equation does in the calculation of the self-diffusion coefficient.

If we now take the Fourier transform of this equation and use the following relations (Problem 21–41)

$$\iiint d\mathbf{r}\, e^{i\mathbf{k}\cdot\mathbf{r}}\nabla p = -ikp(\mathbf{k},\, t)$$

$$\iiint d\mathbf{r}\, e^{i\mathbf{k}\cdot\mathbf{r}}\nabla^2\mathbf{u} = -k^2\mathbf{J}(\mathbf{k},\, t)$$

$$\iiint d\mathbf{r}\, e^{i\mathbf{k}\cdot\mathbf{r}}\nabla(\nabla\cdot\mathbf{u}) = -\mathbf{k}[\mathbf{k}\cdot\mathbf{J}(\mathbf{k},\, t)] \qquad (21\text{–}287)$$

where $p(\mathbf{k},\, t)$ and $\mathbf{J}(\mathbf{k},\, t)$ are the Fourier transforms of the pressure and velocity, respectively, we get

$$m\rho\,\frac{\partial\mathbf{J}(\mathbf{k})}{\partial t} = ikp(\mathbf{k},\, t) - \eta k^2\mathbf{J}(\mathbf{k},\, t) - (\tfrac{1}{3}\eta + \kappa)\mathbf{k}[\mathbf{k}\cdot\mathbf{J}(\mathbf{k},\, t)] \qquad (21\text{–}288)$$

The scalar product in the last term here suggests that we separate $\mathbf{J}(\mathbf{k},\, t)$ into three orthogonal components, $J_\parallel(\mathbf{k},\, t)$, $J_{\perp 1}(\mathbf{k},\, t)$, and $J_{\perp 2}(\mathbf{k},\, t)$, which are the components of $\mathbf{J}(\mathbf{k},\, t)$ parallel and perpendicular to $\mathbf{k}$. Equation (21–288) becomes

$$m\rho\,\frac{\partial J_\parallel(\mathbf{k},\, t)}{\partial t} = ikp(\mathbf{k},\, t) - (\tfrac{4}{3}\eta + \kappa)k^2 J_\parallel(\mathbf{k},\, t) \qquad (21\text{–}289)$$

$$m\rho\,\frac{\partial J_{\perp\alpha}(\mathbf{k},\, t)}{\partial t} = -k^2\eta J_{\perp\alpha}(\mathbf{k},\, t) \qquad \alpha = 1,\, 2 \qquad (21\text{–}290)$$

The second of these equations can be solved immediately to give

$$J_{\perp\alpha}(\mathbf{k},\, t) = J_{\perp\alpha}(\mathbf{k},\, 0)e^{-k^2\eta t/m\rho} \qquad (21\text{–}291)$$

This equation corresponds to Eq. (21–264) in the derivation of D. It is convenient at this point to depart slightly from the procedure used to derive the self-diffusion coefficient and to form the time-correlation function of $J_{\perp\alpha}(\mathbf{k},\, t)$ by multiplying it by $J_{\perp\alpha}^*(\mathbf{k},\, 0) = J_{\perp\alpha}(-\mathbf{k},\, 0)$ and averaging over the equilibrium ensemble. Using Eq. (21–291) for $J_{\perp\alpha}(\mathbf{k},\, t)$, we get

$$C_{\perp\alpha}(\mathbf{k},\, t) \equiv \frac{\langle J_{\perp\alpha}^*(\mathbf{k},\, 0)J_{\perp\alpha}(\mathbf{k},\, t)\rangle}{\langle J_{\perp\alpha}^*(\mathbf{k},\, 0)J_{\perp\alpha}(\mathbf{k},\, 0)\rangle} = \exp\left(\frac{-k^2\eta t}{m\rho}\right) \qquad (21\text{–}292)$$

The velocity $\mathbf{u}(\mathbf{r},\, t)$ can be written as

$$\mathbf{u}(\mathbf{r},\, t) = \sum_{j=1}^{N} \mathbf{v}_j\, \delta[\mathbf{r} - \mathbf{r}_j(t)] \qquad (21\text{–}293)$$

where $\mathbf{v}_j$ is the velocity and $\mathbf{r}_j$ the position of the jth particle. It follows from this that

$$\mathbf{J}(\mathbf{k},\, t) = \sum_{j=1}^{N} \mathbf{v}_j e^{i\mathbf{k}\cdot\mathbf{r}_j} \qquad (21\text{–}294)$$

If we choose our coordinate system so that $\mathbf{k}$ points in the z-direction, then

$$J_\parallel(\mathbf{k},\, t) = \sum_{j=1}^{N} \dot{z}_j(t)e^{ikz_j(t)} \qquad (21\text{–}295a)$$

$$J_{\perp 1}(\mathbf{k},\, t) = \sum_{j=1}^{N} \dot{x}_j(t)e^{ikz_j(t)} \qquad (21\text{–}295b)$$

$$J_{\perp 2}(\mathbf{k},\, t) = \sum_{j=1}^{N} \dot{y}_j(t)e^{ikz_j(t)} \qquad (21\text{–}295c)$$

Note from these last two lines that if the fluid is isotropic, then $C_{11}(\mathbf{k}, t) = C_{12}(\mathbf{k}, t)$ and we can drop the α subscripts in Eq. (21–292). By following the same procedure as before, we can find a time-correlation function expression for η by first noting that it follows from Eq. (21–292) that

$$\left(\frac{\partial^2 C_\perp(\mathbf{k}, t)}{\partial k^2}\right)_{k=0} = -\frac{2\eta t}{\rho m} \tag{21–296}$$

Using Eq. (21–295b) for $J_\perp(\mathbf{k}, t)$ in Eq. (21–292) gives

$$\left\langle \sum_{l=1}^{N} \sum_{j=1}^{N} \dot{x}_l(t)\dot{x}_j(0)e^{ik[z_l(t) - z_j(0)]} \right\rangle = \left\langle \sum_{l=1}^{N} \sum_{j=1}^{N} \dot{x}_l(0)\dot{x}_j(0)e^{ik[z_l(0) - z_j(0)]} \right\rangle e^{-k^2\eta t/m\rho} \tag{21–297}$$

Instead of differentiating this equation according to Eq. (21–296), it is simpler to expand the exponential functions in the two ensemble averages in powers of k and then compare the coefficients of k^2 on the two sides of the equation. This is equivalent to using Eq. (21–296).

We first note that since the velocities and coordinates are independent in an equilibrium ensemble, the ensemble average on the right-hand side of Eq. (21–297) becomes

$$\left\langle \sum_{l=1}^{N} \sum_{j=1}^{N} \left(\frac{k_B T}{m}\right) \delta_{lj} e^{ik[z_l(0) - z_j(0)]} \right\rangle = \frac{Nk_B T}{m} \tag{21–298}$$

We have subscripted the Boltzmann constant with B in this equation to avoid confusion between it and the wave vector appearing in the exponential function.

The evaluation of the ensemble average on the left-hand side of Eq. (21–297) requires some amount of manipulation. Expanding this ensemble average in powers of k gives

$$\left\langle \sum_{l=1}^{N} \sum_{j=1}^{N} \dot{x}_l(t)\dot{x}_j(0) \right\rangle + ik\left\langle \sum_{l=1}^{N} \sum_{j=1}^{N} \dot{x}_l(t)\dot{x}_j(0)[z_l(t) - z_j(0)] \right\rangle$$
$$- \frac{k^2}{2}\left\langle \sum_{l=1}^{N} \sum_{j=1}^{N} \dot{x}_l(t)\dot{x}_j(0)[z_l(t) - z_j(0)]^2 \right\rangle + \cdots \tag{21–299}$$

The first term can be evaluated by noting that because of conservation of momentum,

$$\sum_{l=1}^{N} \dot{x}_l(t) = \sum_{l=1}^{N} \dot{x}_l(0) \tag{21–300}$$

Therefore the first term becomes simply

$$\left\langle \sum_{l=1}^{N} \sum_{j=1}^{N} \dot{x}_l(t)\dot{x}_j(0) \right\rangle = \left\langle \sum_{l=1}^{N} \sum_{j=1}^{N} \dot{x}_l(0)\dot{x}_j(0) \right\rangle = \frac{Nk_B T}{m} \tag{21–301}$$

To evaluate the second term in Eq. (21–299), we write it as the difference between two terms. For the first of these, write (Problem 21–60)

$$\left\langle \sum_{l=1}^{N} \sum_{j=1}^{N} \dot{x}_l(t)\dot{x}_j(0)z_l(t) \right\rangle = \left\langle \sum_{l=1}^{N} \sum_{j=1}^{N} \dot{x}_l(t)\dot{x}_j(t)z_l(t) \right\rangle$$
$$= \left\langle \sum_{l=1}^{N} \sum_{j=1}^{N} \dot{x}_l(0)\dot{x}_j(0)z_l(0) \right\rangle \tag{21–302}$$

The first step follows from conservation of momentum, and the second follows from the stationarity condition for an equilibrium ensemble average. By applying the same argument to the other part of the second term in Eq. (21–299), we see that we get the same result, and so the coefficient of k in Eq. (21–299) is equal to zero.

Putting all of these results in Eq. (21–297) gives then

$$\frac{Nk_BT}{m} - \frac{k^2}{2}\left\langle \sum_{i=1}^{N}\sum_{j=1}^{N} \dot{x}_i(t)\dot{x}_j(0)[z_i(t) - z_j(0)]^2 \right\rangle + \cdots = \frac{Nk_BT}{m}\left(1 - \frac{k^2\eta t}{\rho m} + \cdots\right)$$

(21–303)

from which we find that the coefficient of shear viscosity is

$$\eta = \frac{1}{2Vk_BTt}\left\langle \sum_{i=1}^{N}\sum_{j=1}^{N} p_{xi}(t)p_{xj}(0)[z_i(t) - z_j(0)]^2 \right\rangle$$

(21–304)

where we have written $m\dot{x}_j$ as p_{xj}. This almost corresponds to Eq. (21–268). In order to convert this into an integral over a time correlation as we did in going from Eq. (21–268) to Eq. (21–276), it is necessary to write this in the equivalent form (Problem 21–42)

$$\eta = \frac{1}{2Vk_BTt}\left\langle \sum_{j=1}^{N} [z_j(t)p_{xj}(t) - z_j(0)p_{xj}(0)]^2 \right\rangle$$

(21–305)

The procedure in going from Eq. (21–304) to Eq. (21–305) involves the same type of manipulations as in Eqs. (21–299) and (21–302). (See Problem 21–60.) Note that Eq. (21–305) is in exactly the same form as Eq. (21–268), namely, the ensemble average of $[A(t) - A(0)]^2$, where $A(t)$ is $\mathbf{r}_j(t)$ in Eq. (21–268) and $z_j(t)p_{xj}(t)$ in the above equation.

By differentiating $z_j p_{xj}$ with respect to time, we see (by working backward) that

$$z_j(t)p_{xj}(t) - z_j(0)p_{xj}(0) = \int_0^t dt'\left\{\left[\frac{p_{zj}(t')p_{xj}(t')}{m}\right] + z_j(t')F_{jx}(t')\right\}$$

(21–306)

where $F_{jx}(t')$ is the x-component of the intermolecular forces acting upon particle j at time t'. By squaring this, taking the ensemble average, and changing variables of integration, we finally get the desired result (Problem 21–43)

$$\eta = \frac{1}{VkT}\int_0^\infty dt\, \langle J(0)J(t)\rangle$$

(21–307)

where

$$J = \sum_{j=1}^{N}\left(\frac{p_{xj}p_{zj}}{m} + z_j F_{jx}\right)$$

(21–308)

It is possible to symmetrize the second summation here and write J as

$$J = m\sum_{j=1}^{N} \dot{x}_j\dot{z}_j + \frac{1}{2}\sum_{i\neq j} z_{ij}F_{ij}{}^x$$

(21–309)

where $z_{ij} = z_i - z_j$ and $F_{ij}{}^x = F_i^x - F_j^x$.

The pattern for deriving expressions for the thermal transport coefficients in terms of integrals of time-correlation functions should be clear. We first select the appropriate macroscopic transport equation and solve it in Fourier transform space to derive an equation for the Fourier transform or its correlation function in terms of the macroscopic transport coefficients [cf. Eqs. (21–264) and (21–292)]. We then identify this same Fourier transform in terms of microscopic variables such as the positions and momenta of the molecules of the system [cf. Eqs. (21–270) and (21–295)]. Lastly we expand both the macroscopic and microscopic representations of the Fourier transform (or its correlation function) in powers of the Fourier transform variable k and equate the coefficients of k^2, thus giving the macroscopic coefficient in terms of an ensemble average of microscopic variables. This is then manipulated into the form

$$\gamma = \langle (A(t) - A(0))^2 \rangle \tag{21–310}$$

where γ is the macroscopic transport coefficient and $A(t)$ is some function of microscopic dynamical variables. Lastly, $A(t)$ is written as the integral of dA/dt, and Eq. (21–310) is transformed into the desired form

$$\gamma = \int_0^\infty dt \, \langle \dot{A}(t)\dot{A}(0) \rangle \tag{21–311}$$

We shall do this briefly for one final example, the thermal conductivity.

We start with the heat conduction equation, Eq. (17–31),

$$\rho C_v \frac{\partial \tilde{E}}{\partial t} = \lambda \, \nabla^2 \tilde{E} \tag{21–312}$$

where we have written $\tilde{E}$ instead of T as the dependent variable. We consider $\tilde{E}(\mathbf{r}, t)$ here to be the excess above the average energy. If we let $L(\mathbf{k}, t)$ be the Fourier transform of $\tilde{E}(\mathbf{r}, t)$, then the above equation yields

$$L(\mathbf{k}, t) = L(\mathbf{k}, 0)e^{-\lambda k^2 t / \rho C_v} \tag{21–313}$$

Now the microscopic representation of $\tilde{E}(\mathbf{r}, t)$ is

$$\tilde{E}(\mathbf{r}, t) = \sum_{j=1}^{N} \tilde{E}_j \, \delta(\mathbf{r} - \mathbf{r}_j) \tag{21–314}$$

where

$$\tilde{E}_j = E_j - \langle E_j \rangle \tag{21–315}$$

The Fourier transform of this gives

$$L(\mathbf{k}, t) = \sum_{j=1}^{N} \tilde{E}_j \, e^{i\mathbf{k} \cdot \mathbf{r}_j(t)} \tag{21–316}$$

Without loss of generality, we can take $\mathbf{k}$ in the z-direction, and so

$$L(\mathbf{k}, t) = \sum_{j=1}^{N} \tilde{E}_j \, e^{ikz_j(t)} \tag{21–317}$$

By multiplying both sides of Eq. (21–313) by $L^*(\mathbf{k}, 0)$ and taking the ensemble average, we get

$$\left\langle \sum_{j=1}^{N} \sum_{l=1}^{N} \tilde{E}_j(0)\tilde{E}_l(t)e^{ik[z_l(t) - z_j(0)]} \right\rangle$$

$$= \left\langle \sum_{j=1}^{N} \sum_{l=1}^{N} \tilde{E}_j(0)\tilde{E}_l(0)e^{ik[z_l(0) - z_j(0)]} \right\rangle \exp\left(\frac{-\lambda k^2 t}{\rho C_v}\right) \tag{21–318}$$

This equation is very similar to Eq. (21–297). The procedure from here is essentially the same as that following Eq. (21–297). To evaluate the leading term in either ensemble average, we use the fact that

$$\left\langle \sum_{i=1}^{N} \sum_{j=1}^{N} \tilde{E}_i \tilde{E}_j \right\rangle = \langle (E - \langle E \rangle)^2 \rangle \tag{21–319}$$

which according to fluctuation theory satisfies [cf. Eq. (3–42)]

$$\langle (E - \langle E \rangle)^2 \rangle = k_B T^2 C_v \tag{21–320}$$

Thus it should be easy to see that we eventually can get (Problem 21–44)

$$\lambda = \frac{1}{2Vk_B T^2 t} \left\langle \sum_{i=1}^{N} \sum_{j=1}^{N} [z_j(t) - z_i(0)]^2 \tilde{E}_i(0)\tilde{E}_j(t) \right\rangle \tag{21–321}$$

This equation closely resembles Eq. (21–304). Using conservation of energy, we can write it as

$$\lambda = \frac{1}{2Vk_B T^2 t} \left\langle \sum_{j=1}^{N} [x_j(t)\tilde{E}_j(t) - x_j(0)\tilde{E}_j(0)]^2 \right\rangle \tag{21–322}$$

from which we get the final result

$$\lambda = \frac{1}{Vk_B T^2} \int_0^{\infty} \langle S(t)S(0) \rangle \, dt \tag{21–323}$$

where

$$S(t) = \frac{d}{dt} \sum_{j=1}^{N} x_j \tilde{E}_i \tag{21–324}$$

One usually sees this equation written in terms of a vectorial flux $\mathbf{S}(t)$:

$$\lambda = \frac{1}{3Vk_B T^2} \int_0^{\infty} \langle \mathbf{S}(t) \cdot \mathbf{S}(0) \rangle \, dt \tag{21–325}$$

where

$$S = \frac{d}{dt} \sum_{j=1}^{N} \mathbf{r}_j \tilde{E}_j$$

$$= \sum_{j=1}^{N} \left\{ \left(\frac{p_j^2}{2m} - \langle h_j \rangle \right) \frac{\mathbf{p}_j}{m} + \frac{1}{2} \sum_{j \neq i} (\mathbf{r}_{ij}\mathbf{F}_{ij} + V(r_{ij})|) \right\} \cdot \frac{\mathbf{p}_i}{m} \tag{21–326}$$

where $\langle h_j \rangle$ is the enthalpy per molecule.

We purposely passed over the expression for the bulk viscosity coefficient since its derivation is fairly subtle (e.g., see Helfand's article). In fact, there was some controversy concerning its correct form for a time This controversy no longer exists, and the accepted expression is*

$$\kappa = \frac{1}{VkT} \int_0^{\infty} dt \, \frac{1}{9} \sum_a \sum_b \langle J^{aa}(0)J^{bb}(t) \rangle \tag{21–327}$$

* R. Zwanzig, *Ann. Rev. Phys. Chem.*, **16**, p. 67, 1965.

where a and b run over x, y, and z and

$$J^{ab} = \sum_{j=1}^{N} \frac{p_{ja} p_{jb}}{m} + \sum_{j=1}^{N} r_{ja} F_{jb} - \langle J^{ab} \rangle \tag{21-328}$$

where r_{ja} is the ath component of $\mathbf{r}_j$ and $\langle J^{ab} \rangle$ is a function of pressure $\bar{p}$, internal energy $\bar{E}$, and particle number N:

$$\langle J^{ab} \rangle = \delta_{ab} V \left[\bar{p} + \frac{\partial \bar{p}}{\partial N} (N - \langle N \rangle) + \frac{\partial \bar{p}}{\partial E} (E - \langle E \rangle) \right] \tag{21-329}$$

This $\langle J^{ab} \rangle$, which does not occur in the case of the shear viscosity coefficient, takes account of the ensemble used for the averages. If a microcanonical ensemble is used, both the second and third terms vanish. For a canonical ensemble, only the second term vanishes, whereas for a grand canonical ensemble, none of them vanish.

In this section we have derived time-correlation function expressions for the thermal transport coefficients. These equations do not possess the rigor or generality of those in the previous sections of this chapter, but they are valid for normal hydrodynamic behavior, i.e., for slow processes and for smoothly varying spatial inhomogeneities. In the next chapter, we shall study the limitations of these equations more carefully.

Clearly, these equations are quite formal since an exact determination of the appropriate time-correlation function is a difficult problem, but the simplicity of the final formulas and the physically appealing nature of a time-correlation function allow one to make simple, intuitive approximations, which can afford good agreement with experiment. In addition, at least we now have equations of the correct formal structure, which can be used to derive general theorems such as a law of corresponding states for transport in dense fluids. We shall discuss both such applications in the next section.

21-9   APPLICATIONS OF THE TIME-CORRELATION FUNCTION FORMULAS FOR THE THERMAL TRANSPORT COEFFICIENTS

As we have stated before, the actual calculation of a time-correlation function is a very difficult task in general. In the dilute gas limit, however, where the Boltzmann equation serves as the exact kinetic theory equation, the various time-correlation functions can be evaluated exactly. This has been done by McLennan and Swenson,* who derive at an intermediate stage an integral equation which is identical with that appearing in the first Chapman-Enskog solution of the Boltzmann equation. Thus we see the not so surprising result that in order to evaluate a time-correlation function exactly in the low-density limit, we are essentially back to the Boltzmann equation. Instead of discussing this exact correspondence, we shall instead present some approximate calculations in this section. One of the appealing aspects of the time-correlation function formalism is that it gives not only a formally exact representation of various transport coefficients, but that it also does this in terms of an intuitively simple function for which one can "derive" approximate forms. In this section we shall first discuss the dilute gas limit, then dense fluids, and finally the law of corresponding states for transport processes.

* J. A. McLennan, Jr., and R. J. Swenson, *J. Math. Phys.*, **4**, p. 1527, 1963.

In the dilute gas limit we can neglect the terms involving the intermolecular potential in Eqs. (21–278), (21–307), (21–325), and (21–327) to write

$$D = \frac{1}{m^2} \int_0^\infty \langle p_{ix}(0) p_{ix}(t) \rangle \, dt \tag{21–330}$$

$$\eta = \frac{1}{VkTm^2} \int_0^\infty \left\langle \sum_{i=1}^N \sum_{j=1}^N p_{ix}(0) p_{iy}(0) p_{jx}(t) p_{jy}(t) \right\rangle \, dt \tag{21–331}$$

$$\kappa = \frac{1}{9VkT} \sum_a \sum_b \int_0^\infty \left\langle \left(\sum_{i=1}^N \frac{p_{ia}{}^2(0)}{m} - \langle J^{aa} \rangle \right) \left(\sum_{j=1}^N \frac{p_{ib}{}^2(t)}{m} - \langle J^{bb} \rangle \right) \right\rangle \, dt \tag{21–332}$$

$$\lambda = \frac{1}{3VkT^2} \int_0^\infty \left\langle \sum_{i=1}^N \sum_{j=1}^N \left(\frac{p_i{}^2(0)}{2m} - \tfrac{5}{2}kT \right) \left(\frac{p_j{}^2(t)}{2m} - \tfrac{5}{2}kT \right) \frac{\mathbf{p}_i(0) \cdot \mathbf{p}_j(t)}{m^2} \right\rangle \, dt \tag{21–333}$$

Note that each of these equations is dependent only upon various functions of the momenta of the particles. Clearly the most accurate way to proceed is to solve the Boltzmann equation for the probability that a molecule has a momentum $\mathbf{p}_j(t)$ at time t given that it had some momentum $\mathbf{p}_j$ initially, i.e., for the singlet momentum distribution function. This is the approach mentioned above that leads to essentially the Chapman-Enskog development.

Instead of carrying out this complicated Chapman-Enskog approach, we shall evaluate Eqs. (21–330) through (21–333) by a mean free path model. Before doing this, however, it is instructive to show that the bulk viscosity of a monatomic ideal gas is zero. As Eq. (21–329) indicates, one must specify the ensemble used to evaluate κ, and we shall use the microcanonical ensemble for simplicity. In this ensemble, Eq. (21–332) becomes

$$\kappa = \frac{1}{9VkT} \sum_a \sum_b \int_0^\infty \langle (2K_a(0) - pV)(2K_b(\tau) - pV) \rangle \, dt \tag{21–334}$$

where we have introduced the total kinetic energy in the ath direction, K_a, and used the fact that $\langle J^{aa} \rangle = pV$ in a microcanonical ensemble. If we now sum this over a and b and use the fact that the total kinetic energy K equals $K_x + K_y + K_z$, then Eq. (21–334) becomes

$$\kappa = \frac{1}{9VkT} \int_0^\infty \langle [2K(0) - 3pV][2K(t) - 3pV] \rangle \, dt \tag{21–335}$$

Now since we are neglecting the intermolecular potential in this dilute gas limit, the kinetic energy is equal to the total energy, and since we are averaging over a microcanonical ensemble, this energy is fixed, so that $K(0) = K(t) = E$. It is easy to show from Eqs. (1–37) and Eq. (3–24) that $pV = 2E/3$, and so we see that both factors in the integrand of Eq. (21–335) vanish. It is a good exercise to prove this in the canonical and grand canonical ensembles as well.

We shall now evaluate Eqs. (21–330), (21–331), and (21–333) in a mean free path approximation.* We assume that the motion of the molecules in the gas can be

---

* W. A. Steele, in *Transport Phenomena in Fluids*, edited by H. J. M. Hanley (New York: Dekker, 1969), Chapter 8.

described by a succession of straight-line paths, interrupted intermittently by collisions with other molecules. We furthermore assume that the momentum vector after a collision has a completely random distribution relative to its previous direction. In other words, we are assuming that the correlation of the momentum, or any function of the momentum, is complete until a collision occurs and is zero afterward. It is possible to account for the correlation between successive collisions by computing the so-called persistence of velocity effect, but we shall not do this here.

If we let $P\{\mathbf{p}_i(0), \mathbf{p}_j(0); t\}$ be the fraction of molecules with momenta $\mathbf{p}_i$ and $\mathbf{p}_j$ which will not collide and hence still be correlated at time t, we can write

$$G(t) = \left\langle \sum_{i=1}^{N} \sum_{j=1}^{N} f[\mathbf{p}_i(0)]g[\mathbf{p}_j(0)]P\{\mathbf{p}_i(0), \mathbf{p}_j(0); t\} \right\rangle \qquad (21\text{–}336)$$

where f and g are arbitrary functions and the angular brackets now denote an average over all initial momenta.

The fraction of the molecules that have not collided by time t can be calculated in the following way. Let the collision frequency of the molecules with momenta $\mathbf{p}$ and $\mathbf{p}'$ be $1/\tau_c(\mathbf{p}, \mathbf{p}')$. Then we can write that

$$\frac{\partial P}{\partial t} = -\frac{P}{\tau_c} \qquad (21\text{–}337)$$

which can be integrated to

$$P(\mathbf{p}, \mathbf{p}'; t) = \exp\left[-\frac{t}{\tau_c(\mathbf{p}, \mathbf{p}')} \right] \qquad (21\text{–}338)$$

We can substitute this into Eq. (21–336) to get

$$D = \frac{1}{m^2} \left\langle p_{ix}^2 \tau_c(\mathbf{p}_i, \mathbf{p}_j) \right\rangle \qquad (21\text{–}339)$$

$$\eta = \frac{1}{VkTm^2} \left\langle \sum_{i=1}^{N} \sum_{j=1}^{N} p_{ix} p_{iy} p_{jx} p_{jy} \tau_c(\mathbf{p}_i, \mathbf{p}_j) \right\rangle \qquad (21\text{–}340)$$

$$\lambda = \frac{1}{VkT^2} \left\langle \sum_{i=1}^{N} \sum_{j=1}^{N} \left(\frac{p_i^2}{2m} - \tfrac{5}{2}kT \right)\left(\frac{p_j^2}{2m} - \tfrac{5}{2}kT \right) \frac{\mathbf{p}_i \cdot \mathbf{p}_j}{m^2} \tau_c(\mathbf{p}_i, \mathbf{p}_j) \right\rangle \qquad (21\text{–}341)$$

If we neglect correlations between different molecules, then all cross terms in these three equations vanish, and we have

$$D = \frac{1}{m^2} \left\langle p_x^2 \tau_c(p) \right\rangle \qquad (21\text{–}342)$$

$$\eta = \frac{\rho}{kTm^2} \left\langle p_x^2 p_y^2 \tau_c(p) \right\rangle \qquad (21\text{–}343)$$

$$\lambda = \frac{\rho}{kT^2} \left\langle \left(\frac{p^2}{2m} - \tfrac{5}{2}kT \right)^2 \frac{p^2}{m^2} \tau_c(p) \right\rangle \qquad (21\text{–}344)$$

If we now introduce the Maxwell-Boltzmann distribution function and the reduced momentum

$$w^2 = \frac{p^2}{2mkT}$$

Eqs. (21–342) through (21–344) become

$$D = \frac{8kT}{3\pi^{1/2}m} \int_0^\infty w^4 e^{-w^2} \tau_c(w) \, dw \tag{21-345}$$

$$\eta = \frac{16\rho kT}{15\pi^{1/2}} \int_0^\infty w^6 e^{-w^2} \tau_c(w) \, dw \tag{21-346}$$

$$\lambda = \frac{8\rho k^2 T}{3\pi^{1/2}m} \int_0^\infty w^4 (w^2 - \tfrac{5}{2})^2 e^{-w^2} \tau_c(w) \, dw \tag{21-347}$$

These equations have been derived by Monchick and Mason* by calculating the first-order deviations from the Maxwell-Boltzmann distribution due to the combined effects of collisions and gradients in the appropriate macroscopic variables.

Certainly the simplest approximation that we can make for $\tau_c(w)$ is to assume that it is independent of w and equal to simply the mean free path divided by the average velocity so that

$$\tau_c = \frac{l}{\bar{v}} = \frac{1}{\sqrt{2}\pi\sigma^2\rho} \left(\frac{\pi m}{8kT}\right)^{1/2} \tag{21-348}$$

If we use this in Eqs. (21–345) through (21–347), we get (Problem 24–46)

$$D = \frac{1}{4\pi^{1/2}} \left(\frac{kT}{m}\right)^{1/2} \frac{1}{\sigma^2\rho} \tag{21-349}$$

$$\eta = \frac{1}{4\pi^{1/2}} \frac{(mkT)^{1/2}}{\sigma^2} \tag{21-350}$$

$$\lambda = \frac{5}{6} \left(\frac{kT}{\pi m}\right)^{1/2} \frac{C_v}{N_0 \sigma^2} \tag{21-351}$$

These results are to be compared with Eqs. (16–36) through (16–39), where it can be seen that the agreement with the simple mean free path results of that chapter is quite good.

Gordon† has presented a more sophisticated kinetic-theory evaluation of time-correlation functions which we present now. This approach allows one to account for higher-order collisions in a clear way. To begin with, we assume that the intermolecular forces have a finite range b_0, beyond which they vanish. At the end of the calculation we shall let $b_0 \to \infty$. We shall evaluate the momentum correlation function as an example. For those molecules that have not yet collided, we have that $\langle \mathbf{p}(0) \cdot \mathbf{p}(t) \rangle = \langle \mathbf{p}(0) \cdot \mathbf{p}(0) \rangle$. If we assume molecular chaos, that is, if we assume that the number of collisions that a molecule suffers in any time interval is a Poisson process, then the fraction of molecules that have suffered exactly n collisions in the interval $(0, t)$ is

$$f_n(t) = \frac{1}{n!} \left(\frac{t}{\tau}\right)^n e^{-t/\tau} \tag{21-352}$$

where τ^{-1} is the mean collision frequency. Therefore the number of molecules that has not collided in that time interval is $\exp(-t/\tau)$, and the contribution to the momentum time-correlation function is $\langle \mathbf{p}(0) \cdot \mathbf{p}(0) \rangle \exp(-t/\tau)$.

* L. Monchick and E. A. Mason, *Phys. Fluids*, **10**, p. 1377, 1967.
† R. G. Gordon, *J. Chem. Phys.*, **44**, p. 228, 1966.

The fraction of molecules that has collided once and only once by time t is $(t/\tau)\exp(-t/\tau)$, and so the contribution to $\langle \mathbf{p}(0) \cdot \mathbf{p}(t) \rangle$ from these molecules is

$$\langle \mathbf{p}(0) \cdot \mathbf{p}(1) \rangle \, (t/\tau)\exp(-t/\tau),$$

where $\mathbf{p}(1)$ is the value of $\mathbf{p}$ after the first collision. Clearly we can continue in this way and write

$$\langle \mathbf{p}(0) \cdot \mathbf{p}(t) \rangle = \exp\left(-\frac{t}{\tau}\right)\left[\langle \mathbf{p}(0) \cdot \mathbf{p}(0) \rangle + \frac{t}{\tau} \langle \mathbf{p}(0) \cdot \mathbf{p}(1) \rangle \right.$$

$$\left. + \frac{1}{2!}\left(\frac{t}{\tau}\right)^2 \langle \mathbf{p}(0) \cdot \mathbf{p}(2) \rangle + \cdots \right] \quad (21\text{--}353)$$

In the previous calculation we neglected all the terms beyond the first term in brackets, but here we shall account for them in the following manner.

We first integrate Eq. (21–353) to write

$$3m^2 D = \int_0^\infty \langle \mathbf{p}(0) \cdot \mathbf{p}(t) \rangle \, dt$$

$$= \tau[\langle \mathbf{p}^2(0) \rangle + \langle \mathbf{p}(0) \cdot \mathbf{p}(1) \rangle + \langle \mathbf{p}(0) \cdot \mathbf{p}(2) \rangle + \cdots] \quad (21\text{--}354)$$

We multiply and divide by $\langle \mathbf{p}^2 - \mathbf{p}(0) \cdot \mathbf{p}(1) \rangle$ and rearrange to get (Problem 21–47)

$$3m^2 D = \frac{\tau}{\langle \mathbf{p}^2 - \mathbf{p}(0) \cdot \mathbf{p}(1) \rangle} \{\langle \mathbf{p}^2 \rangle^2 + [\langle \mathbf{p}^2 \rangle \langle \mathbf{p}(0) \cdot \mathbf{p}(2) \rangle - \langle \mathbf{p}(0) \cdot \mathbf{p}(1) \rangle^2]$$

$$+ [\langle \mathbf{p}^2 \rangle \langle \mathbf{p}(0) \cdot \mathbf{p}(3) \rangle - \langle \mathbf{p}(0) \cdot \mathbf{p}(1) \rangle \langle \mathbf{p}(0) \cdot \mathbf{p}(2) \rangle] + \cdots\} \quad (21\text{--}355)$$

The terms in brackets in this equation represent a correlation between the effects of successive collisions and should be rather small (Problem 21–48). We can neglect these and write

$$3m^2 D = \frac{\tau \langle \mathbf{p}^2 \rangle^2}{\langle \mathbf{p}^2 - \mathbf{p}(0) \cdot \mathbf{p}(1) \rangle} \quad (21\text{--}356)$$

This result can be rearranged into a slightly more convenient form by using the fact that a Boltzmann distribution of $\mathbf{p}$ before a collision remains a Boltzmann distribution after a collision when averaged over all types of collisions, so that

$$\langle \mathbf{p}^2 \rangle \equiv \langle \mathbf{p}^2(0) \rangle = \langle \mathbf{p}^2(1) \rangle$$

But

$$\langle \mathbf{p}^2(1) \rangle = \langle \{\mathbf{p}(0) + [\mathbf{p}(1) - \mathbf{p}(0)]\}^2 \rangle$$

$$= \langle \mathbf{p}^2(0) \rangle + 2\langle \mathbf{p}(0) \cdot [\mathbf{p}(1) - \mathbf{p}(0)] \rangle + \langle [\mathbf{p}(1) - \mathbf{p}(0)]^2 \rangle$$

and so

$$0 = 2\langle \mathbf{p}(0) \cdot [\mathbf{p}(1) - \mathbf{p}(0)] \rangle + \langle [\mathbf{p}(1) - \mathbf{p}(0)]^2 \rangle$$

Thus

$$\langle \mathbf{p}^2 - \mathbf{p}(0) \cdot \mathbf{p}(1) \rangle = \langle \mathbf{p}(0) \cdot [\mathbf{p}(0) - \mathbf{p}(1)] \rangle$$

$$= \tfrac{1}{2}\langle [\mathbf{p}(1) - \mathbf{p}(0)]^2 \rangle \equiv \tfrac{1}{2}\langle (\Delta\mathbf{p})^2 \rangle$$

and so

$$3m^2 D = \frac{2\tau \langle \mathbf{p}^2 \rangle^2}{\langle (\Delta\mathbf{p})^2 \rangle} \quad (21\text{--}357)$$

The average over impact parameter b may be written explicitly as

$$\langle(\Delta\mathbf{p})^2\rangle = \frac{1}{\pi b_0^2} \int_0^{b_0} \langle(\Delta\mathbf{p})^2\rangle 2\pi b \, db \tag{21-358}$$

If we give τ in Eq. (21–357) its kinetic theory value $\tau = 1/\rho\bar{v}_r \pi b_0^2 = 1/2^{1/2}\rho\bar{v}_r\pi b_0^2$, where $\bar{v}_r$ is the mean relative velocity, then Eq. (21–357) becomes

$$3m^2 D = \frac{2\langle\mathbf{p}^2\rangle^2}{\rho\bar{v}_r \int_0^{b_0} \langle(\Delta\mathbf{p})^2\rangle 2\pi b \, db}$$

or

$$D = \frac{2\langle\mathbf{p}^2\rangle^2}{3m^2 \rho\bar{v}_r \int_0^\infty \langle(\Delta\mathbf{p})^2\rangle 2\pi b \, db} \tag{21-359}$$

We now show that this equation is the Chapman-Enskog result for D. The average of $(\Delta\mathbf{p})^2$ in Eq. (21–359) is over all collisions and has already been evaluated in Section 16–3. It is shown there that $(\Delta\mathbf{p})^2 = m^2 v_r^2(1 - \cos\chi)/2$, where χ is the deflection angle and that the fraction of collisions with relative velocity v_r is

$$\frac{1}{8}\left(\frac{m}{kT}\right)^2 v_r^3 e^{-mv_r^2/4kT} \, dv_r$$

Thus we have that

$$\langle(\Delta\mathbf{p})^2\rangle = \frac{m^2}{16}\left(\frac{m}{kT}\right)^2 \int_0^\infty v_r^5(1 - \cos\chi)e^{-mv_r^2/4kT} \, dv_r \tag{21-360}$$

and

$$\int \langle(\Delta\mathbf{p})^2\rangle 2\pi b \, db = \frac{m^2}{16}\left(\frac{m}{kT}\right)^2 \int_0^\infty v_r^5 \sigma(v_r)e^{-mv_r^2/4kT} \, dv_r$$

$$= 4mkT \int_0^\infty g^5 \sigma(g)e^{-g^2} \, dg \tag{21-361}$$

where g is a reduced relative velocity, $g^2 = mv_r^2/4kT$. The coefficient of self-diffusion becomes

$$D = \frac{2(3mkT)^2}{3\rho m^2}\left(\frac{\pi m}{16kT}\right)^{1/2} \frac{1}{4mkT \int_0^\infty g^5\sigma(g)e^{-g^2} \, dg} \tag{21-362}$$

Using the standard definition of the collision integrals [Eqs. (19–19) through (19–21)], we see that

$$D = \frac{3}{8\rho}\left(\frac{kT}{m}\right)\frac{1}{\Omega^{(1,1)}} \tag{21-363}$$

which is Eq. (19–26), the Chapman-Enskog result.

The calculation of the time-correlation function for a dense system is, of course, much more difficult than in the binary collision limit. The exact determination of these correlation functions is a complicated dynamical problem involving the motion of many particles. Molecular dynamics calculations have been carried out for a number of systems and these have shed much light on the behavior of time-correlation functions. For example Fig. 21–11 shows the normalized velocity correlation function for both a system, of hard spheres and a system of Lennard-Jones molecules. This molecular dynamics information can be used to suggest simple analytical expressions, which can be used to approximate the time-correlation function. These approximate expressions contain parameters which can be determined in several ways, most obviously by requiring that the first few terms in the Maclaurin expansion for the correlation function be exact. We shall illustrate this approach for the coefficient of self-diffusion. We start with

$$D = \frac{kT}{m} \int_0^\infty \phi(t)\, dt \tag{21–364}$$

where

$$\phi(t) = \frac{\langle \mathbf{p}(0) \cdot \mathbf{p}(t) \rangle}{\langle \mathbf{p}(0) \cdot \mathbf{p}(0) \rangle} \tag{21–365}$$

We expand $\phi(t)$ in a Maclaurin expansion, giving

$$\phi(t) = \sum_{n=0}^\infty (-1)^n A_{2n}\, t^{2n} \tag{21–366}$$

where

$$A_{2n} = \frac{\langle \mathbf{p}^{(n)}(0) \cdot \mathbf{p}^{(n)}(0) \rangle}{(2n)!\, 3mkT} \tag{21–367}$$

where the notation $\mathbf{p}^{(n)}(0)$ denotes the nth time derivative of $\mathbf{p}$ evaluated at $t = 0$. Clearly $A_0 = 1$. To evaluate A_2, it is convenient to first use the fact that

$$\langle \dot{\mathbf{p}}(0) \cdot \dot{\mathbf{p}}(0) \rangle = -\langle \ddot{\mathbf{p}}(0) \cdot \mathbf{p}(0) \rangle$$

and that in the Poisson bracket notation

$$\dot{\mathbf{p}}_1 = -\{H, p_1\} = -\nabla_1 H = -\nabla_1 U_N$$

where ∇_1 represents the gradient with respect to the spatial coordinates of particle 1. By repeating this and then averaging over all initial momenta and coordinates, we get

$$\langle \mathbf{p}(0) \cdot \ddot{\mathbf{p}}(0) \rangle = -kT \langle \nabla_1^2 U_N \rangle$$

If we assume that U_N is pair-wise additive, then

$$\langle \dot{\mathbf{p}}^2(0) \rangle = NkT \langle \nabla^2 u(r_{12}) \rangle$$

$$= \rho kT \int_0^\infty g(r) \nabla^2 u\, 4\pi r^2\, dr \tag{21–368}$$

which is essentially A_2.

Perhaps the simplest approximation that we can make for $\phi(t)$ is to assume that it is Gaussian,* in which case we have $\phi(t) = \exp(-A_2 t^2)$ and

$$D = \frac{kT}{m} \left(\frac{\pi}{A_2} \right)^{1/2} \tag{21–369}$$

This form for $\phi(t)$ does not give the negative region found from molecular dynamics calculations, but requires only one moment, A_2. Below we shall introduce a more flexible form for $\phi(t)$ that does give the negative region, but, in addition, requires A_4 as well as A_2. The A_{2n} become progressively more difficult to calculate, and A_4 is

---

* S. A. Rice, *J. Chem. Phys.*, **33**, p. 1376, 1960.

probably the practical limit. There are not a great deal of data for the coefficient of self-diffusion in dense fluids, but it has been measured in liquid argon at 90°K and a density of 1.374 g/cm^3. Using a Lennard-Jones 6–12 potential with potential parameters $\varepsilon/k = 120°K$ and $\sigma = 3.4$ Å and the molecular dynamics radial distribution function of Verlet,* Eq. (21–368) can be integrated numerically to give $A_2 = 3.94 \times 10^{25}$ sec^{-2},† which according to Eq. (21–369) gives a value of $D = 3.44 \times 10^{-5}$ cm^2/sec. This is to be compared to experimental values of 2.43×10^{-5}‡ cm^2/sec and 2.06×10^{-5}§ cm^2/sec and Rahman's molecular dynamics value of 2.43×10^{-5} cm^2/sec.**

It is possible to carry this procedure one term further and to use A_4. This would allow one to use an approximate form for $\phi(t)$ that contains two parameters. A particularly convenient form is $\phi(t) = \text{sech}(at)\cos(bt)$. The advantage of this form is that it is even in time, gives the negative region that is found in the molecular dynamics calculations, and decays as $e^{-at}\cos bt$ for long times. Douglass†† and Isbister and McQuarrie† have carried out calculations using this form for $\phi(t)$ and find reasonably good agreement with available experimental and molecular dynamics values for D. The A_{2n} become progressively more difficult to calculate, and so this procedure is not readily extended. Nevertheless, it appears to be a useful method for estimating the transport coefficients, although calculations have been done only for D in liquid argon (Problem 21–59).

We have often stated that the odd powers of t do not appear in the Maclaurin expansion of a classical time-correlation function. In a hard-sphere fluid, however, the discontinuous nature of the intermolecular potential gives impulsive forces so that the time derivatives of the velocity of a molecule are no longer well defined or well behaved. In this case, then, instead of an equation like Eq. (21–202), we write

$$\langle \mathbf{v}(0) \cdot \mathbf{v}(t) \rangle = \langle \mathbf{v}^2 \rangle + t\left(\frac{d}{dt}\right)_0 \langle \mathbf{v}(0) \cdot \mathbf{v}(t) \rangle + \frac{t^2}{2}\left(\frac{d^2}{dt^2}\right)_0 \langle \mathbf{v}(0) \cdot \mathbf{v}(t) \rangle + \cdots \quad (21\text{–}370)$$

where the time differentiation is done after the ensemble averaging. When this procedure is carried out for hard spheres, it is found that the linear term in t in the above equation is nonzero, implying that the autocorrelation function has a discontinuous slope at the origin. For a hard-sphere system, then, we can assume that $\langle \mathbf{v}(0) \cdot \mathbf{v}(t) \rangle$ decays exponentially:

$$\langle \mathbf{v}_1(0) \cdot \mathbf{v}_1(t) \rangle = \langle v_1{}^2 \rangle e^{-\alpha t} \quad (21\text{–}371)$$

and determine α from the linear term in Eq. (21-370):

$$\alpha = -\left(\frac{d}{dt}\right)_0 \frac{\langle \mathbf{v}_1(0) \cdot \mathbf{v}_1(t) \rangle}{\langle v_1{}^2 \rangle}$$

$$= -\lim_{\Delta t \to 0}\left(\frac{\langle \mathbf{v}_1 \cdot \Delta\mathbf{v}_1 \rangle}{\langle v_1{}^2 \rangle \, \Delta t}\right)$$

$$= -\lim_{\Delta t \to 0}\left(\frac{\langle \mathbf{p}_1 \cdot \Delta\mathbf{p}_1 \rangle}{\langle p_1{}^2 \rangle \, \Delta t}\right) \quad (21\text{–}372)$$

* L. Verlet, *Phys. Rev.*, **165**, p. 201, 1968.
† D. J. Isbister and D. A. McQuarrie, *J. Chem. Phys.*, **56**, p. 736, 1972.
‡ J. Naghizadeh and S. A. Rice, *J. Chem. Phys.*, **36**, p. 2710, 1962.
§ J. W. Corbett and J. H. Wang, *J. Chem. Phys.*, **25**, p. 422, 1956.
** A. Rahman, *Phys. Rev.*, **136**, p. A405, 1964.
†† D. C. Douglass, *J. Chem. Phys.*, **35**, p. 81, 1961.

We can evaluate this expression in much the same manner that we evaluated $\langle(\Delta p)^2\rangle$ in Eq. (21–359). From Eqs. (16–50) and (16–64), we have

$$
\begin{aligned}
\mathbf{p}_1 \cdot \Delta \mathbf{p}_1 &= m_1\mu(\mathbf{v}_1 \cdot \mathbf{v}_r' - \mathbf{v}_1 \cdot \mathbf{v}_r) \\
&= \mu^2 v_r^2(\cos \chi - 1) + m_1\mu(\mathbf{v}_c \cdot \mathbf{v}_r' - \mathbf{v}_c \cdot \mathbf{v}_r)
\end{aligned}
\tag{21–373}
$$

We are going to average this over collisions, in which case the second and third terms here will vanish. If we average this over the *rate* of collisions of some "tagged" molecule with the others, we get, using Eq. (16–68) with $\rho_1 = 1$ and $\rho_2 = \rho$,

$$
\left\langle \mathbf{p}_1 \cdot \frac{\Delta \mathbf{p}_1}{\Delta t} \right\rangle = \frac{\rho}{2\pi^{1/2}} \frac{m^2\pi\sigma^2}{4} \left(\frac{m}{kT}\right)^{3/2} \int_0^\infty v_r^5 e^{-mv_r^2/4kT}\, dv_r = 8\pi\sigma^2\rho\left(\frac{m}{\pi}\right)^{1/2}(kT)^{3/2}
\tag{21–374}
$$

This gives

$$
\alpha = \frac{8\pi\sigma^2\rho}{3}\left(\frac{kT}{\pi m}\right)^{1/2}
\tag{21–375}
$$

and a coefficient of self-diffusion

$$
D = \frac{kT}{m\alpha} = \frac{3}{8\pi^{1/2}}\left(\frac{kT}{m}\right)^{1/2}\frac{1}{\rho\sigma^2}
\tag{21–376}
$$

which is the Chapman-Enskog result for hard spheres. Thus we see that for a hard-sphere system, an exponentially decaying time-correlation function gives the Chapman-Enskog result. More interestingly, if we multiply the collision rate in Eq. (21–374) by $g(1)$, the hard-sphere radial distribution function at contact, α is increased by $g(1)$ and D is decreased by $g(1)$, giving the Enskog dense hard-sphere fluid result for D [cf. Eq. (19–45)]. So we see that the Enskog theory is in some sense equivalent to assuming an exponentially decaying time-correlation function.*

This has been shown also for the viscosity of a hard-sphere fluid. Wainwright† has shown that if one assumes that the terms in the autocorrelation function giving the viscosity [cf. Eqs. (21–307) and (21–308)] are in the form of a decaying exponential, then one gets

$$
\frac{\eta}{\eta^0 b_0 \rho} = \frac{c_1}{y} + c_2 + c_3 y
$$

where the c_j are quantities similar to α in Eq. (21–372). The integrals here are more complicated than in the case of the coefficient of self-diffusion above since the autocorrelation function giving the viscosity is more complicated, but by using a detailed analysis of hard-sphere collisions, Wainwright is able to show that $c_1 = 1$, $c_2 = \frac{4}{5}$, and $c_3 = 0.7649$, giving

$$
\frac{\eta}{\eta^0 b_0 \rho} = \frac{1}{y} + \frac{4}{5} + 0.7649 y
$$

as compared to the Enskog result [cf. Eq. (19–41)]

$$
\frac{\eta}{\eta^0 b_0 \rho} = \frac{1}{y} + \frac{4}{5} + 0.76 y
$$

---

* This was apparently first shown by H. C. Longuet-Higgins and J. A. Pople, *J. Chem. Phys.*, **25**, p. 884, 1956.

† T. E. Wainwright, *J. Chem. Phys.*, **40**, p. 2932, 1964.

Alder, Gass, and Wainwright* have made a detailed analysis of molecular dynamics calculations and the Enskog theory.

The last topic that we shall discuss in this rather gigantic chapter is the law of corresponding states for the thermal transport processes.† Consider Eq. (21–307) for the viscosity coefficient:

$$\eta = \frac{1}{VkT} \int_0^\infty \langle J(0)J(t) \rangle \, dt \tag{21–377}$$

where

$$J = \sum_{j=1}^N \left(\frac{p_{xj}p_{zj}}{m} + z_j F_{jx} \right) \tag{21–378}$$

We assume that the intermolecular potential is pair-wise additive and of the form $u(r) = \varepsilon f(r/\sigma)$ and introduce the reduced variables

$$r^* = \frac{r}{\sigma}, \quad T^* = \frac{kT}{\varepsilon}, \quad u^* = \frac{u}{\varepsilon}$$

From this basic set, we can see that since time has units of $[\text{mass-(distance)}^2/\text{energy}]^{1/2}$, then t is reduced by $(\varepsilon/m\sigma^2)^{1/2}$, i.e., $t^* = t\varepsilon^{1/2}/m^{1/2}\sigma$. Similarly we see that $v^* = v(m/\varepsilon)^{1/2}$, etc. This all allows us to write J in Eq. (21–377) in the reduced form $J^* = J/\varepsilon$ so that we get

$$\eta = \frac{m^{1/2}\varepsilon^{1/2}}{\sigma^2} \eta^* \tag{21–379}$$

where

$$\eta^* = \eta^*(T^*, \rho^*) = \frac{1}{V^*T^*} \int_0^\infty \langle J^*(0)J^*(t) \rangle \, dt^* \tag{21–380}$$

It is easy to show that the other thermal transport coefficients are (Problem 21–51)

$$\lambda^* = \frac{\lambda k \varepsilon^{1/2}}{m^{1/2}\sigma^2} \tag{21–381}$$

$$D^* = \frac{D m^{1/2}}{\varepsilon^{1/2}\sigma} \tag{21–382}$$

$$\kappa^* = \frac{\kappa \sigma^2}{m^{1/2}\varepsilon^{1/2}} \tag{21–383}$$

Tham and Gubbins‡ have made an extensive test of these relations; the results are shown in Figs. 21–12 through 21–14. The data plotted in Figs. 21–12 and 21–13 cover densities ranging from dilute gas to dense liquid, and pressures from 0 to 2000 atm. Not enough data are available to test Eq. (21–383) for κ^*.

* B. J. Alder, D. M. Gass, and T. E. Wainwright, *J. Chem. Phys.*, **53**, p. 3813, 1970.

† E. Helfand and S. A. Rice, *J. Chem. Phys.*, **32**, p. 1642, 1960.

‡ M. J. Tham and K. E. Gubbins, *I and EC Fundamentals*, **8**, p. 791, 1969; **9**, p. 63, 1970.

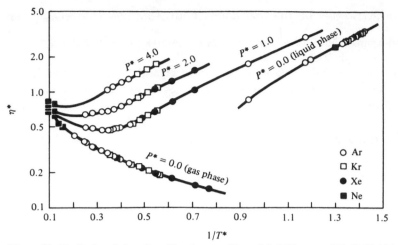

Figure 21–12. **Reduced viscosity of inert gases.** (From M. J. Tham and K. E. Gubbins, *Ind. Eng. Chem.*, *Fundam.*, **8**, 791, 1969.)

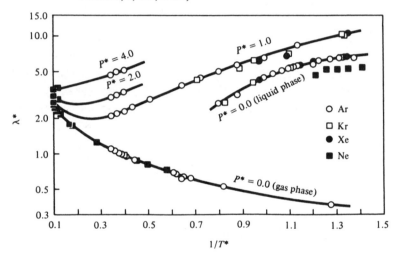

Figure 21–13. **Reduced thermal conductivity of inert gases.** (From M. J. Tham and K. E. Gubbins, *Ind. Eng. Chem. Fundam.*, **8**, 791, 1969.)

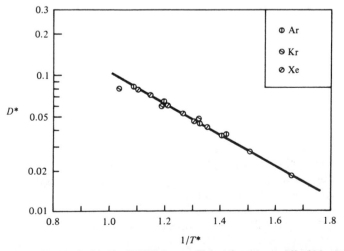

Figure 21–14. **Reduced self-diffusion coefficient for saturated liquid inert gases.** (From M. J. Tham and K. E. Gubbins, *Ind. Eng. Chem. Fundam.*, **8**, 791, 1969.)

ADDITIONAL READING

BERNE, B. J., and HARP, G. D. *Adv. Chem. Phys.*, **17**, p. 63, 1970.

———, 1971. In *Physical chemistry, an advanced treatise*, Vol. 8B. ed. by D. Henderson. New York: Academic.

GORDON, R. G. *Adv. Mag. Resonance*, **3**, p. 1, 1968.

KUBO, R. 1958. In *Lectures in theoretical physics*, Vol. 1, p. 120. New York: Wiley.

———. 1965. In *Statistical mechanics of equilibrium and nonequilibrium*, ed. by J. Meixner. Amsterdam: North-Holland Publ.

———, *Repts. Prog. Phys.*, **29**, p. 255, 1966.

STEELE, W. A. 1969. In *Transport phenomena in fluids*, ed. by H. J. M. Hanley. New York: Dekker.

ZWANZIG, R. W. 1961. In *Lectures in theoretical physics*, Vol. 3, p. 135. New York: Wiley.

———, *Ann. Rev. Phys. Chem.*, **16**, p. 67, 1965.

PROBLEMS

21–1. Derive Eq. (21–12) from Eq. (21–11).

21–2. Prove that if $\mathcal{H}_0|n\rangle = E_n|n\rangle$, then $e^{-i\mathcal{H}_0 t/\hbar}|n\rangle = e^{-iE_n t/\hbar}$ and that $\langle n|e^{i\mathcal{H}_0 t/\hbar} = \langle n|e^{iE_n t/\hbar}$.

21–3. Show that

$$\langle \boldsymbol{\varepsilon} \cdot \mathbf{M}(0)\boldsymbol{\varepsilon} \cdot \mathbf{M}(t)\rangle = \tfrac{1}{3}\langle \mathbf{M}(0) \cdot \mathbf{M}(t)\rangle$$

for an isotropic fluid.

21–4. Referring to Fig. 21–3, show that if we let α_{AB} be the components of the polarizability tensor in a molecule-fixed coordinate system, then the components in an arbitrary system are given by Eq. (21–52).

21–5. Prove that the direction cosines satisfy

$$C_{XJ}{}^2 + C_{YJ}{}^2 + C_{ZJ}{}^2 = 1$$

and

$$C_{Xi}C_{XJ} + C_{Yi}C_{YJ} + C_{Zi}C_{ZJ} = 0 \qquad i \neq j$$

21–6. Prove Eq. (21–57).

21–7. Show that if the three unprimed axes in Fig. 21–3 are principal axes of the molecule, then Eq. (21–52) reduces to Eq. (21–53).

21–8. Derive Eqs. (21–62) to (21–64).

21–9. Show that the square of the anisotropic part of the polarizability is

$$\beta^2 = \tfrac{1}{3}[(\alpha_1 - \alpha_2)^2 + (\alpha_2 - \alpha_3)^2 + (\alpha_3 - \alpha_1)^2]$$

21–10. Show that

$$\bar{\alpha} = \tfrac{1}{3}(\alpha_\| + 2\alpha_\perp)$$

and

$$\beta^2 = \tfrac{2}{3}(\alpha_\| - \alpha_\perp)^2$$

for a linear or symmetric top molecule.

21–11. Derive Eqs. (21–70) through (21–72).

21–12. In Section 21–3, it is necessary to evaluate averages such as $\langle \varepsilon_\alpha{}^\| \varepsilon_\beta{}^\| \varepsilon_\gamma{}^\perp \varepsilon_\delta{}^\perp \rangle_{\text{sphere}}$ where $\varepsilon_i{}^\|$ is the ith component of a unit vector, $\varepsilon_j{}^\perp$ is the jth component of a unit vector perpendicular to the first, and the angular brackets denote a spherical average. The main difficulty here is to write out the components of the unit vector $\varepsilon^\perp$. There are several ways to do this, but one is the following.

We start with a unit vector along the z-axis and rotate the coordinate system about the

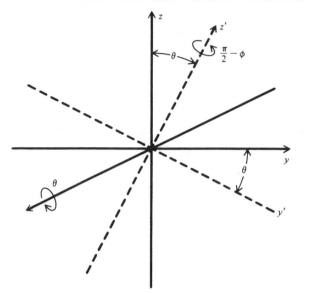

Figure 21–15. **The rotations used to find a unit vector $\epsilon^{\perp}$ perpendicular to an arbitrary unit vector ϵ.**

x-axis by θ degrees (cf. Fig. 21–15). Show that the matrix that gives the unit vector in the new coordinate system is

$$R_\theta = \begin{pmatrix} 1 & 0 & 0 \\ 0 & \cos\theta & \sin\theta \\ 0 & -\sin\theta & \cos\theta \end{pmatrix}$$

Show that the components of the unit vector in this coordinate system are $(0, \sin\theta, \cos\theta)$.

Now rotate this new coordinate system clockwise about the z'-axis by $(\pi/2) - \phi$ radians (cf. Fig. 21–15). Show that the matrix that expresses the vector $(0, \sin\theta, \cos\theta)$ in this coordinate system is

$$R_\phi = \begin{pmatrix} \sin\phi & \cos\phi & 0 \\ -\cos\phi & \sin\phi & 0 \\ 0 & 0 & 1 \end{pmatrix}$$

Show that the unit vector in this new coordinate system is given by $(\sin\theta\cos\phi, \sin\theta\sin\phi, \cos\theta)$, which is the conventional result for a unit vector in a spherical coordinate system. Thus the matrix

$$R_\phi R_\theta = \begin{pmatrix} \sin\phi & \cos\theta\cos\phi & \sin\theta\cos\phi \\ -\cos\phi & \cos\theta\sin\phi & \sin\theta\sin\phi \\ 0 & -\sin\theta & \cos\theta \end{pmatrix}$$

acts upon a vector in the original coordinate system and expresses it in the new coordinate system.

Now choose a unit vector in the original coordinate system that is perpendicular to $(0, 0, 1)$ there. This is $\epsilon^{\perp} = (\cos\alpha, \sin\alpha, 0)$, where α is arbitrary. Show that the new vector $R_\phi R_\theta \epsilon^{\perp}$ is perpendicular to $(\sin\theta\cos\phi, \sin\theta\sin\phi, \cos\theta)$. This gives the components of a vector perpendicular to a unit vector ϵ. Verify that the results given in Eq. (21–91) are independent of the arbitrary angle α.

21–13. Another way to derive the results given in Eq. (21–91) is to begin with a coordinate system in which there is a unit vector along the z-axis and a unit vector perpendicular to that, for example, with components $(\cos\alpha, \sin\alpha, 0)$. We now rotate the coordinate system through the three Euler angles ϕ, θ, ψ. An excellent discussion of the Euler angles is given in *Classical Mechanics* by H. Goldstein (Reading, Mass.: Addison-Wesley, 1950). It is shown there that the matrix corresponding to a rotation through the three Eulerian angles $0 < \theta < \pi$, $0 < \phi < 2\pi$, and $0 < \psi < 2\pi$ is

$$A = \begin{pmatrix} \cos\psi\cos\phi - \cos\theta\sin\phi\sin\psi & -\sin\psi\cos\phi - \cos\theta\sin\phi\cos\psi & \sin\theta\sin\phi \\ \cos\psi\sin\phi + \cos\theta\cos\phi\sin\psi & -\sin\psi\sin\phi + \cos\theta\cos\phi\cos\psi & -\sin\theta\cos\phi \\ \sin\theta\sin\psi & \sin\theta\cos\psi & \cos\theta \end{pmatrix}$$

Now start with a unit vector along the z-axis, $(0, 0, 1)$, and one perpendicular to this $(\cos\alpha, \sin\alpha, 0)$; rotate the coordinate system through the Euler angles; find $A\varepsilon$ and $A\varepsilon^{\perp}$; and then verify Eq. (21–91) by averaging over the three Euler angles, remembering that the "volume element" is $\sin\theta \, d\theta \, d\phi \, d\psi$.

21–14. Derive Eqs. (21–92) and (21–93).

21–15. Derive Eqs. (21–99) and (21–100).

21–16. Prove that

$$i\hbar(\mathbf{J})_{mn} = (E_n - E_m)(\mathbf{M})_{mn} + O(E)$$

where $\mathbf{J} = \dot{M}$; i.e., derive Eq. (21–136).

21–17. Prove that

$$[\mathbf{J}(t)]_{mn} = e^{-i(\omega_n - \omega_m)t}(\mathbf{J})_{mn} + O(E)$$

i.e., derive Eq. (21–137).

21–18. Prove that the trace of a product of matrices is invariant to a cyclic permutation of the order of the matrices.

21–19. Prove that both

$$\phi_{ba}(t) = \frac{1}{i\hbar} \text{Tr}\{\rho[M_a, J_b(t)]\}$$

and

$$\phi_{ba}(t) = \frac{1}{i\hbar} \text{Tr}\{[\rho, M_a]J_b(t)\}$$

go to the correct classical limit.

21–20. Prove that the Kubo transform of an operator A becomes simply the classical expression for A in the classical limit.

21–21. Derive Eq. (21–237).

21–22. Let

$$\phi(t) = \frac{1}{i\hbar} \text{Tr}\{\rho[A, B(t)]\}$$

be designated by $\phi_{BA}(t)$. Then show that $\phi_{BA}(t)$ is real and that $\phi_{BA}(-t) = \phi_{AB}(t)$.

21–23. Prove that

$$\chi''(\omega) = \frac{2}{\pi} \int_0^\infty \chi'(\omega') \frac{\omega \, d\omega'}{\omega^2 - \omega'^2}$$

21–24. Show that the Debye equations, Eqs. (21–177) and (21–178), satisfy the Kramers-Kronig relations.

21–25. Prove that

$$\frac{1}{\pi} \lim_{\alpha \to 0} \frac{\alpha}{\alpha^2 + x^2} = \delta(x)$$

21–26. Equation (21–144) follows from Eq. (21–143) in the following way. Write out the trace operation appearing in Eq. (21–143) to get

$$\text{Tr}\{[\rho,\, M_a]J_b(t)\} = \sum_i \sum_j \sum_k (\rho_{ij} M_{jk} J_{ki} - M_{ij}\rho_{jk}J_{ki})$$

Now show that

$$\rho_{ij} = \frac{\delta_{ij}e^{-\beta E_j}}{Q}$$

Substitute this into the above triple summation to get

$$\sum_j \sum_k (\rho_j M_{jk}J_{kj} - M_{jk}\rho_k J_{kj})$$

Now use the fact that

$$\rho_k = \rho_j e^{-\beta(E_k - E_j)}$$

and Eq. (21–136) to derive Eq. (21–144).

21–27. Prove that Eq. (21–127) is the first-order solution to the time-dependent Schrödinger equation

$$i\hbar\,\frac{\partial \Psi}{\partial t} = [\mathscr{H}_0 + \mathscr{H}_1(t)]\Psi(t)$$

21–28. Prove that the matrix elements of the Kubo transform of an operator A are given by

$$\beta \tilde{A}_{ij} = \left\{\frac{e^{\beta(E_i - E_j)} - 1}{E_i - E_j}\right\} A_{ij}$$

21–29. Following through the derivation given in Section 21–4, prove that if

$$\mathscr{H}_1(t) = -AF(t)$$

and the response is given by B, then

$$\langle B(t)\rangle = \int_0^t dt'\, \phi(t - t')F(t') + O(F^2)$$

where

$$\phi(t) = \frac{1}{i\hbar}\,\text{Tr}\{[\rho,\, A]B(t)\}$$

$$= \frac{1}{i\hbar}\,\text{Tr}\{\rho[A,\, B(t)]\}$$

$$= \beta\,\text{Tr}\{\rho\tilde{A}B(t)\}$$

where

$$\tilde{A} = \frac{1}{\beta}\int_0^\beta d\lambda\, e^{\lambda\mathscr{H}}\,Ae^{-\lambda\mathscr{H}}$$

21–30. In Dirac notation the completeness relation is

$$\sum_m |m\rangle \langle m| = 1$$

It is not really necessary to use Dirac notation in Section 21–1. Prove that the completeness relation can also be written as

$$\sum_m \psi_m^*(x)\psi_m(x') = \delta(x - x')$$

where the ψ_m form a complete, orthonormal set. Hint: This can be proved *formally* by expanding $\delta(x - x')$ in terms of the set $\{\psi_m(x)\}$.

21–31. Prove that the Liouville operator is Hermitian.

21–32. Prove that the operator $\exp(\pm it\mathscr{L})$ is a unitary operator.

21–33. Show that the solution to

$$\frac{\partial\,\Delta f}{\partial t} = -i\mathscr{L}_0\,\Delta f - F(t)(A, f_0)$$

is

$$\Delta f(t) = -\int_0^t e^{i(t'-t)\mathscr{L}_0}(A, f_0)F(t')\,dt'$$

where $\Delta f = 0$ at $t = 0$.

21–34. In this problem we shall discuss the density matrix and the quantum-mechanical Liouville equation. We expand the time-dependent state function in terms of some complete orthonormal set $\{\psi_n(x)\}$ according to

$$\Psi(x, t) = \sum_n c_n(t)\psi_n(x)$$

The density matrix ρ is defined by

$$\rho_{nm} = \overline{c_m{}^*c_n}$$

where the bar represents an ensemble average. The elements of the density matrix clearly depend upon the set $\{\psi_n(x)\}$, i.e., the representation used, but the final result of an ensemble average must be independent of the representation used.

First prove that the trace of ρ is unity. Now prove that the ensemble average of some observable B is given by

$$\overline{(\Psi, \mathscr{B}\Psi)} = \sum_{m,n} B_{mn}\,\overline{c_m{}^*c_n} = \sum_{m,n} B_{mn}\rho_{nm}$$

$$= \sum_m (B\rho)_{mm}$$

$$= \mathrm{Tr}\{\mathscr{B}\rho\}$$

where $\mathscr{B}$ is the operator corresponding to the observable B. Since a trace is independent of the representation or basis, we see that the ensemble average is also.

Lastly, use the fact that

$$\Psi(x, t) = \sum_m c_m(t)\psi_m(x)$$

satisfies the time-dependent Schrödinger equation

$$i\hbar\,\frac{\partial\Psi}{\partial t} = \mathscr{H}\Psi$$

to show that

$$i\hbar\,\frac{\partial\rho_{nm}}{\partial t} = -\sum_k (H_{mk}{}^*\rho_{nk} - \rho_{km}H_{nk})$$

$$= -(\rho\mathscr{H} - \mathscr{H}\rho)_{nm}$$

or

$$i\hbar\,\frac{\partial\rho}{\partial t} = [\mathscr{H}, \rho]$$

which is the quantum-mechanical Liouville equation, Eq. (21–242).

21–35. Prove that

$$\langle \alpha | \rho(t) \rangle = \langle \alpha(t) | \rho \rangle$$

where α is a quantum-mechanical operator and ρ is the density matrix.

21–36. Show that

$$\frac{\partial F_s(\mathbf{k}, t)}{\partial t} = -k^2 D F_s(\mathbf{k}, t)$$

where $F_s(\mathbf{k}, t)$ is the Fourier transform of $G(\mathbf{r}, t)$, where

$$\frac{\partial G}{\partial t} = D \nabla^2 G$$

21–37. Prove that

$$\int_0^t dt' \int_0^t dt'' f(t'' - t') = 2t \int_0^t \left(1 - \frac{\tau}{t} \right) f(\tau) \, d\tau$$

by letting $\tau = t'' - t'$, changing the order of integration, and carrying out one obvious integration.

21–38. Derive the linearized Navier-Stokes equation [Eq. (21–283)] from the Navier-Stokes equation.

21–39. Prove that

$$\nabla \cdot (\text{sym } \nabla \cdot \mathbf{u}) = \tfrac{1}{2} \nabla^2 \mathbf{u} + \tfrac{1}{2} \nabla(\nabla \cdot \mathbf{u})$$

and

$$\nabla \cdot (\nabla \cdot \mathbf{u})\mathsf{I} = \nabla(\nabla \cdot \mathbf{u})$$

21–40. Derive Eq. (21–286).

21–41. Prove that

$$\iiint d\mathbf{r} \, e^{i\mathbf{k} \cdot \mathbf{r}} \nabla p = i\mathbf{k} p(\mathbf{k}, t)$$

$$\iiint d\mathbf{r} \, e^{i\mathbf{k} \cdot \mathbf{r}} \nabla^2 \mathbf{u} = -k^2 \mathbf{J}(\mathbf{k}, t)$$

$$\iiint d\mathbf{r} \, e^{i\mathbf{k} \cdot \mathbf{r}} \nabla(\nabla \cdot \mathbf{u}) = -\mathbf{k}[\mathbf{k} \cdot \mathbf{J}(\mathbf{k}, t)]$$

21–42. Derive Eq. (21–305) from Eq. (21–304).

21–43. Derive Eq. (21–307).

21–44. Derive Eq. (21–321).

21–45. Prove that the bulk viscosity of a monatomic ideal gas is zero.

21–46. Derive Eqs. (21–349) through (21–351).

21–47. Derive Eq. (21–355) from Eq. (21–354).

21–48. Prove that the terms in brackets in Eq. (21–355) represent a correlation between the effects of successive collisions.

21–49. Show that the impulsive nature of the hard-sphere potential leads to the presence of odd powers of t in the Maclaurin expansion of $\langle \mathbf{v}(0) \cdot \mathbf{v}(t) \rangle$.

21–50. Derive Eq. (21–374).

21–51. Using the time-correlation function expressions, show that

$$\lambda^* = \frac{\lambda k \varepsilon^{1/2}}{m^{1/2}\sigma^2}$$

$$D^* = \frac{Dm^{1/2}}{\varepsilon^{1/2}\sigma}$$

$$\kappa^* = \frac{\kappa\sigma^2}{m^{1/2}\varepsilon^{1/2}}$$

21–52. Prove that

$$\int_{-\infty}^{\infty} \langle \mu\mu(t) + \mu(t)\mu \rangle e^{-i\omega t}\, dt = (1 + e^{-\beta\hbar\omega}) \int_{-\infty}^{\infty} \langle \mu\mu(t) \rangle e^{-i\omega t}\, dt$$

and that

$$\varepsilon''(\omega) = \frac{4\pi}{\hbar} \tanh \tfrac{1}{2}\beta\hbar\omega \int_{-\infty}^{\infty} C_+(t)e^{-i\omega t}\, dt$$

where

$$C_+(t) = \tfrac{1}{2}\langle \mu\mu(t) + \mu(t)\mu \rangle$$
$$= \langle \tfrac{1}{2}[\mu, \mu(t)]_+ \rangle$$

where $[a, b]_+ = ab + ba$ is the anticommutator of a and b.

21–53. In Problem 21–52 we introduced the correlation function

$$C_+(t) = \langle \tfrac{1}{2}[A(0), A(t)]_+ \rangle$$

Prove that this more symmetric time-correlation function is a real, even function of time.

21–54. Prove that an autocorrelation function such as $\langle \mathbf{v}(0) \cdot \mathbf{v}(t) \rangle$ is a stationary function; i.e., $\langle \mathbf{v}(t) \cdot \mathbf{v}(t + \tau) \rangle$ is independent of t.

21–55. Use the Schwartz inequality

$$\int f^*g\, dx \le \left(\int f^*f\, dx \right)^{1/2} \left(\int g^*g\, dx \right)^{1/2}$$

to prove that

$$-\langle A^2 \rangle \le \langle A(0)A(t) \rangle \le \langle A^2 \rangle$$

21–56. Since the after-effect function is a real function, the susceptibility $\chi(\omega)$ must be complex. Prove that

$$\chi'(\omega) = \chi'(-\omega)$$
$$\chi''(\omega) = -\chi''(-\omega)$$
$$\chi^*(\omega) = \chi(-\omega)$$

21–57. Prove that if the external field is harmonic, i.e., if

$$F(t) = Fe^{i\omega t}$$

then

$$\langle B(t) \rangle = \chi(\omega)e^{i\omega t}$$

Since the applied field must be real, we represent a monochromatic force by

$$F(t) = \tfrac{1}{2}[Fe^{i\omega t} + F^*e^{-i\omega t}]$$

Show that

$$\langle B(t) \rangle = \tfrac{1}{2}[\chi(\omega)Fe^{i\omega t} + \chi(-\omega)F^*e^{-i\omega t}]$$

Show that this is equivalent to

$$\langle B(t) \rangle = \mathscr{R}e\chi(\omega)Fe^{i\omega t}$$

If $F(t)$ adds a term $-AF(t)$ to the Hamiltonian, this is indicated in the above equation by writing explicitly

$$\langle B(t) \rangle = \mathscr{R}e\chi_{BA}(\omega)Fe^{i\omega t}.$$

21–58. Consider a system that interacts with an external field according to

$$H_1(t) = -AF(t)$$

and that this field induces a response $B(t)$. Since the external field also induces the system to undergo transitions from one state to another, there is an absorption and emission of energy. The difference between the energy absorbed and emitted is the *energy dissipation.*

The energy dissipation per unit time per unit volume $Q(\omega)$ can be related to the susceptibility of the system. The rate of change of the energy of the system is $\partial H_1/\partial t$, and $Q(\omega)$ is obtained by averaging the expectation value of this over one period of the monochromatic field. Show that

$$Q(\omega) = \frac{\omega}{2\pi V} \int_0^{2\pi/\omega} dt \, \langle A(\mathbf{r}, t) \rangle \frac{\partial F}{\partial t}$$

Using the final equation of the previous problem, i.e.,

$$\langle A(t) \rangle = \operatorname{Re} \chi_{AA}(\omega)Fe^{i\omega t}$$

show that

$$Q(\omega) = + \frac{\omega}{2V} \chi_{AA}''(\omega)|F|^2$$

This important equation shows that the energy dissipation is directly related in the imaginary part of the susceptibility.

But from Eq. (21–159), one can also show that (do this)

$$\chi_{AA}''(\omega) = \frac{2\pi}{\hbar} \int_{-\infty}^{\infty} \operatorname{Tr}\{\rho[A, A(t)]\}e^{-i\omega t} \, dt$$

Show that this is equivalent to

$$\chi_{AA}''(\omega) = \frac{4\pi}{\hbar} \tanh \tfrac{1}{2}\beta\hbar\omega \int_{-\infty}^{\infty} C_+(t)e^{-i\omega t} \, dt$$

where $C_+(t)$ is the symmetric autocorrelation function

$$C_+(t) = \langle \tfrac{1}{2}[A(0), A(t)]_+ \rangle$$

where $[\ ]_+$ denotes the anticommutator.

If we call the Fourier transform of $C_+(t)$, $G_{AA}(\omega)$, we have that

$$\chi_{AA}''(\omega) = \frac{4\pi}{\hbar} \tanh(\tfrac{1}{2}\beta\hbar\omega)G_{AA}(\omega)$$

Since the dissipation is linearly related to $G_{AA}(\omega)$, this is often called the power spectrum of the process $A(t)$. In addition, note that the energy dissipated by a system exposed to some external disturbance can be directly related to the autocorrelation function, which describes how spontaneous fluctuations regress to the equilibrium state. The relation between $Q(\omega)$ and $G_{AA}(\omega)$, namely,

$$Q(\omega) = \frac{2\pi\omega}{\hbar V} \tanh(\tfrac{1}{2}\beta\hbar\omega) |F|^2 G_{AA}(\omega)$$

is called the *fluctuation-dissipation theorem*. (See Berne, 1971, "Additional Reading.")

21–59. A simple function that qualitatively approximates the velocity autocorrelation function was apparently first suggested by Douglas.*

$$\phi(t) = \text{sech}(at)\cos(bt)$$

One can evaluate a and b by comparing the Maclaurin expansion for this to the exact expansion

$$\phi(t) = \sum_{n=0}^{\infty} (-1)^n A_{2n} t^{2n}$$

Show that according to this approximation

$$D = \frac{\pi kT}{2ma} \text{sech}\left(\frac{\pi b}{2a}\right)$$

$$= \frac{kT}{m} \left(\frac{\pi}{[2(c-1)A_2]^{1/2}}\right) \text{sech}\left\{\frac{\pi}{2} \left[\frac{(5-c)}{(c-1)}\right]^{1/2}\right\}$$

where $c = 6A_4/A_2^2$. Isbister and McQuarrie† have compared the predictions of this formula to experimental data.

21–60. Prove that‡

$$\left\langle \sum_{i,j} x_i^2(t)p_{iy}(t)p_{jy}(0) \right\rangle = \left\langle \sum_{i,j} x_i^2(t)p_{iy}(t)p_{jy}(t) \right\rangle$$

$$= \left\langle \sum_{i,j} x_i^2(0)p_{iy}(0)p_{jy}(0) \right\rangle$$

$$= \left\langle \sum_{i} x_i^2(0)p_{iy}^2(0) \right\rangle$$

$$\left\langle \sum_{i,j} x_i^2(0)p_{iy}(0)p_{jy}(t) \right\rangle = \left\langle \sum_{i} x_i^2(0)p_{iy}^2(0) \right\rangle$$

$$\left\langle \sum_{i,j} x_j(t)x_i(t)p_{jy}(t)p_{iy}(t) \right\rangle = \left\langle \sum_{i,j} x_j(0)x_i(0)p_{jy}(0)p_{iy}(0) \right\rangle$$

$$= \left\langle \sum_{i} x_i^2(0)p_{iy}^2(0) \right\rangle$$

21–61. Prove that $\phi(t)$ in Eq. (21–187) is an odd function of t.

21–62. Verify Eqs. (21–215).

21–63. Brenner and McQuarrie (*Can. J. Phys.*, **49**, p. 837, 1971) have presented an analysis of collision-induced absorption in rare gas mixtures by considering the Maclaurin expansion

---

* *J. Chem. Phys.*, **35**, p. 81, 1960.
† *J. Chem. Phys.*, **56**, p. 736, 1972.
‡ Cf. E. Helfand, *Phys. Rev.*, **119**, p. 1, 1960.

of the dipole moment autocorrelation function $\phi(t) = \langle \boldsymbol{\mu}(0) \cdot \boldsymbol{\mu}(t) \rangle$. First show that if $I(\omega)$ is the line shape function, then

$$\phi(t) = \sum_{n=0}^{\infty} \frac{(it)^n}{n!} \int_{-\infty}^{\infty} \omega^n I(\omega) \, d\omega$$

$$= \sum_{n=0}^{\infty} \frac{(it)^n}{n!} (-i)^n \left| \frac{d^n}{dt^n} \langle \boldsymbol{\mu}(0) \cdot \boldsymbol{\mu}(t) \rangle \right|_{t=0}$$

$$= \sum_{n=0}^{\infty} \frac{(it)^n}{n!} \gamma_n$$

Show that

$$\gamma_{2n} = \left\langle \left| \frac{d^n \boldsymbol{\mu}}{dt^n} \right|^2 \right\rangle$$

$$\gamma_{2n+1} = 0$$

and in particular that

$$\gamma_0 = 4\pi \int_0^{\infty} e^{-\beta u(r)} |\boldsymbol{\mu}(r)|^2 r^2 \, dr$$

$$\gamma_2 = \frac{4\pi kT}{m} \int_0^{\infty} e^{-\beta u(r)} \left\{ \left| \frac{d\boldsymbol{\mu}}{dr} \right|^2 + \frac{2|\boldsymbol{\mu}|^2}{r^2} \right\} r^2 \, dr$$

where m is the *reduced mass* and $u(r)$ is the intermolecular potential.

Brenner and McQuarrie have evaluated γ_0, γ_2, and γ_4 for a Lennard-Jones potential and a dipole moment function of the form

$$\mu(r) = \mu_0 e^{-r/\zeta} + \frac{c}{r^7}$$

where μ_0, ζ, and c are constants. They then use these three moments in a so-called Gram-Charlier series to derive an approximate line shape $I(\omega)$.

A Gram-Charlier series is a series of the form

$$f(x) = \varphi(x) + \frac{c_3}{3!} \varphi^{(3)}(x) + \frac{c_4}{4!} \varphi^{(4)}(x) + \cdots$$

where $f(x)$ is some distribution function, $\varphi(x)$ is a standardized Gaussian distribution function (i.e., one with zero mean and unit variance), and $\varphi^{(j)}(x)$ is its jth derivative. Show that $\langle \omega \rangle = 0$ and $\langle \omega^2 \rangle = \gamma_2/\gamma_0^2$ so that $x = \omega/\langle \omega^2 \rangle^{1/2}$ for collision-induced absorption in rare gas mixtures. Show also that

$$c_j = (-1)^j \int_{-\infty}^{\infty} H_j(x) f(x) \, dx$$

where $H_j(x)$ is a Hermite polynomial.

We see that a Gram-Charlier series replaces the function $f(x)$ by a Gaussian distribution and correction terms involving the c_j. Since the line shape shown in Fig. 28–9 is symmetric about $\omega = 0$, the first correction, c_3 is zero. Show that the first nonzero correction to the simple Gaussian approximation involving c_4 is given by

$$c_4 = \frac{\gamma_4 \gamma_0}{\gamma_2^2} - 3$$

and that the Gram-Charlier approximation to $I(\omega)$ is

$$I(\omega) = \frac{\gamma_0 \tau}{2\pi^{1/2}} e^{-\omega^2 \tau^2/4} \left\{ 1 + \frac{c_4}{4!} H_4\left(\frac{\omega \tau}{\sqrt{2}} \right) + \cdots \right\}$$

where $\tau = (2\gamma_0/\gamma_2)^{1/2}$. Brenner and McQuarrie show that this gives excellent agreement with experiment.

CHAPTER 22

THE TIME-CORRELATION FUNCTION FORMALISM, II

In this final chapter we shall continue our discussion of the time-correlation function formalism. In Section 22–1 we shall discuss the use of inelastic neutron scattering to probe the structure and dynamics of molecules in condensed media. The theoretical formalism was first presented by van Hove* in 1954. Of particular importance in this section is the introduction of the van Hove space-time correlation function $G(\mathbf{r}, t)$. In the second section we present some of the basic concepts and definitions of random functions and stochastic processes, and in particular, we derive and discuss the Wiener-Khintchine theorem, which relates a correlation function to its associated spectrum. The Wiener-Khintchine theorem is used in the next section to derive the basic equations for the spectrum of light scattered from a solution or a fluid. With the advent of the laser, this has become a powerful experimental tool. We shall see in that section that one can determine the diffusion coefficient and the electrophoretic mobility of macromolecules in solution, the hydrodynamic properties of liquids, and chemical kinetic rate constants from the spectrum of the scattered radiation. Then in Section 22–4 we shall introduce the so-called memory function associated with an autocorrelation function. It turns out that the memory function has both conceptual and computational advantages over the autocorrelation function itself, and we shall illustrate this with a calculation of the velocity autocorrelation function of a monatomic fluid. Lastly, in Section 22–5 we shall derive formulas for the thermal transport coefficients from within the memory function framework. This will more clearly elucidate the small-frequency and wave vector limit that was tacitly assumed in Section 21–8, where we derived such expressions from the appropriate macroscopic transport equations. We shall also be led to the concept of generalized thermal transport coefficients, which depend upon frequency ω and wave vector $\mathbf{k}$, and to generalized hydrodynamics, the extension of hydrodynamics to arbitrary time scale and spatial inhomogeneities, i.e., to arbitrary ω and $\mathbf{k}$.

* L. van Hove, *Phys. Rev.*, **95**, p. 249, 1954.

22–1 INELASTIC NEUTRON SCATTERING

We shall see in this section that an analysis of the scattering of slow or thermal neutrons from fluids has proved to be a powerful technique in obtaining information concerning their dynamical behaviour and microscopic structure. The theoretical development of this technique is due originally to van Hove,* whose derivation is presented here.

van Hove starts with the first Born approximation (Marshall and Lovesey, "Additional Reading"), which says that the differential scattering cross section per unit solid angle and unit interval of outgoing energy ε of the scattered particle (the neutron) is given by

$$\frac{d^2\sigma}{d\Omega\,d\varepsilon} = \frac{m^3}{2\pi^2\hbar^6}\frac{k}{k_0}\,W(\kappa)\sum_i \rho_i \sum_f |\langle i|\sum_{j=1}^N e^{i\kappa\cdot r_j}|f\rangle|^2\,\delta\left[k^2 - k_0{}^2 + \frac{2m}{\hbar^2}(E_f - E_i)\right]$$

(22–1)

where m, $\mathbf{k}_0$, and $\mathbf{k} = \mathbf{k}_0 - \boldsymbol{\kappa}$ are the mass and the initial and final wave vectors of the scattered particle and $\boldsymbol{\kappa}$ is defined by $\boldsymbol{\kappa} = \mathbf{k}_0 - \mathbf{k}$. The operators $\mathbf{r}_j$ represent the position vectors of the N particles of the scattering system (the fluid), whose initial and final states are labeled by i and f, and have energies E_i and E_f. The quantity ρ_i is the statistical probability of the initial state and

$$W(\kappa) = \left|\int \exp(i\boldsymbol{\kappa}\cdot\mathbf{r})V(r)\,d\mathbf{r}\right|^2$$

(22–2)

where $V(r)$ is the interaction between the neutron and the nuclei of the fluid. We have also assumed (for now) that all the nuclei of the fluid interact in the same way with the incident neutrons.

If, besides the momentum transfer $\hbar\boldsymbol{\kappa}$, we introduce the energy transfer through

$$\hbar\omega = \frac{\hbar^2(k_0{}^2 - k^2)}{2m}$$

(22–3)

then Eq. (22–1) can be written in the form

$$\frac{d^2\sigma}{d\Omega\,d\varepsilon} = AS(\boldsymbol{\kappa}, \omega)$$

(22–4)

where

$$A = \frac{Nm^2}{4\pi^2\hbar^5}\frac{k}{k_0}\,W(\kappa)$$

(22–5)

and

$$S(\boldsymbol{\kappa}, \omega) = N^{-1}\sum_i \rho_i \sum_f |\langle i|\sum_{j=1}^N e^{i\kappa\cdot r_j}|f\rangle|^2\,\delta\left(\omega + \frac{E_i - E_f}{\hbar}\right)$$

(22–6)

By introducing the Fourier transform representation of the delta function and proceeding much as we did in Section 21–1, this expression can be rewritten as (Problem 22–1)

$$S(\boldsymbol{\kappa}, \omega) = \frac{1}{2\pi N}\int_{-\infty}^{\infty} dt\, e^{-i\omega t}\sum_{i,\,j}\langle e^{-i\kappa\cdot r_i(0)}e^{i\kappa\cdot r_j(t)}\rangle$$

(22–7)

$$= \frac{1}{2\pi}\int_{-\infty}^{\infty} dt\, e^{-i\omega t}F(\boldsymbol{\kappa}, t)$$

(22–8)

* L. van Hove, *Phys. Rev.*, **95**, p. 249, 1954.

where

$$F(\mathbf{\kappa}, t) = \frac{1}{N} \sum_{i, j} \langle e^{-i\mathbf{\kappa} \cdot \mathbf{r}_i(0)} e^{i\mathbf{\kappa} \cdot \mathbf{r}_j(t)} \rangle \tag{22-9}$$

The quantity $S(\mathbf{\kappa}, \omega)$ is called the scattering function.

We shall now express $S(\mathbf{\kappa}, \omega)$ in terms of a time-dependent extension of the radial distribution function $G(\mathbf{r}, t)$, which is called the van Hove space-time correlation function. This is defined through

$$S(\mathbf{\kappa}, \omega) = (2\pi)^{-1} \int e^{i(\mathbf{\kappa} \cdot \mathbf{r} - \omega t)} G(\mathbf{r}, t) \, d\mathbf{r} \, dt \tag{22-10}$$

$$G(\mathbf{r}, t) = (2\pi)^{-3} \int e^{-i(\mathbf{\kappa} \cdot \mathbf{r} - \omega t)} S(\mathbf{\kappa}, \omega) \, d\mathbf{\kappa} \, d\omega \tag{22-11}$$

Introducing Eq. (22-7) for $S(\mathbf{\kappa}, \omega)$ into Eq. (22-11) and integrating over t, we get (Problem 22-2)

$$G(\mathbf{r}, t) = (2\pi)^{-3} N^{-1} \sum_{l, j}^{N} \int d\mathbf{\kappa} \, e^{-i\mathbf{\kappa} \cdot \mathbf{r}} \langle e^{-i\mathbf{\kappa} \cdot \mathbf{r}_l(0)} e^{i\mathbf{\kappa} \cdot \mathbf{r}_j(t)} \rangle \tag{22-12}$$

We shall see that the factor of $(2\pi)^{-3} N^{-1}$ in this expression makes $G(\mathbf{r}, t)$ independent of N and asymptotically equal to the number density in the thermodynamic limit.

The functions $F(\mathbf{\kappa}, t)$ and $G(\mathbf{r}, t)$ are Fourier-related by

$$G(\mathbf{r}, t) = (2\pi)^{-3} \int d\mathbf{\kappa} \, e^{-i\mathbf{\kappa} \cdot \mathbf{r}} F(\mathbf{\kappa}, t) \tag{22-13}$$

$$F(\mathbf{\kappa}, t) = \int d\mathbf{r} \, e^{i\mathbf{\kappa} \cdot \mathbf{r}} G(\mathbf{r}, t) \tag{22-14}$$

Since $S(\mathbf{\kappa}, \omega)$ is an experimentally observable quantity and hence real, it is easy to see from Eq. (22-11) that $G(\mathbf{r}, t)$ has the symmetry (Problem 22-3)

$$G(-\mathbf{r}, -t) = G^*(\mathbf{r}, t) \tag{22-15}$$

In the classical limit; the operators in Eq. (22-12) commute, and so we may write

$$G(\mathbf{r}, t) = N^{-1} \left\langle \sum_{l, j} \delta[\mathbf{r} + \mathbf{r}_l(0) - \mathbf{r}_j(t)] \right\rangle \tag{22-16}$$

which is a real quantity. Thus we see that complex values of $G(\mathbf{r}, t)$ reflect quantum properties of the system. For simplicity, we shall discuss only the classical limit, although the complete quantal formalism is well developed (cf. Marshall and Lovesey, in "Additional Reading").

Equation (22-16) shows that in the classical limit $G(\mathbf{r}, t)$ splits naturally into two parts: one that describes the correlation of one molecule at different times and one that describes the correlation between distinct particles at different times. Denote these correlation functions by $G_s(\mathbf{r}, t)$ (s stands for same) and $G_d(\mathbf{r}, t)$ (d stands for distinct). Thus we can write

$$G(\mathbf{r}, t) = G_s(\mathbf{r}, t) + G_d(\mathbf{r}, t) \tag{22-17}$$

where

$$G_s(\mathbf{r}, t) = N^{-1} \left\langle \sum_{j=1}^{N} \delta[\mathbf{r} + \mathbf{r}_j(0) - \mathbf{r}_j(t)] \right\rangle \tag{22-18}$$

$$G_d(\mathbf{r}, t) = N^{-1} \left\langle \sum_{j \neq l = 1}^{N} \delta[\mathbf{r} + \mathbf{r}_l(0) - \mathbf{r}_j(t)] \right\rangle \tag{22-19}$$

For $t = 0$ these reduce to

$$G_s(\mathbf{r}, 0) = \delta(\mathbf{r}) \tag{22-20}$$

$$G_d(\mathbf{r}, 0) = \rho_0 g(r) \tag{22-21}$$

where $g(\mathbf{r})$ is the radial distribution function of Chapter 13.

Thus we see that neutron scattering allows one to experimentally probe both the spatial and temporal behavior of the molecules in a fluid. On the other hand, if the incident energy is sufficiently large compared to the energy transfers (as it is in the case of X-ray scattering), the momentum transfer for a given scattering angle is independent of the outgoing energy, and the differential cross section per unit solid angle becomes

$$\frac{d\sigma}{d\Omega} = \int \frac{d^2\sigma}{d\Omega \, d\varepsilon} \, d\varepsilon = \hbar A \int S(\mathbf{\kappa}, \omega) \, d\omega \tag{22-22}$$

$$= \hbar A \int e^{i\mathbf{\kappa} \cdot \mathbf{r}} \, \delta(t) \, G(\mathbf{r}, t) \, d\mathbf{r} \, dt$$

$$\propto [1 + \rho_0 \int e^{i\mathbf{\kappa} \cdot \mathbf{r}} \, g(\mathbf{r}) \, d\mathbf{r}] \tag{22-23}$$

in agreement with the X-ray scattering formulas in Chapter 13.

In what we have discussed so far, we have assumed that all the nuclei in the fluid interact in exactly the same way with the incident neutrons. Generally, however, this is not the case. The scattering usually depends either on isotopic effects or nuclear internal variables, and so the scattering is not the same from all the nuclei. In particular, we wish to extend the above results to the case of spin-dependent scattering. If a_j is an operator that depends upon the spin of the jth particle, we define a spin-dependent space-time correlation function by

$$\Gamma(\mathbf{r}, t) = N^{-1} \left\langle \sum_{l, j=1}^{N} \int d\mathbf{r}' \, a_l(0) \, \delta[\mathbf{r} + \mathbf{r}_l(0) - \mathbf{r}'] a_j(t) \, \delta[\mathbf{r}' - \mathbf{r}_j(t)] \right\rangle \tag{22-24}$$

where

$$a_j(t) = \exp\left(\frac{it \, \mathcal{H}}{\hbar}\right) a_j \exp\left(-\frac{it \, \mathcal{H}}{\hbar}\right)$$

Note that if $a_j = 1$ and we take the classical limit, in which the operators commute, then $\Gamma(\mathbf{r}, t)$ simply reduces to the $G(\mathbf{r}, t)$ given by Eq. (22–16) (Problem 22–4).

For a system of Boltzmann particles with a spin-independent Hamiltonian and described by a canonical distribution function $\rho_i = e^{-\beta E_i}/Q$, there is no correlation between spins nor between spins and positions, and we can write (Problem 22–5)

$$\Gamma(\mathbf{r}, t) = \langle a^2 \rangle \, G_s(\mathbf{r}, t) + \langle a \rangle^2 \, G_d(\mathbf{r}, t) \tag{22-25}$$

where a is any of the a_j and $\langle \cdots \rangle$ denotes an average over the spin states of the corresponding particle. The interaction between a slow neutron and the nuclei of the scattering system is often replaced by the so-called Fermi pseudopotential (cf. Marshall and Lovesey, in "Additional Reading," Chapter 1)

$$V(r) = \left(\frac{2\pi a \hbar^2}{m}\right) \delta(r) \tag{22-26}$$

where m is the mass of the neutron and a is called the scattering length of the nucleus (a may depend upon the nuclear spin state). Thus we may write (Problem 22–6)

$$\frac{d^2\sigma}{d\Omega\,d\varepsilon} = \frac{d^2\sigma_{\text{coh}}}{d\Omega\,d\varepsilon} + \frac{d^2\sigma_{\text{inc}}}{d\Omega\,d\varepsilon} \tag{22–27}$$

where

$$\frac{d^2\sigma_{\text{coh}}}{d\Omega\,d\varepsilon} = \frac{\langle a\rangle^2 N}{2\pi\hbar}\frac{k}{k_0}\int e^{i(\mathbf{\kappa}\cdot\mathbf{r}-\omega t)}G(\mathbf{r},t)\,d\mathbf{r}\,dt \tag{22–28}$$

$$\frac{d^2\sigma_{\text{inc}}}{d\Omega\,d\varepsilon} = \frac{(\langle a^2\rangle - \langle a\rangle^2)N}{2\pi\hbar}\frac{k}{k_0}\int e^{i(\mathbf{\kappa}\cdot\mathbf{r}-\omega t)}G_s(\mathbf{r},t)\,d\mathbf{r}\,dt \tag{22–29}$$

Equations (22–28) and (22–29) are called *coherent* and *incoherent* scattering cross sections.

Note that in the static approximation discussed around Eq. (22–23), we have that (Problem 22–7)

$$\frac{d\sigma_{\text{coh}}}{d\Omega} = \langle a\rangle^2 N\left[1 + \rho_0\int e^{i\mathbf{\kappa}\cdot\mathbf{r}}g(\mathbf{r})\,d\mathbf{r}\right] \tag{22–30}$$

$$\frac{d\sigma_{\text{inc}}}{d\Omega} = (\langle a^2\rangle - \langle a\rangle^2)N \tag{22–31}$$

Equation (22–25) assumes that there is no correlation in the positions of different scattering lengths. The distinction between coherent and incoherent scattering can be seen through the following simple example. First, assume that all the nuclei in the sample interact in exactly the same way with a neutron beam. There will be interference between the scattered waves from each nucleus, just as in X-ray scattering. This is what is referred to as coherent scattering. Suppose now that the sample consists of two types of scatterers and that one interacts with a neutron with an interaction $+b$, the other with energy $-b$, and that there are equal numbers of these species randomly distributed through the sample. Now since there is no correlation of the scattering amplitudes in space, on different nuclei, the spectrum does not contain any information on the collective modes, only on the motion of the individual scatterers, or the so-called single particle motions, or self-correlation functions. This is incoherent scattering. Coherent scattering contains both self-correlation and contributions from other particles being at the space-time point $(\mathbf{r}, t)$.

Some samples are almost purely coherent scatterers—all the nuclei have the same scattering amplitude b. (The cross section is proportional to b^2). A good example of this is argon-36. Some, however, have nuclei spin effects or nuclear internal variables. Hydrogen is the prime example of this. A proton has vastly different scattering amplitudes b, depending on the orientation of the nuclear spin. Thus it has a large incoherent cross section (around 80 barns, 1 barn $= 10^{-24}$ cm^2), while the coherent cross section is less by a factor 40. Thus it is usually difficult to observe the collective modes but easy to see single-particle motion in H-containing samples.

We see from these equations that an experimental determination of the incoherent scattering function $S_{\text{inc}}(\kappa, \omega)$ over a sufficiently broad region in κ–ω space would allow one to determine $G_s(\mathbf{r}, t)$ by Fourier inversion. There are few experimental studies that are extensive enough to allow this to be done. We shall discuss a few such studies shortly, but before doing this we shall discuss some approximations to $G_s(\mathbf{r}, t)$ and its Fourier transform $F_s(\boldsymbol{\kappa}, t)$.

Consider a classical system, for which we can write

$$F_s(\boldsymbol{\kappa}, t) = \langle e^{i\boldsymbol{\kappa} \cdot [\mathbf{r}_1(t) - \mathbf{r}_1(0)]} \rangle \tag{22-32}$$

Take the direction of $\boldsymbol{\kappa}$ to be the x-axis and write

$$F_s(\boldsymbol{\kappa}, t) = \langle e^{i\kappa \int_0^t \dot{x}_1(t') \, dt'} \rangle \tag{22-33}$$

Now using the method of cumulants (Problem 22–8), this can be written as

$$F_s(\kappa, t) = \exp[-\kappa^2 \gamma_1(t) + \kappa^4 \gamma_2(t) + \cdots] \tag{22-34}$$

where

$$\cdot \; \gamma_1(t) = \int_0^t (t - t') \langle x_1(0) x_1(t') \rangle \, dt' \tag{22-35}$$

$$= \tfrac{1}{6} \langle [\mathbf{r}_1(t) - \mathbf{r}_1(0)]^2 \rangle \tag{22-36}$$

and where expressions for $\gamma_2(t)$, etc., have been given by Rahman, Singwi, and Sjölander* in terms of higher-order velocity correlation functions. They also show that for small times as well as long times, these higher terms can be neglected, giving

$$F_s(\kappa, t) = \exp[-\kappa^2 \gamma_1(t)] \tag{22-37}$$

as the often-used Gaussian approximation for $F_s(\kappa, t)$. The self–space-time correlation function corresponding to this is

$$G_s(r, t) = [4\pi\gamma_1(t)]^{-3/2} \exp\left[-\frac{r^2}{4\gamma_1(t)}\right] \tag{22-38}$$

Equation (22–37) can be extended to give non-Gaussian corrections (Problem 22–9)

$$F_s(\kappa, t) = e^{-\kappa^2 \gamma_1(t)} \{1 + \tfrac{1}{2}\alpha_2(t)[\kappa^2 \gamma_1(t)]^2 + \cdots\} \tag{22-39}$$

where

$$\alpha_2(t) = \frac{\langle r^4 \rangle}{3\langle r^2 \rangle^2} - 1 \tag{22-40}$$

Note that $\alpha_2(t) = 0$ for a Gaussian form for $G_s(\mathbf{r}, t)$.

Equation (22–36) shows that $\gamma_1(t) = Dt$ for a diffusing particle, and so we see from Eq. (22–38) that

$$G_s(\mathbf{r}, t) = (4\pi Dt)^{-3/2} \exp\left(-\frac{r^2}{4Dt}\right) \tag{22-41}$$

and that

$$S_{\text{inc}}(\kappa, \omega) = \frac{D\kappa^2/\pi}{\omega^2 + (D\kappa^2)^2} \tag{22-42}$$

We expect that this approximation to be most valid at small frequency since the form of $G_s(\mathbf{r}, t)$ is most valid at long times. Figure 22–1 shows the experimental $S_{\text{inc}}(\kappa, 0)$

* *Phys. Rev.*, **126**, p. 986, 1962.

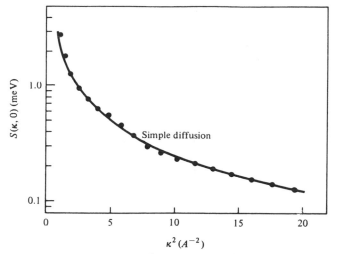

Figure 22–1. $S(\kappa, 0)$ versus κ^2. **The solid line shows the simple diffusion result.** (From K. Sköld *et al.*, *Phys. Rev.*, **6A**, p. 1107, 1972.)

of liquid argon plotted versus κ^2 along with Eq. (22–42) with $D = 1.94 \times 10^{-5}$ cm^2/sec. The agreement is seen to be very good.

Although an exact calculation of $\gamma_1(t)$ requires a knowledge of the velocity auto-correlation function, Eq. (22–36) shows that for short times

$$\gamma_1(t) = \tfrac{1}{6}\langle \mathbf{v} \cdot \mathbf{v}\rangle t^2 = \frac{kT}{2m}\, t^2 \qquad t \to 0 \tag{22–43}$$

This displays the initial free motion that the particle undergoes before colliding with another particle. Thus although the exact behavior of $\gamma_1(t)$ is difficult to calculate for all times, we see that it goes as t^2 for short times and t for long times.

Since

$$F_s(\kappa, t) = \int_{-\infty}^{\infty} e^{i\omega t} S_{\text{inc}}(\kappa, \omega)\, d\omega \tag{22–44}$$

it can be determined by a Fourier inversion of the experimentally determined inco-herent cross section. Then using the form of $F_s(\kappa, t)$ given by Eq. (22–39), one can determine $\gamma_1(t)$ and $\alpha_2(t)$ numerically from the data. This requires high-quality data, but Sköld *et al.*[*] have carried out a thorough study of liquid argon. They present a detailed discussion of the experimental procedure, and their paper is highly recom-mended. Figure 22–2, which comes from their paper, shows the numerically determined values of $\gamma_1(t)$ and $\alpha_2(t)$ as functions of time. It can be seen that $\gamma_1(t)$ goes as t^2 for short times and Dt for long times as one expects. [The term $\hbar^2/8mkT$ that occurs along with the t^2-term in the figure arises from the fact that the authors have Fourier-inverted a symmetric scattering function $\tilde{S}(\kappa, \omega) = e^{-\hbar\omega/2kT}S(\kappa, \omega)$ instead of just $S(\kappa, \omega)$ itself.] Figure 22–3, taken from Dasannacharya and Rao,[†] shows a plot of $G_s(r, t)$ versus r for various values of t.

* *Phys. Rev.*, **6A**, p. 1107, 1972.
† *Phys. Rev.*, **137**, p. A417, 1965.

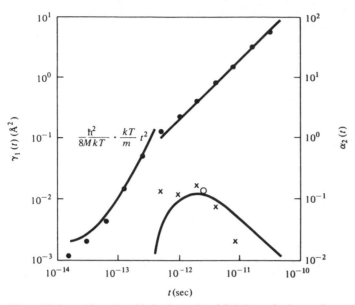

Figure 22–2. $\gamma_1(t)$ and $\alpha_2(t)$ (scale to the right) determined experimentally for liquid argon by Sköld *et al.* The limiting behavior of $\gamma_1(t)$ at small and large times as obtained for a free gas and for simple diffusion, respectively, is also shown. The solid line in the $\alpha_2(t)$ plot shows the result obtained by Levesque and Verlet, and the circle shows the result obtained by Rahman for $\alpha_2(t)$ at $t = 2.5 \times 10^{-12}$ sec for argon at 94.4°K. (From K. Sköld *et al.,* *Phys., Rev.,* **6A**, p. 1107, 1972.)

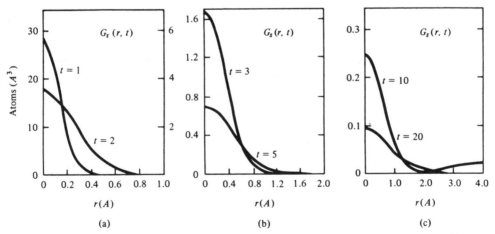

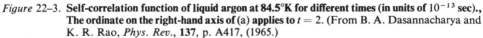

Figure 22–3. **Self-correlation function of liquid argon at 84.5°K for different times (in units of 10^{-13} sec).,** **The ordinate on the right-hand axis of (a) applies to $t = 2$. (From B. A. Dasannacharya and** K. R. Rao, *Phys. Rev.,* **137**, p. A417, (1965.)

It was mentioned in Chapter 11 that neutron scattering can be used to probe the distribution of vibrational frequencies in a crystal. First let us calculate $\gamma_1(t)$ for a one-dimensional harmonic oscillator. In this case (Problem 22–14)

$$\langle \dot{x}_1(0)\dot{x}_1(t) \rangle = \langle \dot{x}_1^2 \rangle \cos \omega_0 t$$

$$= \frac{kT}{m} \cos \omega_0 t \tag{22–45}$$

and

$$\gamma_1(t) = \frac{kT}{m} \frac{(1 - \cos \omega_0 t)}{\omega_0{}^2} \tag{22-46}$$

For a harmonic lattice with a distribution of normal modes $g(\omega)\, d\omega$, then

$$\gamma_1(t) = \frac{kT}{m} \int_0^\infty \left(\frac{1 - \cos \omega t}{\omega^2} \right) g(\omega)\, d\omega \tag{22-47}$$

But if we define a quantity $f(\omega)$ through

$$f(\omega) = \frac{m}{3\pi kT} \int_{-\infty}^\infty dt\, e^{i\omega t} \langle \mathbf{v}(0) \cdot \mathbf{v}(t) \rangle \tag{22-48}$$

and use the fact

$$\gamma_1(t) = \frac{1}{3} \int_0^t dt'\, (t - t') \langle \mathbf{v}(0) \cdot \mathbf{v}(t') \rangle \tag{22-49}$$

then

$$\gamma_1(t) = \frac{1}{3} \int_0^t dt'(t - t') \frac{3kT}{2m} \int_{-\infty}^\infty d\omega\, e^{-i\omega t'} f(\omega)$$

$$= \frac{kT}{m} \int_0^\infty \left(\frac{1 - \cos \omega t}{\omega^2} \right) f(\omega)\, d\omega \tag{22-50}$$

On comparing Eqs. (22–50) and (22–47) we see that $f(\omega)$ [as defined by Eq. (22–48)], is the analog of the distribution of vibrational frequencies in a harmonic lattice.

It is possible to determine $f(\omega)$ in terms of the incoherent scattering function $S_{inc}(\kappa, \omega)$. From Eq. (22–34), we have that

$$F_s(\kappa, t) = 1 - \kappa^2 \gamma_1(t) + O(\kappa^4) \tag{22-51}$$

from which we write

$$\frac{\partial^2 F_s}{\partial t^2} = -\kappa^2 \ddot{\gamma}_1(t) + O(\kappa^4) \tag{22-52}$$

and that

$$\ddot{\gamma}_1(t_1) = -\lim_{\kappa \to 0} \frac{1}{\kappa^2} \frac{\partial^2 F_s}{\partial t^2} \tag{22-53}$$

But Eq. (22–44) shows that

$$F_s(\kappa, t) = \int_{-\infty}^\infty e^{i\omega t} S_{inc}(\kappa, \omega)\, d\omega \tag{22-54}$$

and so

$$-\frac{1}{\kappa^2} \frac{\partial^2 F_s}{\partial t^2} = \frac{1}{\kappa^2} \int_{-\infty}^\infty d\omega\, e^{i\omega t} \omega^2 S_{inc}(\kappa, \omega) \tag{22-55}$$

From Eq. (22–36) we see that

$$\ddot{\gamma}_1(t) = \tfrac{1}{3} \langle \mathbf{v}(0) \cdot \mathbf{v}(t) \rangle \tag{22-56}$$

and so

$$\ddot{\gamma}_1(t) = \frac{kT}{2m} \int_{-\infty}^{\infty} d\omega \, e^{i\omega t} f(\omega) \tag{22-57}$$

By comparing this equation with Eq. (22–55), we see that

$$\frac{kT}{2m} f(\omega) = \lim_{\kappa \to 0} \left[\frac{\omega^2}{\kappa^2} S_{\text{inc}}(\kappa, \omega) \right] \tag{22-58}$$

Thus $S_{\text{inc}}(\kappa, \omega)$ contains information concerning the analog of the distribution of normal modes in liquid. Rahman* has calculated a normalized version of $f(\omega)$ by molecular dynamics, and the result is shown in Fig. 22–4. Also shown in this figure is $f(\omega)$ according to the Langevin equation. In this case

$$\langle \mathbf{v}(0) \cdot \mathbf{v}(t) \rangle = \frac{3kT}{m} e^{-\zeta t} \tag{22-59}$$

where $\zeta = kT/mD$ and so

$$f(\omega) = \frac{2}{\pi} \int_{0}^{\infty} e^{-\zeta t} \cos \omega t \, dt$$

$$= \frac{2}{\pi} \frac{\zeta}{\zeta^2 + \omega^2} \tag{22-60}$$

Note also that

$$f(0) = \frac{2mD}{\pi kT} \tag{22-61}$$

which gives a direct relation between D and $f(0)$.

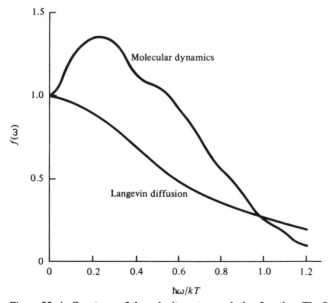

Figure 22–4. **Spectrum of the velocity autocorrelation function. The Lorentzian spectrum of a Langevin-type correlation is also shown.** (From A. Rahman, *Phys. Rev.*, **136**, p. A405, 1964.)

* *Phys. Rev.*, **136**, p. A405, 1964.

Inelastic neutron scattering has proved to be a powerful tool to probe the motion of molecules in both fluids and solids. We have discussed only classical, monatomic fluids in this section, but a great deal of both theoretical and experimental work has been done on quantum fluids and polyatomic fluids. The works by Marshall and Lovesey and by Larsson, Dahlborg, and Sköld ("Additional Reading") discuss these extensions in some detail.

We shall see in Section 22–3 that the spectrum of scattered light also can be used to probe molecular motion (although for different magnitudes of κ and ω) and that the theoretical formalism is quite similar to that of neutron scattering. Before discussing light scattering, however, we must digress to discuss a well-known theorem of stochastic processes known as the Wiener-Khintchine theorem since this theorem plays a central role in the derivation of the basic light-scattering equations.

22–2   THE WIENER-KHINTCHINE THEOREM

Consider some random process described by the function $x(t)$. This might be the trajectory of a particle undergoing Brownian motion, the output of some "noisy" electronic device, the voltage across a biological membrane, etc. If we repeat this process N times, we obtain a set of functions $x(t, \alpha)$, $\alpha = 1, \ldots, N$. We can think of these as N strip recordings as shown in Fig. 22–5.

The mathematical abstraction of this is to consider a family of functions $\{x(t, \alpha)\}$ where α may be continuous as well as discrete.

Now if α is taken to be a random variable with a probability density $p(\alpha)\,d\alpha$, then the family of functions $\{x(t, \alpha)\}$ is a *stochastic process*. A stochastic process can be thought of as a function of two variables t and α, where t is time and α is a random variable. Roughly speaking, the random variable $X(t)$ does not depend upon time in a completely definite way, but only in some probabilistic sense.

More precisely, the random process $X(t)$ is defined or described by a set of probability distributions. At any given time, find the fraction of the total number of functions

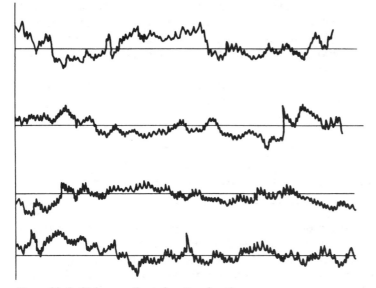

Figure 22–5. **Strip recordings of random functions.**

that have a value between x and $x + dx$. This fraction, $w_1(x, t)\, dx$, is called the *first probability distribution*. Now define the second probability distribution, $w_2(x_1, t_1; x_2, t_2)\, dx_1\, dx_2$, as the joint probability of finding X between x_1 and $x_1 + dx_1$ at time t_1 *and* between x_2 and $x_2 + dx_2$ at time t_2. This process can be continued on through the third, fourth, and all subsequent probability distributions. This set of functions completely characterizes the random process in a statistical sense. If we know the functions w_j for all j, we know all that can be known about the random process. The functions w_j must satisfy certain obvious conditions:

(1) $w_n \geq 0$

(2) $w_n(x_1, t_1; x_2, t_2; \cdots; x_n, t_n)$ is a symmetric function in the set of variables (x_j, t_j). This follows because w_n is a joint probability.

(3) $w_k(x_1, t_1; \ldots; x_k, t_k) = \int \cdots \int dx_{k+1} \cdots dx_n\, w_n(x_1, t_1; \cdots; x_n, t_n)$.

In general, the complete determination of the above set of probability distributions is not feasible. For example, the determination of just $w_1(x, t)$ would require that we determine the number of observations that X lie within x and $x + dx$ for all values of t. Thus we would have to observe the time evolution of a large number of similarly prepared systems. Fortunately, however, there are several reasonable assumptions that can be made that greatly simplify matters. The first and least restrictive assumption that we shall discuss is to assume that the random process is stationary in time. By *stationary* we mean that the form of the probability distribution functions does not depend on a shift of the origin of time. In a sense we assume that the underlying probabilistic mechanism of the process does not change with time. More precisely, we say that a random process is stationary when the probability distributions of $\{x(t, \alpha)\}$ and $\{x(t + \tau, \alpha)\}$ are the same for any τ.

For a stationary random process then, we may, in principle at least, determine the various probability distributions from the experimental observation of $x(t)$ for one system over a long period of time. This long-time record can be cut up into pieces of length T (where T is much longer than any " periodicities " occurring in the process), and these pieces may be treated as observations of different systems in an ensemble of similarly prepared systems. The underlying assumption here is the so-called *ergodic hypothesis*, which states that for a stationary random process, a large number of observations made on a single system at N arbitrary instants of time have the same statistical properties as observing N arbitrarily chosen systems at the same time from an ensemble of similar systems.

Since a shift in the origin of time does not affect any of the statistical properties of a stationary random process, the probability density functions simplify to

General	Stationary
$w_1(x, t)\, dx$	$w_1(x)\, dx$
$w_2(x_1, t_1; x_2, t_2)\, dx_1\, dx_2$	$w_2(x_1, x_2; t_2 - t_1)\, dx_1\, dx_2$
$w_3(x_1, t_1; x_2, t_2; x_3, t_3)\, dx_1\, dx_2\, dx_3$	$w_3(x_1, x_2, x_3; t_2 - t_1, t_3 - t_1)\, dx_1\, dx_2\, dx_3$
etc.	

Lastly, we note one more property that follows from the ergodic hypothesis and the assumption of stationarity. In dealing with general random processes, there are two types of mean values that we encounter. One is obtained by observations made on many systems at some fixed time t (denote this average by $\langle x \rangle$), and the other is the time average made on one system as a function of time (denote this average by $\bar{x}$).

For a stationary random process, both averages yield the same result. In terms of equations, we have

$$\bar{x} = \lim_{T \to \infty} \frac{1}{2T} \int_{-T}^{T} x(t)\, dt \tag{22-62}$$

$$\langle x \rangle = \int_{-\infty}^{\infty} x w_1(x)\, dx \tag{22-63}$$

and for a stationary random process

$$\bar{x} = \langle x \rangle \tag{22-64}$$

according to the ergodic hypothesis.

The probability distributions of stationary random processes are indeed simpler than the general case, but it is necessary, nevertheless, to make further restrictions as well. This leads us to an important classification of random processes. A random process is said to be purely random when values of x at different times are completely uncorrelated. The probability distributions in this case become

$$w_2(x_1, t_1; x_2, t_2) = w_1(x_1, t_1)w_1(x_2, t_2)$$
$$w_3(x_1, t_1; x_2, t_2; x_3, t_3) = w_1(x_1, t_1)w_1(x_2, t_2)w_1(x_3, t_3) \tag{22-65}$$

etc. The random process is completely specified by w_1. Purely random processes do not occur often in physical applications since in most real situations $x(t_1)$ and $x(t_2)$ will be correlated at least for small values of $|t_2 - t_1|$.

The next more complicated case, and one that turns out to be a reasonable abstraction for a large number of physical processes, is to assume that the process is a *Markov process*. In order to define a Markov process, first consider that the time axis is divided into small intervals of length δ, and let $t_j = j\delta$. We now introduce the conditional probability $p_n(x_n, t_n | x_0, t_0; x_1, t_1; \ldots; x_{n-1}, t_{n-1})$, $t_0 < t_1 < \cdots < t_n$, that X lie in the interval $(x_n, x_n + dx_n)$ at time t_n given that $X(t_0) = x_0$, $X(t_1) = x_1$, etc. We define a Markov process by the requirement that

$$p_n(x_n, t_n | x_0, t_0; \ldots; x_{n-1}, t_{n-1}) = p_2(x_n, t_n | x_{n-1}, t_{n-1}) \tag{22-66}$$

In other words, the probability that the system "is in the state n" at time t_n depends only upon its state directly preceding t_n and not upon the previous history of the process. Since all the p_n reduce to p_2, we see that a Markov process is completely characterized by the second distribution function.

There are two functions associated with a continuous stationary random process that are central to the theory of stochastic processes. These two functions are the correlation function and the spectral density. We shall see the stationary random processes can be well characterized by either of these two functions. The *correlation function* of a continuous stationary random process is given by

$$C(\tau) = \lim_{T \to \infty} \frac{1}{2T} \int_{-T}^{T} x(t + \tau)x(t)\, dt \tag{22-67}$$

According to the ergodic hypothesis, $C(\tau)$ also is equal to $\langle x(t + \tau)x(t) \rangle$, i.e., an ensemble average of $x(t + \tau)x(t)$ instead of a time average. We write this as

$$C(\tau) = \iint_{-\infty}^{\infty} x_1 x_2 w_2(x_1, x_2; \tau)\, dx_1\, dx_2 \tag{22-68}$$

Note that if x_1 and x_2 are uncorrelated, then $w_2(x_1, x_2; \tau) = w_1(x_1)w_1(x_2)$ and $C(\tau) = 0$ if $\langle x_j \rangle = 0$.

On physical grounds, we expect that if $x(t)$ is indeed a random process, x_1 and x_2 will be more correlated for small τ than for large τ. In fact, we can prove that

$$-C(0) \leq C(\tau) \leq C(0) \tag{22-69}$$

The proof goes as follows: Start with the obvious inequality

$$[x(t) \pm x(t + \tau)]^2 \geq 0$$

or

$$x^2(t) + x^2(t + \tau) \pm 2x(t)x(t + \tau) \geq 0$$

If we take either the ensemble or time average of both sides of this inequality, we get $2C(0) \geq \pm 2C(\tau)$, which is Eq. (22–69).

One can also show that $C(\tau)$ is an even function of τ since

$$C(\tau) = \overline{x(t)x(t + \tau)} = \overline{x(t - \tau)x(t)}$$

$$= \overline{x(t)x(t - \tau)} = C(-\tau) \tag{22-70}$$

The other central function in the theory of stochastic processes is the *spectral density* of $x(t)$. Let

$$x_T(t) = x(t) \qquad \text{for } -T \leq t \leq T$$
$$= 0 \qquad \text{otherwise} \tag{22-71}$$

Let the Fourier transform of $x_T(t)$ be $A(\omega)$, i.e.,

$$A(\omega) = \int_{-\infty}^{\infty} x_T(t)e^{-i\omega t}\, dt = \int_{-T}^{T} x(t)e^{-i\omega t}\, dt \tag{22-72}$$

so that by Fourier inversion

$$x_T(t) = \frac{1}{2\pi} \int_{-\infty}^{\infty} A(\omega)e^{i\omega t}\, d\omega \tag{22-73}$$

By Parseval's theorem (Problem 22–16) we can write

$$\int_{-\infty}^{\infty} x_T^{\,2}(t)\, dt = \frac{1}{2\pi} \int_{-\infty}^{\infty} |A(\omega)|^2\, d\omega \tag{22-74}$$

This can be proved by writing

$$\int_{-\infty}^{\infty} |A(\omega)|^2\, d\omega = \int_{-\infty}^{\infty} A^*(\omega)A(\omega)\, d\omega = \int_{-\infty}^{\infty} A^*(\omega) \int_{-\infty}^{\infty} x_T(t)e^{i\omega t}\, dt\, d\omega \tag{22-75}$$

and interchanging orders of integration. By dividing Eq. (22–74) by $2T$ and taking the limit, we get

$$\overline{x^2} = \lim_{T \to \infty} \frac{1}{2T} \int_{-T}^{T} x_T^{\,2}(t)\, dt \tag{22-76}$$

$$= \lim_{T \to \infty} \frac{1}{2\pi} \int_{-\infty}^{\infty} \frac{|A(\omega)|^2}{2T}\, d\omega \tag{22-77}$$

Now we define the *spectral density* $S(\omega)$ of $x(t)$ by

$$S(\omega) = \lim_{T \to \infty} \frac{1}{2T} |A(\omega)|^2 \qquad (22\text{--}78)$$

so that

$$\overline{x^2} = \frac{1}{2\pi} \int_{-\infty}^{\infty} S(\omega)\, d\omega \qquad (22\text{--}79)$$

By the ergodic hypothesis, $S(\omega)$ is the same for all $x(t)$ in the ensemble. Note that since $x(t)$ is real, $A^*(\omega) = A(-\omega)$, and $S(\omega)$ is an even function of ω.

This relation has an important and useful physical interpretation. If we consider $x_T(t)$ to be an electric current, then $\overline{x^2}$ is the average power dissipated as the current passes through a unit resistor during the interval $(-T, T)$. Thus $S(\omega)\, d\omega/2\pi$ is the average power dissipated with frequencies between ω and $\omega + d\omega$; hence $S(\omega)$ is often called the *power spectrum* of the random process $x(t)$.

It turns out that the correlation function and the spectral density are connected by the so-called *Wiener-Khintchine theorem*, which states that

$$C(\tau) = \frac{1}{2\pi} \int_{-\infty}^{\infty} S(\omega) e^{i\omega\tau}\, d\omega = \frac{1}{2\pi} \int_{-\infty}^{\infty} S(\omega) \cos \omega\tau\, d\omega \qquad (22\text{--}80)$$

and

$$S(\omega) = \int_{-\infty}^{\infty} C(\tau) e^{-i\omega\tau}\, d\tau = 2 \int_0^{\infty} C(\tau) \cos \omega\tau\, d\tau \qquad (22\text{--}81)$$

Thus according to the Wiener-Khintchine theorem, the correlation function and the spectral density are simply Fourier or cosine transforms of each other. This is a useful relationship since $S(\omega)$ is often experimentally measurable and this spectrum can be inverted to determine the correlation function.

We shall prove the Wiener-Khintchine theorem in two ways, one using the notation of Fourier series and the other using the notation of Fourier transforms.

Let $x(t)$ be represented by a Fourier series over the interval $(-T, T)$:

$$x(t) = \sum_{n=1}^{\infty} a_n \cos \frac{n\pi t}{T} + \sum_{n=1}^{\infty} b_n \sin \frac{n\pi t}{T} \qquad (22\text{--}82)$$

There is no constant term a_0 here since

$$a_0 = \frac{1}{2T} \int_{-T}^{T} x(t)\, dt = \overline{x(t)} \qquad (22\text{--}83)$$

and we assume that the average value of $x(t)$ is zero. Using the fact that $[\cos(n\pi t/T)]$ and $[\sin(n\pi t/T)]$ are orthogonal in $(-T, T)$, we get that

$$\overline{x^2} = \frac{1}{2T} \int_{-T}^{T} \left(\sum_{n=1}^{\infty} a_n \cos \frac{n\pi t}{T} + \sum_{n=1}^{\infty} b_n \sin \frac{n\pi t}{T} \right)^2 dt$$

$$= \sum_{n=1}^{\infty} (a_n^2 + b_n^2) \qquad (22\text{--}84)$$

We can interpret $a_n^2 + b_n^2$ to be the power dissipated at the frequency $\omega_n = n\pi/2T$. From Eq. (22–79), we can make the correspondence

$$\sum_{n=1}^{\infty} (a_n^2 + b_n^2) \to \int_0^{\infty} \frac{S(\omega)\, d\omega}{\pi} \qquad \text{as} \quad T \to \infty$$

Similarly, we can evaluate the correlation function (Problem 22–18)

$$C(\tau) = \overline{x(t)x(t+\tau)} = \frac{1}{2T} \int_{-T}^{T} x(t)x(t+\tau)\, dt$$

$$= \sum_{n=1}^{\infty} (a_n{}^2 + b_n{}^2) \cos \frac{n\pi\tau}{T} \tag{22-85}$$

Now as $T \to \infty$, we can write this as

$$C(\tau) = \frac{1}{\pi} \int_0^{\infty} S(\omega) \cos \omega\tau\, d\tau \tag{22-86}$$

$$= \frac{1}{2\pi} \int_{-\infty}^{\infty} S(\omega) e^{i\omega\tau}\, d\omega \tag{22-87}$$

where we have used the fact that $S(\omega)$ is an even function of ω. By Fourier inversion we get

$$S(\omega) = \int_{-\infty}^{\infty} C(\tau) e^{-i\omega\tau}\, d\tau \tag{22-88}$$

$$= 2 \int_0^{\infty} C(\tau) \cos \omega\tau\, d\tau \tag{22-89}$$

since $C(\tau)$ is an even function of τ.

For the second proof of the Wiener-Khintchine theorem, we start with

$$C_T(\tau) = \frac{1}{2T} \int_{-T}^{T} x_T(t) x_T(t+\tau)\, dt \tag{22-90}$$

so that

$$C(\tau) = \lim_{T \to \infty} C_T(\tau) \tag{22-91}$$

The Fourier transform of $C_T(\tau)$ is

$$\int_{-\infty}^{\infty} C_T(\tau) e^{-i\omega\tau}\, d\tau = \frac{1}{2T} \int_{-\infty}^{\infty} e^{-i\omega\tau} \int_{-\infty}^{\infty} x_T(t) x_T(t+\tau)\, dt\, d\tau$$

$$= \frac{1}{2T} \int_{-\infty}^{\infty} \int_{-\infty}^{\infty} x_T(t) x_T(t+\tau) e^{-i\omega(t+\tau)} e^{i\omega t}\, dt\, d\tau$$

$$= \frac{1}{2T} \int_{-\infty}^{\infty} x_T(t) e^{i\omega t} \int_{-\infty}^{\infty} x_T(t+\tau) e^{-i\omega(t+\tau)}\, dt\, d\tau$$

$$= \frac{1}{2T} A(\omega) A(-\omega) = \frac{1}{2T} A(\omega) A^*(\omega) \tag{22-92}$$

Passing to the limit $T \to \infty$ then and using Eq. (22–78), this becomes

$$\int_{-\infty}^{\infty} C(\tau) e^{-i\omega\tau}\, d\tau = S(\omega) \tag{22-93}$$

or

$$S(\omega) = 2 \int_0^{\infty} C(\tau) \cos \omega\tau\, d\tau \tag{22-94}$$

In many physical applications it is a good approximation that all the probability distributions are Gaussian. A process in which they are all normal is called a Gaussian or normal process.

The discussion of Gaussian processes is most conveniently started with the Fourier series representation of a stationary random function $x(t)$. Let T be a sufficiently long time and suppose that $x(t)$ is periodic with a period $2T$. Then we can write

$$x(t) = \sum_{j=1}^{\infty} (a_j \cos \omega_j t + b_j \sin \omega_j t) \tag{22-95}$$

where $\omega_j = \pi j / T$. Again, there is no constant term since we assume for convenience that the mean value of $x(t)$ is zero.

The coefficients a_j and b_j are random variables, and we assume that they are all mutually independent, have normal distributions with zero means, and that the variances of a_j and b_j are both equal to $\sigma_j{}^2$. Thus we can write

$$p(a_1, a_2, \ldots, b_1, b_2, \ldots) = \prod_{j=1}^{\infty} \frac{1}{2\pi\sigma_j{}^2} \exp\left[-\frac{(a_j{}^2 + b_j{}^2)}{2\sigma_j{}^2}\right] \tag{22-96}$$

Note that

$$\langle a_j \rangle = \langle b_j \rangle = 0$$
$$\langle a_j a_k \rangle = \langle b_j b_k \rangle = \sigma_j{}^2 \delta_{jk} \tag{22-97}$$
$$\langle a_j b_k \rangle = 0$$

In addition, we have

$$\langle x^2(t) \rangle = \sum_{j=1}^{\infty} (\langle a_j{}^2 \rangle \cos^2 \omega_j t + \langle b_j{}^2 \rangle \sin^2 \omega_j t)$$

$$= \sum_{j=1}^{\infty} \sigma_j{}^2 \equiv \sigma^2 \approx \frac{1}{2\pi} \int_{-\infty}^{\infty} S(\omega) \, d\omega \tag{22-98}$$

where the next to the last step follows as $T \to \infty$.

In Problem 22-19 one proves that since $x(t)$ is a linear combination of the a_j and b_j, $x(t)$ itself is a normal process with distribution

$$w_1(x) = \frac{1}{\sqrt{2\pi\sigma^2}} \exp\left(-\frac{x^2}{2\sigma^2}\right) \tag{22-99}$$

Note that time does not appear in this distribution since $X(t)$ is assumed to represent a stationary random process.

For the second probability distribution, we consider

$$x(t_1) = \sum_{j=1}^{\infty} (a_j \cos \omega_j t_1 + b_j \sin \omega_j t_1)$$

$$x(t_2) = \sum_{j=1}^{\infty} (a_j \cos \omega_j t_2 + b_j \sin \omega_j t_2)$$

from which we obtain (Problem 22-20)

$$\langle x(t_1)x(t_2) \rangle = \sum_{j=1}^{\infty} (\langle a_j{}^2 \rangle \cos \omega_j t_1 \cos \omega_j t_2 + \langle b_j{}^2 \rangle \sin \omega_j t_1 \sin \omega_j t_2)$$

$$= \sum_{j=1}^{\infty} \sigma_j{}^2 \cos \omega_j \tau \approx \frac{1}{2\pi} \int_{-\infty}^{\infty} S(\omega) \cos \omega\tau \, d\omega \tag{22-100}$$

$$= C(\tau) = \sigma^2 \rho(\tau) \tag{22-101}$$

where $\tau = t_2 - t_1$ as before and $\rho(\tau)$ is a normalized correlation function defined by $\langle x(0)x(\tau)\rangle/\langle x^2(0)\rangle$. Note that $\langle x(t_1)x(t_2)\rangle$ depends only upon τ, as it should for a stationary process.

According to Problem 22–21 the second probability distribution function is

$$w_2(x_1, x_2; \tau) = \frac{1}{2\pi\sigma^2(1-\rho^2)^{1/2}} \exp\left[-\frac{(x_1^2 + x_2^2 - 2\rho x_1 x_2)}{2\sigma^2(1-\rho^2)}\right] \tag{22–102}$$

The third probability distribution function is a three-dimensional normal distribution which depends on x_1, x_2, and x_3 and $t_2 - t_1$ and $t_3 - t_1$. In fact, all the distribution functions are normal and depend only upon the x_j's, σ^2, and $\rho(\tau_j)$ where $\tau_j = t_j - t_1, j \geq 2$.

We see then that for a Gaussian random process, all the distribution functions can be determined in terms of the correlation function. In many important physical applications, the spectral density is readily determined, and then in turn the correlation function, and then the probability distribution functions.

In the next section we shall need to discuss the correlation function

$$C(t) = \langle x^*(t)x(t)x^*(0)x(0)\rangle \tag{22–103}$$

for Gaussian random process. In order to discuss this, it is best to write $x(t)$ in a Fourier representation of complex exponentials rather than sines and cosines; i.e., write

$$X(t) = \sum_{n=-\infty}^{\infty} c_n e^{i\omega_n t} \tag{22–104}$$

For a Gaussian process

$$\langle c_n^* c_m\rangle = \sigma_n^2 \delta_{nm} \tag{22–105}$$

and (Problem 22–24)

$$\langle c_i^* c_j c_k^* c_l\rangle = \sigma_i^2 \sigma_k^2 \delta_{ij}\delta_{kl} + \sigma_i^2 \sigma_j^2 \delta_{il}\delta_{jk} \tag{22–106}$$

Using these results, it is easy to see that

$$\langle x^*(t)x(0)\rangle = \sum_{j=-\infty}^{\infty} \sigma_j^2 e^{-i\omega_j t} \tag{22–107}$$

and that (Problem 22–25)

$$\langle x^*(t)x(t)x^*(0)x(0)\rangle = \sum_j \sum_k \sigma_j^2 \sigma_k^2 + \sum_j \sum_k \sigma_j^2 \sigma_k^2 e^{i(\omega_k - \omega_j)t}$$

$$= |\langle x^*(0)x(0)\rangle|^2 + |\langle x^*(t)x(0)\rangle|^2 \tag{22–108}$$

This expression is used in the interpretation of certain light-scattering measurements to be discussed in the next section.

There is an important theorem in the theory of Gaussian random processes that says that a one-dimensional Gaussian random process will be Markovian only when the correlation function is of the form $e^{-\alpha\tau}$. This is known as Doob's theorem (for proof of Doob's theorem, see Wang and Uhlenbeck in "Additional Reading"). It also follows from Doob's theorem that the spectrum must be of the form const./ $(\omega^2 + \alpha^2)$. This is often called a Markovian spectrum and is one of the most common power spectra that occur in practice. Doob's theorem can be generalized to n-dimensions. Instead of a single correlation function, one must deal with a correlation matrix whose elements are given by

$$C_{ij}(\tau) = \langle x_i(t)x_j(t + \tau)\rangle$$

and which can be written as

$$\mathbf{C}(\tau) = \exp\left(\mathbf{Q}\tau\right) \tag{22–109}$$

where $\mathbf{Q}$ is some (not generally symmetric) matrix. In this case, the power spectrum need not be of the form const.$/(\omega^2 + \beta^2)$, but instead different forms of $\mathbf{Q}$ will yield different forms for the spectra.

Now that we have discussed some of the more important concepts of random variables and stochastic processes, we can derive the basic formulas of light-scattering spectroscopy by appealing directly to the Wiener-Khintchine theorem.

22–3 LASER LIGHT SCATTERING

It is a standard technique to use light scattering to determine the shape and size of polymer molecules in solution. This is done through a measurement of the intensity and the angular dependence of the scattering. With the development of lasers, it has become possible to determine the spectrum of the scattered light, and we shall see in this section that the spectrum contains information related to the hydrodynamic properties of the scatterers, such as the translational and rotational diffusion coefficients. The theoretical analysis of the scattered spectrum was first presented by Komarov and Fisher[*] and Pecora[†] independently in 1963. In 1964, the spectrum of laser light scattered by dilute solutions of polystyrene latex spheres was observed and was found to exhibit a lineshape in good quantitive agreement with the theory.[‡]

In this section we shall present a derivation due to Cummins et al.,[§] which utilizes the Wiener-Khintchine theorem. Consider a volume v, which contains N identical scatterers in a nonscattering solvent. If we irradiate this scattering volume with a monochromatic plane wave of frequency ω_0, the light scattered through an angle θ by the jth scatterer is given in complex notation by (Problem 13–47)

$$\mathscr{E}_j = A_j(t)e^{i2\pi\phi_j}e^{-i\omega_0 t} \tag{22–110}$$

where $A_j(t)$, the amplitude of $\mathscr{E}_j$, may depend upon the orientation of the scatterer. According to Eq. (13–97)

$$2\pi\phi_j = \frac{2\pi n}{\lambda_0}(\mathbf{u}_0 - \mathbf{u})\cdot\mathbf{r}_j = \boldsymbol{\kappa}\cdot\mathbf{r}_j \tag{22–111}$$

where n is the index of refraction and λ_0 is the wavelength of the radiation in the solution (mostly solvent). The magnitude of κ is given by Eq. (13–98) to be $\kappa = (4\pi n/\lambda_0)\sin(\theta/2)$.

[*] L. I. Komarov and I. Z. Fisher, *Soviet Phys., JETP*, **16**, p. 1358, 1963.
[†] R. Pecora, *J. Chem. Phys.*, **40**, p. 1604, 1964.
[‡] H. Z. Cummins, N. Knable, and Y. Yeh, *Phys. Rev. Lett.*, **12**, p. 150, 1964.
[§] H. Z. Cummins et al., *Biophys. J.*, **9**, p. 518, 1969.

For the total scattered field observed at a large distance R from the scattering volume, we have

$$\mathscr{E}_s = \sum_{j=1}^{N} A_j(t) e^{i\mathbf{\kappa} \cdot \mathbf{r}_j} e^{-i\omega_0 t} \tag{22-112}$$

The total average scattered intensity is given by $I_s = \langle |\mathscr{E}_s|^2 \rangle$, where the angular brackets denotes a time average as in Problem 13–47.

It is perhaps appropriate at this point to present a brief digression on the light-scattering experiment itself. There are two experimental techniques that are used to determine the extremely small-frequency shifts that result. These involve the so-called optical mixing techniques,* in which the scattered spectrum is mixed with itself (*homodyne method*) or mixed with part of the incident beam (*heterodyne method*). Since the interpretation of experimental results depends upon which method is used to determine the scattered spectrum, it is important to be aware of the difference.

In a measurement of the homodyne type, the intensity of the scattered radiation is observed at the cathode of a photomultiplier. The output of the photomultiplier is usually analyzed by either a real-time autocorrelator or a spectrum analyzer.

If a real-time autocorrelator is used, the time autocorrelation function of the output current of the photomultiplier

$$C(t) = \langle i(t) i(0) \rangle$$

is measured. If a spectrum analyzer is used, the Fourier transform of $C(t)$

$$C(\omega) = \frac{1}{2\pi} \int_{-\infty}^{\infty} e^{-i\omega t} C(t) \, dt$$

is measured. Since the photomultiplier current $i(t)$ is proportional to the intensity of the scattered radiation, we have that

$$C(t) \propto \langle \mathscr{E}_s^*(t) \mathscr{E}_s(t) \mathscr{E}_s^*(0) \mathscr{E}_s(0) \rangle$$
$$\propto \langle |\mathscr{E}_s(t)|^2 |\mathscr{E}_s(0)|^2 \rangle \tag{22-113}$$

It is customary at this point to assume that the scattered field is a Gaussian random process. According to Eq. (22–108), then, we have

$$C(t) \propto |I(0)|^2 + |I(t)|^2 \tag{22-114}$$

where

$$I(0) = \langle |\mathscr{E}_s(0)|^2 \rangle \tag{22-115}$$

and

$$I(t) = \langle \mathscr{E}_s(t) \mathscr{E}_s(0) \rangle \tag{22-116}$$

The term involving $I(0)$ represents a dc background that can be subtracted out, so that the $I(t)$ term which contains the spectral information of interest, results. Thus in the homodyne method it is the quantity $|I(t)|^2 = |\langle \mathscr{E}_s^*(t) \mathscr{E}_s(0) \rangle|^2$ that is measured experimentally.

In the heterodyne method, a portion of the incident radiation is mixed with the

---

* H. Z. Cummins and H. L. Swinney ("Additional Reading"); H. Z. Cummins *et al.*, *Biophys. J.* **9**, p. 518, 1969.

scattered radiation before being led into the photomultiplier. The relevant correlation function in this case is

$$C(t) \propto \langle |\mathscr{E}_0(t) + \mathscr{E}_s(t)|^2 \; |\mathscr{E}_0(0) + \mathscr{E}_s(0)|^2 \rangle \tag{22–117}$$

When this is written out, there are 16 terms, 10 of which vanish, 3 of which are time-independent terms which can be subtracted out as in the homodyne method, and the remaining 3 terms are proportional to (Problem 22–26)

$$\langle |\mathscr{E}_0|^2 \rangle \; [e^{i\omega_0 t} \langle \mathscr{E}_s^*(0)\mathscr{E}_s(t) \rangle + e^{-i\omega_0 t} \langle \mathscr{E}_s(0)\mathscr{E}_s^*(t) \rangle] + O(\mathscr{E}_s^4) \tag{22–118}$$

We have assumed here that $|\mathscr{E}_0| \gg |\mathscr{E}_s|$. Thus we see that in the heterodyne method, the correlation function is essentially that of the scattered field.

In order to compute the spectrum of the scattered light, we use the Wiener-Khintchine theorem:

$$I(\omega) = \frac{1}{2\pi} \int_{-\infty}^{\infty} C(\tau)e^{i\omega\tau} \, d\tau \tag{22–119}$$

where $C(-\tau) = C^*(\tau)$ since $I(\omega)$ is real. In the heterodyne method, the autocorrelation function $C(\tau)$ is [(cf. Eq. 22–118)]

$$C(\tau) = e^{i\omega_0\tau}\langle \mathscr{E}_s^*(t)\mathscr{E}_s(t+\tau) \rangle + e^{-i\omega_0\tau}\langle \mathscr{E}_s(t)\mathscr{E}_s^*(t+\tau) \rangle \tag{22–120}$$

where the angular brackets denote a time average or, by the ergodic hypothesis, an ensemble average as well.

If we substitute Eq. (22–112) into $\langle \mathscr{E}_s^*(t)\mathscr{E}_s(t+\tau) \rangle$ we get

$$\langle \mathscr{E}_s^*(t)\mathscr{E}_s(t+\tau) \rangle = \left\langle \sum_{j=1}^{N} A_j^*(t)e^{-i\kappa \cdot r_j(t)}e^{i\omega_0 t} \sum_{l=1}^{N} A_l(t+\tau)e^{i\kappa \cdot r_l(t+\tau)}e^{-i\omega_0(t+\tau)} \right\rangle \tag{22–121}$$

If we now assume that the scatterers are independent, that the position and orientation factors are independent, that the N scatterers are identical, and that the correlation function is stationary, we get

$$\langle \mathscr{E}_s^*(t)\mathscr{E}_s(t+\tau) \rangle = Ne^{-i\omega_0\tau}\langle A^*(0)A(\tau) \rangle \; \langle e^{-i\kappa \cdot r(0)}e^{i\kappa \cdot r(t)} \rangle \tag{22–122}$$

and (Problem 22–27)

$$I(\omega) = \frac{N}{\pi} \int_{-\infty}^{\infty} e^{i\omega\tau}C_A(\tau)C_R(\tau) \, d\tau \tag{22–123}$$

where

$$C_A(\tau) = \langle A^*(0)A(\tau) \rangle \tag{22–124}$$

$$C_R(\tau) = \langle e^{-i\kappa \cdot r(0)}e^{i\kappa \cdot r(\tau)} \rangle \tag{22–125}$$

The separation of the orientation and position is not rigorously correct and should be considered as a simplifying assumption.

If the scatterers are spherical, then the scattering amplitude is independent of time and

$$I(\omega) = \frac{NA^2}{\pi} \int_{-\infty}^{\infty} e^{i\omega\tau}C_R(\tau) \, d\tau \tag{22–126}$$

Note that $C_R(\tau)$ is simply the function $F_s(\kappa, \tau)$ defined in Section 22–1. In terms of the self-part of the van Hove space-time correlation function, we have

$$F_s(\kappa, \tau) = \int e^{i\kappa \cdot r} G_s(r, \tau) \, dr \tag{22–127}$$

$G_s(r, \tau)$ is the ensemble averaged conditional probability that a particle located initially at the origin will be in a volume dr at r at time τ. If the scattering volume is a dilute solution of macromolecules dissolved in a (nonscattering) solvent, $G_s(r, \tau)$ is given by the diffusion equation

$$\frac{\partial G_s}{\partial \tau} = D \, \nabla^2 G_s \tag{22–128}$$

with the initial condition $G_s(r, 0) = \delta(r)$. Thus

$$G_s(r, \tau) = (4\pi D\tau)^{-3/2} \exp\left(-\frac{r^2}{4D\tau}\right) \tag{22–129}$$

$$C_R(\tau) = \exp(-\kappa^2 D\tau) \tag{22–130}$$

and

$$C(\tau) = NA^2 e^{-\kappa^2 D\tau} \tag{22–131}$$

and

$$I(\omega) = 2NA^2 \left[\frac{D\kappa^2/\pi}{\omega^2 + (D\kappa^2)^2}\right] \tag{22–132}$$

Thus the Rayleigh scattered light has a Lorentzian lineshape centered at $\omega = 0$ with a half-width $\Delta\omega_{1/2} = D\kappa^2$. One can determine the value of D by plotting $\Delta\omega_{1/2}$ versus $\kappa^2 = 4k_0^2 \sin^2 \theta/2$, and this is now a fairly standard experimental procedure.

In 1971, Ware and Flygare* presented an interesting extension of the above experiment. They show that one can determine the electrophoretic mobility of a charged macromolecule by observing the spectral distribution of light scattered from solutions of charged macromolecules under the influence of an electric field. The electrophoretic mobility is defined by $v = uE$, where v is the terminal velocity of the macroion in the presence of the field. Consider now the time dependence of $G_s(r, \tau)$ under the condition of a drift in the direction of the electric field. If we take the direction of the field to be the x-direction, then the diffusion-type equation for $G_s(r, \tau)$ becomes

$$\frac{\partial G_s}{\partial \tau} = D \, \nabla^2 G_s + uE \frac{\partial G_s}{\partial x} \tag{22–133}$$

where u is the electrophoretic mobility of the charged macromolecule. This equation can be readily solved by using Fourier transforms to give (Problem 22–28)

$$G_s(r, \tau) = (4\pi D\tau)^{-3/2} \exp\left\{-\frac{[(x + uE\tau)^2 + y^2 + z^2]}{4D\tau}\right\} \tag{22–134}$$

and

$$C(\tau) = NA^2 e^{-i\kappa_x uE\tau} e^{-\kappa^2 D\tau} \tag{22–135}$$

* *Chem. Phys. Lett.*, **12**, p. 81, 1971.

If we choose the x-axis to be perpendicular to the incident radiation, we find that $\kappa_x = \kappa \cos \theta/2$, and the power spectrum is

$$I(\omega) = 2NA^2 \left[\frac{D\kappa^2/\pi}{(\omega - \kappa u E \cos \frac{1}{2}\theta)^2 + (\kappa^2 D)^2} \right] \qquad (22\text{--}136)$$

Thus we see that the Lorentzian lineshape of the Rayleigh scattered light is shifted in frequency by $\Delta\omega = \kappa u E \cos \frac{1}{2}\theta$ under the influence of the electric field. The half-width is still related to the diffusion constant, however, through $\Delta\omega_{1/2} = \kappa^2 D$. Notice that Eqs. (22–130), (22–131), (22–134), and (22–135) do not contain ω_0. This is because in the heterodyne detection method, the scattered light is mixed with a strong component of the incident radiation in a nonlinear detector such as a photomultiplier tube, and the measured spectrum contains only the low-frequency beats between ω_0 and the scattered light, and hence the spectrum is centered at $\omega = 0$. One can use an autocorrelator to observe the correlation function directly, and such a measurement is shown in Fig. 22–6, which is from the paper by Ware and Flygare. It shows the autocorrelation function at zero field and several other fields in a solution of the protein, bovine serum albumin (BSA). The autocorrelation function in the absence of the electric field is an exponential decay with a decay constant, $\tau = 1/\kappa^2 D$. In the presence of the electric field, the autocorrelation function is an exponentially damped cosine whose decay constant is also $1/\kappa^2 D$ and whose period is equal to

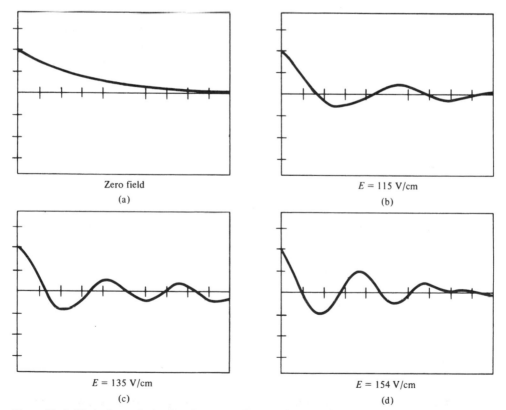

Zero field

(a)

$E = 115$ V/cm

(b)

$E = 135$ V/cm

(c)

$E = 154$ V/cm

(d)

Figure 22–6. **The autocorrelation function at zero field and fields of 115, 135, and 154 V/cm, respectively in a 5 percent solution of BSA in 0.004 M NaCl titrated to pH 9.2 with n-butylamine.** $\lambda_0 = 5145$ Å, $T = 10°C$, **and** $\theta = 0.079$ rad $= 4.5°$. (From B. R. Ware and W. H. Flygare, *Chem. Phys. Lett.*, **12**, p. 81, 1971.)

$\Delta\tau = 2\pi/\kappa_x uE$, which can be reexpressed in the form $\Delta\tau = \lambda_0/uEn \sin\theta$ (Problem 22–30). Thus a measurement of the autocorrelation function in the presence of an electric field provides a simultaneous determination of the diffusion coefficient and the electrophoretic mobility. By determining the period of the autocorrelation function as a function of the field E, one can determine the electrophoretic mobility by plotting $1/\Delta\tau E$ versus $\sin\theta$. Figure 22–7 shows such a plot obtained from the autocorrelation functions in Fig. 22–6.

Note that if the homodyne method of detection had been used, the appropriate correlation function would be

$$C(\tau) = |\langle \mathscr{E}_s^*(0)\mathscr{E}_s(\tau)\rangle|^2 \tag{22–137}$$

which, according to the above, is

$$C(\tau) = NA^2 e^{-2\kappa^2 D\tau} \tag{22–138}$$

Thus the homodyne method yields a frequency spectrum of the form

$$I(\omega) = \frac{NA^2}{2\pi} \int_{-\infty}^{\infty} e^{i\omega\tau} e^{-2\kappa^2 D\tau}\, d\tau \tag{22–139}$$

$$= \frac{NA^2}{\pi} \frac{2D\kappa^2}{\omega^2 + (2D\kappa^2)^2} \tag{22–140}$$

Notice that this is also a Lorentzian frequency distribution, but the half-width in this case is $2D\kappa^2$, rather than $D\kappa^2$ as in the heterodyne method. Figure 22–8 shows the measured half-width versus $\sin^2(\theta/2)$ $[\kappa^2 = (4\pi n_0/\lambda) \sin^2(\theta/2)]$ for polystyrene latex spheres from both the homodyne and heterodyne methods. This figure is taken from

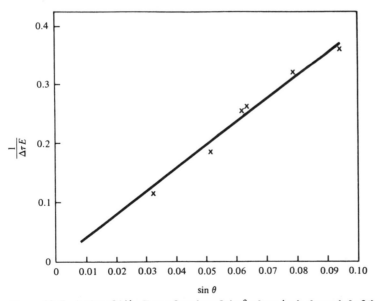

Figure 22–7. **A plot of $1/\Delta\tau E$ as a function of $\sin\theta$ where $\Delta\tau$ is the period of the oscillation of the autocorrelation function (see also Fig. 22–6). The expected κ dependence in the sinusoidal oscillation of the autocorrelation function is evident. The slope of the line leads to an average value for the mobility of 14×10^{-5} cm^2/sec $\cdot$ V which is corrected to $20°C$ to give 18×10^{-5} as shown.** (From B. R. Ware and W. H. Flygare, *Chem. Phys. Lett.*, **12**, p. 81, 1971.)

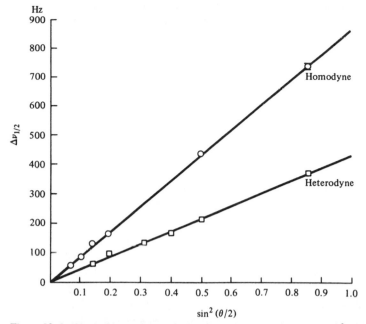

Figure 22–8. **Single Lorentzian half-widths at 25°C as a function of $\sin^2\theta/2$ obtained from homodyne and heterodyne spectra of polystyrene latex spheres, 0.126 $\pm$ 0.004 μ diameter, 0.01 mg/mg; $\bigcirc$: homodyne spectra; $\square$: heterodyne spectra.** (From Cummins *et al.*, *Biophys. J.*, **9**, p. 518, 1969.)

the paper by Cummins *et al.** who give an excellent discussion of the experimental and theoretical details of the homodyne and heterodyne detection methods.

Our results up to now have been for a system of identical scatterers. For a solution of s different kinds of spherical macromolecules, the spectrum is given by

$$I(\omega) = \sum_{j=1}^{s} 2c_j A_j^2 \left[\frac{D_j \kappa^2/\pi}{\omega^2 + (\kappa^2 D_j)^2} \right] \qquad (22\text{–}141)$$

where c_j is the number concentration of the jth scattering species. This is a superposition of Lorentzian curves, all centered at $\omega = 0$, and, in general, it is impossible to extract the various diffusion constants D_j from the experimental data.

If the s species have different electrophoretic mobilities, however, the s Lorentzians will be centered at different frequencies and

$$I(\omega) = \sum_{j=1}^{s} 2c_j A_j^2 \left[\frac{D_j \kappa^2/\pi}{(\omega - \kappa u_j E \cos \frac{1}{2}\theta)^2 + (\kappa^2 D_j)^2} \right] \qquad (22\text{–}142)$$

Thus if the frequency shifts are suitably different and larger than the widths $\kappa^2 D_j$, the separate Lorentzians can be resolved. Ware and Flygare† have studied solutions of bovine serum albumin (BSA) and BSA-dimers and BSA-BSA dimers-fibrinogen solutions.

Equation (22–140) is valid only if the various species are totally noninteracting and,

* *Biophys. J.*, **9**, p. 518, 1969.
† *J. Coll. Interface Sci.*, **39**, p. 670, 1972.

in particular, are not reacting chemically. Let us now consider the case in which the scatterers can interconvert by isomerization or charge transfer, according to

$$A \underset{k_b}{\overset{k_a}{\rightleftharpoons}} B \tag{22–143}$$

Equation (22–133) now becomes two coupled simultaneous equations

$$\frac{\partial G_A}{\partial t} = D_A \nabla^2 G_A + u_A E \frac{\partial G_A}{\partial x} - k_a G_A + k_b G_B$$

$$\frac{\partial G_B}{\partial t} = D_B \nabla^2 G_B + u_B E \frac{\partial G_B}{\partial x} - k_b G_B + k_a G_A \tag{22–144}$$

where a separate $G_s(\mathbf{r}, t)$ has been defined for each species. These two coupled equations can be solved analytically by using Fourier transforms. The general result has been obtained by Haas and Ware* and is an extremely lengthy expression. The numerical results, however, have a straightforward physical interpretation. For very slow kinetics, the spectrum appears as two separate peaks as though there were no reaction. In the other extreme, where the reaction is very fast, the spectrum consists of a single peak centered at an intermediate frequency. For kinetics that occur on approximately the same time scale as either the shift $\kappa u_j E \cos(\theta/2)$ or the width $\kappa^2 D_j$ of the non-reacting spectra, the peaks tend to broaden and move toward each other as the rate constants increase. This is shown in Fig. 22–9, which are computer-simulated spectra calculated from the lengthy equations of Haas and Ware for a hypothetical system with transport coefficients typical of globular proteins. Clearly the magnitude of the rate constants has a dramatic effect on the spectrum, suggesting that accurate determination of kinetics may be possible by this method.

The spectra associated with other types of kinetic schemes have also been calculated, and the reader is referred to the work by Berne and Pecora ("Additional Reading"). Although not many experiments of this sort have yet been carried out, it appears that laser light scattering will become a powerful method of probing reaction kinetics.

So far we have discussed light scattering only from independent macromolecules or scatterers. One can also observe light scattering from liquids, and clearly in this case one cannot assume that the molecules doing the scattering are independent. An important approach to light scattering from such dense systems is to assume that the density fluctuations reponsible for the scattering can be described by linear hydrodynamics. Mountain† whose paper we follow here, was apparently the first to present calculations from a hydrodynamic model.

If the scattering centers are not independent, we cannot ignore cross terms as we did in going from Eq. (22–121) to Eq. (22–122), and so Eq. (22–123) is modified to read (Problem 22–31)

$$S(\mathbf{\kappa}, \omega) = \frac{A^2}{\pi} \int_{-\infty}^{\infty} e^{-i\omega\tau} F(\mathbf{\kappa}, \tau) \, d\tau \tag{22–145}$$

where we have used the notation $S(\mathbf{\kappa}, \omega)$ in place of $I(\omega)$ and

$$F(\mathbf{\kappa}, \tau) = \frac{1}{N} \sum_{i,j} \langle e^{-i\mathbf{\kappa}\cdot\mathbf{r}_i(0)} e^{i\mathbf{\kappa}\cdot\mathbf{r}_j(t)} \rangle \tag{22–146}$$

---

* Unpublished results.
† *Rev. Mod. Phys.*, **38**, p. 205, 1966.

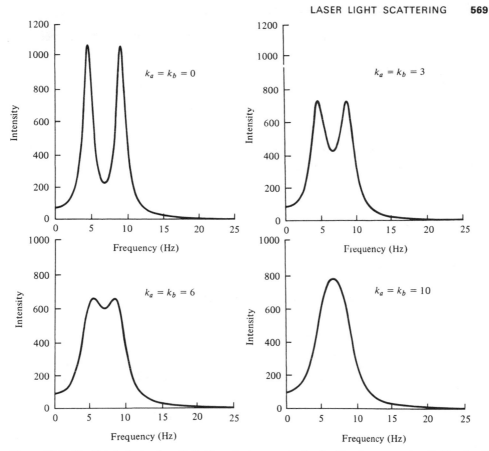

Figure 22-9. **Predicted electrophoretic light-scattering spectra for the kinetic system** $A \rightleftarrows B$. **Simulated experimental conditions are** $D_A = D_B = 6 \times 10^{-7}$ cm²/sec, $\mu_A = 1 \times 10^{-4}$ cm²/sec · V, $\mu_B - 2 \times 10^{-4}$ cm²/sec · V, $E = 100$ V/cm, $\theta = 1°$, $\lambda_0 = 5.145 \times 10^{-5}$ cm. (From D. D. Haas and B. R. Ware, private communication; see B. R. Ware, in " Additional Reading.")

as in Eq. (22–9). Since the phase space operator that corresponds to the density is

$$\rho(\mathbf{r}, t) = \sum_{j=1}^{N} \delta[\mathbf{r} - \mathbf{r}_j(t)] \qquad (22\text{–}147)$$

we see that the operators in the ensemble average in Eq. (22–146) are the Fourier transforms of $\rho(\mathbf{r}, t)$, namely,

$$\hat{\rho}(\mathbf{\kappa}, 0) = \int d\mathbf{r}\ e^{-i\mathbf{\kappa} \cdot \mathbf{r}} \rho(\mathbf{r}, 0)$$

$$\hat{\rho}^*(\mathbf{\kappa}, t) = \int d\mathbf{r}\ e^{i\mathbf{\kappa} \cdot \mathbf{r}} \rho(\mathbf{r}, t) \qquad (22\text{–}148)$$

Thus Eq. (22–146) can be written in the form

$$F(\mathbf{\kappa}, t) = \frac{1}{N} \langle \hat{\rho}^*(\mathbf{\kappa}, 0)\hat{\rho}(\mathbf{\kappa}, t) \rangle$$

$$= \frac{1}{N} \langle \hat{\rho}_{-\mathbf{\kappa}}(0)\hat{\rho}_{\mathbf{\kappa}}(t) \rangle \qquad (22\text{–}149)$$

which shows that $F(\kappa, t)$ and hence the scattered spectrum $S(\kappa, \omega)$ can be determined from the correlation function of $\hat{\rho}(\kappa, t)$. Mountain determines $\hat{\rho}(\kappa, t)$ from the linearized hydrodynamics equations in the following manner.

According to Eq. (21–286), the linearized Navier-Stokes equation is

$$m\rho \frac{\partial \mathbf{u}}{\partial t} = -\nabla p + \eta_s \nabla^2 \mathbf{u} + (\tfrac{1}{3}\eta_s + \eta_v) \nabla(\nabla \cdot \mathbf{u}) \tag{22–150}$$

where we write η_s and η_v for the shear and bulk viscosities, respectively. The mass continuity equation

$$\frac{\partial \rho}{\partial t} + \operatorname{div} \rho\mathbf{u} = 0 \tag{22–151}$$

can be linearized by writing $\rho = \rho_0 + \rho_1$, where ρ_0 is the equilibrium density:

$$\frac{\partial \rho_1}{\partial t} + \rho_0 \operatorname{div} \mathbf{u} = 0 \tag{22–152}$$

The linearized energy balance can be written as [cf. Eq. (17–30)]

$$\rho_m \frac{\partial \tilde{E}}{\partial t} = -p \operatorname{div} \mathbf{u} + \lambda \nabla^2 T_1 \tag{22–153}$$

We can use Eq. (22–152) to eliminate div $\mathbf{u}$ in Eqs. (22–150) and (22–153) and use straightforward thermodynamic manipulations to get (Problem 22–32),

$$m\rho_0 \frac{\partial \mathbf{u}}{\partial t} + \frac{c_0^2}{\gamma} \operatorname{grad} \rho_1 + \frac{c_0^2 \beta m \rho_0}{\gamma} \operatorname{grad} T_1 - (\tfrac{4}{3}\eta_s + \eta_v) \operatorname{grad} \operatorname{div} \mathbf{u} = 0 \tag{22–154}$$

and

$$m\rho_0 c_v \frac{\partial T_1}{\partial t} - \frac{c_v(\gamma - 1)}{\beta} \frac{\partial \rho_1}{\partial t} - \lambda \nabla^2 T_1 = 0 \tag{22–155}$$

where c_v is the specific heat, $\gamma = c_p/c_v$, β is the thermal expansion coefficient, c_0 is the zero-frequency speed of sound, and η_s and η_v are the shear and bulk viscosity, respectively.

By taking the divergence of Eq. (22–154) to eliminate $\mathbf{u}$ through Eq. (22–152), we get two simultaneous partial differential equations in the two independent thermodynamic variables $\rho_1(\mathbf{r}, t)$ and $T_1(\mathbf{r}, t)$. By taking the spatial Fourier transform and a Laplace transform with respect to time, these may be transformed into two simultaneous algebraic equations for

$$\hat{\rho}(\kappa, s) = \int d\mathbf{r} \int_0^\infty dt \, e^{-i\kappa \cdot \mathbf{r}} e^{-st} \rho_1(\mathbf{r}, t) \tag{22–156}$$

$$\hat{T}(\kappa, s) = \int d\mathbf{r} \int_0^\infty dt \, e^{-i\kappa \cdot \mathbf{r}} e^{-st} T_1(\mathbf{r}, t) \tag{22–157}$$

We can then solve these equations for $\hat{\rho}(\kappa, t)$ in terms of $\hat{\rho}(\kappa, 0)$, multiply by $\hat{\rho}(-\kappa, 0)$ and average over an equilibrium ensemble of initial conditions. The algebra involved is quite lengthy, but one eventually gets (see Mountain, in " Additional Reading," for details)

$$\langle \rho_{-\kappa}(0)\rho_\kappa(t) \rangle = \langle |\rho_\kappa(0)|^2 \rangle \Psi(\kappa, t) \tag{22–158}$$

where

$$\Psi(\kappa, t) = \left(\frac{\gamma - 1}{\gamma}\right)\exp\left(-\frac{\kappa^2 \lambda t}{\rho_0 c_p}\right) + \frac{1}{\gamma}\exp(-\kappa^2 \Gamma t)[\cos c_0 \kappa t + b(\kappa)\sin c_0 \kappa t]$$

(22–159)

where Γ is the acoustic attenuation coefficient

$$\Gamma = \frac{\frac{1}{2}(\frac{4}{3}\eta_s + \eta_v)}{m\rho_0} + \left(\frac{\lambda}{\rho_0 c_v}\right)(\gamma - 1)$$

(22–160)

and

$$b(\kappa) = \left[\Gamma + \left(\frac{\lambda}{\rho_0 c_p}\right)(\gamma - 1)\right]\frac{\kappa}{c_0}$$

(22–161)

Equation (22–145) shows that $\langle |\rho_\kappa(0)|^2 \rangle = 2A^2 S(\kappa, 0)$, and so we can finally write (Problem 22–33)

$$\frac{S(\kappa, \omega)}{S(\kappa, 0)} = (1 - \gamma)\frac{\kappa^2 \lambda / \rho_0 c_p}{\omega^2 + (\kappa^2 \lambda / \rho_0 c_p)^2}$$

$$+ \frac{1}{\gamma}\left[\frac{\kappa^2 \Gamma}{(\omega + c_0 \kappa)^2 + (\kappa^2 \Gamma)^2} + \frac{\kappa^2 \Gamma}{(\omega - c_0 \kappa)^2 + (\kappa^2 \Gamma)^2}\right]$$

$$+ \frac{b(\kappa)}{\gamma}\left[\frac{c_0 \kappa + \omega}{(\omega + c_0 \kappa)^2 + (\kappa^2 \Gamma)^2} + \frac{c_0 \kappa - \omega}{(\omega - c_0 \kappa)^2 + (\kappa^2 \Gamma)^2}\right]$$

(22–162)

In practice, this last term is difficult to observe.

We see that the spectrum consists of a central (Rayleigh) component and two shifted (Brillouin) components. This is shown schematically in Fig. 22–10. The width of the Rayleigh peak is $\kappa^2 \lambda / \rho_0 c_p$ (thermal diffusion), and the width of the Brillouin peak is $\kappa^2 \Gamma$. The ratio of the intensity of the Rayleigh component to the intensity of the Brillouin component is $\gamma - 1$ (Problem 22–34). This is known as the Landau-Placzek ratio. Fleury and Boon* have observed the Brillouin scattering from liquid

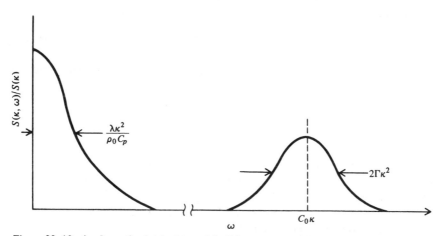

Figure 22–10. **A schematic plot (not to scale) of the spectrum of light scattered from a fluid.** (From R. D. Mountain, *Chem. Rubber Co. Crit. Rev. Solid State*, 1, p. 5, 1970.)

* *Phys. Rev.*, **186**, p. 244, 1969.

argon. In their experiment, $\kappa \approx 2.1 \times 10^7$ m^{-1}, and the observed Brillouin shift was $\approx 1.8 \times 10^{10}$ sec^{-1}, leading to a value of $(1.8 \times 10^{10})/(2.1 \times 10^7) = 857$ m/sec for the speed of sound in liquid argon under the conditions of the experiment, 85°K and 592.5 mmHg. The low-frequency speed of sound measured acoustically is 853 m/sec. Mountain ("Additional Reading") also gives an excellent discussion of light scattering from polyatomic fluids. The work by Fleury and Boon ("Additional Reading") provides a thorough and recent review of light scattering from fluids.

22–4   THE MEMORY FUNCTION

The direct evaluation of a time-correlation function is a prohibitively difficult task for an N-body system. Consequently it is not surprising that there have been many attempts to derive differential or integral equations whose solution would yield a correlation function. No simple or convenient equation (or set of equations) has yet been found, but one such attempt, due originally to Zwanzig ("Additional Reading"), has led to a function, associated with a correlation function, which turns out to be very useful for both conceptual and computational purposes. This function, called the *memory function*, is the central topic of this section. We shall follow a derivation due to Berne, Boon, and Rice ("Additional Reading") although the two works by Berne as well as Berne and Harp ("Additional Reading") are also highly recommended.

Let $\psi(t)$ be a normalized autocorrelation function of the phase space function $U(p, q)$ which for convenience we assume to be real; i.e., let

$$\psi(t) = \langle U(0)U(t) \rangle \tag{22–163}$$

with

$$\psi(0) = \langle U(0)U(0) \rangle = 1 \tag{22–164}$$

Clearly U can be chosen in this way by defining $U(t) = \alpha(t)/\langle \alpha^2 \rangle^{1/2}$, where $\alpha(t) = \alpha(p, q; t)$ is any suitable space function. For convenience, we also assume that

$$\langle U \rangle = 0 \tag{22–165}$$

If the angular brackets denote a canonical ensemble average, then we can write $\psi(t)$ more explicitly as

$$\psi(t) = \frac{1}{Q} \iint dp\, dq\, U(p, q)\, [\exp(it\mathscr{L})U(p, q)]\exp[-\beta H(p, q)] \tag{22–166}$$

where as usual (p, q) represents all the momenta and coordinates necessary to describe the N-body system, $\mathscr{L}$ is the Liouville operator, and Q is simply the canonical partition function. We have used Eq. (21–227) to write $U(t)$ as $\exp(it\mathscr{L})U(0)$. These brackets about $\exp(it\mathscr{L})U(p, q)$ are really unnecessary since $H(p, q)$ is a constant of the motion.

If we differentiate $\psi(t)$ twice with respect to t, we get

$$\frac{d^2\psi}{dt^2} = Q^{-1} \int dp\, dq\, U(p, q)[i\mathscr{L}\exp(it\mathscr{L})i\mathscr{L}U(p, q)]\exp(-\beta H) \tag{22–167}$$

We can now integrate the right-hand side of this equation by parts or simply use the Hermitian property of $\mathscr{L}$ to get (Problem 22–36)

$$\frac{d^2\psi}{dt^2} = -Q^{-1} \int dp \, dq \, [i\mathscr{L}U]\exp(it\mathscr{L})[i\mathscr{L}U]\exp(-\beta H)$$

$$= -\langle \Phi(p, q)\exp(it\mathscr{L})\Phi(p, q) \rangle \tag{22–168}$$

where

$$\Phi(p, q) = i\mathscr{L}U = \dot{U}$$

since the time evolution of U satisfies

$$\frac{dU}{dt} = i\mathscr{L}U \tag{22–169}$$

Since $\exp(it\mathscr{L})\dot{U}(p, q) = \dot{U}(p, q; t)$, Eq. (22–168) can be written as

$$\frac{d^2\psi}{dt^2} = -\langle \dot{U}(0)\dot{U}(t) \rangle \equiv -\phi(t) \tag{22–170}$$

where the last equality defines $\phi(t)$. Note that the initial conditions for this equation are

$$\psi(0) = 1$$
$$\dot{\psi}(0) = \langle U(0)\dot{U}(0) \rangle = 0 \tag{22–171}$$

The Laplace transform of Eq. (22–170) gives

$$s^2\hat{\psi}(s) - s = -\hat{\phi}(s) \tag{22–172}$$

where s is the Laplace transform variable. If we substitute this into the identity, for $s \neq 0$,

$$[s^2\hat{\psi}(s) - s]\hat{\psi}(s) = \left\{ 1 + \left(\frac{1}{s}\right) [s^2\hat{\psi}(s) - s] \right\} [s\hat{\psi}(s) - 1] \tag{22–173}$$

we find

$$-\hat{\phi}(s)\hat{\psi}(s) = \left[1 - \left(\frac{1}{s}\right)\hat{\phi}(s) \right][s\hat{\psi}(s) - 1] \tag{22–174}$$

For values of s such that $1 - (1/s)\hat{\phi}(s) \neq 0$, this can be written as

$$s\hat{\psi}(s) - 1 = -\left[1 - \left(\frac{1}{s}\right)\hat{\phi}(s) \right]^{-1} \hat{\phi}(s)\hat{\psi}(s)$$

which is readily inverted by using the convolution theorem of Laplace transforms (Problem 22–37) to give

$$\frac{d\psi}{dt} = -\int_0^t d\tau \, K(\tau)\psi(t - \tau) \tag{22–175}$$

where

$$\hat{K}(s) = \left[1 - \left(\frac{1}{s}\right)\hat{\phi}(s) \right]^{-1} \hat{\phi}(s) \tag{22–176}$$

Notice that this is an integrodifferential equation for $\psi(t)$, Unfortunately the kernel $K(\tau)$ is not a known function. Nevertheless, we can manipulate $K(\tau)$ into an interesting form which will, in fact, turn out to be the memory function associated with $\psi(t)$. Since by Eq. (22–170)

$$\phi(t) = \langle \dot{U}(0)\dot{U}(t) \rangle$$
$$= \langle \dot{U}(0)\exp(it\mathscr{L})\dot{U}(0) \rangle$$

its Laplace transform $\hat{\phi}(s)$ can be written in terms of the so-called resolvent operator $(s - i\mathscr{L})^{-1}$:

$$\hat{\phi}(s) = \langle \dot{U}(s - i\mathscr{L})^{-1}\dot{U} \rangle \tag{22–177}$$

At this point we define a projection operator $\mathscr{P}$, first introduced into statistical mechanics by Zwanzig,* such that if $G(p, q)$ is a suitably well-behaved phase space function, then

$$\mathscr{P}G(p, q) = U(p, q)f_{eq} \int dp'\, dq'\, U(p', q')G(p', q') \tag{22–178}$$

where f_{eq} is simply

$$f_{eq} = Q^{-1} \exp\left(-\beta H\right)$$

The operator "projects" the phase space function $G(p, q)$ onto $U(p, q)$. It is easy to prove that $\mathscr{P}^2 G = \mathscr{P}G$, which is one of the requirements of a projection operator (Problem 22–38).

By using the operator identity (Problem 22–39)

$$\mathscr{A}^{-1} = \mathscr{B}^{-1} + \mathscr{A}^{-1}(\mathscr{B} - \mathscr{A})\mathscr{B}^{-1} \tag{22–179}$$

with

$$\mathscr{A} = s - i\mathscr{L}$$

and

$$\mathscr{B} = s - i(1 - \mathscr{P})\mathscr{L}$$

we can rewrite Eq. (22–177) for $\hat{\phi}(s)$ in the form

$$\hat{\phi}(s) = \langle \dot{U}[s - i(1 - \mathscr{P})\mathscr{L}]^{-1}\dot{U} \rangle + \langle \dot{U}[s - i\mathscr{L}]^{-1}i\mathscr{P}\mathscr{L}[s - i(1 - \mathscr{P})\mathscr{L}]^{-1}\dot{U} \rangle \tag{22–180}$$

Now we can use the definition of $\mathscr{P}$ [Eq. (22–178)] to rewrite the second term in this equation. First write this second term as $\int dp\, dq\, \dot{U}[s - i\mathscr{L}]^{-1}i\mathscr{P}G$, where

$$G = \mathscr{L}[s - i(1 - \mathscr{P})\mathscr{L}]^{-1}\dot{U}f_{eq}$$

An application of Eq. (22–178) gives then

$$\mathscr{P}G = Uf_{eq} \int dp\, dq\, U\mathscr{L}[s - i(1 - \mathscr{P})\mathscr{L}]^{-1}\dot{U}f_{eq}$$
$$= Uf_{eq}\langle U\mathscr{L}[s - i(1 - \mathscr{P})\mathscr{L}]^{-1}\dot{U} \rangle$$
$$= iUf_{eq}\langle \dot{U}[s - i(1 - \mathscr{P})\mathscr{L}]^{-1}\dot{U} \rangle \tag{22–181}$$

---

*J. Chem. Phys., **33**, p. 1338, 1960.

where to derive the last step we have used the Hermitian property of $\mathscr{L}$; i.e., we have used the fact that [cf. Eq. (21–230) and Problem 22–40]

$$\langle \alpha^* \mathscr{L} \beta \rangle = \langle (\mathscr{L}\alpha)^* \beta \rangle = i \langle \dot{\alpha} \beta \rangle \qquad (22\text{–}182)$$

We might point out at this point that these manipulations are very similar to those involving the Dirac notation (Appendix G), and, in fact, the articles by Berne and Harp (1970) and Berne (1971) given in "Additional Reading" exploit this analogy to develop an elegant notation for the time-correlation function and memory function formalism.

Using Eq. (22–181), then, in Eq. (22–180), we get

$$\hat{\phi}(s) = \langle \dot{U}[s - i(1 - \mathscr{P})\mathscr{L}]^{-1}\dot{U} \rangle - \langle \dot{U}(s - i\mathscr{L})^{-1}U \rangle \langle \dot{U}[s - i(1 - \mathscr{P})\mathscr{L}]^{-1}\dot{U} \rangle$$

$$(22\text{–}183)$$

Again if we use the operator identity given by Eq. (22–179) with $\mathscr{A} = s - i\mathscr{L}$ and $\mathscr{B} = s$, we see that (Problem 22–22)

$$\langle \dot{U}(s - i\mathscr{L})^{-1}U \rangle = \frac{1}{s}\langle U\dot{U} \rangle + \frac{1}{s}\langle \dot{U}(s - i\mathscr{L})^{-1}i\mathscr{L}U \rangle$$

$$= \frac{1}{s}\langle \dot{U}(s - i\mathscr{L})^{-1}\dot{U} \rangle \qquad (22\text{–}184)$$

But note that this is simply $\hat{\phi}(s)/s$ according to Eq. (22–177). Thus Eq. (22–183) reads

$$\hat{\phi}(s) = \langle \dot{U}[s - i(1 - \mathscr{P})\mathscr{L}]^{-1}\dot{U} \rangle - \frac{1}{s}\hat{\phi}(s)\langle \dot{U}[s - i(1 - \mathscr{P})\mathscr{L}]^{-1}\dot{U} \rangle$$

and if this is substituted into Eq. (22–176) for $\hat{K}(s)$, we get that

$$\hat{K}(s) = \langle \dot{U}[s - i(1 - \mathscr{P})\mathscr{L}]^{-1}\dot{U} \rangle \qquad (22\text{–}185)$$

which by inverse transformation becomes (Problem 22–41)

$$K(t) = \langle \dot{U} \exp[it(1 - \mathscr{P})\mathscr{L}]\dot{U} \rangle \qquad (22\text{–}186)$$

This is the desired equation for the memory function $K(t)$. Note that the operator in this equation contains the projection operator.

We can summarize this algebraically long derivation quite simply by writing that if $\psi(t)$ is a normalized correlation function of U and $\langle U \rangle = 0$, then $\psi(t)$ satisfies the integrodifferential equation

$$\frac{d\psi}{dt} = -\int_0^t K(\tau)\psi(t - \tau)\,d\tau \qquad (22\text{–}187)$$

where the kernel of this equation is given by

$$K(\tau) = \langle \dot{U} \exp[i\tau(1 - \mathscr{P})\mathscr{L}]\dot{U} \rangle \qquad (22\text{–}188)$$

and is called the memory function.

We shall see shortly that the memory function can be very useful for computational purposes, but before doing so, we shall discuss some of its general properties. For example, using the fact that $(1 - \mathscr{P})\mathscr{L}$ is Hermitian (Problem 22–42), i.e.,

$$\langle g(1 - \mathscr{P})\mathscr{L}f \rangle = \langle [(1 - \mathscr{P})\mathscr{L}g]^*f \rangle$$

for any two suitably well-behaved phase space functions f and g, it follows that

$$\langle g \exp[it(1 - \mathscr{P})\mathscr{L}]f \rangle = \langle f \exp[-it(1 - \mathscr{P})\mathscr{L}]g^* \rangle \tag{22–189}$$

and that a memory function is stationary. Furthermore, $K(t)$ is an even, real function of time. To prove this, start with

$$K(t) = \langle \dot{U} \exp[it(1 - \mathscr{P})\mathscr{L}]\dot{U} \rangle$$

$$= \sum_{n=0}^{\infty} \frac{t^n}{n!} \langle (i\mathscr{L}U)[i(1 - \mathscr{P})\mathscr{L}]^n (i\mathscr{L}U) \rangle \tag{22–190}$$

Consider the term arising from $n = 1$. This is

$$\langle (i\mathscr{L}U)[i(1 - \mathscr{P})\mathscr{L}](i\mathscr{L}U) \rangle = \langle (i\mathscr{L}U)[i\mathscr{L}(i\mathscr{L}U)] \rangle - \langle (i\mathscr{L}U)[i\mathscr{P}\mathscr{L}(i\mathscr{L}U)] \rangle$$

Clearly

$$\langle (i\mathscr{L}U)[i\mathscr{L}(i\mathscr{L}U)] = \langle \dot{U}\ddot{U} \rangle = 0$$

Also

$$\langle (i\mathscr{L}U)[i\mathscr{P}\mathscr{L}(i\mathscr{L}U)] \rangle = \langle (i\mathscr{L}U)U \rangle \langle U[i\mathscr{L}(i\mathscr{L}U)] \rangle$$
$$= 0$$

since $\langle (i\mathscr{L}U)U \rangle = \langle \dot{U}U \rangle = 0$. By iteration, one can show that all terms arising from odd values of n vanish, and therefore $K(t)$ is a real even function of time (Problem 22–43). One can also use the Schwartz inequality to prove that (cf. Problem 21–55)

$$-\langle \dot{U}\dot{U} \rangle \leq K(t) \leq \langle \dot{U}\dot{U} \rangle \tag{22–191}$$

One of the most important properties of the memory function, however, is that it seems to lend itself very readily to simple approximations. For example, Eq. (22–187) shows that if we assume that $K(\tau)$ decays extremely rapidly, e.g., $K(\tau) \approx \delta(\tau)$, then the correlation function decays exponentially. Although this form of a correlation function is not rigorously correct, we have seen in these last two chapters that such an approximate form leads to classic physical theories such as the Debye theory of dielectric relaxation and the Enskog hard-sphere transport theory. Since such a simple form for $K(\tau)$ leads to useful expressions for the correlation function $\psi(\tau)$, one might hope that more detailed assumptions for $K(\tau)$ would lead to an even better $\psi(t)$. Such an approach has been used by a number of authors, but here we shall continue to follow Berne, Boon, and Rice,[*] who were one of the first to explore this idea. A more recent paper by Harp and Berne[†] gives a good summary of the various approaches.

We shall consider the memory function associated with the normalized velocity autocorrelation function

$$\psi(t) = \frac{\langle \mathbf{v}_1(0) \cdot \mathbf{v}_1(t) \rangle}{\langle \mathbf{v}_1{}^2 \rangle} \tag{22–192}$$

Since this correlation function involves a vector dot product, we must generalize some of the above concepts. For example, the projection operator becomes

$$\mathscr{P}\mathbf{G}(p, q) = \mathbf{U}(p, q)f_{eq} \int dp \, dq \, \mathbf{U}(p, q) \cdot \mathbf{G}(p, q) \tag{22–193}$$

---

[*] *J. Chem. Phys.*, **45**, p. 1086, 1966.
[†] *Phys. Rev.*, **A2**, p. 975, 1970.

It is straightforward to show that since

$$i\mathscr{L}\mathbf{v}_1 = \frac{\mathbf{F}_1}{m}$$

then the kernel function $\hat{K}(s)$ becomes (Problem 22–44)

$$\hat{K}(s) = \left[1 - s^{-1}\frac{\langle\mathbf{F}_1\cdot\hat{\mathbf{F}}_1(s)\rangle}{m^2\langle v_1{}^2\rangle}\right]^{-1}\frac{\langle\mathbf{F}_1\cdot\hat{\mathbf{F}}_1(s)\rangle}{m^2\langle v_1{}^2\rangle} \tag{22–194}$$

where $\mathbf{F}_1$ is the force on molecule 1 and $\hat{\mathbf{F}}_1(s)$ is the Laplace transform of $\mathbf{F}_1(t)$. The value of $\hat{K}(0)$ is easily determined from the Einstein relation for the diffusion coefficient. For

$$D = \frac{kT}{m}\int_0^\infty dt\,\psi(t)$$

$$= \frac{kT}{m}\lim_{s\to 0}\int_0^\infty dt\,e^{-st}\,\psi(t)$$

$$= \frac{kT}{m}\hat{\psi}(0) \tag{22–195}$$

But according to Eqs. (22–174) and (22–176),

$$s\hat{\psi}(s) = 1 - \hat{K}(s)\hat{\psi}(s)$$

and so we see that

$$D = \left(\frac{kT}{m}\right)[\hat{K}(0)]^{-1} \tag{22–196}$$

Thus $\hat{K}(0)$ is directly related to the diffusion coefficient.

Berne *et al.* argue on several grounds that a good approximation for $K(t)$ is an exponentially decaying function

$$K(\alpha, \gamma, t) = \gamma\exp(-\alpha|t|) \tag{22–197}$$

where α and γ are adjustable constants. One can readily show that this form for the memory function leads to a correlation function of the form (Problem 22–45)

$$\psi(t) = \frac{1}{s_+ - s_-}(s_+ e^{s_- t} - s_- e^{s_+ t}) \tag{22–198}$$

where

$$s_\pm = -\frac{\alpha}{2}\left\{1 \mp \left[1 - \left(\frac{4\gamma}{\alpha^2}\right)\right]^{1/2}\right\} \tag{22–199}$$

This approximate $\psi(t)$ is tested against the molecular dynamics data of Rahman.* The two parameters α and γ are determined from Rahman's "data." From Eq. (22–196), we see that

$$\frac{\gamma}{\alpha} = \frac{kT}{mD}$$

---

* *Phys. Rev.*, **136**, p. A405, 1964.

In addition, since (Problem 22–56)

$$K(0) = \frac{\langle F_1{}^2 \rangle}{m^2 \langle v_1{}^2 \rangle}$$

we find that

$$\gamma = \frac{\langle F_1{}^2 \rangle}{m^2 \langle v_1{}^2 \rangle} = \frac{\langle F_1{}^2 \rangle}{3mkT} = \frac{\langle \nabla_1{}^2 U_N \rangle}{3m}$$

where U_N is the total intermolecular potential and $\nabla_1{}^2$ denotes the Laplacian with respect to particle 1. Using Rahman's data, Berne *et al.* find that $\langle \nabla_1{}^2 U_N \rangle = 11.0 \times 10^3 \text{ erg} \cdot \text{cm}^{-2}$. Using Rahman's value of $D = 2.43 \times 10^{-5} \text{ cm}^2/\text{sec}$, it is found that $\alpha = 8.06 \times 10^{12} \text{ sec}^{-1}$ and $\gamma = 6.5 \times 10^{25} \text{ sec}^{-2}$. With these values of α and γ, the $s_{\pm}$ are complex and

$$\psi(t) = \exp\left(-\frac{4.03t}{\tau_0}\right)\left[\cos\left(\frac{4.03\sqrt{3}\,t}{\tau_0}\right) + 3^{-1/2}\sin\left(\frac{4.03\sqrt{3}\,t}{\tau_0}\right)\right] \tag{22-200}$$

where $\tau_0 = 10^{-12}$ sec. This autocorrelation function is plotted in Fig. 22–11 along with Rahman's molecular dynamics value and a simple exponential approximation for $\psi(t)$. One sees that the qualitative features of the autocorrelation function are reproduced. Although the agreement is not perfect, it is important to note that the theoretical function correctly predicts a negative region for $\psi(t)$, despite the very simple form of the trial memory function.

Berne *et al.* also calculate a normalized power spectrum, defined by

$$G(\omega) = \frac{kT}{mD} \int_0^\infty dt\, \psi(t) \cos \omega t \tag{22-201}$$

This is essentially the function $f(\omega)$ defined in Eq. (22–48), but here

$$G(0) = 1$$

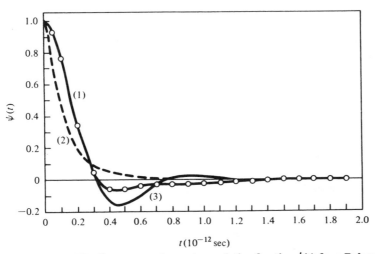

Figure 22–11. **The linear momentum autocorrelation function** $\psi(t)$ **from Rahman's molecular dynamics calculations (1), from the Markovian approximation** $\psi(t) = \exp(-\beta|t|)$ **(2), and Eq. (22–200) (3). (From B. J. Berne, J. P. Boon, and S. A. Rice,** *J. Chem. Phys.,* **45, p. 1086, 1966.)**

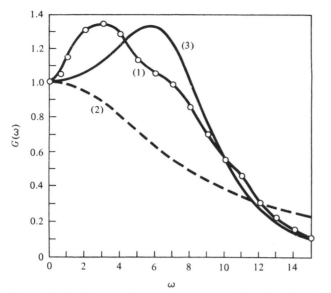

Figure 22–12. **The normalized power spectrum** $G(\omega)$ **from Rahman's molecular dynamics calculations (1), from the Markovian approximation (2), and from Eq. (22–202) (3).** (From B. J. Berne, J. P. Boon, and S. A. Rice, *J. Chem. Phys.*, **45**, p. 1086, 1966.)

It is easy to show that Eq. (22–197) leads to (Problem 22–46)

$$G(\omega) = \frac{\gamma^2}{\gamma^2 + \alpha^2(1 - 2\gamma/\alpha^2)\omega^2 + \omega^4}$$
$$= \frac{0.420}{0.420 - 6.5 \times 10^{-27}\omega^2 + 10^{-52}\omega^4} \qquad (22\text{--}202)$$

This theoretical power spectrum is plotted in Fig. 22–12 along with the Markovian approximation (which leads to a Lorentzian power spectrum) and the molecular dynamics data of Rahman. Again the agreement is seen to be fairly good.

It appears then that the memory function offers a convenient computational approach to linear response theory calculations.

22–5 DERIVATION OF THERMAL TRANSPORT COEFFICIENTS

In Section 21–8 we derived time-correlation function expressions for the thermal transport coefficients by starting with the appropriate macroscopic transport equation. Since these derivations are based on the classical macroscopic transport equations, the resulting expressions are in some sense restricted to long-time, small wavelengths. In this section we shall rederive these results within a memory function formalism and investigate the long-time, small wavelength limit in some detail. We shall follow the 1971 review article of Berne ("Additional Reading").

We start with the diffusion coefficient. In Section 21–8 we showed that [Eq. (21–264)]

$$F_s(\mathbf{k}, t) = \exp(-k^2 Dt) \qquad (22\text{--}203)$$

where this results by taking the Fourier transform of the macroscopic diffusion equation. We also saw in Section 21–8 [Eq. (21–271)] that

$$F_s(\mathbf{k}, t) = \langle e^{-i\mathbf{k}\cdot\mathbf{r}(0)}e^{i\mathbf{k}\cdot\mathbf{r}(t)}\rangle \qquad (22\text{--}204)$$

This correlation function obeys the memory function equation

$$\frac{\partial F_s(\mathbf{k}, t)}{\partial t} = -\int_0^t d\tau \, K(\mathbf{k}, \tau) F_s(\mathbf{k}, t - \tau) \tag{22–205}$$

where

$$K(\mathbf{k}, t) = \langle i\mathscr{L}U \exp\left[it(1 - \mathscr{P})\mathscr{L}\right] i\mathscr{L}U \rangle \tag{22–206}$$

with

$$U = \exp(i\mathbf{k} \cdot \mathbf{r}) \tag{22–207}$$

Since $i\mathscr{L}U = \dot{U} = i\mathbf{k} \cdot \mathbf{v} \exp[i\mathbf{k} \cdot \mathbf{r}]$, it follows that

$$K(\mathbf{k}, t) = \mathbf{k} \cdot \langle \mathbf{v} \exp(-i\mathbf{k} \cdot \mathbf{r}) \exp[it(1 - \mathscr{P})\mathscr{L}] \mathbf{v} \exp(i\mathbf{k} \cdot \mathbf{r}) \rangle \cdot \mathbf{k}$$
$$= \mathbf{k} \cdot \mathsf{D}_k(t) \cdot \mathbf{k} \tag{22–208}$$

where the tensor $\mathsf{D}_k(t)$ is defined by

$$\mathsf{D}_k(t) = \langle \mathbf{v} \exp(-i\mathbf{k} \cdot \mathbf{r}) \exp[it(1 - \mathscr{P})\mathscr{L}] \mathbf{v} \exp(i\mathbf{k} \cdot \mathbf{r}) \rangle \tag{22–209}$$

Equation (22–205) becomes

$$\frac{\partial F_s}{\partial t} = -\int_0^t d\tau \, [\mathbf{k} \cdot \mathsf{D}_k(\tau) \cdot \mathbf{k}] F_s(\mathbf{k}, t - \tau) \tag{22–210}$$

By taking the Laplace transform of this equation, we get

$$\hat{F}_s(\mathbf{k}, s) = [s + \mathbf{k} \cdot \hat{\mathsf{D}}_k(s) \cdot \mathbf{k}]^{-1} \tag{22–211}$$

where the carets denote the Laplace transform of that function. This should be compared to the Laplace transform of $F_s(\mathbf{k}, t)$ given by Eq. (22–203), namely,

$$\hat{F}_s(\mathbf{k}, s) = (s + k^2 D)^{-1} \tag{22–212}$$

We see that $\mathsf{D}_k(s)$ is a complicated k- and s-dependent diffusion coefficient. We wish to see under what conditions, i.e., in what limit, Eq. (22–211) goes over into Eq. (22–212).

Using Eq. (22–203) as a guide, we redefine the time scale by introducing

$$\tau = k^2 t \tag{22–213}$$

We now consider the double limit

$$\lim_{\substack{t \to \infty \\ \tau = \text{const}}} \lim_{k \to 0} F_s(k, t) = \lim_{\substack{t \to \infty \\ \tau = \text{const}}} \lim_{k \to 0} e^{-Dk^2 t} = e^{-D\tau} \tag{22–214}$$

This limit selects out from the exact $F_s(\mathbf{k}, t)$ given in Eq. (22–211) the small k and long-time behavior; i.e., it selects out the hydrodynamic limit, where the wave vectors are small (smoothly spatially varying properties) and the time is long (compared to molecular times). To take this limit of the exact $F_s(\mathbf{k}, t)$, we formally take the inverse of Eq. (22–211),

$$F_s(\mathbf{k}, t) = \frac{1}{2\pi i} \oint ds \, e^{st} \, [s + \mathbf{k} \cdot \hat{\mathsf{D}}_k(s) \cdot \mathbf{k}]^{-1} \tag{22–215}$$

introduce the new time variable $\tau = k^2 t$ and a corresponding new Laplace transform variable $s = k^2 x$ to get

$$F_s(\mathbf{k}, t) = \frac{1}{2\pi i} \oint dx \, e^{x\tau} \left[x + \frac{\mathbf{k} \cdot \hat{\mathsf{D}}_k(k^2 x) \cdot \mathbf{k}}{k^2} \right]^{-1} \tag{22–216}$$

We now take the double limit of this equation

$$\lim_{\substack{t \to \infty \\ \tau = \text{const}}} \lim_{k \to 0} F_s(\mathbf{k}, t) = \frac{1}{2\pi i} \oint dx \, e^{xt} \left[x + \lim_{k \to 0} \mathbf{k} \cdot \hat{D}_k(k^2 x) \cdot \frac{\mathbf{k}}{k^2} \right]^{-1}$$

$$= \exp\left[-\left(\lim_{s \to 0} \lim_{k \to 0} \frac{\mathbf{k} \cdot \hat{D}_k(s) \cdot \mathbf{k}}{k^2} \right) \tau \right] \tag{22-217}$$

By comparing this with Eq. (22–214) we see that

$$D = \lim_{s \to 0} \lim_{k \to 0} \frac{\mathbf{k} \cdot \hat{D}_k(s) \cdot \mathbf{k}}{k^2} \tag{22-218}$$

This is not yet a useful expression for D since $\hat{D}_k(s)$ is expressed in terms of a projection operator in Eq. (22–209). We can obtain a more useful expression for D by starting with Eq. (22–177), which in the present notation becomes

$$\hat{\phi}(\mathbf{k}, s) = \mathbf{k} \cdot \langle \mathbf{v} \exp(-i\mathbf{k} \cdot \mathbf{r})[s - i\mathcal{L}]^{-1} \mathbf{v} \exp(i\mathbf{k} \cdot \mathbf{r}) \rangle \cdot \mathbf{k} \tag{22-219}$$

$$= \frac{\hat{K}(\mathbf{k}, s)}{[1 + (1/s)\hat{K}(\mathbf{k}, s)]}$$

$$= \frac{k^2 \tilde{D}_k(s)}{[1 + (k^2/s)\tilde{D}_k(s)]} \tag{22-220}$$

where we have defined

$$\tilde{D}_k(s) = \frac{\mathbf{k} \cdot \hat{D}_k(s) \cdot \mathbf{k}}{k^2} \tag{22-221}$$

so that

$$\frac{1}{k^2} \hat{\phi}(\mathbf{k}, s) = \frac{\tilde{D}_k(s)}{[1 + (k^2/s)\tilde{D}_k(s)]} \tag{22-222}$$

Now according to Eqs. (22–218) and (22–221)

$$D = \lim_{s \to 0} \lim_{k \to 0} \tilde{D}_k(s)$$

and so

$$\lim_{s \to 0} \lim_{k \to 0} \frac{1}{k^2} \hat{\phi}(\mathbf{k}, s) = D$$

Using Eq. (22–219) for $\hat{\phi}(\mathbf{k}, s)$, we see that

$$D = \lim_{s \to 0} \lim_{k \to 0} \mathbf{k} \cdot \langle \mathbf{v} \exp(-i\mathbf{k} \cdot \mathbf{r})(s - i\mathcal{L})^{-1} \mathbf{v} \exp(i\mathbf{k} \cdot \mathbf{r}) \rangle \cdot \mathbf{k} \tag{22-223}$$

In this limit, the exponentials in the above expression may be replaced by unity, so that the ensemble average becomes $\langle \mathbf{v}(s - i\mathcal{L})^{-1} \mathbf{v} \rangle$ which is the Laplace transform of $\langle \mathbf{v}(0)\mathbf{v}(t) \rangle$. It is straightforward to show that for an isotropic fluid that (Problem 22–51)

$$\langle \mathbf{v}(0)\mathbf{v}(t) \rangle = \tfrac{1}{3} \langle \mathbf{v}(0) \cdot \mathbf{v}(t) \rangle \mathbf{I} \tag{22-224}$$

where $\mathbf{I}$ is the unit tensor. Thus

$$\langle \mathbf{v}(s - i\mathcal{L})^{-1} \mathbf{v} \rangle = \tfrac{1}{3} \langle \mathbf{v} \cdot (s - i\mathcal{L})^{-1} \mathbf{v} \rangle \mathbf{I}$$

We now substitute this into Eq. (22–223) and use the fact that $\mathbf{k} \cdot \mathbf{l} \cdot \mathbf{k}/k^2 = 1$, we finally get

$$D = \lim_{s \to 0} \tfrac{1}{3} \langle \mathbf{v} \cdot (s - i\mathscr{L})^{-1} \mathbf{v} \rangle$$

$$= \lim_{s \to 0} \frac{1}{3} \int_0^\infty dt \, e^{-st} \langle \mathbf{v}(0) \cdot \mathbf{v}(t) \rangle \tag{22–225}$$

which is essentially the result derived in Section 21–8 by starting with the diffusion equation. Equation (22–225), however, was derived as the limit of a general expression. If s is replaced by $i\omega$ in the above equation, we see that the "ordinary" diffusion equation can be thought of as a low-frequency limit. In fact, we can define a frequency-dependent diffusion coefficient through Eqs. (22–223) through (22–225). Let

$$D(\omega) = \lim_{k \to 0} \tilde{D}_k(\omega)$$

where $i\omega = s$. Equation (22–223) becomes

$$D(\omega) = \lim_{k \to 0} \mathbf{k} \cdot \langle \mathbf{v} \exp(-i\mathbf{k} \cdot \mathbf{r})(i\omega - i\mathscr{L})^{-1} \mathbf{v} \exp(i\mathbf{k} \cdot \mathbf{r}) \rangle \cdot \mathbf{k}$$

and Eq. (22–225) becomes

$$D(\omega) = \frac{1}{3} \int_0^\infty dt \, e^{-i\omega t} \langle \mathbf{v}(0) \cdot \mathbf{v}(t) \rangle \tag{22–226}$$

Zwanzig* has shown that $D(\omega)$ and $D(\mathbf{k}, \omega)$ can be defined through a generalization of Fick's first law of diffusion to arbitrary space-time variations of the driving force:

$$\mathbf{j}(\mathbf{k}, \omega) = -D(\mathbf{k}, \omega)(\nabla c)_{k, \omega} \tag{22–227}$$

where $\mathbf{j}(\mathbf{k}, \omega)$ is the Fourier component of the flux of the diffusing species, as a response to the $(\mathbf{k}, \omega)$th Fourier component of the gradient of the concentration of the diffusing species.

In Section 21–8 we also derived a time-correlation function expression for the shear viscosity by starting with the linearized Navier-Stokes equation. Equation (21–307) shows that the shear viscosity is given by

$$\eta = \frac{1}{VkT} \int_0^\infty dt \, \langle J(0)J(t) \rangle \tag{22–228}$$

where

$$J = \sum_{j=1}^N \left(\frac{p_{xj} p_{zj}}{m} + z_j F_{jx} \right) \tag{22–229}$$

We defined a correlation function $C_\perp(\mathbf{k}, t)$ [Eq. (21–292)]

$$C_\perp(\mathbf{k}, t) = \frac{\langle J_\perp^*(\mathbf{k}, 0) J_\perp(\mathbf{k}, t) \rangle}{\langle J_\perp^*(\mathbf{k}, 0) J_\perp(\mathbf{k}, 0) \rangle} = \exp(-k^2 v t) \tag{22–230}$$

where $v = \eta/m\rho$ and

$$J(\mathbf{k}, t) = \sum_{j=1}^N \dot{x}_j(t) e^{ikz_j(t)} \tag{22–231}$$

* *Phys. Rev.*, **133**, p. A50, 1964.

Since $C_\perp(\mathbf{k}, t)$ is a normalized autocorrelation function of the phase space function

$$U = \frac{J_\perp(\mathbf{k}, t)}{\langle J_\perp(\mathbf{k})J_\perp(\mathbf{k})\rangle^{1/2}} \qquad (22\text{-}232)$$

then it satisfies a memory function equation

$$\frac{\partial C_\perp}{\partial t} = -\int_0^t d\tau \, K_\perp(\mathbf{k}, \tau)C_\perp(\mathbf{k}, t - \tau) \qquad (22\text{-}233)$$

where

$$K_\perp(\mathbf{k}, t) = \langle i\mathscr{L}U \exp[it(1 - \mathscr{P})\mathscr{L}]i\mathscr{L}U\rangle \qquad (22\text{-}234)$$

where U is given by Eq. (22–232). The analog of Eq. (22–211) is

$$\hat{C}_\perp(\mathbf{k}, s) = [s + \hat{K}_\perp(\mathbf{k}, s)]^{-1} \qquad (22\text{-}235)$$

One proceeds in much the same way as we did above for the diffusion coefficient, and one can show that (Problem 22–52)

$$v = \lim_{s \to 0} \lim_{k \to 0} \frac{1}{k^2} \hat{K}_\perp(\mathbf{k}, s) \qquad (22\text{-}236)$$

and that (cf. Berne (1971) "Additional Reading"],

$$\eta = \lim_{s \to 0} \frac{1}{VkT} \int_0^\infty dt \, e^{-st}\langle J(0)J(t)\rangle \qquad (22\text{-}237)$$

Again we can define a frequency-dependent shear viscosity by

$$\eta(\omega) = \frac{1}{VkT} \int_0^\infty dt \, e^{-i\omega t}\langle J(0)J(t)\rangle \qquad (22\text{-}238)$$

Zwanzig and Mountain* have presented an interesting application of this formula. When a mechanical force is applied suddenly to a fluid, the fluid responds elastically at first, just as if it were a solid body. The initial response may be described by two quantities, the high-frequency limit G_∞ of the shear modulus (or modulus of rigidity) and the high-frequency limit K_∞ of the bulk modulus (or modulus of compression). Zwanzig and Mountain argue quite generally that

$$G(\omega) = i\omega\eta(\omega) \qquad (22\text{-}239)$$

and

$$K(\omega) = K(0) + i\omega\kappa(\omega) \qquad (22\text{-}240)$$

where $\kappa(\omega)$ is the frequency-dependent bulk viscosity. We shall consider only the shear modulus here. We wish to find the high-frequency limit of $\eta(\omega)$. This can be done by generating a power series in $1/\omega$ in the following way. Let $\omega t = \tau$ in Eq. (22–238) to get

$$\eta(\omega) = \frac{1}{VkT\omega} \int_0^\infty d\tau \, e^{-i\tau} \left\langle J(0)J\left(\frac{\tau}{\omega}\right)\right\rangle \qquad (22\text{-}241)$$

* *J. Chem. Phys.*, **43**, p. 4464, 1965.

Assuming the time-correlation function has an expansion in powers of t with a non-zero radius of convergence, we may write (Problem 22–53)

$$\eta(\omega) = \left(\frac{\eta_1}{i\omega}\right) + \frac{\eta_2}{(i\omega)^2} + \cdots \tag{22–242}$$

where

$$\eta_1 = \frac{1}{VkT} \langle J(0)J(0) \rangle$$

$$\eta_2 = \frac{1}{VkT} \langle J(0)J'(0) \rangle$$

and so on. Thus we see that the high-frequency shear modulus G_∞ is given by

$$G_\infty = \frac{1}{VkT} \langle J(0)J(0) \rangle \tag{22–243}$$

which is just an equilibrium average of J in Eq. (22–229). After some manipulation, one finds that

$$G_\infty = \rho kT + \frac{2\pi}{15} \rho^2 \int_0^\infty dr\, g(r) \frac{d}{dr}\left(r^4 \frac{du}{dr}\right) \tag{22–244}$$

for a pair-wise additive intermolecular potential (Problem 22–54). Zwanzig and Mountain give an excellent discussion of high-frequency elastic moduli of simple fluids.

We have seen in this section that it is possible to extend the formulas for the thermal transport coefficients to include frequency dependence and wave vector dependence. The extension of the hydrodynamic equations themselves is presently a very active and interesting area of the statistical mechanical theory of transport and is called generalized hydrodynamics. Several of the references in the "Additional Reading" should serve as an introduction to this field.

ADDITIONAL READING

General

BERNE, B. J. 1971. In *Physical chemistry, an advanced treatise*, Vol. 8B, ed. by D. Henderson. New York: Academic.
———, and FORSTER, D. *Ann. Rev. Phys. Chem.*, **22**, p. 563, 1971.
———, and HARP, G. D. *Adv. Chem. Phys.*, **17**, p. 63, 1970.
HARP, G., and BERNE, B. J. *Phys. Rev.*, **2A**, p. 975, 1970.
MOUNTAIN, R. D. *Chem. Rubber Co. Crit. Rev. Solid State*, **1**, p. 5, 1970.
ZWANZIG, R. W. 1961. In *Lectures in theoretical physics*. Vol. 3. New York: Interscience.

Neutron Scattering

EGELSTAFF, P. A. 1965. *Thermal neutron scattering*. New York: Academic.
LARSSON, K. E., DAHLBORG, V., and SKÖLD, K. 1968. In *Simple dense fluids*, ed. by H. L. Frisch and Z. W. Salsburg. New York: Academic.
MARSHALL, W., and LOVESEY, S. W. 1971. *Theory of thermal neutron scattering*. Oxford: Oxford University Press.
SKÖLD, K., et al. *Phys. Rev.*, **6A**, p. 1107, 1972.

Stochastic Processes

McQuarrie, D. A. 1975. In *Physical chemistry, an advanced treatise,* Vol. 11B, ed. by D. Henderson. New York: Academic.

Papoulis, A. 1965. *Probability, random variables, and stochastic processes.* New York: McGraw-Hill.

Light Scattering

Berne, B. J. *Accts. Chem. Res.*, **6**, p. 318, 1973.

———, and Pecora, R. 1976. *Introduction to the molecular theory of light scattering.* New York: Wiley.

Cummins, H. Z., and Swinney, H. L. *Prog. Optics*, **8**, p. 133, 1970.

Fleury, P. A., and Boon, J. P. *Adv. Chem. Phys.*, **24**, p. 1, 1973.

Pecora, R. *Ann. Rev. Biophys., and Bioengineering*, **1**, p. 257, 1972.

Ware, B. *Adv. Colloid and Interface Sci.*, **4**, p. 1, 1974.

Yeh, Y., and Keeler, R. N. *Quart. Rev. Biophys.*, **2**, p. 315, 1969.

Memory Functions

Berne, B. J. 1971. In *Physical chemistry, an advanced treatise*, Vol. 8B, ed. by D. Henderson. New York: Academic.

———, Boon, J. P., and Rice, S. A. *J. Chem. Phys.*, **45**, p. 1086, 1966.

———, and Harp, G. D. *Adv. Chem. Phys.*, **17**, p. 63, 1970.

Generalized Hydrodynamics

Ailawadi, N. K., Rahman, A., and Zwanzig, R. W. *Phys. Rev.*, **A4**, p. 1616, 1971.

Kadanoff, L. P., and Martin, P. C. *Ann. Phys.*, **24**, p. 419, 1963.

Oppenheim, I. *Berichte der Bunsengellschaft fur Phys. Chem.*, **75**, p. 385, 1971.

Selwyn, P. A., and Oppenheim, I. *Physics*, **54**, p. 161, 1971.

PROBLEMS

22–1. Derive Eq. (22–7) from Eq. (22–6).

22–2. Derive Eq. (22–12).

22–3. Using the fact that $S(\kappa, \omega)$ is real, prove that $G(-\mathbf{r}, -t) = G^*(\mathbf{r}, t)$.

22–4. Prove that Eq. (22–24) reduces to Eq. (22–16) when $a_J = 1$ and the classical limit is used.

22–5. Derive Eq. (22–25).

22–6. Derive Eq. (22–27).

22–7. Derive Eqs. (22–30) and (22–31).

22–8. Prove that $F_s(\kappa, t)$ can be written in the form

$$F_s(\kappa, t) = \exp[-\kappa^2 \gamma_1(t) + \kappa^4 \gamma_2(t) + \cdots]$$

where

$$\gamma_1(t) = \tfrac{1}{6}\langle (\mathbf{r}_1(t) - \mathbf{r}_1(0))^2 \rangle$$

22–9. Derive the non-Gaussian corrections to $F_s(\kappa, t)$.

22–10. Prove that $\gamma_1(t)$ [Eq. (22–36)] has the expansion

$$\gamma_1(t) = \frac{t^2}{2\beta m} - \langle \ddot{x}_1(0)^2 \rangle \frac{t^4}{4!} + \cdots$$

and that

$$S_{\text{inc}}(\kappa, \omega) = (4\pi\alpha)^{-1/2} e^{-\omega^2/4\alpha} \left\{ 1 + \frac{\beta^2 m^2}{4! \kappa^2} \langle \ddot{x}_1(0)^2 \rangle H_4\left(\frac{\omega}{\sqrt{2}\alpha}\right) + \cdots \right\}$$

where $\alpha = \kappa^2/2\beta m$ and $H_n(x) = (-1)^n e^{x^2/2}(d^n/dx^n)e^{-x^2/2}$ is the nth Hermite polynomial.

22–11. Prove that $G_s(r, t)$ can be written in the form

$$G_s(r, t) = [4\pi\rho(t)]^{-3/2} \exp\left[-\frac{r^2}{4\rho(t)} \right] \times [1 + b_6(t)H_6(\alpha r) + \cdots]$$

where

$$b_6 = \frac{\alpha_2(t)}{24}$$

and

$$\alpha_2 = \frac{\langle r^4 \rangle}{3 \langle r^2 \rangle^2} - 1$$

Rahman* has calculated non-Gaussian corrections to $G_s(r, t)$ by molecular dynamics.

22–12. Define the moments of either $S_{coh}(\varkappa, \omega)$ or $S_{inc}(\varkappa, \omega)$ by

$$\langle \omega^n \rangle = \int_{-\infty}^{\infty} \omega^n S(\varkappa, \omega) \, d\omega$$

Prove that

$$\langle \omega^0 \rangle_{inc} = 1$$

$$\langle \omega^0 \rangle_{coh} = 1 + \int dr \, e^{i\varkappa \cdot r} g(r)$$

$$\langle \omega^2 \rangle_{inc} = kT \frac{\varkappa^2}{m}$$

$$\langle \omega^2 \rangle_{coh} = kT \frac{\varkappa^2}{m}$$

$$\langle \omega^4 \rangle_{inc} = \frac{\varkappa^4}{\beta m^2} \left[\frac{3}{4} + \frac{4\pi}{3\varkappa^2} \int_0^\infty dr \, r^2 g(r) \left(\frac{\partial^2 u}{\partial r^2} + \frac{2}{r} \frac{\partial u}{\partial r} \right) \right]$$

$$\langle \omega^4 \rangle_{coh} = \frac{\varkappa^4}{\beta m} \left[\frac{3}{\beta} + \frac{2\pi}{15} \int_0^\infty dr \, g(r) \left(3 \frac{\partial^2 u}{\partial r^2} + \frac{2}{r} \frac{\partial u}{\partial r} \right) \right] + O(\varkappa^6)$$

22–13. In 1958, Vineyard† introduced an interesting approximate expression for $G_d(r, t)$ in terms of $G_s(r, t)$ and the equilibrium radial distribution function. First define $P(\mathbf{r}', t | \mathbf{r}, t)$ to be the conditional probability that a different particle is at $(\mathbf{r}', t)$ if the first, which had been at $\mathbf{r} = 0$ at $t = 0$, is at $(\mathbf{r}, t)$. Show that $P(\mathbf{r}', t | \mathbf{r}, t)$ satisfies the relations

$$\int d\mathbf{r}' \, P(\mathbf{r}', t | \mathbf{r}, t) = N - 1$$

$$\lim_{t \to 0} P(\mathbf{r}', t | \mathbf{r}, t) = g(\mathbf{r}')$$

$$\lim_{t \to \infty} P(\mathbf{r}', t | \mathbf{r}, t) = g(\mathbf{r}' - \mathbf{r})$$

where g is the equilibrium radial distribution function. Show that

$$\int d\mathbf{r} \, P(\mathbf{r}', t | \mathbf{r}, t) G_s(\mathbf{r}, t) = G_d(\mathbf{r}', t)$$

Unfortunately, $P(\mathbf{r}', t | \mathbf{r}, t)$ is not known. Vineyard approximated it by its long-time value $g(\mathbf{r}' - \mathbf{r})$ to derive the approximation

$$\int d\mathbf{r} \, g(\mathbf{r}' - \mathbf{r}) G_s(\mathbf{r}, t) = G_d^c(\mathbf{r}', t)$$

where the superscript c denotes "convolution" approximation for G_d. Show that Vineyard's convolution approximation can also be written as

$$F_d(\varkappa, t) = G(\varkappa) F_s(\varkappa, t)$$

* *Phys. Rev.*, **136**, p. A405, 1964.
† *Phys. Rev.*, **110**, p. 999, 1958.

where $G(\varkappa)$ is the Fourier transform of $g(\mathbf{r})$. Show that this is equivalent to

$$F(\varkappa, t) = S(\varkappa)F_s(\varkappa, t)$$

where $S(\varkappa) = 1 + G(\varkappa)$ is the structure factor. Rahman* has given an interesting molecular dynamics discussion of the convolution approximation.

22–14. Show that

$$\gamma_1(t) = \frac{kT}{m} \frac{(1 - \cos \omega_0 t)}{\omega_0^2}$$

for a one-dimensional harmonic oscillator.

22–15. Give an example of a stochastic process in which α is a continuous random variable.

22–16. Prove Parseval's theorem.

22–17. What form does Parseval's theorem take on if $x(t)$ is written as a Fourier series:

$$x(t) = \sum_{n=0}^{\infty} a_n \cos \omega_n t + \sum_{n=1}^{\infty} b_n \sin \omega_n t$$

22–18. Derive Eq. (22–85).

22–19. Prove that if X_1 and X_2 are Gaussian random variables; then $Y = X_1 + X_2$ is also Gaussian.

22–20. Derive Eq. (22–100).

22–21. Prove that if $X_1(t)$ and $X_2(t)$ are Gaussian random variables, then the second probability distribution function is given by Eq. (22–102).

22–22. Derive Eq. (22–184).

22–23. Calculate the power spectrum associated with a correlation function of the form

$$C(\tau) = e^{-|\tau|/\tau_c}$$

Plot $S(\omega)$ versus $\ln \omega$. Notice that $S(\omega)$ is fairly flat out to $\omega \approx 1/\tau_c$, at which point it decreases as ω^{-2}. One calls this flat region a "white" region since it contains all frequencies equally. What form of $C(\tau)$ would lead to a white spectrum for all frequencies?

22–24. Prove that if $X(t)$ is a Gaussian random variable, then it can be written as

$$X(t) = \sum_{n=-\infty}^{\infty} c_n e^{i\omega_n t}$$

with

$$\langle c_m^* c_n \rangle = \sigma_n^2 \, \delta_{nm}$$

and

$$\langle c_i^* c_j c_k^* c_l \rangle = \sigma_i^2 \sigma_k^2 \delta_{ij} \delta_{kl} + \sigma_i^2 \sigma_j^2 \delta_{il} \delta_{jk}$$

22–25. Prove that if $X(t)$ is a Gaussian random variable, then

$$\langle x^*(t)x(t)x^*(0)x(0) \rangle = |\langle x^*(0)x(0) \rangle|^2 + |\langle x^*(t)x(0) \rangle|^2$$

22–26. Derive Eq. (22–118).

22–27. Derive Eq. (22–123).

22–28. Solve Eq. (22–133).

22–29. In an electrophoretic light-scattering experiment, prove that if the direction of the electric field is perpendicular to the incident radiation, then $\kappa_x = \kappa \cos(\theta/2)$.

22–30. In an electrophoretic light-scattering experiment, prove that the autocorrelation has a period of oscillation given by $\Delta\tau = \lambda_0/uEn \sin \theta$.

* *Phys. Rev.*, **136**, p. A405, 1964.

22–31. Show that if the scattering molecules are not independent, then

$$I(\omega) = \frac{A^2}{\pi} \int_{-\infty}^{\infty} e^{-i\omega\tau} F(\varkappa, \tau) \, d\tau$$

where

$$F(\varkappa, t) = \frac{1}{N} \langle \rho(-\varkappa, 0)\rho(\varkappa, t) \rangle$$

where

$$\rho(\varkappa, t) = \sum_j \exp[i\varkappa \cdot \mathbf{r}_j(t)]$$

22–32. Derive Eqs. (22–154) and (22–155).

22–33. Derive Eq. (22–162) from Eq. (22–159).

22–34. Prove that the ratio of the intensity of the Rayleigh component to the Brillouin component is $\gamma - 1$, where $\gamma = C_p/C_v$. This ratio is called the Landau-Placzek ratio.

22–35. Derive the basic light-scattering formulas starting from Eqs. (J–27) and (J–28).

22–36. Derive Eq. (22–168) from Eq. (22–167).

22–37. Derive Eq. (22–175).

22–38. Prove that $\mathscr{P}^2 G = \mathscr{P} G$, where $\mathscr{P}$ is the projection operator defined in Eq. (22–178).

22–39. Verify the operator identity

$$\mathscr{A}^{-1} = \mathscr{B}^{-1} + \mathscr{A}^{-1}(\mathscr{B} - \mathscr{A})\mathscr{B}^{-1}$$

22–40. Prove that

$$\langle \alpha^* \mathscr{L} \beta \rangle = i\langle \dot\alpha \beta \rangle$$

22–41. Show that the inverse Laplace transform of

$$\hat{K}(s) = \langle \dot{U}[s - i(1 - \mathscr{P})\mathscr{L}]^{-1}\dot{U} \rangle$$

is

$$K(t) = \langle \dot{U} \exp[it(1 - \mathscr{P})\mathscr{L}]\dot{U} \rangle$$

22–42. Prove that $\mathscr{P}$ and $\mathscr{L}$, and hence $(1 - \mathscr{P})\mathscr{L}$, are Hermitian operators.

22–43. Prove that the classical memory function $K(t)$ is a real, even function of time.

22–44. Derive Eq. (22–194).

22–45. Prove that a memory function of the form

$$K(t) = \gamma e^{-\alpha|t|}$$

leads to a correlation function of the form

$$\psi(t) = \frac{1}{s_+ - s_-} (s_+ e^{s_- t} - s_- e^{s_+ t})$$

where

$$s_\pm = -\frac{\alpha}{2}\left\{1 \mp \left[1 - \left(\frac{4\gamma}{\alpha^2}\right)\right]^{1/2}\right\}$$

22–46. Prove that if the memory function associated with the velocity autocorrelation function has the form $K(t) = \gamma \exp(-\alpha|t|)$, then

$$G(\omega) = \frac{kT}{mD} \int_0^{\infty} dr \, \psi(t) \cos \omega t$$

$$= \frac{\gamma^2}{\omega^4 + \alpha^2(1 - 2\gamma/\alpha^2)\omega^2 + \gamma^2}$$

22–47. Berne, Pechukas, and Harp* have used the ideas of information theory to derive an approximate expression for $G_s(\mathbf{r}, t)$ (see also Problem 22–13). Suppose we know $\langle r^2(t) \rangle$, where

$$\langle r^2(t) \rangle = \int d\mathbf{r} \; r^2 G_s(\mathbf{r}, t)$$

The information functional associated with the conditional probability distribution $G_s(\mathbf{r}, t)$ is

$$I[G_s(\mathbf{r}, t)] = -\int d\mathbf{r} \; G_s(\mathbf{r}, t) \ln G_s(\mathbf{r}, t)$$

We wish to maximize this information subject to the two constraints

$$\int d\mathbf{r} \; G_s(\mathbf{r}, t) = 1$$

$$\int d\mathbf{r} \; r^2 G_s(\mathbf{r}, t) = \langle r^2(t) \rangle$$

Use the method of Lagrange multipliers to show that

$$G_s(\mathbf{r}, t) = [\tfrac{2}{3}\pi\langle r^2(t) \rangle]^{-3/2} \exp\left[-\frac{r^2}{\tfrac{2}{3}\langle r^2(t) \rangle} \right]$$

which is the well-known Gaussian approximation for $G_s(\mathbf{r}, t)$. Show that

$$\langle r^2(t) \rangle = 2 \int_0^t d\tau \; (t - \tau) \langle \mathbf{v}(0) \cdot \mathbf{v}(\tau) \rangle$$

so that $G_s(\mathbf{r}, t)$ is completely determined under this approximation if $\langle \mathbf{v}(0) \cdot \mathbf{v}(\tau) \rangle$ is known.

22–48. Starting with Eq. (22–101), show that if $\mathbf{v}(t)$ can be considered to be a Gaussian random variable, then the joint probability of $\mathbf{v}(t)$ and $\mathbf{v}_0$ is

$$P(\mathbf{v}(t), \mathbf{v}_0) = \left[\frac{m}{2\pi kT(1 - \Psi^2(t))} \right]^{3/2} \exp\left[-\frac{m}{2kT} \frac{(v^2 + v_0^2 - 2\mathbf{v} \cdot \mathbf{v}_0 \Psi(t))}{(1 - \Psi^2(t))} \right]$$

where $\Psi(t)$ is the normalized velocity autocorrelation function.

Defining a transition probability $K(\mathbf{v}, t | \mathbf{v}_0, 0)$ through

$$P(\mathbf{v}(t), \mathbf{v}_0) = K(\mathbf{v}, t | \mathbf{v}_0, 0) f(\mathbf{v}_0)$$

where $f(\mathbf{v}_0)$ is a Maxwell distribution, show that the Gaussian transition probability has the form

$$K(\mathbf{v}, t | \mathbf{v}_0, 0) = \left[\frac{m}{2\pi kT(1 - \Psi^2(t))} \right]^{3/2} \exp\left[-\frac{m}{2kT} \frac{(\mathbf{v} - \mathbf{v}_0 \Psi(t))^2}{(1 - \Psi^2(t))} \right]$$

22–49. Berne, Pechukas, and Harp* have used the ideas of information theory to approximate the probability that an atom will have a velocity $\mathbf{v}$ at time t given that it had the velocity $\mathbf{v}_0$ initially. Denote this conditional probability by $K(\mathbf{v}, t | \mathbf{v}_0, 0)$. Since the joint probability of $(\mathbf{v}, t)$ and $(\mathbf{v}_0, 0)$ is

$$P(\mathbf{v}(t), \mathbf{v}_0) = K(\mathbf{v}, t | \mathbf{v}_0, 0) f(\mathbf{v}_0)$$

where $f(\mathbf{v}_0)$ is the Maxwell distribution, the information functional that is used is

$$I[P] = -\int d\mathbf{v} \int d\mathbf{v}_0 \; P(\mathbf{v}(t), \mathbf{v}_0) \ln P(\mathbf{v}(t), \mathbf{v}_0)$$

* *J. Chem. Phys.*, **49**, p. 3125, 1968.

Show that the maximum of this subject to the two constraints

$$\iint dv \, dv_0 \, P(\mathbf{v}(t), \mathbf{v}_0) = 1$$

$$\iint dv \, dv_0 \, (\mathbf{v} \cdot \mathbf{v}_0) P(\mathbf{v}(t)\mathbf{v}_0) = \langle v_0{}^2 \rangle \Psi(t)$$

gives

$$K(\mathbf{v}, t | \mathbf{v}_0, 0) = \left\{ \frac{m}{2\pi kT(1 - \Psi^2(t))} \right\}^{3/2} \exp\left\{ -\frac{m}{2kT} \frac{(\mathbf{v} - \mathbf{v}_0 \Psi(t))^2}{(1 - \Psi^2(t))} \right\}$$

which is a Gaussian transition probability, i.e., the transition probability that one would obtain by assuming that the variable $\mathbf{v}(t)$ can be treated as a Gaussian random variable (see the previous problem). The next problem discusses a very interesting test of this assumption using molecular dynamics calculations.

22–50. Harp and Berne* have presented molecular dynamics calculations of the linear and angular momentum autocorrelation functions and have made a thorough analysis of the assumption that $\mathbf{v}(t)$ can be treated as a Gaussian random variable. If, in fact, $\mathbf{v}(t)$ were a Gaussian random variable, the joint distribution of $\mathbf{v}(t)$ and $\mathbf{v}(0)$ would be (cf. Problems 22–48 and 22–49)

$$P[\mathbf{v}(t), \mathbf{v}(0)] = \left\{ \frac{m}{2\pi kT(1 - \Psi^2(t))} \right\}^{3/2} \exp\left\{ -\frac{m}{2kT} \frac{(v^2 + v_0{}^2 - 2\mathbf{v} \cdot \mathbf{v}_0 \Psi(t))}{(1 - \Psi^2(t))} \right\}$$

where $\Psi(t) = \langle \mathbf{v}(t) \cdot \mathbf{v}(0) \rangle / \langle v^2 \rangle$. The key point here is that a knowledge of $\Psi(t)$ is all that is required to calculate any autocorrelation function involving a higher power of $\mathbf{v}$. For example, show that the normalized translational kinetic energy autocorrelation function $\mathscr{E}_2(t)$, where

$$\mathscr{E}_2(t) = \frac{\langle v^2(0) v^2(t) \rangle}{\langle v^4 \rangle}$$

is given in terms of $\Psi(t)$ under the Gaussian approximation as

$$\mathscr{E}_{2G}(t) = \tfrac{3}{5}[1 + \tfrac{2}{3}\Psi^2(t)]$$

Show also that for $\mathscr{E}_4(t) = \langle v^4(0) v^4(t) \rangle / \langle v^8 \rangle$,

$$\mathscr{E}_{4G}(t) = \frac{1}{945}[225 + 600\Psi^2(t) + 120\Psi^4(t)]$$

Harp and Berne have evaluated $\mathscr{E}_2(t)$, $\mathscr{E}_4(t)$, and $\mathscr{E}_8(t)$ by molecular dynamics and have compared these to the Gaussian approximations, and the results for $\mathscr{E}_2(t)$ and $\mathscr{E}_4(t)$ are shown in Fig. 22–13. It can be seen that the agreement is moderately good. This type of calculation is an excellent illustration of the utility and power of molecular dynamics calculations.

22–51. Prove that

$$\langle \mathbf{v}(0)\mathbf{v}(t) \rangle = \tfrac{1}{3}\langle \mathbf{v}(0) \cdot \mathbf{v}(t) \rangle I$$

in an isotropic medium.

22–52. Derive Eq. (22–236).

22–53. Derive Eq. (22–242).

22–54. Derive Eq. (22–244).

22–55. Zwanzig† has given an interesting relation between the incoherent neutron scat-

---

* *J. Chem. Phys.*, **49**, p. 1249, 1968
† *Phys. Rev.*, **133**, p. A50, 1964.

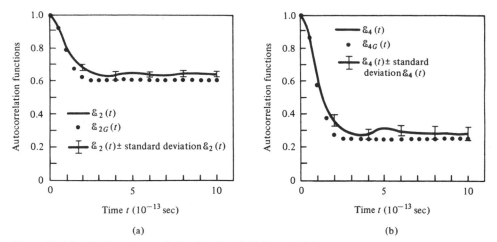

Figure 22–13. (a) **The autocorrelation function $\mathscr{E}_2(t)$ from a Molecular dynamics simulation for liquid CO and the autocorrelation function $\mathscr{E}_{2G}(t)$ from the Gaussian approximation. (b) The autocorrleation function $\mathscr{E}_4(t)$ from a molecular dynamics simulation for liquid CO and the autocorrelation function $\mathscr{E}_{4G}(t)$ from the Gaussian approximation.** (From G. Harp and B. J. Berne, *J. Chem. Phys.*, **49**, p. 1249, 1968.)

tering function $S_{\text{inc}}(\mathbf{k}, \omega)$ and the generalized diffusion coefficient $D(\mathbf{k}, \omega)$. $S_{\text{inc}}(\mathbf{k}, \omega)$ is given by

$$S_{\text{inc}}(\mathbf{k}, \omega) = \frac{1}{2\pi} \int_{-\infty}^{\infty} dt\, e^{-i\omega t} \langle \rho(-\mathbf{k}, 0)\rho(\mathbf{k}, t)\rangle$$

where

$$\rho(\mathbf{k}, t) = \exp[i\mathbf{k} \cdot \mathbf{r}(t)]$$

First prove that

$$\omega^2 \int_{-\infty}^{\infty} dt\, e^{-i\omega t} \langle \rho(-\mathbf{k}, 0)\rho(\mathbf{k}, t)\rangle = \int_{-\infty}^{\infty} dt\, e^{-i\omega t} \langle \dot{\rho}(-\mathbf{k}, 0)\dot{\rho}(\mathbf{k}, t)\rangle$$

Now define the generalized diffusion coefficient tensor $\mathbf{D}(\mathbf{k}, \omega)$ by

$$D_{ab}(\mathbf{k}, \omega) = \frac{1}{3} \int_{0}^{\infty} dt\, e^{-i\omega t} \langle v_b(0)e^{-i\mathbf{k}\cdot\mathbf{r}(0)}v_a(t)e^{i\mathbf{k}\cdot\mathbf{r}(t)}\rangle$$

and prove that

$$S_{\text{inc}}(\mathbf{k}, \omega) = \frac{1}{\pi\omega^2} \mathbf{k} \cdot \text{Re}\, \mathbf{D}(\mathbf{k}, \omega) \cdot \mathbf{k}$$

Show that for an isotropic medium

$$S_{\text{inc}}(\mathbf{k}, \omega) \rightarrow \frac{k^2 D}{\pi\omega^2} \quad \text{as } k \rightarrow 0 \quad \text{and} \quad \text{as } \omega \rightarrow 0$$

in agreement with Eq. (22–42).

Zwanzig actually does not assume the classical limit and derives the general result

$$S_{\text{inc}}(\mathbf{k}, \omega) = \left[\frac{\beta\hbar\omega}{\pi\omega^2(1 - e^{-\beta\hbar\omega})}\right] \mathbf{k} \cdot \text{Re}\, \mathbf{D}(\mathbf{k}, \omega) \cdot \mathbf{k}$$

Derive this.

22–56. Prove that if the memory function associated with the velocity autocorrelation function has the form $K(t) = \gamma e^{-\alpha|t|}$, then

$$K(0) = \frac{\langle F_1^2\rangle}{m^2 \langle v_1^2\rangle}$$

VALUES OF SOME PHYSICAL CONSTANTS AND ENERGY CONVERSION FACTORS

Table A–1. **Values of some physical constants**

quantity	symbol	value
Avogadro's number	N_0	6.0222×10^{23}
Planck constant	h	6.6262×10^{-27} erg-sec
	$\hbar$	1.0546×10^{-27} erg-sec
Boltzmann constant	k	1.3806×10^{-16} erg/molecule-deg K
gas constant	R	8.3143×10^{7} ergs/mole-deg K
		1.9872 cal/mole-deg K
speed of light	c	2.9979×10^{10} cm/sec
proton charge	e	4.8032×10^{-10} esu
electron mass	m_e	9.1096×10^{-28} g
atomic mass unit	amu	1.6605×10^{-24} g
Bohr magneton	μ_B	9.2741×10^{-21} erg/gauss
nuclear magneton	μ_N	5.0509×10^{-24} erg/gauss

Source: B. N. Taylor, W. H. Parker, and D. N. Langenberg, *Rev. Mod. Phys.*, **41**, p. 375, 1969.

Table A–2. **Energy conversion factors**

	ergs	eV	cm^{-1}	$°K$	kcal	kcal/mole	atomic units
1 erg	1	6.2420×10^{11}	5.0348×10^{15}	7.2441×10^{15}	2.3901×10^{-11}	1.4394×10^{13}	2.294×10^{10}
1 eV	1.6021×10^{-12}	1	8.0657×10^{3}	1.1605×10^{4}	3.8390×10^{-23}	2.3119×10^{1}	3.675×10^{-2}
1 cm^{-1}	1.9862×10^{-16}	1.2398×10^{-4}	1	1.4388	4.7471×10^{-27}	2.8588×10^{-3}	4.556×10^{-6}
1°K	1.3804×10^{-16}	8.6167×10^{-5}	6.9502×10^{-1}	1	3.2993×10^{-27}	1.9869×10^{-3}	3.116×10^{-6}
1 kcal	4.1840×10^{10}	2.6116×10^{22}	2.1066×10^{26}	3.3009×10^{26}	1	6.0222×10^{23}	9.597×10^{20}
1 kcal/mole	6.9446×10^{-14}	4.3348×10^{-2}	3.4964×10^{2}	5.0307×10^{2}	1.6598×10^{-24}	1	1.594×10^{-3}
1 atomic unit	4.360×10^{-11}	27.21	2.195×10^{5}	3.158×10^{5}	1.042×10^{-21}	6.275×10^{2}	1

FOURIER INTEGRALS AND THE DIRAC DELTA FUNCTION

The Fourier transform $g(k)$ of the function $f(x)$ is defined by

$$g(k) = \frac{1}{(2\pi)^{1/2}} \int_{-\infty}^{\infty} f(x)e^{-ikx}\,dx \tag{B-1}$$

The Fourier integral theorem states that if $f(x)$ is single valued and periodic with at most a finite number of finite discontinuities, maxima, and minima, and

$$\int_{-\infty}^{\infty} |f(x)|^2\,dx$$

is finite, then

$$f(x) = \frac{1}{(2\pi)^{1/2}} \int_{-\infty}^{\infty} g(k)e^{ikx}\,dk \tag{B-2}$$

Equations (B-1) and (B-2) are said to be a Fourier transform pair. Other definitions of the Fourier transforms are in use, differing by numerical factors and by the use of complex conjugates of the integrands given here.

In vector notation, if $V(\mathbf{r})$ is a function of the vector $\mathbf{r} = (x, y, z)$ and $\overline{V}(\mathbf{k})$ its Fourier transform, then

$$\overline{V}(\mathbf{k}) = \frac{1}{(2\pi)^{3/2}} \iiint_{-\infty}^{\infty} V(\mathbf{r})e^{-i\mathbf{k}\cdot\mathbf{r}}\,d\mathbf{r} \tag{B-3}$$

and

$$V(\mathbf{r}) = \frac{1}{(2\pi)^{3/2}} \iiint_{-\infty}^{\infty} \overline{V}(\mathbf{k})e^{i\mathbf{k}\cdot\mathbf{r}}\,d\mathbf{k} \tag{B-4}$$

Fourier transforms can be used in a formal manner to obtain a useful representation for the Dirac delta function. The Dirac delta function is defined by

$$\int_{-\infty}^{\infty} \delta(x - a)\Phi(x)\, dx = \Phi(a) \tag{B-5}$$

$$\int_{-\infty}^{\infty} \delta(x)\, dx = 1 \tag{B-6}$$

The Fourier transform of $\delta(x)$ is

$$\frac{1}{(2\pi)^{1/2}} \int_{-\infty}^{\infty} \delta(x)e^{-iux}\, dx = \frac{1}{(2\pi)^{1/2}} \tag{B-7}$$

The formal inverse gives the desired relation:

$$\delta(x) = \frac{1}{2\pi} \int_{-\infty}^{\infty} e^{iux}\, du \tag{B-8}$$

More generally, in three dimensions,

$$\frac{1}{(2\pi)^{3/2}} \iiint_{-\infty}^{\infty} \delta(\mathbf{k})e^{-i\mathbf{k}\cdot\mathbf{r}}\, d\mathbf{k} = \frac{1}{(2\pi)^{3/2}} \tag{B-9}$$

$$\delta(\mathbf{k}) = \frac{1}{(2\pi)^3} \iiint_{-\infty}^{\infty} e^{i\mathbf{k}\cdot\mathbf{r}}\, d\mathbf{r} \tag{B-10}$$

where $\delta(\mathbf{k}) = \delta(k_x)\delta(k_y)\delta(k_z)$.

A triple integral such as

$$\overline{V}(|\mathbf{k}|) = \frac{1}{(2\pi)^{3/2}} \iiint_{-\infty}^{\infty} V(|\mathbf{r}|)e^{-i\mathbf{k}\cdot\mathbf{r}}\, d\mathbf{r} \tag{B-11}$$

may be converted to a single integral, and hence more easily handled, by converting to spherical coordinates:

$$d\mathbf{r} = dx\, dy\, dz \to r^2 \sin\theta\, dr\, d\theta\, d\phi$$

$$\mathbf{k}\cdot\mathbf{r} \to kr \cos\theta$$

and so

$$\overline{V}(|\mathbf{k}|) = \frac{1}{(2\pi)^{3/2}} \int_0^\infty \int_0^\pi \int_0^{2\pi} V(|\mathbf{r}|)e^{-ikr\cos\theta} r^2\, dr \sin\theta\, d\theta\, d\phi$$

The integral over ϕ gives 2π. The integral over θ gives

$$\overline{V}(|\mathbf{k}|) = \left(\frac{2}{\pi}\right)^{1/2} \int_0^\infty V(r) \frac{r \sin kr}{k}\, dr \tag{B-12}$$

In order to evaluate Fourier transforms, the identity $e^{iz} = \cos z + i \sin z$ is often useful if one remembers that

$$\int_{-\infty}^{\infty} f(x)\, dx = 2 \int_0^\infty f(x)\, dx \qquad \text{if } f(x) \text{ is even}$$

$$\int_{-\infty}^{\infty} f(x)\, dx = 0 \qquad \text{if } f(x) \text{ is odd}$$

For example, consider the Fourier transform of $1/(x^2 + \alpha^2)$.

$$\phi(k) = (2\pi)^{-1/2} \int_{-\infty}^{\infty} \frac{e^{-ikx}}{x^2 + \alpha^2} \, dx$$

$$= 2(2\pi)^{-1/2} \int_{0}^{\infty} \frac{\cos kx \, dx}{x^2 + \alpha^2} = \left(\frac{\pi}{2}\right)^{1/2} \alpha^{-1} e^{-k|\alpha|}$$

You should be able to invert $\phi(k)$ to get $(x^2 + \alpha^2)^{-1}$.

PROBLEMS

B–1. Show that the Fourier transform of $f(x) = e^{-\alpha|x|}$ is

$$\left(\frac{2}{\pi}\right)^{1/2} \frac{\alpha}{k^2 + \alpha^2}$$

B–2. Show that the Fourier transform of $e^{-\lambda x^2}$ is $(2\lambda)^{-1/2} e^{-k^2/4\lambda}$.

B–3. Solve the equation

$$e^{-|\alpha|} = \int_{0}^{\infty} \cos \alpha t f(t) \, dt$$

for $f(x)$ given that $f(x)$ is an even function of x.

B–4. Given that $f(\mathbf{r}) = Ne^{-r^2/a^2}$ where $r = |\mathbf{r}|$ (three dimensions), show that its Fourier transform $\phi(\mathbf{k}) = a^3 N e^{-k^2 a^2/4}/2^{3/2}$.

B–5. The delta function is not a function in the strict sense, but has meaning only if it occurs under an integral sign. For example, we can assign a meaning to $x\delta(x)$ by multiplying by an arbitrary but continuous function $f(x)$ and integrating to get

$$\int_{-\infty}^{\infty} f(x)x\delta(x) \, dx = 0$$

and thus one often sees $x\delta(x) = 0$. In a similar way, show that

$$x\delta'(x) = -\delta(x)$$

and

$$\delta(ax) = a^{-1}\delta(x)$$

B–6. Show that

$$\int_{-\infty}^{\infty} \delta(x - a) \cos x \, dx = \cos a$$

by first evaluating

$$I(\sigma) = (2\pi\sigma^2)^{-1/2} \int_{-\infty}^{\infty} \cos x e^{-(x-a)^2/2\sigma^2} \, dx$$

and then taking the limit of $\sigma \to 0$. This shows that Eq. (1–72) with $\sigma \to 0$ is one representation of the delta function. When in doubt, this is always a sure way of handling $\delta(x)$.

B–7. Another representation of $\delta(x)$ was used in deriving Eq. (10–131). Let $u_k(x)$ be an orthonormal complete set. Let $\psi(x)$ be a suitably arbitrary function which can be expanded in terms of the $\{u_k(x)\}$:

$$\psi(x) = \sum_{k} a_k u_k(x)$$

First find the a_k by multiplying by $u_k*(x)$ and integrating and then substituting the result back into the above equation to get

$$\psi(x) = \int \left[\sum_k u_k*(x')u_k(x) \right] \psi(x')\, dx'$$

Conclude from this that

$$\delta(x - x') = \sum_k u_k*(x')u_k(x)$$

Generalize this result to n-dimensions.

DEBYE HEAT CAPACITY FUNCTION

Table C–1. Debye heat capacity function, $C_V/3R$ as a function of Θ_D/T

Θ_D/T	0.0	0.1	0.2	0.3	0.4	0.5	0.6	0.7	0.8	0.9	1.0
0.0	1.0000	0.9995	0.9980	0.9955	0.9920	0.9876	0.9822	0.9759	0.9687	0.9606	0.9517
1.0	0.9517	0.9420	0.9315	0.9203	0.9085	0.8960	0.8828	0.8692	0.8550	0.8404	0.8254
2.0	0.8254	0.8100	0.7943	0.7784	0.7622	0.7459	0.7294	0.7128	0.6961	0.6794	0.6628
3.0	0.6628	0.6461	0.6296	0.6132	0.5968	0.5807	0.5647	0.5490	0.5334	0.5181	0.5031
4.0	0.5031	0.4883	0.4738	0.4595	0.4456	0.4320	0.4187	0.4057	0.3930	0.3807	0.3686
5.0	0.3686	0.3569	0.3455	0.3345	0.3237	0.3133	0.3031	0.2933	0.2838	0.2745	0.2656
6.0	0.2656	0.2569	0.2486	0.2405	0.2326	0.2251	0.2177	0.2107	0.2038	0.1972	0.1909
7.0	0.1909	0.1847	0.1788	0.1730	0.1675	0.1622	0.1570	0.1521	0.1473	0.1426	0.1382
8.0	0.1382	0.1339	0.1297	0.1257	0.1219	0.1182	0.1146	0.1111	0.1078	0.1046	0.1015
9.0	0.1015	0.09847	0.09558	0.09280	0.09011	0.08751	0.08500	0.08259	0.08025	0.07800	0.07582
10.0	0.07582	0.07372	0.07169	0.06973	0.06783	0.06600	0.06424	0.06253	0.06087	0.05928	0.05773
11.0	0.05773	0.05624	0.05479	0.05339	0.05204	0.05073	0.04946	0.04823	0.04705	0.04590	0.04478
12.0	0.04478	0.04370	0.04265	0.04164	0.04066	0.03970	0.03878	0.03788	0.03701	0.03617	0.03535
13.0	0.03535	0.03455	0.03378	0.03303	0.03230	0.03160	0.03091	0.03024	0.02959	0.02896	0.02835
14.0	0.02835	0.02776	0.02718	0.02661	0.02607	0.02553	0.02501	0.02451	0.02402	0.02354	0.02307
15.0	0.02307	0.02262	0.02218	0.02174	0.02132	0.02092	0.02052	0.02013	0.01975	0.01938	0.01902

Source: K. S. Pitzer, Quantum Chemistry (Englewood Cliffs, N.J.: Prentice-Hall, 1953).

HARD-SPHERE RADIAL DISTRIBUTION FUNCTION

A listing of a FORTRAN program that gives essentially the exact hard-sphere radial distribution function is shown below. It numerically inverts the Laplace transform of the Percus-Yevick $rg(r)$ (cf. G. J. Throop and R. J. Bearman, *J. Chem. Phys.*, **42**, p. 2408, 1965) and adds on a correction due to Verlet and Weis (*Phys. Rev.*, **A5**, p. 939, 1972) that brings $g(r)$ into agreement with the computer (Monte Carlo and molecular dynamics) results. The author wishes to thank D. Henderson for permission to present this program here.

```
      FUNCTION GR(RHO,R1)

C     CALCULATION OF GO(R) USING HENDERSONS PROGRAM              3R
C     CORRECTION IS MADE TO P.Y GO(R) USING VERLET AND WEIS METHOD
C     RHO = DENISTY*SIGMA**3      R1 = R/SIGMA                   GR

      COMPLEX EX12,IG102,IG202,IG212,IG222,IG302,IG312,IG322,IG332,  3R
     1 IG402,IG412,IG422,IG432,IG442,IT2,EX2,                        GR
     3IXL1C,IXL2C,IXL3C,IXL4C,IXL5C,TLTC,JX,ISPC,ISDPC               GR
      REAL LT,LT1,LT2,LP                                             GR
      DATA PIE,K,RHOS/3.14159265,0,0.0/

      IF( K .NE. 0) GO TO 1                                          GR
      X = 2.*PIE/3.                                                  GR
      JX = CEXP(CMPLX(0.0,X))                                        GR
      K = 1                                                          3R
    1 IF(ABS(RHO - RHOS) .LT. 1.E-4)GO TO 10                         GR

      RHOS = RHO                                                     GR
      ETAC = PIE/6.**RHO                                            GR
      ETA = ETAC*(1. - ETAC/16.)                                    3R
      ETAM = 1. - ETA                                               GR
      BOT1 = ETAM**4                                                GR
      TOP = 1. + 2.*ETA                                             GR
      TOP1 = TOP*TOP                                                3R
      C1 = TOP1/BOT1                                                3R
      BOT2 = 4.*BOT1                                                GR
      TOPP = 2. + ETA                                               GR
      TOP2 = TOPP*TOPP                                              3R
      C2 = -TOP2/BOT2                                               3R
      C2 = 6.*ETA*C2                                                GR
      C3 = ETA*C1/2.                                                3R
      A1 = 0.75*ETA*ETA*(1. - ETA*(0.7117 + 0.114*ETA))/BOT1        3R
      A2 = 24.*A1/ETA*ETAM*ETAM/(1. + 0.5*ETA)                      GR
      A3=CUBER(ETAC/ETA)
      ETA12 = 12.*ETA                                               GR
      LT1 = 1. + ETA/2.
      LT2 = 1 + 2.*ETA                                             GR
      LP = LT1                                                      GR
      XETA = 1./(1. - ETA)                                          3R
      ETA3 = (1. - ETA)**2                                          GR
      SP1 = 1.                                                      GR
      SP2 = 4.*ETA*XETA                                            GR
      SP3 = 6.*(ETA*XETA)**2                                        3R
      SDP1 = 2.                                                     GR
      SDP2 = SP2                                                    GR
      STP = 2.                                                      GR
      FF = (-ETA + 3.)*ETA + 3.                                     GR
      PAR=SQRT(1.0+2.0*(ETA**2/FF)**2)
      YP=CUBER(1.0+PAR)
      YM=CUBER(1.0-PAR)
      PAR=CUBER(2.0*ETA*FF)
      T1 = XETA*(-2.*ETA + PAR*(YP + YM))                           3R
      IT2 = XETA*(-2.*ETA + PAR*(YP*JX + YM/JX))                    GR
      XF1 = 3.*ETA    3                                             3R
      PAR = ETA3**2/ETA3                                            GR
      XF2 = PAR*3./4.                                               GR
      PAR = PAR*ETA3/ETA                                            GR
      XF3 = PAR*3./8.                                               3R
      PAR = PAR*ETA3/ETA                                            3R
      XF4 = PAR*9./32.                                              GR
      LT = LT1*T1 + LT2                                             GR
      SDP = SDP1*T1 + SDP2                                          GR
      SP = (SP1*T1 + SP2)*T1 + SP3                                  GR
      EX11 = EXP(-T1)                                               GR
      XL1 = (LT/SP)**2                                             3R
      XL2 = 15.*SDP**2 - 4.*SP*STP                                  GR
      XL3 = LT + 4.*T1*LP                                           GR
      XL4 = LT*SDP/SP                                               3R
      XL5 = 2.*LT + 3.*T1*LP                                        GR
      G101 = T1*LT/XF1/SP                                           3R
      G201 = -LT/XF2/ SP**2                                         GR
      G211 = LT*(1. - T1*(2. + SDP/SP)) + 2.*LP*T1                  GR
      G221 = LT*T1                                                  3R
```

```
      G301 = LT/XF3/SP**3                                                   GR
      G311 = LT**2*T1/SP**2*(3.*SDP**2 - SP*STP) - 3.*LT*SDP/SP*(           GR
     1 LT + 3.*T1*LP) + 6.*LP*(LT + LP*T1)                                  GR
      G321 = LT*(6.*LP*T1 + LT*(2. - 3.*SDP*T1/SP))                         GR
      G331 = G221*LT                                                        GR
      G401 = -LT/XF4/SP**4                                                  GR
      G411 = 5.*T1*XL1*LT/SP*SDP*(2.*SP*STP - 3.*SDP**2) + XL1*             GR
     1 XL2*XL3 - 24.*LP*XL4*XL5 + 12.*LP**2*(3.*LT + 2.*T1*LP)              GR
      G421 = (T1*XL1*XL2 - 12.*(XL4*XL3 - LP*XL5))*LT                       GR
      G431 =(-6.*T1*XLP + 3.*XL3)*LT**2                                     GR
      G441 = G331 *LT                                                       GR
      ILTC = LT1*IT2 + LT2                                                  GR
      ISDPC = SDP1*IT2 + SDP2                                               GR
      ISPC = (SP1*IT2 + SP2)*IT2 + SP3                                      GR
      EX12 = CEXP(-IT2)                                                     GR
      IXL1C = (ILTC/ISPC)**2                                               GR
      IXL2C = 15.*ISDPC**2 - 4.*ISPC*STP                                    GR
      IXL3C = ILTC + 4.*IT2*LP                                             GR
      IXL4C = ILTC*ISDPC/ISPC                                              GR
      IXL5C = 2.*ILTC + 3.*IT2*LP                                          GR
      IG102 = IT2*ILTC/XF1/ISPC                                            GR
      IG202 = -ILTC/XF2/ISPC**2                                            GR
      IG212 = ILTC*(1. - IT2*(2. + ISDPC/ISPC)) + 2.*LP*IT2                 GR
      IG222 = ILTC*IT2                                                     GR
      IG302 = ILTC/XF3/ISPC**3                                             GR
      IG312 = ILTC**2*IT2/ISPC**2*(3.*ISDPC**2 - ISPC*STP) - 3.*ILTC*       GR
     1 ISDPC/ISPC*(ILTC + 3.*IT2*LP) + 6.*LP*(ILTC + LP*IT2)               GR
      IG322 = ILTC*(6.*LP*IT2 + ILTC*(2. - 3.*ISDPC*IT2/ISPC))             GR
      IG332 = IG222*ILTC                                                   GR
      IG402 = -ILTC/XF4/ISPC**4                                            GR
      IG412 = 5.*IT2*IXL1C*ILTC/ISPC*ISDPC*(2.*ISPC*STP - 3.*ISDPC**2)      GR
     1 + IXL1C*IXL2C*IXL3C - 24.*LP*IXL4C*IXL5C + 12.*LP**2*(3.*ILTC        GR
     2 + 2.*IT2*LP)                                                        GR
      IG422 = (IT2*IXL1C*IXL2C - 12.*(IXL4C*IXL3C - LP*IXL5C))*ILTC         GR
      IG432 = (-6.*IT2*IXL4C + 3.*IXL3C)*ILTC**2                            GR
      IG442 = IG332*ILTC                                                   GR

   10 R = R1*A3                                                            GR
      IF (R .GE. .9999999 .AND. R .LE. 5.) GO TO 1010                      GR
      IF (R .GT. 5.)GO TO 1011                                             GR
      R3 = R*R*R                                                           GR
      GR = C1 + C2*R + C3*R3                                               GR
      RETURN                                                               GR

 1011 GR = 1.                                                              GR
      RETURN                                                               GR
 1010 RRG = 0.                                                             GR
C                                                                          GR
C     FIRST SHELL CONTRIBUTION                                             GR
C                                                                          GR
      EX1 = EX11*EXP(R*T1)                                                 GR
      EX2 = EX12*CEXP(R*IT2)                                               GR
      RRG = RRG + G101*EX1 + 2.*IG102*EX2                                  GR
      IF(R .LE. 2.)GO TO 99                                                GR
C                                                                          GR
C     SECOND SHELL CONTRIBUTION                                            GR
C                                                                          GR
      EX1 = EX1*EX11                                                       GR
      EX2 = EX2*EX12                                                       GR
      RRG = RRG + G201*EX1*(G211 + G221*R) + 2.*IG202*EX2*(IG212 +         GR
     1 IG222*R)                                                            GR
      IF(R .LE. 3.)GO TO 99                                                GR
C                                                                          GR
C     THIRD SHELL CONTRIBUTION                                             GR
C                                                                          GR
      RX = R - 3.                                                          GR
      EX1 = EX1*EX11                                                       GR
      EX2 = EX2*EX12                                                       GR
      RRG = RRG + G301*EX1*((G331*RX + G321)*RX + G311) + 2.*IG302*        GR
     1 EX2*((IG332*RX + IG322)*RX + IG312)                                 GR
      IF (R .LE. 4.)GO TO 99                                               GR
C                                                                          GR
C     FOURTH SHELL CONTRIBUTION                                            GR
C                                                                          GR
```

```
      RX = R - 4.                                                        GR
      EX1 = EX1*EX11                                                     GR
      EX2 = EX2*EX12                                                     GR
      RRG = RRG + G401*EX1*(((G441*RX + G431)*RX + G421)*RX + G411) +    GR
     1 2.*EX2*(G402*(((G442*RX + G432)*RX + G422)*RX + G412)             GR
   99 Z = A2*(R1 - 1.)                                                   GR
      GR = RRG/R + A1/R1*EXP(-Z)*COS(Z)                                  GR
      RETURN
      END




      FUNCTION CUBER(X)

      IF(X.LT.0.0) GO TO 10
      CUBER=X**(1.0/3.0)
      RETURN
   10 AX=-X
      AA=AX**(1.0/3.0)
      CUBER=-AA
      RETURN
      END
```

TABLES FOR THE *m*–6–8 POTENTIAL

The second virial coefficient, collision integrals, and Chapman-Enskog correction factors for the *m*–6–8 potential,

$$\frac{u(r)}{\varepsilon} = \frac{1}{m-6}(6+2\gamma)\left(\frac{d}{r^*}\right)^m - \frac{1}{m-6}[m-\gamma(m-8)]\left(\frac{d}{r^*}\right)^6 - \gamma\left(\frac{d}{r^*}\right)^8$$

where ε is the depth of the well, σ is the distance at which $u = 0$, i.e., $u(\sigma) = 0$, r_m is the distance at which u is a minimum, i.e., $u(r_m) = -\varepsilon$, and $d = r_m/\sigma$. In the tables, $m = 11$, $\gamma = 3$, and $T^* = kT/\varepsilon$. The author wishes to thank Howard Hanley for permission to present these tables here.

Second virial coefficients for the 11–6–8 ($\gamma = 3.0$) **potential**

T^*	B^*
0.60000	−4.87068
0.65000	−4.17866
0.70000	−3.63207
0.75000	−3.19016
0.80000	−2.82595
0.85000	−2.52090
0.90000	−2.26188
0.95000	−2.03934
1.00000	−1.84619
1.10000	−1.52774
1.20000	−1.27637
1.30000	−1.07314
1.40000	−0.90561
1.50000	−0.76526
1.60000	−0.64607
1.70000	−0.54368
1.80000	−0.45482
1.90000	−0.37705
2.00000	−0.30844
2.20000	−0.19307
2.40000	−0.10001
2.60000	−0.02353
2.80000	0.04033
3.00000	0.09434
3.20000	0.14054
3.40000	0.18044
3.60000	0.21519
3.80000	0.24568
4.00000	0.27260
4.50000	0.32768
5.00000	0.36984
5.50000	0.40287
6.00000	0.42925
6.50000	0.45064
7.00000	0.46820
7.50000	0.48277
8.00000	0.49495
8.50000	0.50522
9.00000	0.51392
9.50000	0.52134
10.00000	0.52768
12.00000	0.54528
14.00000	0.55495
16.00000	0.56008
18.00000	0.56244
20.00000	0.56304
26.00000	0.55939
30.00000	0.55484

Collision integrals for the 11-6-8 ($\gamma = 3.0$) **potential**

T^*	$\Omega^{(1, 1)*}$	$\Omega^{(1, 2)*}$	$\Omega^{(2, 2)*}$	$\Omega^{(1, 3)*}$	$\Omega^{(2, 3)*}$	$\Omega^{(3, 3)*}$
0.6000	1.77495683	1.50044068	1.97293436	1.31799182	1.75119807	1.62664569
0.6500	1.71029244	1.44430297	1.90393332	1.27150018	1.68427961	1.56699361
0.7000	1.65268780	1.39511086	1.83999062	1.23134857	1.62310690	1.51434932
0.7500	1.59908326	1.34980755	1.77844390	1.19570368	1.56853486	1.46629105
0.8000	1.55313888	1.31078609	1.72505031	1.16477532	1.52060206	1.42331600
0.8500	1.50943774	1.27685162	1.67759146	1.13794402	1.47708942	1.38516896
0.9000	1.46922464	1.24690235	1.63261115	1.11429982	1.43791883	1.35111649
0.9500	1.43430228	1.21885599	1.59074373	1.09308629	1.40277761	1.32011258
1.0000	1.40190117	1.19315885	1.55245548	1.07387989	1.37099892	1.29166006
1.1000	1.34357947	1.14923890	1.48592018	1.04096448	1.31621990	1.24211144
1.2000	1.29400805	1.11350836	1.42835730	1.01376097	1.27020089	1.20067018
1.3000	1.25350469	1.08291541	1.37936665	0.99076132	1.23131087	1.16550068
1.4000	1.21753570	1.05644518	1.33706527	0.97101082	1.19832065	1.13518176
1.5000	1.18421312	1.03359443	1.30037160	0.95379473	1.16989256	1.10879871
1.6000	1.15504628	1.01369579	1.26748280	0.93866542	1.14515079	1.08571844
1.7000	1.12997253	0.99605801	1.23842116	0.92523500	1.12343252	1.06539195
1.8000	1.10801605	0.98032225	1.21281020	0.91321359	1.10414466	1.04735142
1.9000	1.08807262	0.96617115	1.19000803	0.90233838	1.08693482	1.03119717
2.0000	1.06958556	0.95330369	1.16943369	0.89242579	1.07150251	1.01661888
2.2000	1.03769859	0.93098908	1.13394880	0.87496953	1.04494356	0.99132506
2.4000	1.01099595	0.91222000	1.10430496	0.86001408	1.02282001	0.97012749
2.6000	0.98812353	0.89605171	1.07911862	0.84700946	1.00399848	0.95198017
2.8000	0.96833164	0.88188385	1.05760250	0.83545818	0.98777679	0.93639700
3.0000	0.95078230	0.86935931	1.03897395	0.82511265	0.97355794	0.92267975
3.2000	0.93533311	0.85817805	1.02262673	0.81576952	0.96097407	0.91051120
3.4000	0.92165145	0.84808145	1.00809869	0.80725371	0.94971433	0.89959636
3.6000	0.90944407	0.83888754	0.99507016	0.79943579	0.93955025	0.88972610
3.8000	0.89830455	0.83048268	0.98332895	0.79220960	0.93029732	0.88074277
4.0000	0.88805170	0.82273302	0.97266670	0.78549248	0.92181381	0.87251447
4.5000	0.86583179	0.80570453	0.94972898	0.77063955	0.90331730	0.85454060
5.0000	0.84748180	0.79123150	0.93087726	0.75764824	0.88775363	0.83942074
5.5000	0.83178498	0.77866684	0.91494074	0.74632705	0.87436348	0.82639695
6.0000	0.81824763	0.76758447	0.90118200	0.73623864	0.86261104	0.81497356
6.5000	0.80636877	0.75767422	0.88909357	0.72714416	0.85214217	0.80481054
7.0000	0.79575232	0.74872313	0.87833681	0.71886917	0.84270646	0.79565537
7.5000	0.78617885	0.74056466	0.86866131	0.71128089	0.83411929	0.78732605
8.0000	0.77746627	0.73306709	0.85986201	0.70427535	0.82623927	0.77968899
8.5000	0.76948072	0.72613592	0.85180140	0.69777133	0.81895853	0.77263782
9.0000	0.76211483	0.71969252	0.84436560	0.69170305	0.81219147	0.76608977
9.5000	0.75528526	0.71367522	0.83746685	0.68601729	0.80587034	0.75997739
10.0000	0.74893100	0.70803109	0.83103386	0.68066998	0.79993999	0.75424598
12.0000	0.72717121	0.68842664	0.80892213	0.66200350	0.77931652	0.73434495
14.0000	0.70961693	0.67234483	0.79097716	0.64660102	0.76234894	0.71800629
16.0000	0.69491929	0.65873431	0.77587443	0.63351590	0.74794007	0.70415536
18.0000	0.68228800	0.64695238	0.76283719	0.62216028	0.73542548	0.69214297
20.0000	0.67122838	0.63657807	0.75137094	0.61214403	0.72437144	0.68154551
26.0000	0.64454920	0.61136970	0.72350146	0.58775833	0.69736849	0.65570675
30.0000	0.63045621	0.59797285	0.70865817	0.57478102	0.68293315	0.64192048

Collision integral combinations for the 11–6–8 ($\gamma = 3.0$) **potential**

T^*	f_η	f_λ	f_D
0.600	1.0002	1.0002	1.0001
0.650	1.0001	1.0001	1.0001
0.700	1.0000	1.0001	1.0001
0.750	1.0000	1.0001	1.0001
0.800	1.0000	1.0001	1.0001
0.850	1.0000	1.0000	1.0001
0.900	1.0000	1.0001	1.0001
0.950	1.0000	1.0001	1.0002
1.000	1.0001	1.0001	1.0002
1.100	1.0001	1.0002	1.0003
1.200	1.0002	1.0003	1.0005
1.300	1.0003	1.0005	1.0006
1.400	1.0004	1.0007	1.0007
1.500	1.0006	1.0009	1.0010
1.600	1.0008	1.0012	1.0012
1.700	1.0010	1.0016	1.0015
1.800	1.0012	1.0019	1.0017
1.900	1.0014	1.0022	1.0019
2.000	1.0017	1.0026	1.0021
2.200	1.0021	1.0033	1.0026
2.400	1.0026	1.0040	1.0030
2.600	1.0030	1.0047	1.0034
2.800	1.0034	1.0053	1.0038
3.000	1.0038	1.0059	1.0041
3.200	1.0041	1.0064	1.0044
3.400	1.0044	1.0069	1.0047
3.600	1.0047	1.0073	1.0050
3.800	1.0049	1.0077	1.0052
4.000	1.0052	1.0081	1.0054
4.500	1.0057	1.0088	1.0059
5.000	1.0061	1.0094	1.0063
5.500	1.0064	1.0099	1.0066
6.000	1.0066	1.0103	1.0069
6.500	1.0068	1.0106	1.0071
7.000	1.0070	1.0109	1.0072
7.500	1.0071	1.0111	1.0074
8.000	1.0072	1.0112	1.0075
8.500	1.0073	1.0114	1.0076
9.000	1.0074	1.0115	1.0077
9.500	1.0074	1.0116	1.0078
10.000	1.0075	1.0117	1.0078
12.000	1.0077	1.0119	1.0080
14.000	1.0077	1.0120	1.0081
16.000	1.0078	1.0121	1.0082
18.000	1.0078	1.0121	1.0082
20.000	1.0078	1.0121	1.0082

DERIVATION OF THE GOLDEN RULE OF PERTURBATION THEORY

Consider a system whose Hamiltonian can be written in the form

$$\mathcal{H} = \mathcal{H}_0 + \lambda V(t) \tag{F–1}$$

where λ is small, $\mathcal{H}_0$ is independent of time, and $V(t)$ is a function of time and the spatial coordinates of the system. The system obeys the time-dependent Schrödinger equation

$$i\hbar \frac{\partial \Psi}{\partial t} = \mathcal{H}\Psi \tag{F–2}$$

The eigenfunctions ψ_m and eigenvalues E_m of $\mathcal{H}_0$ are given by

$$\mathcal{H}_0 \psi_m = E_m \psi_m \tag{F–3}$$

We assume that these are known and that the eigenfunctions form a complete orthonormal set.

Now if $\mathcal{H}$ in Eq. (F–2) were independent of time, then $\Psi(t) = \psi_m e^{-iE_m t/\hbar}$ would be a solution. This suggests that we expand $\Psi(t)$ as

$$\Psi(t) = \sum_m a_m(t)\psi_m e^{-iE_m t/\hbar} \tag{F–4}$$

Substitute this into Eq. (F–2) to get

$$i\hbar \sum_m \frac{da_m}{dt} \psi_m e^{-iE_m t/\hbar} + \sum_m a_m \psi_m E_m e^{-iE_m t/\hbar}$$
$$= \sum_m a_m \psi_m E_m e^{-iE_m t/\hbar} + \lambda \sum_m V a_m \psi_m e^{-iE_m t/\hbar}$$

Now multiply by ψ_k^*, integrate over all space, and use the orthonormal property of the ψ_j to get

$$i\hbar \frac{da_k}{dt} = \lambda \sum_n H_{kn}^{(1)} e^{i\omega_{kn} t} a_n \tag{F–5}$$

where $H_{kn}^{(1)}$ is the kn matrix element of $V(t)$, i.e.,

$$H_{kn}^{(1)} = \int \psi_k^* V(t) \psi_n \, dq \tag{F-6}$$

and $\hbar\omega_{kn} = E_k - E_n$.

The set of equations for the $a_k(t)$, Eq. (F–5), is equivalent to the time-dependent Schrödinger equation, Eq. (F–2). This set of equations can be solved by successive approximation by writing

$$a_n = \sum_{i=0}^{\infty} \lambda^i a_n^{(i)} \tag{F-7}$$

The substitution of this into Eq. (F–5) gives to linear powers in λ

$$a_n^{(0)} = \text{const} \tag{F-8}$$

and

$$i\hbar \frac{da_k^{(1)}}{dt} = \sum_n H_{kn}^{(1)} e^{i\omega_{kn}t} a_n^{(0)} \tag{F-9}$$

Suppose now that initially the system is known to be in the stationary state ψ_m, so that $\Psi(0) = \psi_m$, and that the time-dependent perturbation $\lambda V(t)$ is independent of time except for being turned on at time 0. Then $a_k^{(0)} = \delta_{km}$ and Eq. (F–9) becomes

$$i\hbar \frac{da_k^{(1)}}{dt} = H_{km}^{(1)} e^{i\omega_{km}t} \qquad t > 0 \tag{F-10}$$

or

$$a_k^{(1)} = (i\hbar)^{-1} \int_0^t H_{km}^{(1)} e^{i\omega_{km}t'} \, dt'$$

$$= -\frac{H_{km}^{(1)}}{\hbar} \frac{e^{i\omega_{km}t} - 1}{\omega_{km}} \qquad k \neq m \tag{F-11}$$

Now according to Eq. (F–4) (assuming Ψ normalized),

$$\int \Psi^*(t)\Psi(t) \, dq = \sum_{m=0}^{\infty} |a_m(t)|^2 = 1$$

and so $|a_m(t)|^2$ has the interpretation of being the probability that the system is in the state m at time t. To first order in λ then

$$|a_k^{(1)}|^2 = 4|H_{mk}^{(1)}|^2 \frac{\sin^2 \tfrac{1}{2}\omega_{km}t}{\hbar^2 \omega_{km}^2} \tag{F-12}$$

This is the (first-order) probability of transition from state m to state $k \neq m$. This probability is shown as a function of E_k in Fig. F–1. As t increases, the height of the central portion of the curve in Fig. F–1 increases as t^2 and its width decreases as $1/t$. The other peaks simply oscillate in time. Thus the most likely transitions are those to states whose energies lie under the central peak in Fig. F–1. Since this peak is given by the first zeros of $\sin x$, then $x = \pm\pi$, and so the energies of the most probable states satisfy

$$|E_m - E_k| < \frac{2\pi\hbar}{t} \tag{F-13}$$

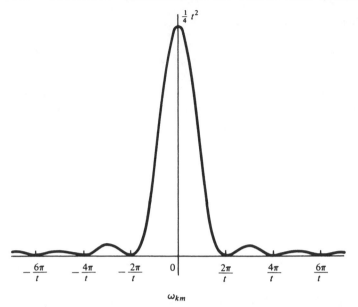

Figure F–1. **A plot of** $\sin^2 \frac{1}{2}\omega_{km} t / \omega_{km}^2$ **versus** ω_{km}, **showing how the central peak grows as** t^2 **and narrows as** $1/t$.

Suppose now that there is actually a continuum of states about the state k. Then we would wish to know the probability that there is a transition into this group of states, rather than to just the one state k. If we assume that $|H_{km}^{(1)}|^2$ is a smooth, slowly varying function of k and let $\rho(E_k)$ be the density of states about E_k, then the probability of making a transition from m to a state about k is

$$P = |H_{km}^{(1)}|^2 \int_* dE_k\, \rho(E_k) \left[\frac{\sin(\omega_{km} t/2)}{(\omega_{km}/2)} \right]^2 \tag{F–14}$$

where the asterisk indicates that we are integrating only over a small region around E_k. Now for large enough t, the central peak must include all the states around E_k. If we assume that $\rho(E_k)$ is not a strong function of E_k, we can take $\rho(E_k)$ out from under the integral sign to get

$$P = |H_{km}^{(1)}|^2 \rho(E_k) \int_* \left[\frac{\sin(\omega_{km} t/2)}{(\omega_{km}/2)} \right]^2 dE_k \tag{F–15}$$

But for large t the area under the central peak is essentially all the area, so we can extend the limits here to $\pm\infty$ and use the fact that

$$\int_{-\infty}^{\infty} \frac{\sin^2 x}{x^2}\, dx = \pi$$

to get finally

$$P = \Gamma t \tag{F–16}$$

where Γ, the transition rate, is given by

$$\Gamma = \frac{2\pi}{\hbar} |H_{km}^{(1)}|^2 \rho(E_k) \tag{F–17}$$

where $|E_m - E_k| < 2\pi\hbar/t$. This result is known as *Fermi's golden rule* and is a well-known famous result of first-order time-dependent perturbation theory.

A similar result can be obtained by noticing that as t becomes large, $[\sin(\omega_{km} t/2)/(\omega_{km}/2)]^2$ becomes more and more peaked about $E_k = E_m$, has total area $2\pi t/\hbar$, and hence approaches $(2\pi t/\hbar)\,\delta(E_k - E_m)$, except for the negligible oscillations in the wings. Thus we again have

$$P = \Gamma t$$

with

$$\Gamma = \frac{2\pi}{\hbar}\,|H_{km}^{(1)}|^2\,\delta(E_k - E_m) \tag{F–18}$$

Remember that to use this formula numerically, we must sum P over a continuous group of final states. Baym (cf. "References") gives a thorough discussion of the limit of the validity of the golden rule.

In deriving the golden rule, we have assumed that the perturbation is turned on suddenly at $t = 0$. It is often more convenient to "turn on" the perturbation very slowly. This can be done by turning on the perturbation

$$V(t) = e^{\eta t} V \qquad 0 < t$$

$$= V \qquad t > 0$$

where $\eta > 0$ in the infinite past, and then letting $\eta \to 0$ in the final results. Since the derivation of the golden rule requires that t be large, the result must be independent of how the perturbation was initially introduced.

In Section 21–1,

$$\lambda V(t) = -E_0\,\mathbf{M} \cdot \boldsymbol{\varepsilon}\cos\omega t$$

$$= -\frac{E_0\,\mathbf{M} \cdot \boldsymbol{\varepsilon}}{2}\,(e^{i\omega t} + e^{-i\omega t})$$

where E_0 is the amplitude of an electric field of frequency ω, $\boldsymbol{\varepsilon}$ is a unit vector along its direction, and $\mathbf{M}$ is the dipole moment of the system. It is straightforward to show in this case that

$$a^{(1)} = -\frac{E_0\langle f|\mathbf{M} \cdot \boldsymbol{\varepsilon}|i\rangle}{2\hbar}\left[\frac{e^{i(\omega_{fi} - \omega)t} - 1}{\omega_{fi} - \omega} + \frac{e^{i(\omega_{fi} + \omega)t} - 1}{\omega_{fi} + \omega}\right]$$

and

$$|a_f^{(1)}|^2 = E_0^2|\langle f|\boldsymbol{\varepsilon} \cdot \mathbf{M}|i\rangle|^2\left[\frac{\sin^2\frac{1}{2}(\omega_{fi} - \omega)t}{\hbar^2(\omega_{fi} - \omega)^2} + \frac{\sin^2\frac{1}{2}(\omega_{fi} + \omega)t}{\hbar^2(\omega_{fi} + \omega)^2}\right.$$

$$\left. + \frac{8\cos\omega t\,\sin\frac{1}{2}(\omega + \omega_{fi})t\,\sin\frac{1}{2}(\omega - \omega_{fi})t}{(\omega - \omega_{fi})(\omega + \omega_{fi})}\right]$$

and that

$$P_{i \to f}(\omega) = \frac{\pi E_0^2}{2\hbar^2}\,|\langle f|\boldsymbol{\varepsilon} \cdot \mathbf{M}|i\rangle|^2[\delta(\omega_{fi} - \omega) + \delta(\omega_{fi} + \omega)] \tag{F–19}$$

REFERENCES

BAYM, G. 1969. *Lectures in quantum mechanics*. New York: Benjamin.
BETHE, H. A., and JACKIW, R. W. 1968. *Intermediate quantum mechanics*, 2nd ed. New York: Benjamin.
MERZBACHER, E. 1970. *Quantum mechanics*, 2nd ed. New York: Wiley.

THE DIRAC BRA AND KET NOTATION

It is common in quantum-mechanical derivations to have integrals such as

$$\int \Psi_n^* \Psi_m \, dq \quad \text{or} \quad \int \Psi_n^* M \Psi_m \, dq$$

where M is some operator. It is convenient to define a scalar or inner product (a, b) of two functions a and b by

$$(a, b) = \int a^* b \, dq \tag{G-1}$$

where we assume that both a and b are suitably well behaved over the range of integration. In this notation, the above two integrals are (Ψ_n, Ψ_m) and $(\Psi_n, M\Psi_m)$.

If λ is a scalar (possibly complex), the inner product has the easily proved properties (Problem G–1)

$$(\Psi_n, \Psi_m)^* = (\Psi_m, \Psi_n) \tag{G-2}$$

$$(\Psi_n, \lambda\Psi_m) = \lambda(\Psi_n, \Psi_m) \tag{G-3}$$

$$(\lambda\Psi_n, \Psi_m) = \lambda^*(\Psi_n, \Psi_m) \tag{G-4}$$

The inner product defined here is similar to the inner or dot product of vectors. Consider two three-dimensional vectors, $\mathbf{a}$ and $\mathbf{b}$

$$\begin{aligned} \mathbf{a} &= a_x \varepsilon_x + a_y \varepsilon_y + a_z \varepsilon_z \\ \mathbf{b} &= b_x \varepsilon_x + b_y \varepsilon_y + b_z \varepsilon_z \end{aligned} \tag{G-5}$$

Since ε_x, ε_y, and ε_z form an orthonormal set, i.e., $\varepsilon_i \cdot \varepsilon_j = \delta_{ij}$, we see that

$$\mathbf{a} \cdot \mathbf{b} = a_x b_x + a_y b_y + a_z b_z \tag{G-6}$$

If we extend this to an N-dimensional space, for example, still with $\varepsilon_i \cdot \varepsilon_j = \delta_{ij}$, we get

$$\mathbf{a} \cdot \mathbf{b} = \sum_{j=1}^{N} a_j b_j \tag{G-7}$$

It is possible to extend this further to an infinitely dimensional function space. Instead of a simple dot product, we use the above definition of an inner product as as an integral. Thus consider a complete orthonormal set $\{\phi_j(x)\}$. By orthonormal we mean that

$$\int \phi_i^*(x)\phi_j(x)\,dx = \delta_{ij}$$

and by complete we mean that this set spans the space; i.e., there are no other functions orthonormal to the set $\{\phi_j(x)\}$. This means that all functions defined in the "space" spanned by the set $\{\phi_j(x)\}$ can be expanded as

$$f(x) = \sum f_j \phi_j(x) \tag{G-8}$$

Note the analogy with Eq. (G–5). If $h(x) = \sum h_j \phi_j(x)$, then their inner product is

$$
\begin{aligned}
(h(x), f(x)) &= \sum_{i,j} h_i^* f_j \int \phi_i^* \phi_j \, dx \\
&= \sum_{i,j} h_i^* f_j \delta_{ij} = \sum_j h_j^* f_j
\end{aligned}
\tag{G-9}
$$

in analogy with Eq. (G–7). Just as the length of a three-dimensional vector is given by

$$|\mathbf{a}| = (\mathbf{a} \cdot \mathbf{a})^{1/2} = (a_x^2 + a_y^2 + a_z^2)^{1/2}$$

the "length" of $f(x)$ is given by

$$\|f(x)\| = (f, f)^{1/2} = \left(\sum_j f_j^* f_j \right)^{1/2} \tag{G-10}$$

Similarly, just as two three-dimensional vectors are orthogonal (perpendicular) if their dot product is zero, we say that the two functions $h(x)$ and $f(x)$ are orthogonal if their inner product vanishes. These ideas are a brief introduction to the beautiful mathematical concepts of vector space, Hilbert space, etc. The works by Friedman, Merzbacher, and Jackson ("References") discuss these in more detail.

We are now ready to discuss an elegant notation due to Dirac. The state of a system is described by a wave function, which may be considered to be a vector in some space. Hence we call the wave function a state vector and represent it by $|m\rangle$. The set of all state vectors forms a vector space. The set of all possible prefactors $\langle m|$ forms another vector space, said to be the dual space. We set up a set of rules for operations in these two spaces such as

$$|a\rangle + |b\rangle \longleftrightarrow \langle a| + \langle b| \tag{G-11}$$

$$\lambda|a\rangle \longleftrightarrow \lambda^*\langle a| \tag{G-12}$$

The two spaces are connected by defining the product of a prefactor with a state vector by

$$\langle a|b\rangle = (\Psi_a, \Psi_b) \tag{G-13}$$

Since this represents a bracket notation for an inner product, the quantity $\langle a|$ is called a bra vector and $|b\rangle$ is called a ket vector. The two vector spaces are said to be dual vector spaces. Using this definition, it is easy to show that (Problem G–3)

$$\langle a|b\rangle = \langle b|a\rangle^* \tag{G-14}$$

$$\langle a|\lambda b\rangle = \lambda\langle a|b\rangle \tag{G-15}$$

$$\langle \lambda a|b\rangle = \lambda^*\langle a|b\rangle \tag{G-16}$$

This last line shows the motivation behind the definition in Eq. (G–12).

We can define an operator in the state space through

$$|b\rangle = A|a\rangle$$

If it so happens that

$$A|a\rangle = \alpha|a\rangle \tag{G-17}$$

we say that $|a\rangle$ and α are an eigenfunction and an eigenvalue, respectively, of A. We can also introduce a set of orthonormal vectors, $|m\rangle$, that span the state space and expand a state vector $|a\rangle$ by (Problem G–4)

$$|a\rangle = \sum_m \langle m|a\rangle |m\rangle \tag{G-18}$$

Note the analogy of this equation with Eq. (G–8), where $f_j = \int \phi_j*(x) f(x)\, dx$.

Since $\langle m|a\rangle$ is a scalar, Eq. (G–18) can be written as

$$|a\rangle = \sum_m |m\rangle \langle m|a\rangle \tag{G-19}$$

which shows that $|m\rangle \langle m|$ can be thought of as an operator that projects $|a\rangle$ onto $|m\rangle$. Equation (G–19) also shows that

$$\sum_m |m\rangle \langle m| = 1 \tag{G-20}$$

which is called the completeness or closure relation. Generally the combination $|a\rangle \langle b|$ is an operator since it operates on a ket to give another ket.

REFERENCES

DIRAC, P. A. M. 1958. *The principles of quantum mechanics*, 4th ed. Oxford: Oxford University Press.
FRIEDMAN, B. 1956. *Principles and techniques of applied mathematics*. New York: Wiley.
JACKSON, J. D. 1962. *Mathematics for quantum mechanics*. New York: Benjamin.
MERZBACHER, E. 1970. *Quantum mechanics*, 2nd ed. New York: Wiley.
SHILOV, G. E. 1961. *An introduction to the theory of linear vector spaces*. Englewood Cliffs, N.J.: Prentice-Hall.

PROBLEMS

G–1. Prove that

$$(\Psi_n, \Psi_m)* = (\Psi_m, \Psi_n)$$
$$(\Psi_n, \lambda\Psi_m) = \lambda(\Psi_n, \Psi_m)$$
$$(\lambda\Psi_n, \Psi_m) = \lambda*(\Psi_n, \Psi_m)$$

G–2. Discuss the analogy between a simple three-dimensional vector space spanned by ε_x, ε_y, ε_z and the space of all functions that can be expanded in terms of $\{\sin mx\}$.

G–3. Prove that

$$\langle b|a\rangle* = \langle a|b\rangle$$
$$\langle a|\lambda b\rangle = \lambda \langle a|b\rangle$$
$$\langle \lambda a|b\rangle = \lambda* \langle a|b\rangle$$

Compare these equations to those in Problem G–1.

G–4. Derive Eq. (G–18).

G–5. Write out Eq. (G–20) in terms of functions Ψ_n and integrals; i.e., what does this equation look like if one does not use Dirac notation?

THE HEISENBERG TIME-DEPENDENT REPRESENTATION

Before discussing the title of this appendix, we must first introduce a number of definitions and notions of operators and functions of operators. Let A and B be operators and c_1 and c_2 be (possibly complex) scalars. Then we have

$$(c_1 A + c_2 B)\Psi = c_1 A\Psi + c_2 B\Psi \tag{H–1}$$

The successive application of two operators is defined through

$$AB\Psi = A(B\Psi) \tag{H–2}$$

Generally the order of application is important since often $AB \neq BA$. We define the commutator of A and B by

$$[A, B] = (AB - BA) \tag{H–3}$$

We can readily construct a polynomial function of an operator A by forming

$$P_N(A) = c_0 + c_1 A + c_2 A^2 + \cdots + c_N A^N \tag{H–4}$$

By considering the limit $N \to \infty$, we can define more general functions of A through power series. For example,

$$e^A = 1 + A + \frac{1}{2!}A^2 + \frac{1}{3!}A^3 + \cdots \tag{H–5}$$

Note that (Problem H–1)

$$e^{A+B} \neq e^A e^B \quad \text{if} \quad [A, B] \neq 0$$

If $A\Psi = a\Psi$, then it is clear that $A^n\Psi = a^n\Psi$ and $f(A)\Psi = f(a)\Psi$ if f is an analytic function.

There are a number of special kinds of operators. If $(\phi, A\Psi) = (A\phi, \Psi)$, then A is said to be a *Hermitian* or *self-adjoint* operator. It is easy to prove that a Hermitian operator has real eigenvalues (Problem H–2). For this reason, the quantum-mechanical operators that correspond to physical, observable properties are Hermitian. Generally an operator will not be self-adjoint. We define an operator $A^\dagger$, called the *adjoint* of A, by

$$(\phi, A\Psi) = (A^\dagger \phi, \Psi) \tag{H-6}$$

Clearly $(A^\dagger)^\dagger = A$ (Problem H–3). If A is Hermitian, $A = A^\dagger$. It is also easy to prove that (Problem H–4)

$$(c_1 A + c_2 B)^\dagger = c_1{}^* A^\dagger + c_2{}^* B^\dagger \tag{H-7}$$

and

$$(AB)^\dagger = B^\dagger A^\dagger \tag{H-8}$$

A *unitary* operator is defined by

$$UU^\dagger = U^\dagger U = I \tag{H-9}$$

where I is the unit operator. A unitary operator corresponds to a rotation in a vector space. This is proved in the following manner: The length of a vector is $(\phi, \phi)^{1/2}$. But $(\phi, \phi) = (\phi, U^\dagger U \phi) = (U\phi, U\phi)$, which implies that $\|\phi\| = \|U\phi\|$, where $\|\phi\|$ denotes the length of ϕ. Also, it is easy to show that if $(\phi, \Psi) = 0$, then $(U\phi, U\Psi) = 0$, and so we see that the operation U must correspond to a rotation.

If A is Hermitian and $f(A)$ is a real function, then (Problem H–5) $U = \exp[if(A)]$ is a unitary operator. From $U^\dagger U = UU^\dagger = I$, it follows that if $\Psi = U\phi$, then $\phi = U^\dagger \Psi$. Thus for a unitary operator, the adjoint $U^\dagger$ has the property if being an inverse operation as well, so that

$$U^\dagger = U^{-1} \tag{H-10}$$

If an operator depends continuously on a continuous parameter t, then the derivative of A is defined by

$$\frac{dA}{dt} = \lim_{\varepsilon \to 0} \frac{A(t + \varepsilon) - A(t)}{\varepsilon} \tag{H-11}$$

Problem H–6 has you prove that

$$\frac{d}{dt}(AB) = \frac{dA}{dt} B + A \frac{dB}{dt} \tag{H-12}$$

and

$$\frac{d}{dt} e^{iAt} = iAe^{iAt} \tag{H-13}$$

We are now ready to introduce the Heisenberg picture of time-dependent quantum mechanics. There are a number of different so-called representations of quantum mechanics. The two most common are probably the Schrödinger representation and the Heisenberg representation. In the Schrödinger picture, the temporal variation of a system is given by time-dependent wave functions, and the operators corresponding to observables are independent of time. In the Heisenberg picture the states of the system are represented by time-independent wave functions, and the operators are time dependent.

Since the physically significant quantities of quantum mechanics are inner products, we require that

$$\int \chi_0{}^* A(t)\Psi_0 \, dq = \int \chi^*(t) A_0 \Psi(t) \, dq \tag{H-14}$$

The first is the Heisenberg representation, and the second is the Schrödinger representation. Now for a time-independent Hamiltonian, the time-dependent Schrödinger equation

$$ih \frac{\partial \Psi}{\partial t} = \mathcal{H}\Psi \tag{H-15}$$

can be solved formally by

$$\Psi(t) = U(t)\Psi(0) \tag{H-16}$$

where

$$U(t) = \exp\left(-\frac{i\mathcal{H}t}{\hbar}\right) \tag{H-17}$$

This operator is defined through its power series.

Substitute this into Eq. (H–14) to get

$$(\chi_0, A(t)\Psi_0) = (U(t)\chi_0, A_0 U(t)\Psi_0) = (\chi_0, U_t^\dagger A_0 U_t \Psi_0)$$
$$= (\chi_0, U_t^* A_0 U_t \Psi_0) \tag{H-18}$$

from which we see that

$$A(t) = U_t^* A_0 U_t = e^{i\mathcal{H}t/\hbar} A_0 e^{-i\mathcal{H}t/\hbar} \tag{H-19}$$

This gives the time dependence of the operator $A(t)$. We can differentiate this to get

$$ih \frac{dA}{dt} = ih \frac{\partial U_t^*}{dt} A_0 U_t + ih U_t^* A_0 \frac{dU_t}{dt}$$

$$= -\mathcal{H}U_t^* A_0 U_t + U_t^* A_0 U_t \mathcal{H}$$

$$= -\mathcal{H}A_t + A_t \mathcal{H} = [A_t, \mathcal{H}]$$

Thus,

$$ih \frac{dA}{dt} = [A_t, \mathcal{H}] \tag{H-20}$$

where we have assumed that A_0 has no explicit time dependence. This equation is called the Heisenberg equation of motion. We have shown that Eq. (H–19) is a solution to Eq. (H–20).

REFERENCES

JACKSON, J. D. 1962. *Mathematics for quantum mechanics.* New York: Benjamin.
MERZBACHER, E. 1970. *Quantum mechanics*, 2nd ed. New York: Wiley.
MESSIAH, A. 1966. *Quantum mechanics*, Vol. 1. New York: Wiley.

PROBLEMS

H–1. Prove that $e^{A+B} \neq e^A e^B$ if $[A, B] \neq 0$.

H–2. Prove that the eigenvalues of a Hermitian operator are real.

H–3. Prove that $(A^\dagger)^\dagger = A$ and that $(A^{-1})^{-1} = A$.

H–4. Prove that $(AB)^\dagger = B^\dagger A^\dagger$ and that $(AB)^{-1} = B^{-1}A^{-1}$.

H–5. Prove that if A is Hermitian and f is a real function, then $U = \exp[if(A)]$ is unitary.

H–6. Prove that

$$\frac{d(AB)}{dt} = \frac{dA}{dt}B + A\frac{dB}{dt}$$

and

$$\frac{de^{iAt}}{dt} = iAe^{iAt}$$

THE POYNTING FLUX VECTOR

In this appendix, we shall introduce the concept of the Poynting flux vector and then derive Eq. (21–16). Consider a single particle of charge q moving with velocity $\mathbf{v}$. The rate of work done by an external electromagnetic field is $q\mathbf{v} \cdot \mathbf{E}$. There is no contribution here from the magnetic part $\mathbf{B}$ since the magnetic force is perpendicular to $\mathbf{v}$. For a continuous distribution of charge and current, then the total rate of work on the volume V is

$$\int_V \mathbf{J} \cdot \mathbf{E} \, d\mathbf{r} \tag{I–1}$$

where $\mathbf{J}$ is the macroscopic current density.

Now recall that Maxwell's equations are

$$\text{div } \mathbf{D} = 4\pi\rho \qquad\qquad \text{curl } \mathbf{H} = \frac{4\pi}{c}\mathbf{J} + \frac{1}{c}\frac{\partial \mathbf{D}}{\partial t}$$

$$\text{div } \mathbf{B} = 0 \qquad \text{curl } \mathbf{E} + \frac{1}{c}\frac{\partial B}{\partial t} = 0 \tag{I–2}$$

with the constitutive relations $\mathbf{D} = \varepsilon\mathbf{E}$, $\mathbf{J} = \sigma\mathbf{E}$, and $\mathbf{B} = \mu\mathbf{H}$ for an isotropic, conducting medium.

We use the Maxwell equation for curl $\mathbf{H}$ to eliminate $\mathbf{J}$ in Eq. (I–1) to get

$$\int_V \mathbf{J} \cdot \mathbf{E} \, d\mathbf{r} = \frac{1}{4\pi} \int_V \left[c\mathbf{E} \cdot (\nabla \times \mathbf{H}) - \mathbf{E} \cdot \frac{\partial \mathbf{D}}{\partial t} \right] d\mathbf{r} \tag{I–3}$$

We now use the vector identity

$$\nabla \cdot (\mathbf{E} \times \mathbf{H}) = \mathbf{H} \cdot (\nabla \times \mathbf{E}) - \mathbf{E} \cdot (\nabla \times \mathbf{H}) \tag{I–4}$$

and the Maxwell equation for curl $\mathbf{E}$ (Faraday's law) to get

$$-\int_V \mathbf{J} \cdot \mathbf{E} \, d\mathbf{r} = \frac{1}{4\pi} \int_V \left[c\nabla \cdot (\mathbf{E} \times \mathbf{H}) + \mathbf{E} \cdot \frac{\partial \mathbf{D}}{\partial t} + \mathbf{H} \cdot \frac{\partial \mathbf{B}}{\partial t} \right] d\mathbf{r} \tag{I–5}$$

If we assume that the medium is linear, i.e., that ε and μ are constants, then we can express the time derivatives in Eq. (I–5) in a more conventional form. Using the fact that $\mathbf{D} = \varepsilon\mathbf{E}$ and $\mathbf{B} = \mu\mathbf{H}$, we can write

$$\frac{1}{4\pi}\left(\mathbf{E}\cdot\frac{\partial\mathbf{D}}{\partial t} + \mathbf{H}\cdot\frac{\partial\mathbf{B}}{\partial t}\right) = \frac{1}{8\pi}\frac{\partial}{\partial t}(\mathbf{E}\cdot\mathbf{D} + \mathbf{B}\cdot\mathbf{H})$$

and if we identify $(\mathbf{E}\cdot\mathbf{D} + \mathbf{B}\cdot\mathbf{H})/8\pi$ as the energy density u, Eq. (I–5) can be written in the form

$$-\int_V \mathbf{J}\cdot\mathbf{E}\,d\mathbf{r} = \int_V \left[\frac{\partial u}{\partial t} + \frac{c}{4\pi}\nabla\cdot(\mathbf{E}\times\mathbf{H})\right]d\mathbf{r} \tag{I–6}$$

and since the volume V is arbitrary, this can be written as

$$\frac{\partial u}{\partial t} + \nabla\cdot\mathbf{S} = -\mathbf{J}\cdot\mathbf{E} \tag{I–7}$$

This is the (energy) continuity equation. By comparing this with the usual continuity equations, we see that the vector $\mathbf{S}$ represents the flux of energy and is called the *Poynting vector*

$$\mathbf{S} = \frac{c}{4\pi}(\mathbf{E}\times\mathbf{H}) \tag{I–8}$$

and that the quantity

$$u = \frac{1}{8\pi}(\mathbf{E}\cdot\mathbf{D} + \mathbf{H}\cdot\mathbf{B}) \tag{I–9}$$

can, in fact, be identified as the energy density of the electromagnetic field. Equation (I–9) can also be derived more directly by considering the work done by electric and magnetic fields separately, but the derivation here is quite valid.

We now wish to derive Eq. (21–16) from Eq. (I–8). Maxwell's equations for a nonconducting medium with no free charges are

$$\operatorname{div}\mathbf{E} = 0 \qquad \operatorname{curl}\mathbf{E} + \frac{1}{c}\frac{\partial\mathbf{B}}{\partial t} = 0$$

$$\operatorname{div}\mathbf{B} = 0 \qquad \operatorname{curl}\mathbf{B} - \frac{\mu\varepsilon}{c}\frac{\partial\mathbf{E}}{\partial t} = 0 \tag{I–10}$$

It is a standard procedure to show from these equations that each Cartesian component of $\mathbf{E}$ and $\mathbf{B}$ satisfies the wave equation (Problem I–1)

$$\nabla^2\zeta = \frac{1}{v^2}\frac{\partial^2\zeta}{\partial t^2} \tag{I–11}$$

where ζ represents any of the components of $\mathbf{E}$ and $\mathbf{B}$ and

$$v = \frac{c}{(\varepsilon\mu)^{1/2}} \tag{I–12}$$

Clearly v is the velocity of propagation of the electromagnetic wave through the medium characterized by ε and μ. The solutions to Eq. (I–11) are the plane wave solutions, which have the form (Problem I–2)

$$\zeta = e^{i\mathbf{k} \cdot \mathbf{r} - i\omega t} \tag{I–13}$$

where

$$k = |\mathbf{k}| = \frac{\omega}{v} = (\mu\varepsilon)^{1/2} \frac{\omega}{c} \tag{I–14}$$

By studying the properties of the solution Eq. (I–13), one sees that the wave propagates in the direction of the wave vector $\mathbf{k}$ (Problem I–3). Although ζ does satisfy the wave equation, since it represents components of $\mathbf{E}$ and $\mathbf{B}$, it must also satisfy Maxwell's equations as well.

Let us assume that $\mathbf{E}$ and $\mathbf{B}$ are plane waves of the form

$$\mathbf{E}(\mathbf{r}, t) = \boldsymbol{\varepsilon}_1 E_0 \, e^{i\mathbf{k} \cdot \mathbf{r} - i\omega t}$$
$$\mathbf{B}(\mathbf{r}, t) = \boldsymbol{\varepsilon}_2 B_0 \, e^{i\mathbf{k} \cdot \mathbf{r} - i\omega t} \tag{I–15}$$

where $\boldsymbol{\varepsilon}_1$ and $\boldsymbol{\varepsilon}_2$ are constant unit vectors, and E_0 and B_0 are (possibly complex) amplitudes. The Maxwell equations div $\mathbf{E} = 0$ and div $\mathbf{B} = 0$ require that (Problem I–4)

$$\boldsymbol{\varepsilon}_1 \cdot \mathbf{k} = 0 \quad \text{and} \quad \boldsymbol{\varepsilon}_2 \cdot \mathbf{k} = 0 \tag{I–16}$$

which mean that $\mathbf{E}$ and $\mathbf{B}$ are both perpendicular to the direction of propagation $\mathbf{k}$. We call such a wave *transverse*.

In addition, if we substitute Eqs. (I–15) into the Maxwell equation for curl $\mathbf{E}$, we find that (Problem I–5)

$$(\mathbf{k} \times \boldsymbol{\varepsilon}_1)E_0 = \frac{\omega}{c} \boldsymbol{\varepsilon}_2 B_0 \tag{I–17}$$

We write this as

$$\boldsymbol{\varepsilon}_2 = \frac{(\mathbf{k} \times \boldsymbol{\varepsilon}_1)}{\dfrac{\omega B_0}{c E_0}} \tag{I–18}$$

Since $\boldsymbol{\varepsilon}_1$ and $\boldsymbol{\varepsilon}_2$ are unit vectors and $\boldsymbol{\varepsilon}_1$, $\boldsymbol{\varepsilon}_2$, and $\mathbf{k}$ are mutually orthogonal (Problem I–6), we must have

$$k = |\mathbf{k}| = \frac{\omega B_0}{c E_0} \tag{I–19}$$

and since $k = \omega/v = (\mu\varepsilon)^{1/2}\omega/c$, it follows that

$$B_0 = (\mu\varepsilon)^{1/2} E_0 \tag{I–20}$$

The Poynting vector is

$$\mathbf{S} = \frac{c}{4\pi} (\mathbf{E} \times \mathbf{H})$$

and so for the case of a plane wave

$$\mathbf{S} = (\boldsymbol{\varepsilon}_1 \times \boldsymbol{\varepsilon}_2) \frac{cE_0 B_0}{4\pi\mu} \cos^2 (\mathbf{k} \cdot \mathbf{r} - \omega t)$$

Now it is easy to show (Problem I–7) that

$$\boldsymbol{\varepsilon}_1 \times \boldsymbol{\varepsilon}_2 = \frac{\mathbf{k}}{k} \tag{I–21}$$

and so if we average $\mathbf{S}$ over one cycle, one finds that (Problem I–8)

$$\mathbf{S}_{\mathrm{av}} = \frac{cE_0 B_0}{8\pi\mu} \frac{\mathbf{k}}{k}$$

$$= \frac{c}{8\pi} \left(\frac{\varepsilon}{\mu}\right)^{1/2} E_0^2 \frac{\mathbf{k}}{k} \tag{I–22}$$

Since the index of refraction n is defined as the ratio of the speed of light in a vacuum to the speed of light in the material medium, then Eq. (I–14) shows that

$$n = (\mu\varepsilon)^{1/2} \tag{I–23}$$

Thus Eq. (I–22) can be written as

$$S_{\mathrm{av}} = \frac{cn}{8\pi\mu} E_0^2 = \frac{v\varepsilon E_0^2}{8\pi} \tag{I–24}$$

which equals Eq. (21–16) for the nonmagnetic medium ($\mu = 1$) discussed there.

REFERENCES

FEYNMAN, R. P., LEIGHTON, R. B., and SANDS, M. 1965. *The Feynman lectures on physics.* Reading, Mass: Addison-Wesley.

JACKSON, J. D. 1962. *Classical electrodynamics.* New York: Wiley.

PROBLEMS

I–1. Derive the wave equation from Maxwell's equations.

I–2. Show that Eq. (I–13) is a solution to the wave equation.

I–3. Show that the function

$$\zeta = e^{i\mathbf{k} \cdot \mathbf{r} - i\omega t}$$

is a plane wave propagating in the direction of $\mathbf{k}$.

I–4. Derive Eqs. (I–16), showing that an electromagnetic wave is transverse.

I–5. Derive Eq. (I–17).

I–6. Prove that the vectors $\boldsymbol{\varepsilon}_1$, $\boldsymbol{\varepsilon}_2$, and $\mathbf{k}$ are mutually orthogonal.

I–7. Prove that $\boldsymbol{\varepsilon}_1 \times \boldsymbol{\varepsilon}_2 = \mathbf{k}/k$.

I–8. Derive Eq. (I–22).

THE RADIATION EMITTED BY AN OSCILLATING DIPOLE

In this appendix we shall discuss the electromagnetic radiation emitted by an oscillating electric dipole, and, in particular, we shall derive Eqs. (21–46) and (21–47).

We start with Maxwell's equations

$$\text{div } \mathbf{D} = 4\pi\rho \qquad\qquad \text{curl } \mathbf{H} = \frac{4\pi}{c}\mathbf{J} + \frac{1}{c}\frac{\partial \mathbf{D}}{\partial t} \tag{J–1}$$

$$\text{div } \mathbf{B} = 0 \qquad \text{curl } \mathbf{E} + \frac{1}{c}\frac{\partial \mathbf{B}}{\partial t} = 0$$

with the constitutive equations $\mathbf{D} = \varepsilon\mathbf{E}$, $\mathbf{J} = \sigma\mathbf{E}$, and $\mathbf{B} = \mu\mathbf{H}$. Since div $\mathbf{B} = 0$ and div curl $\mathbf{V} = 0$ for any vector $\mathbf{V}$, we define a vector $\mathbf{A}$, called the vector potential, by

$$\mathbf{B} = \text{curl } \mathbf{A} \tag{J–2}$$

With this form for $\mathbf{B}$, the Maxwell equation for curl $\mathbf{E}$ reads

$$\text{curl}\left(\mathbf{E} + \frac{1}{c}\frac{\partial \mathbf{A}}{\partial t}\right) = 0 \tag{J–3}$$

Since curl grad $\zeta = 0$ for any function ζ, the terms in parentheses here can be written as the gradient of some scalar function, e.g., ϕ, to get

$$\mathbf{E} + \frac{1}{c}\frac{\partial \mathbf{A}}{\partial t} = -\text{grad } \phi \tag{J–4}$$

In terms of $\mathbf{A}$ and ϕ, the two inhomogeneous Maxwell equations become

$$\text{curl curl } \mathbf{A} + \frac{1}{c^2}\frac{\partial^2 \mathbf{A}}{\partial t^2} + \frac{1}{c}\text{grad}\frac{\partial \phi}{\partial t} = \frac{4\pi}{c}\mathbf{J} \tag{J–5}$$

$$\text{div grad } \phi + \frac{1}{c}\frac{\partial}{\partial t}\text{div } \mathbf{A} = -4\pi\rho \tag{J–6}$$

or (Problem J–1)

$$\nabla^2\phi + \frac{1}{c}\frac{\partial}{\partial t}(\nabla \cdot \mathbf{A}) = -4\pi\rho \tag{J–7}$$

$$\nabla^2\mathbf{A} - \frac{1}{c^2}\frac{\partial^2\mathbf{A}}{\partial t^2} - \nabla\left(\nabla \cdot \mathbf{A} + \frac{1}{c}\frac{\partial\phi}{\partial t}\right) = -\frac{4\pi}{c}\mathbf{J} \tag{J–8}$$

Since $\mathbf{B} = \text{curl } \mathbf{A}$, the vector potential is arbitrary to within an added gradient of some scalar function ζ. Thus $\mathbf{B}$ is unchanged by

$$\mathbf{A} \to \mathbf{A} + \nabla\zeta \tag{J–9}$$

Similarly $\mathbf{E}$ is unchanged by (Problem J–2)

$$\phi \to \phi - \frac{1}{c}\frac{\partial\zeta}{\partial t} \tag{J–10}$$

The two transformations (J–9) and (J–10) allow us to choose $\mathbf{A}$ and ϕ such that

$$\nabla \cdot \mathbf{A} + \frac{1}{c}\frac{\partial\phi}{\partial t} = 0 \tag{J–11}$$

Notice that this uncouples Eqs. (J–7) and (J–8) to

$$\nabla^2\phi - \frac{1}{c^2}\frac{\partial^2\phi}{\partial t^2} = -4\pi\rho \tag{J–12}$$

$$\nabla^2\mathbf{A} - \frac{1}{c^2}\frac{\partial^2\mathbf{A}}{\partial t^2} = -\frac{4\pi}{c}\mathbf{J} \tag{J–13}$$

Equations (J–11) through (J–13) are equivalent to Maxwell's equations.

We remark in passing that transformations (J–9) and (J–10) are called gauge transformations, and the invariance of the fields under such a transformation is called gauge invariance. The particular relation given by Eq. (J–11) is called the Lorentz gauge. Although this gauge is not unique, it is the most commonly used since it leads to the wave equations, Eqs. (J–12) and (J–13). See the work by Jackson (" References ") for a detailed discussion of this problem.

Both Eqs. (J–12) and (J–13) are inhomogeneous wave equations of the form

$$\nabla^2\Psi - \frac{1}{c^2}\frac{\partial^2\Psi}{\partial t^2} = -4\pi f(\mathbf{r}, t) \tag{J–14}$$

where $f(\mathbf{r}, t)$ is a known source distribution. Such an equation can be solved by the method of Green's functions (cf. Jackson, Section 6.6) to give

$$\Psi(r, t) = \int \frac{\delta\left(t' + \dfrac{|\mathbf{r} - \mathbf{r}'|}{c} - t\right)}{|\mathbf{r} - \mathbf{r}'|} f(\mathbf{r}', t')\, d\mathbf{r}'\, dt' \tag{J–15}$$

The quantity

$$G(\mathbf{r}, t; \mathbf{r}', t') \equiv \frac{\delta\left(t' + \dfrac{|\mathbf{r} - \mathbf{r}'|}{c} - t\right)}{|\mathbf{r} - \mathbf{r}'|} \tag{J–16}$$

is called a (retarded) Green's function.

Of direct interest to this appendix is the special case of Eq. (J–15)

$$\mathbf{A}(\mathbf{r}, t) = \frac{1}{c} \int d\mathbf{r}' \int dt' \, \frac{\mathbf{J}(\mathbf{r}', t')}{|\mathbf{r} - \mathbf{r}'|} \, \delta\left(t' + \frac{|\mathbf{r} - \mathbf{r}'|}{c}, - t\right) \tag{J–17}$$

Assume now that $\mathbf{J}(\mathbf{r}, t) = \mathbf{J}(\mathbf{r})e^{-i\omega t}$. Equation (J–17) becomes then

$$\mathbf{A}(\mathbf{r}, t) = \mathbf{A}(\mathbf{r})e^{-i\omega t} \tag{J–18}$$

with

$$\mathbf{A}(\mathbf{r}) = \frac{1}{c} \int \mathbf{J}(\mathbf{r}') \, \frac{e^{ik|\mathbf{r}-\mathbf{r}'|}}{|\mathbf{r} - \mathbf{r}'|} \, d\mathbf{r}' \tag{J–19}$$

where as usual $k = \omega/c$.

If the region in which $\mathbf{J}$ is nonzero is small and $\mathbf{r}$ is far from this small region, then $|\mathbf{r} - \mathbf{r}'| \approx r$, and Eq. (J–19) becomes

$$\mathbf{A}(\mathbf{r}) = \frac{e^{ikr}}{cr} \int \mathbf{J}(\mathbf{r}') \, d\mathbf{r}' \tag{J–20}$$

More precisely, one expands the plane wave $e^{ik|\mathbf{r}-\mathbf{r}'|}/|\mathbf{r} - \mathbf{r}'|$ in a series of Bessel functions and retains only the first term (cf. Jackson). Equation (J–20) can be integrated by parts to give

$$\int \mathbf{J} \, d\mathbf{r}' = - \int \mathbf{r}'[\nabla \cdot \mathbf{J}(\mathbf{r}')] \, d\mathbf{r}' \tag{J–21}$$

but from the equation for continuity of charge, namely,

$$\frac{\partial \rho}{\partial t} + \text{div } \mathbf{J} = 0 \tag{J–22}$$

we see that if $\rho(\mathbf{r}, t) = \rho(\mathbf{r})e^{-i\omega t}$, then

$$i\omega\rho = \text{div } \mathbf{J}$$

and so

$$\mathbf{A}(\mathbf{r}) = -ik\boldsymbol{\mu} \, \frac{e^{ikr}}{r} \tag{J–23}$$

where $\boldsymbol{\mu}$ is the electric dipole moment

$$\boldsymbol{\mu} = \int \mathbf{r}'\rho(\mathbf{r}') \, d\mathbf{r}' \tag{J–24}$$

Equation (J–23) is a central equation of this appendix, since once $\mathbf{A}(\mathbf{r}, t) = \mathbf{A}(\mathbf{r})e^{-i\omega t}$ is known, $\mathbf{E}$ and $\mathbf{B}$ are known through Eqs. (J–2) and (J–1), namely (Problem J–3),

$$\mathbf{B} = \text{curl } \mathbf{A} \tag{J–25}$$

$$\mathbf{E} = \frac{i}{k} \text{ curl } \mathbf{B} \tag{J–26}$$

Equation (J–26) is valid only away from the source. Straightforward application of these two equations to Eq. (J–23) yields (Problem J–4)

$$\mathbf{B} = k^2(\boldsymbol{\varepsilon}_r \times \boldsymbol{\mu}) \, \frac{e^{ikr}}{r} \tag{J–27}$$

and

$$\mathbf{E} = \mathbf{B} \times \boldsymbol{\varepsilon}_r \tag{J–28}$$

where $\boldsymbol{\varepsilon}_r$ is a unit vector in the direction of $\mathbf{r}$.

In order to derive Eq. (21–47), the rate of radiation emitted per unit solid angle in the direction of X, we let $\varepsilon_r = \varepsilon_X$ and $\mu = (\varepsilon_X \mu_{0X} + \varepsilon_Y \mu_{0Y} + \varepsilon_Z \mu_{0Z}) \cos \omega t$. Equations (J–27) and (J–28) become (Problem J–5)

$$\mathbf{B} = k^2(\varepsilon_Z \mu_{0Y} - \varepsilon_Y \mu_{0Z}) \cos \omega t \, \frac{e^{ikr}}{r} \tag{J–29}$$

$$\mathbf{E} = k^2(\varepsilon_Y \mu_{0Y} + \varepsilon_Z \mu_{0Z}) \cos \omega t \, \frac{e^{ikr}}{r} \tag{J–30}$$

Note that $\mathbf{E}$ and $\mathbf{B}$ are perpendicular to each other as they should be (cf. Appendix I). The flux of energy is given by the Poynting vector $\mathbf{S}$ (cf. Appendix I)

$$\mathbf{S} = \frac{c}{4\pi} (\mathbf{E} \times \mathbf{H}) \tag{J–31}$$

or

$$\mathbf{S} = \frac{c}{4\pi} (\mathbf{E} \times \mathbf{H}^*) \tag{J–32}$$

if $\mathbf{E}$ and $\mathbf{H}$ are given in complex notation.

Using Eqs. (J–29) and (J–30), we see that

$$\mathbf{S} = \varepsilon_X \frac{ck^4}{4\pi} (\mu_{0Y}{}^2 + \mu_{0Z}{}^2) \cos^2 \omega t \tag{J–33}$$

which is Eq. (21–47) if we average over one vibrational cycle.

REFERENCES

JACKSON, J. D. 1962. *Classical electrodynamics*. New York: Wiley.

LARRAIN, P., and CORSO, D. R. 1970. *Electromagnetic fields and waves*. San Francisco: Freeman.

PROBLEMS

J–1. Show that Eqs. (J–5) and (J–6) are equivalent to Eqs. (J–7) and (J–8).

J–2. Show that E is unchanged by the transformation $\phi \to \phi - (1/c)(\partial\zeta/\partial t)$ where ζ is a scalar function.

J–3. Derive Eq. (J–26).

J–4. Derive and discuss geometrically Eqs. (J–27) and (J–28). Remember that an electromagnetic wave is transverse.

J–5. Derive Eqs. (J–29) and (J–30).

DIELECTRIC CONSTANT AND ABSORPTION

In this appendix, we introduce the idea of representing the frequency-dependent dielectric constant as a complex number, and then interpret the real and imaginary parts. First recall the static equations

$$D = 4\pi\sigma \qquad D = E + 4\pi P$$

$$D = \varepsilon_s E \qquad P = \chi E \tag{K-1}$$

$$D = (1 + 4\pi\chi)E \tag{K-2}$$

where ε_s is the static (zero-frequency) dielectric constant and χ is the susceptibility. These equations become inapplicable for high-frequency fields, or if we wish to keep these equations, we must generalize and reinterpret ε and χ. This is what we shall do.

We could look at how $P(t)$ behaves when $E(t)$ varies, but it is more convenient to look at $D(t)$ instead of $P(t)$. Since $D = (1 + 4\pi\chi)E$, this amounts to essentially the same thing. Consider the case in which $E(t) = E_0 \cos \omega t$. If this field has persisted for a time, $D(t)$ must also vary with the same frequency, but must not necessarily be in phase with $E(t)$ since, in general, there is a time lag between $P(t)$ and $E(t)$. This can be expressed by writing

$$D(t) = D_0 \cos(\omega t - \phi) = D_1 \cos \omega t + D_2 \sin \omega t \tag{K-3}$$

with

$$D_1 = D_0 \cos \phi \qquad D_2 = D_0 \sin \phi \tag{K-4}$$

The phase angle ϕ is certainly a function of ω and should go to zero as ω goes to zero, which gives $D(t) = D_1 \cos \omega t = \varepsilon E_0 \cos \omega t$ as $\omega \to 0$. In general, however, we can define two "dielectric constants" $\varepsilon'(\omega)$ and $\varepsilon''(\omega)$ by

$$D_1 = \varepsilon'(\omega)E_0 \quad \text{and} \quad D_2 = \varepsilon''(\omega)E_0 \tag{K-5}$$

The dielectric constant we studied before then was $\varepsilon'(0)$. Again we say that $\varepsilon''(\omega) \to 0$

as $\omega \to 0$. We shall show shortly that $\varepsilon''(\omega)$ is directly related to the absorption coefficient of the system. Since this represents energy lost from the incident electric field, this is often called dielectric loss.

It is convenient to define a complex dielectric constant by

$$\hat{\varepsilon}(\omega) = \varepsilon'(\omega) - i\varepsilon''(\omega) \qquad (K-6)$$

since if we write $E = E_0\, e^{i\omega t}$, we then have

$$D = \hat{\varepsilon}E \qquad (K-7)$$

even for time-dependent fields. We, of course, take only the real part of this equation. Realize that writing $\hat{\varepsilon}$ is a convenient formal device that allows us to write $D = \hat{\varepsilon}E$. It does not mean that ε becomes complex at high frequencies. We shall see how Maxwell's equations can similarly be reinterpreted to accommodate a complex formalism for ε, but before doing this let us look at the relation between $E(t)$ and $D(t)$ once again.

We can represent the time lag between $D(t)$ and $E(t)$, or $P(t)$ and $E(t)$, by means of an after-effect or response function:

$$D(t) = E(t) + \int_0^t E(t')\alpha(t - t')\, dt' \qquad (K-8)$$

where the part of D that can follow E instantaneously is written as a separate term. Introducing $E(t) = E_0 \cos \omega t$, we get

$$D(t) - E_0 \cos \omega t = E_0 \int_0^t \alpha(t - t') \cos \omega t'\, dt'$$

$$= E_0 \int_0^t \alpha(x) \cos \omega(t - x)\, dx \qquad (K-9)$$

We assume now that E has persisted long enough to make $D(t)$ a periodic function with frequency ω. This means that t is larger than the time t_0 it takes for $\alpha(t)$ to vanish. In view of this then, we can set the upper limit of the above integral equal to ∞, and so Eq. (K–9) becomes

$$D(t) - E_0 \cos \omega t = E_0 \int_0^\infty \alpha(x) \cos \omega(t - x)\, dx$$

or using Eqs. (K–3) and (K–5),

$$(\varepsilon'(\omega) - 1) \cos \omega t + \varepsilon''(\omega) \sin \omega t$$

$$= \cos \omega t \int_0^\infty \alpha(x) \cos \omega x\, dx + \sin \omega t \int_0^\infty \alpha(x) \sin \omega x\, dx$$

from which we see

$$\varepsilon'(\omega) - 1 = \int_0^\infty \alpha(t) \cos \omega t\, dt \qquad (K-10)$$

$$\varepsilon''(\omega) = \int_0^\infty \alpha(t) \sin \omega t\, dt \qquad (K-11)$$

These equations are similar to the equations that are derived in Section 21–5. They are derived here without giving $\alpha(t)$ any molecular significance, whereas in Section 21–5, $\alpha(t)$ is a dipole moment correlation function. It is from a pair of equations just like this that the Kramers-Kronig relations between $\varepsilon'(\omega)$ and $\varepsilon''(\omega)$ are derived [cf. Eqs. (21–182 and 21–183)].

Now we wish to show how the introduction of $\hat{\varepsilon}$ affects some other electromagnetic results, particularly Maxwell's equations. One result we wish to preserve is the low-frequency relation between the dielectric constant ε and the index of refraction n. This relation is $\varepsilon = n^2$. We shall accordingly define a complex index of refraction by

$$\hat{\varepsilon} = \hat{n}^2 \tag{K–12}$$

As in all the generalizations we shall make, we assume that the real part of our complex quantities is just the ordinary property and that the pure imaginary part vanishes as $\omega \to 0$. This is simply to assure consistency with all our ordinary low-frequency equations.

Consider now Maxwell's equations

$$\text{curl } \mathbf{H} - \frac{1}{c}\frac{\partial \mathbf{D}}{\partial t} = \frac{4\pi \mathbf{J}}{c} \tag{K–13}$$

$$\text{curl } \mathbf{E} + \frac{1}{c}\frac{\partial \mathbf{B}}{\partial t} = 0 \tag{K–14}$$

$$\text{div } \mathbf{D} = 4\pi\rho \tag{K–15}$$

$$\text{div } \mathbf{B} = 0 \tag{K–16}$$

along with the material equations $\mathbf{J} = \sigma\mathbf{E}$, $\mathbf{D} = \varepsilon\mathbf{E}$, $\mathbf{B} = \mu\mathbf{H}$. We shall confine ourselves to nonmagnetic materials ($\mu = 1$) and no currents or free charges ($\mathbf{J} = 0$; $\rho = 0$). Now eliminate $\mathbf{B} = \mu\mathbf{H}$ from the first two Maxwell equations by taking $\partial/\partial t$ of the first and taking curl of the second (curl curl = grad div $-\nabla^2$). One eventually gets

$$\nabla^2 \mathbf{E} = \frac{\varepsilon\mu}{c^2}\frac{\partial^2 \mathbf{E}}{\partial t^2} \tag{K–17}$$

i.e., a wave equation for $\mathbf{E}$. It was from this equation incidentally that Maxwell predicted that electromagnetic radiation was wavelike and propagated with a velocity $c/(\mu\varepsilon)^{1/2}$ (ε and $\mu = 1$ in a vacuum).

Now substitute $E_x = E_0\, e^{i(\omega t - kx)}$ into the wave equation to get

$$\nabla^2 E_x + \frac{\varepsilon\mu\omega^2}{c^2}\, E_x = 0 \tag{K–18}$$

This tells us that the wave vector k is

$$k^2 = \frac{\varepsilon\mu\omega^2}{c^2} = \frac{\varepsilon\omega^2}{c^2} \quad \text{(nonmagnetic)} \tag{K–19}$$

Now in order to preserve the form of this equation, we define

$$\hat{k}^2 = \frac{\omega^2\hat{\varepsilon}}{c^2} \tag{K–20}$$

or

$$\hat{k} = \frac{\omega}{c}\,\hat{n} \tag{K–21}$$

Set

$$\hat{n} = n - i\kappa = (\hat{\varepsilon})^{1/2} \tag{K-22}$$

and so

$$\varepsilon'(\omega) = n^2 - \kappa^2 \tag{K-23}$$

$$\varepsilon''(\omega) = 2\kappa n \tag{K-24}$$

Our equation for E_x is

$$E_x = E_0 e^{i\omega t} e^{-ikx} = E_0 e^{-\omega\kappa x/c} e^{i\omega t} e^{-ikx} \tag{K-25}$$

or the intensity is $|E_x|^2 = E_0^2 e^{-2\omega\kappa x/c}$. Comparing this to Lambert's law, $I = I_0 e^{-\alpha x}$, we finally see that

$$\alpha(\omega) = \frac{2\omega\kappa}{c} = \frac{\varepsilon''(\omega)\omega}{nc} \approx \frac{\omega\varepsilon''(\omega)}{c} \qquad \text{(gases)} \tag{K-26}$$

The quantities without the $\frown$ denote the real part, and we have used the fact that $n \approx 1$ for gases. Thus we see that the absorption coefficient $\alpha(\omega)$ and the imaginary part of the complex dielectric constant are directly related.

REFERENCES

BORN, M., and WOLF, E. 1959. *Principles of optics*. London: Pergamon.
FROHLICH, H. 1958. *Theory of Dielectrics*. Oxford: Oxford University Press.

INDEX